VISCOSITY

THERMOPHYSICAL PROPERTIES OF MATTER
The TPRC Data Series

A Comprehensive Compilation of Data by the
Thermophysical Properties Research Center (TPRC), Purdue University

Y. S. Touloukian, Series Editor
C. Y. Ho, Series Technical Editor

New data on thermophysical properties are being constantly accumulated at TPRC. Contact TPRC and use its interim updating services for the most current information.

THERMOPHYSICAL PROPERTIES OF MATTER
VOLUME 11

VISCOSITY

Y. S. Touloukian

Director
Thermophysical Properties Research Center
and
Distinguished Atkins Professor of Engineering
School of Mechanical Engineering
Purdue University
and
Visiting Professor of Mechanical Engineering
Auburn University

S. C. Saxena

Professor of Energy Engineering
University of Illinois
Chicago Circle
and
Consultant
Thermophysical Properties Research Center
Purdue University

P. Hestermans

Director
Belgian Institute of High Pressure
Sterrebeek, Belgium
Formerly
Affiliate Senior Researcher
Thermophysical Properties Research Center
Purdue University

IFI/PLENUM • NEW YORK-WASHINGTON

Library of Congress Catalog Card Number 73-129616

ISBN (13-Volume Set) 0-306-67020-8
ISBN (Volume 11) 0-306-67031-3

IFI/Plenum Data Company is a division of
Plenum Publishing Corporation
227 West 17th Street, New York, N.Y. 10011

Distributed in Europe by Heyden & Son, Ltd.
Spectrum House, Alderton Crescent
London NW4 3XX, England

Printed in the United States of America

"In this work, when it shall be found that much is omitted, let it not be forgotten that much likewise is performed..."

SAMUEL JOHNSON, A.M.

From last paragraph of Preface to his two-volume *Dictionary of the English Language*, Vol. I, page 5, 1755, London, Printed by Strahan.

Foreword

In 1957, the Thermophysical Properties Research Center (TPRC) of Purdue University, under the leadership of its founder, Professor Y. S. Touloukian, began to develop a coordinated experimental, theoretical, and literature review program covering a set of properties of great importance to science and technology. Over the years, this program has grown steadily, producing bibliographies, data compilations and recommendations, experimental measurements, and other output. The series of volumes for which these remarks constitute a foreword is one of these many important products. These volumes are a monumental accomplishment in themselves, requiring for their production the combined knowledge and skills of dozens of dedicated specialists. The Thermophysical Properties Research Center deserves the gratitude of every scientist and engineer who uses these compiled data.

The individual nontechnical citizen of the United States has a stake in this work also, for much of the science and technology that contributes to his well-being relies on the use of these data. Indeed, recognition of this importance is indicated by a mere reading of the list of the financial sponsors of the Thermophysical Properties Research Center; leaders of the technical industry of the United States and agencies of the Federal Government are well represented.

Experimental measurements made in a laboratory have many potential applications. They might be used, for example, to check a theory, or to help design a chemical manufacturing plant, or to compute the characteristics of a heat exchanger in a nuclear power plant. The progress of science and technology demands that results be published in the open literature so that others may use them. Fortunately for progress, the useful data in any single field are not scattered throughout the tens of thousands of technical journals published throughout the world. In most fields, fifty percent of the useful work appears in no more than thirty or forty journals. However, in the case of TPRC, its field is so broad that about 100 journals are required to yield fifty percent. But that other fifty percent! It is scattered through more than 3500 journals and other documents, often items not readily identifiable or obtainable. Over 75,000 references are now in the files.

Thus, the man who wants to use existing data, rather than make new measurements himself, faces a long and costly task if he wants to assure himself that he has found all the relevant results. More often than not, a search for data stops after one or two results are found—or after the searcher decides he has spent enough time looking. Now with the appearance of these volumes, the scientist or engineer who needs these kinds of data can consider himself very fortunate. He has a single source to turn to; thousands of hours of search time will be saved, innumerable repetitions of measurements will be avoided, and several billions of dollars of investment in research work will have been preserved.

However, the task is not ended with the generation of these volumes. A critical evaluation of much of the data is still needed. Why are discrepant results obtained by different experimentalists? What undetected sources of systematic error may affect some or even all measurements? What value can be derived as a "recommended" figure from the various conflicting values that may be reported? These questions are difficult to answer, requiring the most sophisticated judgment of a specialist in the field. While a number of the volumes in this Series do contain critically evaluated and recommended data, these are still in the minority. The data are now being more intensively evaluated by the staff of TPRC as an integral part of the effort of the National Standard Reference Data System (NSRDS). The task of the National Standard Reference Data System is to organize and operate a comprehensive program to prepare compilations of critically evaluated data on the properties of substances. The NSRDS is administered by the National Bureau of Standards under a directive from the Federal Council for Science and Technology, augmented by special legislation of the Congress of the United States. TPRC is one of the national resources participating in the National Standard Reference Data System in a united effort to satisfy the needs of the technical community for readily accessible, critically evaluated data.

As a representative of the NBS Office of Standard Reference Data, I want to congratulate Professor Touloukian and his colleagues on the accomplishments represented by this Series of reference data books. Scientists and engineers the world over are indebted to them. The task ahead is still an awesome one and I urge the nation's private industries and all concerned Federal agencies to participate in fulfilling this national need of assuring the availability of standard numerical reference data for science and technology.

EDWARD L. BRADY
Associate Director for Information Programs
National Bureau of Standards

Preface

Thermophysical Properties of Matter, the TPRC Data Series, is the culmination of seventeen years of pioneering effort in the generation of tables of numerical data for science and technology. It constitutes the restructuring, accompanied by extensive revision and expansion of coverage, of the original *TPRC Data Book*, first released in 1960 in loose-leaf format, 11″ × 17″ in size, and issued in June and December annually in the form of supplements. The original loose-leaf *Data Book* was organized in three volumes: (1) metallic elements and alloys; (2) nonmetallic elements, compounds, and mixtures which are solid at N.T.P., and (3) nonmetallic elements, compounds, and mixtures which are liquid or gaseous at N.T.P. Within each volume, each property constituted a chapter.

Because of the vast proportions the *Data Book* began to assume over the years of its growth and the greatly increased effort necessary in its maintenance by the user, it was decided in 1967 to change from the loose-leaf format to a conventional publication. Thus, the December 1966 supplement of the original *Data Book* was the last supplement disseminated by TPRC.

While the manifold physical, logistic, and economic advantages of the bound volume over the loose-leaf oversize format are obvious and welcome to all who have used the unwieldy original volumes, the assumption that this work will no longer be kept on a current basis because of its bound format would not be correct. Fully recognizing the need of many important research and development programs which require the latest available information, TPRC has instituted a *Data Update Plan* enabling the subscriber to inquire, by telephone if necessary, for specific information and receive, in many instances, same-day response on any new data processed or revision of published data since the latest edition. In this context, the TPRC Data Series departs drastically from the conventional handbook and giant multivolume classical works, which are no longer adequate media for the dissemination of numerical data of science and technology without a continuing activity on contemporary coverage. The loose-leaf arrangements of many works fully recognize this fact and attempt to develop a combination of bound volumes and loose-leaf supplement arrangements· as the work becomes increasingly large. TPRC's *Data Update Plan* is indeed unique in this sense since it maintains the contents of the TPRC Data Series current and live on a day-to-day basis between editions. In this spirit, I strongly urge all purchasers of these volumes to complete in detail and return the *Volume Registration Certificate* which accompanies each volume in order to assure themselves of the continuous receipt of annual listing of corrigenda during the life of the edition.

The TPRC Data Series consists initially of 13 independent volumes. The first seven volumes were published in 1970, Volumes 8 and 9 in 1972, and Volume 10 in 1973. Volumes 11, 12, and 13 are planned for 1975. It is also contemplated that subsequent to the first edition, each volume will be revised, up-dated, and reissued in a new edition approximately every fifth year. The organization of the TPRC Data Series makes each volume a self-contained entity available individually without the need to purchase the entire Series.

The coverage of the specific thermophysical properties represented by this Series constitutes the most comprehensive and authoritative collection of numerical data of its kind for science and technology.

Whenever possible, a uniform format has been used in all volumes, except when variations in presentation were necessitated by the nature of the property or the physical state concerned. In spite of the wealth of data reported in these volumes, it should be recognized that all volumes are not of the same degree of completeness. However, as additional data are processed at TPRC on a continuing basis, subsequent editions will become increasingly more complete and up to date. Each volume in the Series basically comprises three sections, consisting of a text, the body of numerical data with source references, and a material index.

The aim of the textual material is to provide a complementary or supporting role to the body of numerical data rather than to present a treatise on the subject of the property. The user will find a basic theoretical treatment, a comprehensive presentation of selected works which constitute reviews, or compendia of empirical relations useful in estimation of the property when there exists a paucity of data or when data are completely lacking. Established major experimental techniques are also briefly reviewed.

The body of data is the core of each volume and is presented in both graphical and tabular formats for convenience of the user. Every single point of numerical data is fully referenced as to its original source and no secondary sources of information are used in data extraction. In general, it has not been possible to critically scrutinize all the original data presented in these volumes, except to eliminate perpetuation of gross errors. However, in a significant number of cases, such as for the properties of liquids and gases and the thermal conductivity and thermal diffusivity of all the elements, the task of full evaluation, synthesis, and correlation has been completed. It is hoped that in subsequent editions of this continuing work, not only new information will be reported but the critical evaluation will be extended to increasingly broader classes of materials and properties.

The third and final major section of each volume is the material index. This is the key to the volume, enabling the user to exercise full freedom of access to its contents by any choice of substance name or detailed alloy and mixture composition, trade name, synonym, etc. Of particular interest here is the fact that in the case of those properties which are reported in separate companion volumes, the material index in each of the volumes also reports the contents of the other companion volumes.* The sets of companion volumes are as follows:

Thermal conductivity:	Volumes 1, 2, 3
Specific heat:	Volumes 4, 5, 6
Radiative properties:	Volumes 7, 8, 9
Thermal expansion:	Volumes 12, 13

The ultimate aims and functions of TPRC's Data Tables Division are to extract, evaluate, rec-

*For the first edition of the Series, this arrangement was not feasible for Volumes 7 and 8 due to the sequence and the schedule of their publication. This situation will be resolved in subsequent editions.

oncile, correlate, and synthesize all available data for the thermophysical properties of materials with the result of obtaining internally consistent sets of property values, termed the "recommended reference values." In such work, gaps in the data often occur, for ranges of temperature, composition, etc. Whenever feasible, various techniques are used to fill in such missing information, ranging from empirical procedures to detailed theoretical calculations. Such studies are resulting in valuable new estimation methods being developed which have made it possible to estimate values for substances and/or physical conditions presently unmeasured or not amenable to laboratory investigation. Depending on the available information for a particular property and substance, the end product may vary from simple tabulations of isolated values to detailed tabulations with generating equations, plots showing the concordance of the different values, and, in some cases, over a range of parameters presently unexplored in the laboratory.

The TPRC Data Series constitutes a permanent and valuable contribution to science and technology. These constantly growing volumes are invaluable sources of data to engineers and scientists, sources in which a wealth of information heretofore unknown or not readily available has been made accessible. We look forward to continued improvement of both format and contents so that TPRC may serve the scientific and technological community with ever-increasing excellence in the years to come. In this connection, the staff of TPRC is most anxious to receive comments, suggestions, and criticisms from all users of these volumes. An increasing number of colleagues are making available at the earliest possible moment reprints of their papers and reports as well as pertinent information on the more obscure publications. I wish to renew my earnest request that this procedure become a universal practice since it will prove to be most helpful in making TPRC's continuing effort more complete and up to date.

It is indeed a pleasure to acknowledge with gratitude the multisource financial assistance received from over fifty sponsors which has made the continued generation of these tables possible. In particular, I wish to single out the sustained major support received from the Air Force Materials Laboratory–Air Force Systems Command, the Defense Supply Agency, the Office of Standard Reference Data–National Bureau of Standards, and the Office of Advanced Research and Technology–National Aeronautics and Space Administration. TPRC is indeed proud to have been designated as a National

Information Analysis Center for the Department of Defense as well as a component of the National Standard Reference Data System under the cognizance of the National Bureau of Standards.

While the preparation and continued maintenance of this work is the responsibility of TPRC's Data Tables Division, it would not have been possible without the direct input of TPRC's Scientific Documentation Division and, to a lesser degree, the Theoretical and Experimental Research Divisions. The authors of the various volumes are the senior staff members in responsible charge of the work. It should be clearly understood, however, that many have contributed over the years and their contributions are specifically acknowledged in each volume. I wish to take this opportunity to personally thank those members of the staff, assistant researchers, graduate research assistants, and supporting graphics and technical typing personnel without whose diligent and painstaking efforts this work could not have materialized.

Y. S. TOULOUKIAN

Director
Thermophysical Properties Research Center
Distinguished Atkins Professor of Engineering

Purdue University
West Lafayette, Indiana
October 1974

Introduction to Volume 11

This volume of *Thermophysical Properties of Matter, the TPRC Data Series*, presents the data and information on the viscosity of fluids and fluid mixtures and follows the general format of Volume 3 of this Series.

The volume comprises three major sections: the front text on *theory, estimation, and measurement* together with its bibliography, the main body of *numerical data* with its references, and the *material index*.

The text material is intended to assume a role complementary to the main body of numerical data, the presentation of which is the primary purpose of this volume. It is felt that a moderately detailed discussion of the theoretical nature of the property under consideration together with an overview of predictive procedures and recognized experimental methods and techniques will be appropriate in a major reference work of this kind. The extensive reference citations given in the text should lead the interested reader to sufficient literature for a more comprehensive study. It is hoped, however, that enough detail is presented for this volume to be self-contained for the practical user.

The main body of the volume consists of the presentation of numerical data compiled over the years in a most meticulous manner. The coverage includes 59 pure fluids, most of which are identical to those covered in Volumes 3 and 6 of this Series, and 129 systems of fluid mixtures which are felt to be of greatest engineering importance. The extraction of all data directly from their original sources ensures freedom from errors of transcription. Furthermore, a number of gross errors appearing in the original source documents have been corrected. The organization and presentation of the data together with other pertinent information on the use of the tables and figures is discussed in detail in the introductory material to the section entitled *Numerical Data*.

The data on pure fluids have been critically evaluated, analyzed, and synthesized, and "recommended reference values" are presented, with the available experimental data given in departure plots.

The recommended values are those that were considered to be the most probable when assessments were made of the available data and information. It should be realized, however, that these recommended values are not necessarily the final true values and that changes directed toward this end will often become necessary as more data become available. Future editions will contain these changes.

The data on fluid mixtures have been smoothed graphically and the smoothed values as well as the experimental data are presented in both graphical and tabular forms. Furthermore, the experimental data for binary mixtures have been fitted with equations of the Sutherland type and the Sutherland coefficients have been calculated and are presented.

As stated earlier, all data have been obtained from their original sources and each data set is so referenced. TPRC has in its files all data-source documents cited in this volume. Those that cannot readily be obtained elsewhere are available from TPRC in microfiche form.

This volume has grown out of activities made possible principally through the support of the Air Force Materials Laboratory–Air Force Systems Command, the Defense Supply Agency, and the American Society of Heating, Refrigerating and Air-Conditioning Engineers, Inc., all of which are gratefully acknowledged.

Inherent to the character of this work is the fact that in the preparation of this volume we have drawn most heavily upon the scientific literature and feel a debt of gratitude to the authors of the referenced articles. While their often discordant results have caused us much difficulty in reconciling their findings, we consider this to be our challenge and our contribution to negative entropy of information, as an effort is made to create from the randomly distributed data a condensed, more orderly state.

While this volume is primarily intended as a reference work for the designer, researcher, experimentalist, and theoretician, the teacher at the graduate level may also use it as a teaching tool to point out to his students the topography of the state of knowledge on the viscosity of fluids. We believe there is also much food for reflection by the specialist and the academician concerning the meaning of "original" investigation and its "information content."

The authors are keenly aware of the possibility of many weaknesses in a work of this scope. We hope that we will not be judged too harshly and that we will receive the benefit of suggestions regarding references omitted, additional material groups needing more detailed treatment, improvements in pre-

sentation or in recommended values, and, most important, any inadvertent errors. If the *Volume Registration Certificate* accompanying this volume is returned, the reader will assure himself of receiving annually a list of corrigenda as possible errors come to our attention.

West Lafayette, Indiana
October 1974

Y. S. TOULOUKIAN
S. C. SAXENA
P. HESTERMANS

Contents

Theory, Estimation, and Measurement

Numerical Data

Material Index

GROUPING OF MATERIALS AND
LIST OF FIGURES AND TABLES

* L = saturated liquid, V = saturated vapor, G = gas.

* L = saturated liquid, V = saturated vapor, G = gas.

---

* L = saturated liquid, G = gas.

* L = saturated liquid, G = gas.

* L = saturated liquid, V = saturated vapor, G = gas.

---

* L = saturated liquid, V = saturated vapor, G = gas.

----

* G = gas.

Theory, Estimation, and Measurement

Notation

a	Root-mean-square radius in equations (50 and 51); Numerical constant
a'	Proportionality constant
A	Atomic weight; Work function for melting point; Numerical constant
A'	Numerical constant
A_{ij}	Parameter [equation (41)]
b	Impact parameter; Van der Waals constant; Numerical constant
B	Numerical constant
c	Numerical constant
c'	Numerical constant
C	Numerical constant
C^1	Numerical constant
C'	Numerical constant
C_{ij}	Parameter [equation (41)]
C_p	Molar specific heat at constant pressure
C_v	Molar specific heat at constant volume
d	Displacement; Diameter
D	Self-diffusion coefficient; Numerical constant
D_{ij}	Diffusion coefficient
E	Total energy; Numerical constant
E_s	Energy of sublimation
E_μ	Numerical constant
ΔE_{vap}	Energy of vaporization
ΔE_{act}	Activation energy
$f(\ )$	Function [equation (64)]
f_0	Resonant frequency
$f_\mu^{(n)}$	Correction factor
F	Numerical constant; Resistance force
F_a^*	Partition function
F_n	Partition function
g	Gravitational acceleration; Initial relative speed [equation (10)]
$g^{(2)}$	Pair correlation function; Equilibrium radial distribution function
G	Force constant of potential energy; Numerical constant
h	Planck's constant
H	Numerical constant
ΔH_{vap}	Enthalpy of vaporization

ΔH_{vb}	Latent heat of vaporization
I	Moment of inertia
k	Coefficient of thermal conductivity [equation (1)]; Boltzmann's constant; Wave vector [equation (105)]
$k°$	Translational thermal conductivity
k_1	Adiabatic compressibility
K	Transmission coefficient; Numerical constant; Bulk modulus
l	Length
L_1, L_2, L_3	Mean absolute deviation, root-mean-square deviation, and maximum absolute deviation from smoothed values [defined in equations (47)–(49)]
m	Mass of a molecule; Numerical constant; Molecular weight
M	Molecular weight
n	Numerical constant; Number of Molecules
N	Avogadro's number; Number of data points
p	Dipole moment [equation (8)]
P	Pressure
P_c	Critical pressure
P_R	Reduced pressure
Q	Numerical constant
r	Radius
R	Neighborhood of the resonant frequency; Radius; Universal gas constant; Resistance [equation (136)]; Numerical constant
s	Displacement
S	Numerical constant
S_i	Collision cross section
t	Time; Temperature, C
T	Absolute temperature, K
T_b	Boiling temperature
T_c	Critical temperature
T_m	Melting temperature
T_R	Reduced temperature
T^*	Reduced temperature
u_s	Speed of sound

U	Numerical constant	λ	Mean free path; Logarithmic decrement; Distance
v	Specific volume; Volume of an atom; Velocity	Λ^*	Reduced de Brogie wavelength
$\bar{v}$	Mean speed	μ	Coefficient of viscosity
V	Molar volume	μ^*	Reduced viscosity
V_f	Free volume	μ°	Viscosity at atmospheric pressure
V_A	Volume of a gram atom	ν	Coefficient of kinematic viscosity
w	Parameter [equation (71)]	ν_0	Molecular vibrational frequency
W	Activation energy; Viscous drag; Apparent weight	ξ	Parameter [equations (31) and (32)]
W^d	Energy dissipated per cycle	π	3.14159...
W^v	Vibrational energy	ρ	Density
x	Displacement	$\bar{\rho}$	Average gas density
x_i	Mole fraction of the ith component	ρ_c	Critical density
x_t^{11}	Double Fourier transform of transverse current–current correlation function	ρ_i	Density of the ith component
		ρ_R	Reduced density
Z	Number of moles of a component; Compressibility coefficient	ρ^*	Reduced density
		σ	Size parameter
α	Molecular mobility; Numerical constant	σ_0	Potential parameter [equations (8) and (9)]
α'	Coefficient of thermal expansion	τ	Period of vibration; Mean life [equations (70) and (71)]
α_{ij}	Interaction parameter		
β	Friction constant; Numerical constant	ϕ	Azimuthal angle
β'	Coefficient [equation (129)]	Φ	Angular deflection [equations (54) and (55)]
γ	Parameter [equation (10)]		
δ	Deviation function; Correction factor; Potential parameter	χ	Deflection angle in a binary collision
		Ψ_2	Coefficient of the Legendre polynomial of order 2
Δ	Logarithmic decrement; Differential increment	Ψ_{ij}	Sutherland coefficient
		Ψ'_{ij}	Parameter [equation (40)]
ϵ	Small correction factor; Measure of intermolecular depth; Potential parameter; Difference in energy	ψ_{ij}	Parameter [equation (4)]
		ω	Angular frequency, angular velocity
		ω_c	Collision frequency
ζ	Orientation factor [equation (8)]	ω_L	Larmor frequency
θ	Einstein characteristic temperature	$\Omega^{(l,n)}$	Viscosity collision integral
θ_i	Mass rate of flow; Angle [equation (8)]	$\Omega^{(l,n)*}$	Reduced viscosity collision integral

Viscosity of Gases and Gas Mixtures

1. INTRODUCTION

An adequate knowledge of viscosity plays a very important role in a variety of interesting engineering problems involving fluid flow and momentum transfer. This much-needed information is scattered throughout the literature, as may be seen from an examination of the many sources cited in [1] for a limited number of materials, either as obtained from an experimental measurement or as values computed according to a certain theoretical procedure. The probability of finding even an approximate value of viscosity decreases considerably as the molecular complexity of the material increases and/or the interest shifts toward extremes in such environmental conditions as temperature, density, magnetic fields, electric fields, etc. The information available for multicomponent systems is meager in comparison with that for pure substances, and in general the theoretical understanding of the phenomenon is less developed for the liquid state than for the gaseous state. Measurements of the viscosity of liquids and their mixtures are quite scarce. In the absence of elaborate experimental information and adequate theoretical understanding of the coefficient of viscosity for fluids and their mixtures, it would be most desirable to critically evaluate the available information and by a judicious interplay of theory and experiment develop, as well as possible, both the standard data and reliable procedures for theoretical calculations. This volume is an initial effort in this broad and general direction. In the first part we review the present state of the art of theory, estimation, and measurement techniques of gases and gas mixtures, and then of liquids and liquid mixtures. The second part deals with the critical evaluation of viscosity data obtained by different workers and different techniques, and lists the recommended values for pure and mixed materials in the gaseous and liquid states. In this entire volume we have implied by the word fluid its traditional meaning, the gaseous and liquid states.

2. THEORETICAL METHODS

A. Introduction

The history of the development of the kinetic theory of gases is both long and interesting. Chapman and Cowling [2] in their classic book give a brief description of this long development of several centuries. Brush, in a series of articles [3–9], has referred in a very original fashion to the contribution of Herapath, Waterson, Clausius, Maxwell, and others. Chapman [10] has delivered a very interesting lecture on the history of development of kinetic theory. The kinetic theory of transport processes is described in different detail and with varying degrees of rigor in a number of textbooks by Kennard [11], Jeans [12, 13], Loeb [14], Saha and Srivastava [15], Present [16], Herzfeld and Smallwood [17], Cowling [18], Knudsen [19], Guggenheim [20], Kauzmann [21], Golden [22], etc. Desloge [23–27] has written a number of articles presenting a pedagogical approach to the theoretical expressions for the transport properties coefficients starting from the Boltzmann transport equation. In their treatises, Chapman and Cowling [2] and Hirschfelder, Curtiss, and Bird [28] have presented a detailed rigorous treatment of the derivation of transport coefficients. Additional works which must be mentioned in this context are those of Mintzer [29], Mazo [3], Liboff [31], Cercignani [32], Waldmann [33, 34], Hochstim [35], and DeGroot [36]. The general theory of irreversible processes is also developed to derive transport coefficients [36–38].

We briefly refer below to the kinetic theory expressions for the coefficient of viscosity as obtained by simple and by more rigorous theories. The simple mean-free-path and the rigorous Chapman–Enskog theories lead to quite different theoretical expressions, but Monchick [40, 41] has successfully developed the interconnection between the two theories and their equivalence.

In Volume 3 of this series, *Thermal Conductivity of Nonmetallic Liquids and Gases* [42], we have described the various theories and the theoretical expressions for the coefficient of thermal conductivity. As the mechanisms of transport of energy and momentum are similar in many ways there is an inherent interconnection between the coefficients of thermal conductivity and viscosity. We will, therefore, when discussing the latter, omit at places certain basic details which have already been given in connection with thermal conductivity [42]. Furthermore, the scope of our present text is to reproduce most of the practical results and refer to all major and relevant works so that consulting the widely scattered literature becomes easier. Many similar efforts of varying scope are referred to later, but mention must be made here of a series of survey articles by Liley [43–46] reviewing the work on transport properties of gases.

B. The Mean-Free-Path Theories

The transport of momentum is considered in a homogeneous gas which is spherically symmetric and monatomic, so that no inelastic collisions occur, and the pressure and density are such that only binary collisions between the gas molecules occur and the collisions between the gas and wall are negligible in comparison to gas–gas collisions. If the temperature is high enough so that the quantum effects are negligible and classical mechanics is adequate, if there is only a small velocity gradient so that $v_{x+dx} = v_x + (\partial v/\partial x)\Delta x$ accurately describes the velocity variation over Δx, and if the temperature is low enough so that the gas is un-ionized, undissociated, and not electronically excited, the simple kinetic theory predicts that

$$\mu = \tfrac{1}{3}\rho\bar{v}\lambda = k/C_v \tag{1}$$

Here μ is the coefficient of viscosity, ρ the density of the molecules, $\bar{v}$ the mean speed, λ the mean free path, k the coefficient of thermal conductivity, and C_v the specific heat at constant volume.

Different numerical factors are found in equation (1) if consideration is given to the dependence of mean free path and collision rate on molecular velocity. A more rigorous calculation gives

$$\mu = \frac{5\pi}{32}\rho\bar{v}\lambda \tag{2}$$

or more precisely

$$\mu = \frac{5\pi}{32}(1 + \epsilon)\rho\bar{v}\lambda \tag{3}$$

where ϵ is a small number whose value depends upon the nature of the intermolecular force field. Thus, ϵ is zero for a Maxwellian gas and increases to 0.016 for a

gas composed of rigid impenetrable spheres. The variation in the numerical coefficient of these relations for viscosity is mainly due to the tendency of the molecules to continue moving in their original direction even after a collision.

If the simple mean-free-path arguments are applied to a mixture consisting of n different gases, the resulting expression for the coefficient of viscosity, μ_{mix}, in terms of the viscosities of the pure components and other quantities, is [11, 47]

$$\mu_{\text{mix}} = \sum_{i=1}^{j} \mu_i \bigg/ \left(1 + \sum_{\substack{j=1 \\ j \neq i}}^{n} \psi_{ij}\frac{x_j}{x_i}\right) \tag{4}$$

where

$$\psi_{ij} = \frac{S_{ij}}{S_i}\frac{[1 + (M_i/M_j)]^{1/2}}{\sqrt{2}} \tag{5}$$

Here μ_i, x_i, and M_i are the coefficients of viscosity, mole fraction, and molecular weight of component i in the mixture, respectively; S_i and S_{ij} are the collision cross sections for molecules of type i and types i and j, respectively. This general form of equation (4) has been extensively studied, both to determine the physical significance of ψ_{ij}, and in the development of methods based on equation (4) which can be used for the estimation of μ_{mix} and which offer different alternatives for equation (5). These will be dealt with later at appropriate places in this chapter.

These results of simple kinetic theory are only of historical importance because estimates based on these expressions are in crude agreement with the directly observed values even for simple systems. The principal limitation of this approach consists in neglecting the effect of intermolecular forces during molecular collisions. In the rigorous approach of Chapman and Enskog this feature is considered and the theoretical expressions for viscosity are derived for a pure gas as well as for multicomponent gas mixtures. These expressions have been further refined in more recent years, as will be briefly described in the next section.

C. The Rigorous (Chapman–Enskog) Theories

The pioneer work of Enskog and Chapman is described in the treatise on the kinetic theory of nonuniform gases by Chapman and Cowling [2]. Many notable efforts have been made since then to reformulate the problem in different ways by adopting different approaches, developing more general and sometimes equivalent and alternative approaches for solving the Boltzmann equation, and deriving the expressions for transport coefficients. It will be in order to refer to some of these efforts: Kirkwood [48, 49], Grad [50, 51], Kumar [52, 53], Green [54–56], Green and Piccirelli [57], Hoffman and Green [58], Snider [59], Mazur and Biel [60], Su [61], McLennan [62], Garcia-Coling, Green, and Chaos [63], Fujita [64], Bogoliubov [65, 66], Desai and Ross [67], and Tip [1172]. Montroll and Green [68] have reviewed various efforts aimed at developing the statistical mechanics of transport processes. Grad [69–71] has introduced a very strong approach to the formulation of transport coefficients of dilute gases. Zwanzig [72] reviewed the formulation of transport coefficients in terms of time-correlation functions. Model calculations have also been used in kinetic theory to simplify many of the complicated aspects while retaining all the essential features: see Bhatnagar, Gross, and Krook [73], Welander [74], Gross and Krook [75], Gross and Jackson [76], Sirovich [77], Enoch [78], Hamel [79], Willis [80], and Holway [81]. We refer to studies which have derived expressions for the coefficient of viscosity for pure gases and their mixtures of increasing molecular complexity and under different environmental conditions of temperature, pressure, etc. It is also appropriate to mention a recent article by Mason [82], who has reviewed the present art of calculation of transport coefficients in neutral gases and their mixtures.

a. Pure Monatomic Gases

The theoretical first-approximation Chapman–Cowling expression for the coefficient of shear viscosity of a pure monatomic gas under the same assumptions as mentioned above is [2, 28]

$$[\mu]_1 = \frac{a\sqrt{MT}}{\sigma^2 \Omega^{(2,2)*}} \tag{6}$$

Here $\sigma^2 \Omega^{(2,2)*}$ is the viscosity collision cross section, σ is a size parameter, and $\Omega^{(2,2)*}$ is a function of the reduced temperature $T^* = kT/\epsilon$. ϵ is a measure of the depth of the attractive part of the intermolecular potential, T the temperature, and k the Boltzmann constant. The quantity a is a numerical factor and if μ be expressed in g cm^{-1} sec^{-1}, σ in Å (10^{-8} cm), T in degrees K, its value is 266.93×10^{-7}.

The higher approximations to μ are represented in terms of $[\mu]_1$, the nth approximation being

$$[\mu]_n = [\mu]_1 f_\mu^{(n)} \tag{7}$$

$f_\mu^{(n)}$ has been evaluated up to $n = 3$ and found to be very feebly dependent on the nature of the intermolecular potential for moderate temperature ranges and not much different from unity [28]. The expression

for $f_\mu^{(3)}$ according to the procedure of Chapman and Cowling [2] is quite complicated, and Kihara [83] has developed an alternative scheme for representing the transport coefficients as an infinite series. The latter procedure approximates the actual intermolecular potential as a perturbation to the Maxwellian model. Joshi [85, 86] on the other hand has developed another approximation scheme in which the actual potential energy function is regarded as a perturbation over the rigid-sphere model and has derived the expressions for $f_\mu^{(2)}$ and $f_\mu^{(3)}$. In either formulation the higher-order approximation correction factors are simpler than those derived by the method of Chapman and Cowling [2, 28], and a tabulation of $f_\mu^{(2)}$ is available for the Lennard-Jones (12-6) potential on the Kihara approximation scheme [87].

b. Multicomponent Systems of Monatomic Gases

The general expression for the first approximation to viscosity of a multicomponent mixture is derived by Curtiss and Hirschfelder [88]. The higher second and third Chapman–Cowling approximations have been derived by Saxena and Joshi [89, 90] and Joshi [91], respectively. The Kihara approximation procedure has been extended by Mason [92], and the theoretical expression for a binary gas system on the Kihara–Mason scheme is derived by Joshi and Saxena [93]. The general characteristics of a gas mixture have been discussed by Waldmann [94] on the basis of the first-approximation Chapman–Cowling theoretical expression for the viscosity coefficient. Hirschfelder, Taylor, Kihara, and Rutherford [95] have theoretically examined the conditions under which the viscosity of a binary mixture will exhibit either a maximum or a minimum in the plot of viscosity *versus* composition of the mixture. They [95] have based their studies on the first-approximation Chapman–Cowling expression. Kessel'man and Litvinov [1158] have described the calculation of multicomponent viscosity from the first-approximation theoretical expression in conjunction with a Lennard-Jones (12–6) intermolecular potential with parameters regarded as depending on temperature. Barbe [1160] has developed automatic computer calculation procedure for multicomponent viscosity based on the kinetic theory expression.

c. Nonpolar Polyatomic Gases and Multicomponent Systems

The transport theory of polyatomic gases is much more complicated than that of monatomic gases, for two reasons. First, the intermolecular potential is not central for polyatomic systems and due consideration must be given to its orientation or direction dependence. Second, the collisions are not all elastic and various complications associated with inelastic collisions must be properly considered. Consistent with the general style and scope of this text we refer briefly below to the various efforts made to resolve the overall understanding of the momentum transfer process in the above two categories.

Curtiss and co-workers [96–100] have developed the classical theory of nonspherical molecules by suitably modifying the Boltzmann equation and considering only the rotational motion. Curtiss [96] applied the perturbation technique of Chapman–Enskog and solved the Boltzmann equation to derive expressions for the transport coefficients which may be regarded as referring to rigid convex nonspherical bodies in which the center of mass is a center of symmetry. Curtiss and Muckenfuss [97] specialized the calculations [96] to a *spherocylindrical* model and presented results for shear viscosity as a function of two parameters characterizing the shape and mass distribution of the molecule. These calculations have also been extended to multicomponent mixtures [98] and further examined in detail including rigid convex nonspherical molecules with symmetric-top mass distributions [99, 100]. Others who have considered this molecular model are Sandler and Dahler [101] and Kagan and Afana'sev [102]. Another molecular model which has been studied in detail and for which the coefficient of viscosity is derived is the loaded sphere [103, 104]. Historically, the molecular model having internal energy, first studied by Pidduck [105], consisted of perfectly rough, elastic, rigid spherical molecules. For such molecules the energy of translation and the energy of rotation are interconvertible [2]. In more recent years the kinetic description of such a dilute gas of perfectly rough spheres was developed in considerable detail by Condiff, Lu, and Dahler [106], McLaughlin and Dahler [107], and Waldmann [108]. Dahler [109] made some interesting comments concerning the developments in the transport theory of polyatomic fluids. Pople [110, 111] has treated the interaction between nonspherical molecules as consisting of a central part and directional terms of various angular symmetries. He considered in particular the axially symmetric molecules. Attempts [28, 112, 113] have been made to further extend such an approach, but mainly equilibrium thermodynamic properties have been computed. The transport properties of gases with

rotational states have also been examined by McCourt and Snider [114, 115] and Kagan and Maksimov [116]. Studies have been made of transport phenomena in diatomic gases [117], the probability for rotational energy transfer in a collision [118], the relation between angular distribution and transport cross sections [119], etc. The subject of molecular friction in dilute gases has been discussed by Dahler and co-workers [120–122]. Bjerre [123] has derived the expressions for shear viscosity starting from the theory of Curtiss and Muckenfuss [96–98] and specializing them for a model appropriate for planar molecules. Other molecular models have been developed by Morse [124] and Brau [125] to account for the collision term in the kinetic equation for polyatomic gases.

The topic of molecular collisions in polyatomic molecules has received considerable attention both theoretically and experimentally. Here we refer only to a series of articles written by Curtiss and co-workers [126–135, 1164–1170] on this subject, which deals with collisions between diatomic and polyatomic molecules and considers both rotational and vibrational excitations. Wang Chang and Uhlenbeck [136, 137] developed a formal theory of transport phenomena in dilute polyatomic gases. They treated the problem semi-quantum-mechanically, treating the translational motion of the molecules classically and the internal motion quantum-mechanically. This enabled them to assume the existence of quantum inverse collisions. Furthermore, they considered two cases: one in which the energy exchange between the translational and internal degrees of freedom is easy [136], and the other extreme case in which such an energy transfer is quite rare [137]; see also Wang Chang, Uhlenbeck, and de Boer [138]. However, the Wang Chang–Uhlenbeck equation is much more complicated than the Boltzmann equation, and an attempt by Finkelstein and Harris [139] to linearize the former is interesting. They used the geometrical technique of Finkelstein [140]. Hanson and Morse [141] have developed the kinetic model equations for a gas with internal structure by employing a modified diagonal approximation and the Wang Chang–Uhlenbeck equation. A classical theory of transport phenomena in dilute polyatomic gases is developed by Taxman [142] as an extension of the Chapman–Enskog theory for monatomic gases [2]. This theory [142] is also the classical limit of the work of Wang Chang and Uhlenbeck [137].

The formal theory of Wang Chang and Uhlenbeck [136–138] and of Taxman [142] has been very cleverly simplified by Mason and Monchick [143] and Monchick, Yun, and Mason [144, 145] to derive expressions for transport coefficients. They have neglected terms arising from considerations of inelastic collisions which are small and expressed the others in terms of measurable quantities. The potential of this procedure is also successfully tested in predicting the other transport properties [146–148]. A similar success is demonstrated for the loaded sphere model calculations of the thermal diffusion factor [149, 150]. Alievskii and Zhdanov [151] have discussed the transport phenomena in mixtures of polyatomic gases. Curtiss [1171, 1193] has recently derived an explicit classical expression for the viscosity of a low-density gas of rotating and nonvibrating diatomic molecules. Stevens [1173] performed calculations for methane including inelastic collisions and introducing approximations in the calculation of transport cross sections. He found that viscosity is hardly influenced by inelastic effects.

d. Pure Polar Gases and Multicomponent Systems

The properties of polar gases are hard to calculate because the interaction between two molecules depends on their relative orientations and the calculation of molecular trajectories for angle-dependent potentials is not easy. The occurrence of inelastic collisions and resonant transfer of internal energy complicates the analysis considerably. The nonspherical shape of the molecules gives rise to short-range orientation-dependent overlapping repulsive forces. The attractive force between polar molecules arises from three different sources: dispersion, the interaction between permanent electrostatic distributions (dipoles and higher multipoles), and interactions arising from electric moments induced by the permanent moments of other molecules. A detailed discussion of this topic is given by Buckingham and Pople [156, 157], Saxena and Joshi [158], and Hirschfelder, Curtiss, and Bird [28].

Krieger [159] assumed the following type of Stockmayer potential [160, 161] to correlate and estimate the viscosities of polar gases:

$$\Phi(r) = 4\epsilon\left[\left(\frac{\sigma_0}{r}\right)^{12} - \left(\frac{\sigma_0}{r}\right)^6\right] - \left(\frac{p^2}{r^3}\right)\zeta \qquad (8)$$

where

$$\zeta = 2\cos\theta_1\cos\theta_2 - \sin\theta_1\sin\theta_2\cos\phi$$

Here p is the dipole moment of the molecule, ζ is an orientation factor in which θ_1 and θ_2 are the angles of inclination of the two dipole axes to the line joining the centers of the molecules, and ϕ is the azimuthal angle between them. In the limit when $p \to 0$, $\phi(r)$ is just the Lennard-Jones (12–6) potential, and ϵ and σ_0 are the potential parameters. Krieger [159] further assigned a constant value of 2 to ζ, which implies that the dipoles maintain an attractive end-on position, corresponding to the maximum attractive orientation, throughout their encounter. This assumption transforms the above angle-dependent potential into the following central potential:

$$\Phi(r) = 4\epsilon[(\sigma_0/r)^{12} - (\sigma_0/r)^6 - \delta(\sigma_0/r)^3] \qquad (9)$$

where

$$\delta = p^2/2\epsilon\sigma_0^3$$

Krieger [159] evaluated the viscosity collision integral for the reduced temperature range, T^*, from 1.0 to 512 and for nine equally spaced δ values from 0.00 to 2.00. He [159] correlated the viscosity data for twelve polar gases and determined the values of the potential which he found inadequate for highly polar gases. Liley [163] made certain comments concerning the accuracy of the tabulated viscosity collision integral by Krieger [159] and presented a retabulation for the low temperature range, $T^* = 0.70$ to 5.00. More detailed calculations of Itean, Glueck, and Svehla [164] confirmed an error in the original calculations of Krieger [159]. However, the Itean *et al.* [164] corrected calculations give only unreasonable values for the potential parameters if experimental data are fitted with the theoretical predictions on this model.

Monchick and Mason [165] argued that in Krieger's model all repulsive orientations are neglected, and the orientation of aligned dipoles of maximum attraction and rotational energy is the one in which the molecules spend the least amount of time; hence this model may be unrealistic. They suggested a model in which all relative orientations are accounted for but still the dipole field is replaced by a central field. The Monchick and Mason [165] model assumes that the molecular trajectories are insignificantly affected by the inelastic collisions even when they occur quite frequently. They justify this on the consideration of energy grounds because the rotational energy at ordinary temperatures is much smaller than the translational kinetic energy, which is of the order of kT. This assumption is likely to be reasonable for shear viscosity because of the small contribution of inelastic collisions to momentum transport [145]. This assumption simplifies the theoretical expression of μ given by Taxman [142] so that

$$\frac{1}{\mu} = \frac{8}{5(\pi m k T)^{1/2}} \int [(1 - \cos^2 \chi)b \, db \, d\phi]$$
$$\times \exp(-\gamma^2)\gamma^7 \, d\gamma \qquad (10)$$

where

$$\gamma^2 = \left(\frac{m}{4kT}\right)g^2$$

Here m is the mass of a molecule, k the Boltzmann constant, T the temperature, χ the deflection angle in a binary collision, b the impact parameter, ϕ the azimuthal angle, and g the initial relative speed. Equation (10) is the same as that obtained for no internal degrees of freedom.

Monchick and Mason [165] further argued that the relative orientation of the molecules over a small range around the distance of closest approach remains almost constant, and the angle of deflection is primarily and mainly controlled by this particular relative orientation rather than by all the possible orientations assumed along the entire trajectory from $t = -\infty$ to $t = +\infty$. The work of Horn and Hirschfelder [166] also supports this point of view. The idea of a fixed relative orientation during a collision leads one, in actual calculation, to treat ζ as a constant ζ_0 (value of ζ at the distance of closest approach) and thus replace Φ by a multiplicity of central field potentials corresponding to all values of ζ_0 between -2 and $+2$. The collision integrals are then calculated for each of these potentials and average values are determined by giving the proper weight of the potential. The latter is essentially the probability of the collision taking place along that potential. The viscosity is then computed by the same expression as that for nonpolar gases except that $\Omega^{(2,2)*}$ is replaced by the average value, $\langle \Omega^{(2,2)*} \rangle_{av}$, obtained according to the above procedure. This is a valid approach for all orders of the kinetic-theory approximations as shown by Mason, Vanderslice, and Yos [167]. Mason and Monchick [168] have extended this model with reasonable success for the computation of the viscosities of mixtures. Singh and Das Gupta [1162] have analyzed the data on polar gases according to a simple preaveraged 12–6–6 intermolecular potential. They [1163] have also studied the properties of binary mixtures of polar gases where one component has a predominance of dipole moment while the other has a quadrupole moment only.

e. Quantum Effects

The calculation of viscosity of light gases at low temperatures is complicated because of the appearance of quantum-mechanical diffraction and statistical effects [28]. The collision cross sections must now be computed using quantum mechanics instead of classical mechanics [169, 170]. It also becomes imperative to work through the quantum-mechanical version of the Boltzmann equation as given by Uehling and Uhlenbeck [171]. Considerable progress has been made in both of these directions, and an excellent review on the subject by Buckingham and Gal [172] has appeared. Here we refer to some of the pertinent works which may prove specially useful in the art of computing viscosities of gases at low temperatures.

Detailed discussions of derivations of the Boltzmann equation using different quantum-mechanical approaches are available in two recent review articles by de Boer [173] and Mori, Oppenheim, and Ross [174]. Other interesting derivations have appeared since then: Waldman [175], Snider [176, 177], Hoffman, Mueller, and Curtiss [178], and Hoffman [179]. Mention may be made of the diagram technique of Prigogine and co-workers [180–182] in handling the transport equation in quantum gases. Quantum-mechanical kinetic theory has been worked out in detail by Mueller and Curtiss [183, 184] for a gas of loaded spheres. de Boer and Bird [185, 186] have derived correction factors to be applied to the classical collision integrals to estimate the quantum effects. Their calculations are valid for relatively high temperatures (above the reduced temperature, T^*, of five) and for a monotonic decreasing intermolecular potential function [187]. Choi and Ross [188] have calculated the first-order quantum correction by solving without any approximation the equation of motion of a two-particle system and have estimated the magnitude by assuming a simple model for molecular interactions. Buckingham and Gal [172] have computed the quantum corrections assuming the Buckingham–Corner [189] intermolecular potential. Imam-Rahajoe, Curtiss, and Bernstein [190] and Munn, Smith, Mason, and Monchick [191] have determined the contribution of quantum effects to the transport cross sections assuming a Lennard-Jones (12–6) intermolecular potential function. More detailed calculations of the phase shifts and quantum corrections to transport corrections have been made in recent years by Curtiss and Powers [192], Wood and Curtiss [193], Munn, Mason, and Smith [194], Smith, Mason, and Vanderslice [195], Bernstein,

Curtiss, Imam–Rahajoe, and Wood [196], and Aksarailian and Cerceau [1161].

A number of calculations have been made on the isotopic varieties of lighter gases (helium and hydrogen) and their mixtures. This is because quantum corrections are expected to be large for such systems and many of these have been experimentally studied. We mention here several such efforts. Assuming the interaction model to be of rigid-sphere type, Massey and Mohr [197] calculated the quantum collision cross sections and collision integrals. This work followed a series of investigations for He^4 assuming different types of molecular interactions. Massey and Mohr [198] and Massey and Buckingham [199] did calculations using the Slater interaction potential [201]; Buckingham, Hamilton, and Massey [202] for six different potentials; de Boer [203], Keller [204], Monchick, Mason, Munn, and Smith [205], and Larsen, Witte, and Kilpatrick [206] for the Lennard-Jones (12–6) potential [207]. Keller [204] has considered the modified exp-six potential derived by Mason and Rice [208]. Similar calculations have been made for He^3 by Buckingham and Temperley [209], de Boer and Cohen [210], Buckingham and Scriven [211], Cohen, Offerhaus, and de Boer [212], Halpern and Buckingham [213], Keller [204], and Monchick *et al.* [205]. Some of these authors have also discussed the properties of the mixtures of He^3 and He^4 [214].

A number of interesting calculations have been made on the isotopes of hydrogen. Cohen, Offerhaus, Leeuwen, Roos, and de Boer [215] computed the viscosities of ortho- and para-hydrogen assuming a spherically symmetric Lennard-Jones (12–6) type of interaction potential [207]. A similar investigation is due to Buckingham, Davies, and Gilles [216], who approximated the force field by a Buckingham–Corner type potential [189]. Takayanagi and Ohno [217] and Niblett and Takayanagi [218] have further extended the scope of these calculations by considering the nonspherical potential. Waldmann [219] has discussed the kinetic theory of para–ortho-hydrogen mixtures, for which Hartland and Lipsicas [220] have made some interesting comments. Diller and Mason [221] have calculated the transport properties of H_2, D_2, HD, and some of their mixtures employing a Lennard-Jones (12–6) potential.

Calculations of the viscosity of atomic hydrogen at low temperatures have also been made by several workers: Buckingham and Fox [222], Buckingham, Fox, and Gal [223], Buckingham and Gal [172], Browing and Fox [224], etc. Konowalow,

Hirschfelder, and Linder [225] have computed the viscosity of oxygen and sulfur atoms from the potential energy curves at large separations. It may be pointed out that the low-temperature viscosity studies help in the understanding of the operation of low-density free jets such as those which occur in space vehicles and low-density wind tunnels [226].

f. High-Temperature Calculations

The calculation of viscosity at high temperatures is of particular interest to design engineers and to the outer space exploration program. The computation is tedious because with increasing temperature, internal energy excitations, electronic excitations, dissociation, and various degrees of ionization must be considered. Multiplicity of intermolecular potentials, nonequilibrium between the electron and heavy-particle temperatures, appearance of quantum corrections for high-density plasmas and, at extremely high temperatures (above 10^6 K), for low-density plasmas, and resonant charge exchange between ions are the main factors making the calculation of transport properties at high temperatures difficult. However, many significant improvements have been made in recent years, and in many cases reliable estimates of viscosity are possible up to high temperatures of practical need. Many review articles and books, differing in scope and emphasis, summarize these developments, e.g., Chapman and Cowling [2], Hochstim [35], Spitzer [227], Ahtye [228, 229], and Brokaw [230].

The kinetic equations and the calculation of transport properties of ionized gases and plasmas have been recently reviewed in a series of articles by Tchen [249], Lewis [250], and Hochstim and Massel [251]. Here we will refer very briefly to some of the work which is of direct relevance to the calculation of viscosity of gases under partial or complete ionization.

The calculation of viscosity at high temperatures is easy if the contributions of internal degrees of freedom, electronic excitations, dissociation, and ionization are ignored. Under such assumptions the theory of Chapman and Enskog [2, 28] may be used if the molecular interactions and corresponding viscosity collision integrals are known. Amdur and Mason [231], Kamnev and Leonas [232], and Balyaev and Leonas [233] adopted this approach and predicted properties of rare gases and homonuclear diatomic gases, hydrogen, nitrogen, and oxygen, up to 15,000 K. In each case the interaction potentials were determined by experiments on the elastic scattering of fast molecular beams. Amdur, in a series of articles

[234–236], has explained the limitations of such an approach and their effect on the calculated values of transport coefficients. Brokaw [237] has discussed the role of viscosity in calculating the convective heat transfer in high-temperature gases. Yos [238, 239] has computed the viscosity of hydrogen, nitrogen, oxygen, and air in the temperature range 1000–30,000 K and for pressures from 1 to 30 atm. The values for the fully ionized case were made to agree with those of Spitzer and Harm [240]. The viscosity of dissociating gases has been computed by Mason and co-workers with the assumption of no ionization and no electronic excitation for hydrogen [241, 242], nitrogen and oxygen [243], and air [244, 245]. Krupenie, Mason, and Vanderslice [246] have computed the viscosity of Li + Li, Li + H, and O + H systems in the temperature range 1000–10,000 K. Belov and Klyuchnikov [247] have also considered the viscosity of the weakly ionized LiH plasma in the temperature range 1000–10,000 K and at five pressure levels. The viscosity values of alkali metal vapors have been computed by Davies, Mason, and Munn [248]. Belov [1156] has computed the viscosity of partially ionized hydrogen in the temperature range of 6000–30,000 K and for pressures of 0.001, 0.01, 0.1, 1, and 10 atm. The effect of charge transfer is included.

It was observed by Ahtye [229] that for ionized gases higher Chapman–Enskog approximations are needed because the convergence of the infinite series representing the transport coefficients is poor due to the small mass of the electron. Devoto extended the formulation of viscosity to include second [252], third [253], and even higher approximations [254, 255]. In view of the great complexity of these expressions, Devoto [256] has also attempted to simplify them, and has assessed the adequacy of these simple expressions by performing actual calculations for partially ionized argon.

A number of other interesting developments have been made which facilitate the calculation of viscosity at high temperatures in general. Mason and Sherman [257] have made estimates of the cross sections for symmetric resonant charge exchange between ions differing by one electronic charge. Chmieleski and Ferziger [258] have presented a modified Chapman–Enskog approach for an ionized gas where heavy particle and electron temperatures are allowed to differ, though up to zero order all species have the same macroscopic velocity. This inequality of temperature is caused mainly by the fact that the relaxation time for energy exchange

between the heavy and light species is much larger than the time for each individual species to acquire equilibrium with itself. In the limit of equal temperature for electron and ion, these expressions are identical with the results obtained adopting the Chapman–Enskog approach. Sandler and Mason [259] have considered a scheme for the solution of the Boltzmann equation which converges more rapidly than the usual Chapman–Cowling procedure [2]. They considered a particular gas system called an almost-Lorentzian mixture, where the mass of one component is far greater than the other and the proportion of the lighter component in the mixture is smaller than that of the heavier component. A partially ionized gas mixture constitutes such a system. Hahn, Mason, Miller, and Sandler [1192] have made calculations to determine the contributions of dynamic shielding to the transport properties of partially ionized argon both at low and high degrees of ionization. Meador [260] has discussed a collision model, which is similar in many respects to a Lorentz gas, for an ionized gas plasma.

A number of calculations have been made of the transport properties in general and viscosity in particular of ionized gases as a function of temperature and pressure. Some of these will be quoted here. Devoto and Li [261] have tabulated the viscosity of partially ionized helium in chemical equilibrium at pressures of 0.01, 0.1, 1, and 5 atm and for temperatures ranging from 4000 to 30,000 K. Kulik, Panevin, and Khvesyuk [262] have reported the computed values of viscosity of ionized argon in the temperature range 2000–30,000 K and for pressure levels of 1, 0.1, 0.01, 0.001, and 0.0001 kg/cm^2. Devoto [263] has graphically reported the viscosity values of equilibrium partially ionized krypton and xenon covering temperatures between 2000 and 20,000 K at pressures of 0.01, 0.1, 1, and 10 atm. Devoto [264, 265] has also tabulated the viscosity values for partially ionized hydrogen at these four pressure levels but for temperatures ranging up to 50,000 K. Grier [266] has given tabulations of transport properties of ionizing atomic hydrogen.

Mason, Munn, and Smith [267] have used repulsive and attractive screened coulomb potentials to represent interactions among charged particles in an ionized gas. They have computed the classical Chapman–Enskog collision integrals for these potentials over a wide range of reduced temperatures, the latter being equivalent to a wide range of electron densities and temperatures. This work has also included a discussion of quantum effects at high densities and temperatures. This work supersedes the earlier computation of collision integrals for repulsive screened coulomb potentials by Smith, Mason, and Munn [268]. Beshinske and Curtiss [269] have recently initiated the study of a dense fluid of molecules composed of nuclei and electrons with purely coulomb interaction potentials.

Dalgarno and Smith [270] have calculated the viscosity of atomic hydrogen for temperatures up to 10^5 K and estimated that the classical calculations are adequate for temperatures above 100 K; below this temperature quantum corrections are important. Dalgarno [271] has also shown that the effect of quantum symmetry on viscosity cross section is small for the collision of two similar particles. It is also appropriate to mention the calculations of momentum transfer and total and differential cross sections for scattering from a coulomb potential with exponential screening by Everhart and co-workers [272, 273].

g. High-Density (or Pressure) Calculations

The calculation of viscosity of a dense gas becomes very complicated because of the possibility of occurrence of more than two particle collisions and the transfer of momentum from the mass center of one particle to another through the action of intermolecular forces [2, 28]. These two effects are briefly referred to as "higher-order collisions" and "collisional transfer of momentum," respectively. David Enskog's [2] efforts are pioneering contributions to the study of dense gases. He modified the Boltzmann equation and applied it to a dense gas of rigid spherical molecules. Since then this molecular model has been extensively studied because for such molecules the probability of multiple collisions is negligible and the collisions are instantaneous [2]. Curtiss [274] and Cohen [275–277] have briefly referred to the various efforts made to understand the transport behavior of a dense gas, and a more detailed review on the subject by Ernst, Haines, and Dorfman [278] has recently appeared. We now cite the different works which have helped in the understanding of this difficult subject and may also help in the prediction of viscosity of moderately dense or dense gases in general. A few attempts to examine the individual gases are also mentioned.

As in the case of a theory for dilute gas, here also for a dense gas an appropriate development of transport theory involves the formulation of an alternative or modification to the Boltzmann equation. Many attempts have been made in this direction by Bogoliubov [65], Cohen [280], Sengers and Cohen [281], Cohen [282, 285], Green and Piccirelli [57], Piccirelli [286], García-Colin [287], and others, as discussed below. It may be pointed out that an interesting question concerning the appropriate definition of temperature arises in the kinetic theory of dense gases. Two temperature definitions are possible, based either on the kinetic or total energy densities. The latter includes the molecular-interaction potential energy. This is discussed by García-Colin and Green [288] and Ernst [289]. The two definitions are equivalent as far as the coefficient of shear viscosity is concerned, but only the second definition is consistent with the irreversible thermodynamics [289].

We now mention some simple kinetic-theory approaches which have been developed to understand the transport processes in dense fluids, in certain cases for specialized molecular interactions. Dymond and Alder [290] developed a theory for transport coefficients on the basis of the van der Waals concept of a dense fluid. Making certain simplifying assumptions about the pair distribution functions, Longuet-Higgins and Pople [291] and Longuet-Higgins and Valleau [292] have derived an expression for the shear viscosity of a dense fluid of hard spheres, and Valleau [293] for rough spheres exerting no attractive forces. Longuet-Higgins and Valleau [294] developed the theory for a dense gas whose molecules attract each other according to a square-well potential, and Valleau [295], Naghizadeh [296], and McLaughlin and Davis [297] extended the theory to mixtures. McCoy, Sandler, and Dahler [298] have also worked out the theory of a dense gas of perfectly rough spheres including the effect of rotational degrees of freedom. Sandler and Dahler [299] have computed from their theory the shear viscosity for a dense gas of loaded spheres. Sather and Dahler [300] have considered a dense polyatomic fluid whose molecules interact with impulsive forces and derived, among other transport coefficients, the expression for shear viscosity. Some other authors who have used statistical mechanics to study the kinetic theory of a dense gas composed of rigid spherical molecules are O'Toole and Dahler [301] and Livingston and Curtiss [302]. Ono and Shizume [303] discuss the transport coefficients of a moderately dense gas on the basis of the statistical mechanics of irreversible processes.

Snider and Curtiss [304] developed the kinetic theory of moderately dense gases by ignoring the effect of three-body collisions and considering the collisional transfer of momentum arising from the distortion of the radial distribution function [305].

Their expressions when evaluated for the limiting case of a rigid-sphere gas give the same results as those of Enskog [2, 28]. These expressions were simplified by Snider and McCourt [307] and evaluated for a case where molecules interact according to an inverse power potential. Curtiss, McElroy, and Hoffman [308] have performed the numerical calculations of the first- and second-order density corrections to the transport coefficients of a gas, assuming a Lennard-Jones (12–6) interaction potential. Starting from a generalized Boltzmann equation valid to all orders in density [57] and adopting a method similar to that of García-Colin, Green, and Chaos [63], García-Colin and Flores [309, 310] have derived the expressions for shear viscosity to terms linear in density for a moderately dense gas.

Stogryn and Hirschfelder [312] have developed a theory to compute the initial pressure dependence of viscosity. They approximated the three-body collisions effectively by a two-body collision between a monomer and a dimer. The fractions of molecules in bound and metastable states are calculated according to procedures outlined by Hill [313, 314] and Stogryn and Hirschfelder [312, 315]. The contribution of collisional transfer is obtained by a semiempirical modification of the Enskog theory [2, 28]. This theory has been applied to explain many experimentally observed facts with reasonable success [316, 320].

Singh and Bhattacharyya [321] have derived the relation for computing the viscosity of moderately dense gases with appreciable quadrupole moments. Their approach is similar to that developed by Stogryn and Hirschfelder [312]; they assumed equal probability for all the relative orientations of the interacting quadrupoles and employed equilibrium constants for dimerization for quadrupolar gases as evaluated by Singh and Das Gupta [322]. Singh and Manna [323] have presented a similar formulation for moderately dense dipolar gases using the equilibrium concentrations of dimers as evaluated by Singh, Deb, and Barua [324]. Kim and Ross [325], on the other hand, have developed a theory for moderately dense gases in which, though the contribution of collisional transfer is neglected, a more complicated picture of a triple collision is considered by including in the calculation what they call quasi-dimers due to orbiting collisions, in addition to bound and metastable dimer states.

Curtiss and co-workers have developed the theory for dense gases as an improvement of their theory for moderately dense gases [304] by including the contribution of three-body collisions, as have Hollinger and Curtiss [326], Hollinger [327], and Hoffman and Curtiss [328, 330]. Bennett and Curtiss [331] have recently derived the transport coefficients for mixtures on the basis of a modified Boltzmann equation, considering the effects from both collisional transfer and three-body collisions. The various collision integrals which appear in this formulation are evaluated numerically for the Lennard-Jones potential. In this formulation the effect of bound pairs is not included; it is probably small at higher temperatures. Sengers, in a series of papers [332–336, 1174], has discussed how the expressions for transport coefficients change if details of collisions are properly accounted for. On including certain types of recollisions and cyclic collisions he finds a divergence in the density expansion of the transport coefficients. This particular topic has been discussed in recent years by Dorfman and Cohen [337, 338], Dorfman [339, 340], Stecki [341], Andrews [342, 343], Fujita [344, 345], and Ernst, Haines, and Dorfman [278] in considerable detail. Sengers [346, 347], Hanley, McCarty, and Sengers [348], and Kestin, Paykoc, and Sengers [1175] have considered the experimental data on viscosity of gases and their parametric dependence on the density of the gas. Hoffman, Mueller, and Curtiss [178], Imam-Rahajoe and Curtiss [349], Grossmann [350–353], Grossman and Baerwinkel [354], Fujita [355], and Morita [357] have discussed the various features of dense gases from the viewpoint of quantum mechanics.

Another approach used to study the density dependence of transport coefficients in a moderately dense gas is based on expressions in terms of time-correlation functions. The developments of this approach and the various methods used in recent years have been reviewed by Zwanzig [72], Helfand [358], Ernst, Haines, and Dorfman [278], and Ernst [279]. Reference is made to the efforts of Kawasaki and Oppenheim [359–362], Frisch and Berne [363], Storer and Frisch [364], Prigogine [365], Ernst, Dorfman, and Cohen [366, 367], Ernst [368, 369], Zwanzig [371, 372], Weinstock [373–378], and Goldman [379], whose work has helped very much in the development of the theory of dense gases.

The various procedures used to derive the theoretical expressions for the transport coefficients of a moderately dense gas, based either on a generalized Boltzmann equation and the distribution function approach or the correlation function approach, have been compared by a number of workers such as García-Colin and Flores [380], Chaos and García-Colin [381], Stecki and Taylor [382], Prigogine and

Resibois [383], Resibois [384–387], Brocas and Resibois [388], and Nicolis and Severne [389]. Mo, Gubbins, and Dufty [1187] have developed a perturbation theory for predicting the transport properties of pure fluids and their mixtures. Good agreement is reported between the calculated and experimental viscosity values of both pure and mixed dense gases and liquids. Attempts have also been made in recent years by Tham and Gubbin [1188] and Wakeham, Kestin, Mason, and Sandler [1189] to extend the Enskog theory of dense gases to multicomponent mixtures. The theory is found to agree with the available experimental data.

h. Magnetic- and Electric-Field Effects

A good way of determining the contribution of the nonspherical shape of polyatomic molecules to the transport processes is to study the effects of magnetic and electric fields. In 1930 Senftleben [390] experimentally examined the effect of magnetic field on the thermal conductivity of paramagnetic diatomic gases. A similar investigation was made in relation to shear viscosity [391–394] and a number of other studies were made about the same time [395–400]. A simple mean-free-path kinetic theory to explain this magnetic-field dependence in paramagnetic gases was developed by Gorter [401] and Zernike and Van Lier [402]. In the externally applied magnetic field, the magnetic moment causes the molecular axis to precess around it with a Larmor frequency, ω_L. Thus, the changing orientation of the axis between collisions alters the effective collision cross section, and the net effect of the external field is to introduce an additional averaging over different orientation. It is also evident in this picture that collision frequency ω_c, and hence pressure, should be a controlling factor, and indeed this effect is found to be dependent upon the ratio of the field to the pressure of the gas. Thirty-two years later Beenakker, Scoles, Knaap, and Jonkman [403] showed that the transport properties of any polyatomic gas are influenced by the presence of an external magnetic field; hence in recent literature this phenomenon has been referred to as the "Senftleben–Beenakker" effect. The first measurement [403] was confined to nitrogen up to 21 kOe at pressures of 12.2 and 5.4 mm Hg. Since this preliminary work, the viscosity of many other gases has been studied. For example, O_2, NO, CO, normal H_2 and D_2, para-H_2, ortho-D_2, HD, CH_4, CF_4, and CO_2 have been studied by Korvig, Hulsman, Knaap, and Beenakker [384, 406]. In a smilar fashion the thermal conductivity of nonspherical gases (H_2, D_2,

O_2, N_2, CO, and CO_2) is altered in the presence of an external magnetic field, as shown by the experiments of Gorelik and Sinitsyn [407] and Gorelik, Redkoborodyi, and Sinitsyn [408].

Efforts to develop a more rigorous theory to explain the effects of external field, starting from a rigorous Boltzmann equation [108, 114, 409] and adopting a procedure somewhat parallel to that of Chapman and Enskog, have been made by Kagan and Maksimov [116, 410], McCourt and Snider [411], Knapp and Beenakker [412], Tip [413], Levi and McCourt [414], Tip, Levi, and McCourt [415], Tip [416], and Hooyman, Mazur, and de Groot [417]. These theoretical studies also established that energy and momentum transport will also occur perpendicular to the directions of external field and gradient. Korvig, Hulsman, Knaap, and Beenakker [418] have reported experimental results of this transverse effect in the case of viscosity for O_2, N_2, and HD at room temperature. The experimental work of Kikoin, Balashov, Lazarev, and Neushtadt [419, 420] on oxygen and nitrogen has shown the necessity of more detailed study of this transverse effect. In the last few years many additional investigations have been made to understand the effect of external magnetic field on the transport properties of gases: Tip [421], Korvig, Knapp, Gordon, and Beenakker [422], Korvig, Honeywell, Bose, and Beenakker [423], Gorelik and Sinitsyn [424], Levi, McCourt, and Hajdu [425], Levi, McCourt, and Beenakker [426], McCourt, Knapp, and Moraal [427], Gorelik, Nikolaevskii, and Sinitsyn [428], Hulsman and Burgmans [1180], Moraal, McCourt, and Knaap [1181], Korving [1182], Tommasini, Levi, Scoles, de Groot, van den Brocke, van den Meigdenberg, and Beenakker [1183], Hulsman, van Waasdijk, Burgmans, Knaap, and Beenakker [1184], Hulsman and Knaap [1185], and Beenakker and McCourt [1186]. Studies have also been made to determine the effect of the magnetic field on the properties of mixtures: viscosity [429], diffusion [430], and thermal diffusion [431].

Similar studies have been conducted to investigate the effect of an external electric field on the transport properties of gases: Senftleben [432], Amme [433], Borman, Gorelik, Nikolaev, and Sinitsyn [434], Borman, Nikolaev, and Nikolaev [435], Gallinaro, Meneghetti, and Scoles [436], and Levi, McCourt, and Tip [437].

i. Critical and Rarefied Gas Regions

Our understanding of the properties of fluids near the critical point is far from being satisfactory

[438], and much theoretical and experimental work needs to be done. The status of knowledge concerning viscosity is reviewed in recent articles by Sengers [439, 1176, 1177], Sengers and Sengers [440], Deutch and Zwanzig [441], Fixman [442], and Teague and Pings [443]. Cercignani and Sernagiotto [444] have recently discussed the Poiseuille flow of a rarefied gas in a cylindrical tube and solved the integro-differential equation numerically for the Bhatnagar, Gross, and Krook model. Because of the limited present understanding of these topics, we refer to them only briefly here.

3. ESTIMATION METHODS

A. Introduction

A number of methods have been developed to compute the viscosity of gases and their multicomponent mixtures under conditions of temperature and composition where directly measured values are not available. Many ways have emerged from the framework of Chapman–Enskog theory [2] to estimate the collision integrals either through a simplified adjusted potential or a more complicated potential whose parameters are obtained from critical constants or boiling point constants, or from viscosity data over a limited temperature range. Attempts have been made to arrange the rigorous theory expression in such a form that various groups of quantities depend only in an insensitive way on the temperature, composition, etc., so that once the expression is adjusted for one or two observed values of viscosity, the reliable estimation for other conditions is possible with great ease. Many sources list methods with various viewpoints and consequently with varying degrees of rigor. Reid and Sherwood [445] in their book describe correlation procedures for the viscosity data of gases as a function of temperature, and methods of calculation for pure gases and mixtures. Westenberg [446] and Brokaw [740] have discussed the calculation of viscosity of gases and multicomponent mixtures on the basis of rigorous kinetic theory for polar and nonpolar gases, labile atoms, and radicals. Hilsenrath and Touloukian [447] and Hilsenrath, Beckett, Benedict, Fano, Hoge, Masi, Nuttal, Touloukian, and Woolley [448] have recommended viscosity data for a number of gases based on various empirical or kinetic theory expressions. Svehla [449, 450] and Simon, Liu, and Hartnett [451, 773] have tabulated the estimated values of viscosities of a number of gases and mixtures as a function of temperature. Because of the interest of

the petroleum industry, viscosities have been computed for natural gases [452], light hydrocarbons [453, 777], and lubricants [454]. Some other articles will be referred to later while discussing the individual estimation procedures.

B. Pure Gases

The rigorous kinetic theory expression given earlier can be used to compute the viscosity of the desired gas under specified conditions if all the necessary related information is known; this view is supported by a large number of studies [28, 809]. For simple molecules in the predissociation and pre-ionization range at ordinary pressures, the basic information necessary is the intermolecular potential, and hence, the computed viscosity collision cross section. Much effort has been devoted to determining the nature of intermolecular forces as well as in the computation of collision integrals. We refer to many such studies here, for they are of prime importance in the calculation of viscosities of gases and gaseous mixtures.

Various books [2, 28] discuss the subject of intermolecular forces, but it will be sufficient here to mention two recent publications [455, 456] which exclusively deal with this complicated subject from different points of views. Some other exhaustive reviews on the subject are due to Margenau [457], Fitts [458], Pauly and Toennies [459], Lichten [460], Buckingham [461], Dalgarno [462], Walker, Monchick, Westenberg, and Fowin [463], Treanor and Skinner [464], and Certain and Bruch [370]. Some papers deal with particular features in detail, e.g., zero-point energy [465], long-range intermolecular forces [466–469], moderately long-range intermolecular forces [470, 471], short-range intermolecular forces [472–474], exchange forces [475, 476], additivity of intermolecular forces [477–479], quasi-spherical [480, 481] and polar [482] molecule interactions, and resonant charge exchange [483, 484]. The determination of short-range intermolecular forces from measurements of elastic scattering of high-energy beams has been discussed by Amdur [485] and Amdur and Jordan [486]. In spite of all such studies, the understanding of intermolecular forces is still quite primitive [487], and the qualitative features thus derived are combined with experimental data to determine the unknown parameters which are adjusted in this process to values depending upon the property and the temperature range used. Here again extensive work has been done, and we briefly review below the various semiempirical potential forms so far used and the effort to determine their unknown parameters.

Various semiempirical potential forms used for computing transport properties are reviewed in a number of articles [2, 488–491] and in many more; some of these will be referred to later. The simple inverse (or exponential) attractive (or repulsive) potentials have been considered to compute transport property collision integrals [492–496]. The more complicated potential forms are square-well [776], various Lennard-Jones (12–6) [497–499], (9–6) and (28–7) [500], (m–6) for m = 9, 12, 15, 18, 21, 24, 30, 40, 50, and 75 [501], modified Buckingham exp-six [502, 503], Morse [504, 505], and the Lennard-Jones (12–6) with an added quadrupole–quadrupole term [506]. Barker, Fock, and Smith [507] have computed the viscosity collision integral for the Kihara spherical-core potential [84] and for another particular potential derived by Guggenheim and McGlashan [508]. Some other forms used for polar gases or for gases at low and high temperatures have been referred to earlier in the text.

Mention may also be made of other potential forms which have been studied in connection with the various equilibrium properties but their use in the calculation of viscosity still remains to be explored. Some such references are: Pollara and Funke [509], Saxena and Joshi [510, 511], Saxena, Joshi, and Ramaswamy [512], Saksena and Saxena [113, 513], Saxena and Saksena [514], Saksena, Nain, and Saxena [515], Varshni [516], Dymond, Rigby, and Smith [517, 1206], Nain and Saxena [518, 529], Feinberg and de Rocco [519], de Rocco and Hoover [520], de Rocco, Spurling, and Storvick [521], Spurling and de Rocco [522], Storvick, Spurling, and de Rocco [523], McKinley and Reed [524], Lawley and Smith [525], Dymond and Smith [526], Spurling and Mason [527], Carra and Konowalow [528], Nain and Saksena [530], Konowalow [531], and Dymond and Alder [1207].

A considerable amount of work has been done to determine the potential parameters of the different above-mentioned semiempirical potential functions—from theory as well as from experimental data. In reference [456] there are review articles by Mason and Monchick [532], Bernstein and Muckerman [533], Birnbaum [534], Bloom and Oppenheim [535]; some others have been referred to earlier in this section. Potential parameters are also well estimated on the basis of critical or boiling-point constants [28, 536–539, 735] and from densities in the liquid phase [540]. The independent calculation of long-range dispersion forces is also possible from experimental data [541, 542], somewhat in the same manner as repulsive forces are determined from the scattering measurements on molecular beams [485, 486, 543, 544]. A series of articles discuss and demonstrate the limitations associated with the choice of proper data if appropriate values of the parameters are to be obtained. Some of these are by Zwimino and Keller [545], Munn [546], Munn and Smith [547], Klein [548], Hanley and Klein [549, 550], Klein and Hanley [551], Mueller and Brackett [552], and Hogervorst [1196].

The experimental data on viscosity as a function of temperature have been used extensively to determine the parameters of the intermolecular potentials. Such methods are developed by Hirschfelder, Curtiss, and Bird [28], Bird, Hirschfelder, and Curtiss [553], Srivastava and Madan [554], Hawksworth [555], Mason and Rice [208], Whalley and Schneider [556], and Robinson and Ferron [557]. Using these methods or their minor modifications, many workers have determined the potential parameters from the viscosity data, for example, Mason and Rice [558], Hanley [559, 560], Hanley and Childs [561, 917], Childs and Hanley [775], de Rocco and Halford [562], Milligan and Liley [563], Saran [564], Pal [565], and Chakraborti [566]. In a somewhat analogous fashion the experimental data giving the temperature dependence of thermal conductivity have been used to determine the intermolecular potentials [567, 568]. Similarly the measurements on self-diffusion [569, 571] and the isotopic thermal diffusion factor [572–575] are used to determine intermolecular forces between similar molecules of a gas. Next to viscosity, the second virial coefficient data as a function of temperature have been employed most extensively to determine force fields. Some of these investigations were conducted by Yntema and Schneider [576], Whalley and Schneider [577], Schamp, Mason, Richardson, and Altman [578], Schamp, Mason, and Su [579], Barua [580, 581], Srivastava [582], Srivastava and Barua [583], Barua and Saran [584], and Mason, Amdur, and Oppenheim [585]. Zero-pressure Joule–Thomson data have also been used to determine potential parameters [28, 586–588]. Combination of these two properties to determine the potential parameters is also suggested [589]. Parameters are also evaluated from the properties of the molecules in the solid state [590–594] and from x-ray scattering data [595]. Theoretical calculations of intermolecular forces between rare gas atoms are still common [596–598]. Indeed, many workers have employed simultaneously the data on various properties to get the best overall adjusted potential parameters, for example, Fender [599], Bahethi and Saxena [600], Barua and Chakraborti [601], Chakraborti [602, 603], Srivastava and Saxena [604], Konowalow, Taylor, and Hirschfelder [605], Konowalow and Hirschfelder [606], Bahethi and Saxena [607, 608], Konowalow and Carra [609, 610], Konowalow [611, 612], and Saxena and Bahethi [613].

Semitheoretical combination rules have been suggested to determine the interaction potential between unlike molecules from the knowledge of potentials between like molecules. Such semiempirical combination rules have been given for Lennard-Jones (12–6) [614, 1197, 1198], modified Buckingham exp-six [615, 616], and Morse [617, 1198] potentials and have been extensively tested against the experimental data on different properties of mixtures [28, 554, 555, 558, 559, 614, 615, 618, 619]. It was soon realized that an alternative and maybe a better approach would be to determine the interaction potential parameters from the experimental data on the properties of mixtures themselves. The data on viscosity of binary mixtures have been used to determine unlike interactions by Srivastava [620], but now it is well understood that the appropriate properties are only those which are sensitive to such interactions, such as diffusion and thermal diffusion. Data on viscosity and thermal conductivity [621–628] have nevertheless been used as a good check for the appropriateness of the potential. Recently Alvarez-Rizzatti and Mason [1199] have given a perturbation and a variation method for the calculation of dipole–quadrupole dispersion coefficients. They have thus derived the combination rules.

A number of workers have used the experimental data on the interdiffusion coefficient of gas mixtures as a function of temperature to determine the parameters of the potential, for example, Amdur, Ross, and Mason [629], Amdur and Shuler [630], Amdur and Beatty [631], Amdur and Malinauskas [632], Mason, Annis, and Islam [633], Srivastava [634], Srivastava and Barua [635], Paul and Srivastava [636], Srivastava and Srivastava [637], Srivastava [638], Walker and Westenberg [639–642], Saxena and Mathur [643], and Mathur and Saxena [644]. Srivastava and Madan [645] suggested the use of thermal diffusion data as a function of temperature to determine the unlike potential parameters. Saxena [646, 647] and Srivastava [648] have discussed and refined this method. Calculations by Madan [649] and Saxena [650] of other transport properties and comparison with the ob-

served values revealed that the technique has a great potential in experimentally determining the forces between molecules. Srivastava and Srivastava [651] and Srivastava [652] have used the thermal diffusion data to determine the three parameters of the modified exp-six potential. In recent years thermal diffusion measurements have been used extensively to probe into the nature of intermolecular-force laws [653–658]. Simultaneous use of diffusion and thermal diffusion data has also been made to determine the potential functions [659–661].

The determination of potential functions on the basis of any type of experimental data is limited primarily because of the scarcity of accurate measurements. Consequently, theoretical calculation have turned out to be very useful and attempts are being continuously made to refine the theoretical approaches or develop new ones; for example, McQuarrie and Hirschfelder [662], Kim and Hirschfelder [663], and Certain, Hirschfelder, Kolos, and Wolniewicz [664]. Some other calculations of specific interaction potentials for atoms and molecules in their ground and excited states have been made by Mason, Ross, and Schatz [665], Ross and Mason [666], Mason and Hirschfelder [667, 668], Mason and Vanderslice [669], Vanderslice and Mason [670, 671], and Fallon, Mason, and Vanderslice [672]. The interaction energies have been computed between ions and neutral atoms by Mason and Vanderslice [673–678] using the ion-scattered measurements. Binding energies of He_2^+, Ne_2^+, and Ar_2^+ have also been computed by Mason and co-workers [679–681] on the basis of ion-scattering data. A number of calculations of potential energy from spectroscopic data have been made in recent years for ground and excited states of atomic and molecular diatomic gases by Vanderslice, Mason, Maisch, and Lippincott [682], Vanderslice, Mason, and Lippincott [683], Vanderslice, Mason, and Maisch [684, 685], Fallon, Vanderslice, and Mason [686, 687], Tobias and Vanderslice [691], Vanderslice [692], Krupenie, Mason, and Vanderslice [693], Weissman, Vanderslice; and Battino [694], Knof, Mason, and Vanderslice [695], Krupenie and Weissman [696], and Benesch, Vanderslice, Tilford, and Wilkinson [697–699].

As already pointed out [28, 536–539], the potential parameters are also obtained from the knowledge of critical constants through semiempirical relations. We refer here to a number of papers which deal with the determination of critical constants of complicated gases and their multicomponent mixtures. They are: Stiel and Thodos [700] for saturated aliphatic hydrocarbons; Thodos for naphthenic hydrocarbons [701], aromatic hydrocarbons [702], and unsaturated [703] and saturated [704] aliphatic hydrocarbons; Forman and Thodos for hydrocarbons [705] and organic compounds [706]; Ekiner and Thodos for binary mixtures of aliphatic hydrocarbons [707], ethane–*n*-heptane system [708], and ethane–*n*-pentane system [709]; and Grieves and Thodos [710, 711] for binary systems of gases and hydrocarbons. Grieves and Thodos have also studied the critical temperatures [712] and pressures [713] of multicomponent mixtures of hydrocarbons. Many ternary systems [714], methane–propane–*n*-pentane systems [715], methane–ethane–*n*-butane systems [716, 717], ethane–*n*-pentane–*n*-heptane systems [718], ethane–propane–*n*-butane systems [719], ethane–*n*-butane–*n*-pentane systems [720] have been investigated and their critical constants determined by Thodos and co-workers. Ekiner and Thodos [721–723] have proposed an interaction model for representing the critical temperatures and pressures of methane-free aliphatic hydrocarbon mixtures. Rastogi and Girdhar [724] have proposed a semiempirical relationship between the critical constants and the chain length of saturated hydrocarbons. Gunn, Chuch, and Prausnitz [725] have recently determined the effective critical constants for light gases which exhibit appreciable quantum effects, and Gambill [726–728] has reviewed the methods for estimating critical properties.

A number of attempts have been made to develop semitheoretical correlating expressions for the viscosity of pure gases based on the theoretical equations (6) and (7). Thus, Keyes [729] suggested that for the Lennard-Jones (12–6) potential $f^{(3)}/\Omega^{(2,2)*}$ be replaced by a three-term equation involving only the independent parameter T^*. Gambill [730] has tabulated the ratio as a function of T^*, Westenberg [731] and Sutten and Klimov [732, 733] have represented the viscosity collision integral, $\Omega^{(2,2)*}$, by different polynomials involving T^*, and recently Kim and Ross [734] have suggested the following three expressions for the different reduced temperature ranges:

$$\Omega^{(2,2)*} \simeq 1.604(T^*)^{-1/2}, \quad 0.4 < T^* < 1.4$$

$$\Omega^{(2,2)*} \simeq 0.7616[1 + (1.09)T^*)], \quad 1 < T^* < 5$$

$$\Omega^{(2,2)*} \simeq 1.148T^* - 0.145, \quad 20 < T^* < 100$$

$$(11)$$

These formulas lead to values which agree with the directly calculated values within maximum deviations of 0.7, 0.1, and 0.1 %, respectively. Hattikudur and

Thodos [1201] have represented the reduced viscosity integral by the following relation:

$$\Omega^{(2,2)*}(T^*) = \frac{1.155}{T^{*0.1462}} + \frac{0.3945}{e^{0.6672T^*}} + \frac{2.05}{e^{2.168T^*}} \quad (12)$$

This equation produces the original computed values in the $T^* = 0.30$ to $T^* = 400$ range within an average deviation of 0.13 % and a maximum deviation of 0.54 % at $T^* = 0.30$. For $T^* \geq 1.15$, the average deviation is 0.09 %, with a maximum deviation of 0.15 % at $T^* = 1.15$. Neufeld, Janzen, and Aziz [1202] employed the following twelve-adjustable-parameter equation:

$$\Omega^{(2,2)*}(T^*) = (A/T^{*B}) + [C/\exp(DT^*)]$$
$$+ [E/\exp(FT^*)] + [G/\exp(HT^*)]$$
$$+ RT^{*B} \sin(ST^{*W} - P) \quad (13)$$

They found that this relation reproduces the actual values within an average deviation of 0.050 % and a maximum deviation of 0.16 % at $T^* = 100$. Klimov [733] has also reported the polynomials representing the viscosity collision integral for polar gases [28]. Brokaw [735] has expressed the collision integral for polar gases, $\Omega p^{(2,2)*}$, in terms of its value for nonpolar gases, $\Omega np^{(2,2)*}$, by the simple relation

$$\Omega p^{(2,2)*} \simeq \Omega np^{(2,2)*} + \frac{0.2\delta^2}{T^*} \quad (14)$$

This result is based on the collision integral tabulations of Monchick and Mason [165]. Brokaw [735] has given alignment charts for $\Omega np^{(2,2)*}$ as a function of T^* to obtain quick estimates of viscosity with fair accuracy.

Bromley and Wilke [736] wrote the theoretical expression in a slightly modified form and presented nomographs for rapid calculations. This procedure has been extensively used and recommended by Holmes and Baerns [737], and an interesting comment is made by Weintraub and Corey [738] which facilitates the estimation of viscosity at high temperatures. More recently, Brokaw [739] has presented alignment charts similar to those of Bromley and Wilke [736].

Many semiempirical forms have been used to represent the temperature dependence of viscosity. Licht and Stechert [741] considered the data for twenty-five gases and discussed the following four forms:

$$\mu = aT^n \quad (15)$$

$$\mu = \frac{KT^{3/2}}{T + S} \quad \text{or} \quad \frac{KT^{1/2}}{1 + (S/T)} \quad (16)$$

$$\mu = \frac{bT^{1/2}}{\exp(c/T)} \tag{17}$$

$$\mu = dT(T^{3/4} + T^{-3/4})^{-m} \tag{18}$$

These are all two-constant equations, these being a and n, K and S, b and c, and d and m in the four cases, respectively. Sutherland [1200] derived the form of the second equation for the coefficient of viscosity of a gas whose molecules are spherical and attract each other. More complicated relations have also been used. These are in many cases modified forms of the above relations, for instance [741, 875],

$$\mu = \frac{AT^{1/2}}{1 + C/T + D/T^2} \tag{19}$$

$$\mu = (A + BT + CT^2 + DT^3)T^{1/2} \tag{20}$$

$$\mu = \frac{BT^{1/2}}{\exp[C'/(T + \alpha)]} \tag{21}$$

$$\mu = \frac{QT^{[1/2 + 2/(s-1)]}}{1 + UT^{(m-s)/(s-1)}} \tag{22}$$

For the empirical choice of $m = 5$ and $s = 9$, this equation reduces to

$$\mu = \frac{QT^{5/4}}{T^{1/2} + U} \tag{23}$$

In the following relation the value 3 has been used for S, as well as many other empirical choices:

$$\mu = \frac{KT^n}{1 + (S/T)} \tag{24}$$

The unknown constants are A, C, and D, A, B, C, and D, B, C', and α, Q, U, m, and s, and K, n, and S in equations (19), (20), (21), (22), and (24), respectively. The simple polynomial expansion in temperature as well as many other semiempirical forms have been used for individual or groups of gases [447, 453, 729, 742, 746, 749, 754, 774, 778], but these will not be enumerated here.

The principle of corresponding states has also been applied to develop procedures for correlating viscosity data [28]. Smith and Brown [747] and Whalley [748] have discussed extensively the form of this law and analyzed the data on viscosity of a large number of gases. Comings and Egly [1153] developed a graphical correlation on the basis of available data to predict viscosity of gases at high pressures. Tham and Gubbins have correlated the available experimental dense-gas viscosity data of rare gases [1190] and nonpolar polyatomic gases [1191] by applying the principle of corresponding states. Licht and Stechert [741] used the same principle to develop a universal equation for predicting viscosities of gases. They even presented a nomograph [741] to be used along with their proposed equation. Bromley and Wilke [736] suggested a simple relation for the prediction of viscosity based on the rigorous theory expression in which the potential parameters were eliminated in favor of critical temperature and volume. The use of this equation is further facilitated by the presentation of two curves by Gegg and Purchas [755]. Shimotake and Thodos [756] and more recently Trappeniers, Botzen, Ten Seldam, Van Den Berg, and Van Oosten [757] have given the corresponding states correlations for the viscosity of rare gases. Thodos and co-workers have developed similar relations for diatomic gases [758], para-hydrogen [760], air [761], carbon dioxide [762], sulfur dioxide [1154], ammonia [763], and gaseous water [764]. Recently more ambitious efforts have been made in employing the principle of corresponding states in correlating the viscosity data of spherical molecules with a high degree of accuracy over a wide temperature range by Dymond [1203], Kestin, Ro, and Wakeham [1204], and Neufeld and Aziz [1205].

Stiel and Thodos [765] analyzed the viscosity data at atmospheric pressure for fifty-two nonpolar gases on the basis of a dimensional analysis approach, to develop a correlation involving reduced temperature. This approach has been successfully extended to dissociated and undissociated gases up to 10,000 K [766], to polar gases [767], and to hydrocarbon gases [1155]. Lefrancois [1159] has outlined a procedure for the computation of the viscosity of pure gases as a function of pressure based on the numerous measurements of the compressibility factors for gases.

Many of the above-mentioned works also include a discussion on the correlation of viscosity of dense gases, but reference may be made now to some other papers which deal exclusively with this aspect, for example, Starling and Ellington [768], Lennert and Thodos [769], Elzinga and Thodos [770], Jossi, Stiel, and Thodos [771], and Stiel and Thodos [772]. Viscosities of pure gases are also generated from the experimental data on other transport properties through the framework of kinetic theory [2, 28]. In particular, thermal conductivity data have been used, and the relation between μ and k has been confirmed from direct experimental work [827]. Saxena and Saxena [828], Saxena, Gupta, and Saxena [829], and Saxena and Gupta [830] have in this way generated the viscosity values for rare and diatomic gases from their measurements on k as a function of temperature.

C. Multicomponent Gas Systems

A number of empirical and semiempirical relations have been used to estimate the viscosity of multicomponent gas mixtures. Some of these procedures can be justified to a large extent as simplifications of the rigorous theory expression. To assess the methods one needs to evaluate the simplifying limitations and the nature of the gas molecules involved. We outline below the various methods used so far for estimating viscosities of mixtures and point out their basis and probable degree of success.

Many of the earlier semiempirical relations employed for computing viscosities of mixtures are given by Partington [778]. One such relation is due to Enskog [779] and has been recently reexamined by Keyes [729]. Gambill [780] has reviewed the prediction methods. We list below some of the major methods which have proved useful and have been tested extensively in many cases. Hirschfelder, Curtiss, and Bird [28] found that to a good approximation the viscosity of a binary mixture of heavy isotopes is given by

$$\mu_{\text{mix}}^{-1/2} = x_1[\mu_1]^{-1/2} + x_2[\mu_2]^{-1/2} \qquad (25)$$

The well-known Sutherland form [47] and the simple quadratic form

$$\mu_{\text{mix}} = \mu_1 x_1^2 + \mu_{12}x_1x_2 + \mu_2 x_2^2 \qquad (26)$$

for the viscosity of mixtures have been mentioned [11] though never sufficiently tested. Not too much is known about reliable prediction procedures for dense gas mixtures [780, 789] at the present time, and this development will have to await our theoretical understanding of the dense gases and more experimental work on such systems.

a. Method of Buddenberg and Wilke

Buddenberg and Wilke [781] showed that the viscosity data on mixtures are adequately correlated by the following Sutherland [47] type relation:

$$\mu_{\text{mix}} = \sum_{i=1}^{n} \mu_1 \bigg/ \left[\left(1 + \frac{1.385\mu_i}{x_i\rho_i}\right)\sum_{\substack{j=1\\j\neq i}}^{n} x_j/D_{ij}\right] \qquad (27)$$

here D_{ij} is the diffusion coefficient and ρ_i is the density of the ith component. Wilke [782] further simplified this relation to

$$\mu_{\text{mix}} = \sum_{i=1}^{n} \mu_1 \bigg/ \left(1 + \frac{1}{x_i}\sum_{\substack{j=1\\j\neq i}}^{n} x_j\Psi_{ij}\right) \qquad (28)$$

where

$$\Psi_{ij} = \frac{[1 + (\mu_i/\mu_j)^{1/2}(M_j/M_i)^{1/4}]^2}{(4/\sqrt{2})[1 + (M_i/M_j)]^{1/2}}$$

Hirschfelder, Curtiss, and Bird [28] have discussed the assumptions under which a relation of the type given by Buddenberg and Wilke [781] is derived from the rigorous kinetic theory expression. Bromley and Wilke [736] and more recently Brokaw [739] have given alignment charts which facilitate the computation of Ψ_{ij} as given by the above equation. Saxena and Narayanan [783] and Mathur and Saxena [784] have examined the method of Wilke for nonpolar multicomponent mixtures up to about 1300 K with reasonable success. These workers have also suggested that Ψ_{ij} computed at a lower temperature may be used for computation of μ_{mix} at higher temperatures. That similar conclusions are valid for mixtures involving polar gases is established by the calculations of Mathur and Saxena [785].

b. Method of Saxena and Narayanan

Saxena and Narayanan [783] suggested that Ψ_{ij} in the μ_{mix} expression of Wilke may be regarded as disposable parameters independent of composition and temperature and may thus be determined from two experimental mixture viscosities. Their [783] checks against data at higher temperatures, as well as for the mixtures of three gases, demonstrated the potential and promise of the proposed method. Mathur and Saxena [786] successfully examined this method for binary systems of polar and nonpolar gases.

c. Method of Herning and Zipperer

Herning and Zipperer [787] suggested that μ_{mix} may be estimated from a still simpler relation than that of Wilke [782]:

$$\mu_{\text{mix}} = \sum_{i=1}^{n} (x_i\mu_i M_i^{1/2}) \bigg/ \sum_{i=1}^{n} (x_i M_i^{1/2}) \qquad (29)$$

This form is equivalent to Wilke's if

$$\Psi_{ij} = (M_j/M_i)^{1/2} = \Psi_{ji}^{-1} \qquad (30)$$

This formula has been tested extensively for hydrocarbon and other mixtures with an uncertainty of better than 2% [780]. Recently Tondon and Saxena [788] tested it for mixtures involving polar gases, and found that the method is particularly good for such binary mixtures where the mass ratio for the two components is small. For 174 mixtures of 11 systems the average absolute deviation between theory and experiment is 6.1%, and this improves to 2.7% for 89 mixtures when three systems involving gases of large mass ratio are excluded.

d. Method of Dean and Stiel

Dean and Stiel [789] developed a relationship to estimate the viscosity of nonpolar gases at ordinary pressures in terms of the pseudocritical constants of the mixture. Their recommended expression is

$$\mu_{mix}\xi = 34.0 \times 10^{-5}T_R^{8/9}, \qquad T_R < 1.5 \qquad (31)$$

and

$$\mu_{mix}\xi = 166.8 \times 10^{-5}(0.1338T_R - 0.0932)^{5/9}$$
$$T_R \geq 1.5 \tag{32}$$

where

$$\xi = T_{cm}^{1/6} \left/ \left[\left(\sum_i x_i M_i\right)^{1/2} P_{cm}^{2/3} \right]\right.$$

Here μ_{mix} is centipoises, $T_R = T/T_c$, and the defining relations for pseudocritical constants of the mixture as recommended by these authors [789] are

$$T_{cm} = \sum_i x_i T_{ci}$$

$$V_{cm} = \sum_i x_i V_{ci}$$

$$Z_{cm} = \sum_i x_i Z_{ci}$$

and

$$P_{cm} = Z_{cm}RT_{cm}/V_{cm}$$

They [789] have examined 339 experimental mixtures in twenty-two binary systems and reproduced the μ_{mix} values on the basis of the above relations within an overall average of 1.7%.

e. Method of Strunk, Custead, and Stevenson

Strunk, Custead, and Stevenson [790] suggested on the basis of approximate theoretical analysis that the viscosity of a binary mixture of nonpolar gases may be computed on the basis of an expression similar to that given by the Chapman–Enskog rigorous kinetic theory [28]:

$$\mu_{mix} = \frac{266.93 \times 10^{-7}(TM_{mix})^{1/2}}{\sigma_{mix}^2 \Omega_{mix}^{(2,2)*}} \tag{33}$$

where

$$M_{mix} = \sum_{i=1}^{n} x_i M_i$$

$$\sigma_{mix} = \sum_{i=1}^{n} x_i \sigma_i$$

and $\Omega_{mix}^{(2,2)*}$ is a function of the reduced temperature T^*, where

$$T^* = \frac{kT}{\epsilon_{mix}}$$

and

$$\frac{\epsilon_{mix}}{k} = \sum_{i=1}^{n} x_i \left(\frac{\epsilon_i}{k}\right)\sigma_i^3 \left/ \sigma_{mix}^3\right.$$

Thus, all one needs in the calculation are the parameters of the Lennard-Jones (12–6) potential for the pure components, and the mixture composition. These authors examined 201 binary mixtures of eleven different nonpolar gases. Strunk and Fehsenfeld [791] also evaluated the potential of these equations to predict viscosity of multicomponent mixtures of nonpolar gases. Their [791] detailed calculations on 136 mixtures containing three to seven components from sixteen different gases indicated that the experimental viscosities could be reproduced within −0.3 to −6.7% for 95% of the time. This led them to suggest that the numerical coefficient in equation (33) be replaced by 276.27 for ternary and higher-order mixtures. With this modification the viscosities could be reproduced to lie within +3.2 and −3.2% of the actual values 95% of the time.

f. Method of Ulybin

Ulybin [792] has suggested an empirical method in which the viscosity of a mixture at a temperature T_1 is related to its value at a lower temperature T_2 according to the following equation:

$$\mu_{mix}(T_2) = \mu_{mix}(T_1) \sum_{i=1}^{n} x_i[\mu_i(T_2)/\mu_i(T_1)] \tag{34}$$

His detailed calculations on binary and ternary mixtures did reproduce the experimental value in most of the cases within the uncertainty in the latter. The somewhat remarkable success of this empirical relation is not surprising, in the light of the work of Saxena [793]. He [793] has given a theoretical basis to this formula; hence this relation is not to be regarded as empirical, but as an approximate theoretical expression. The discussion by Saxena [793] deals with the case of thermal conductivity but an exactly parallel argument can be given for the case of viscosity.

g. Sutherland Form and Rigorous Kinetic Theory

The success of the Sutherland form [47] in representing the experimental data on viscosity of gas mixtures is already evident from some of the work referred to above. This led to a large number of investigations which will be mentioned in this section, they form the basis of the many methods of calculation of viscosities of multicomponent gas mixtures described later.

Cowling [794] and Cowling, Gray, and Wright [795] gave a simple physical interpretation to the coefficient Ψ_{ij} as the ratio of the efficiencies with which molecules j and molecules i separately impede the transport of momentum by molecules i. On the basis of this interpretation [794], they [795] have been able to develop the physical significance of the rigorous theory expression for viscosity [2]. Francis [796], Brokaw [797, 798], Hansen [799], Wright and Gray [800], Burnett [801], and Yos [802] made notable attempts to interpret the rigorous theory expression for μ_{mix} and in this process derived relations for Ψ_{ij}. Various approximations have been made by different workers resulting in different explicit expressions for Ψ_{ij}, the Sutherland coefficients. Some of these expressions of the interrelation between Ψ_{ij} and Ψ_{ji} have been used to develop methods for the predictions of μ_{mix}. These will be described now.

h. Method of Saxena and Gambhir

Following the analysis of Wright and Gray [800], Saxena and Gambhir [803] suggested the following relation connecting Ψ_{ij} with Ψ_{ji}:

$$\frac{\Psi_{ij}}{\Psi_{ji}} = \frac{\mu_i}{\mu_j}\left(\frac{M_j}{M_i}\right)^{0.85} \tag{35}$$

Thus, if the μ_{mix} value is known at one composition, equations (28) and (35), together with the knowledge of pure component viscosities, serve to obtain Ψ_{ij} and Ψ_{ji}. Detailed calculations by Saxena and Gambhir [804] on the binary and ternary mixtures of nonpolar gases indicated that this scheme is capable of reproducing the viscosity values to greater accuracy than the experimental uncertainties. Their [804] calculations also revealed that Ψ_{ij} and Ψ_{ji} may be regarded as independent of composition, so that the same set correlates the data over the entire range, and may also be used for multicomponent mixtures. They [804] also found that these Sutherland coefficients are feebly dependent on temperature; the experimental data over the temperature range 300–1300 K could be adequately represented by the Ψ_{ij}'s calculated at 300 K. Mathur and Saxena [805] applied the method to binary mixtures of nonpolar–polar gases and found the same conclusion to be valid. Their [805] calculations covering 79 binary mixtures reproduced

the experimental values within an average absolute deviation of 0.4%.

i. Method of Gambhir and Saxena

Gambhir and Saxena [806] examined the temperature and composition dependence of Ψ_{ij} and Ψ_{ji} on the basis of the theoretical expression for μ_{mix}. After making certain reasonable assumptions, they [806] found that if the mass of the one gas is sufficiently larger than the other in the binary mixture, the following simple relation connects Ψ_{ij} with Ψ_{ji}:

$$\frac{\Psi_{ij}}{\Psi_{ji}} = \frac{\mu_i M_j}{\mu_j M_i} \frac{50 M_i + 33 M_j}{33 M_i + 50 M_j} \tag{36}$$

Numerical calculations of Saksena and Saxena [807] established that this procedure, where the above relation and one μ_{mix} experimental value are used to compute the Sutherland coefficients, is completely satisfactory. Experimental data on ten binary systems could be reproduced within an overall average absolute deviation of 0.7%, whereas for a ternary system this number improved to 0.5%. These calculations on mixtures of nonpolar gases also established that the assumption of the temperature and composition independence of Sutherland coefficients is a good and practical one. Mathur and Saxena [808] made a detailed study of a similar nature for mixtures of polar and nonpolar gases and found that the method and above conclusions are also valid for these gas systems.

j. Method of Saxena and Gambhir

Saxena and Gambhir [810] suggested that Ψ_{ij} may be calculated in the Sutherland equation with the help of translational or frozen thermal conductivity data (i.e., the thermal conductivity of monatomic gases and in polyatomic gases that part of total thermal conductivity which is due to translational degrees of freedom only) so that

$$k_{\mathrm{mix}}^{\circ} = \sum_{i=1}^{n} k_i \bigg/ \left[1 + \sum_{j=1}^{n} \Psi_{ij}(x_j/x_i)\right] \tag{37}$$

Here Ψ_{ij} is computed according to the formula derived by Mason and Saxena [812]:

$$\Psi_{ij} = \frac{1}{2\sqrt{2}}\left(1 + \frac{M_i}{M_j}\right)^{-1/2}\left[1 + \left(\frac{k_i^{\circ}}{k_j^{\circ}}\right)^{1/2}\left(\frac{M_i}{M_j}\right)^{1/4}\right]^2 \tag{38}$$

Ψ_{ji} is obtained from Ψ_{ij} by interchanging the subscripts referring to the molecular species. Numerical calculations of Saxena and Gambhir [810], and

Gandhi and Saxena [811] on the binary mixtures of rare gases showed good reliability for the method, particularly when one recalls that the knowledge of thermal conductivity is employed to predict the values for viscosity.

k. Method of Brokaw

Brokaw [797, 798] manipulated the expression for the multicomponent mixture into the Sutherland form and derived the increasingly complicated expressions for Ψ_{ij}. In approximations other than the first the expression for Ψ_{ij} is quite complicated and requires knowledge of the interaction potential and different collision integrals, so that the actual calculation of μ_{mix} becomes as difficult as the kinetic-theory expression. The first-approximation expressions for the Sutherland coefficient suggest that

$$\frac{\Psi_{ij}}{\Psi_{ji}} = \frac{\mu_i}{\mu_j}\frac{M_j}{M_i} \quad (39)$$

Gupta and Saxena [815] employed this relation and one value of μ_{mix} in the Sutherland form to compute Ψ_{ij} and Ψ_{ji}. On this basis they [815] successfully correlated the data on twenty-two binary systems and twelve ternary mixtures of argon–neon–helium. They also confirmed that, treating these Ψ_{ij} as temperature independent, the high temperature viscosities could be reproduced within an average absolute deviation of 0.8%.

Brokaw [798] also suggested a simplified form for Ψ_{ij}

$$\Psi_{ij} = \Psi'_{ij} + \frac{M_i\sqrt{\Psi'_{ij}} - M_j\sqrt{\Psi'_{ji}}}{2(M_i + M_j) + M_j\sqrt{\Psi'_{ji}}}\sqrt{\Psi'_{ij}} \quad (40)$$

where

$$\Psi'_{ij} = \frac{\mu_i}{\mu_j}\frac{2M_j}{M_i + M_j}$$

$$\mu_i \times 10^7 = \frac{266.93\sqrt{M_i T}}{\sigma_{ii}^2\Omega_{ii}^{(2,2)}}$$

and

$$\mu_{ij} \times 10^7 = \frac{266.93\sqrt{2TM_iM_j/(M_i + M_j)}}{\sigma_{ij}^2\Omega_{ij}^{(2,2)*}}$$

Brokaw's [798] limited calculations on three binary and one ternary systems of nonpolar gases indicated a very good accuracy for this procedure. Tondon and Saxena [788, 813], however, made detailed calculations on 224 binary mixtures of nonpolar–polar gases and found an average absolute diagreement of

3.0%. On the other hand the rigorous theory reproduced these results within an average absolute deviation of 1.0%.

Tondon and Saxena [788, 813] suggested a modification to the above procedure of Brokaw [798]. It consisted in using the experimental values for the viscosity of the pure components instead of the theoretically calculated ones. This reproduced the data on 95 mixtures at the lower temperatures within an average absolute deviation of 1.2%. They [788, 813] also suggested that these computed values of Ψ_{ij} at the lower temperatures may be used in computing viscosities at the higher temperatures. This procedure led to the reproduction of 174 experimental data points within an average absolute deviation of 1.8%. It is to be noted that the simplicity does not impair the accuracy seriously; these computed values are in better agreement with the experiments than the original suggestion of Brokaw [798].

Brokaw [814] has simplified his complicated expressions for Ψ_{ij} and suggested [735] that

$$\Psi_{ij} = S_{ij}A_{ij}(\mu_i/\mu_j)^{1/2} \quad (41)$$

where

$$S_{ij} = \frac{\sigma_{ij}^2\Omega_{ij}^{(2,2)*}}{(\sigma_{ii}^2\Omega_{ii}^{(2,2)*}\sigma_{jj}^2\Omega_{jj}^{(2,2)*})^{1/2}}$$

$$A_{ij} \equiv \left(C_{ij}\frac{M_j}{M_i}\right)^{1/2}$$

$$\times \left[1 + \frac{(M_i/M_j) - (M_i/M_j)^{0.45}}{2\left(1 + \frac{M_i}{M_j}\right) + \frac{1 + (M_i/M_j)^{0.45}}{1 + C_{ij}}C_{ij}}\right]$$

and

$$C_{ij} \equiv \left[\frac{4M_iM_j}{(M_i + M_j)^2}\right]^{1/4}$$

For mixtures of nonpolar gases $S_{ij} = 1$, while for polar–nonpolar gas mixtures

$$S_{ij} = S_{ji} \cong \frac{1 + (T_i^*T_j^*)^{1/2} + (\delta_i\delta_j/4)}{[1 + T_i^* + (\delta_i^2/4)]^{1/2}[1 + T_j^* + (\delta_j^2/4)]^{1/2}} \quad (42)$$

In the limit when $\delta_i = \delta_j = 0$, as for nonpolar gases, the above relation does not reduce to $S_{ij} = 1$, and hence Brokaw [814] suggested that when δ_i and δ_j are both less than 0.1, S_{ij} should be taken to be unity. A_{ij} is a function of molecular-weight ratio and Brokaw [735] has given a scale giving A_{ij} and A_{ji} in terms of M_i/M_j to facilitate numerical calculations. Pal and Bhattacharyya [1194] and Brokaw [1195] have performed calculations on binary polar gas mixtures to check the accuracy of this procedure [735, 814].

l. Viscosity from Thermal Conductivity Data

Saxena and Agrawal [816] employed the framework of the transport theory [2], and computed viscosities of seven binary systems of rare gases from thermal conductivity data. Their [816] indirectly generated values of μ_{mix} were found to be in good agreement with the directly measured values. Since then this approach has been used to estimate the viscosities of binary systems for rare gases by Saxena and Tondon [817] and for mixtures involving polyatomic gases by Saxena and Gupta [628, 818]. The various assumptions involved in these interrelating expressions and their consequences for the generated data are also discussed by Gupta [819], Gupta and Saxena [820], Gandhi and Saxena [821], and Mathur and Saxena [822].

m. Viscosity from Interdiffusion Data

Data on interdiffusion coefficients can be used to generate reliable values of viscosities on the basis of the Chapman and Enskog theory [2] as illustrated by Mathur and Saxena [644] and Nain and Saxena [823]. The reverse of this approach, the determination of diffusion coefficients from viscosity data, has been more common in recent years [824].

D. Sutherland Coefficients

It is clear from the discussion in the previous section that the Sutherland form is a very successful one for correlating the data on binary systems, for predicting the values at high temperatures, and for multicomponent systems. The determination of these coefficients, Ψ_{ij}, is not a straightforward job and many suggestions have been made [825, 826, 1218]. Saxena [1218] found from an extensive numerical analysis on sixty-six binary systems involving both polar and nonpolar gases that the following Sutherland form:

$$\mu_{mix} = \frac{\mu_1}{1 + \Psi_{12}(x_2/x_1)} + \frac{\mu_2}{1 + \Psi_{21}(x_1/x_2)} \quad (43)$$

is satisfactory when two different procedures were employed to determine Ψ_{ij}. In the *first method* Ψ_{ij} and Ψ_{ji} were assumed to be interrelated by

$$\frac{\Psi_{ij}}{\Psi_{ji}} = \frac{\mu_i}{\mu_j} \frac{M_j}{M_i} \quad (44)$$

while in the *second method* this relation was modified to

$$\frac{\Psi_{ij}}{\Psi_{ji}} = \frac{\mu_i}{\mu_j} \left(\frac{M_j}{M_i}\right)^{0.85} \quad (45)$$

In both procedures the values of μ_i, μ_j, and μ_{mix} at one mixture composition must be known to correlate the data of μ_{mix} over the entire composition range at the specified temperature.

Tables 1, on pages 47a to 86a, shows how the calculated values of Ψ_{12} and Ψ_{21} obtained by one-parameter fits to the available experimental data reported in the next section using equations (43) and (44) (the first method) and equations (43) and (45) (the second method), depend on the value of μ_{mix} for the particular mixture composition used in making the fit and also on the temperature. The last column gives the viscosity values of the pure component on which the calculations are based. The relative constancy in the values of Ψ_{ij} for a given gas pair and temperature indicates the accuracy with which equation (43) represents the data.

Table 2, on pages 87a to 92a, contains recommended values of Ψ_{ij} for these mixtures, picked from the values in Table 1, along with three measures (L_1, L_2, and L_3) of the deviations of experimental data from the smoothed values computed with these Ψ_{ij}. If $\Delta\mu$ is the percent deviation from the smoothed value

$$\Delta\mu = \frac{\mu_{exp} - \mu_{smoothed}}{\mu_{smoothed}} \times 100 \quad (46)$$

then L_1, the mean absolute deviation, is given by

$$L_1 = \frac{1}{N} \sum_{i=1}^{N} \Delta\mu_i \quad (47)$$

here N is the number of data points. L_2, the root-mean-square deviation, is given by

$$L_2 = \frac{1}{N} \sum_{i=1}^{N} (\Delta\mu_i)^2 \quad (48)$$

L_3, the maximum absolute deviation, is given by

$$L_3 = \Delta\mu_{max} \quad (49)$$

At each temperature, values of Ψ_{ij} obtained by each method were selected to give the generally most favorable set of values of L_i (usually the smallest values). The relative effectiveness of the two methods is evident from comparison of the two sets of L_i; for practical interpolation one would pick the set of Ψ_{ij} that gives the more satisfactory L_i.

The presentation of all the Ψ_{ij} values calculated from the available experimental data (in Table 1) in addition to presenting the recommended sets of Ψ_{ij} values (in Table 2) is believed to be justified. First, the selected values given in Table 2 show mainly the temperature dependence, whereas the full values of Ψ_{ij} in Table 1 show both the composition and temperature dependences. Thus the extensive tabulation in Table 1 provides a general basis for data correlation and analysis and should be useful for further studies on these dependences. Second, the fact that Ψ_{ij} are weakly dependent upon composition and temperature is true only for mixtures of simple molecules, and it is not true for mixtures of complex molecules such as highly polar and polyatomic molecules, for which the full values in Table 1 are needed. Third, the full values of Ψ_{ij} in Table 1 are useful for the estimation of viscosity values at high temperatures and for multicomponent systems.

4. EXPERIMENTAL METHODS

A. Introduction

Historically, the early interest in the measurement of viscosity was directed more to liquids than to gases. This is obviously because of the practical thrust and everyday interest in the general problem of the flow of a liquid through a pipe. Dunstan and Thole [831] in their monograph briefly review the measurement done on pure liquids prior and subsequent to 1895 through about 1912. This work [831] also includes a brief reference to the viscosity of liquid mixtures, electrolytic solutions, and colloidal solutions. In 1928, Hatschek [832] published a more detailed account of the work done on the viscosity of liquids, similar in scope to that of Dunstan and Thole [831]. A more detailed description of the techniques of measurement of viscosity of gases and liquids is given by Barr [833]. Through these years the increasing interest in the viscosity of non-Newtonian fluids has led to the development of special techniques for such materials. Van Wazer, Lyons, Kim, and Colwell [834] have given an excellent description of the various viscometers developed and commercially available pertinent to the rheological studies. They [834] also append a list of 100 selected books on rheology. In this section, consistent with the scope of this monograph, we will describe and refer to more recent work and to techniques which have resulted in a large body of data of reasonable accuracy. No claim can be made concerning its completeness, though it is hoped that this will constitute a fairly comprehensive survey of

the work done during the last three to four decades. Gases will be discussed specially here and liquids in a subsequent chapter. Very briefly, Partington [778], Kestin [835], and Westenberg [446] have discussed the major methods of determining the viscosity of gases. Experimental measurements of viscosity fall in two general categories, absolute and relative. Absolute viscosity measurements differ from relative measurements in that the latter lead to viscosity values in terms of the viscosity of a known substance.

B. Various Methods of Measurement

a. The Capillary-Flow Method

The foundation of this method was laid in 1839 by the work of Hagen [836], who measured the flow rates of water through capillaries of varying bore and length. Poiseuille [837] in 1840 published a note, and his subsequent work describes in detail the theory of fluid flow through thin glass capillaries. It is on these pioneer investigations that a large number of efforts are based. Viscosity determinations, made with various variations of the same simplifying assumptions, do have to include many corrections before accurate values of viscosity can be computed from direct measurements. These will be discussed below, but in passing it may be mentioned that Fryer [838] has recently considered the theory of gas flow through capillaries, covering all the three pressure regimes when the mean free path is smaller than, comparable to, and greater than the diameter of the tube.

In the simple case of an incompressible Newtonian fluid flowing steadily through a capillary which is a perfect cylinder and in which the flow is everywhere laminar, with no slip at the wall, the mass rate of flow at the inlet, θ_i, is given by

$$\theta_i = \frac{\pi a^4 \bar{\rho}(P_i - P_o)}{8l\mu} \tag{50}$$

Here a is the root-mean-square radius of the tube, l its length, $\bar{\rho}$ the gas density evaluated at the capillary temperature and average pressure between inlet and outlet, P_i and P_o are the pressures at the inlet and outlet, respectively. For a compressible fluid flowing through a capillary of mean radius, a, with slip at the wall, and including the kinetic-energy correction, the above equation is given by [833]

$$\mu = \frac{\pi a^4 \bar{\rho}(P_i - P_o)}{8l\theta_i(1 + \delta)}\left(1 + \frac{4\zeta}{b}\right) - c\frac{\theta_i}{8\pi l} \tag{51}$$

Here δ is a small correction for nonuniformity of the bore, $(1 + 4\zeta/b)$ accounts for the slip at the wall, and the last term arises because of the departure of the flow patterns at the inlet and outlet of the capillary from true parabolic velocity distribution. The detailed form of the equation depends on the nature of the experimental arrangement and the procedure being adopted in taking the data; see for example Shimotake and Thodos [839], Flynn, Hanks, Lemaire, and Ross [840], Giddings, Kao, and Kobayashi [841], Kao, Ruska, and Kobayashi [1146], and Carr, Parent, and Peck [842]. It is found that stable laminar flow exists as long as the Reynolds number is less than 2000 [835, 840].

In one variant of this general capillary flow method, the constant-volume gas viscometer, the gas transpires from a bulb containing the test gas through the capillary into a constant low-pressure region. In many cases the latter is just atmospheric pressure or a very low pressure obtained by continuous pumping. The fall in gas pressure of the bulb is noted over a known period of time. Since the historical work of Graham [843] frequent use of this general technique is made in determining the viscosities of gases and gaseous mixtures. Edwards [844] employed this principle and measured the viscosity of air between 15 and 444.5 C. This work resolved the controversy over the applicability of the Sutherland model to predict the temperature dependence of viscosity arising out of the experimental work of Williams [845] and the comment of Rankine [846]. Kenney, Sarjant, and Thring [847] built a similar apparatus with emphasis on design for work at high temperatures. They [847] measured the viscosity of nitrogen–carbon dioxide gas mixtures up to about 900 C with an estimated accuracy of 2%. Bonilla, Brooks, and Walker [848] employed this type of apparatus with a platinum capillary coiled in the form of a helix and made measurements on steam and nitrogen at atmospheric pressure. They went up to the maximum temperature of 1102.2 C for nitrogen and 1205.6 C for steam. They corrected their data for coiling of the capillary as outlined by White [849]. Bonilla, Wang, and Weiner [850] built another apparatus and measured the viscosity of steam, heavy-water vapor, and argon relative to the known values for nitrogen. The measurements at atmospheric pressure extend up to as high as about 1500 C. McCoubrey and Singh [851, 852] employed a glass constant-volume gas viscometer and maintained a much lower pressure at the exit end of the capillary by continuously pumping, and thus determined the relative values of viscosity within an uncertainty of about 1%. They worked with a number of polyatomic

quasi-spherical molecules and pentanes in the temperature range 20 to about 200 C. A similar viscometer has been used by Raw and co-workers [871–873] to measure the viscosity of binary gas mixtures in the temperature range 0–400 C with an overall accuracy of ±1%. Smith and co-workers [874–877] have devised a modified viscometer of this type and made relative measurements on pure gases over a wide temperature range, 77–1500 K, with an estimated accuracy of about 1%. Recently this group has reported data on inert gases [1151] and three gases each composed of quasi-spherical molecules [1152]. Pena and Esteban [1148, 1149] have employed a constant volume capillary viscometer and determined viscosities of organic vapors in the temperature range from −10 to 150 C. It is, thus, clear that this arrangement of capillary-flow viscometers is appropriate for moderate-accuracy absolute or relative measurements on gases at pressures around one atmosphere. The marked simplicity and convenience of operation of such a viscometer has made it attractive for undergraduate laboratory experimentation [853].

Trautz and Weizel [854] initiated a different variant of this general principle of transpiration of gas through a capillary to determine viscosity. They allowed the gas to flow through the capillary into the atmosphere from a reservoir whose volume was not kept constant; instead a known volume of gas from it is pushed by increasing the pressure and the time is recorded. Thus, both pressure and volume of the gas at the inlet side of the capillary change with time. The integration of the basic flow equation thus becomes somewhat difficult because of the variation in both pressure and volume, and consequently this procedure has been preferred for relative measurements.

Rankine [855, 856] devised a very clever capillary transpiration viscometer which is simple, employs a very small quantity of gas, and can be readily adopted for relative measurements. It consists of a closed glass loop of which one vertical side is wide while the other is a capillary. A mercury pellet descending in the wide leg exerts a known force and forces the gas up through the capillary. The pressure difference across the capillary remains constant because it is due only to the mercury pellet. At high pressures it is necessary to account for the buoyancy effect for the pellet. The volume rate of gas flow through the capillary is computed by timing the descent of the pellet between two masks on the wide tube. The viscometer is symmetrical about a horizontal axis and can be rotated to allow the movement of the pellet in the opposite direction. The surface tension of the mercury pellet plays a very important role, particularly if the gases are not quite inert. Rankine and Smith [859] corrected for such a possibility by taking observations for each case both with the pellet intact and then broken into two or three segments. It is assumed that the capillary effect is doubled and tripled in a pellet broken into two and three segments, respectively. Rankine [860, 861] has used this technique extensively to determine the viscosity of gases and vapors as a function of temperature at ordinary pressures, in order to determine molecular sizes.

Comings and Egly [862] and Baron, Roof, and Wells [863] suitably modified the original design of the Rankine viscometer so that measurements at elevated pressures and temperatures may be made. Comings and Egly's [862] work covers ethylene and carbon dioxide at 40 C and extends up to a maximum pressure of 137.1 atm. They claim a maximum probable uncertainty of 2% for measurements below 89 atm, and 4% above this pressure. Baron, Roof, and Wells [863], on the other hand, took measurements on nitrogen, methane, ethane, and propane in the pressure range 100–8000 psi and at temperatures of 125, 175, 225, and 275 F. The precision of their data is better than 1%.

Heath [864] used a glass Rankine viscometer and made relative measurements at 18 C and 70 cm Hg pressure for various mixtures of helium–argon, helium–nitrogen, helium–carbon dioxide, hydrogen–argon, hydrogen–nitrogen, and hydrogen–carbon dioxide. A similar viscometer was used to measure the viscosity of rare gas mixtures within an accuracy of ±1.0% at about 18 C and 70 cm Hg pressure [865–867].

Williams [845], in his experiment, displaced a known volume of gas but controlled the flow rate so that the gas inlet pressure and the pressure difference across the capillary were constant throughout the experiment. Anfilogoff and Partington [778] have described in detail the design of such a viscometer and in recent years Raw and co-workers [868–870] have employed an apparatus of the same general principle and measured viscosities of gases and gaseous mixtures up to a maximum temperature of 1000 C with an estimated uncertainty of 1%.

A number of capillary viscometers have been designed to obtain viscosity values (relative in most cases) of gases over wide temperature and pressure ranges through the basic Hagen–Poiseuille equation. The pressure difference across the capillary is kept constant and the flow rate of the gas transpiring through the capillary is measured accurately. Some

important efforts of this type are by Timrot [878], Makavetskas, Popov, and Tsederberg [879, 880], Vasilesco [881], Lazarre and Vodar [882, 883], Luker and Johnson [884], Andreev, Tsederberg, and Popov [885], Rivkin and Levin [886], Lee and Bonilla [887], Masiá, Paniego, and Pinto [1147], etc. Flynn, Hanks, Lemaire, and Ross [840] and Giddings, Kao, and Kobayashi [841] have developed very accurate absolute viscometers, and reported data on gases as a function of temperature and pressure with an accuracy of a few tenths of a percent. The measurements of Ross *et al.* cover a maximum and a minimum temperature of 150 C [888] and −100 C [889], respectively, and pressures up to a maximum of 250 atm. The measurements of Kobayashi *et al.* [841, 890] cover the temperature range −90 to 137.78 C and the pressure range 6.8–544.4 atm.

A very important variation in the general capillary method was introduced by Michels and Gibson [891] in 1931 while engaged in measurements at high pressures. A known pressure difference is imposed across the capillary and the flow rate is determined under the decreasing pressure head. Several alternative procedures have been developed to obtain this type of operation and these unsteady state viscometers will be mentioned below. Careful interpretation of the observed data leads to very accurate absolute values of viscosity. Michels and Gibson's [891] measurements on nitrogen at 25, 50, and 75 C and up to 1000 atm have been extended up to 2000 atm on hydrogen and deuterium [892], argon [893], and carbon dioxide [894]. Trappeniers, Botzen, Van Den Berg, and Van Oosten [895] have recently revived this work and measured the viscosity of neon at 25, 50, and 75 C and at pressures up to 1800 atm, for krypton [896] at these temperatures and pressures up to 2050 atm, and at 125 C at pressures between 1300 and 1900 atm. Some other workers who have employed this general principle to measure viscosity over a limited temperature range at ordinary pressures are: Bond [897], Rigden [898], Thacker and Rowlinson [899], Chakraborti and Gray [900, 901], and Lambert *et al.* [902]. In most cases these measurements are relative.

Shimotake and Thodos [839] developed a viscometer and, based on this unsteady-state method, determined the viscosity of ammonia. Their [839] relative measurements cover the pressure range 250–5000 psia and temperatures of 100, 150, and 200 C. Thodos and co-workers have also done careful measurements on sulfur dioxide [903], argon, krypton, and xenon [904], and helium, neon, and nitrogen [907]. Eakin and Ellington [908] and Starling, Eakin, and Ellington [909] developed another design for a viscometer on this very principle, and reported data on the viscosity of propane within an estimated accuracy of $\pm 0.5\%$ for nine temperatures between 77 and 280 F and for pressures in the range 100–8000 psia. On the basis of this viscometer a large body of data was developed which is of special practical interest to the petroleum industry [753, 910–916]. Guevara, McInteer, and Wageman [1208] determined relative values of viscosity employing a capillary viscometer in the temperature range 1100–2150 K at atmospheric pressure with an accuracy of $\pm 0.4\%$ and precision of $\pm 0.1\%$. The data are reported on viscosity ratios for hydrogen, helium, argon, and nitrogen [1208], krypton [1209], neon [1210], and xenon [1211].

b. The Oscillating-Disk (Solid-Body) Method

This method, like the capillary-flow method, has a long history following the pioneer work of Maxwell [918] in 1870. This method in many respects is the opposite of the capillary-flow method. Here the test fluid is kept stationary while a solid body oscillates and the effect of shearing stresses on the oscillations makes possible, if properly analyzed, the determination of viscosity. It may be recalled that in the capillary-flow method it is the test fluid which moves and the knowledge of flow rate and associated pressure difference permit the calculation of viscosity. The principle of the solid-body method involves the measurement of the period and amplitudes of the damped oscillations of a suitable solid body suspended from an elastic wire in the test fluid and then in vacuum. The latter makes possible correction for the damping due to the torsion of the suspension wire in a straightforward manner. However, the exact theoretical description of the velocity field around the oscillating body in the test fluid is not simple; this is the major limiting feature of this method. These complications and their theoretical resolution for various shapes of the oscillating body have been understood only in recent years; this is reviewed by Kestin [835]. In particular, the shapes which have been adopted are a sphere or a thin cylindrical disk oscillating freely in the fluid, or a thin disk oscillating between two fixed parallel disks with finite spacing. This latter alternative has received wide use for the determination of viscosity both relative and absolute. Craven and Lambert [919] employed a sealed quartz bulb pendulum drawn out from a 1-cm-diameter tubing. The lower end was drawn out to form a pointer. The pendulum was set into oscillations and

the damping time was measured as a function of pressure of the gas. The measurements were taken relative to air with an estimated error of 1%, as the pressure independent damping times were taken to be directly proportional to the viscosity of the gas. A detailed discussion of the various efforts made to theoretically and experimentally examine this method is beyond our scope, and we refer the reader to the article of Kestin [835] and to the number of original articles referred in it. We will briefly review below some of the recent efforts and point out developments which have helped considerably in improving the potential of the technique and work which has produced a large body of data.

The Kammerling Onnes Laboratory at Leiden ~~Kammerling~~ Kamnerlingh initiated experimental and theoretical studies of this oscillating-disk-type apparatus: Van Itterbeek and Claes [920, 921], Van Itterbeek and Keesom [922, 929], Van Itterbeek and Van Paemel [923, 924, 930], Keesom and Macwood [925, 926], and Macwood [927, 928]. In more recent years Van Itterbeek and his co-workers [931–933] have also measured the viscosity of binary mixtures of monatomic and diatomic gases in the temperature range 72.0–291.1 K with an estimated error of 1%. The viscosity calculation was made from the equation

$$\mu = C\left(\frac{\lambda}{\tau} - \frac{\lambda_0}{\tau_0}\right) \tag{52}$$

where C, a constant of the apparatus, is obtained from

$$C = \frac{4I}{\pi R^4} \frac{d_1 d_2}{d_1 + d_2} \tag{53}$$

Here I is the moment of inertia of the oscillating disk, R the radius of the oscillating disk, d_1 and d_2 the distances between the oscillating and fixed disks, λ and λ_0 the logarithmic decrements of the oscillations in the test fluid and vacuum, respectively, and τ and τ_0 the periods of the oscillations in the test fluid and vacuum, respectively. Two types of oscillation systems have been employed. In one the distance between the fixed disks could not be changed, while in the second it was adjustable. These authors [934–937] have also measured the viscosity of light gases and their mixtures down to temperatures as low as 14 K.

Mason and Maass [939] developed a design of the oscillating-disk viscometer somewhat similar to that of Sutherland and Maass [938], to measure the viscosity of gases in the critical region. They [938] claim a differential accuracy of 1 in 3000 and an absolute accuracy of 1 in 1000 in measurements over a temperature range 0–100 C and for pressures up to 150 atm. The calculation procedure is the same as described above. Johnston and McCloskey [940] also built a viscometer of the same general pattern [938] and measured the viscosity of a number of gases [940, 941] between room and liquid-oxygen temperature with an accuracy of 0.3% at 300 K to about 0.8% at 90 K.

Kestin and Pilarczyk [942] measured the viscosity of gases by an accurately built oscillating-disk viscometer and pointed out the necessity of improving the theory of this apparatus if highly precise values are to be obtained. Kestin and Wang [943] succeeded in semiempirically developing the edge correction factor arising because of the finite size of the disk and re-evaluated [944] the earlier measurements [942]. Kestin, Leidenfrost, and Liu [945] further examined the edge correction factor and verified experimentally the procedure of relative measurements in such a viscometer for moderate spacings. This provides considerable confidence in the measurements of Kestin and Leidenfrost [946, 947] on pure gases, which were taken on a modified version of the apparatus of Kestin and Moszynski [948].

Around this time a number of additional improvements in the theory of such a viscometer appeared: Mariens and Van Paemel [949], Dash and Taylor [950], and Newell [951]. These made it possible to evaluate the experimental information on an absolute basis to get very accurate values of viscosity. Kestin and Leidenfrost [952, 953] thus succeeded in determining the absolute values of viscosity of gases and gas mixtures at 20 C over a range of pressure values, using their earlier viscometer [947] with a very high degree of accuracy. Kestin and co-workers [954–961] have reported data at 20 and 30 C for a large number of binary systems and pure gases as a function of pressure from 1 to about 50 atm with an estimated accuracy of the order of 0.2%, and an uncertainty of no more than 0.04% for the relative values of the mixtures in comparison with the pure gases. Di Pippo, Kestin, and Whitelaw [962] have also designed an absolute high-temperature viscometer appropriate at atmospheric pressure in the temperature range 20–950 C. In recent years Kestin and co-workers [1213–1215] have employed an oscillating-disk viscometer and reported the relative measurements of the viscosity of pure gases and their binary mixtures in the temperature range 25–700 C and at atmospheric pressure with a precision of $\pm 0.1\%$.

Clifton [963] measured the viscosity of krypton in the temperature range 297 to 666 K and calibrated his viscometer with helium. He also found that the rigorous theory [951], with approximate geometrical dimensions of the viscometer, gave the calibration factor within about 3%. Thus he provided another very much needed experimental proof of the theory of viscometer as well as the calibration procedure which forms the basis of all relative measurements. Pal and Barua [964] constructed a metal viscometer and determined the viscosity of H_2–N_2 and H_2–NH_3 gas mixtures in the temperature range 33–206 C at one atmosphere pressure. They calibrated their apparatus according to the procedure pointed out by Clifton [963] employing the viscosity data for H_2 and N_2 of Barua *et al.* [888] and Kestin and Whitelaw [965]. Pal and Barua [966–969] have reported data on a number of other pure gases and binary gas systems in this temperature range. A similar approach has been adopted by Gururaja, Tirunarayanan, and Ramachandran [970] who have reported data on binary and ternary mixtures at ambient temperature and pressure.

c. The Rotating-Cylinder (Sphere or Disk) Method

The uniform rotation of a sphere, disk, or cylinder in concentric spherical shells, fixed parallel planes, or a fixed concentric cylinder, respectively, is used to determine the viscosity of the fluid enclosed between the two surfaces. A historical account of this method is to be found in reference [833]. Because of practical convenience, the coaxial cylinder geometry has been preferred by most of the workers with this method. A brief review of such efforts will be given here, with special reference to work which has appeared since the review of Barr [833]. In its most commonly used variant, the angular deflection, Φ, of the inner cylinder is noted when the outer cylinder is rotated with a constant angular velocity of ω. Let r_i and r_o be the radii of the inner and outer cylinders, respectively, and l the length of the inner cylinder where the test fluid is enclosed between the two cylinders. If the end effects which arise because of the finite length of the inner cylinder are ignored, the viscosity is obtained from a rather simple relation

$$\mu = \frac{\pi \Phi I (r_o^2 - r_i^2)}{r_i^2 r_o^2 \tau^2 \omega l} \qquad (54)$$

Here I and τ are the moment of inertia and period of vibration of the inner cylinder and Φ is obtained by noting the steady-state deflection as read on a straight scale located at a distance d from the mirror and attached to the suspension system of the inner cylinder so that

$$\tan \Phi = s/2d \qquad (55)$$

It may also be remarked that the speed of the rotating cylinder must be so chosen that the fluid flow remains viscous and radial or eddy motion does not occur [833]. The mathematical theory for the correction of end effects has not yet been developed, but these are reduced by providing "guard rings" above and below the suspended cylinder. These are the major considerations which limit the absolute nature of this method and impair the accuracy. In principle, either of the two cylinders can be rotated with a constant angular velocity, though consideration of the instability of motion suggests a preference for the outer cylinder to be rotated [835].

Gilchrist [971] built a constant deflection type coaxial cylinder apparatus, having guard cylinders both at the top and bottom, and measured the viscosity of air. He used a bifilar phosphor bronze strip for suspension. Later Harrington [972] tried to improve upon this design. He used quartz fibers instead of phosphor bronze and very accurately determined the geometrical constants of the apparatus and the moment of the inertia of the inner cylinder. His results on air at about 23 C are claimed to be accurate within a maximum uncertainty of 0.04%. He also claimed that for his apparatus at ordinary pressures the correction amounts to about 2 parts in 100,000. Yen [973] and Van Dyke [974] used this apparatus to determine the viscosities of oxygen, nitrogen, hydrogen, and carbon dioxide. The adaption of this apparatus for operation at low pressures and the theory of slip are discussed by Millikan [975], Stacy [976], Van Dyke [974], States [977], and Blankenstein [978]. Several other efforts have been made to build improved versions of the basic Harrington–Gilchrist apparatus to measure viscosities of normal pentane and isopentane [979] and air [980, 981].

Reamer, Cokelet, and Sage [982] built a rotating cylinder viscometer for measurements at pressures up to 25,000 psia in the temperature range 0–500 F. They reported data on n-pentane with an estimated accuracy of 0.4%. Additional measurements have been reported on this apparatus for ethane [983] and ammonia [984] and mixtures of nitrogen–n-heptane, nitrogen–n-octane [985], and methane–n-butane [986].

d. The Falling-Sphere (Body) Method

The principle of this method, its scope and limitations, and many of the experimental attempts made are described in references [833] and [835]. The basis for this method is in Stokes' law, according to which the viscous drag, W, on a rigid sphere of radius a, falling in an infinite homogeneous fluid which has attained a uniform velocity of v (free from accelerations) is

$$W = 6\pi\mu a v \qquad (56)$$

Furthermore, under these conditions, W is equal to the apparent weight of the sphere so that

$$W = \tfrac{4}{3}\pi a^3 (\rho_s - \rho_f) g \qquad (57)$$

Here ρ_s and ρ_f are the densities of the sphere and the fluid respectively, and g the acceleration due to gravity. Combining these two equations

$$\mu = \frac{2}{9} \frac{(\rho_s - \rho_f) g}{v} a^2 \qquad (58)$$

This relation is valid only for extremely low Reynolds numbers, though modifications to this law have been proposed for higher Reynolds numbers [833, 835]. For bodies other than spheres Stokes' law is modified so that

$$W = 6\pi\mu a v/\delta \qquad (59)$$

where the value of δ depends upon the shape of the body [833].

Ishida [987] employed this principle and by observing the rate of fall of charged droplets in the test gas determined the viscosity of the latter. It is necessary to consider the effect of slip in view of the small size of the drops, and further, it is implied that the electric field of the drops does not alter the viscosity of the test gas.

Hawkins, Solberg, and Potter [988] described a falling-body viscometer similar to that which Lawaczeck developed in 1919. It consists of a metal cylindrical weight falling through the test fluid contained in a vertical tube closed at the lower end and having a diameter slightly greater than that of the weight. Under certain conditions the simple measurement of the time t needed for the weight to fall through a fixed distance is a measure of the viscosity so that

$$\mu = C(\rho_s - \rho_f) t \qquad (60)$$

Here C is a constant dependent on the dimensions of the apparatus and can be determined if an experiment is made with a fluid of known viscosity. These workers [988] described a viscometer appropriate for measurements up to pressures of 3500 psi and temperatures of 1000 F. The viscometer was rotated through 180° to permit the body to fall in the tube in the opposite sense and the measurements repeated.

A combination of an inclined tube and a rolling ball has been used as a convenient, simple empirical method for the last fifty years to determine the viscosity of fluids. Hubbard and Brown [989] derived general relations, through the use of dimensional analysis, between the variables involved and the simple calibration for the rolling ball viscometer, in the streamline region of fluid flow. An empirical correlation is also given which enables viscosity to be estimated from data taken in the turbulent region of flow. The correlating functions were evaluated from data taken on a viscometer consisting of a precision-bore inclined glass tube, and times to traverse a known distance were determined with an automatic photo-electric device. This design was further modified by Smith and Brown [747].

Bicher and Katz [990] employed a rolling-ball inclined-tube viscometer and measured the viscosities of methane, propane, and their mixtures with an average error of 3.2%. The ranges of pressure and temperature examined were 400–5000 psia and 77–473 F, respectively.

Swift, Christy, Heckes, and Kurata [991] designed a falling-body viscometer and have reported viscosities of liquid methane, ethane, propane, and *n*-butane [992]. Huang, Swift, and Kurata [993] modified the design of the viscometer [992] so that measurements were possible up to as high a pressure as 12,000 psia. They [993] reported measurements on methane and propane at pressures to 5000 psia and went down to the lowest temperature of -170 C with an estimated precision of $\pm 1.2\%$. These authors have also extended the measurements to the mixtures of methane and propane [994].

Stefanov, Timrot, Totskii, and Chu Wen-hao [1150] have employed an improved falling-weight viscometer to measure the viscosity of the vapors of sodium and potassium as a function of temperature and pressure.

e. The Less-Developed Methods: Based on Ultrasonic, Shock Tube, and Electric Arc Measurements

Recent interest in the exploration and understanding of outer space have led to the development of methods which may give viscosity values at high temperatures up to about 15,000 K. A very limited amount of experimental work has been done and many difficulties are not resolved, the techniques are not entirely satisfactory. A considerable amount of theoretical and experimental work is needed to establish the techniques so that reliable data may be obtained. In view of the unsatisfactory state of the art only a brief account of the efforts made so far will be sufficient.

Measurements of the velocity of sound in a gas permit its temperature to be determined [995]. Carnevale *et al.* [996–998] employed this principle and measured the viscosity at high temperatures from the knowledge of the velocity and absorption of ultrasonic waves in the test gas. In particular, they [998] determined the viscosity of helium up to 1300 K and of argon up to 8000 K at one atmosphere. This attempt has been extended to include polyatomic gases and temperatures as high as 17,000 K [999], and high pressures up to 100 atm [1000, 1001]. Besides experimental difficulties, there still remain many theoretical questions to be answered. A critical evaluation of this ultrasonic technique has been given by Ahtye [1002], who has included in the theory of ultrasonic absorption, in addition to components due to viscosity and thermal conductivity, also terms which arise due to chemical relaxation and radiative heat transfer. Madigosky [1003], while discussing his results of ultrasonic attenuation in gases at high densities, has pointed out the need for considering a significant absorption resulting from the bulk viscosity, in addition to shear viscosity, thermal conductivity, etc.

Measurement of the heat transfer to the side wall of a shock tube is used in conjunction with a suitable equilibrium boundary layer theory to determine viscosity of shock heated gases. Carey, Carnevale, and Marshall [1004] thus determined the viscosities of argon, oxygen, nitrogen, and carbon dioxide up to 4000 K. Hartunian and Marrone [1005] used this principle to determine the viscosity of dissociated oxygen with an estimated accuracy of $\pm 4\%$.

Theoretical understanding and experimental techniques have been developed to the point that measurements on a confined electric arc are capable of yielding fairly accurate data on viscosity and other properties of the gas [1006]. Schreiber, Schumaker, and Benedetto [1007] have recently described the details of an argon-plasma source and related instrumentation, along with some preliminary measurements of a continuing program. Schreiber, Hunter, and Benedetto [1144] have measured the viscosity of an argon plasma at one atmosphere and in the temperature range 10,000–13,000 K.

Dedit, Galperin, Vermesse, and Vodar [1145]

have described an apparatus in which the record of displacements of a column of mercury as a function of time is employed to determine viscosity of a gas compressed to varying pressures.

Viscosity of Liquids and Liquid Mixtures

1. INTRODUCTION

In the preceding sections a brief discussion is given of the theoretical status, estimation procedures, and experimental techniques for gases and gas mixtures at ordinary as well as at high pressures before condensation occurs. We will now review the similar art in relation to pure liquids and their mixtures. Many of the ideas developed in connection with the studies on gases are still valid, either as such or with appropriate modifications, and consequently, our present discussion will be essentially a continuation specialized for liquids and consistent with our over-all plan to be brief but relatively complete in references. The work on liquids is less extensive than that on gases, though in recent years more attention has been paid to the former.

Many monographs are available which describe the different theories developed to explain the liquid state and the different thermodynamic and transport properties. Some of these are by Frenkel [1008], Green [1009], Rice and Gray [1010], Kirkwood [1011], and Hirschfelder, Curtiss, and Bird [28]. Many excellent review articles have also appeared, e.g., Rice [1012], Kimball [1013], Lebelt and Cohen [1014], Brush [1015], Partington [1016], Hildebrand [1017], and deBoer [1064]. These describe the status of the current theory and its ability to explain the observed experimental facts. In the next section we mention the theoretical efforts made to describe the mechanism of momentum transfer in liquids, and hence, the coefficient of viscosity. The next two sections describe the empirical approach to estimating and experimentally measuring the viscosity of liquids. It may be pointed out that very often the term fluidity is used in literature to represent the reciprocal of viscosity. The reason for this is that for liquids the fact to explain is not their viscosity, i.e., their tendency to offer resistance under the influence of a shearing stress, but their fluidity, i.e., their capability of yielding to such a stress [1008].

2. THEORETICAL METHODS

A. Introduction

Although the liquid state is intermediate between the solid and gaseous states, most materials have properties in the liquid state which are close to those of one or the other of these two states. For a simple example, liquids, like gases, adopt the shape of the container—they lack rigidity. Similarly, liquids, like solids, are hard to compress, in sharp contrast with gases. From the molecular point of view, the molecules are closely packed in solids and in liquids, while in gases the intermolecular separations are so large that the molecular motion is random and free from the influence of the other molecules for most of the time. In liquids, on the other hand, molecules are so closely packed that the molecular motion is much more limited in space and is controlled by the influence of many neighboring molecules. Thus, the transport of momentum in liquids takes place, in sharp contrast with gases at ordinary pressures, not by the actual movement of molecules, but by the intense influence of intermolecular force fields. It is this basic difference in the mechanism of momentum transfer which is responsible for the opposite qualitative dependence of viscosity on temperature for gases and liquids. The viscosity of gases increases with temperature, while that of liquids decreases with temperature. This simple concept can be developed to give an appreciation of the mechanism of transport of momentum, and hence, of the coefficient of viscosity. We will now discuss the various theories developed to explain the phenomenon of viscosity in liquids.

B. The Simple Theories

It seems from the above brief description of the viscous nature of liquids that formulation of a simple theory to explain it has very little promise. Nevertheless, some efforts at the early stages of the development of the subject were made by ingeniously interpreting the motion of molecules and by associating special mechanisms of momentum transfer during collision, as reviewed by Frenkel [1008] and Andrade [1018]. By considering the forces of collision to be the only important factor, J. D. van der Waals derived the following expression for the coefficient of viscosity μ [1018]:

$$\mu = \frac{8\sqrt{\pi}}{15} n^2 d^4 m^{1/2} k^{1/2} \frac{v}{v-b} T^{1/2} e^{-\epsilon/RT} \quad (61)$$

Here n is the number of molecules of mass m and diameter d per square centimeter, ϵ is the difference between the amount of potential energy that the molecules of the liquid possess on an average and the amount which they possess at the moment of a collision, v represents volume, and b is the van der Waals constant. This theory predicts $(1/\mu)(d\mu/dT)_v$ to be positive, although experiments lead to negative values for this factor.

The theory of Andrade [1018, 1019] may be mentioned because many of its predictions have survived the experimental checks to some extent. He attempted to develop the theory from the solid state point of view. Assuming that at the melting point the frequency of vibration is equal to that in the solid state, and that one-third of the molecules are vibrating along each of the three directions normal to one another, Andrade [1018] showed that

$$\mu = 5.1 \times 10^{-4}(AT_m)^{1/2}(V_A)^{-2/3} \quad (62)$$

Here A is the atomic weight, T_m is the melting point, and V_A is the volume of a gram atom at temperature T_m. The above formula checked well against the data on monatomic metals at the melting point. The predictions were less satisfactory for liquid halogens, oxygen, and hydrogen.

Andrade [1019] also extended his theory to explain the temperature and pressure dependence of viscosity. Assuming the frequency of vibration of the liquid molecules, v, to be constant, Andrade [1019] showed that the temperature dependence of viscosity is given by

$$\mu = A \exp(c/T) \quad (63)$$

where A and c are constants. By including the temperature dependence of volume he found [1019], instead of the above expression, a more complicated result,

$$\mu v^{1/3} = A \exp[cf(v)/T] \quad (64)$$

Here v is the specific volume. When the molecular interaction potential is approximated by the van der

Waals relation, the above relation becomes

$$\mu v^{1/3} = A \exp(c/vT) \quad (65)$$

If the temperature dependence of the frequency v is also considered, equation (64) becomes

$$\mu v^{1/6} = (A'/\sqrt{k_1}) \exp(c'/vT) \quad (66)$$

Here A' and c' are constants and k_1 is the adiabatic compressibility. Checks against the experimental data showed that equation (64) leads to values which are in better agreement with the experimental results than equation (66). This is interpreted as indicating that some compensating effect is responsible for the superiority of equation (65) in representing the observed data. Andrade [1019] also argued that equation (66) will give the pressure dependence of μ if k_1 and v are given appropriate values corresponding to the pressure under consideration. Consequently,

$$\frac{\mu_p}{\mu_1} = \left(\frac{v_1}{v_p}\right)^{1/6} \sqrt{\frac{k_{1,1}}{k_{1,p}}} \left[\exp\left\{\frac{c}{T}\left(\frac{1}{v_p} - \frac{1}{v_1}\right)\right\}\right] \quad (67)$$

Here the subscripts on μ, v, and k_1 refer to the pressure, p, or the pressure at one atmosphere at which these quantities are to be interpreted. Andrade [1019] found the above relation to be satisfactory up to about 3000 atm. Andrade also suggested that in the absence of adiabatic compressibility, isothermal compressibility values may be used. The constant c is to be obtained from equation (64). Andrade [1066] has given additional comments on the scope of these formulas and assessed them against experimental data.

Frenkel [1008] has discussed simple approaches to derive expressions for μ. Considering the molecules of a liquid to be spheres of radius a, he takes the resistance F suffered by a molecule as it moves with an average velocity $\bar{v}$ with respect to the surrounding molecules, on the basis of Stokes' law to be

$$F = 6\pi a\mu\bar{v} = \alpha^{-1}\bar{v} \quad (68)$$

where α is the mobility of the molecule. α is related to the self-diffusion coefficient D by Einstein's relation

$$\alpha = D/kT \quad (69)$$

Here k is the Boltzmann constant. The dependence of the mean life of an atom τ in an equilibrium position on temperature is given by

$$\tau = \tau_0 e^{W/kT} \quad (70)$$

where W is the activation energy and τ_0 is a constant. The average velocity of translation of the molecules through the whole volume of the liquid is

$$w = \delta/\tau = (\delta/\tau_0) e^{-W/kT} \qquad (71)$$

and the self-diffusion coefficient, which determines the rate of their mixing together is

$$D = \delta^2/\sigma\tau = (\delta^2/\sigma\tau_0) e^{-W/kT} \qquad (72)$$

Substituting these relations one gets

$$\mu = (kT\tau_0/\pi a\delta^2) e^{W/kT} \cong A\, e^{W/kT} \qquad (73)$$

The above relation successfully accounts for the experimentally observed temperature trend of μ, though the absolute computed values are 10^2 to 10^3 times greater than the experimental values. This disagreement is explained by the decrease of W with increasing T. If this dependence is assumed in terms of a parameter γ, such that

$$W = W_0 - \gamma kT \qquad (74)$$

the value of A then changes to

$$A = \frac{kT\tau_0}{\pi a\delta^2} e^{-\gamma} \qquad (75)$$

The μ values are thus reduced by a factor of e^γ. Similarly, if the pressure dependence of W is included according to the relation

$$W = W_0 + (\beta v_0 P/K) \qquad (76)$$

where if v is the volume of an atom, v_0 is the value of v for $P = 0$, and K is the bulk modulus, then the factor A comes out to be an exponential function of pressure

$$A = A_0\, e^{P\gamma/\alpha' KT} = A_0\, e^{P/P_0} \qquad (77)$$

Here A_0 is the value of A for $P = 0$, and α' is the coefficient of thermal expansion, and P_0 is that characteristic pressure where viscosity has increased by a factor of e. This exponential increase of viscosity with pressure is in accord with the experimental data. The above analytical treatment is valid only for moderate values of pressures where $\gamma = v_0\alpha'\beta/k$.

Furth [1038] derived a formula for the viscosity of a liquid by assuming the momentum transfer to take place by the irregular Brownian movement of the "holes" [1039]. These "holes" were likened to clusters in a gas and thus, in analogy with the gas theory of viscosity and with the assumption of the equipartition law of energy, he [1038] showed that

$$\mu = 0.915 \frac{RT}{V} \sqrt{\frac{m}{\sigma}}\, e^{A/RT} \qquad (78)$$

where R is the universal gas constant, σ the surface tension, and A the work function at the melting point.

He [1038] compared his theory with experiments as well as with the theories of Andrade [1019] and Ewell and Eyring [1022]. Furth [1039] developed the concepts of the hole theory of liquids from basic principles of classical statistical mechanics and found he was able to quantitatively reproduce the thermodynamic properties. Auluck, De, and Kothari [1106] further refined the theory and successfully explained the variation of viscosity with pressure.

A good critical review of these simple theories and their abilities to explain momentum transport in liquid is given by Eisenschitz [1065].

C. The Reaction-Rate Theory

Eyring [1020] developed an interesting pictorial description of the liquid state and derived an explanation for the phenomenon of viscosity by the application of the theory of absolute reaction rates [1021]. In a liquid, if a molecule is assumed to be bound to others by bonds of total energy E, then to vaporize a single molecule will require an energy equal to $E/2$ provided no hole is left behind in the liquid. This is because each bond is shared between two molecules. However, if a hole is created in the liquid while vaporizing a molecule, an energy of E will be required. Now, if we return this gas molecule to the liquid we get back an energy $E/2$ only. Using this picture of a liquid, Eyring [1020] concluded that it takes just the same energy to create a hole in a liquid the size of a molecule as to vaporize a single molecule without leaving a hole. Like a gas molecule in empty space, a hole in the liquid can take up a great number of different positions. Whenever a hole is created in the liquid, a neighboring molecule jumps into it leaving behind an empty lattice point, and this process goes on. Consequently, each hole contributes essentially a new degree of translation to the liquid [1020], by permitting the relative motion of molecules near the hole with a minimum of disturbance to other molecules.

Viscous flow was considered as a chemical reaction in which a molecule moving in a plane occasionally acquires the activation energy necessary to slip over the potential barrier to the next equilibrium position in the same plane. The average distance between these equilibrium positions in the direction of motion is λ while the distance between neighboring molecules in the same direction is λ_2, which may or may not be equal to λ. The distance from molecule to molecule in the plane normal to the direction of motion is λ_3. λ_1 is the perpendicular distance between two neighboring layers of molecules in relative motion. Eyring [1020] showed that the viscosity of the liquid is given by

$$\mu = \frac{\lambda_1 h F_n}{K \lambda^2 \lambda_2 \lambda_3 F_a^*} \exp \frac{\Delta E_{act}}{kT} \tag{79}$$

Here K is the transmission coefficient and is the measure of the chance that a molecule having once crossed the potential barrier will react and not recross in the reverse direction. K is usually unity for chemical reactions and will be given this value in the present work. F_n is the partition function of the normal molecule and F_a^* that of the activated molecule with a degree of freedom corresponding to flow. ΔE_{act} is the activation energy for the flow process and h is Planck's constant. Further simplification results if $\lambda = \lambda_1$, for then

$$\lambda_1 / \lambda^2 \lambda_2 \lambda_3 = N/V \tag{80}$$

Here N is Avogadro's number and V is the molar volume. If the degree of freedom corresponding to flow is assumed to be a translational one, while the other degrees of freedom are the same for the initial and activated states, the ratio of the partition functions [1022, 1023] is

$$F_n / F_a^* = (2\pi m k T)^{1/2} (V_f^{1/3}/h) \tag{81}$$

where V_f is the free volume. Eyring and Hirschfelder [1023] have shown that

$$V_f^{1/3} = \frac{bRT}{V^{2/3} N^{1/3} (P + a/V^2)} \text{ per molecule} \tag{82}$$

Here a and b are constants. If ΔE_{vap} is the energy of vaporization,

$$\frac{a}{V^2} = \frac{\Delta E_{vap}}{V} \gg P \tag{83}$$

so that

$$V_f^{1/3} = \frac{bRTV^{1/3}}{N^{1/3} \Delta E_{vap}} \tag{84}$$

$b = 2$ for simple cubic packing and varies weakly with temperature and for other types of packing.

Ewell and Eyring [1022] argued that for a molecule to flow into a hole, it is not necessary that the latter be of the same size as the molecule. Consequently, they write $\Delta E_{act} = \Delta E_{vap} n^{-1}$ for viscous flow, because ΔE_{vap} is the energy required to make a hole in a liquid of the size of a molecule. Combining all these relations one finally gets

$$\mu = \frac{Nh}{V} \frac{(2\pi m k T)^{1/2}}{h} \frac{bRTV^{1/3}}{N^{1/3} \Delta E_{vap}} \exp \frac{\Delta E_{vap}}{nRT} \tag{85}$$

The above relation is used by Ewell and Eyring [1022] to analyze the viscosity data as a function of

temperature for a number of liquids with choices for n varying between 2 and 5. It was found that the theory could reproduce the trend in the temperature dependence of μ but the computed values are greater than the observed ones by a factor of 2 or 3 for most liquids. Many possibilities exist which may be responsible for this discrepancy. Any departure of K from unity will further worsen the agreement between theory and experiment. The packing factor cannot explain this large discrepancy. A good possibility is advanced in the "persistence of velocity theory," that a moving molecule after acquiring the necessary activation energy may move more than one intermolecular distance, so that λ may be equal to $\lambda_1, 2\lambda_1, 3\lambda_1, \ldots$, for any individual elementary process. A strong possibility is that the flow process is bimolecular rather than a unimolecular one [28, 1022, 1024]. Thus, two molecules in adjacent layers which are in relative motion temporarily form a pair, rotate through approximately 90°, and then separate. During the rotation the two molecules will sweep out an extra volume which would be of the order of one-third of the molecular volume.

In order to account for the pressure dependence, Ewell and Eyring [1022] argued that in the above formula one should substitute

$$\Delta E_{vap} = V(P_{int} + P_{ext}) \tag{86}$$

$P_{int} = (\partial E/\partial V)_T$ must therefore be known to account for the pressure dependence of μ. These authors [1022] used the μ data to compute a consistent set of P_{int} values and compared them with those obtained from the thermodynamic relation

$$P_{int} = (\partial E/\partial V)_T = T(\partial P/\partial T)_v - P \tag{87}$$

ΔE_{vap} is related with the more familiar enthalpy of vaporization, ΔH_{vap}, such that [28]

$$\Delta H_{vap} = \Delta E_{vap} + RT \tag{88}$$

Furthermore, the energy of vaporization can be estimated according to the Trouton's rule [28]

$$\Delta E_{vap} = 9.4 R T_b \tag{89}$$

where T_b is the boiling point at one atmosphere.

Kincaid, Eyring, and Stearn [1143] have summarized all the working relations and the underlying theory needed to calculate the viscosity of any normal liquid as a function of temperature and pressure.

D. The Significant-Structure Theory

Eyring and co-workers [1026–1029] improved the "holes in solid" model theory [1024, 1025] to picture the liquid state by identifying three significant structures: (i) solid-like degrees of freedom because of the confinement of a molecule to an equilibrium position as a result of its binding by its neighbors; (ii) positional degeneracy in the solid-like structure due to the availability of vacant sites to a molecule, in addition to its equilibrium position; and (iii) gas-like degrees of freedom for a molecule which escapes from the solid lattice. A liquid molecule, according to significant-structure theory, possesses both solid-like and gas-like degrees of freedom, the relative contribution of the two types being V_s/V and $(V - V_s)/V$ respectively. Here V_s is the molar volume of the solid at the melting point and V is the molar volume of the liquid at the temperature of interest. In brief, a molecule has solid-like properties for the short time it vibrates about an equilibrium position and then it assumes instantly the gas-like behavior on jumping into the neighboring vacancy.

The above method of significant structures leads to the following relation for the viscosity of a liquid [1030, 1031]:

$$\mu = \frac{V_s}{V}\mu_s + \left(1 - \frac{V_s}{V}\right)\mu_g \qquad (90)$$

Here μ_s and μ_g are the viscosity contributions from the solid-like and gas-like degrees of freedom, respectively. The expressions for μ_s and μ_g are given by Carlson, Eyring, and Ree [1031]. Eyring and Ree [1032] have discussed in detail the evaluation of μ_s from the reaction rate theory of Eyring [1020], assuming that a solid molecule can jump into all neighboring empty sites. They [1032] give an expression for μ which in a more general form is [1033]

$$\mu = \frac{Nh}{Zk}\frac{6}{\sqrt{2}}\frac{\Psi}{V - V_s}\exp\left[\frac{a'E_sV_s}{(V - V_s)RT}\right]$$
$$\times \exp\frac{-P(V - V_s)}{RT} + \frac{V - V_s}{V}\frac{2}{3d^2}\left(\frac{mkT}{\pi^3}\right)^{1/2} \qquad (91)$$

Here N is the number of nearest neighbors, E_s is the energy of sublimation, Ψ is the partition function for the oscillator under consideration, a' is the proportionality constant, m is the molecular mass, and d is the molecular diameter, $a'E_sV_s/(V - V_s)$ is the activation energy for jumping. The second exponential is introduced in order to take care of the effect of pressure. At higher pressures, the kinetic energy of molecules becomes correspondingly large and thus the activation free energy is reduced by the kinetic energy.

Lu, *et al.* [1034] have extended the scope of the significant-structure theory to include the molten salts also. The μ expression is of the general form (90), where

$$\mu_s = \frac{Nh}{Zk}\frac{V}{V_s}\frac{6}{\sqrt{2}}(V - V_s)^{-1}[1 - \exp(-\theta/T)]^{-1}$$
$$\times \exp\left[-\frac{a'E_s(V/V_s)^{1/3}}{2RT(V - V_s)/V_s}\right]\exp\left[-\frac{P(V - V_s)}{RT}\right] \qquad (92)$$

$$\mu_g = [n_1/(n_1 + n_2)]\mu_{g_1} + [n_2/(n_1 + n_2)]\mu_{g_2} \qquad (93)$$

$$u_{g_1} = d_1^{-2}(m_1kT/\pi^3)^{1/2}$$

and

$$\mu_{g_2} = d_2^{-2}(m_2kT/\pi^3)^{1/2}$$

μ_{g_1} and μ_{g_2} are the viscosities contributed by monomer and dimer gas-like molecules respectively, d_1 and d_2 are the diameters of the monomer and dimer gas-like molecules respectively, m_1 and m_2 are the molecular weights of monomer and dimer species, n_1 and n_2 are the number of molecules of monomer and dimer species respectively, and θ is the Einstein characteristic temperature.

E. The Cell or Lattice Theory

Lennard-Jones and Devonshire [1035, 1036] introduced a simple model to describe the critical phenomena in gases [1035] and in liquids [1036], which is referred to in the literature by various names such as cell, lattice, cage, free-volume, or one-particle model. In this model each particle is confined to a cell or cage by its nearest neighbors. These cells are assumed to be spherical in shape, and the particles remain in their mean lattice positions, except the one under consideration which roams or wanders under the influence of a spherically symmetric potential in the cage. Thus, the mathematical formulation was made tractable on intuitive grounds by effectively reducing the description to a one-particle model. This concept was regarded as an improvement over the empirical hole theory of Eyring [1020] in as much as a more quantitative description was given in the model, in the size of the cell, the motion of each molecule within its cell, the distribution of lattice sites, etc. Pople [1037] further expanded these ideas by considering the influence of noncentral forces. He considered the polar liquids HCl, H_2S, and PH_3, and assumed that the rotational and translational motions of the molecules can be treated separately. The molecules were regarded to be fixed in position at the center of their cells, but at the same time free to rotate in the field of the others.

Eisenschitz [1040] employed the cell model and developed a theory for viscosity by considering the motion of the representative molecule to be Brownian and their distribution according to the Smoluchowski equation. The force within the cell was assumed to be proportional to the distance from the center and increasing from the center to the surface of the cell, but to remain constant outside the surface, the final expression being

$$\mu = \frac{27}{40\sqrt{2\pi}}[m\beta(kT)^{5/2}/R^6G^{5/2}]\exp(GR^2/2kT) \tag{94}$$

Here β is the friction constant, m is the molecular mass, G is the force constant of potential energy, and R is the cell radius. If the friction constant, β, is assumed to depend weakly on temperature, the above formula gives a good representation of the temperature dependence of μ on T in spite of the fact that a somewhat unrealistic parabolic potential-energy form is assumed in the formulation. Many of the shortcomings of this derivation have been overcome by the author in a subsequent publication [1068] which, however, does substantiate the final results of his earlier work [1040].

Mention may be made of some efforts to extend and modify the cell theory to give a better appreciation of the properties of liquids. Wentorf *et al.* [1041] showed that the theory of Lennard-Jones and Devonshire is not adequate for fluid densities below and near the critical point but improves at higher densities. Kirkwood [1042] developed a formulation of the free-volume theory from the general principles of statistical mechanics under well-defined approximations. This theory [1042] leads to the results of Lennard-Jones and Devonshire [1035, 1036] in the first approximation. The assumption of empty and multiple occupancy of the cells, and the calculation of their volume, etc., are discussed by a number of workers in relation to the thermodynamic properties, which lie outside the scope of our present effort. Good discussion and reviews of many such efforts are given in the articles of Rowlinson and Curtiss [1043] and Buehler *et al.* [1044].

Dahler, Hirschfelder, and Thacher [1045] started with the nonlinear integral equation for the free volume of a liquid given by Kirkwood [1042] and numerically solved it for the Lennard-Jones (12–6) potential [1046]. In order to achieve this solution they [1046] spherically symmetrized the free volume and employed a Boltzmann type of averaging for the pair interaction. However, the quantitative predictions of thermodynamic properties were unsatisfactory [1046]. This deficiency of the improved theory was attributed to the neglect of spatial correlations between the motions of the molecules in neighboring cells. Chung and Dahler [1047] have given an approximate theory of molecular correlations in liquids. De Boer and co-workers [1014, 1048, 1049] have made extensive studies of this nature, which resulted in a theory for the liquid state which is referred to as the "cell-cluster theory." Dahler and Cohen [1050] have developed the cell-cluster theory for a binary liquid solution. These theories have not been employed to formulate the transport properties. A possible check of the cell model is provided by the work of Dahler [1076] who computed the radial distribution function for liquids on such an approach. Levelt and Hurst [1083] have developed a quantum-mechanical treatment for the cell model but considered calculations of only the macroscopic thermodynamic properties. Collins and Raffel [1051] presented an approximate treatment of the viscosity of a liquid of rigid sphere molecules employing simple ideas of the free volume theory and concerning themselves with the collisional transport of momentum. They have introduced a correction for the blocking effect of third neighbors. Their final result for the collisional contribution to shear viscosity is

$$\mu_c = \frac{2d(mkT)^{1/2}}{5\sqrt{\pi}v[1-(v_0/v)^{1/3}]} \tag{95}$$

Here d is the diameter of the molecule; the quantity v_0/v, the ratio of the incompressible volume to the molecular volume, is recommended by the authors [1051] to be computed from the following relation [1067]:

$$u_s = \frac{1-\frac{2}{3}(v_0/v)^{1/3}}{1-(v_0/v)^{1/3}}\left\{\frac{C_pRT/M}{C_p[1-\frac{2}{3}(v_0/v)^{1/3}]-R}\right\}^{1/2} \tag{96}$$

C_p is the molar specific heat, M is the molecular weight, and u_s is the velocity of sound in the liquid. The calculated μ_c values are found to be of the order of a quarter to a half of the experimental viscosity values for various low-molecular-weight liquids [1051].

F. The Statistical-Mechanical Theory

The foundation of the statistical-mechanical theory of liquids was laid by the efforts of Kirkwood [48, 1011], Mayer and Montroll [1052], Mayer [1053], Born and Green [1054], and others. These workers have derived integral equations, the solutions of which give the distribution functions for the molecules in the liquid. The functions involve the position, velocity of the molecules, derivatives of these quantities with respect to time, and intermolecular potentials. We will now refer briefly to some of the specific work in the following.

Born and Green [1054, 1055] developed from general kinetic theory an expression for the coefficient of viscosity as

$$\mu = \tfrac{1}{30} \int v(r)\Phi'(r)r^3 \, dr - \tfrac{1}{15}m \int \Phi_2(v)v^4 \, dv \quad (97)$$

Here $\Phi'(r)$ is the interaction potential at a separation distance r, v and Φ_2 are functions of r, v is the velocity, and m the molecular mass. The first term in the above expression is due to the intermolecular forces and is much greater than the second term due to the thermal motion of the molecules. In an effort to derive a simple expression for μ, Born and Green [1055] dropped the second term and through a series of approximations found for a face-centered-cubic structure and for a Lennard-Jones (12–6) intermolecular potential that

$$\mu = \frac{\pi^2}{315}(42\pi)^{1/2}\left(\frac{r_0}{r_1}\right)^5 \frac{mv_0}{r_0} e^{-\Phi(r_1)/kT} \quad (98)$$

Here v_0 is the molecular vibrational frequency near the equilibrium point r_0, and r_1 is the distance of nearest neighbors from a given molecule. Thus the work of Born and Green [1055] provided an explanation from kinetic theory of the empirical expressions for μ discussed before [1018, 1019, 1038]. However, Born and Green's work [1054–1056] did not include explicit expressions for the distribution functions, and the difficulty of numerical computations for liquids prevented any theoretical estimation of μ.

Kirkwood, Buff, and Green [1058] derived the following general expression for the coefficient of viscosity based on the statistical mechanical theory of transport processes developed by Kirkwood [48]:

$$\mu = \rho_m \frac{kT}{2\zeta} + \frac{\pi\zeta}{15kT}\frac{N^2}{V^2}\int_0^\infty R^3 \frac{d\Phi(R)}{dR}\Psi_2(R)g_0^{(2)}(R) \, dR \tag{99}$$

Here $\Phi(R)$ is the intermolecular pair potential, N is the Avogadro number, V is the molar volume, ζ is the Brownian motion friction constant arising from the total force acting on a molecule, ρ_m is the mass density at a point R in a fluid, $g_0^{(2)}(R)$ is the equilibrium radial distribution or pair correlation function, and $\Psi_2(R)$ is obtained from the solution of a differential

equation. Implicit in the determination of these functions is the knowledge of the intermolecular potential. The general statistical-mechanical theory of distribution functions in liquids is given by Kirkwood [1011, 1058] and Kirkwood and Salsburg [1059] and an integral equation is formulated, the solution of which gives the radial distribution function [1060]. Explicit solutions of the integral equation for nonpolar liquids composed of rigid spherical molecules are obtained by Kirkwood and Boggs [1061] and Kirkwood, Maun, and Alder [1062]. In the latter work, the theory of Kirkwood [1058] and the slightly different formulation of Born and Green [1054] are considered, to bring out the relative differences in the two theories. Kirkwood, Lewinson, and Alder [1063] further extended the work of Kirkwood, Maun, and Alder [1062] by considering a more realistic intermolecular force field of the Lennard-Jones type.

Kirkwood, Buff, and Green [1057] computed μ for liquid argon at its normal boiling point on the basis of the above expression, the Lennard-Jones interaction potential, and an approximate radial distribution function obtained from the intensity measurements of x-ray scattering. Their [1057] result involving the friction constant is

$$\mu = \frac{8.53 \times 10^{-15}}{\zeta} + 2.63 \times 10^6\zeta \quad (100)$$

Here μ is in poises and they estimated $\zeta = 4.84 \times 10^{-10}$ g sec^{-1}. The above result clearly shows that the contribution to μ arising from the momentum transport (first term) is of less importance than the contribution of intermolecular forces (second term). This result is valid for liquids and is in sharp contrast to that for gases. Zwanzig *et al.* [1082] further improved the calculation by employing a more accurate equilibrium-radial distribution function and the friction constant.

Rice and Allnatt [1010, 1012, 1069, 1070] developed a model from dense-fluid kinetic theory in which it is no more necessary to assume, as Kirkwood's theory [48] does, that the momentum transfer during collision between particles is small. They approximated the pair-interaction potential by an impenetrable rigid core and a soft attraction. In such a model liquid, a moving molecule undergoes a collision similar to that between two rigid cores, followed by a Brownian motion under the influence of the soft potential of the neighboring molecules. The singlet and doublet distribution functions are calculated for this model [1069–1071]. The shear viscosity has a kinetic component given by [1069, 1072]

$$\mu_k = \frac{5kT}{8g^{(2)}(\sigma)} \frac{[1 + \frac{4}{15}(\pi\rho\sigma^3)g^{(2)}(\sigma)]}{[\Omega^{(2,2)} + \{5\zeta_s/4\rho m g^{(2)}(\sigma)\}]} \quad (101)$$

where

$$\Omega^{(2,2)} = (4\pi kT/m)^{1/2}\sigma^2$$

Here σ is the hard-core diameter, ζ_s is the friction constant arising from the autocorrelation of the soft force on a molecule, ρ the number density, and $g^{(2)}$ is the pair correlation function.

The intermolecular-force contribution to viscosity for $R_{12} = \sigma$ (collisional contribution) is given by [1072]

$$\mu_v = \mu_v^{(1)}(\sigma) + \mu_v^{(2)}(\sigma) + \mu_v^{(3)}(\sigma) \quad (102)$$

The expressions for $\mu_v^{(1)}(\sigma)$, $\mu_v^{(2)}(\sigma)$, and $\mu_v^{(3)}(\sigma)$ are complicated and will not be reproduced. For the region $R_{12} > \sigma$, the soft-potential contribution to viscosity is [1072]

$$\mu_v = \frac{\pi\zeta_s\rho^2}{15kT} \int_\sigma^\infty R_{12}^3 \frac{d\Phi}{dR_{12}} g_0^{(2)}(R_{12})\Psi_2(R_{12}) dR_{12}$$

$$(103)$$

Here $\Psi_2(R_{12})$ is the coefficient of Legendre polynomials of order two arising from the shear components of the rate of strain.

Wei and Davis [1073] extended the theory of Rice and Allnatt to mixtures. They [1073] derived the singlet distribution functions and obtained the kinetic contribution to shear viscosity. In a subsequent paper these authors [1074] report the doublet distribution functions and a complete expression for the shear viscosity involving kinetic, collisional, and soft-potential contributions. A comparison of their results [1074] with the corresponding formulation of Rice and Allnatt [1069, 1070] is also given. For further details, the original papers must be consulted.

Longuet-Higgins and Valleau [294] and Davis, Rice, and Sengers [1077] have worked out the theory of shear viscosity for a square-well potential. This theory is further discussed by Davis and Luks [1078], who also present numerical results for liquid argon. The theoretical expression is [1078]

$$\mu = \frac{5}{16\sigma_1^2}\left(\frac{mkT}{\pi}\right)^{1/2}\left\{ \frac{[1 + \frac{2}{5}b\rho(g(\sigma_1) + R^3 g(\sigma_2)\Psi)]}{g(\sigma_1) + R^2 g(\sigma_2)[E + \frac{1}{6}(\epsilon/kT)^2]}\right.$$

$$\left. + \frac{48}{25\pi}(b\rho)^2(g(\sigma_1) + R^4 g(\sigma_2)E)\right\}$$

$$(104)$$

Here

$$\Psi = 1 - e^{\epsilon/kT} + \frac{\epsilon}{2kT}\left[1 + \frac{4}{\pi}e^{\epsilon/kT}\int_{\sqrt{\epsilon/kT}}^\infty e^{-x^2} x^2\, dx\right]$$

$$E = e^{\epsilon/kT} - \frac{\epsilon}{2kT} - 2\int_0^\infty x^2(x^2 + \epsilon/kT)^{1/2} e^{-x^2}\, dx$$

$$b = (2/3)\pi\sigma_1^3, \qquad R = \sigma_2/\sigma_1$$

where σ_1, σ_2, and ϵ are the potential parameters of the square-well intermolecular potential and $g(\sigma_1)$ and $g(\sigma_2)$ are the equilibrium radial distribution functions. These authors suggest that one determine the repulsive and attractive radii and the depth of the attractive square-well potential from the gaseous virial coefficient data. Furthermore, $g(\sigma_1)$ and $g(\sigma_2)$ were determined from the experimental thermal conductivity and equation of state data by fitting against the theoretical expressions. The agreement between the computed and experimental values for liquid argon was found to be satisfactory [1078]. However, these authors [1078] also outline an entirely theoretical procedure for computing the pair correlation functions. The numerical results for viscosity are given for argon, krypton, and xenon [1078, 1079] and the authors claim that a "square-well" fluid is an adequate first approximation to a real fluid [1084].

G. Correlation Function Theories

In this section, a brief reference is made to the use of the time-dependent correlation functions as a tool to determine viscosity. Kadanoff and Martin [1080] have given a good account of the state of the art and have pointed out the complications associated with such an approach. Their paper [1080] must be referred to for details and for references to some of the other work in this area. Forster, Martin, and Yep [1081] have described a moment method to calculate shear viscosity from the long-range (small wavevector k) and long-time (small angular frequency ω) part of the correlation function. In particular, their starting relation is

$$\mu = \lim_{\omega \to 0}\left[\lim_{k \to 0}(\omega/k^2)x_t^{11}(k, \omega)\right] \quad (105)$$

where x_t^{11} is the double Fourier transform of the transverse current–current correlation function. They have evaluated the various parameters of this relation assuming a Gaussian spectral function, and have computed numerical results for argon which are found to be in reasonable agreement with the experimental data.

H. Theories for Liquids of Complicated Molecular Structures

In the above sections we have dealt with theories which have been developed for normal or simple liquids composed of spherically symmetric monatomic molecules. Even for such simple liquids these theories predict viscosity values correct in most cases only within an order of magnitude. The viscosity of polyatomic, nonspherical, polar, and association liquids is harder to calculate and the task becomes increasingly harder as complicated organic and inorganic liquids, fused salts, glasses, polymers, etc. are considered. However, the practical engineering interest in such liquids is amazing. The present scope of our effort does not permit us to undertake a comprehensive review of the state of the art. Frenkel [1008] has referred to some earlier work in this field and many recent publications [1085–1089] include a good account of the present ability to deal with such nonideal liquids of special shaped molecules. Much remains to be done in both the theoretical and experimental areas.

3. ESTIMATION METHODS

A. Introduction

The inadequate state of the development of the theory of liquids has led to the generation of a number of correlative and predictive procedures for viscosity of liquids and their mixtures. Unfortunately, in almost all cases these are based on rather empirical or semiempirical approaches. We will refer to some of these below rather briefly because the domains of their applicability and the estimate of the extent of their uncertainties are still not known with enough reliance. What is conspicuously lacking is a good correlation of the existing data and its critical evaluation against procedures which at least appear to have been logically developed. Our efforts indeed are directed towards such an ultimate goal, but one must be content here with a brief statement of the procedures and a limited statement concerning their appropriateness to reproduce the available data. The data, in many cases, are taken at face value and are not representative of the entire stock of available information. For convenience in presentation, we have artificially divided the various procedures into three categories. This may be regarded as appropriate because of the provisional and to some extent incomplete nature of this section.

B. Procedures Based on the Principle of the Corresponding States

The principle of the corresponding states has been applied to liquids in the same way as to gases [28], the basic assumption being that the intermolecular potential between two molecules is a universal function of the reduced intermolecular separation. This assumption is a good approximation for spherically symmetric monatomic nonpolar molecules. For complicated molecules the principle becomes increasingly crude and many modified versions have very often been used with varying degrees of success. In general, more parameters are introduced in the corresponding state correlations on somewhat empirical grounds in the hope that this modification in some way compensates for the lack of fulfillment of the above stated assumption. We may quote the work of Helfand and Rice [1090] and Rogers and Brickwedde [1091], who have discussed the classical and quantum versions of the principle of corresponding states in relation to the viscosity. Very briefly, the classical viscosity is

$$\mu = \mu(T, \rho, \epsilon, \sigma, k) \tag{106}$$

Here T is the temperature, ρ the density, ϵ the potential-well depth, σ the collision parameter, and k the Boltzmann constant. The reduced viscosity, $\mu^* = \mu\sigma^2/\sqrt{m\epsilon}$, is a different universal function of reduced temperature, T^*, and reduced density, ρ^*, so that

$$\mu^* = \mu^*(T^*, \rho^*) \tag{107}$$

In quantum fluids we have

$$\mu = \mu(T, \rho, \epsilon, \sigma, k, h) \tag{108}$$

where h is Planck's constant. In reduced dimensionless form equation (104) becomes

$$\mu^* = \mu^*(T^*, \rho^*, \Lambda^*) \tag{109}$$

here Λ^* is a sort of reduced de Broglie wavelength associated with the molecule of a certain kinetic energy. In the limit of $\Lambda^* \to 0$ the quantum-mechanical equations reduce to the corresponding classical equations.

Rogers and Brickwedde [1091] have investigated the saturated-liquid viscosity of ^{3}He, ^{4}He, H_2, D_2, T_2, Ne, N_2, and Ar on the basis of the above equations. They [1091] correlate the properties of the heteronuclear isotopic molecules with the effective value of Λ^* obtained for the homonuclear molecules by the following relation:

$$\Lambda_{eff}^* = \Lambda^* \left[1 + \frac{1}{6} \frac{(m_1 - m_2)^2}{m_1 m_2} \right] \quad (110)$$

where m_1 and m_2 are the atomic masses of the two atoms of the heteronuclear molecule.

Boon and Thomaes [1092] and Boon, Legros, and Thomaes [1093] examined the validity of the principle of corresponding states in conjunction with the data on viscosity of many such simple liquids as Ar, Kr, Xe, O_2, N_2, CO, CH_4, and CD_4. Along the liquid–vapor equilibrium curve μ^* is a unique function of T^*. They found that plots of $\ln \mu^*$ against $1/T^*$ are approximately linear, although the data do not lie on one line for all liquids. Ar, Kr, and Xe data lie on one curve and the data points for N_2 and CO fall very close on the same reduced curve. Surprisingly, the oxygen viscosity data lie on a different curve, as do the data for CH_4 and CD_4. These authors [1094–1095] have also extended the principle to mixtures of two liquids and examined it against their own data. The logarithm of the relative kinematic viscosity, $v_R = (v/v_0)$, was plotted against $1/T$ for each binary mixture. Here, $v = \mu/\rho$ and the reference value v_0 was taken as that of argon at 88.98 K. The systems examined were Ar–Kr, Ar–CH_4, Kr–CH_4, Ar–O_2, and CH_4–CD_4. The principle of corresponding states for binary mixtures of more complicated molecules, such as the normal alkane series, is discussed by Holleman and Hijmans [1097], though they do not consider the particular case of viscosity.

C. Semitheoretical or Empirical Procedures for Pure Liquids

Gambill [1098, 1099] in two review articles has referred to a large body of effort which has gone into the development of a number of correlating expressions to predict liquid viscosities and their variations with temperature and pressure. We recommend that readers consult his articles and the sixty-nine references quoted in them [1098, 1099]. Thodos and co-workers [759, 760, 762–764] in a series of articles have exmined the viscosity data of a number of substances in the gaseous and liquid states and have presented smooth plots of excess or residual viscosity, $\mu - \mu^*$, as a function of reduced density, ρ/ρ_c. μ^* is the viscosity of the fluid at one atmosphere pressure at the temperature of interest, and ρ_c is the value of ρ at the critical temperature, the critical density. Jossi, Stiel, and Thodos [771], from dimensional-analysis arguments, showed that $(\mu - \mu^*)\zeta$ is a function of ρ/ρ_c, where

$\zeta = T_c^{1/6}/M^{1/2}P_c^{2/3}$ for nonpolar and polar liquids [771, 772]. Lennart and Thodos [1100] also related $(\mu - \mu^*)\zeta$ to $(\partial P_R/\partial T_R)_{\rho_R}$ for simple fluids, argon, krypton, and xenon. Here $P_R = P/P_c$, $T_R = T/T_c$, and $\rho_R = \rho/\rho_c$. Dolan et al. [1101] and Lee and Ellington [1102] have also employed the principle of a unique plot between $\mu - \mu^*$ and density to correlate their own and other available data on n-butane and n-decane.

Swift et al. [992], while correlating their data on methane, ethane, propane, and n-butane suggested plotting

$$\frac{\mu}{\sqrt{M}} \frac{P_{c(x)}}{P_c} \frac{\rho_c}{\rho} P_R \text{ versus } T_R$$

Here $P_{c(x)}$ is the critical pressure, P_c, of the reference substance x. This was intended to be an improvement on an earlier practice where plots of

$$\frac{\mu}{\sqrt{M}} \frac{P_{c(x)}}{P_c} P_R \text{ versus } T_R$$

were employed to synthesize data. These authors [992] also confirm the relation

$$\mu_c = K \left(\frac{\rho_c^{2/3}}{M^{1/6}} \right) T_c^{1/2} \quad (111)$$

where K is a constant independent of the fluid which Swift et al. [992] found to be equal to 0.00569. ρ_c is in g/cc, T_c is in degrees Kelvin, and μ_c is in centipoises. Swift et al. [992] chose ethane as the reference substance x, and their correlation predicts saturated liquid viscosities for normal paraffins from methane to n-octane within $\pm 5\%$ over the reduced temperature range from 0.65 to 0.95.

Othmer and Conwell [1103] suggested a linear correlation for viscosity of liquids as a function of temperature. They found that a log–log plot of viscosity against the vapor pressure of a reference material at the same temperature is linear. They [1103] have presented a semitheoretical analysis justifying such a correlation. Choosing the reference material as water, they have analyzed the data for eleven representative liquids. The plot using the vapor pressure of water at the same reduced temperature (T/T_c) instead of T is suggested by them as still more promising. Othmer and Silvis [1104] extended the approach to solutions of solids in liquids or of mixtures of liquids, and examined the case of caustic soda solutions in which the plots of the log of the mixture viscosity against the viscosity of water at the same temperature were found to be linear for different concentrations of the solutions.

Thomas [1105] found that the viscosity of a large number of liquids to be adequately correlated by

$$\mu = (0.1167 \, \rho^{0.5})10^{\alpha} \qquad (112)$$

where

$$\alpha = B(1 - T_R)/T_R$$

Here μ is in centipoises, ρ in g/cc, $T_R = T/T_c$, and B is a constant which depends upon the structure of the liquid and is tabulated by Thomas [1105]. This is based on an average correlation of the data, though in many cases the error can be almost an order of magnitude. The range of applicability of this equation is limited to $T_R \leq 0.7$.

Gambill [1098] suggested

$$\mu = \frac{17.05 \, \rho^{1.333} T_b}{\mu_{gas}^{1.333} \, \Delta H_{vb}} 10^{\alpha} \qquad (113)$$

where

$$\alpha = (M \, \Delta H_{vb}/17.85T) - 1.80$$

Here μ is in centipoises, ρ is in g/cc, T_b is the normal boiling point in degrees Kelvin, M is the molecular weight, and ΔH_{vb} is the latent heat of vaporization at T_b in Btu/lb. For 12 different organic liquids in the temperature range 0–40 C, he found the average and maximum deviations between experimental and calculated viscosity values as 33 % and 94 %, respectively.

Gambill [1098, 1099] has given some other forms and generalized charts which have proven useful in representing the viscosity of liquids as a function of temperature and pressure. He particularly recommends the expressions of Andrade [1019] which are given earlier. Dunstan and Thole [831] also list many forms connecting the viscosity at a temperature, t, to that at a lower temperature, t_0 and the empirical constants:

$$\mu_t = \mu_{t_0}/(1 + \beta t)^n \qquad (114)$$

or in a simplified form

$$\mu_t = \frac{A}{1 + \alpha t + \beta t^2} \qquad (115)$$

or

$$\mu_t = \frac{A(T_c - t)}{t - t_1} \qquad (116)$$

where t_1 is a temperature below the melting point. A more complicated version is

$$\mu = A \sqrt{T} \frac{(t - t_1)^2 + C}{(t - t_0)^2 + C^1} \qquad (117)$$

Here α, β, A, C, and C^1 are constants.

Recently Das, Ibrahim, and Kuloor [1107] have suggested that the kinematic viscosity at 20 C and the atmospheric pressure of organic liquids is correlated well by molecular weight and the two empirical constants A and B by the following form:

$$(\mu_{20}/\rho) = AM^B \qquad (118)$$

D. Semitheoretical or Empirical Procedures for Mixtures of Liquids

Gambill [1108] and Dunstan and Thole [831] have listed many forms which have been used to compute viscosities of miscible liquids at a fixed temperature and pressure. Some of these for binary mixtures are:

$$\mu_{mix}^{-1} = \mu_1^{-1}x_1 + \mu_2^{-1}x_2$$
$$\mu_{mix} = \mu_1 x_1 + \mu_2 x_2 \qquad (119)$$
$$\mu_{mix} = \mu_1 x_1 - \mu_2 x_2$$

and

$$\log \mu_{mix} = x_1 \log \mu_1 + x_2 \log \mu_2 \qquad (120)$$

Here μ_{mix} is computed from the knowledge of pure components viscosities and composition only. If one value of μ_{mix} is known, relations with one adjustable parameter have been tried such as:

$$\log \mu_{mix} = x_1 \log \mu_1 + x_2 \log \mu_2 + x_1 x_2 \, d \qquad (121)$$

$$\log \mu_{mix} = x_1^2 \log(\mu_1\mu_2/\mu_{12}^2) + 2x_1 \ln(\mu_{12}/\mu_2) + \ln \mu_2 \qquad (122)$$

Katti and Chaudhri [1109] suggest that

$$\log \mu_{mix} V_{mix} = x_1 \log \mu_1 V_1 + x_2 \log \mu_2 V_2$$
$$+ x_1 x_2 (W\mu/RT) \qquad (123)$$

Here V is the molar volume and $W\mu$ is referred to as the interaction energy for the activation of flow; it is suggested that it be determined from the known value of μ_{mix} for an equimolar mixture at one temperature. These authors have confirmed the validity of such a procedure for a number of systems [1110–1112].

Heric [1113] suggested the following generalization for the kinematic viscosity, v, of an n-component system:

$$\log v_{mix} = \sum_{i=1}^{n} x_i \log v_i + \sum_{i=1}^{n} x_i \log M_i$$
$$- \log \sum_{i=1}^{n} x_i M_i + \delta_{i...n} \qquad (124)$$

where

$$\delta_{i\ldots n} = \frac{1}{2} \sum_{i=1}^{n} \sum_{j=1}^{n} x_i x_j \alpha_{ij}$$

Here α_{ij} is an interaction parameter, with $\alpha_{ij} = \alpha_{ji}$ and $\alpha_{ii} = \alpha_{jj} = 0$. $\delta_{i\ldots n}$ is a deviation function, representing departure from a noninteracting system. For a binary system

$$\delta_{12} = x_1 x_2 \alpha_{12} = x_1 x_2 (W\mu/RT) \qquad (125)$$

For a multicomponent system, assuming binary interactions only, Heric [1113] suggested an improved relation,

$$\delta_{i\ldots n} = \sum_{\substack{i=1 \\ i<j}}^{n} x_i x_j [\alpha_{ij} + \alpha_{ij}^1 (x_i - x_j)] \qquad (126)$$

as an example,

$$\delta_{12} = x_1 x_2 [\alpha_{12} + \alpha_{12}' + \alpha_{12}'(x_1 - x_2)] \qquad (127)$$

α_{12} and α_{12}' are to be determined from the experimental data as explained by Heric [1113]. Heric further suggested that inclusion of a term representing ternary interactions will be essential so that

$$\delta_{123} = \sum_{\substack{i=1 \\ i<j}}^{3} x_i x_j [\alpha_{ij} + \alpha_{ij}'(x_i - x_j)] + x_1 x_2 x_3 \beta \qquad (128)$$

where β may be regarded as concentration independent or its variation may be accounted by the form

$$\beta = \beta_{123} + \beta_{123}'(x_1 - x_2) \qquad (129)$$

Numerical calculations could not suggest which procedure is better, because composition-dependent β improved the reproduction only within the limits of uncertainty of the data.

Kalidas and Laddha [1114] simplified the following relation for the kinematic viscosity of a ternary mixture:

$$\log v = x_1^3 \log v_1 + 3x_1^2 x_2 \log v_{12} + 3x_1 x_2^2 \log v_{21}$$

$$+ x_2^3 \log v_2 - \log\left(x_1 + x_2 \frac{M_2}{M_1}\right)$$

$$+ 3x_1^2 x_2 \log\left(\frac{2 + M_2/M_1}{3}\right) \qquad (130)$$

$$+ 3x_1 x_2^2 \log\left(\frac{1 + 2M_2/M_1}{3}\right) + x_2^3 \log\left(\frac{M_2}{M_1}\right)$$

By considering a simplified model for ternary molecular interactions these authors [1114] derived from the above equation, due to McAllister [1216], an

explicit expression with seven unknown constants for the kinematic viscosity of a three-component mixture. Six of these constants were obtained by analyzing the experimental data for the three binary systems possible with a three-component system. The seventh unknown parameter was adjusted while fitting the experimental data on a ternary system to the theoretical expression. Their [1114] experimental data on acetone–methanol–ethylene glycol mixtures at 30 C were found to be adequately correlated by their proposed theoretical expression.

Huang, Swift, and Kurata [1115] correlated their data on binary systems at higher pressures by plotting residual viscosity $\mu_{mix} - \mu_{mix}^{\circ}$ versus molar density. μ_{mix}°, the viscosity of the mixture at the atmospheric pressure, was obtained from the relation

$$\mu_{mix}^{\circ} = (x_1\sqrt{M_1}\mu_1^{\circ} + x_2\sqrt{M_2}\mu_2^{\circ})/(x_1\sqrt{M_1} \\ + x_2\sqrt{M_2}) \qquad (131)$$

Saxena [1217] suggested an expression of the Sutherland–Wassiljewa form to correlate the data on viscosity of multicomponent mixtures, in analogy to the parallel work on gaseous mixtures. He found that the data on binary systems is very well represented by the following relation:

$$\mu_{mix} = \frac{\mu_1}{1 + \Psi_{12}(x_2/x_1)} + \frac{\mu_2}{1 + \Psi_{21}(x_1/x_2)} \qquad (132)$$

where

$$\frac{\Psi_{12}}{\Psi_{21}} = \frac{M_2}{M_1} \cdot \frac{\mu_1}{\mu_2}$$

4. EXPERIMENTAL METHODS

A. Introduction

The viscosity of liquids is simpler to measure than that of gases primarily because of the convenience of handling; furthermore, fairly accurate values are determined with relative ease as liquids are much more viscous than gases. The technological interest in lubrication has encouraged detailed study of the subject as early as almost a century ago [1116]. Historically, more detailed attention is given to the determination of viscosity of liquids than to that of gases as is evident from the review accounts given in the monographs of Dunstan and Thole [831], Hatschek [832], Barr [833], and others. In addition to the development of different absolute methods already mentioned in connection with gases, many relative methods have been developed as quick and fairly accurate alternatives in compliance with the practical demands. Partington [1016] has given a detailed reference to the various efforts made until almost twenty years back; in our brief review here we will mention some of the more recent work on the viscosity determination of Newtonian fluids. The survey here is unfortunately incomplete and constitutes what may be called a stray sampling of recent efforts in the literature. As the basic principles of the methods are already given while dealing with gases, a straightforward approach is followed below.

B. The Capillary-Flow Viscometers

A large variety of viscometers (or more appropriately viscosimeters) are developed on the general principle of liquid flow through a capillary. The designs of a large number of such viscometers in historical sequence are given by Hatschek [832] and Partington [1016]. We have referred to some work in connection with gases, and we will not repeat any reference to these efforts here. Many capillary viscometers have been developed to obtain data on liquid hydrocarbons. Lipkin, Davison, and Kurtz [1117] have described two such viscometers for work at low and high temperatures and pressures. They [1117] reported data on propane, butane, and iso-butane with an accuracy of $\pm 2\%$. Lee and co-workers, whose work has been described earlier [453, 908–916], have measured the viscosity of liquid n-butane [1101] and n-decane [1102]. A number of workers have used an Ostwald-type capillary viscometer. Boon and Thomaes [1092–1096, 1118, 1119] have measured the kinematic viscosity of a number of liquids and their mixtures at saturation vapor pressure over a range of temperatures with a stated precision of 1%. Katti and Chaudhri [1109] measured viscosity of binary mixtures with an Ostwald viscometer having an accuracy of 0.5%. The measurements have been extended to many more binary systems [1110–1112]. Denny and Ferenbaugh [1120] developed a capillary-tube viscometer for superheated liquids and reported results for CCl_4. An Ostwald viscometer is used by Mullin and Osman [1121] for viscosity of solutions; they reported results for nickel ammonium sulfate aqueous solutions in the temperature ranges 10–35 C with an estimated precision of $\pm 0.3\%$.

Swindells, Coe, and Godfrey [1122] determined the viscosity of water at 20 C with a high degree of accuracy with a capillary-flow viscometer, to provide a standard value for relative measurements. They found the value to be 0.010019 ± 0.000003 poise, which is appreciably different from the value 0.01005

poise taken so far as standard. Following this work, the National Bureau of Standards in the USA has adopted the absolute viscosity of water at 20 C as 0.01002 poise. Agaev and Yusibova [1157] have reported measurements of the viscosity of heavy water in the pressure range of 1–1200 kg/cm^2, and temperature range of 4–100 C.

C. The Oscillating-Disk Viscometers

Van Itterbeek, Zink, and Van Paemel [1123] measured the viscosity of liquid oxygen, nitrogen, argon, and hydrogen as a function of temperature using an oscillating-disk absolute viscometer. The viscosity is determined from the record of the logarithmic decrement of the amplitude of the oscillation. The measurements on liquids were further extended to pressures up to 100 atm [1124, 1125] and it was found that the viscosity increases linearly with pressure.

D. The Falling-Body Viscometers

Hubbard and Brown [1126] determined the viscosity of liquid n-pentane with a high pressure rolling-ball viscometer in the temperature range 25–250 C and at pressures up to 1000 psi. The measurements were relative and estimated to have a varying uncertainty of 5–10%. The data above 150 C are less accurate. As already mentioned while discussing measurements on gases, Swift et al. [991, 992] have employed a falling-cylinder viscometer to determine the viscosity of liquid hydrocarbons. Using a falling-ball viscometer Chacon-Tribin, Loftus, and Satterfield [1127] have determined the viscosity of vanadium pentoxide–potassium sulfate eutectic mixture at 461, 505, and 586 C. Riebling [1128] described a variant of this general type of viscometer, which is especially useful at high temperatures up to 1750 C. In this design, the ball does not freely fall, but its motion is controlled by attaching it to an analytical balance, and thus its effective weight and therefore its velocity can be suitably varied. The details of this improved counterbalanced sphere viscometer, along with its related instrumentation and necessary corrections, are described by the author.

E. The Coaxial-Cylinder Viscometers

Moynihan and Cantor [1129] measured the viscosity of molten BeF_2 by the fixed-cup rotating-cylinder method using Brookfield Synchro-Lectric viscometers. The temperature range covered is 573.7–979 C and the uncertainty in the viscosity value at any temperature level is estimated to be less than

$\pm 3\%$. Cantor, Ward, and Moynihan [1130] determined the viscosity of molten BeF_2–LiF solutions covering the concentration range 36–99 mole% of BeF_2. The overall temperature range was 367–967 C, though for each mixture the temperature range was less extensive. The data at each composition was fitted to the form:

$$\mu = A \exp(E_\mu/RT) \qquad (133)$$

and the constants A and E_μ are tabulated. The equation for pure BeF_2 is

$$\mu = 7.603 \times 10^{-9} \exp[(52590/RT) \\ + (1.471 \times 10^6/T^2)] \qquad (134)$$

Here μ is in poises and T in degrees Kelvin. It is shown that the viscosity of the mixtures at a fixed temperature, as well as the activation energy, decreases exponentially for this system.

F. Other Types of Viscometers

Cottingham [1131] described a viscometer suitable for relative measurements of viscosity of low melting point metals in the temperature range 20–600 C. Measured values for methanol, bismuth, and lead are compared with the existing values in the literature. The viscometer consists of a tank filled with the test liquid. The two flat end faces of the drum are in light contact with the sides of the tank, and only a small clearance separates the bottom of the drum and the tank. A scraper lightly pressed against the top of the drum forms two compartments in the tank and prevents any liquid flow from one compartment into the other as the drum is rotated. However, liquid is dragged through the narrow duct at the bottom and a head of liquid builds up in one compartment, which in turn forces a part of the liquid to flow back. A measure of the viscosity is the equilibrium value of the liquid head at the steady state, i.e., when equal volumes of liquid flow in opposite directions through the duct per unit time. The viscometer is designed to measure viscosities between one and more than a thousand centipoises, and the influence of the various variables on the viscosity measurement is analyzed.

Welber [1132] and Welber and Quimby [1133] have described in detail the principle and operation of a simple viscometer in which the electrical characteristics of a piezoelectric cylinder of quartz oscillating in a torsional mode are measured. The logarithmic decrement Δ of the system is defined as

$$\Delta = W^d/2W^v \qquad (135)$$

Here W^d is the energy dissipated per cycle and W^v is the vibrational energy of the system. The resistance R in the neighborhood of the resonant frequency f_0 is given in terms of Δ by

$$R = KMf_0\Delta \qquad (136)$$

where M is the mass of the crystal and the constant K, dependent on the electrode geometry, is obtained experimentally. The product $\mu\rho$ is related to $(\Delta - \Delta_0)^2$ as in the oscillating-disk viscometers. Δ_0 is the value of Δ in vacuum and is referred to as nuisance decrement. Webeler and Hammer [1134–1136] have used this technique to measure viscosity of liquid helium at low temperatures. DeBock *et al.* [1137, 1138] have reported data on liquid argon as a function of pressure (0–200 kg/cm^2) and temperature (between the boiling and critical points) with an estimated accuracy of better than 3%.

Solov'ev and Kaplun [1139] describe a vibration viscometer for the measurement of viscosity of liquids within fractions of a percent and of a moving liquid within 1.5%. The design is appropriate for high temperatures and pressures and requires only a small quantity of the test fluid. A thin plate attached to a rod and suspended through an elastic element executes plane oscillations under the influence of a harmonic force. The equation of motion is analyzed for the frequency–phase and frequency–amplitude modes of operation, and it has been pointed out that the selection of the mode is dependent on the viscosity of the test liquid.

Krutin and Smirnitskii [1140] describe the theory of what they refer to as a vibrating-rod or probe viscometer. The forced longitudinal and torsional vibration characteristics of a slender rod (or probe) in a liquid are shown to depend upon the viscosity and density of the liquid, the density of the probe, the modulus of elasticity and internal loss coefficient in the probe material, the configuration of the probe cross section, and the driving frequency. By introducing the damping coefficient, a measure of the influence of damping of the fluid on the vibrational characteristics of the probe, appropriate analytical treatment is developed to guide proper selection of the various quantities for accurate viscosity measurement.

Andrade and Dodd [1141, 1142] used a rectangular channel formed between two plane steel surfaces as a viscometer for detecting small relative changes in viscosity (a few parts in a million) while investigating the influence of an electric field on viscosity.

TABLE 1. COMPOSITION AND TEMPERATURE DEPENDENCE OF Ψ_{ij} ON DIFFERENT SCHEMES OF COMPUTATION

Gas Pair [Reference]	Temp. (K)	Mole Fraction of Heavier Component	First Method Ψ_{12}	Ψ_{21}	Second Method Ψ_{12}	Ψ_{21}	Viscosity (N s m^{-2} x 10^{-6})
Ar–He [165]	72.0	0.0000					7.94
		0.1590	0.2086	2.603	0.2496	2.206	
		0.2580	0.1905	2.377	0.2316	2.047	
		0.3570	0.1948	2.431	0.2400	2.121	
		0.3910	0.1924	2.401	0.2381	2.104	
		0.4585	0.1933	2.412	0.2406	2.126	
		0.5380	0.1900	2.371	0.2384	2.107	
		0.5570	0.1892	2.361	0.2378	2.102	
		0.6570	0.1900	2.371	0.2402	2.122	
		0.8280	0.1807	2.255	0.2316	2.047	
		1.0000					6.35
Ar–He [165]	81.1	0.0000					8.59
		0.1590	0.2166	2.634	0.2588	2.228	
		0.2580	0.1968	2.394	0.2387	2.055	
		0.3570	0.2005	2.438	0.2462	2.120	
		0.3910	0.2005	2.439	0.2472	2.129	
		0.4585	0.1974	2.400	0.2451	2.110	
		0.5380	0.1983	2.411	0.2476	2.132	
		0.5570	0.1935	2.353	0.2424	2.088	
		0.6570	0.1928	2.344	0.2432	2.094	
		0.8280	0.2177	2.647	0.2727	2.349	
		1.0000					7.05
Ar–He [165]	90.2	0.0000					9.08
		0.1590	0.2129	2.539	0.2536	2.141	
		0.2580	0.2050	2.444	0.2480	2.094	
		0.3570	0.2008	2.394	0.2462	2.079	
		0.3910	0.2018	2.407	0.2484	2.097	
		0.5380	0.1956	2.333	0.2444	2.064	
		0.5570	0.1953	2.329	0.2443	2.063	
		0.6570	0.1904	2.271	0.2404	2.030	
		0.8280	0.1551	1.849	0.2020	1.705	
		1.0000					7.60
Ar–He [165]	192.5	0.0000					14.60
		0.1055	0.2619	2.481	0.3039	2.039	
		0.2000	0.2577	2.441	0.3027	2.031	
		0.3030	0.2527	2.394	0.3003	2.015	
		0.4110	0.2507	2.375	0.3008	2.018	
		0.4650	0.2559	2.425	0.3079	2.066	
		0.4940	0.2515	2.382	0.3034	2.036	
		0.6220	0.2551	2.416	0.3095	2.077	
		0.7110	0.2476	2.346	0.3027	2.031	
		0.8010	0.2414	2.287	0.2972	1.994	
		0.8055	0.2376	2.251	0.2931	1.967	
		0.8870	0.2317	2.195	0.2877	1.930	
		1.0000					15.38
Ar–He [165]	229.5	0.0000					16.35
		0.1050	0.2707	2.498	0.3138	2.051	
		0.1990	0.2645	2.441	0.3100	2.026	
		0.3010	0.2567	2.369	0.3042	1.988	
		0.4090	0.2578	2.379	0.3083	2.015	
		0.5640	0.2581	2.382	0.3098	2.025	
		0.6210	0.2532	2.337	0.3072	2.008	
		0.7100	0.2473	2.282	0.3021	1.975	
		0.8000	0.2358	2.177	0.2910	1.902	
		0.8050	0.2390	2.206	0.2945	1.925	
		0.8865	0.2341	2.161	0.2903	1.897	
		1.0000					17.68

TABLE 1.   COMPOSITION AND TEMPERATURE DEPENDENCE OF Ψ_{ij} ON DIFFERENT SCHEMES OF COMPUTATION (continued)

Gas Pair [Reference]	Temp. (K)	Mole Fraction of Heavier Component	First Method Ψ_{12}	Ψ_{21}	Second Method Ψ_{12}	Ψ_{21}	Viscosity (N s m^{-2} x 10^{-6})
Ar–He [211]	288.2	0.0000					19.66
		0.1922	0.2888	2.552	0.3376	2.113	
		0.2915	0.2842	2.512	0.3347	2.095	
		0.5337	0.2791	2.467	0.3336	2.088	
		0.6119	0.2723	2.407	0.3274	2.049	
		0.6846	0.2852	2.521	0.3420	2.141	
		0.7705	0.2770	2.448	0.3343	2.092	
		0.8074	0.2684	2.373	0.3256	2.038	
		0.8572	0.2697	2.384	0.3274	2.049	
		0.9093	0.2653	2.345	0.3233	2.024	
		0.9507	0.2706	2.391	0.3292	2.060	
		1.0000					22.20
Ar–He [165]	291.1	0.0000					19.35
		0.1590	0.2854	2.532	0.3326	2.090	
		0.2580	0.2795	2.479	0.3285	2.064	
		0.3570	0.2758	2.447	0.3267	2.052	
		0.3910	0.2697	2.393	0.3205	2.014	
		0.4585	0.2689	2.386	0.3211	2.017	
		0.5380	0.2610	2.315	0.3138	1.971	
		0.5570	0.2578	2.287	0.3106	1.951	
		0.6570	0.2620	2.324	0.3169	1.991	
		0.8280	0.2673	2.282	0.3141	1.973	
		1.0000					21.77
Ar–He [165]	291.1	0.0000					19.13
		0.1050	0.2891	2.540	0.3349	2.084	
		0.1990	0.2744	2.411	0.3202	1.992	
		0.3010	0.2678	2.353	0.3158	1.965	
		0.4090	0.2650	2.328	0.3154	1.962	
		0.4640	0.2608	2.292	0.3120	1.941	
		0.6210	0.2532	2.225	0.3067	1.908	
		0.7100	0.2447	2.150	0.2989	1.860	
		0.8000	0.2506	2.014	0.3066	1.907	
		0.8050	0.2465	2.166	0.3023	1.881	
		0.8865	0.2325	2.043	0.2884	1.795	
		1.0000					21.73
Ar–He [213]	291.2	0.0000					19.40
		0.0610	0.2987	2.629	0.3456	2.154	
		0.2080	0.2863	2.520	0.3349	2.087	
		0.2990	0.2811	2.474	0.3312	2.064	
		0.4380	0.2809	2.472	0.3339	2.081	
		0.5200	0.2751	2.421	0.3289	2.050	
		0.5740	0.2760	2.429	0.3307	2.061	
		0.6450	0.2751	2.421	0.3308	2.061	
		0.7200	0.2661	2.342	0.3221	2.007	
		0.7820	0.2698	2.374	0.3267	2.036	
		0.8440	0.2647	2.330	0.3220	2.007	
		0.9140	0.2687	2.365	0.3269	2.037	
		1.0000					22.00
Ar–He [223]	293.0	0.0000					19.73
		0.5094	0.2782	2.478	0.3324	2.096	
		0.6180	0.2733	2.434	0.3286	2.072	
		1.0000					22.11
Ar–He [223]	373.0	0.0000					23.20
		0.5094	0.2878	2.483	0.3423	2.091	
		0.6180	0.2862	2.469	0.3421	2.090	
		1.0000					26.84

TABLE 1. COMPOSITION AND TEMPERATURE DEPENDENCE OF Ψ_{ij} ON
DIFFERENT SCHEMES OF COMPUTATION (continued)

Gas Pair [Reference]	Temp. (K)	Mole Fraction of Heavier Component	First Method Ψ_{12}	Ψ_{21}	Second Method Ψ_{12}	Ψ_{21}	Viscosity (N s m^{-2} x 10^{-6})
Ar-He [211]	373.2	0.0000					23.55
		0.1922	0.3046	2.598	0.3554	2.146	
		0.2015	0.2852	2.432	0.3323	2.007	
		0.5337	0.2984	2.545	0.3541	2.138	
		0.6119	0.2910	2.482	0.3470	2.096	
		0.6846	0.2954	2.519	0.3525	2.129	
		0.7706	0.2941	2.508	0.3519	2.125	
		0.8074	0.2850	2.431	0.3428	2.070	
		0.8572	0.2841	2.423	0.3423	2.067	
		0.9093	0.2897	2.470	0.3485	2.104	
		0.9507	0.3161	2.696	0.3755	2.268	
		1.0000					27.56
Ar-He [211]	456.2	0.0000					26.91
		0.1922	0.3158	2.629	0.3680	2.169	
		0.6119	0.3043	2.532	0.3609	2.127	
		0.6846	0.3019	2.512	0.3591	2.116	
		0.8074	0.2958	2.462	0.3539	2.086	
		0.8572	0.2939	2.446	0.3523	2.076	
		0.9093	0.3074	2.558	0.3665	2.160	
		0.9507	0.3485	2.900	0.4079	2.404	
		1.0000					32.27
Ar-He [223]	473.0	0.0000					27.15
		0.6180	0.2960	2.500	0.3523	2.108	
		1.0000					32.08
Ar-He [223]	523.0	0.0000					29.03
		0.6180	0.2975	2.500	0.3539	2.106	
		1.0000					34.48
Ar-Kr [278]	291.2	0.0000					22.10
		0.1090	0.7172	1.341	0.7551	1.263	
		0.2280	0.7221	1.350	0.7606	1.272	
		0.3300	0.7241	1.354	0.7629	1.276	
		0.4430	0.7256	1.356	0.7646	1.279	
		0.5460	0.7234	1.352	0.7624	1.275	
		0.6730	0.7228	1.351	0.7619	1.274	
		0.7770	0.7133	1.333	0.7524	1.258	
		0.8650	0.7354	1.375	0.7748	1.296	
		1.0000					24.80
Ar-Ne [180]	72.3	0.0000					11.72
		0.1613	0.4854	1.765	0.5179	1.700	
		0.3231	0.4858	1.767	0.5190	1.704	
		0.5011	0.4863	1.769	0.5202	1.707	
		0.6707	0.4858	1.767	0.5201	1.707	
		0.8300	0.4919	1.789	0.5267	1.729	
		1.0000					6.38
Ar-Ne [180]	90.3	0.0000					13.52
		0.1634	0.4900	1.692	0.5219	1.627	
		0.3265	0.5014	1.732	0.5350	1.668	
		0.4828	0.5044	1.742	0.5387	1.679	
		0.6713	0.4989	1.723	0.5335	1.663	
		0.8390	0.4772	1.648	0.5114	1.594	
		1.0000					7.75
Ar-Ne [180]	193.4	0.0000					23.52
		0.1698	0.5408	1.647	0.5747	1.580	
		0.3292	0.5414	1.649	0.5760	1.583	
		0.5024	0.5432	1.654	0.5783	1.590	
		0.6690	0.5552	1.691	0.5911	1.625	
		0.8298	0.5391	1.642	0.5748	1.580	
		1.0000					15.29

TABLE 1.   COMPOSITION AND TEMPERATURE DEPENDENCE OF Ψ_{ij} ON DIFFERENT SCHEMES OF COMPUTATION (continued)

Gas Pair [Reference]	Temp. (K)	Mole Fraction of Heavier Component	First Method Ψ_{12}	Ψ_{21}	Second Method Ψ_{12}	Ψ_{21}	Viscosity (N s m^{-2} x10^{-6})
Ar-Ne [180]	229.0	0.0000					26.70
		0.1654	0.5496	1.614	0.5835	1.547	
		0.3348	0.5470	1.606	0.5813	1.541	
		0.4308	0.5430	1.594	0.5774	1.531	
		0.5017	0.5494	1.613	0.5844	1.549	
		0.6507	0.5451	1.601	0.5804	1.539	
		0.8320	0.5408	1.588	0.5764	1.528	
		1.0000					18.00
Ar-Ne [180]	291.1	0.0000					31.29
		0.1693	0.5611	1.569	0.5949	1.502	
		0.3227	0.5673	1.587	0.6022	1.520	
		0.4970	0.5627	1.574	0.5979	1.509	
		0.6757	0.5665	1.584	0.6023	1.520	
		0.8323	0.5688	1.591	0.6050	1.527	
		1.0000					22.15
Ar-Ne [213]	291.2	0.0000					30.70
		0.1570	0.5795	1.601	0.6146	1.532	
		0.2210	0.5716	1.579	0.6062	1.512	
		0.3280	0.5573	1.540	0.5913	1.475	
		0.4360	0.5616	1.552	0.5964	1.487	
		0.5410	0.5738	1.585	0.6095	1.520	
		0.6380	0.5732	1.583	0.6090	1.519	
		0.7260	0.5801	1.603	0.6164	1.537	
		0.8030	0.5548	1.533	0.5906	1.473	
		0.9000	0.5544	1.532	0.5904	1.472	
		1.0000					22.00
Ar-Ne [221]	293.0	0.0000					30.92
		0.2680	0.5782	1.600	0.6136	1.532	
		0.6091	0.5758	1.593	0.6117	1.527	
		0.7420	0.5735	1.586	0.6096	1.522	
		1.0000					22.13
Ar-Ne [221]	373.0	0.0000					36.23
		0.2680	0.5973	1.591	0.6332	1.522	
		0.6091	0.5965	1.589	0.6329	1.521	
		0.7420	0.5939	1.582	0.6304	1.515	
		1.0000					26.93
Ar-Ne [221]	473.0	0.0000					42.20
		0.2680	0.6068	1.573	0.6429	1.505	
		0.6091	0.6118	1.586	0.6486	1.518	
		0.7420	0.6132	1.590	0.6501	1.521	
		1.0000					32.22
Ar-Ne [221]	523.0	0.0000					45.01
		0.2680	0.6117	1.575	0.5481	1.507	
		0.6091	0.6096	1.570	0.6462	1.502	
		0.7420	0.6164	1.588	0.6533	1.519	
		1.0000					34.60
Ar-Xe [324]	291.2	0.0000					22.10
		0.1090	0.5377	1.736	0.5861	1.583	
		0.2130	0.5252	1.696	0.5727	1.547	
		0.3000	0.5257	1.697	0.5738	1.550	
		0.4050	0.5281	1.705	0.5771	1.559	
		0.4980	0.5227	1.687	0.5719	1.544	
		0.5980	0.5254	1.696	0.5753	1.554	
		0.7010	0.5195	1.677	0.5696	1.538	
		0.7920	0.5186	1.674	0.5691	1.537	
		0.9050	0.5281	1.705	0.5793	1.564	
		1.0000					22.50

TABLE 1. COMPOSITION AND TEMPERATURE DEPENDENCE OF Ψ_{ij} ON
DIFFERENT SCHEMES OF COMPUTATION (continued)

Gas Pair [Reference]	Temp. (K)	Mole Fraction of Heavier Component	First Method Ψ_{12}	Ψ_{21}	Second Method Ψ_{12}	Ψ_{21}	Viscosity ($N\ s\ m^{-2} \times 10^{-6}$)
He–Kr [325]	283.2	0.0000					19.52
		0.1021	0.1925	3.223	0.2302	2.442	
		0.2046	0.1866	3.124	0.2281	2.419	
		0.3086	0.1820	3.047	0.2265	2.403	
		0.4995	0.1790	2.996	0.2280	2.418	
		0.7098	0.1755	2.938	0.2275	2.414	
		0.8100	0.1723	2.884	0.2252	2.389	
		0.8845	0.1683	2.817	0.2217	2.352	
		0.9454	0.1498	2.508	0.2022	2.145	
		1.0000					24.41
He–Kr [278]	291.2	0.0000					19.40
		0.0690	0.1858	3.043	0.2182	2.264	
		0.1510	0.1826	2.991	0.2197	2.279	
		0.2720	0.1760	2.882	0.2175	2.257	
		0.3530	0.1740	2.849	0.2181	2.263	
		0.4390	0.1739	2.848	0.2205	2.288	
		0.6000	0.1708	2.796	0.2205	2.288	
		0.6980	0.1618	2.650	0.2122	2.202	
		0.7970	0.1698	2.780	0.2223	2.307	
		0.8910	0.1587	2.599	0.2113	2.193	
		1.0000					24.80
He–Kr [325]	373.2	0.0000					23.35
		0.1021	0.2072	3.301	0.2473	2.497	
		0.2046	0.1968	3.135	0.2390	2.413	
		0.3086	0.1957	3.118	0.2415	2.438	
		0.4995	0.1936	3.085	0.2436	2.459	
		0.7098	0.1890	3.012	0.2418	2.441	
		0.8100	0.1834	2.922	0.2369	2.392	
		0.8845	0.1835	2.924	0.2379	2.401	
		0.9454	0.1962	3.125	0.2515	2.539	
		1.0000					30.68
He–Ne [179]	20.4	0.0000					3.50
		0.2560	0.4051	2.036	0.4565	1.800	
		0.4920	0.3952	1.986	0.4483	1.768	
		0.7200	0.3976	1.999	0.4530	1.786	
		1.0000					3.51
He–Ne [179]	65.8	0.0000					7.45
		0.2580	0.4686	1.684	0.5165	1.456	
		0.5090	0.4627	1.663	0.5144	1.450	
		0.7610	0.4601	1.653	0.5150	1.452	
		1.0000					10.45
He–Ne [179]	90.2	0.0000					9.12
		0.2510	0.4884	1.663	0.5366	1.434	
		0.4910	0.4841	1.649	0.5358	1.431	
		0.7550	0.4802	1.635	0.5351	1.430	
		1.0000					13.50
He–Ne [179]	194.0	0.0000					14.93
		0.2440	0.5167	1.648	0.5658	1.416	
		0.4820	0.5148	1.642	0.5670	1.419	
		0.7590	0.5121	1.633	0.5670	1.419	
		1.0000					23.60
He–Ne [325]	284.2	0.0000					19.29
		0.0340	0.5318	1.753	0.5824	1.506	
		0.2801	0.5199	1.714	0.5713	1.477	
		0.4995	0.5159	1.700	0.5691	1.472	
		0.6804	0.5238	1.727	0.5789	1.497	
		0.7850	0.5228	1.723	0.5785	1.496	
		0.9091	0.5061	1.668	0.5625	1.455	
		0.9461	0.5062	1.669	0.5629	1.456	
		0.9900					25.50

TABLE 1. COMPOSITION AND TEMPERATURE DEPENDENCE OF Ψ_{ij} ON DIFFERENT SCHEMES OF COMPUTATION (continued)

Gas Pair [Reference]	Temp. (K)	Mole Fraction of Heavier Component	First Method Ψ_{12}	Ψ_{21}	Second Method Ψ_{12}	Ψ_{21}	Viscosity (N s m^{-2} x 10^{-6})
He-Ne [213]	291.2	0.0000					19.20
		0.1580	0.5234	1.645	0.5714	1.409	
		0.2500	0.5227	1.643	0.5720	1.410	
		0.3930	0.5205	1.635	0.5715	1.409	
		0.5650	0.5222	1.641	0.5753	1.418	
		0.6550	0.5000	1.571	0.5532	1.364	
		0.7830	0.5027	1.580	0.5575	1.374	
		0.8940	0.4540	1.427	0.5097	1.257	
		1.0000					30.80
He-Ne [221]	293.0	0.0000					19.41
		0.2379	0.5260	1.664	0.5758	1.430	
		0.4376	0.5211	1.649	0.5730	1.423	
		0.7341	0.5061	1.602	0.5606	1.392	
		1.0000					30.92
He-Ne [179]	293.1	0.0000					19.61
		0.2620	0.5263	1.680	0.5769	1.445	
		0.4980	0.5204	1.661	0.5731	1.435	
		0.7520	0.5166	1.649	0.5716	1.431	
		1.0000					30.97
He-Ne [221]	373.0	0.0000					22.81
		0.2379	0.5271	1.673	0.5773	1.437	
		0.4376	0.5198	1.650	0.5716	1.423	
		0.7341	0.5076	1.611	0.5622	1.400	
		1.0000					36.23
He-Ne [325]	373.2	0.0000					23.35
		0.0340	0.5331	1.768	0.5844	1.521	
		0.2801	0.5183	1.719	0.5697	1.482	
		0.4995	0.5167	1.714	0.5701	1.484	
		0.6804	0.5146	1.707	0.5695	1.482	
		0.7850	0.5165	1.713	0.5722	1.489	
		0.9091	0.4791	1.589	0.5355	1.393	
		0.9461	1.299	4.307	1.339	3.484	
		0.9900					35.49
He-Ne [221]	473.0	0.0000					26.72
		0.2379	0.5253	1.677	0.5755	1.441	
		0.4376	0.5180	1.653	0.5699	1.427	
		0.7341	0.5059	1.615	0.5605	1.404	
		1.0000					42.20
He-Ne [221]	523.0	0.0000					28.53
		0.2379	0.5237	1.673	0.5737	1.438	
		0.7341	0.5140	1.642	0.5688	1.426	
		1.0000					45.01
He-Xe [324]	291.2	0.0000					19.40
		0.0630	0.1296	3.683	0.1568	2.638	
		0.1690	0.1293	3.673	0.1644	2.767	
		0.2010	0.1251	3.555	0.1609	2.708	
		0.3040	0.1224	3.477	0.1619	2.725	
		0.4010	0.1201	3.412	0.1621	2.727	
		0.4940	0.1185	3.367	0.1623	2.731	
		0.5940	0.1130	3.210	0.1575	2.651	
		0.6870	0.1135	3.225	0.1594	2.682	
		0.7920	0.1142	3.245	0.1614	2.716	
		0.8980	0.0994	2.824	0.1450	2.440	
		1.0000					22.40

TABLE 1. COMPOSITION AND TEMPERATURE DEPENDENCE OF Ψ_{ij} ON DIFFERENT SCHEMES OF COMPUTATION (continued)

Gas Pair [Reference]	Temp. (K)	Mole Fraction of Heavier Component	First Method Ψ_{12}	Ψ_{21}	Second Method Ψ_{12}	Ψ_{21}	Viscosity (N s m^{-2} x 10^{-6})
Kr-Ne [278]	291.2	0.0000					31.30
		0.0650	0.3892	2.032	0.4344	1.832	
		0.1110	0.3898	2.035	0.4360	1.838	
		0.2290	0.3891	2.031	0.4372	1.843	
		0.3390	0.3916	2.044	0.4413	1.861	
		0.4380	0.3899	2.035	0.4404	1.857	
		0.5330	0.3892	2.032	0.4405	1.857	
		0.6470	0.3747	1.956	0.4258	1.796	
		0.7970	0.3702	1.932	0.4222	1.780	
		0.8890	0.3974	2.074	0.4512	1.902	
		1.0000					24.90
Kr-Xe [324]	291.2	0.0000					24.70
		0.1150	0.7470	1.285	0.7735	1.244	
		0.2010	0.7500	1.290	0.7767	1.249	
		0.2960	0.7567	1.302	0.7838	1.260	
		0.3930	0.7554	1.299	0.7824	1.258	
		0.4910	0.7477	1.286	0.7745	1.245	
		0.5950	0.7590	1.306	0.7863	1.264	
		0.6930	0.7419	1.276	0.7688	1.236	
		0.7860	0.7600	1.307	0.7873	1.266	
		0.8960	0.7368	1.267	0.7639	1.228	
		1.0000					22.50
Ne-Xe [324]	291.2	0.0000					31.00
		0.1030	0.2787	2.510	0.3241	2.204	
		0.1990	0.2734	2.462	0.3203	2.178	
		0.2850	0.2699	2.431	0.3182	2.164	
		0.3930	0.2711	2.442	0.3216	2.186	
		0.5040	0.2655	2.391	0.3167	2.154	
		0.5940	0.2672	2.406	0.3197	2.173	
		0.7940	0.2649	2.386	0.3190	2.169	
		0.9030	0.2568	2.312	0.3109	2.114	
		1.0000					22.40
Ar-H$_2$ [226]	293.0	0.0000					8.75
		0.3485	0.2787	2.186	0.3189	1.598	
		0.5543	0.2708	2.124	0.3170	1.588	
		0.7058	0.2627	2.060	0.3126	1.566	
		1.0000					22.11
Ar-H$_2$ [226]	373.0	0.0000					10.29
		0.3485	0.2817	2.140	0.3212	1.559	
		0.5543	0.2732	2.075	0.3189	1.548	
		0.7058	0.2702	2.053	0.3199	1.553	
		1.0000					26.84
Ar-H$_2$ [226]	473.0	0.0000					21.11
		0.3485	0.3266	4.258	0.3870	3.225	
		0.5543	0.2987	3.895	0.3546	2.954	
		0.7058	0.2909	3.794	0.3462	2.884	
		1.0000					32.08
Ar-H$_2$ [226]	523.0	0.0000					12.96
		0.3485	0.2929	2.182	0.3332	1.585	
		0.5543	0.2840	2.115	0.3299	1.570	
		1.0000					34.48
He-H$_2$ [74]	273.2	0.0000					8.41
		0.1881	1.098	0.9690	1.134	0.9029	
		0.3986	1.094	0.9655	1.130	0.8993	
		0.5972	1.101	0.9715	1.136	0.9046	
		0.7509	1.095	0.9661	1.130	0.8993	
		0.8640	1.093	0.9644	1.128	0.8977	
		0.8957	1.171	1.033	1.205	0.9595	
		0.9609	1.054	0.9298	1.089	0.8665	
		1.0000					18.92

TABLE 1.   COMPOSITION AND TEMPERATURE DEPENDENCE OF Ψ_{ij} ON
DIFFERENT SCHEMES OF COMPUTATION (continued)

Gas Pair [Reference]	Temp. (K)	Mole Fraction of Heavier Component	First Method Ψ_{12}	Ψ_{21}	Second Method Ψ_{12}	Ψ_{21}	Viscosity (N s m^{-2} x 10^{-6})
He-H$_2$ [74]	288.2	0.0000					8.78
		0.1881	1.105	0.9820	1.142	0.9154	
		0.3986	1.108	0.9843	1.144	0.9171	
		0.5972	1.112	0.9877	1.147	0.9197	
		0.7509	1.113	0.9888	1.148	0.9202	
		0.8640	1.128	1.002	1.162	0.9319	
		0.8957	1.100	0.9772	1.134	0.9094	
		0.9609	1.089	0.9681	1.124	0.9011	
		1.0000					19.61
He-H$_2$ [327]	291.7	0.0000					8.81
		0.1890	1.113	0.9886	1.150	0.9218	
		0.3530	1.145	1.017	1.183	0.9484	
		0.5030	1.169	1.039	1.207	0.9677	
		0.5650	1.193	1.060	1.231	0.9868	
		0.6830	1.192	1.059	1.229	0.9851	
		0.8110	1.228	1.091	1.263	1.013	
		1.0000					19.69
He-H$_2$ [221]	293.0	0.0000					8.75
		0.3082	1.127	0.9921	1.164	0.9246	
		0.3931	1.129	0.9937	1.166	0.9258	
		0.4480	1.118	0.9837	1.154	0.9162	
		1.0000					19.74
He-H$_2$ [221]	373.0	0.0000					10.29
		0.3082	1.107	0.9751	1.143	0.9085	
		0.3931	1.120	0.9863	1.156	0.9189	
		0.4480	1.114	0.9807	1.150	0.9135	
		1.0000					23.20
He-H$_2$ [74]	373.2	0.0000					10.45
		0.1881	1.089	0.9653	1.125	0.8995	
		0.3986	1.096	0.9720	1.132	0.9054	
		0.5972	1.082	0.9591	1.117	0.8931	
		0.7509	1.090	0.9665	1.125	0.8999	
		0.8640	1.061	0.9408	1.096	0.8765	
		0.8957	1.046	0.9273	1.081	0.8643	
		0.9609	1.114	0.9876	1.148	0.9184	
		1.0000					23.41
He-H$_2$ [221]	473.0	0.0000					12.11
		0.3082	1.111	0.9840	1.148	0.9170	
		0.3931	1.125	0.9964	1.162	0.9285	
		0.4480	1.114	0.9864	1.150	0.9189	
		1.0000					27.15
He-H$_2$ [221]	523.0	0.0000					12.96
		0.3082	1.111	0.9845	1.147	0.9175	
		0.3931	1.121	0.9933	1.157	0.9256	
		0.4480	1.118	0.9913	1.155	0.9235	
		1.0000					29.03
Ne-H$_2$ [221]	290.4	0.0000					8.78
		0.1610	0.5615	1.584	0.6017	1.201	
		0.3470	0.5420	1.529	0.5838	1.166	
		0.5050	0.5898	1.664	0.6373	1.272	
		0.6570	0.5216	1.471	0.5689	1.136	
		0.7950	0.4940	1.3935	0.5441	1.086	
		1.0000					31.16
Ne-H$_2$ [221]	293.0	0.0000					8.75
		0.2285	0.5482	1.553	0.5882	1.179	
		0.5391	0.5413	1.533	0.5870	1.177	
		0.7480	0.5319	1.507	0.5811	1.165	
		1.0000					30.92

TABLE 1. COMPOSITION AND TEMPERATURE DEPENDENCE OF Ψ_{ij} ON
DIFFERENT SCHEMES OF COMPUTATION (continued)

Gas Pair [Reference]	Temp. (K)	Mole Fraction of Heavier Component	First Method		Second Method		Viscosity $(N\ s\ m^{-2} \times 10^{-6})$
			Ψ_{12}	Ψ_{21}	Ψ_{12}	Ψ_{21}	
Ne-H$_2$ [221]	373.0	0.0000					10.29
		0.2285	0.5450	1.549	0.5848	1.177	
		0.5391	0.5409	1.538	0.5867	1.181	
		0.7480	0.5242	1.490	0.5734	1.154	
		1.0000					36.23
Ne-H$_2$ [221]	473.0	0.0000					12.11
		0.2285	0.5424	1.558	0.5825	1.184	
		0.5391	0.5357	1.539	0.5815	1.182	
		0.7480	0.5249	1.508	0.5743	1.168	
		1.0000					42.20
Ne-H$_2$ [221]	523.0	0.0000					12.96
		0.2285	0.5422	1.563	0.5824	1.188	
		0.5391	0.5395	1.555	0.5856	1.195	
		0.7480	0.5295	1.526	0.5790	1.181	
		1.0000					45.01
Ar-NH$_3$ [134]	298.2	0.0000					10.16
		0.0540	1.080	1.142	1.129	1.050	
		0.1720	0.9952	1.052	1.036	0.9638	
		0.2740	0.9876	1.044	1.028	0.9562	
		0.3860	0.9701	1.026	1.009	0.9390	
		0.5010	0.9735	1.029	1.013	0.9425	
		0.5950	0.9786	1.035	1.018	0.9475	
		0.6910	0.9793	1.035	1.019	0.9482	
		0.7850	0.9880	1.045	1.028	0.9564	
		0.8520	0.9800	1.036	1.020	0.9489	
		1.0000					22.54
Ar-NH$_3$ [134]	308.2	0.0000					10.49
		0.0380	1.162	1.238	1.221	1.145	
		0.1680	1.001	1.066	1.042	0.9771	
		0.2950	0.9818	1.046	1.022	0.9578	
		0.3990	0.9734	1.037	1.013	0.9495	
		0.5190	0.9590	1.022	0.9981	0.9356	
		0.6190	0.9468	1.009	0.9859	0.9241	
		0.7020	0.9461	1.008	0.9854	0.9237	
		0.7950	0.9367	0.9978	0.9763	0.9152	
		0.8600	0.9331	0.9940	0.9730	0.9121	
		1.0000					23.10
Ar-NH$_3$ [134]	353.2	0.0000					11.98
		0.0530	1.017	1.111	1.062	1.021	
		0.1840	0.9696	1.060	1.010	0.9714	
		0.2780	0.9703	1.061	1.011	0.9721	
		0.3810	0.9646	1.054	1.005	0.9662	
		0.4910	0.9539	1.043	0.9934	0.9555	
		0.5940	0.9472	1.035	0.9868	0.9491	
		0.6840	0.9425	1.030	0.9822	0.9447	
		0.8600	0.9392	1.027	0.9794	0.9420	
		1.0000					25.71
Ar-SO$_2$ [35]	298.2	0.0000					22.45
		0.1910	0.5862	1.602	0.6127	1.560	
		0.2500	0.5839	1.596	0.6104	1.554	
		0.3140	0.5893	1.611	0.6161	1.569	
		0.4040	0.5918	1.618	0.6188	1.576	
		0.5000	0.5918	1.618	0.6189	1.576	
		0.6120	0.5991	1.638	0.6265	1.595	
		0.7200	0.6012	1.643	0.6287	1.601	
		0.8300	0.6159	1.683	0.6436	1.639	
		0.9540	0.6868	1.877	0.7152	1.821	
		1.0000					13.17

TABLE 1.   COMPOSITION AND TEMPERATURE DEPENDENCE OF Ψ_{ij} ON DIFFERENT SCHEMES OF COMPUTATION (continued)

Gas Pair [Reference]	Temp. (K)	Mole Fraction of Heavier Component	First Method Ψ_{12}	Ψ_{21}	Second Method Ψ_{12}	Ψ_{21}	Viscosity (N s m^{-2} x 10^{-6})
Ar-SO$_2$ [35]	308.2	0.0000					23.10
		0.0240	0.5374	1.499	0.5602	1.456	
		0.1500	0.5189	1.447	0.5414	1.407	
		0.2540	0.5124	1.429	0.5352	1.391	
		0.3620	0.5166	1.441	0.5403	1.404	
		0.4640	0.5088	1.419	0.5327	1.384	
		0.5810	0.4958	1.383	0.5198	1.351	
		0.6660	0.4871	1.359	0.5113	1.329	
		0.7620	0.4751	1.325	0.4993	1.298	
		0.8720	0.4373	1.220	0.4610	1.198	
		0.8930	0.4228	1.179	0.4461	1.159	
		1.0000					13.28
Ar-SO$_2$ [35]	353.2	0.0000					25.71
		0.0430	0.4984	1.349	0.5181	1.306	
		0.1630	0.5192	1.405	0.5413	1.365	
		0.2640	0.5162	1.397	0.5389	1.359	
		0.3870	0.5116	1.385	0.5349	1.349	
		0.4830	0.5074	1.374	0.5311	1.339	
		0.5860	0.4978	1.348	0.5218	1.316	
		0.6870	0.4896	1.325	0.5138	1.296	
		0.7810	0.4916	1.331	0.5163	1.302	
		0.8850	0.4838	1.310	0.5088	1.283	
		0.9200	0.4671	1.264	0.4918	1.240	
		1.0000					15.23
C$_6$H$_6$-C$_6$H$_{12}$ (Liquid) [355]	298.2	0.0000					605.90
		0.0967	1.582	1.188	1.591	1.182	
		0.2186	1.598	1.201	1.607	1.194	
		0.3530	1.595	1.198	1.603	1.191	
		0.5126	1.599	1.202	1.607	1.194	
		0.6636	1.598	1.201	1.605	1.192	
		0.7826	1.618	1.216	1.624	1.207	
		0.8718	1.662	1.248	1.667	1.239	
		1.0000					869.00
C$_6$H$_6$-CH$_3$(CH$_2$)$_4$CH$_3$ (Liquid) [355]	298.2	0.0000					605.90
		0.1189	0.9164	2.036	0.9266	2.029	
		0.2784	0.9296	2.066	0.9393	2.057	
		0.4296	0.8862	1.969	0.8948	1.960	
		0.5950	0.8681	1.929	0.8761	1.919	
		0.7335	0.8536	1.897	0.8611	1.886	
		0.8719	0.8231	1.829	0.8303	1.818	
		1.0000					300.80
C$_6$H$_6$-OMCTS (Liquid) [360]	291.2	0.0000					670.30
		0.0881	1.518	1.533	1.626	1.344	
		0.3511	1.450	1.464	1.524	1.260	
		0.5997	1.445	1.460	1.501	1.241	
		0.7738	1.442	1.456	1.488	1.230	
		0.8529	1.436	1.450	1.478	1.222	
		0.9369	1.433	1.447	1.471	1.216	
		1.0000					2520.00
C$_6$H$_6$-OMCTS (Liquid) [360]	298.2	0.0000					602.40
		0.0341	1.477	1.543	1.590	1.359	
		0.0699	1.451	1.516	1.555	1.330	
		0.1407	1.426	1.489	1.519	1.299	
		0.2235	1.410	1.472	1.493	1.277	
		0.2938	1.403	1.466	1.481	1.266	
		0.3751	1.392	1.454	1.462	1.250	
		0.4689	1.394	1.455	1.457	1.246	
		0.6211	1.391	1.452	1.445	1.236	
		0.6777	1.392	1.454	1.444	1.234	
		0.7510	1.349	1.409	1.396	1.194	
		0.8434	1.407	1.469	1.450	1.240	
		0.8753	1.397	1.459	1.439	1.230	
		0.9028	1.427	1.490	1.467	1.254	
		0.9291	1.390	1.452	1.430	1.222	
		1.0000					2190.00

TABLE 1. COMPOSITION AND TEMPERATURE DEPENDENCE OF Ψ_{ij} ON DIFFERENT SCHEMES OF COMPUTATION (continued)

Gas Pair [Reference]	Temp. (K)	Mole Fraction of Heavier Component	First Method Ψ_{12}	Ψ_{21}	Second Method Ψ_{12}	Ψ_{21}	Viscosity (N s m^{-2} x 10^{-6})
C_6H_6-OMCTS (Liquid) [360]	308.2	0.0000					523.50
		0.0886	1.349	1.485	1.442	1.300	
		0.3517	1.322	1.455	1.392	1.254	
		0.6020	1.323	1.456	1.378	1.241	
		0.7741	1.306	1.437	1.353	1.219	
		0.8544	1.330	1.463	1.373	1.237	
		0.9373	1.324	1.458	1.364	1.229	
		1.0000					1806.00
C_6H_6-OMCTS (Liquid) [360]	318.2	0.0000					460.30
		0.0888	1.301	1.502	1.393	1.316	
		0.3526	1.259	1.454	1.327	1.254	
		0.6036	1.262	1.457	1.317	1.244	
		0.7763	1.263	1.458	1.310	1.238	
		0.8562	1.256	1.450	1.300	1.229	
		0.9134	1.170	1.351	1.213	1.146	
		1.0000					1514.00
CO_2-H_2 [234]	300.0	0.0000					8.91
		0.1112	0.2057	2.679	0.2373	1.946	
		0.2150	0.2024	2.636	0.2394	1.964	
		0.4054	0.1988	2.590	0.2432	1.995	
		0.5871	0.1951	2.541	0.2440	2.002	
		0.8006	0.1984	2.585	0.2514	2.063	
		0.8821	0.1892	2.465	0.2431	1.994	
		1.0000					14.93
CO_2-H_2 [234]	400.0	0.0000					10.81
		0.1112	0.2163	2.626	0.2478	1.895	
		0.2150	0.2169	2.633	0.2545	1.945	
		0.4054	0.2111	2.562	0.2554	1.952	
		0.5871	0.2072	2.515	0.2561	1.958	
		0.8006	0.2084	2.529	0.2614	1.998	
		0.8821	0.1977	2.399	0.2516	1.923	
		1.0000					19.44
CO_2-H_2 [234]	500.0	0.0000					12.56
		0.1112	0.2261	2.634	0.2583	1.896	
		0.2150	0.2242	2.613	0.2618	1.921	
		0.4054	0.2191	2.553	0.2635	1.933	
		0.5871	0.2142	2.496	0.2631	1.930	
		0.8006	0.2043	2.381	0.2570	1.886	
		0.8821	0.1998	2.328	0.2536	1.861	
		1.0000					23.53
CO_2-H_2 [234]	550.0	0.0000					13.41
		0.1112	0.2314	2.650	0.2643	1.906	
		0.2150	0.2289	2.621	0.2668	1.924	
		0.4054	0.2076	2.378	0.2502	1.805	
		0.5871	0.2201	2.520	0.2691	1.941	
		0.8006	0.2217	2.540	0.2748	1.982	
		0.8821	0.2144	2.456	0.2686	1.937	
		1.0000					25.56
CO_2-N_2 [337]	297.7	0.0000					17.80
		0.2260	0.7307	1.363	0.7581	1.321	
		0.2770	0.7236	1.350	0.7506	1.308	
		0.3260	0.7285	1.359	0.7558	1.317	
		0.5800	0.7188	1.341	0.7458	1.300	
		0.7500	0.7114	1.327	0.7384	1.287	
		0.8000	0.7591	1.416	0.7868	1.371	
		1.0000					14.99
CO_2-N_2O [234]	300.0	0.0000					14.93
		0.1087	0.9896	0.9929	0.9896	0.9929	
		0.1903	1.003	1.006	1.003	1.006	
		0.3967	0.9927	0.9961	0.9927	0.9961	
		0.5976	0.9928	0.9962	0.9928	0.9962	
		1.0000					14.88

TABLE 1.   COMPOSITION AND TEMPERATURE DEPENDENCE OF Ψ_{ij} ON DIFFERENT SCHEMES OF COMPUTATION (continued)

Gas Pair [Reference]	Temp. (K)	Mole Fraction of Heavier Component	First Method Ψ_{12}	Ψ_{21}	Second Method Ψ_{12}	Ψ_{21}	Viscosity (N s m^{-2} x 10^{-6})
CO_2-N_2O [234]	400.0	0.0000					19.44
		0.1087	0.9968	0.9974	0.9968	0.9974	
		0.1903	1.004	1.005	1.004	1.005	
		0.3967	0.9929	0.9934	0.9929	0.9934	
		0.5976	0.9927	0.9933	0.9927	0.9933	
		0.8003	0.9920	0.9926	0.9920	0.9926	
		1.0000					19.43
CO_2-N_2O [234]	500.0	0.0000					23.53
		0.1087	0.9900	0.9892	0.9900	0.9892	
		0.1903	0.9940	0.9933	0.9940	0.9933	
		0.3967	0.9905	0.9897	0.9905	0.9897	
		0.5976	0.9909	0.9901	0.9909	0.9901	
		0.8003	0.9972	0.9964	0.9972	0.9964	
		1.0000					23.55
CO_2-N_2O [234]	550.0	0.0000					25.65
		0.1087	1.016	1.020	1.016	1.020	
		0.1903	1.013	1.017	1.013	1.017	
		0.3967	0.9956	0.9996	0.9956	0.9996	
		0.5976	0.9956	0.9996	0.9956	0.9996	
		0.8003	1.000	1.004	1.000	1.004	
		1.0000					25.55
CO_2-O_2 [337]	300.0	0.0000					20.80
		0.1950	0.7674	1.464	0.7895	1.436	
		0.3060	0.7239	1.382	0.7443	1.354	
		0.3390	0.7189	1.372	0.7392	1.345	
		0.5600	0.6722	1.283	0.6914	1.258	
		0.7100	0.6920	1.321	0.7119	1.295	
		0.8000	0.7131	1.361	0.7333	1.334	
		0.9170	0.6989	1.334	0.7191	1.308	
		1.0000					14.99
CO_2-C_3H_8 [234]	300.0	0.0000					14.93
		0.2117	0.7177	1.314	0.7179	1.314	
		0.4224	0.7173	1.313	0.7174	1.313	
		0.5975	0.7159	1.311	0.7160	1.311	
		0.8106	0.7154	1.310	0.7155	1.310	
		1.0000					8.17
CO_2-C_3H_8 [234]	400.0	0.0000					19.44
		0.2117	0.7182	1.307	0.7184	1.307	
		0.4224	0.7188	1.309	0.7190	1.308	
		0.5975	0.7173	1.306	0.7174	1.306	
		0.8106	0.7144	1.301	0.7146	1.300	
		1.0000					10.70
CO_2-C_3H_8 [234]	500.0	0.0000					23.53
		0.2117	0.7273	1.311	0.7275	1.311	
		0.4224	0.7282	1.313	0.7283	1.312	
		0.5975	0.7332	1.322	0.7333	1.321	
		0.8106	0.7479	1.348	0.7481	1.348	
		1.0000					13.08
CO_2-C_3H_8 [234]	550.0	0.0000					25.56
		0.2117	0.7342	1.322	0.7344	1.322	
		0.4224	0.7335	1.321	0.7336	1.321	
		0.5975	0.7335	1.321	0.7336	1.321	
		0.8106	0.7293	1.313	0.7294	1.313	
		1.0000					14.22
CO-C_2H_4 [227]	300.0	0.0000					17.76
		0.2632	0.7446	1.282	0.7447	1.282	
		0.4354	0.7624	1.313	0.7625	1.313	
		0.8062	0.7897	1.360	0.7898	1.360	
		1.0000					10.33

TABLE 1. COMPOSITION AND TEMPERATURE DEPENDENCE OF Ψ_{ij} ON DIFFERENT SCHEMES OF COMPUTATION (continued)

Gas Pair [Reference]	Temp. (K)	Mole Fraction of Heavier Component	First Method Ψ_{12}	Ψ_{21}	Second Method Ψ_{12}	Ψ_{21}	Viscosity (N s m^{-2} x 10^{-6})
CO-C$_2$H$_4$ [227]	400.0	0.0000					21.83
		0.2632	0.3555	2.011	0.3986	1.870	
		0.4354	0.3605	2.039	0.4061	1.906	
		0.8062	0.3667	2.074	0.4152	1.948	
		1.0000					13.42
CO-C$_2$H$_4$ [227]	500.0	0.0000					25.48
		0.2632	0.7817	1.230	0.7818	1.230	
		0.4354	0.7932	1.248	0.7933	1.248	
		0.8062	0.8129	1.279	0.8130	1.279	
		1.0000					16.22
CO-C$_2$H$_4$ [227]	550.0	0.0000					27.14
		0.2632	0.7906	1.226	0.7907	1.226	
		0.4354	0.8054	1.249	0.8055	1.249	
		0.8062	0.8144	1.263	0.8145	1.263	
		1.0000					17.53
CO-H$_2$ [327]	293.3	0.0000					8.84
		0.1190	0.3210	2.230	0.3596	1.683	
		0.1910	0.3212	2.231	0.3628	1.698	
		0.2740	0.3159	2.194	0.3596	1.683	
		0.3860	0.3088	2.145	0.3552	1.663	
		0.4940	0.3081	2.140	0.3573	1.673	
		0.6130	0.3046	2.116	0.3564	1.668	
		1.0000					17.68
CO-N$_2$ [227]	300.0	0.0000					17.76
		0.1629	1.007	1.004	1.007	1.004	
		0.3432	1.005	1.002	1.005	1.002	
		0.6030	0.9990	0.9963	0.9990	0.9963	
		0.8154	0.9978	0.9951	0.9978	0.9951	
		1.0000					17.81
CO-N$_2$ [227]	400.0	0.0000					21.83
		0.1629	1.002	0.9987	1.002	0.9987	
		0.3432	0.9959	0.9928	0.9959	0.9928	
		0.6030	1.006	1.002	1.006	1.002	
		0.8154	1.006	1.003	1.006	1.003	
		1.0000					21.90
CO-N$_2$ [227]	500.0	0.0000					25.48
		0.1629	1.001	0.9962	1.001	0.9962	
		0.3432	1.005	1.000	1.005	1.000	
		0.6030	1.000	0.9955	1.000	0.9954	
		0.8154	0.9994	0.9948	0.9994	0.9948	
		1.0000					25.60
CO-N$_2$ [227]	550.0	0.0000					27.14
		0.1629	0.9984	0.9938	0.9985	0.9938	
		0.3432	0.9994	0.9948	0.9995	0.9948	
		0.6030	1.005	0.9998	1.005	0.9998	
		0.8154	1.007	1.002	1.007	1.002	
		1.0000					27.27
CO-O$_2$ [227]	300.0	0.0000					17.76
		0.2337	1.007	0.9936	1.017	0.9830	
		0.4201	1.000	0.9863	1.009	0.9758	
		0.7733	0.9994	0.9858	1.009	0.9753	
		1.0000					20.57
CO-O$_2$ [227]	400.0	0.0000					21.83
		0.2337	1.020	0.9901	1.029	0.9795	
		0.4201	1.015	0.9858	1.024	0.9752	
		0.7733	1.012	0.9828	1.021	0.9723	
		1.0000					25.68

TABLE 1.   COMPOSITION AND TEMPERATURE DEPENDENCE OF Ψ_{ij} ON
DIFFERENT SCHEMES OF COMPUTATION (continued)

Gas Pair [Reference]	Temp. (K)	Mole Fraction of Heavier Component	First Method		Second Method		Viscosity (N s m^{-2} x 10^{-6})
			Ψ_{12}	Ψ_{21}	Ψ_{12}	Ψ_{21}	
CO-O$_2$ [227]	500.0	0.0000					25.48
		0.2337	1.024	0.9884	1.034	0.9777	
		0.4201	1.020	0.9837	1.029	0.9731	
		0.7733	1.019	0.9835	1.029	0.9729	
		1.0000					30.17
CCl$_4$-OMCTS (Liquid) [360]	291.2	0.0000					1001.00
		0.1780	1.112	0.8515	1.142	0.7927	
		0.3227	1.143	0.8752	1.175	0.8155	
		0.5718	1.171	0.8970	1.204	0.8359	
		0.7258	1.183	0.9063	1.216	0.8443	
		0.8618	1.185	0.9074	1.218	0.8451	
		0.9815	1.142	0.8749	1.175	0.8157	
		1.0000					2520.00
CCl$_4$-OMCTS (Liquid) [360]	298.2	0.0000					901.00
		0.1089	1.058	0.8392	1.086	0.7808	
		0.1965	1.102	0.8743	1.133	0.8147	
		0.2890	1.120	0.8886	1.153	0.8285	
		0.4288	1.139	0.9034	1.172	0.8425	
		0.5841	1.156	0.9171	1.190	0.8552	
		0.6590	1.158	0.9187	1.192	0.8566	
		0.8443	1.172	0.9296	1.205	0.8662	
		0.9264	1.179	0.9352	1.212	0.8711	
		0.9773	1.279	1.014	1.310	0.9420	
		1.0000					2190.00
CCl$_4$-OMCTS (Liquid) [360]	308.2	0.0000					781.00
		0.1756	1.047	0.8727	1.076	0.8134	
		0.3239	1.079	0.8998	1.111	0.8397	
		0.5732	1.106	0.9226	1.140	0.8613	
		0.7290	1.129	0.9418	1.163	0.8789	
		0.8636	1.125	0.9383	1.159	0.8755	
		0.9817	1.084	0.9041	1.118	0.8447	
		1.0000					1806.00
CCl$_4$-OMCTS (Liquid) [360]	318.2	0.0000					686.60
		0.1779	1.009	0.8820	1.038	0.8226	
		0.3249	1.041	0.9105	1.073	0.8503	
		0.5816	1.082	0.9460	1.116	0.8840	
		0.7307	1.094	0.9565	1.128	0.8936	
		0.8652	1.113	0.9733	1.147	0.9087	
		0.9821	1.074	0.9392	1.108	0.8779	
		1.0000					1514.00
CF$_4$-SF$_6$ [339]	303.1	0.0000					17.67
		0.2460	0.8117	1.497	0.8461	1.446	
		0.5090	0.7815	1.441	0.8131	1.390	
		0.7430	0.7738	1.427	0.8045	1.375	
		1.0000					15.90
CF$_4$-SF$_6$ [339]	313.1	0.0000					18.17
		0.2460	0.8129	1.498	0.8473	1.447	
		0.5090	0.7839	1.445	0.8156	1.393	
		0.7430	0.7727	1.424	0.8034	1.372	
		1.0000					16.36
CF$_4$-SF$_6$ [339]	329.1	0.0000					18.94
		0.2460	0.8149	1.501	0.8495	1.451	
		0.5090	0.7838	0.444	0.8154	1.392	
		0.7430	0.7719	1.422	0.8026	1.371	
		1.0000					17.06
CF$_4$-SF$_6$ [339]	342.0	0.0000					19.57
		0.2460	0.8144	1.504	0.8490	1.453	
		0.5090	0.7798	1.440	0.8113	1.388	
		0.7430	0.7713	1.424	0.8019	1.372	
		1.0000					17.59

TABLE 1.   COMPOSITION AND TEMPERATURE DEPENDENCE OF Ψ_{ij} ON
DIFFERENT SCHEMES OF COMPUTATION (continued)

Gas Pair [Reference]	Temp. (K)	Mole Fraction of Heavier Component	First Method Ψ_{12}	Ψ_{21}	Second Method Ψ_{12}	Ψ_{21}	Viscosity (N s m^{-2} x 10^{-6})
C_6-H_{12}- $CH_3(CH_2)_4CH_3$ (Liquid) [355]	298.2	0.0000					869.00
		0.0966	0.7736	2.288	0.7758	2.287	
		0.2480	0.7637	2.259	0.7658	2.257	
		0.4127	0.7513	2.222	0.7533	2.220	
		0.5502	0.7410	2.192	0.7428	2.190	
		0.7258	0.7218	2.135	0.7235	2.133	
		0.8286	0.7079	2.094	0.7096	2.092	
		1.0000					300.80
D_2-H_2 [179]	14.4	0.0000					0.79
		0.2690	0.8117	1.272	0.8496	1.202	
		0.5040	0.8126	1.274	0.8502	1.203	
		0.7600	0.8493	1.331	0.8872	1.255	
		1.0000					1.00
D_2-H_2 [179]	20.4	0.0000					1.08
		0.3340	0.8020	1.254	0.8392	1.184	
		0.6770	0.7995	1.250	0.8366	1.181	
		1.0000					1.37
D_2-H_2 [179]	71.5	0.0000					3.24
		0.2480	0.8316	1.204	0.8683	1.134	
		0.5020	0.8301	1.202	0.8669	1.133	
		0.7490	0.8430	1.220	0.8801	1.150	
		1.0000					4.44
D_2-H_2 [179]	90.1	0.0000					3.86
		0.2620	0.8294	1.192	0.8658	1.123	
		0.5020	0.8285	1.191	0.8651	1.122	
		0.7450	0.8361	1.201	0.8730	1.132	
		1.0000					5.33
D_2-H_2 [179]	196.0	0.0000					6.75
		0.2510	0.8327	1.191	0.8691	1.122	
		0.4970	0.8347	1.194	0.8714	1.125	
		0.7530	0.8355	1.195	0.8724	1.126	
		1.0000					9.36
D_2-H_2 [179]	229.0	0.0000					7.57
		0.2480	0.8335	1.200	0.8703	1.131	
		0.5050	0.8322	1.198	0.8690	1.129	
		0.7550	0.8448	1.217	0.8819	1.146	
		1.0000					10.43
D_2-H_2 [179]	293.1	0.0000					8.86
		0.2460	0.8336	1.191	0.8701	1.122	
		0.5070	0.8392	1.199	0.8761	1.130	
		0.7530	0.8363	1.195	0.8732	1.126	
		1.0000					12.30
D_2-HD [179]	14.4	0.0000					0.91
		0.2610	0.8980	1.086	0.9164	1.062	
		0.4970	0.8846	1.070	0.9028	1.046	
		0.7160	0.8761	1.059	0.8944	1.036	
		1.0000					1.00
D_2-HD [179]	20.4	0.0000					1.27
		0.2420	0.9074	1.086	0.9258	1.062	
		0.5030	0.9189	1.100	0.9377	1.075	
		0.7510	0.9092	1.088	0.9278	1.064	
		1.0000					1.41
D_2-HD [179]	71.5	0.0000					3.93
		0.2540	0.9362	1.091	0.9552	1.067	
		0.5070	0.9348	1.090	0.9536	1.065	
		0.7550	0.9342	1.089	0.9529	1.065	
		1.0000					4.48

TABLE 1.   COMPOSITION AND TEMPERATURE DEPENDENCE OF Ψ_{ij} ON
DIFFERENT SCHEMES OF COMPUTATION (continued)

Gas Pair [Reference]	Temp. (K)	Mole Fraction of Heavier Component	First Method Ψ_{12}	Ψ_{21}	Second Method Ψ_{12}	Ψ_{21}	Viscosity (N s m^{-2} x 10^{-6})
D_2-HD [179]	90.1	0.0000					4.74
		0.2380	0.9290	1.084	0.9478	1.059	
		0.4920	0.9286	1.083	0.9473	1.059	
		0.7490	0.9227	1.076	0.9414	1.052	
		1.0000					5.40
D_2-HD [179]	196.0	0.0000					8.22
		0.2490	0.9306	1.081	0.9493	1.057	
		0.5000	0.9281	1.079	0.9468	1.054	
		0.7500	0.9280	1.078	0.9467	1.054	
		1.0000					9.40
D_2-HD [179]	229.0	0.0000					9.10
		0.2490	0.9315	1.075	0.9502	1.051	
		0.4950	0.9324	1.076	0.9511	1.052	
		0.7550	0.9309	1.074	0.9495	1.050	
		1.0000					10.48
D_2-HD [179]	293.1	0.0000					10.75
		0.2580	0.9375	1.080	0.9563	1.056	
		0.5090	0.9347	1.077	0.9534	1.053	
		0.7360	0.9310	1.073	0.9496	1.048	
		1.0000					12.40
C_2H_6-H_2 [229]	293.0	0.0000					8.76
		0.1485	0.2067	2.971	0.2490	2.386	
		0.5500	0.1912	2.748	0.2422	2.321	
		1.0000					9.09
C_2H_6-H_2 [229]	373.0	0.0000					10.33
		0.1485	0.2186	2.949	0.2617	2.354	
		0.5500	0.2087	2.816	0.2610	2.348	
		1.0000					11.42
C_2H_6-H_2 [229]	473.0	0.0000					12.13
		0.1485	0.2286	2.936	0.2725	2.333	
		0.5500	0.2197	2.821	0.2726	2.333	
		1.0000					14.09
C_2H_6-H_2 [229]	523.0	0.0000					12.96
		0.1485	0.2322	2.942	0.2766	2.336	
		0.5500	0.2223	2.816	0.2752	2.324	
		1.0000					15.26
C_2H_3-CH_4 [229]	293.0	0.0000					10.87
		0.1884	0.6594	1.478	0.6940	1.416	
		0.5126	0.6570	1.473	0.6917	1.411	
		0.8097	0.6543	1.466	0.6892	1.406	
		1.0000					9.09
C_2H_6-CH_4 [229]	373.0	0.0000					13.31
		0.1884	0.6690	1.462	0.7037	1.399	
		0.5126	0.6652	1.453	0.6999	1.391	
		0.8097	0.6627	1.448	0.6976	1.387	
		1.0000					11.42
C_2H_6-CH_4 [229]	473.0	0.0000					16.03
		0.1884	0.6751	1.440	0.7096	1.377	
		0.5126	0.6749	1.439	0.7097	1.377	
		0.8097	0.6733	1.436	0.7083	1.374	
		1.0000					14.09
C_2H_6-CH_4 [229]	523.0	0.0000					17.25
		0.1884	0.6789	1.438	0.7136	1.376	
		0.5126	0.6788	1.438	0.7136	1.376	
		0.8097	0.6759	1.432	0.7109	1.371	
		1.0000					15.26

TABLE 1.   COMPOSITION AND TEMPERATURE DEPENDENCE OF Ψ_{ij} ON DIFFERENT SCHEMES OF COMPUTATION (continued)

Gas Pair [Reference]	Temp. (K)	Mole Fraction of Heavier Component	First Method		Second Method		Viscosity ($N\ s\ m^{-2} \times 10^{-6}$)
			Ψ_{12}	Ψ_{21}	Ψ_{12}	Ψ_{21}	
C_2H_6-C_3H_8 [229]	293.0	0.0000					9.09
		0.5673	0.7754	1.290	0.7995	1.256	
		0.7437	0.7637	1.271	0.7876	1.238	
		0.8474	0.7719	1.285	0.7959	1.251	
		1.0000					8.01
C_2H_6-C_3H_8 [229]	373.0	0.0000					11.42
		0.5673	0.7755	1.288	0.7996	1.254	
		0.7437	0.7699	1.279	0.7939	1.245	
		0.8474	0.7739	1.286	0.7979	1.252	
		1.0000					10.08
C_2H_6-C_3H_8 [229]	473.0	0.0000					14.09
		0.5673	0.7764	1.280	0.8004	1.246	
		0.7437	0.7580	1.250	0.7817	1.217	
		0.8474	0.7792	1.285	0.8033	1.251	
		1.0000					12.53
C_2H_6-C_3H_8 [229]	523.0	0.0000					15.26
		0.5673	0.7797	1.280	0.8037	1.246	
		0.7437	0.7749	1.272	0.7988	1.238	
		0.8474	0.7832	1.286	0.8072	1.251	
		1.0000					13.63
C_2H_4-H_2 [230]	195.2	0.0000					6.70
		0.2501	0.2224	2.888	0.2696	2.359	
		0.5087	0.2176	2.825	0.2699	2.362	
		0.6444	0.2129	2.765	0.2668	2.334	
		0.8082	0.2276	2.956	0.2842	2.487	
		1.0000					7.18
C_2H_4-H_2 [230]	233.2	0.0000					7.40
		0.1638	0.2218	2.792	0.2646	2.244	
		0.2501	0.2268	2.855	0.2740	2.324	
		0.5129	0.2231	2.808	0.2757	2.338	
		0.6444	0.2205	2.776	0.2747	2.330	
		0.8082	0.2126	2.676	0.2681	2.274	
		1.0000					8.18
C_2H_4-H_2 [230]	272.2	0.0000					8.30
		0.1638	0.2266	2.775	0.2695	2.224	
		0.2501	0.2248	2.753	0.2708	2.235	
		0.5129	0.2188	2.679	0.2705	2.232	
		0.6444	0.2169	2.656	0.2706	2.233	
		0.8082	0.2293	2.809	0.2858	2.358	
		1.0000					9.43
C_2H_4-H_2 [230]	293.2	0.0000					8.73
		0.2160	0.2238	2.687	0.2678	2.166	
		0.5173	0.2204	2.646	0.2721	2.201	
		0.7033	0.2132	2.560	0.2673	2.162	
		0.8107	0.2095	2.514	0.2646	2.140	
		1.0000					10.12
C_2H_4-H_2 [230]	328.2	0.0000					9.43
		0.2100	0.2279	2.666	0.2716	2.140	
		0.5173	0.2258	2.640	0.2777	2.188	
		0.7033	0.2152	2.517	0.2692	2.121	
		0.8107	0.2057	2.406	0.2605	2.052	
		1.0000					11.22
C_2H_4-H_2 [230]	373.2	0.0000					10.30
		0.2114	0.2335	2.648	0.2774	2.119	
		0.5173	0.2306	2.615	0.2827	2.159	
		0.7033	0.2276	2.581	0.2823	2.157	
		0.8107	0.2191	2.484	0.2746	2.098	
		1.0000					12.64

TABLE 1.   COMPOSITION AND TEMPERATURE DEPENDENCE OF Ψ_{ij} ON
DIFFERENT SCHEMES OF COMPUTATION (continued)

Gas Pair [Reference]	Temp. (K)	Mole Fraction of Heavier Component	First Method Ψ_{12}	Ψ_{21}	Second Method Ψ_{12}	Ψ_{21}	Viscosity (N s m^{-2} x 10^{-6})
C_2H_4-H_2 [230]	423.2	0.0000					11.23
		0.2114	0.2358	2.617	0.2795	2.090	
		0.5197	0.2282	2.533	0.2797	2.091	
		0.7201	0.2299	2.551	0.2848	2.130	
		0.8043	0.2287	2.539	0.3847	2.129	
		1.0000					14.08
C_2H_4-H_2 [230]	473.2	0.0000					12.11
		0.2114	0.2397	2.611	0.2836	2.081	
		0.5197	0.2371	2.583	0.2893	2.123	
		0.7201	0.2379	2.592	0.2932	2.152	
		0.8043	0.2356	2.566	0.2918	2.142	
		1.0000					15.47
C_2H_4-H_2 [230]	523.2	0.0000					12.94
		0.2114	0.2487	2.664	0.2938	2.120	
		0.5116	0.2443	2.617	0.2967	2.141	
		0.7201	0.2474	2.651	0.3032	2.188	
		0.8043	0.2479	2.655	0.3046	2.198	
		1.0000					16.81
C_2H_4-N_2 [227]	300.0	0.0000					17.81
		0.2405	0.7445	1.286	0.7446	1.285	
		0.5695	0.7589	1.310	0.7950	1.310	
		0.7621	0.7744	1.337	0.7745	1.337	
		1.0000					10.33
C_2H_4-N_2 [227]	400.0	0.0000					21.90
		0.2405	0.7751	1.261	0.7752	1.261	
		0.5695	0.7900	1.285	0.7901	1.285	
		0.7621	0.8164	1.328	0.8165	1.328	
		1.0000					13.48
C_2H_4-N_2 [227]	500.0	0.0000					25.60
		0.2405	0.7963	1.259	0.7964	1.258	
		0.5695	0.8046	1.272	0.8047	1.272	
		0.7621	0.8229	1.301	0.8230	1.301	
		1.0000					16.22
C_2H_4-N_2 [227]	550.0	0.0000					27.27
		0.2405	0.7995	1.246	0.7996	1.245	
		0.5695	0.8107	1.263	0.8108	1.263	
		0.7621	0.8318	1.296	0.8320	1.296	
		1.0000					17.53
C_2H_4-O_2 [227]	293.0	0.0000					10.10
		0.2297	1.316	0.7508	1.324	0.7410	
		0.5855	1.327	0.7572	1.336	0.7473	
		0.8694	1.316	0.7511	1.325	0.7413	
		1.0000					20.19
C_2H_4-N_2 [227]	323.0	0.0000					11.07
		0.2297	1.308	0.7572	0.317	0.7474	
		0.5855	1.323	0.7659	1.332	0.7559	
		0.8694	1.314	0.7608	1.323	0.7509	
		1.0000					21.81
C_2H_4-O_2 [227]	373.0	0.0000					12.62
		0.2297	1.304	0.7712	1.312	0.7613	
		0.5855	1.310	0.7749	1.319	0.7649	
		0.8694	1.297	0.7672	1.305	0.7573	
		1.0000					24.33

TABLE 1. COMPOSITION AND TEMPERATURE DEPENDENCE OF Ψ_{ij} ON DIFFERENT SCHEMES OF COMPUTATION (continued)

Gas Pair [Reference]	Temp. (K)	Mole Fraction of Heavier Component	First Method		Second Method		Viscosity (N s m^{-2} x 10^{-6})
			Ψ_{12}	Ψ_{21}	Ψ_{12}	Ψ_{21}	
CH$_3$(CH$_2$)$_5$CH$_3$-(CH$_3$)$_2$CHCH$_2$C(CH$_3$)$_3$ [354]	303.2	0.0000					4.79
		0.1550	0.6812	1.339	0.7264	1.188	
		0.3658	0.6801	1.337	0.7270	1.189	
		0.4830	0.6656	1.308	0.7125	1.166	
		0.6992	0.6521	1.282	0.7006	1.146	
		0.8941	0.6127	1.204	0.6630	1.085	
		1.0000					8.29
CH$_3$(CH$_2$)$_5$CH$_3$-(CH$_3$)$_2$CHCH$_2$C(CH$_3$)$_3$ [354]	323.2	0.0000					5.13
		0.1550	0.7162	1.418	0.7661	1.263	
		0.3658	0.6906	1.368	0.7386	1.217	
		0.4830	0.6511	1.289	0.6973	1.149	
		0.6992	0.6351	1.258	0.6832	1.126	
		0.8941	0.6235	1.235	0.6739	1.111	
		1.0000					8.82
CH$_3$(CH$_2$)$_5$CH$_3$-(CH$_3$)$_2$CHCH$_2$C(CH$_3$)$_3$ [354]	333.2	0.0000					5.32
		0.4830	0.6623	1.331	0.7096	1.187	
		1.0000					9.01
H$_2$-HD [179]	14.4	0.0000					0.79
		0.2540	0.8522	1.142	0.8766	1.106	
		0.5010	0.8604	1.153	0.8852	1.117	
		0.7570	0.8352	1.119	0.8598	1.085	
		1.0000					0.88
H$_2$-HD [179]	20.4	0.0000					1.11
		0.2450	0.8649	1.147	0.8897	1.111	
		0.5050	0.8768	1.163	0.9020	1.126	
		0.7540	0.8881	1.178	0.9133	1.140	
		1.0000					1.25
H$_2$-HD [179]	71.5	0.0000					3.26
		0.2500	0.8985	1.107	0.9233	1.071	
		0.4990	0.9020	1.111	0.2969	1.075	
		0.7490	0.9009	1.110	0.9257	1.074	
		1.0000					3.95
H$_2$-HD [179]	90.1	0.0000					3.92
		0.2530	0.8888	1.095	0.9131	1.059	
		0.4990	0.8991	1.108	0.9239	1.072	
		0.7410	0.9131	1.125	0.9381	1.088	
		1.0000					4.75
H$_2$-HD [179]	196.0	0.0000					6.70
		0.2360	0.1082	17.90	0.1462	10.92	
		0.4960	0.1335	22.09	0.1671	12.48	
		0.7460	0.1645	27.23	0.1935	14.45	
		1.0000					8.16
H$_2$-HD [179]	229.0	0.0000					7.45
		0.1960	0.8916	1.090	0.9158	1.054	
		0.4970	0.9048	1.106	0.9296	1.070	
		0.7480	0.9029	1.104	0.9277	1.068	
		1.0000					9.10
H$_2$-HD [179]	293.1	0.0000					8.83
		0.2410	0.9089	1.121	0.9340	1.085	
		0.4980	0.9014	1.112	0.9263	1.076	
		0.7980	0.9457	1.166	0.9709	1.128	
		1.0000					10.69
H$_2$-CH$_4$ [229]	293.0	0.0000					8.76
		0.0777	0.3411	2.187	0.3855	1.811	
		0.3978	0.3331	2.136	0.3849	1.808	
		0.5145	0.3309	2.122	0.3847	1.807	
		0.7192	0.3306	2.120	0.3873	1.819	
		1.0000					10.87

TABLE 1. COMPOSITION AND TEMPERATURE DEPENDENCE OF Ψ_{ij} ON DIFFERENT SCHEMES OF COMPUTATION (continued)

Gas Pair [Reference]	Temp. (K)	Mole Fraction of Heavier Component	First Method Ψ_{12}	Ψ_{21}	Second Method Ψ_{12}	Ψ_{21}	Viscosity (N s m^{-2} x 10^{-6})
H_2-HC_4 [1]	293.2	0.0000					9.24
		0.2083	0.3483	2.276	0.3990	1.910	
		0.3909	0.3713	2.427	0.4280	2.049	
		0.4904	0.3530	2.307	0.4088	1.958	
		0.6805	0.3494	2.284	0.4069	1.948	
		1.0000					11.25
H_2-HC_4 [1]	333.2	0.0000					10.08
		0.2083	0.3586	2.292	0.4102	1.921	
		0.3909	0.3745	2.394	0.4310	2.018	
		0.4909	0.3546	2.267	0.4101	1.920	
		0.6805	0.3485	2.228	0.4056	1.899	
		1.0000					12.55
H_2-HC_4 [229]	373.0	0.0000					10.33
		0.0777	0.3501	2.162	0.3947	1.786	
		0.3978	0.3434	2.121	0.3955	1.789	
		0.5145	0.3400	2.100	0.3938	1.782	
		0.7192	0.3400	2.100	0.3968	1.795	
		1.0000					13.31
H_2-HC_4 [1]	373.2	0.0000					10.90
		0.2083	0.3569	2.244	0.4076	1.877	
		0.3909	0.3704	2.328	0.4260	1.962	
		0.4909	0.3519	2.212	0.4068	1.873	
		0.6805	0.3467	2.179	0.4036	1.858	
		1.0000					13.80
H_2-CH_4 [229]	473.0	0.0000					12.13
		0.0777	0.3527	2.124	0.3966	1.750	
		0.3978	0.3521	2.120	0.4045	1.784	
		0.5145	0.3457	2.082	0.3996	1.763	
		0.7192	0.3477	2.094	0.4046	1.785	
		1.0000					16.03
H_2-CH_4 [229]	523.0	0.0000					12.96
		0.0777	0.3610	2.158	0.4066	1.781	
		0.3978	0.3548	2.121	0.4073	1.784	
		0.5145	0.3503	2.094	0.4044	1.771	
		0.7192	0.3534	2.113	0.4104	1.798	
		1.0000					17.25
H_2-NO [340]	273.2	0.0000					8.49
		0.1975	0.2780	1.955	0.3118	1.462	
		0.2299	0.2844	2.000	0.3208	1.505	
		0.2835	0.3083	2.168	0.3507	1.645	
		0.4508	0.3178	2.235	0.3662	1.717	
		0.7045	0.3176	2.233	0.3709	1.740	
		0.8503	0.3486	2.451	0.4037	1.893	
		1.0000					17.97
H_2-NO [334]	293.2	0.0000					8.88
		0.0510	0.3220	2.287	0.3577	1.694	
		0.1002	0.3213	2.282	0.3592	1.701	
		0.1499	0.3190	2.266	0.3585	1.698	
		0.1931	0.3060	2.173	0.3448	1.633	
		0.2500	0.3204	2.275	0.3637	1.723	
		0.2944	0.3186	2.263	0.3631	1.720	
		0.3425	0.3078	2.186	0.3524	1.669	
		0.3926	0.3202	2.274	0.3677	1.742	
		0.4423	0.3150	2.237	0.3632	1.720	
		0.4891	0.3197	2.270	0.3693	1.749	
		0.5393	0.3104	2.204	0.3605	1.707	
		0.6204	0.3269	2.322	0.3793	1.797	
		0.6416	0.3125	2.219	0.3647	1.728	
		0.6900	0.3246	2.305	0.3780	1.790	
		0.7453	0.3077	2.185	0.3616	1.713	
		0.7932	0.3307	2.349	0.3855	1.826	
		0.8430	0.2607	1.851	0.3157	1.495	
		0.8947	0.3236	2.298	0.3795	1.798	
		0.9524	0.2519	1.789	0.3091	1.464	
		1.0000					18.61

TABLE 1. COMPOSITION AND TEMPERATURE DEPENDENCE OF Ψ_{ij} ON DIFFERENT SCHEMES OF COMPUTATION (continued)

Gas Pair [Reference]	Temp. (K)	Mole Fraction of Heavier Component	First Method Ψ_{12}	Ψ_{21}	Second Method Ψ_{12}	Ψ_{21}	Viscosity ($N\ s\ m^{-2} \times 10^{-6}$)
H_2-N_2 [252]	82.2	0.0000					3.62
		0.1600	0.2666	2.466	0.3080	1.919	
		0.3510	0.2803	2.592	0.3304	2.059	
		0.4410	0.2823	2.610	0.3344	2.083	
		0.6200	0.2814	2.602	0.3362	2.095	
		0.7590	0.2880	2.663	0.3445	2.147	
		1.0000					5.44
H_2-N_2 [252]	90.2	0.0000					3.92
		0.1600	0.2939	2.459	0.3369	1.899	
		0.3510	0.2775	2.322	0.3244	1.829	
		0.4410	0.2775	2.322	0.3269	1.843	
		0.6200	0.2956	2.473	0.3498	1.972	
		0.7590	0.2948	2.466	0.3507	1.978	
		0.8660	0.2993	2.504	0.3564	2.010	
		1.0000					6.51
H_2-N_2 [252]	291.1	0.0000					8.77
		0.1600	0.3302	2.297	0.3724	1.746	
		0.4410	0.3227	2.244	0.3719	1.743	
		0.6200	0.3152	2.192	0.3676	1.723	
		0.7590	0.3514	2.444	0.4065	1.905	
		0.8660	0.2839	1.975	0.3401	1.594	
		1.0000					17.52
H_2-N_2 [252]	291.1	0.0000					8.77
		0.1360	0.3250	2.268	0.3654	1.718	
		0.1600	0.3286	2.294	0.3707	1.743	
		0.1870	0.3223	2.250	0.3642	1.713	
		0.2960	0.3159	2.205	0.3605	1.696	
		0.4000	0.3136	2.189	0.3610	1.698	
		0.4410	0.3198	2.232	0.3689	1.735	
		0.5170	0.3134	2.188	0.3637	1.711	
		0.6200	0.3107	2.168	0.3629	1.707	
		0.6900	0.3094	2.159	0.3629	1.707	
		0.7590	0.3433	2.396	0.3984	1.874	
		0.8660	0.2711	1.892	0.3272	1.539	
		1.0000					17.46
H_2-N_2 [252]	291.2	0.0000					8.82
		0.1360	0.3266	2.292	0.3676	1.739	
		0.1870	0.3235	2.271	0.3658	1.730	
		0.2960	0.3166	2.222	0.3615	1.710	
		0.4000	0.3141	2.205	0.3617	1.711	
		0.5170	0.3137	2.202	0.3641	1.722	
		0.6900	0.3095	2.172	0.3631	1.717	
		1.0000					17.46
H_2-N_2 [341]	307.2	0.0000					9.07
		0.2000	0.3178	2.206	0.3590	1.680	
		0.3991	0.3151	2.188	0.3622	1.695	
		0.5100	0.3156	2.191	0.3657	1.711	
		0.5794	0.3185	2.211	0.3703	1.732	
		0.7977	0.3231	2.243	0.3785	1.771	
		1.0000					18.16
H_2-N_2 [341]	325.4	0.0000					9.94
		0.2000	0.3395	2.458	0.3866	1.886	
		0.3991	0.3374	2.443	0.3882	1.894	
		0.5100	0.3327	2.408	0.3849	1.878	
		0.5794	0.3310	2.397	0.3842	1.874	
		0.7977	0.3401	2.462	0.3959	1.931	
		1.0000					19.09

TABLE 1.   COMPOSITION AND TEMPERATURE DEPENDENCE OF Ψ_{ij} ON
DIFFERENT SCHEMES OF COMPUTATION (continued)

Gas Pair [Reference]	Temp. (K)	Mole Fraction of Heavier Component	First Method Ψ_{12}	Ψ_{21}	Second Method Ψ_{12}	Ψ_{21}	Viscosity (N s m^{-2} x 10^{-6})
H_2-N_2 [341]	373.2	0.0000					10.42
		0.2000	0.3512	2.421	0.3982	1.850	
		0.3991	0.3332	2.297	0.3822	1.775	
		0.5100	0.3312	2.283	0.3823	1.776	
		0.5794	0.3282	2.262	0.3804	1.767	
		0.7977	0.3112	2.145	0.3664	1.702	
		1.0000					21.01
H_2-N_2 [341]	422.7	0.0000					11.49
		0.2005	0.3618	2.511	0.4111	1.922	
		0.3988	0.3483	2.417	0.3991	1.866	
		0.4996	0.3376	2.343	0.3891	1.819	
		0.5988	0.3353	2.326	0.3882	1.815	
		0.8002	0.3465	2.404	0.4019	1.879	
		1.0000					23.01
H_2-N_2 [341]	478.2	0.0000					12.64
		0.2005	0.3799	2.641	0.4327	2.027	
		0.3988	0.3491	2.427	0.4000	1.874	
		0.4996	0.3498	2.432	0.4022	1.884	
		0.5988	0.3540	2.460	0.4077	1.910	
		0.8002	0.3547	2.465	0.4101	1.921	
		1.0000					25.27
H_2-N_2O [234]	300.0	0.0000					8.91
		0.2143	0.2108	2.756	0.2496	2.055	
		0.4039	0.2084	2.725	0.2540	2.091	
		0.6011	0.2089	2.731	0.2592	2.134	
		1.0000					14.88
H_2-N_2O [234]	400.0	0.0000					10.81
		0.2143	0.2250	2.733	0.2642	2.020	
		0.4039	0.2209	2.683	0.2663	2.037	
		0.6011	0.2220	2.697	0.2722	2.082	
		1.0000					19.43
H_2-N_2O [234]	500.0	0.0000					12.56
		0.2143	0.2335	2.719	0.2728	2.000	
		0.4039	0.2295	2.672	0.2750	2.016	
		0.6011	0.2303	2.681	0.2804	2.056	
		1.0000					23.55
H_2-N_2O [234]	550.0	0.0000					13.41
		0.2143	0.2371	2.717	0.2766	1.996	
		0.4039	0.2338	2.679	0.2793	2.016	
		0.6011	0.2343	2.685	0.2844	2.052	
		1.0000					25.55
H_2-O_2 [334]	293.2	0.0000					8.78
		0.0520	0.3551	2.445	0.3964	1.803	
		0.1000	0.3344	2.302	0.3724	1.694	
		0.1530	0.3297	2.270	0.3690	1.678	
		0.2060	0.3209	2.209	0.3605	1.640	
		0.2550	0.3163	2.178	0.3570	1.624	
		0.2780	0.2600	1.790	0.2937	1.336	
		0.3590	0.3054	2.103	0.3484	1.584	
		0.4060	0.2966	2.042	0.3403	1.548	
		0.4470	0.2975	2.049	0.3427	1.558	
		0.4930	0.2940	2.024	0.3402	1.547	
		0.5430	0.2902	1.998	0.3376	1.536	
		0.5910	0.2771	1.908	0.3252	1.479	
		0.6510	0.2844	1.958	0.3345	1.521	
		0.7000	0.2816	1.939	0.3329	1.514	
		0.7480	0.2791	1.922	0.3315	1.508	
		0.7950	0.2715	1.869	0.3249	1.478	
		0.8470	0.2711	1.866	0.3256	1.481	
		0.8950	0.2555	1.759	0.3110	1.414	
		0.9550	0.2385	1.642	0.2952	1.343	
		1.0000					20.24

TABLE 1. COMPOSITION AND TEMPERATURE DEPENDENCE OF Ψ_{ij} ON DIFFERENT SCHEMES OF COMPUTATION (continued)

Gas Pair [Reference]	Temp. (K)	Mole Fraction of Heavier Component	First Method Ψ_{12}	Ψ_{21}	Second Method Ψ_{12}	Ψ_{21}	Viscosity (N s m^{-2} x 10^{-6})
H_2-O_2 [327]	293.6	0.0000					8.85
		0.1610	0.3094	2.131	0.3452	1.570	
		0.2730	0.3073	2.116	0.3475	1.580	
		0.3800	0.3096	2.132	0.3538	1.609	
		0.5270	0.3042	2.094	0.3521	1.601	
		0.6700	0.2949	2.031	0.3459	1.573	
		1.0000					20.40
H_2-O_2 [337]	297.37	0.2500					15.60
		0.3670	0.2910	6.854	0.3506	4.937	
		0.5750	0.2932	6.926	0.3473	4.890	
		0.6500	0.2824	6.671	0.3350	4.717	
		0.7450	0.3064	7.239	0.3570	5.026	
		0.8170	0.3696	8.731	0.4163	5.862	
		1.0000					20.80
H_2-O_2 [227]	300.0	0.0000					8.89
		0.2192	0.3278	2.248	0.3689	1.672	
		0.3970	0.3030	2.079	0.3469	1.572	
		0.6055	0.3064	2.102	0.3563	1.615	
		0.8165	0.2935	2.013	0.3475	1.575	
		1.0000					20.57
H_2-O_2 [227]	400.0	0.0000					10.87
		0.2192	0.3272	2.198	0.3674	1.630	
		0.3970	0.3147	2.114	0.3592	1.594	
		0.6055	0.3182	2.138	0.3682	1.634	
		0.8165	0.3095	2.080	0.3635	1.613	
		1.0000					25.68
H_2-O_2 [227]	500.0	0.0000					12.59
		0.2192	0.3329	2.205	0.3734	1.634	
		0.3970	0.3199	2.119	0.3645	1.595	
		0.6055	0.3212	2.128	0.3712	1.624	
		0.8165	0.3045	2.017	0.3583	1.568	
		1.0000					30.17
H_2-O_2 [227]	550.0	0.0000					13.81
		0.2192	0.3416	2.326	0.3847	1.730	
		0.3970	0.3204	2.181	0.3658	1.645	
		0.6055	0.3220	2.192	0.3725	1.675	
		0.8165	0.3063	2.085	0.3603	1.620	
		1.0000					32.20
H_2-C_3H_8 [340]	273.2	0.0000					8.60
		0.0313	0.1583	0.3960	0.1942	3.059	
		0.0785	0.1434	3.588	0.1772	2.791	
		0.0891	0.1400	3.502	0.1734	2.731	
		0.1500	0.1333	3.334	0.1687	2.657	
		0.2218	0.1310	3.277	0.1695	2.670	
		0.3271	0.1326	3.318	0.1753	2.760	
		0.5182	0.1255	3.139	0.1707	2.689	
		0.6978	0.1331	3.330	0.1819	2.865	
		0.8037	0.1671	4.180	0.2207	3.476	
		1.0000					7.52
H_2-C_3H_8 [229]	300.0	0.0000					8.91
		0.0775	0.1537	3.666	0.1895	2.846	
		0.1250	0.1504	3.588	0.1883	2.828	
		0.2118	0.1477	3.522	0.1891	2.839	
		0.4182	0.1507	3.595	0.1983	2.978	
		0.6296	0.1523	3.633	0.2030	3.048	
		0.8179	0.1644	3.923	0.2178	3.271	
		1.0000					8.17

TABLE 1.   COMPOSITION AND TEMPERATURE DEPENDENCE OF Ψ_{ij} ON DIFFERENT SCHEMES OF COMPUTATION (continued)

Gas Pair [Reference]	Temp. (K)	Mole Fraction of Heavier Component	First Method Ψ_{12}	Ψ_{21}	Second Method Ψ_{12}	Ψ_{21}	Viscosity (N s m^{-2} x 10^{-6})
H_2-C_3H_8 [229]	400.0	0.0000					10.81
		0.0775	0.1629	3.600	0.1993	2.773	
		0.1250	0.1606	3.549	0.1993	2.773	
		0.2118	0.1576	3.482	0.1996	2.777	
		0.4182	0.1636	3.615	0.2124	2.955	
		0.6296	0.1616	3.571	0.2129	2.962	
		0.8179	0.1708	3.775	0.2245	3.123	
		1.0000					10.70
H_2-C_3H_8 [229]	500.0	0.0000					12.56
		0.0775	0.1723	3.618	0.2101	2.778	
		0.1250	0.1689	3.547	0.2085	2.757	
		0.2118	0.1655	3.475	0.2082	2.753	
		0.4182	0.1674	3.516	0.2161	2.857	
		0.6296	0.1705	3.581	0.2224	2.941	
		0.8179	0.1855	3.897	0.2400	3.173	
		1.0000					13.08
H_2-C_3H_8 [229]	550.0	0.0000					13.47
		0.0775	0.1778	3.684	0.2171	2.831	
		0.1250	0.1725	3.573	0.2127	2.774	
		0.2118	0.1692	3.507	0.2125	2.772	
		0.4182	0.1708	3.540	0.2198	2.867	
		0.6296	0.1746	3.617	0.2268	2.959	
		0.8179	0.1875	3.886	0.2421	3.158	
		1.0000					14.22
CH_4-O_2 [334]	293.2	0.0000					11.12
		0.0510	0.9550	1.057	0.9908	0.9887	
		0.0990	0.9166	1.014	0.9495	0.9475	
		0.1420	0.9227	1.021	0.9563	0.9542	
		0.1980	0.9140	1.012	0.9471	0.9451	
		0.2510	0.9051	1.002	0.9380	0.9360	
		0.2960	0.9107	1.008	0.9441	0.9421	
		0.3490	0.9266	1.026	0.9611	0.9590	
		0.5010	0.9243	1.023	0.9591	0.9571	
		0.5490	0.9296	1.029	0.9647	0.9627	
		0.5970	0.9257	1.025	0.9609	0.9588	
		0.6470	0.9256	1.024	0.9609	0.9589	
		0.7020	0.9292	1.028	0.9648	0.9628	
		0.7650	0.9018	0.9980	0.9372	0.9352	
		0.7990	0.9367	1.037	0.9725	0.9705	
		0.8490	0.9237	1.022	0.9597	0.9576	
		0.8980	0.9099	1.007	0.9460	0.9440	
		0.9510	0.8927	0.9880	0.9291	0.9271	
		1.0000					20.04
CH_4-C_3H_8 [229]	293.0	0.0000					10.87
		0.3684	0.5042	1.881	0.5502	1.764	
		0.6383	0.4992	1.862	0.5454	1.748	
		0.8341	0.5072	1.892	0.5540	1.776	
		1.0000					8.01
CH_4-C_3H_8 [229]	373.0	0.0000					13.31
		0.3684	0.5063	1.838	0.5520	1.722	
		0.6383	0.5014	1.820	0.5475	1.707	
		0.8341	0.5159	1.872	0.5628	1.755	
		1.0000					10.08
CH_4-C_3H_8 [229]	473.0	0.0000					16.03
		0.3684	0.5179	1.821	0.5640	1.704	
		0.6383	0.5120	1.800	0.5582	1.687	
		0.8341	0.5238	1.842	0.5708	1.725	
		1.0000					12.53

TABLE 1.   COMPOSITION AND TEMPERATURE DEPENDENCE OF Ψ_{ij} ON DIFFERENT SCHEMES OF COMPUTATION (continued)

Gas Pair [Reference]	Temp. (K)	Mole Fraction of Heavier Component	First Method		Second Method		Viscosity (N s m^{-2} x 10^{-6})
			Ψ_{12}	Ψ_{21}	Ψ_{12}	Ψ_{21}	
CH$_4$-C$_3$H$_8$ [229]	523.0	0.0000					17.25
		0.3684	0.5247	1.825	0.5711	1.707	
		0.6383	0.5193	1.807	0.5658	1.691	
		0.8341	0.5261	1.830	0.5731	1.713	
		1.0000					13.63
N$_2$-NO [315]	293.0	0.0000					17.47
		0.2674	1.010	1.004	1.015	0.9989	
		0.5837	1.001	0.9957	1.006	0.9904	
		0.6948	1.013	1.007	1.018	1.002	
		1.0000					18.82
N$_2$-NO [315]	373.0	0.0000					20.84
		0.2674	1.011	0.9935	1.016	0.9882	
		0.5837	0.9941	0.9767	0.9990	0.9715	
		0.6948	1.000	0.9830	1.005	0.9777	
		1.0000					22.72
N$_2$-O$_2$ [337]	298.7	0.0000					17.80
		0.1320	1.093	1.068	1.104	1.057	
		0.2560	1.027	1.003	1.036	0.9927	
		0.4100	1.030	1.006	1.039	0.9955	
		0.5100	1.035	1.011	1.044	1.001	
		0.6600	1.026	1.002	1.035	0.9917	
		0.7600	1.059	1.035	1.068	1.023	
		1.0000					20.80
N$_2$-O$_2$ [227]	300.0	0.0000					17.81
		0.2178	1.002	0.9912	1.012	0.9807	
		0.4107	1.006	0.9945	1.015	0.9839	
		0.7592	0.9988	0.9878	1.008	0.9774	
		1.0000					20.57
N$_2$-O$_2$ [227]	400.0	0.0000					21.90
		0.2178	1.009	0.9827	1.018	0.9721	
		0.4107	1.012	0.9862	1.022	0.9756	
		0.7592	1.009	0.9826	1.018	0.9721	
		1.0000					25.68
N$_2$-O$_2$ [227]	500.0	0.0000					25.60
		0.2178	1.016	0.9849	1.025	0.9743	
		0.4107	1.020	0.9882	1.029	0.9776	
		0.7592	1.012	0.9813	1.022	0.9708	
		1.0000					30.17
N$_2$-O$_2$ [227]	550.0	0.0000					17.53
		0.2178	1.231	0.9080	1.241	0.8973	
		0.4107	1.031	0.7609	1.039	0.7515	
		0.7592	1.193	0.8805	1.203	0.8698	
		1.0000					27.14
N$_2$O-C$_3$H$_8$ [234]	300.0	0.0000					14.88
		0.2018	0.7255	1.324	0.7256	1.324	
		0.4171	0.7291	1.330	0.7292	1.330	
		0.7984	0.7323	1.336	0.7324	1.336	
		1.0000					8.17
N$_2$O-C$_3$H$_8$ [234]	400.0	0.0000					19.43
		0.2018	0.7362	1.339	0.7363	1.339	
		0.4171	0.7309	1.330	0.7310	1.330	
		0.7984	0.7316	1.331	0.7317	1.331	
		1.0000					10.70
N$_2$O-C$_3$H$_8$ [234]	500.0	0.0000					23.55
		0.2018	0.7517	1.356	0.7518	1.356	
		0.4171	0.7348	1.325	0.7349	1.325	
		0.7984	0.7380	1.331	0.7381	1.331	
		1.0000					13.08

TABLE 1.   COMPOSITION AND TEMPERATURE DEPENDENCE OF Ψ_{ij} ON
DIFFERENT SCHEMES OF COMPUTATION (continued)

Gas Pair [Reference]	Temp. (K)	Mole Fraction of Heavier Component	First Method Ψ_{12}	Ψ_{21}	Second Method Ψ_{12}	Ψ_{21}	Viscosity ($N\,s\,m^{-2} \times 10^{-6}$)
$N_2O-C_3H_8$ [234]	550.0	0.0000					25.56
		0.2018	0.7429	1.338	0.7431	1.338	
		0.4171	0.7366	1.327	0.7367	1.326	
		0.7984	0.7335	1.321	0.7336	1.321	
		1.0000					14.22
$HCl-CO_2$ [346]	291.0	0.0000					14.26
		0.2000	0.1053	0.1238	0.1053	0.1204	
		0.4000	0.1190	0.1399	0.1190	0.1361	
		0.6000	0.1451	0.1706	0.1452	0.1660	
		0.8000	0.1967	0.2313	0.1970	0.2251	
		1.0000					14.64
$HCl-CO_2$ [346]	291.16	0.0000					14.44
		0.1000	0.8876	1.043	0.8997	1.028	
		0.2000	0.8862	1.042	0.8983	1.026	
		0.3000	0.8845	1.040	0.8967	1.025	
		0.4000	0.8824	1.037	0.8946	1.022	
		0.5000	0.8796	1.034	0.8920	1.019	
		0.6000	0.8783	1.032	0.8907	1.018	
		0.7000	0.8755	1.029	0.8880	1.015	
		0.8000	0.8734	1.027	0.8861	1.012	
		0.9000	0.8658	1.018	0.8786	1.004	
		1.0000					14.83
SO_2-CO_2 [346]	289.0	0.0000					14.58
		0.2000	0.7311	1.248	0.7531	1.215	
		0.4000	0.7318	1.249	0.7543	1.217	
		0.6000	0.7278	1.243	0.7506	1.211	
		0.8000	0.7216	1.232	0.7446	1.202	
		1.0000					12.43
SO_2-CO_2 [346]	289.0	0.0000					14.77
		0.1000	0.7297	1.245	0.7515	1.212	
		0.2000	0.7310	1.247	0.7531	1.215	
		0.3000	0.7301	1.246	0.7523	1.213	
		0.4000	0.7315	1.248	0.7540	1.216	
		0.5000	0.7323	1.249	0.7550	1.218	
		0.6000	0.7294	1.244	0.7522	1.213	
		0.7000	0.7298	1.245	0.7528	1.214	
		0.8000	0.7230	1.234	0.7461	1.203	
		0.9000	0.7243	1.236	0.7476	1.206	
		1.0000					12.60
SO_2-CO_2 [35]	298.2	0.0000					14.80
		0.0800	0.7357	1.203	0.7569	1.170	
		0.1520	0.7399	1.210	0.7615	1.177	
		0.1790	0.7395	1.210	0.7612	1.177	
		0.2770	0.7380	1.207	0.7599	1.175	
		0.3890	0.7304	1.195	0.7523	1.163	
		0.4240	0.7394	1.209	0.7617	1.178	
		0.5030	0.7324	1.198	0.7547	1.167	
		0.5960	0.7323	1.198	0.7548	1.167	
		0.6550	0.7228	1.182	0.7453	1.152	
		0.7120	0.7190	1.176	0.7417	1.147	
		0.7830	0.7233	1.183	0.7461	1.154	
		0.8220	0.7246	1.185	0.7476	1.156	
		0.9720	0.8163	1.335	0.8400	1.299	
		1.0000					13.17

TABLE 1. COMPOSITION AND TEMPERATURE DEPENDENCE OF Ψ_{ij} ON DIFFERENT SCHEMES OF COMPUTATION (continued)

Gas Pair [Reference]	Temp. (K)	Mole Fraction of Heavier Component	First Method Ψ_{12}	Ψ_{21}	Second Method Ψ_{12}	Ψ_{21}	Viscosity (N s m^{-2} x 10^{-6})
SO_2-CO_2 [35]	308.2	0.0000					15.38
		0.0410	0.7178	1.210	0.7385	1.177	
		0.1770	0.7231	1.219	0.7445	1.186	
		0.2690	0.7227	1.218	0.7444	1.186	
		0.3960	0.7327	1.235	0.7551	1.203	
		0.5090	0.7270	1.225	0.7495	1.194	
		0.6080	0.7252	1.222	0.7478	1.192	
		0.6970	0.7160	1.207	0.7387	1.177	
		0.7820	0.7153	1.206	0.7382	1.176	
		0.8660	0.6988	1.178	0.7217	1.150	
		1.0000					13.28
SO_2-CO_2 [35]	353.2	0.0000					17.30
		0.0480	0.7669	1.268	0.7899	1.235	
		0.1820	0.7531	1.245	0.7757	1.212	
		0.2880	0.7507	1.241	0.7733	1.209	
		0.3880	0.7483	1.237	0.7710	1.205	
		0.5000	0.7478	1.236	0.7706	1.204	
		0.5980	0.7476	1.236	0.7706	1.204	
		0.6940	0.7443	1.231	0.7674	1.199	
		0.7920	0.7431	1.229	0.7663	1.198	
		0.8780	0.7437	1.230	0.7671	1.199	
		1.0000					15.23
CCl_4-CH_2Cl_2 [292]	293.15	0.0000					10.25
		0.1575	0.7167	1.355	0.7498	1.297	
		0.2015	0.7234	1.368	0.7570	1.309	
		0.4986	0.7085	1.339	0.7416	1.282	
		0.6886	0.7131	1.348	0.7467	1.291	
		0.8616	0.7101	1.342	0.7438	1.286	
		1.0000					9.82
CCl_4-CH_2Cl_2 [292]	353.26	0.0000					12.02
		0.2261	0.6974	1.309	0.7291	1.252	
		0.6351	0.7015	1.316	0.7345	1.261	
		1.0000					11.60
CCl_4-CH_2Cl_2 [292]	413.43	0.0000					14.27
		0.1615	0.7085	1.343	0.7411	1.285	
		0.2882	0.7275	1.379	0.7614	1.321	
		0.4738	0.7060	1.339	0.7390	1.282	
		0.7096	0.7199	1.365	0.7536	1.307	
		0.8739	0.7295	1.383	0.7633	1.324	
		1.0000					13.63
$(CH_3)_2CHOH$-CCl_4 (Liquid) [352]	313.2	0.0000					1330.00
		0.1210	0.5343	2.461	0.5881	2.353	
		0.2550	0.5599	2.579	0.6143	2.458	
		0.3150	0.5667	2.610	0.6206	2.483	
		0.3980	0.5798	2.671	0.6333	2.534	
		0.5000	0.5969	2.749	0.6497	2.599	
		0.5790	0.6168	2.841	0.6691	2.677	
		0.6750	0.6331	2.917	0.6842	2.738	
		0.7800	0.6677	3.076	0.7175	2.871	
		0.8850	0.6987	3.219	0.7465	2.987	
		1.0000					739.00
CH_3OH-CCl_4 (Liquid) [352]	313.2	0.0000					0.46
		0.0900	0.5513	1.633	0.5998	1.404	
		0.2100	0.5271	1.561	0.5738	1.343	
		0.2800	0.5135	1.521	0.5598	1.311	
		0.3200	0.5038	1.492	0.5498	1.287	
		0.4900	0.4629	1.371	0.5093	1.192	
		0.6500	0.4167	1.234	0.4646	1.088	
		0.6970	0.4042	1.197	0.4530	1.060	
		0.8070	0.3646	1.080	0.4149	0.9712	
		0.8950	0.3339	0.9890	0.3854	0.9023	
		1.0000					0.74

TABLE 1.   COMPOSITION AND TEMPERATURE DEPENDENCE OF Ψ_{ij} ON DIFFERENT SCHEMES OF COMPUTATION (continued)

Gas Pair [Reference]	Temp. (K)	Mole Fraction of Heavier Component	First Method Ψ_{12}	Ψ_{21}	Second Method Ψ_{12}	Ψ_{21}	Viscosity (N s m^{-2} x 10^{-6})
$CH_3COOCH_2C_6H_5$-	313.2	0.0000					625.60
$C_4H_8O_2$		0.2000	1.260	0.9936	1.296	0.9430	
(Liquid)		0.3000	1.196	0.9428	1.228	0.8934	
[351]		0.3800	1.190	0.9382	1.221	0.8887	
		0.5200	1.182	0.9319	1.212	0.8824	
		0.6450	1.161	0.9152	1.190	0.8663	
		0.7480	1.142	0.9005	1.171	0.8525	
		0.8750	1.217	0.9593	1.246	0.9067	
		1.0000					1352.50
NH_3-C_2H_4	293.2	0.0000					9.82
[222]		0.1133	0.7402	1.188	0.7667	1.142	
		0.1929	0.7358	1.181	0.7624	1.135	
		0.3039	0.7380	1.184	0.7652	1.139	
		0.4828	0.7370	1.183	0.7649	1.139	
		0.7007	0.7374	1.183	0.7661	1.141	
		0.8904	0.7450	1.196	0.7744	1.153	
		1.0000					10.08
NH_3-C_2H_4	373.2	0.0000					12.79
[222]		0.1133	0.7284	1.221	0.7552	1.175	
		0.1929	0.7279	1.220	0.7550	1.174	
		0.3039	0.7301	1.224	0.7578	1.178	
		0.4828	0.7294	1.222	0.7576	1.178	
		0.7007	0.7272	1.219	0.7560	1.176	
		0.8904	0.7346	1.231	0.7641	1.188	
		1.0000					12.57
NH_3-C_2H_4	473.2	0.0000					16.46
[222]		0.1133	0.7236	1.273	0.7513	1.226	
		0.1929	0.7194	1.266	0.7471	1.220	
		0.3039	0.7225	1.271	0.7506	1.225	
		0.4828	0.7220	1.270	0.7505	1.225	
		0.7007	0.7183	1.264	0.7473	1.220	
		0.8904	0.7209	1.268	0.7504	1.225	
		1.0000					15.41
NH_3-C_2H_4	523.2	0.0000					181.3
[222]		0.1133	0.7174	1.286	0.7452	1.239	
		0.1929	0.7153	1.282	0.7431	1.236	
		0.3039	0.7178	1.287	0.7460	1.241	
		0.4828	0.7190	1.289	0.7477	1.244	
		0.7007	0.7155	1.283	0.7445	1.238	
		0.8904	0.7179	1.287	0.7474	1.243	
		1.0000					16.66
NH_3-H_2	293.2	0.0000					8.77
[222]		0.1082	0.2674	2.018	0.3028	1.659	
		0.2239	0.2627	1.982	0.3030	1.660	
		0.2975	0.2603	1.964	0.3033	1.661	
		0.5177	0.2547	1.922	0.3039	1.665	
		0.7087	0.2505	1.890	0.3034	1.662	
		0.9005	0.2505	1.890	0.3064	1.679	
		1.0000					9.82
NH_3-H_2	306.2	0.0000					9.06
[341]		0.1950	0.2325	1.679	0.2645	1.387	
		0.3990	0.2189	1.581	0.2589	1.358	
		0.5360	0.2104	1.520	0.2543	1.334	
		0.6770	0.1934	1.397	0.2392	1.255	
		0.8550	0.1610	1.163	0.2060	1.080	
		1.0000					10.59

TABLE 1. COMPOSITION AND TEMPERATURE DEPENDENCE OF Ψ_{ij} ON DIFFERENT SCHEMES OF COMPUTATION (continued)

Gas Pair [Reference]	Temp. (K)	Mole Fraction of Heavier Component	First Method Ψ_{12}	Ψ_{21}	Second Method Ψ_{12}	Ψ_{21}	Viscosity $(\text{N s m}^{-2} \times 10^{-6})$
NH_3-H_2 [341]	327.2	0.0000					9.49
		0.1950	0.2365	1.667	0.2685	1.374	
		0.3990	0.2274	1.603	0.2680	1.372	
		0.5360	0.2145	1.512	0.2585	1.323	
		0.6770	0.2019	1.423	0.2484	1.272	
		0.8550	0.1810	1.276	0.2289	1.172	
		1.0000					11.37
NH_3-H_2 [341]	371.2	0.0000					10.40
		0.1950	0.2592	1.751	0.2938	1.441	
		0.3990	0.2389	1.614	0.2802	1.374	
		0.5360	0.2274	1.537	0.2723	1.336	
		0.6770	0.2052	1.386	0.2518	1.235	
		0.8550	0.1533	1.035	0.1968	0.9652	
		1.0000					13.00
NH_3-H_2 [222]	373.2	0.0000					10.30
		0.1082	0.2930	1.993	0.3298	1.629	
		0.2239	0.2875	1.956	0.3289	1.625	
		0.2975	0.2856	1.943	0.3295	1.628	
		0.5177	0.2773	1.886	0.3271	1.616	
		0.7087	0.2767	1.882	0.3307	1.634	
		0.9005	0.2742	1.865	0.3312	1.636	
		1.0000					12.79
NH_3-H_2 [341]	421.2	0.0000					11.46
		0.1400	0.2498	1.628	0.2782	1.317	
		0.4054	0.2658	1.733	0.3100	1.467	
		0.5170	0.2595	1.691	0.3067	1.451	
		0.6005	0.2413	1.573	0.2892	1.369	
		0.8042	0.1975	1.287	0.2462	1.165	
		1.0000					14.85
NH_3-H_2 [222]	473.2	0.0000					12.11
		0.1082	0.3176	1.974	0.3559	1.606	
		0.2239	0.3153	1.960	0.3583	1.617	
		0.2975	0.3132	1.947	0.3585	1.618	
		0.5177	0.3077	1.912	0.3588	1.619	
		0.7087	0.3036	1.887	0.3584	1.618	
		0.9005	0.3025	1.880	0.3604	1.626	
		1.0000					16.46
NH_3-H_2 [341]	479.2	0.0000					12.62
		0.1400	0.2626	1.647	0.2921	1.330	
		0.4054	0.2833	1.777	0.3287	1.497	
		0.5170	0.2778	1.742	0.3262	1.485	
		0.6005	0.2674	1.677	0.3172	1.444	
		0.8042	0.2255	1.414	0.2769	1.261	
		1.0000					17.00
NH_3-H_2 [222]	523.2	0.0000					12.96
		0.2239	0.3250	1.963	0.3685	1.616	
		0.2975	0.3223	1.946	0.3680	1.613	
		0.5177	0.3159	1.908	0.3672	1.610	
		0.7087	0.3120	1.884	0.3670	1.609	
		0.9005	0.3080	1.860	0.3660	1.605	
		1.0000					18.13
H_2-$(C_2H_5)_2O$ [226]	288.16	0.0000					8.68
		0.1330	0.1063	4.654	0.1419	3.617	
		0.2650	0.1023	4.476	0.1418	3.616	
		1.0000					7.29
H_2-$(C_2H_5)_2O$ [226]	373.16	0.0000					10.35
		0.1330	0.1156	4.636	0.1523	3.555	
		0.2650	0.1121	4.494	0.1529	3.570	
		1.0000					9.49

TABLE 1. COMPOSITION AND TEMPERATURE DEPENDENCE OF Ψ_{ij} ON DIFFERENT SCHEMES OF COMPUTATION (continued)

Gas Pair [Reference]	Temp. (K)	Mole Fraction of Heavier Component	First Method Ψ_{12}	Ψ_{21}	Second Method Ψ_{12}	Ψ_{21}	Viscosity (N s m^{-2} x 10^{-6})
H_2-$(C_2H_5)_2O$ [226]	423.15	0.0000					11.34
		0.1330	0.1197	4.665	0.1570	3.563	
		0.2650	0.1132	4.412	0.1539	3.491	
		1.0000					10.70
H_2-$(C_2H_5)_2O$ [226]	486.16	0.0000					12.48
		0.1330	0.1230	4.646	0.1606	3.532	
		0.2650	0.1187	4.484	0.1602	3.523	
		1.0000					12.15
HCl-H_2 [228]	294.16	0.0000					8.81
		0.2031	0.1781	1.975	0.2070	1.486	
		0.5042	0.2003	2.220	0.2466	1.771	
		0.7179	0.1961	2.175	0.2476	1.779	
		0.8220	0.1920	2.129	0.2452	1.761	
		1.0000					14.37
HCl-H_2 [228]	327.16	0.0000					9.41
		0.2031	0.2128	2.257	0.2470	1.697	
		0.5042	0.2069	2.193	0.2532	1.739	
		0.7179	0.2027	2.150	0.2543	1.747	
		0.8220	0.1994	2.114	0.2529	1.737	
		1.0000					16.05
HCl-H_2 [228]	372.16	0.0000					10.36
		0.2031	0.2232	2.288	0.2582	1.714	
		0.5042	0.2140	2.193	0.2604	1.729	
		0.7179	0.2050	2.101	0.2564	1.702	
		0.8220	0.2039	2.090	0.2574	1.709	
		1.0000					18.28
HCl-H_2 [228]	427.16	0.0000					11.42
		0.2409	0.2231	2.254	0.2597	1.700	
		0.5092	0.2114	2.136	0.2575	1.686	
		0.6989	0.1868	1.887	0.2363	1.546	
		0.8417	0.1642	1.660	0.2152	1.409	
		1.0000					20.44
HCl-H_2 [228]	473.16	0.0000					12.24
		0.2409	0.2385	2.292	0.2762	1.719	
		0.5092	0.2287	2.198	0.2755	1.715	
		0.6989	0.2249	2.161	0.2764	1.720	
		0.8417	0.2218	2.131	0.2761	1.719	
		1.0000					23.04
HCl-H_2 [228]	523.16	0.0000					13.15
		0.2991	0.2418	2.275	0.2819	1.718	
		0.5178	0.2386	2.245	0.2861	1.743	
		0.6312	0.2305	2.169	0.2805	1.709	
		0.7947	0.2295	2.159	0.2831	1.725	
		1.0000					25.28
SO_2-H_2 [231]	290.16	0.0000					8.88
		0.1676	0.1354	3.035	0.1661	2.216	
		0.2286	0.1321	2.960	0.1657	2.211	
		0.2963	0.1273	2.854	0.1634	2.180	
		0.5075	0.1230	2.756	0.1655	2.207	
		0.8215	0.1205	2.701	0.1682	2.244	
		1.0000					12.59
SO_2-H_2 [347]	303.2	0.0000					9.00
		0.2005	0.1406	3.023	0.1736	2.221	
		0.4059	0.1506	3.237	0.1938	2.481	
		0.4919	0.1512	3.250	0.1966	2.516	
		0.5957	0.1587	3.412	0.2068	2.646	
		0.8219	0.1506	3.239	0.2010	2.573	
		1.0000					13.30

TABLE 1. COMPOSITION AND TEMPERATURE DEPENDENCE OF Ψ_{ij} ON
DIFFERENT SCHEMES OF COMPUTATION (continued)

Gas Pair [Reference]	Temp. (K)	Mole Fraction of Heavier Component	First Method		Second Method		Viscosity (N s m^{-2} x 10^{-6})
			Ψ_{12}	Ψ_{21}	Ψ_{12}	Ψ_{21}	
SO$_2$-H$_2$ [231]	318.16	0.0000					9.45
		0.1676	0.1386	3.002	0.1691	2.181	
		0.2286	0.1364	2.955	0.1702	2.195	
		0.2963	0.1293	2.801	0.1651	2.129	
		0.5075	0.1291	2.798	0.1722	2.220	
		0.8028	0.1034	2.241	0.1481	1.910	
		1.0000					13.86
SO$_2$-H$_2$ [347]	328.2	0.0000					9.56
		0.2005	0.1406	2.965	0.1730	2.172	
		0.4000	0.1511	3.187	0.1939	2.435	
		0.4863	0.1491	3.145	0.1939	2.435	
		0.5975	0.1522	3.210	0.1996	2.506	
		0.7866	0.1598	3.370	0.2102	2.639	
		1.0000					14.40
SO$_2$-H$_2$ [231]	343.16	0.0000					9.94
		0.1657	0.1401	2.953	0.1701	2.135	
		0.1676	0.1417	2.987	0.1722	2.161	
		0.2366	0.1385	2.921	0.1726	2.166	
		0.2963	0.1325	2.795	0.1684	2.114	
		0.4823	0.1326	2.797	0.1752	2.199	
		0.6175	0.1277	2.694	0.1728	2.169	
		0.6999	0.1276	2.690	0.1741	2.185	
		0.8028	0.1283	2.705	0.1765	2.215	
		1.0000					14.98
SO$_2$-H$_2$ [231]	365.16	0.0000					10.37
		0.1657	0.1445	2.979	0.1750	2.147	
		0.1676	0.1459	3.007	0.1769	2.170	
		0.2306	0.1414	2.914	0.1752	2.149	
		0.4823	0.1362	2.806	0.1789	2.195	
		0.6175	0.1298	2.675	0.1749	2.146	
		0.6999	0.1363	2.810	0.1837	2.254	
		0.8228	0.1296	2.670	0.1781	2.185	
		1.0000					15.99
SO$_2$-H$_2$ [347]	373.2	0.0000					10.47
		0.2005	0.1569	3.090	0.1911	2.241	
		0.4000	0.1735	3.418	0.2182	2.558	
		0.4863	0.1758	3.464	0.2226	2.610	
		0.5975	0.1765	3.476	0.2251	2.639	
		0.7866	0.1853	3.650	0.2363	2.771	
		1.0000					16.89
SO$_2$-H$_2$ [231]	397.16	0.0000					11.02
		0.1636	0.1482	2.985	0.1788	2.143	
		0.3265	0.1435	2.889	0.1817	2.178	
		0.4698	0.1409	2.837	0.1836	2.201	
		0.6760	0.1358	2.735	0.1826	2.189	
		1.0000					17.39
SO$_2$-H$_2$ [347]	423.2	0.0000					11.55
		0.2000	0.1706	3.258	0.2070	2.352	
		0.4018	0.1616	3.085	0.2042	2.321	
		0.5023	0.1661	3.173	0.2119	2.408	
		0.6024	0.1691	3.229	0.2170	2.467	
		0.8110	0.1771	3.381	0.2280	2.592	
		1.0000					19.22
SO$_2$-H$_2$ [231]	432.16	0.0000					11.67
		0.1512	0.1602	3.131	0.1919	2.233	
		0.1676	0.1537	3.004	0.1849	2.1520	
		0.3265	0.1473	2.880	0.1856	2.159	
		0.4698	0.1458	2.850	0.1887	2.196	
		0.6760	0.1453	2.840	0.1928	2.243	
		1.0000					18.97

TABLE 1.   COMPOSITION AND TEMPERATURE DEPENDENCE OF Ψ_{ij} ON DIFFERENT SCHEMES OF COMPUTATION (continued)

Gas Pair [Reference]	Temp. (K)	Mole Fraction of Heavier Component	First Method Ψ_{12}	Ψ_{21}	Second Method Ψ_{12}	Ψ_{21}	Viscosity $(N\ s\ m^{-2} \times 10^{-6})$
SO_2-H_2 [231]	472.16	0.0000					12.87
		0.1512	0.1526	3.013	0.1828	2.148	
		0.3265	0.1526	3.014	0.1920	2.257	
		0.4905	0.1515	2.992	0.1959	2.302	
		0.6760	0.1465	2.893	0.1942	2.283	
		1.0000					20.71
SO_2-H_2 [347]	473.2	0.0000					12.26
		0.2000	0.1699	3.129	0.2050	2.247	
		0.4018	0.1563	2.879	0.1976	2.166	
		0.5023	0.1535	2.827	0.1974	2.164	
		0.6024	0.1546	2.847	0.2010	2.203	
		0.8110	0.1479	2.725	0.1975	2.165	
		1.0000					21.15
NH_3-CH_4 [346]	287.66	0.0000					10.91
		0.1000	0.8311	0.9832	0.8347	0.9787	
		0.2000	0.8243	0.9751	0.8279	0.9707	
		0.3000	0.8230	0.9737	0.8268	0.9694	
		0.4000	0.8242	0.9750	0.8280	0.9708	
		0.5000	0.8211	0.9714	0.8250	0.9673	
		0.6000	0.8206	0.9708	0.8246	0.9668	
		0.7000	0.8156	0.9648	0.8196	0.9610	
		0.8000	0.8123	0.9609	0.8164	0.9572	
		0.9000	0.8264	0.9777	0.8307	0.9739	
		1.0000					9.79
NH_3-CH_4 [134]	298.2	0.0000					11.00
		0.0740	0.8876	1.020	0.8916	1.016	
		0.1970	0.8645	0.9936	0.8684	0.9892	
		0.3020	0.8573	0.9853	0.8612	0.9810	
		0.4040	0.8264	0.9498	0.8302	0.9457	
		0.4970	0.8150	0.9368	0.8189	0.9328	
		0.5910	0.8134	0.9348	0.8173	0.9310	
		0.7000	0.8327	0.9571	0.8368	0.9532	
		0.7950	0.8299	0.9538	0.8340	0.9500	
		0.8980	0.8598	0.9882	0.8641	0.9843	
		1.0000					10.16
NH_3-CH_4 [134]	308.2	0.0000					11.38
		0.0800	0.9124	1.051	0.9166	1.046	
		0.1850	0.8884	1.023	0.8924	1.019	
		0.3040	0.8741	1.007	0.8780	1.002	
		0.4060	0.8732	1.006	0.8773	1.001	
		0.4990	0.8853	1.020	0.8894	1.015	
		0.5980	0.8547	0.9844	0.8588	0.9802	
		0.6970	0.8256	0.9508	0.8297	0.9470	
		0.7980	0.7991	0.9203	0.8032	0.9167	
		0.8710	0.7982	0.9193	0.8024	0.9158	
		1.0000					10.49
NH_3-CH_4 [134]	353.2	0.0000					12.53
		0.0460	0.8558	0.9503	0.8594	0.9457	
		0.1780	0.8653	0.9608	0.8691	0.9563	
		0.2900	0.8637	0.9590	0.8675	0.9546	
		0.3940	0.8612	0.9562	0.8651	0.9520	
		0.4970	0.8575	0.9521	0.8615	0.9480	
		0.5960	0.8556	0.9500	0.8596	0.9459	
		0.6890	0.8487	0.9423	0.8528	0.9384	
		0.7780	0.8439	0.9370	0.8480	0.9332	
		0.8350	0.8491	0.9428	0.8533	0.9390	
		1.0000					11.98

TABLE 1. COMPOSITION AND TEMPERATURE DEPENDENCE OF Ψ_{ij} ON DIFFERENT SCHEMES OF COMPUTATION (continued)

Gas Pair [Reference]	Temp. (K)	Mole Fraction of Heavier Component	First Method Ψ_{12}	Ψ_{21}	Second Method Ψ_{12}	Ψ_{21}	Viscosity ($N\ s\ m^{-2} \times 10^{-6}$)
SO_2-CH_4 [35]	308.2	0.0000					11.38
		0.0850	0.5065	1.733	0.5545	1.542	
		0.2210	0.4884	1.671	0.5355	1.489	
		0.3020	0.4862	1.664	0.5343	1.485	
		0.4330	0.4770	1.632	0.5260	1.462	
		0.5670	0.4660	1.594	0.5160	1.434	
		0.6740	0.4564	1.562	0.5073	1.410	
		0.7910	0.4410	1.509	0.4928	1.370	
		0.8710	0.4214	1.442	0.4737	1.317	
		1.0000					13.28
SO_2-CH_4 [35]	353.2	0.0000					12.53
		0.1460	0.5056	1.663	0.5524	1.476	
		0.2600	0.5279	1.736	0.5791	1.548	
		0.3920	0.4970	1.635	0.5461	1.460	
		0.4780	0.4945	1.627	0.5445	1.455	
		0.5900	0.4901	1.612	0.5412	1.446	
		0.6810	0.4859	1.598	0.5377	1.437	
		0.8710	0.4914	1.616	0.5451	1.457	
		1.0000					15.21
NH_3-N_2 [222]	293.2	0.0000					9.82
		0.1117	0.9367	0.8670	0.9594	0.8242	
		0.2853	0.9316	0.8623	0.9552	0.8206	
		0.4362	0.9284	0.8594	0.9531	0.8188	
		0.7080	0.9207	0.8522	0.9474	0.8139	
		0.8889	0.9225	0.8539	0.9507	0.8167	
		1.0000					17.45
NH_3-N_2 [347]	297.2	0.0000					10.28
		0.2036	0.9701	0.9372	0.9962	0.8931	
		0.4291	0.9783	0.9450	1.005	0.9014	
		0.4973	0.9719	0.9389	0.9991	0.8958	
		0.5980	0.9473	0.9412	1.002	0.8982	
		0.7993	0.8595	0.8303	0.8867	0.7950	
		1.0000					17.50
NH_3-N_2 [347]	327.2	0.0000					11.37
		0.2036	0.8779	0.8584	0.8999	0.8167	
		0.4291	0.9469	0.9259	0.9732	0.8832	
		0.4973	0.9351	0.9144	0.9615	0.8725	
		0.5980	0.9176	0.8972	0.9441	0.8567	
		0.7993	0.9362	0.9154	0.9642	0.8750	
		1.0000					19.13
NH_3-N_2 [347]	373.2	0.0000					13.07
		0.2036	0.8589	0.8791	0.8812	0.8371	
		0.4291	0.9303	0.9522	0.9570	0.9091	
		0.4973	0.9115	0.9330	0.9380	0.8910	
		0.5980	0.9109	0.9324	0.9379	0.8910	
		0.7993	0.9077	0.9291	0.9358	0.8890	
		1.0000					21.01
NH_3-N_2 [222]	373.2	0.0000					12.79
		0.1117	0.9147	0.9229	0.9387	0.8790	
		0.2853	0.9136	0.9218	0.9386	0.8789	
		0.4362	0.9055	0.9136	0.9311	0.8719	
		0.7080	0.9001	0.9081	0.9274	0.8684	
		0.8889	0.8943	0.9012	0.9228	0.8641	
		1.0000					20.85
NH_3-N_2 [347]	423.2	0.0000					14.93
		0.2397	0.8789	0.9363	0.9034	0.8932	
		0.4080	0.8982	0.9569	0.9245	0.9140	
		0.5072	0.8915	0.9496	0.9181	0.9076	
		0.6015	0.9167	0.9765	0.9444	0.9336	
		0.7748	0.9100	0.9694	0.9383	0.9277	
		1.0000					23.05

TABLE 1. COMPOSITION AND TEMPERATURE DEPENDENCE OF Ψ_{ij} ON
DIFFERENT SCHEMES OF COMPUTATION (continued)

Gas Pair [Reference]	Temp. (K)	Mole Fraction of Heavier Component	First Method Ψ_{12}	Ψ_{21}	Second Method Ψ_{12}	Ψ_{21}	Viscosity ($N \cdot s \cdot m^{-2} \times 10^{-6}$)
NH_3-N_2 [222]	473.2	0.0000					16.46
		0.1117	0.8950	0.9843	0.9205	0.9394	
		0.2853	0.8899	0.9786	0.9160	0.9348	
		0.4362	0.8871	0.9755	0.9137	0.9325	
		0.7080	0.8824	0.9704	0.9103	0.9291	
		0.8889	0.8762	0.9635	0.9050	0.9236	
		1.0000					24.62
NH_3-N_2 [222]	523.2	0.0000					18.13
		0.1117	0.8802	0.9992	0.9058	0.9543	
		0.2853	0.8837	1.003	0.9102	0.9589	
		0.4362	0.8814	1.000	0.9084	0.9570	
		0.7080	0.8773	0.9959	0.9054	0.9539	
		0.8889	0.8722	0.9901	0.9011	0.9493	
		1.0000					26.27
NH_3-N_2 [347]	573.2	0.0000					16.80
		0.2397	0.8761	0.9596	0.9011	0.9160	
		0.4080	0.9139	1.001	0.9413	0.9569	
		0.5072	0.9240	1.012	0.9520	0.9677	
		0.6015	0.9316	1.020	0.9600	0.9759	
		0.7748	0.9250	1.013	0.9537	0.9695	
		1.0000					25.23
NH_3-N_2O [35]	298.2	0.0000					10.16
		0.1050	0.5417	0.9570	0.5647	0.8653	
		0.2070	0.7384	1.305	0.7817	1.198	
		0.3030	0.7333	1.296	0.7765	1.190	
		0.4060	0.7213	1.274	0.7641	1.171	
		0.5040	0.7258	1.282	0.7693	1.179	
		0.5980	0.7325	1.294	0.7767	1.190	
		0.7020	0.7381	1.304	0.7827	1.199	
		0.8020	0.7002	1.237	0.7446	1.141	
		0.8990	0.6971	1.232	0.7421	1.137	
		1.0000					14.86
NH_3-N_2O [35]	308.2	0.0000					10.49
		0.1120	0.6490	1.144	0.6823	1.043	
		0.2100	0.6637	1.170	0.7000	1.070	
		0.3130	0.6695	1.180	0.7077	1.082	
		0.4020	0.6732	1.187	0.7128	1.090	
		0.5020	0.6805	1.200	0.7217	1.103	
		0.6020	0.6823	1.203	0.7245	1.108	
		0.7060	0.6860	1.209	0.7293	1.115	
		0.8210	0.6930	1.222	0.7374	1.127	
		0.9510	0.6963	1.227	0.7416	1.134	
		1.0000					15.38
NH_3-N_2O [35]	353.2	0.0000					11.98
		0.1420	0.6510	1.165	0.6856	1.064	
		0.2210	0.6593	1.180	0.6959	1.080	
		0.3200	0.6620	1.185	0.7002	1.087	
		0.4080	0.6638	1.188	0.7032	1.091	
		0.5020	0.6659	1.192	0.7067	1.097	
		0.6060	0.6609	1.183	0.7026	1.090	
		0.7160	0.6577	1.177	0.7006	1.087	
		0.8160	0.6562	1.174	0.7001	1.087	
		0.9190	0.6409	1.147	0.6858	1.064	
		1.0000					17.30

TABLE 1. COMPOSITION AND TEMPERATURE DEPENDENCE OF Ψ_{ij} ON
DIFFERENT SCHEMES OF COMPUTATION (continued)

Gas Pair [Reference]	Temp. (K)	Mole Fraction of Heavier Component	First Method		Second Method		Viscosity (N s m^{-2} x 10^{-6})
			Ψ_{12}	Ψ_{21}	Ψ_{12}	Ψ_{21}	
SO_2-N_2O [35]	298.2	0.0000					14.86
		0.0430	0.7685	1.262	0.7915	1.229	
		0.1780	0.7316	1.201	0.7530	1.169	
		0.2970	0.7360	1.209	0.7579	1.177	
		0.4010	0.7367	1.210	0.7590	1.178	
		0.4930	0.7421	1.219	0.7647	1.187	
		0.5960	0.7460	1.225	0.7689	1.194	
		0.7020	0.7328	1.203	0.7557	1.173	
		0.8000	0.7284	1.196	0.7514	1.166	
		0.9000	0.7488	1.230	0.7722	1.199	
		0.9140	0.8187	1.344	0.8425	1.308	
		1.0000					13.17
SO_2-N_2O [35]	308.2	0.0000					15.38
		0.0420	0.8986	1.515	0.9298	1.481	
		0.1470	0.7606	1.282	0.7838	1.249	
		0.2490	0.7511	1.266	0.7739	1.233	
		0.3980	0.7405	1.248	0.7632	1.216	
		0.4760	0.7378	1.244	0.7605	1.212	
		0.5750	0.7341	1.237	0.7569	1.206	
		0.6720	0.7110	1.198	0.7335	1.169	
		0.7770	0.6931	1.168	0.7156	1.140	
		0.8790	0.6593	1.111	0.6819	1.086	
		1.0000					13.28
SO_2-N_2O [35]	353.2	0.0000					17.30
		0.0350	0.7489	1.238	0.7709	1.205	
		0.1830	0.7463	1.234	0.7685	1.201	
		0.2730	0.7443	1.230	0.7666	1.198	
		0.3750	0.7415	1.226	0.7639	1.194	
		0.4740	0.7429	1.228	0.7655	1.196	
		0.5760	0.7398	1.223	0.7626	1.192	
		0.6750	0.7337	1.213	0.7565	1.182	
		0.7860	0.7299	1.207	0.7529	1.177	
		0.8950	0.7188	1.188	0.7420	1.160	
		1.0000					15.23
NH_3-O_2 [222]	293.2	0.0000					9.82
		0.1245	0.9346	0.8524	0.9603	0.7968	
		0.2921	0.9287	0.8470	0.9556	0.7929	
		0.5214	0.9272	0.8456	0.9563	0.7935	
		0.7014	0.9218	0.8407	0.9527	0.7905	
		0.8649	0.9172	0.8365	0.9500	0.7883	
		1.0000					20.23
NH_3-O_2 [222]	373.2	0.0000					12.79
		0.1245	0.9215	0.9076	0.9490	0.8503	
		0.2921	0.9170	0.9031	0.9456	0.8472	
		0.5214	0.9146	0.9007	0.9449	0.8466	
		0.7014	0.9107	0.8969	0.9425	0.8444	
		0.8649	0.9104	0.8966	0.9436	0.8455	
		1.0000					24.40
NH_3-O_2 [222]	473.2	0.0000					16.46
		0.1245	0.9077	0.9674	0.9372	0.9086	
		0.2921	0.9062	0.9657	0.9365	0.9079	
		0.5214	0.9033	0.9626	0.9349	0.9064	
		0.7014	0.9001	0.9592	0.9328	0.9043	
		0.8649	0.9032	0.9626	0.9370	0.9084	
		1.0000					29.02
NH_3-CH_3NH_2 [348]	273.0	0.0000					9.20
		0.2500	0.7121	1.372	0.7457	1.313	
		0.5000	0.7065	1.361	0.7401	1.303	
		0.7500	0.6988	1.347	0.7325	1.290	
		1.0000					8.71

TABLE 1.   COMPOSITION AND TEMPERATURE DEPENDENCE OF Ψ_{ij} ON DIFFERENT SCHEMES OF COMPUTATION (continued)

Gas Pair [Reference]	Temp. (K)	Mole Fraction of Heavier Component	First Method Ψ_{12}	Ψ_{21}	Second Method Ψ_{12}	Ψ_{21}	Viscosity ($N\,s\,m^{-2} \times 10^{-6}$)
NH_3-CH_3NH_2 [348]	298.0	0.0000					10.09
		0.2500	0.7057	1.378	0.7393	1.319	
		0.5000	0.7016	1.370	0.7352	1.311	
		0.7500	0.6965	1.359	0.7303	1.302	
		1.0000					9.43
NH_3-CH_3NH_2 [348]	323.0	0.0000					10.99
		0.2500	0.7011	1.385	0.7346	1.326	
		0.5000	0.6977	1.378	0.7313	1.320	
		0.7500	0.6925	1.368	0.7262	1.311	
		1.0000					10.15
NH_3-CH_3NH_2 [348]	348.0	0.0000					11.89
		0.2500	0.6979	1.391	0.7313	1.332	
		0.5000	0.6954	1.386	0.7290	1.328	
		0.7500	0.6929	1.381	0.7267	1.324	
		1.0000					10.88
NH_3-CH_3NH_2 [348]	373.0	0.0000					12.79
		0.2500	0.6938	1.394	0.7271	1.335	
		0.5000	0.6922	1.391	0.7258	1.333	
		0.7500	0.6914	1.389	0.7253	1.332	
		1.0000					11.61
NH_3-CH_3NH_2 [348]	423.0	0.0000					14.60
		0.2500	0.6875	1.401	0.7208	1.342	
		0.5000	0.6851	1.396	0.7186	1.338	
		0.7500	0.6858	1.397	0.7196	1.340	
		1.0000					13.07
NH_3-CH_3NH_2 [348]	473.0	0.0000					16.47
		0.2500	0.6848	1.404	0.7181	1.345	
		0.5000	0.6897	1.414	0.7235	1.355	
		0.7500	0.6843	1.402	0.7181	1.345	
		1.0000					14.66
NH_3-CH_3NH_2 [348]	523.0	0.0000					18.25
		0.2500	0.6807	1.407	0.7139	1.348	
		0.5000	0.6822	1.410	0.7158	1.352	
		0.7500	0.6803	1.406	0.7141	1.348	
		1.0000					16.11
NH_3-CH_3NH_2 [348]	573.0	0.0000					20.03
		0.2500	0.6765	1.408	0.7095	1.349	
		0.5000	0.6987	1.454	0.7330	1.394	
		0.7500	0.6734	1.401	0.7071	1.344	
		1.0000					17.56
NH_3-CH_3NH_2 [348]	623.0	0.0000					21.81
		0.2500	0.6782	1.419	0.7114	1.361	
		0.5000	0.6736	1.410	0.7069	1.352	
		0.7500	0.6731	1.409	0.7068	1.352	
		1.0000					19.01
NH_3-CH_3NH_2 [348]	673.0	0.0000					23.60
		0.2500	0.6724	1.413	0.7054	1.355	
		0.5000	0.6708	1.410	0.7041	1.352	
		0.7500	0.6678	1.404	0.7014	1.347	
		1.0000					20.48
$CH_3COOCH_2C_6H_5$- $C_6H_5NH_2$ (Liquid) [351]	303.2	0.0000					3145.70
		0.1250	0.5253	1.613	0.5497	1.571	
		0.3000	0.5264	1.616	0.5516	1.576	
		0.4350	0.5262	1.615	0.5517	1.576	
		0.4950	0.5301	1.627	0.5560	1.589	
		0.6050	0.5327	1.635	0.5589	1.597	
		0.7500	0.5387	1.654	0.5653	1.615	
		0.8500	0.5424	1.665	0.5692	1.626	
		1.0000					1652.40

TABLE 1. COMPOSITION AND TEMPERATURE DEPENDENCE OF Ψ_{ij} ON DIFFERENT SCHEMES OF COMPUTATION (continued)

Gas Pair [Reference]	Temp. (K)	Mole Fraction of Heavier Component	First Method Ψ_{12}	First Method Ψ_{21}	Second Method Ψ_{12}	Second Method Ψ_{21}	Viscosity (N s m^{-2} x 10^{-6})
$CH_3COOCH_2C_6H_5$-$CH_3C_6H_4OH$ (Liquid) [351]	313.2	0.0000					6180.00
		0.1150	0.3628	2.302	0.3775	2.280	
		0.2720	0.3809	2.417	0.3966	2.396	
		0.4350	0.3820	2.424	0.3978	2.403	
		0.6200	0.3840	2.436	0.3999	2.416	
		0.8100	0.3964	2.515	0.4127	2.493	
		1.0000					1352.50
$(CH_3)_2O$-CH_3Cl [349]	308.2	0.0000					9.66
		0.0460	1.010	0.9497	1.016	0.9426	
		0.2220	1.036	0.9742	1.043	0.9671	
		0.2990	1.040	0.9782	1.047	0.9710	
		0.4010	1.042	0.9794	1.048	0.9722	
		0.5080	1.041	0.9790	1.048	0.9718	
		0.6040	1.047	0.9841	1.053	0.9768	
		0.6990	1.047	0.9845	1.054	0.9773	
		0.8020	1.055	0.9914	1.061	0.9841	
		0.8770	1.062	0.9985	1.069	0.9910	
		1.0000					11.26
$(CH_3)_2O$-CH_3Cl [349]	353.2	0.0000					10.98
		0.0630	1.031	0.9705	1.037	0.9634	
		0.1910	1.038	0.9772	1.045	0.9701	
		0.2810	1.043	0.9817	1.049	0.9745	
		0.4000	1.035	0.9750	1.042	0.9678	
		0.4740	1.041	0.9802	1.048	0.9730	
		0.5880	1.040	0.9797	1.047	0.9724	
		0.6690	1.041	0.9799	1.047	0.9727	
		0.7610	1.035	0.9748	1.042	0.9676	
		1.0000					12.78
$(CH_3)_2O$-SO_2 [349]	308.2	0.0000					9.66
		0.0580	1.027	1.039	1.050	1.010	
		0.1840	0.9999	1.011	1.021	0.9828	
		0.2940	1.000	1.012	1.021	0.9832	
		0.3910	0.9980	1.009	1.019	0.9810	
		0.4920	0.9937	1.005	1.015	0.9767	
		0.5910	0.9969	1.008	1.018	0.9799	
		0.6920	0.9881	0.9994	1.009	0.9713	
		0.7820	0.9827	0.9940	1.004	0.9661	
		0.8440	0.9708	0.9819	0.9916	0.9546	
		1.0000					13.28
$(CH_3)_2O$-SO_2 [349]	353.2	0.0000					10.98
		0.0490	1.039	1.042	1.062	1.013	
		0.1900	1.023	1.026	1.045	0.9972	
		0.2790	1.023	1.026	1.045	0.9968	
		0.3890	1.016	1.018	1.037	0.9894	
		0.5040	1.010	1.013	1.031	0.9838	
		0.5700	1.010	1.013	1.031	0.9841	
		0.6480	0.9930	0.9954	1.014	0.9673	
		0.7480	1.011	1.013	1.032	0.9847	
		0.8600	1.006	1.008	1.027	0.9796	
		1.0000					15.23
CH_3Cl-SO_2 [349]	308.2	0.0000					11.26
		0.0450	1.013	1.090	1.031	1.070	
		0.1670	0.9774	1.051	0.9937	1.032	
		0.2860	0.9685	1.042	0.9845	1.022	
		0.3690	0.9579	1.031	0.9737	1.011	
		0.4920	0.9582	1.031	0.9741	1.011	
		0.6040	0.9539	1.026	0.9697	1.007	
		0.6900	0.9529	1.025	0.9687	1.006	
		0.7680	0.9429	1.014	0.9586	0.9951	
		0.8470	0.9276	0.9979	0.9434	0.9793	
		1.0000					13.28

TABLE 1.   COMPOSITION AND TEMPERATURE DEPENDENCE OF Ψ_{ij} ON
DIFFERENT SCHEMES OF COMPUTATION (continued)

Gas Pair [Reference]	Temp. (K)	Mole Fraction of Heavier Component	First Method Ψ_{12}	Ψ_{21}	Second Method Ψ_{12}	Ψ_{21}	Viscosity (N s m^{-2} x 10^{-6})
CH_3Cl-SO_2 [349]	353.2	0.0000					12.78
		0.0510	1.004	1.069	1.021	1.049	
		0.1830	0.9806	1.044	0.9968	1.024	
		0.2850	0.9799	1.043	0.9961	1.023	
		0.3940	0.9682	1.031	0.9841	1.011	
		0.4830	0.9667	1.029	0.9826	1.009	
		0.5890	0.9636	1.026	0.9795	1.006	
		0.6860	0.9558	1.018	0.9716	0.9981	
		0.7930	0.9406	1.001	0.9564	0.9825	
		1.0000					15.23
SO_2-SO_2F_2 [350]	273.0	0.0000					12.26
		0.2500	0.8342	1.153	0.8615	1.110	
		0.5000	0.8301	1.147	0.8575	1.105	
		0.7500	0.8205	1.134	0.8482	1.093	
		1.0000					14.13
SO_2-SO_2F_2 [350]	323.0	0.0000					14.42
		0.2500	0.8152	1.155	0.8420	1.112	
		0.5000	0.8086	1.145	0.8357	1.104	
		0.7500	0.8035	1.138	0.8311	1.098	
		1.0000					16.22
SO_2-SO_2F_2 [350]	373.0	0.0000					16.52
		0.2500	0.8049	1.159	0.8316	1.117	
		0.5000	0.7975	1.148	0.8245	1.107	
		0.7500	0.7972	1.148	0.8248	1.107	
		1.0000					18.28
SO_2-SO_2F_2 [350]	423.0	0.0000					18.62
		0.2500	0.7964	1.164	0.8230	1.122	
		0.5000	0.7904	1.156	0.8174	1.115	
		0.7500	0.7901	1.155	0.8177	1.115	
		1.0000					20.29
SO_2-SO_2F_2 [350]	473.0	0.0000					20.69
		0.2500	0.7950	1.178	0.8218	1.135	
		0.5000	0.7887	1.169	0.8158	1.127	
		0.7500	0.7931	1.175	0.8208	1.134	
		1.0000					22.25
SO_2-SO_2F_2 [350]	523.0	0.0000					22.69
		0.2500	0.7937	1.185	0.8206	1.142	
		0.5000	0.7909	1.180	0.8182	1.139	
		0.7500	0.7942	1.185	0.8220	1.144	
		1.0000					24.22
SO_2-SO_2F_2 [350]	573.0	0.0000					24.68
		0.2500	0.7964	1.198	0.8236	1.155	
		0.5000	0.7939	1.194	0.8213	1.152	
		0.7500	0.7963	1.198	0.8242	1.156	
		1.0000					26.14
SO_2-SO_2F_2 [350]	623.0	0.0000					26.61
		0.2500	0.7990	1.209	0.8265	1.167	
		0.5000	0.7963	1.205	0.8239	1.163	
		0.7500	0.7989	1.209	0.8268	1.167	
		1.0000					28.01
SO_2-SO_2F_2 [350]	673.0	0.0000					28.45
		0.2500	0.7967	1.211	0.8241	1.168	
		0.5000	0.7974	1.212	0.8251	1.169	
		0.7500	0.7994	1.215	0.8274	1.172	
		1.0000					29.83

TABLE 1. COMPOSITION AND TEMPERATURE DEPENDENCE OF Ψ_{ij} ON DIFFERENT SCHEMES OF COMPUTATION (continued)

Gas Pair [Reference]	Temp. (K)	Mole Fraction of Heavier Component	First Method Ψ_{12}	Ψ_{21}	Second Method Ψ_{12}	Ψ_{21}	Viscosity (N s m^{-2} x 10^{-6})
Air-CO$_2$ [346]	290.0	0.0000					14.55
		0.2000	0.7065	1.326	0.7310	1.288	
		0.4000	0.7068	1.326	0.7317	1.289	
		0.6000	0.7059	1.325	0.7311	1.288	
		0.8000	0.7043	1.322	0.7297	1.286	
		1.0000					17.97
Air-CH$_4$ [334]	293.2	0.0000					11.21
		0.1090	0.9244	1.042	0.9557	0.9863	
		0.1990	0.8683	0.9791	0.8965	0.9251	
		0.3020	0.8771	0.9890	0.9064	0.9354	
		0.4050	0.8962	1.011	0.9269	0.9565	
		0.5050	0.8956	1.101	0.9267	0.9563	
		0.6090	0.9027	1.018	0.9344	0.9642	
		0.7130	0.9110	1.027	0.9432	0.9733	
		0.8040	0.9252	1.043	0.9578	0.9884	
		0.9020	0.9244	1.042	0.9572	0.9877	
		1.0000					17.95
Air-CH$_4$ [334]	293.2	0.0000					11.09
		0.0450	0.9088	1.013	0.9384	0.9575	
		0.1500	0.8944	0.9972	0.9236	0.9423	
		0.2530	0.8762	0.9769	0.9048	0.9232	
		0.3540	0.8950	0.9978	0.9252	0.9440	
		0.4410	0.9037	1.007	0.9346	0.9536	
		0.5590	0.7961	0.8876	0.8244	0.8411	
		0.6540	0.9036	1.007	0.9353	0.9543	
		0.7490	0.9060	1.010	0.9381	0.9572	
		0.8540	0.9343	1.042	0.9670	0.9866	
		0.9490	0.9452	1.054	0.9779	0.9978	
		1.0000					17.96
Air-CH$_4$ [334]	293.2	0.0000					11.29
		0.1060	0.8968	1.025	0.9267	0.9691	
		0.1990	0.8623	0.9853	0.8906	0.9313	
		0.3000	0.8611	0.9839	0.8899	0.9306	
		0.3840	0.8191	0.9359	0.8467	0.8854	
		0.5050	0.8850	1.011	0.9160	0.9579	
		0.6010	0.8917	1.019	0.9234	0.9656	
		0.6990	0.8999	1.028	0.9321	0.9747	
		0.7980	0.9080	1.038	0.9406	0.9836	
		0.9010	0.9016	1.030	0.9345	0.9772	
		1.0000					17.84
Air-CH$_4$ [334]	293.2	0.0000					11.28
		0.0480	0.8947	1.014	0.9239	0.9583	
		0.1520	0.8845	1.002	0.9135	0.9475	
		0.2520	0.8630	0.9780	0.8914	0.9245	
		0.3480	0.8873	1.006	0.9174	0.9516	
		0.4420	0.8977	1.017	0.9287	0.9633	
		0.5530	0.9075	1.029	0.9392	0.9742	
		0.6360	0.9007	1.021	0.9325	0.9672	
		0.7470	0.9126	1.034	0.9450	0.9801	
		0.8520	0.9318	1.056	0.9646	1.000	
		0.9460	0.9727	1.102	1.005	1.043	
		1.0000					17.97
NH$_3$-Air [346]	288.7	0.0000					9.88
		0.1000	0.9106	0.8454	0.9329	0.7998	
		0.2000	0.9157	0.8501	0.9391	0.8051	
		0.3000	0.9083	0.8432	0.9322	0.7992	
		0.4000	0.9024	0.8377	0.9270	0.7947	
		0.5000	0.8986	0.8342	0.9241	0.7922	
		0.6000	0.8929	0.8290	0.9193	0.7881	
		0.7000	0.9362	0.8692	0.9644	0.8267	
		0.8000	0.8842	0.8208	0.9126	0.7824	
		0.9000	0.8908	0.8270	0.9205	0.7891	
		1.0000					18.10

TABLE 1. COMPOSITION AND TEMPERATURE DEPENDENCE OF Ψ_{ij} ON
DIFFERENT SCHEMES OF COMPUTATION (continued)

Gas Pair [Reference]	Temp. (K)	Mole Fraction of Heavier Component	First Method		Second Method		Viscosity (N s m^{-2} x 10^{-6})
			Ψ_{12}	Ψ_{21}	Ψ_{12}	Ψ_{21}	
HCl-Air [346]	291.3	0.0000					17.94
		0.2000	0.7194	1.155	0.7329	1.136	
		0.4000	0.7171	1.151	0.7310	1.133	
		0.6000	0.7111	1.141	0.7254	1.125	
		0.8000	0.6829	1.096	0.6974	1.081	
		1.0000					14.07
HCl-Air [346]	289.7	0.0000					18.18
		0.1000	0.7237	1.161	0.7370	1.143	
		0.2000	0.7201	1.156	0.7336	1.137	
		0.3000	0.7190	1.154	0.7327	1.136	
		0.4000	0.7178	1.152	0.7317	1.134	
		0.5000	0.7151	1.148	0.7292	1.130	
		0.6000	0.7109	1.141	0.7252	1.124	
		0.7000	0.7080	1.136	0.7224	1.120	
		0.8000	0.6993	1.122	0.7139	1.107	
		0.9000	0.6935	1.113	0.7083	1.098	
		1.0000					14.26
H$_2$S-Air [346]	290.36	0.0000					18.27
		0.1000	0.6808	1.161	0.6899	1.149	
		0.2000	0.6800	1.160	0.6894	1.148	
		0.3000	0.6785	1.157	0.6881	1.146	
		0.4000	0.6792	1.159	0.6890	1.147	
		0.5000	0.6736	1.149	0.6835	1.138	
		0.6000	0.6788	1.158	0.6890	1.147	
		0.7000	0.6795	1.159	0.6899	1.148	
		0.8000	0.6766	1.154	0.6871	1.144	
		0.9000	0.6806	1.161	0.6913	1.151	
		1.0000					12.60

TABLE 2.   RECOMMENDED SETS OF Ψ_{ij} AND L-VALUES FOR THE VISCOSITY DATA

Gas Pair	Temp. (K)	Mole Fraction of Heavier Component	First Method Ψ_{12}	Ψ_{21}	L_1 (%)	L_2 (%)	L_3 (%)	Second Method Ψ_{12}	Ψ_{21}	L_1 (%)	L_2 (%)	L_3 (%)
Ar-He	72.0	0.3570						0.2400	2.121	0.490	0.736	1.485
	72.0	0.4585	0.1933	2.412	0.711	1.090	2.950					
	81.0	0.5380	0.1983	2.411	0.806	1.269	3.439	0.2476	2.132	0.710	0.933	1.683
	90.2	0.3570	0.2010	2.402	1.177	1.551	3.233	0.2465	2.086	0.709	1.104	2.793
	192.5	0.4650	0.2559	2.448	0.572	0.646	0.967					
	192.5	0.4940						0.3034	2.036	0.312	0.376	0.561
	229.5	0.4090						0.3083	2.015	0.424	0.501	0.880
	229.5	0.5640	0.2581	2.382	0.731	0.898	1.721					
	288.2	0.2915	0.2842	2.512	0.468	0.570	1.125	0.3347	2.095	0.264	0.337	0.651
	291.1	0.3910	0.2697	2.393	0.956	1.191	2.400	0.3205	2.014	0.643	0.806	1.500
	291.1	0.3010	0.2678	2.353	1.140	1.345	2.782	0.3158	1.965	0.698	0.884	1.975
	291.2	0.4380	0.2809	2.472	0.612	0.759	1.668	0.3339	2.081	0.377	0.451	0.834
	293.0	0.5094	0.2782	2.478	0.230	0.326	0.461	0.3324	2.096	0.164	0.233	0.329
	373.0	0.5094	0.2878	2.483	0.070	0.099	0.140	0.3423	2.091	0.009	0.013	0.018
	373.2	0.6119						0.3470	2.096	0.497	0.755	1.935
	373.2	0.6846	0.2954	2.519	0.496	0.723	1.604					
	456.2	0.6119	0.3043	2.532	0.438	0.699	1.722	0.3609	2.127	0.302	0.406	0.874
	473.0	0.6180	0.2960	2.500	0.000	0.000	0.000	0.3523	2.108	0.000	0.000	0.000
	523.0	0.6180	0.2975	2.500	0.000	0.000	0.000	0.3539	2.106	0.000	0.000	0.000
Ar-Kr	291.2	0.2280	0.7221	1.350	0.166	0.203	0.336	0.7606	1.272	0.176	0.208	0.338
Ar-Ne	72.3	0.5011	0.4863	1.769	0.083	0.124	0.264	0.5202	1.707	0.110	0.155	0.299
	90.3	0.6713	0.4989	1.723	0.467	0.574	0.979	0.5335	1.663	0.457	0.580	0.962
	193.4	0.5024	0.5432	1.654	0.264	0.395	0.838	0.5783	1.590	0.283	0.412	0.864
	229.0	0.6507	0.5451	1.601	0.195	0.225	0.367	0.5804	1.539	0.164	0.197	0.330
	291.1	0.4970						0.5979	1.509	0.213	0.245	0.339
	291.1	0.6757	0.5665	1.584	0.162	0.212	0.328					
	291.2	0.2210	0.5716	1.579	0.496	0.615	1.196	0.6062	1.512	0.499	0.600	1.167
	293.0	0.6091	0.5758	1.593	0.106	0.132	0.190	0.6117	1.527	0.082	0.101	0.132
	373.0	0.6091	0.5965	1.589	0.066	0.087	0.138	0.6329	1.521	0.052	0.077	0.130
	473.0	0.6091	0.6118	1.586	0.147	0.217	0.369	0.6486	1.518	0.154	0.227	0.386
	523.0	0.2680	0.6117	1.575	0.130	0.164	0.243	0.5481	1.507	0.130	0.170	0.266
Ar-Xe	291.2	0.5980	0.5254	1.696	0.207	0.292	0.673	0.5753	1.554	0.211	0.255	0.516
He-Kr	283.2	0.2046	0.1866	3.124	0.759	0.844	1.252	0.2281	2.419	0.172	0.220	0.376
	291.2	0.2720	0.1760	2.882	0.770	0.973	1.862					
	291.2	0.3530						0.2181	2.263	0.217	0.263	0.482
	373.2	0.2046	0.1968	3.135	0.709	0.834	1.201					
	373.2	0.4995						0.2436	2.459	0.224	0.281	0.574
He-Ne	20.4	0.7200	0.3976	1.999	0.375	0.522	0.867	0.4530	1.786	0.266	0.329	0.450
	65.8	0.5090	0.4627	1.663	0.236	0.345	0.586					
	65.8	0.7610						0.5150	1.452	0.061	0.083	0.136
	90.2	0.4910	0.4841	1.649	0.196	0.259	0.414	0.5358	1.431	0.034	0.045	0.072
	194.0	0.4820	0.5148	1.642	0.097	0.120	0.173	0.5670	1.419	0.032	0.055	0.095
	284.2	0.2801	0.5199	1.714	0.188	0.217	0.329	0.5713	1.477	0.185	0.221	0.400
	291.2	0.3930						0.5715	1.409	0.440	0.641	1.201
	291.2	0.5650	0.5222	1.641	0.526	0.762	1.326					
	293.0	0.4376	0.5211	1.649	0.377	0.473	0.690	0.5730	1.423	0.263	0.349	0.558
	293.1	0.4980	0.5204	1.661	0.235	0.327	0.544	0.5731	1.435	0.125	0.182	0.309
	373.0	0.4376	0.5198	1.650	0.409	0.503	0.669	0.5716	1.423	0.294	0.360	0.456
	373.2	0.4995	0.5167	1.714	0.950	2.035	5.327	0.5701	1.484	0.911	2.039	5.355
	473.0	0.4376	0.5180	1.653	0.407	0.500	0.663	0.5699	1.427	0.291	0.357	0.449
	523.0	0.2379	0.5237	1.673	0.222	0.313	0.443	0.5737	1.438	0.110	0.156	0.220
He-Xe	291.2	0.2010	0.1251	3.555	1.056	1.152	1.837					
	291.2	0.7920						0.1614	2.716	0.343	0.452	0.901
Kr-Ne	291.2	0.1110						0.4360	1.838	0.395	0.476	0.812
	291.2	0.5330	0.3892	2.032	0.320	0.542	1.232					
Kr-Xe	291.2	0.2010	0.7500	1.290	0.292	0.336	0.544	0.7767	1.249	0.293	0.340	0.564
Ne-Xe	291.2	0.3930	0.2711	2.442	0.447	0.524	0.894					
	291.2	0.5940						0.3197	2.173	0.206	0.250	0.418
Ar-H$_2$	293.0	0.3485	0.2787	2.186	0.534	0.659	0.895					
	293.0	0.5543						0.3170	1.588	0.159	0.194	0.241
	373.0	0.3485						0.3212	1.559	0.090	0.121	0.196
	373.0	0.5543	0.232	2.075	0.443	0.679	1.164					
	523.0	0.3485	0.2929	2.182	0.528	0.646	0.809	0.3332	1.585	0.497	0.724	1.226

TABLE 2.   RECOMMENDED SETS OF Ψ_{ij} AND L-VALUES FOR THE VISCOSITY DATA (continued)

Gas Pair	Temp. (K)	Mole Fraction of Heavier Component	First Method Ψ_{12}	Ψ_{21}	L_1 (%)	L_2 (%)	L_3 (%)	Second Method Ψ_{12}	Ψ_{21}	L_1 (%)	L_2 (%)	L_3 (%)
$He-H_2$	273.2	0.7509	1.095	0.9661	0.244	0.416	1.038	1.130	0.8993	0.246	0.417	1.034
	288.2	0.5972	1.112	0.9877	0.145	0.172	0.286	1.147	0.9197	0.131	0.157	0.268
	291.7	0.5030	1.169	1.039	1.039	1.196	1.978	1.207	0.9677	0.988	1.139	1.883
	293.0	0.3931	1.129	0.9937	0.207	0.315	0.539	1.166	0.9258	0.205	0.320	0.551
	373.0	0.4480	1.114	0.9807	0.196	0.240	0.300	1.150	0.9135	0.192	0.235	0.309
	373.2	0.7509	1.090	0.9665	0.284	0.360	0.629	1.125	0.8999	0.282	0.363	0.632
	473.0	0.4480	1.114	0.9864	0.218	0.315	0.531	1.150	0.9189	0.214	0.317	0.540
	523.0	0.4480	1.118	0.9913	0.151	0.210	0.348	1.155	0.9235	0.146	0.198	0.323
$Ne-H_2$	290.4	0.1610	0.5615	1.584	1.530	1.712	1.975	0.6017	1.201	1.418	1.611	2.317
	293.0	0.2285	0.5482	1.553	0.343	0.424	0.578					
	293.0	0.5391						0.5870	1.177	0.104	0.134	0.208
	373.0	0.2285						0.5840	1.177	0.175	0.246	0.410
	373.0	0.5391	0.5409	1.538	0.332	0.416	0.602					
	473.0	0.5391	0.5357	1.539	0.349	0.442	0.661	0.5815	1.182	0.115	0.156	0.255
	523.0	0.2285						0.5824	1.188	0.108	0.137	0.202
	523.0	0.5391	0.5395	1.555	0.207	0.257	0.359					
$Ar-NH_3$	298.2	0.6910	0.9793	1.035	0.402	0.600	1.519	1.019	0.9482	0.401	0.598	1.513
	308.2	0.3990	0.9734	1.037	0.907	1.043	1.987	1.013	0.9495	0.902	1.038	1.984
	353.2	0.3810	0.9646	1.054	0.504	0.585	0.866	1.005	0.9662	0.501	0.583	0.862
$Ar-SO_2$	298.2	0.5000	0.5918	1.618	0.477	0.599	1.073	0.6189	1.576	0.484	0.604	1.076
	308.2	0.2540	0.5124	1.429	1.373	1.829	3.297	0.5352	1.391	1.294	1.741	3.200
	353.2	0.0430	0.4984	1.349	0.712	0.843	1.389	0.5181	1.306	0.718	0.908	1.547
$C_6H_6-C_6H_{12}$	298.2	0.5126	1.599	1.202	0.262	0.414	0.962	1.607	1.194	0.246	0.395	0.930
$C_6H_6-CH_3(CH_2)_4CH_3$	298.2	0.4296	0.8862	1.969	1.264	1.459	2.329	0.8948	1.960	1.293	1.489	2.357
C_6H_6-OMCTS	291.2	0.3511	1.450	1.464	0.336	0.571	1.357	1.524	1.260	0.739	0.916	1.770
	298.2	0.4689						1.457	1.246	0.594	0.793	1.608
	298.2	0.6211	1.391	1.452	0.378	0.526	1.038					
	308.2	0.6020	1.323	1.456	0.185	0.287	0.568	1.378	1.241	0.433	0.608	1.215
	318.2	0.3526						1.327	1.254	0.618	0.771	1.310
	318.2	0.6036	1.262	1.457	0.342	0.529	0.915					
CO_2-H_2	300.0	0.2150	0.2024	2.636	0.423	0.502	0.762	0.2394	1.964	0.330	0.391	0.547
	400.0	0.112	0.2163	2.626	0.440	0.550	0.906					
	400.0	0.2150						0.2545	1.945	0.283	0.453	1.048
	500.0	0.2150	0.2242	2.613	0.590	0.674	0.960	0.2618	1.921	0.215	0.270	0.538
	550.0	0.1112						0.2643	1.906	0.544	0.829	1.892
	550.0	0.8006	0.2217	2.540	0.946	1.280	2.121					
CO_2-N_2	297.7	0.2260	0.7307	1.363	0.548	0.671	1.057	0.7581	1.321	0.548	0.671	1.058
CO_2-N_2O	300.0	0.3967	0.9927	0.9961	0.093	0.158	0.309	0.9927	0.9961	0.093	0.158	0.309
	400.0	0.3967	0.9929	0.9934	0.094	0.163	0.356	0.9929	0.9934	0.094	0.163	0.356
	500.0	0.5967	0.9909	0.9901	0.068	0.101	0.203	0.9909	0.9901	0.068	0.101	0.203
	550.0	0.8003	1.000	1.004	0.233	0.267	0.397	1.000	1.004	0.233	0.267	0.397
CO_2-O_2	300.0	0.3390	0.7189	1.372	1.131	1.610	3.248	0.7392	1.345	1.130	1.609	3.245
$CO_2-C_3H_8$	300.0	0.4224	0.7173	1.313	0.050	0.064	0.095	0.7174	1.313	0.050	0.064	0.095
	400.0	0.5975	0.7173	1.306	0.069	0.085	0.126	0.7174	1.306	0.069	0.085	0.126
	500.0	0.5975	0.7332	1.322	0.310	0.381	0.626	0.7333	1.321	0.310	0.381	0.626
	550.0	0.5975	0.7335	1.321	0.055	0.092	0.180	0.7336	1.321	0.055	0.092	0.180
$CO_2-C_2H_4$	300.0	0.4354	0.7624	1.313	0.686	0.843	1.122	0.7625	1.313	0.686	0.843	1.122
	400.0	0.4354	0.3605	2.039	0.327	0.416	0.629	0.4061	1.906	0.442	0.560	0.839
	500.0	0.4354	0.7932	1.248	0.450	0.558	0.777	0.7933	1.248	0.450	0.558	0.777
	550.0	0.4354	0.8054	1.249	0.361	0.468	0.730	0.8055	1.249	0.361	0.468	0.730
CO_2-H_2	293.3	0.2740	0.3159	2.194	0.673	0.739	0.889	0.3596	1.683	0.226	0.292	0.500
CO_2-N_2	300.0	0.6030	0.9990	0.9963	0.129	0.171	0.258	0.9990	0.9963	0.129	0.171	0.258
	400.0	0.1629	1.002	0.9987	0.141	0.172	0.267	1.002	0.9987	0.141	0.172	0.267
	500.0	0.1629	1.001	0.9962	0.067	0.099	0.190	1.001	0.9962	0.067	0.099	0.190
	550.0	0.6030	1.005	0.9998	0.115	0.145	0.229	1.005	0.9998	0.115	0.145	0.229
CO_2-O_2	300.0	0.4201	1.000	0.9863	0.098	0.158	0.273	1.009	0.9758	0.100	0.157	0.272
	400.0	0.4201	1.015	0.9858	0.088	0.111	0.163	1.024	0.9752	0.088	0.111	0.163
	500.0	0.4201	1.020	0.9837	0.061	0.102	0.176	1.029	0.9731	0.062	0.102	0.176
CCl_4-OMCTS	291.2	0.5718	1.171	0.8970	0.656	0.980	1.995	1.204	0.8359	0.649	0.971	1.979
	298.2	0.4288	1.139	0.9034	0.778	0.949	1.987	1.172	0.8425	0.776	0.947	1.984
	308.2	0.5732	1.106	0.9226	0.745	1.049	2.090	1.140	0.8613	0.745	1.049	2.089
	318.2	0.5816	1.082	0.9460	0.921	1.361	2.639	1.116	0.8840	0.919	1.360	2.636

TABLE 2. RECOMMENDED SETS OF Ψ_{ij} AND L-VALUES FOR THE VISCOSITY DATA (continued)

Gas Pair	Temp. (K)	Mole Fraction of Heavier Component	First Method Ψ_{12}	Ψ_{21}	L_1 (%)	L_2 (%)	L_3 (%)	Second Method Ψ_{12}	Ψ_{21}	L_1 (%)	L_2 (%)	L_3 (%)
CF_4-SF_6	303.1	0.5090	0.7815	1.441	0.654	0.957	1.623	0.8131	1.390	0.682	0.991	1.676
	313.1	0.5090	0.7839	1.445	0.681	0.941	1.555	0.8156	1.393	0.712	0.978	1.610
	329.1	0.5090	0.7838	1.444	0.729	1.009	1.669	0.8154	1.392	0.759	1.046	1.724
	342.1	0.5090	0.7798	1.440	0.744	1.094	1.858	0.8113	1.388	0.772	1.127	1.910
C_6H_{12}-CH_3 $(CH_2)_4CH_3$	298.2	0.5502	0.7410	2.192	0.958	1.072	1.411	0.7428	2.190	0.964	1.078	1.416
D_2-H_2	14.4	0.5040	0.8126	1.274	0.462	0.773	1.338	0.8502	1.203	0.453	0.767	1.328
	20.4	0.3340	0.8020	1.254	0.061	0.087	0.123	0.8392	1.184	0.062	0.087	0.123
	71.5	0.2480	0.8316	1.204	0.171	0.252	0.427	0.8683	1.134	0.172	0.254	0.433
	90.1	0.5020	0.8285	1.191	0.112	0.168	0.287	0.8651	1.122	0.110	0.172	0.295
	196.0	0.4970	0.8347	1.194	0.045	0.064	0.107	0.8714	1.125	0.049	0.068	0.114
	229.0	0.2480	0.8335	1.200	0.163	0.242	0.412	0.8703	1.131	0.163	0.245	0.418
	293.1	0.7530	0.8363	1.195	0.103	0.127	0.167	0.8732	1.126	0.105	0.128	0.162
D_2-HD	14.4	0.4970	0.8846	1.070	0.326	0.411	0.611	0.9028	1.046	0.319	0.403	0.600
	20.4	0.7510	0.9092	1.088	0.202	0.309	0.530	0.9278	1.064	0.203	0.308	0.527
	71.5	0.5070	0.9348	1.090	0.028	0.038	0.063	0.9536	1.065	0.030	0.041	0.067
	90.1	0.4920	0.9286	1.083	0.080	0.129	0.223	0.9473	1.059	0.081	0.130	0.223
	196.0	0.5000	0.9281	1.079	0.036	0.061	0.106	0.9468	1.054	0.036	0.061	0.106
	229.0	0.2490	0.9315	1.075	0.024	0.031	0.048	0.9502	1.051	0.024	0.031	0.048
	293.1	0.5090	0.9347	1.077	0.089	0.109	0.147	0.9534	1.053	0.090	0.110	0.148
C_6H_6-H_2	293.0	0.1485	0.2067	2.971	1.052	1.488	2.104	0.2490	2.386	0.418	0.591	0.835
	373.0	0.1485	0.2186	2.949	0.623	0.882	1.247	0.2617	2.354	0.041	0.058	0.083
	473.0	0.1485	0.2286	2.936	0.539	0.762	1.077	0.2725	2.333	0.001	0.002	0.003
	523.0	0.1485	0.2322	2.942	0.597	0.844	1.193	0.2766	2.336	0.076	0.107	0.152
C_2H_6-CH_4	293.0	0.5126	0.6570	1.473	0.080	0.099	0.136	0.6917	1.411	0.071	0.088	0.119
	373.0	0.5126	0.6652	1.453	0.101	0.133	0.210	0.6999	1.391	0.092	0.121	0.192
	473.0	0.5126	0.6749	1.439	0.024	0.035	0.059	0.7097	1.377	0.018	0.030	0.051
	523.0	0.5126	0.6788	1.438	0.038	0.061	0.105	0.7136	1.376	0.034	0.057	0.099
C_2H_6-C_3H_8	293.0	0.5673	0.7754	1.290	0.209	0.309	0.525	0.7995	1.256	0.209	0.309	0.525
	373.0	0.8474	0.7739	1.286	0.093	0.118	0.178	0.7979	1.252	0.092	0.118	0.178
	473.0	0.5673	0.7764	1.280	0.303	0.480	0.828	0.8004	1.246	0.303	0.480	0.827
	523.0	0.7437	0.7749	1.272	0.179	0.220	0.296	0.7988	1.238	0.178	0.219	0.293
C_2H_4-H_2	195.2	0.2501	0.2224	2.888	0.466	0.590	0.940	0.2969	2.359	0.247	0.369	0.688
	233.2	0.2501	0.2268	2.855	0.557	0.642	0.771	0.2740	2.324	0.405	0.674	1.464
	272.2	0.2501	0.2248	2.753	0.427	0.530	0.811					
	272.2	0.5129						0.2705	2.232	0.183	0.326	0.713
	293.2	0.2160						0.2678	2.166	0.177	0.272	0.521
	293.2	0.5173	0.2204	2.646	0.451	0.523	0.673					
	328.2	0.2100						0.2716	2.140	0.357	0.455	0.714
	328.2	0.5173	0.2258	2.640	0.566	0.689	1.013					
	373.2	0.2114						0.2774	2.119	0.271	0.354	0.602
	373.2	0.5173	0.2306	2.615	0.333	0.405	0.559					
	423.2	0.5197						0.2797	2.091	0.150	0.207	0.343
	423.2	0.7201	0.2299	2.551	0.346	0.570	1.119					
	473.2	0.5197						0.2893	2.123	0.309	0.454	0.864
	473.2	0.7201	0.2379	2.592	0.135	0.184	0.339					
	523.2	0.5116						0.2967	2.141	0.297	0.344	0.430
	523.2	0.7201	0.2474	2.651	0.157	0.223	0.385					
C_2H_4-N_2	300.0	0.5695	0.7589	1.310	0.487	0.597	0.749	0.7950	1.310	0.487	0.597	0.749
	400.0	0.5695	0.7900	1.285	0.640	0.811	1.211	0.7901	1.285	0.640	0.811	1.211
	500.0	0.5695	0.8046	1.272	0.406	0.529	0.831	0.8047	1.272	0.406	0.529	0.831
	550.0	0.5695	0.8107	1.263	0.489	0.625	0.950	0.8108	1.263	0.489	0.625	0.590
C_2H_4-O_2	293.0	0.2297	1.316	0.7508	0.132	0.227	0.392	1.324	0.7410	0.132	0.228	0.394
	323.0	0.8694	1.314	0.7608	0.169	0.216	0.330	1.323	0.7509	0.169	0.216	0.330
	373.0	0.5855	1.310	0.7749	0.135	0.166	0.225	1.319	0.7649	0.135	0.166	0.226
n-C_7H_{16} - $(CH_3)_2$ $CHCH_2C$ $(CH_3)_3$	303.2	0.4830	0.6656	1.308	0.702	0.797	1.062	0.7125	1.166	0.619	0.709	0.981
	323.2	0.3658	0.6906	1.368	1.580	1.874	2.748	0.7386	1.217	1.537	1.828	2.708
	333.2	0.4830	0.6623	1.331	0.000	0.000	0.000	0.7096	1.187	0.000	0.000	0.000
H_2-HD	14.4	0.2540	0.8522	1.142	0.377	0.469	0.662	0.8766	1.106	0.376	0.465	0.648
	20.4	0.5050	0.8768	1.163	0.329	0.407	0.560	0.9020	1.126	0.326	0.403	0.554
	71.5	0.7490	0.9009	1.110	0.057	0.071	0.106	0.9257	1.074	0.057	0.072	0.107
	90.1	0.4990	0.8991	1.108	0.338	0.414	0.533	0.9239	1.072	0.339	0.416	0.535
	196.0	0.2360	0.1082	17.90	1.775	2.192	3.007	0.1462	10.92	1.537	1.892	2.536
	229.0	0.7480	0.9029	1.104	0.184	0.264	0.444	0.9277	1.068	0.184	0.266	0.448
	293.1	0.2410	0.9089	1.121	0.516	0.698	1.137	0.9340	1.085	0.516	0.696	1.130

TABLE 2.   RECOMMENDED SETS OF Ψ_{ij} AND L-VALUES FOR THE VISCOSITY DATA (continued)

Gas Pair	Temp. (K)	Mole Fraction of Heavier Component	First Method Ψ_{12}	Ψ_{21}	L_1 (%)	L_2 (%)	L_3 (%)	Second Method Ψ_{12}	Ψ_{21}	L_1 (%)	L_2 (%)	L_3 (%)
H_2-CH_4	293.0	0.3978	0.3331	2.136	0.274	0.382	0.713	0.3849	1.808	0.051	0.074	0.141
	293.2	0.4904	0.3530	2.307	0.769	1.154	2.208	0.4088	1.958	0.826	1.177	2.076
	333.2	0.2083	0.3586	2.292	0.750	1.033	1.908	0.4102	1.921	0.633	0.123	0.225
	373.0	0.0777						0.3947	1.786	0.070	0.083	0.117
	373.0	0.3978	0.3434	2.121	0.285	0.355	0.582					
	373.2	0.2083	0.3369	2.244	0.710	0.918	1.616	0.4076	1.877	0.578	0.995	0.972
	473.0	0.3978	0.3521	2.120	0.238	0.347	0.643	0.4045	1.784	0.251	0.353	0.539
	523.0	0.0777						0.4066	1.781	0.124	0.153	0.216
	523.0	0.3978	0.3548	2.121	0.265	0.349	0.527					
H_2-NO	273.2	0.2835	0.3083	2.168	1.867	2.587	4.861	0.3507	1.645	2.233	2.889	5.318
	293.2	0.6416	0.3125	2.219	0.723	0.825	1.565					
	293.2	0.5393						0.3605	1.707	0.630	0.817	2.042
H_2-N_2	82.2	0.3510	0.2803	2.592	0.599	1.042	2.285	0.3304	2.059	0.932	1.430	3.055
	90.2	0.1600	0.2939	2.459	0.785	1.273	2.359	0.3369	1.899	0.811	0.954	1.596
	291.1	0.1600	0.3302	2.297	0.801	0.903	1.151	0.3724	1.746	0.534	0.760	1.451
	291.1	0.4410	0.3198	2.232								
	291.1	0.5170						0.3637	1.711			
	291.2	0.2960	0.3166	2.222	0.571	0.746	1.400					
	291.2	0.5170						0.3641	1.722	0.211	0.254	0.399
	307.2	0.3991						0.3622	1.695	0.339	0.466	0.868
	307.2	0.5100	0.3156	2.191	0.234	0.264	0.326					
	325.4	0.3991	0.3374	2.443	0.275	0.339	0.514	0.3882	1.894	0.217	0.246	0.313
	373.2	0.3991	0.3332	2.297	0.789	1.211	2.546	0.3882	1.775	0.529	0.896	1.912
	422.7	0.3988	0.3483	2.417	0.794	1.056	1.875	0.3991	1.866	0.642	0.831	1.412
	478.2	0.8002	0.3547	2.465	0.915	1.572	3.426	0.4101	1.921	0.903	1.290	2.581
H_2-N_2O	300.0	0.4039						0.2540	2.091	0.409	0.513	0.752
	300.0	0.6011	0.2089	2.731	0.154	0.227	0.387					
	400.0	0.4039						0.2663	2.037	0.289	0.360	0.514
	400.0	0.6011	0.2220	2.697	0.254	0.356	0.592					
	500.0	0.4039						0.2750	2.016	0.270	0.334	0.461
	500.0	0.6011	0.2303	2.681	0.243	0.362	0.617					
	550.0	0.4039						0.2793	2.016	0.290	0.355	0.444
	550.0	0.6011	0.2343	2.685	0.204	0.314	0.538					
H_2-O_2	293.2	0.4470	0.2975	2.049	1.934	2.698	5.846					
	293.2	0.4930						0.3402	1.547	1.479	2.268	6.351
	293.6	0.2730	0.3073	2.116	0.337	0.419	0.779	0.3475	1.580	0.301	0.392	0.714
	300.0	0.6055	0.3064	2.102	0.999	1.602	3.147	0.3563	1.615	0.732	0.963	1.591
	400.0	0.6055	0.3182	2.138	0.507	0.712	1.333					
	400.0	0.8165						0.3635	1.613	0.326	0.382	0.498
	500.0	0.6055	0.3212	2.128	0.598	0.893	1.696	0.3712	1.624	0.353	0.437	0.717
	550.0	0.6055	0.3220	2.192	0.884	1.441	2.830	0.3725	1.675	0.659	0.864	1.520
$H_2-C_3H_8$	273.2	0.1500	0.1333	3.334	1.383	1.823	3.599					
	273.2	0.3271						0.1753	2.760	1.060	1.330	2.661
	300.0	0.0775						0.1895	2.846	0.829	1.096	1.605
	300.0	0.4182	0.1507	3.595	0.431	0.554	0.856					
	400.0	0.1250	0.1606	3.549	0.429	0.514	0.817					
	400.0	0.2118						0.1996	2.777	0.829	1.185	2.191
	500.0	0.0775						0.2101	2.778	0.729	0.897	1.397
	500.0	0.1250	0.1689	3.547	0.480	0.587	0.880					
	550.0	0.0775						0.2171	2.831	0.722	0.819	1.145
	550.0	0.1250	0.1725	3.573	0.533	0.656	1.117					
CH_4-O_2	293.2	0.1420	0.9227	1.021	0.304	0.382	0.853	0.9563	0.9542	0.322	0.386	0.836
$CH_4-C_3H_8$	293.0	0.3684	0.5042	1.881	0.165	0.228	0.378	0.5502	1.764	0.163	0.217	0.349
	373.0	0.3684	0.5063	1.838	0.243	0.297	0.369	0.5520	1.722	0.240	0.295	0.394
	473.0	0.3684	0.5179	1.821	0.218	0.281	0.436	0.5640	1.704	0.215	0.271	0.401
	523.0	0.3684	0.5247	1.825	0.147	0.229	0.393	0.5711	1.707	0.146	0.216	0.369
N_2-NO	293.0	0.2674	1.010	1.004	0.179	0.250	0.414	1.015	0.9989	0.179	0.250	0.415
	373.0	0.6948	1.000	0.9830	0.244	0.302	0.421	1.005	0.9777	0.244	0.302	0.422
N_2-O_2	298.7	0.5100	1.035	1.011	0.515	0.673	1.321	1.044	1.001	0.515	0.674	1.325
	300.0	0.4107	1.006	0.9945	0.118	0.153	0.238	1.015	0.9839	0.118	0.153	0.238
	400.0	0.4107	1.012	0.9862	0.085	0.104	0.129	1.022	0.9756	0.085	0.104	0.129
	500.0	0.4107	1.020	0.9882	0.122	0.158	0.246	1.029	0.9776	0.122	0.158	0.247
	550.0	0.7592	1.193	0.8805	2.814	4.283	7.334	1.203	0.8698	2.814	4.278	7.326
$N_2O-C_3H_8$	300.0	0.4171	0.7291	1.330	0.103	0.126	0.162	0.7292	1.330	0.103	0.126	0.162
	400.0	0.4171	0.7309	1.330	0.091	0.140	0.241	0.7310	1.330	0.091	0.140	0.240
	500.0	0.7984	0.7380	1.331	0.275	0.372	0.607	0.7381	1.331	0.275	0.372	0.607
	550.0	0.4171	0.7366	1.327	0.141	0.182	0.282	0.7367	1.326	0.141	0.182	0.282

TABLE 2.   RECOMMENDED SETS OF Ψ_{ij} AND L-VALUES FOR THE VISCOSITY DATA (continued)

Gas Pair	Temp. (K)	Mole Fraction of Heavier Component	First Method Ψ_{12}	Ψ_{21}	L_1 (%)	L_2 (%)	L_3 (%)	Second Method Ψ_{12}	Ψ_{21}	L_1 (%)	L_2 (%)	L_3 (%)
HCl-CO$_2$	291.00	0.6000	0.1451	0.1206	0.612	0.776	1.340	0.1452	0.1660	0.612	0.776	1.340
	291.16	0.5000	0.8796	1.034	0.171	0.189	0.261	0.8920	1.019	0.162	0.180	0.253
SO$_2$-CO$_2$	289.0	0.6000	0.7278	1.243	0.165	0.195	0.262	0.7506	1.211	0.145	0.175	0.236
	289.0	0.6000	0.7294	1.244	0.093	0.124	0.247	0.7522	1.213	0.080	0.112	0.234
	298.2	0.3890	0.7304	1.195	0.322	0.371	0.592	0.7523	1.163	0.307	0.355	0.599
	308.2	0.6080	0.7252	1.222	0.287	0.373	0.744	0.7478	1.192	0.289	0.365	0.729
	353.2	0.5000	0.7478	1.236	0.132	0.162	0.272	0.7706	1.204	0.119	0.149	0.261
CCl$_4$-CH$_2$Cl$_2$	293.15	0.6886	0.7131	1.348	0.219	0.290	0.536	0.7467	1.291	0.209	0.279	0.505
	353.26	0.6351	0.7015	1.316	0.116	0.164	0.232	0.7345	1.261	0.145	0.205	0.289
	413.43	0.7096	0.7199	1.365	0.437	0.541	0.956	0.7536	1.307	0.439	0.544	0.964
(CH$_3$)$_2$ CHOH-CCl$_4$	313.2	0.5000	0.5969	2.749	2.301	2.546	3.759	0.6497	2.599	2.069	2.272	3.253
CH$_3$OH-CCl$_4$	313.2	0.3200	0.5038	1.492	2.215	2.637	4.127	0.5498	1.287	2.041	2.436	3.835
CH$_3$COOCH$_2$ C$_6$H$_5$- C$_4$H$_8$O$_2$	313.2	0.3800	1.190	0.9382	0.828	1.112	2.233	1.221	0.8887	0.844	1.132	2.255
NH$_3$-C$_2$H$_4$	293.2	0.3039	0.7380	1.184	0.069	0.084	0.150	0.7652	1.139	0.071	0.098	0.195
	373.2	0.4828	0.7294	1.222	0.062	0.075	0.114	0.7576	1.178	0.070	0.086	0.140
	473.2	0.4828	0.7220	1.270	0.070	0.096	0.189	0.7505	1.225	0.059	0.094	0.164
	523.2	0.3039	0.7178	1.287	0.057	0.078	0.123	0.7460	1.241	0.064	0.080	0.136
NH$_3$-H$_2$	293.2	0.2975	0.2603	1.964	0.509	0.602	0.927	0.3033	1.661	0.043	0.052	0.084
	306.2	0.3990	0.2189	1.581	1.882	2.225	3.365					
	306.2	0.5360						0.2543	1.334	1.213	1.489	2.569
	327.2	0.3990	0.2274	1.603	1.634	1.874	2.566					
	327.2	0.5360						0.2585	1.323	1.006	1.139	1.456
	371.2	0.3990	0.2389	1.614	2.535	3.032	4.834					
	371.2	0.5360						0.2723	1.336	1.925	2.363	3.947
	373.2	0.2239						0.3289	1.625	0.092	0.109	0.189
	373.2	0.2975	0.2856	1.943	0.513	0.620	0.961					
	421.2	0.1400	0.2498	1.628	1.502	1.874	3.239					
	421.2	0.6005						0.2892	1.369	1.655	1.900	2.546
	473.2	0.2975	0.3132	1.947	0.378	0.438	0.635	0.3585	1.618	0.057	0.103	0.247
	479.2	0.6005	0.2674	1.677	1.260	1.532	2.321	0.3172	1.444	1.469	1.782	2.970
	523.2	0.2975	0.3223	1.946	0.404	0.474	0.666	0.3680	1.613	0.049	0.056	0.080
H$_2$-(C$_2$H$_5$)$_2$O	288.16	0.1330	0.1063	4.654	0.727	1.029	1.455	0.1419	3.617	0.007	0.009	0.013
	373.16	0.1330	0.1156	4.636	0.588	0.831	1.175	0.1523	3.555	0.080	0.114	0.161
	423.15	0.1330	0.1197	4.665	1.055	1.492	2.110	0.1570	3.563	0.413	0.584	0.826
	486.16	0.1330	0.1230	4.646	0.675	0.954	1.350	0.1609	3.532	0.054	0.076	0.107
HCl-H$_2$	294.16	0.8220	0.1920	2.129	1.226	1.843	3.508	0.2452	1.761	0.112	0.131	0.167
	327.16	0.2031	0.2128	2.257	0.508	0.590	0.756					
	327.16	0.5042						0.2532	1.739	0.276	0.509	1.015
	372.16	0.2031						0.2582	1.714	0.099	0.136	0.243
	372.16	0.5042	0.2140	2.193	0.702	0.960	0.769					
	427.16	0.5092	0.2114	2.136	1.509	1.745	2.164	0.2575	1.686	0.888	1.150	1.685
	473.16	0.2409	0.2385	2.292	0.655	0.783	1.131	0.2762	1.719	0.021	0.035	0.068
	523.16	0.5178	0.2386	2.245	0.399	0.471	0.671	0.2861	1.743	0.293	0.378	0.600
SO$_2$-H$_2$	290.16	0.2286	0.1321	2.960	0.875	1.021	1.553	0.1657	2.211	0.149	0.230	0.487
	303.2	0.4059	0.1506	3.237	0.752	1.292	2.734	0.1938	2.481	1.319	2.125	4.525
	318.16	0.2286	0.1364	2.955	1.161	1.405	2.221	0.1702	2.195	0.595	0.795	1.368
	328.2	0.4863	0.1491	3.145	0.730	1.100	2.332	0.1939	2.435	1.222	2.134	4.657
	343.16	0.2366	0.1385	2.921	0.841	0.953	1.466	0.1726	2.166	0.283	0.400	0.858
	365.16	0.2306	0.1414	2.914	0.778	0.894	1.391	0.1752	2.149	0.264	0.375	0.712
	373.2	0.2005	0.1569	3.090	1.821	2.104	2.919	0.1911	2.241	2.789	3.199	4.217
	397.16	0.3265	0.1435	2.889	0.618	0.780	1.310	0.1817	2.178	0.251	0.351	0.634
	423.2	0.5023	0.1661	3.173	0.541	0.665	1.110	0.2119	2.408	0.676	0.795	1.206
	432.16	0.1676	0.1537	3.004	1.028	1.190	1.720	0.1849	2.1520	0.566	0.771	1.484
	472.16	0.3265	0.1526	3.014	0.183	0.288	0.551	0.1920	2.257	0.679	1.041	2.005
	473.2	0.4018	0.1563	2.879	0.866	1.515	3.334					
	473.2	0.6024						0.2010	2.203	0.394	0.485	0.805
NH$_3$-CH$_4$	287.66	0.5000	0.8211	0.9714	0.149	0.181	0.331	0.8250	0.9673	0.144	0.176	0.322
	298.2	0.7000	0.8327	0.9571	0.703	0.826	1.180	0.8368	0.9532	0.702	0.823	1.164
	308.2	0.4060	0.8732	1.006	1.048	1.430	2.763	0.8773	1.001	1.044	1.426	2.755
	353.2	0.5960	0.8556	0.9500	0.232	0.279	0.452	0.8596	0.9459	0.227	0.273	0.447
SO$_2$-CH$_4$	308.2	0.4330	0.4770	1.632	1.104	1.212	1.576	0.5260	1.462	0.940	1.042	1.445
	353.2	0.3920	0.4970	1.635	0.709	1.133	2.796	0.5461	1.460	0.577	1.039	2.648

TABLE 2. RECOMMENDED SETS OF Ψ_{ij} AND L-VALUES FOR THE VISCOSITY DATA (continued)

Gas Pair	Temp. (K)	Mole Fraction of Heavier Component	First Method Ψ_{12}	Ψ_{21}	L_1 (%)	L_2 (%)	L_3 (%)	Second Method Ψ_{12}	Ψ_{21}	L_1 (%)	L_2 (%)	L_3 (%)
NH_3-N_2	293.2	0.4362	0.9284	0.8594	0.154	0.184	0.299	0.9531	0.8188	0.103	0.129	0.218
	297.2	0.2036						0.9962	0.8931	0.819	1.468	3.238
	297.2	0.4973	0.9719	0.9389	0.771	1.508	3.354					
	327.2	0.4973	0.9351	0.9144	0.769	1.156	2.369	0.9615	0.8725	0.783	1.177	2.432
	373.2	0.7993	0.9077	0.9291	0.722	1.070	2.056	0.9353	0.8890	0.698	1.098	2.198
	373.2	0.4362	0.9055	0.9136	0.210	0.245	0.390	0.9311	0.8719	0.166	0.200	0.345
	423.2	0.4080	0.8982	0.9569	0.504	0.609	0.890	0.9245	0.9140	0.529	0.639	0.938
	473.2	0.4362	0.8871	0.9755	0.149	0.169	0.216	0.9137	0.9325	0.115	0.130	0.174
	523.2	0.4362	0.8814	1.000	0.097	0.120	0.169	0.9084	0.9570	0.082	0.094	0.133
	573.2	0.5072	0.9240	1.012	0.627	1.023	2.192	0.9520	0.9677	0.642	1.038	2.222
NH_3-N_2O	298.2	0.5040						0.7693	1.179	0.522	0.589	1.028
	298.2	0.5980	0.7325	1.294	0.481	0.589	1.043					
	308.2	0.3130	0.6695	1.180	0.519	0.596	0.873	0.7077	1.082	0.637	0.727	0.979
	353.2	0.2210	0.6593	1.180	0.212	0.259	0.445					
	353.2	0.3200						0.7002	1.087	0.211	0.293	0.660
SO_2-N_2O	298.2	0.4930	0.7421	1.219	0.420	0.537	1.312	0.7647	1.187	0.426	0.542	1.326
	308.2	0.4760	0.7378	1.244	1.041	1.297	2.072	0.7605	1.212	1.027	1.284	2.056
	353.2	0.4740	0.7429	1.228	0.236	0.313	0.531	0.7655	1.196	0.223	0.300	0.515
NH_3-O_2	293.2	0.5214	0.9272	0.8456	0.142	0.167	0.227	0.9563	0.7935	0.082	0.099	0.137
	373.2	0.5214	0.9146	0.9007	0.114	0.135	0.213	0.9449	0.8466	0.054	0.070	0.119
	473.2	0.5214	0.9033	0.9626	0.082	0.106	0.147	0.9349	0.9064	0.054	0.061	0.081
$NH_3-CH_3NH_2$	273.0	0.5000	0.7065	1.361	0.226	0.277	0.347	0.7401	1.303	0.216	0.265	0.334
	298.0	0.5000	0.7016	1.370	0.160	0.196	0.248	0.7352	1.311	0.149	0.183	0.229
	323.0	0.5000	0.6977	1.378	0.148	0.181	0.238	0.7313	1.320	0.137	0.169	0.225
	348.0	0.5000	0.6954	1.386	0.088	0.109	0.150	0.7290	1.328	0.078	0.097	0.133
	373.0	0.5000	0.6922	1.391	0.043	0.059	0.095	0.7258	1.333	0.034	0.047	0.077
	423.0	0.7500	0.6858	1.397	0.053	0.069	0.107	0.7196	1.340	0.047	0.058	0.071
	473.0	0.2500	0.6848	1.404	0.123	0.199	0.344	0.7181	1.345	0.123	0.211	0.366
	523.0	0.2500	0.6807	1.407	0.042	0.063	0.107	0.7139	1.348	0.047	0.076	0.131
	573.0	0.2500	0.6765	1.408	0.569	0.907	1.565	0.7095	1.349	0.567	0.921	1.591
	623.0	0.5000	0.6736	1.410	0.104	0.167	0.289	0.7069	1.352	0.091	0.153	0.265
	673.0	0.5000	0.6708	1.410	0.081	0.100	0.142	0.7041	1.352	0.066	0.084	0.125
CH_3COOCH_2 C_6H_5-$C_6H_5NH_2$	303.2	0.4950	0.5301	1.627	0.050	0.135	0.542	0.5560	1.589	0.046	0.147	0.573
CH_3COOCH_2 C_6H_5-$CH_3C_6H_4OH$	313.2	0.2720	0.3809	2.417	0.645	0.860	1.325					
	313.2	0.4350						0.3978	2.403	0.613	0.850	1.414
$(CH_3)_2O$-CH_3Cl	308.2	0.5080	1.041	0.9790	0.199	0.247	0.413	1.048	0.9718	0.198	0.246	0.412
	353.2	0.5880	1.040	0.9797	0.091	0.119	0.233	1.047	0.9724	0.091	0.119	0.232
$(CH_3)_2O$-SO_2	308.2	0.4920	0.9937	1.005	0.269	0.308	0.550	1.015	0.9767	0.268	0.307	0.549
	353.2	0.5040	1.010	1.013	0.276	0.373	0.744	1.031	0.9838	0.277	0.376	0.747
CH_3Cl-SO_2	308.2	0.6040	0.9539	1.026	0.387	0.469	0.732	0.9697	1.007	0.384	0.466	0.726
	353.2	0.5890	0.9636	1.026	0.398	0.469	0.738	0.9795	1.006	0.396	0.467	0.737
SO_2-SO_2F_2	273.0	0.5000	0.8301	1.147	0.195	0.250	0.380	0.8575	1.105	0.184	0.237	0.366
	323.0	0.5000	0.8086	1.145	0.180	0.226	0.334	0.8357	1.104	0.163	0.205	0.305
	373.0	0.5000	0.7975	1.148	0.130	0.219	0.380	0.8245	1.107	0.120	0.200	0.347
	423.0	0.5000	0.7094	1.156	0.107	0.179	0.309	0.8174	1.115	0.096	0.159	0.276
	473.0	0.7500	0.7931	1.175	0.122	0.165	0.269	0.8208	1.134	0.115	0.174	0.297
	523.0	0.5000	0.7909	1.180	0.094	0.116	0.145	0.8182	1.139	0.092	0.114	0.157
	573.0	0.5000	0.7939	1.194	0.077	0.095	0.131	0.8213	1.152	0.075	0.092	0.115
	623.0	0.5000	0.7963	1.205	0.083	0.103	0.142	0.8239	1.163	0.082	0.100	0.127
	673.0	0.5000	0.7974	1.212	0.040	0.052	0.082	0.8251	1.169	0.048	0.061	0.092
Air-CO_2	290.0	0.6000	0.7059	1.325	0.037	0.045	0.063	0.7311	1.288	0.024	0.033	0.053
Air-CH_4	293.2	0.5050	0.8956	1.010	0.575	0.691	1.170	0.9267	0.9563	0.591	0.714	1.224
	293.2	0.4410	0.9037	1.007	0.911	1.914	5.827	0.9346	0.9536	0.928	1.912	5.801
	293.2	0.5050	0.8850	1.011	0.916	1.417	3.794	0.9160	0.9579	0.940	1.439	3.825
	293.2	0.6360	0.9007	1.021	0.547	0.738	1.841	0.9325	0.9672	0.574	0.768	1.907
NH_3-Air	288.7	0.3000	0.9083	0.8432	0.442	0.551	1.088	0.9322	0.7992	0.394	0.532	1.233
Air-HCl	289.7	0.6000	0.7109	1.141	0.335	0.369	0.483	0.7252	1.124	0.311	0.342	0.464
	291.3	0.6000	0.7111	1.141	0.490	0.654	1.188	0.7254	1.125	0.465	0.634	1.170
H_2S-Air	290.36	0.8000	0.6766	1.154	0.133	0.145	0.208	0.6871	1.144	0.112	0.129	0.245

References to Text

1. Touloukian, Y. S., Gerritsen, J. K., and Moore, N. Y., *Thermophysical Properties Research Literature Retrieval Guide*, 3 books, Plenum Press, New York, 2936 pp., 1967.
2. Chapman, S. and Cowling, T. G., *The Mathematical Theory of Non-Uniform Gases*, 3rd Edition, prepared in cooperation with D. Burnett, Cambridge University Press, London, 423 pp., 1970.
3. Brush, S. G., "The Development of the Kinetic Theory of Gases. I. Herapath," *Ann. Sci.*, **13**, 188–98, 1957.
4. Brush, S. G., "The Development of the Kinetic Theory of Gases. II. Waterson," *Ann. Sci.*, **13**, 273–82, 1957.
5. Brush, S. G., "The Development of the Kinetic Theory of Gases. III. Clausius," *Ann. Sci.*, **14**, 185–96, 1958.
6. Brush, S. G., "The Development of the Kinetic Theory of Gases. IV. Maxwell," *Ann. Sci.*, **14**, 243–55, 1958.
7. Brush, S. G., "Development of the Kinetic Theory of Gases. V. The Equation of State," *Am. J. Phys.*, **29**, 593–605, 1961.
8. Brush, S. G., "Development of the Kinetic Theory of Gases. VI. Viscosity," *Am. J. Phys.*, **30**, 269–81, 1962.
9. Brush, S. G., "John James Waterston and the Kinetic Theory of Gases," *Am. Sci.*, **49**, 202–14, 1961.
10. Chapman, S., "The Kinetic Theory of Gases Fifty Years Ago," in *Lectures in Theoretical Physics* (Brittin, W. E., Barut, A. O., and Guenin, M., Editors), Vol. IXC, Kinetic Theory, Gordon and Breach, Science Publishers, Inc., New York, 1–13, 1967.
11. Kennard, E. H., *Kinetic Theory of Gases, with an Introduction to Statistical Mechanics*, McGraw-Hill, New York, 483 pp., 1938.
12. Jeans, J. H., *An Introduction to the Kinetic Theory of Gases*, Cambridge University Press, London, 311 pp., 1946.
13. Jeans, J. H., *The Dynamical Theory of Gases*, Dover Publication reprint, 444 pp., 1954.
14. Loeb, L. B., *The Kinetic Theory of Gases*, Dover Publication reprint, 687 pp., 1961.
15. Saha, M. N. and Srivastava, B. N., *A Treatise on Heat, including Kinetic Theory of Gases, Thermodynamics and Recent Advances in Statistical Thermodynamics*, Indian Press, Calcutta, 935 pp., 1950.
16. Present, R. D., *Kinetic Theory of Gases*, McGraw-Hill, New York, 267 pp., 1958.
17. Herzfeld, K. F. and Smallwood, H. M., "The Kinetic Theory of Ideal Gases," Chapter I, *States Matter*, in Vol. II of *Treatise on Physical Chemistry* (Taylor, H. S. and Glasstone, S., Editors), D. Van Nostrand Co., Inc., New York, 1–185, 1951.
18. Cowling, T. G., *Molecules in Motion*, Anchor Press, Tiptree, Essex, 183 pp., 1950.
19. Knudsen, M., *The Kinetic Theory of Gases: Some Modern Aspects*, Methuen, London, 61 pp., 1950.
20. Guggenheim, E. A., "The Kinetic Theory of Gases," in *Elements of the Kinetic Theory of Gases*, Topic 6 of Vol. 1, *The International Encyclopedia of Physical Chemistry and Chemical Physics*, Pergamon Press, Oxford, 92 pp., 1960.
21. Kauzmann, W., *Kinetic Theory of Gases*, Vol. 1 of *Thermal Properties of Matter*, Benjamin, New York, 248 pp., 1966.
22. Golden, S., *Elements of the Theory of Gases*, Addison-Wesley Publishing Co., Reading, Mass., 154 pp., 1964.
23. Desloge, E. A. and Matthysse, S. W., "Collision Term in the Boltzmann Transport Equation," *Am. J. Phys.*, **28**, 1–11, 1960.
24. Desloge, E. A., "Fokker–Planck Equation," *Am. J. Phys.*, **31**, 237–46, 1963.
25. Desloge, E. A., "Coefficients of Diffusion, Viscosity, and Thermal Conductivity of a Gas," *Am. J. Phys.*, **30**, 911–20, 1962.
26. Desloge, E. A., "Transport Properties of a Simple Gas," *Am. J. Phys.*, **32**, 733–42, 1964.
27. Desloge, E. A., "Transport Properties of a Gas Mixture," *Am. J. Phys.*, **32**, 742–8, 1964.
28. Hirschfelder, J. O., Curtiss, C. F., and Bird, R. B., *Molecular Theory of Gases and Liquids*, John Wiley, & Sons, New York, 1219 pp., 1954; reprinted with Notes added, 1249 pp., 1964.
29. Mintzer, D., "Transport Properties of Gases," Chapter 1 in *The Mathematics of Physics and Chemistry* (Margenau, H. and Murphy, G. S., Editors), 2nd Edition, D. Van Nostrand Co., New York, 49 pp., 1956.
30. Mazo, R. M., "Transport Phenomena," in *Statistical Mechanical Theories of Transport Processes*, Topic 9 of Vol. 1, *The International Encyclopedia of Physical Chemistry and Chemical Physics*, Pergamon Press, Oxford, 166 pp., 1967.
31. Liboff, R. L., *Introduction to the Theory of Kinetic Equations*, John Wiley & Sons, New York, 397 pp., 1969.
32. Cercignani, C., *Mathematical Methods in Kinetic Theory*, Plenum Press, New York, 227 pp., 1969.
33. Waldmann, L., *Statistical Mechanics of Equilibrium and Non-Equilibrium* (Meixner, J., Editor), North-Holland Publishing Co., Amsterdam, 117 pp., 1965.
34. Waldmann, L., *Transporterscheinungen in Gasen von Mittlerem Druck* (Flugge, S., Editor), Handbuch der Physik, Springer-Verlag, Berlin, Band 12, 1958.
35. Hochstim, A. R., Editor, *Kinetic Processes in Gases and Plasmas*, Academic Press, New York, 458 pp., 1969.
36. DeGroot, S. R., *Thermodynamics of Irreversible Processes*, North-Holland Publishing Co., Amsterdam, 242 pp., 1952.
37. Prigogine, I., *Non-Equilibrium Statistical Mechanics*, Interscience Publishers, Inc., New York, 319 pp., 1962.
38. Prigogine, I., Resibois, P., and Severne, G., "Irreversible Processes in Dilute Monatomic Gases," in *Proc. International Seminar on the Transport Properties of Gases*, Brown University, Providence, Rhode Island, 7–38, 1964.
39. Montgomery, D., "The Foundations of Classical Kinetic Theory," in *Lectures in Theoretical Physics* (Brittin, W. E.,

Barut, A. O., and Guenin, M., Editors), Vol. IX C, *Kinetic Theory*, Gordon and Breach, Science Publishers, Inc., New York, 791 pp., 1967.

40. Monchick, L., "Equivalence of the Chapman–Enskog and the Mean-Free-Path Theory of Gases," *Phys. Fluids*, **11**, 1393–8, 1962.

41. Monchick, L. and Mason, E. A., "Free-Flight Theory of Gas Mixtures," *Phys. Fluids*, **10**, 1377–90, 1967.

42. Touloukian, Y. S., Liley, P. E., and Saxena, S. C., *Thermal Conductivity—Nonmetallic Liquids and Gases*, Vol. 3 of *Thermophysical Properties of Matter* (The TPRC Data Series), IFI/Plenum Data Corp., New York, 707 pp., 1970.

43. Liley, P. E., "Survey of Recent Work on the Viscosity, Thermal Conductivity and Diffusion of Gases and Gas Mixtures," in *Thermodynamic and Transport Properties of Gases, Liquids and Solids*, Symposium ASME, New York, 40–69, 1959.

44. Liley, P. E., "Review of Work on the Transport Properties of Gases and Gas Mixtures," Purdue University, TPRC Report 10, 57 pp., 1959.

45. Liley, P. E., "Review of Work on the Transport Properties of Gases and Gas Mixtures," Supplement 1, Purdue University, TPRC Report 12, 14 pp., 1961.

46. Liley, P. E., "Survey of Recent Work on the Viscosity, Thermal Conductivity and Diffusion of Gases and Liquefied Gases Below 500 K," Purdue University, TPRC Report 13, 33 pp., 1961.

47. Sutherland, W., "The Viscosity of Mixed Gases," *Phil. Mag.*, **40**, 421–31, 1895.

48. Kirkwood, J. G., "The Statistical Mechanical Theory of Transport Processes. I. General Theory," *J. Chem. Phys.*, **14**, 180–201, 1946.

49. Kirkwood, J. G., "The Statistical Mechanical Theory of Transport Processes. II. Transport in Gases," *J. Chem. Phys.*, **15**, 72–6, 1947.

50. Grad, H., "Singular and Nonuniform Limits of Solutions of the Boltzmann Equation," Vol. I of *SIAM-AMS Proceedings: Transport Theory*, 269–308, 1967.

51. Grad, H., "Accuracy and Limits of Applicability of Solutions of Equations of Transport: Dilute Monatomic Gases," in *Proc. International Seminar on the Transport Properties of Gases*, Brown University, Providence, R. I., 39–57, 1964.

52. Kumar, K., "Polynomial Expansions in Kinetic Theory of Gases," *Ann. Phys.*, **37**, 113–41, 1966.

53. Kumar, K., "The Chapman–Enskog Solution of the Boltzmann Equation: A Reformation in Terms of Irreducible Tensors and Matrices," *Aust. J. Phys.*, **20**, 205–52, 1967.

54. Green, M. S., "Markoff Random Processes and the Statistical Mechanics of Time-Dependent Phenomena," *J. Chem. Phys.*, **20**, 1281–95, 1952.

55. Green, M. S., "Boltzmann Equation from the Statistical Mechanical Point of View," *J. Chem. Phys.*, **25**, 836–55, 1956.

56. Green, M. S., "The Non-Equilibrium Pair Distributing Function at Low Densities," *Physica*, **29**, 393–403, 1958.

57. Green, M. S. and Piccirelli, R. A., "Basis of the Functional Assumption in the Theory of the Boltzmann Equation," *Phys. Rev.*, **132**, 1388–410, 1963.

58. Hoffmann, D. K. and Green, H. S., "On a Reduction of Liouville's Equation to Boltzmann's Equation," *J. Chem. Phys.*, **43**, 4007–16, 1965.

59. Snider, R. F., "Variational Methods for Solving the Boltzmann Equation," *J. Chem. Phys.*, **41**, 591–5, 1964.

60. Mazur, P. and Biel, J., "On the Derivation of the Boltzmann Equation," *Physica*, **32**, 1633–48, 1966.

61. Su, C. H., "Kinetic Equation of Classical Boltzmann Gases," *Phys. Fluids*, **7**, 1248–55, 1964.

62. McLennan, J. A., "Convergence of the Chapman–Enskog Expansion for the Linearized Boltzmann Equation," *Phys. Fluids*, **8**, 1580–4, 1965.

63. García-Colin, L. S., Green, M. S., and Chaos, F., "The Chapman–Enskog Solution of the Generalized Boltzmann Equation," *Physica*, **32**, 450–78, 1966.

64. Fujita, S., "Boltzmann Equation Approach to Transport Phenomena," in *Lectures in Theoretical Physics* (Britten, W. E., Barut, A. O., and Guenin, M., Editors), Vol. IX C, *Kinetic Theory*, Gordon and Breach, Science Publishers, Inc., New York, 231–63, 1967.

65. Bogoliubov, N. N., "Problems of a Dynamical Theory in Statistical Physics," English translation by Gora, E. K., in *Studies in Statistical Mechanics*, Vol. I (de Boer, J. and Uhlenbeck, E. K., Editors), North-Holland Publishing Co., Amsterdam, 131 pp., 1962.

66. Bogoliubov, N. N., "Kinetic Equations," *J. Phys. (USSR)*, **10**, 265–74, 1946.

67. Desai, R. C. and Ross, J., "Solutions of Boltzmann Equation and Transport Processes," *J. Chem. Phys.*, **49**, 3754–64, 1968.

68. Montroll, E. W. and Green, M. S., "Statistical Mechanics of Transport and Nonequilibrium Processes," *Ann. Rev. Phys. Chem.*, **5**, 449–76, 1954.

69. Grad, H., "On the Kinetic Theory of Rarefied Gases," *Commun. Pure Appl. Math.*, **2**, 331–407, 1949.

70. Grad, H., "Asymptotic Theory of the Boltzmann Equation," *Phys. Fluids*, **6**, 147–81, 1963.

71. Grad, H., "Statistical Mechanics, Thermodynamics, and Fluid Mechanics of Systems with an Arbitrary Number of Integrals," *Commun. Pure Appl. Math.*, **5**, 455–94, 1952.

72. Zwanzig, R., "Time-Correlation Functions and Transport Coefficients in Statistical Mechanics," *Ann. Rev. Phys. Chem.*, **16**, 67–102, 1965.

73. Bhatnagar, P. L., Gross, E. P., and Krook, M., "A Model for Collision Processes in Gases. I. Small Amplitude Processes in Charged and Neutral One-Component Systems," *Phys. Rev.*, **94**, 511–25, 1954.

74. Welander, P., "On the Temperature Jump in a Rarefied Gas," *Ark. Fys.*, **7**(44), 507–53, 1954.

75. Gross, E. P. and Krook, M., "Model for Collision Processes in Gases: Small-Amplitude Oscillations of Charged Two-Component Systems," *Phys. Rev.*, **102**, 593–604, 1956.

76. Gross, E. P. and Jackson, E. A., "Kinetic Models and the Linearized Boltzmann Equation," *Phys. Fluids*, **2**, 432–41, 1959.

77. Sirovich, L., "Kinetic Modeling of Gas Mixtures," *Phys. Fluids*, **5**, 908–18, 1962.

78. Enoch, J., "Kinetic Model for High Velocity Ratio Near Free Molecular Flow," *Phys. Fluids*, **5**, 913–24, 1962.

79. Hamel, B. B., "Kinetic Model for Binary Gas Mixtures," *Phys. Fluids*, **8**, 418–25, 1965.

80. Willis, D. R., "Comparison of Kinetic Theory Analysis of Linearized Conette Flow," *Phys. Fluids*, **5**, 127–35, 1962.

81. Holway, L. H., "New Statistical Models for Kinetic Theory: Methods of Construction," *Phys. Fluids*, **9**, 1958–73, 1966.

82. Mason, E. A., "Transport in Neutral Gases," Chapter 3 of *Kinetic Pressures in Gases and Plasmas* (Hochstim, A. R., Editor), Academic Press, New York, 57–100, 1969.

83. Kihara, T., *Imperfect Gases*, Asakura Press, Tokyo, 334 pp., 1949; English translation by the United States Office of Air Research, Wright–Patterson Air Force Base.

84. Kihara, T., "Virial Coefficients and Models of Molecules in Gases," *Rev. Mod. Phys.*, **25**, 831–43, 1953.

85. Joshi, R. K., "The Rigid Sphere Perturbation Procedure for Theoretical Evaluation of the Transport Coefficients of Pure Gases," *Chem. Phys. Letters*, **1**, 575–8, 1968.

86. Joshi, R. K., "Self-Consistent Approximation Procedure for Theoretical Evaluation of Transport Properties of Binary Gas Mixtures," *Indian J. Pure Appl. Phys.*, **7**, 381–4, 1969.

87. Saxena, S. C., "On the Two Schemes of Approximating the Transport Coefficients (Chapman–Cowling and Kihara), *J. Phys. Soc.* (Japan), **11**, 367–9, 1956.

88. Curtiss, C. F. and Hirschfelder, J. O., "Transport Properties of Multicomponent Gas Mixtures," *J. Chem. Phys.*, **17**, 550–5, 1949.

89. Saxena, S. C. and Joshi, R. K., "Evaluation of the Determinant Elements Occurring in the Second Approximation to the Viscosity of Binary Gas Mixtures," *Physica*, **29**, 870–2, 1963.

90. Saxena, S. C. and Joshi, R. K., "The Chapman–Cowling Second Approximation to the Viscosity Coefficient of Binary Gas Mixtures," *Indian J. Phys.*, **37**, 479–85, 1963.

91. Joshi, R. K., "The Chapman–Cowling Third Approximation to the Viscosity Coefficient of Binary Gas Mixtures," *Phys. Letters*, **15**, 32–4, 1965.

92. Mason, E. A., "Higher Approximations for the Transport Properties of Binary Gas Mixtures. I. General Formulas," *J. Chem. Phys.*, **27**, 58–84, 1957.

93. Joshi, R. K. and Saxena, S. C., "A Second Approximation Formula for the Viscosity Coefficient of Binary Gas Mixtures," *Physica*, **31**, 762–3, 1965.

94. Waldmann, L., "Remarks on the Transport Properties of Gaseous Isobar Mixtures," *Physica*, **30**, 914–20, 1964.

95. Hirschfelder, J. O., Taylor, M. H., Kihara, T., and Rutherford, R., "Viscosity of Two-Component Gaseous Mixtures," *Phys. Fluids*, **4**, 663–8, 1961.

96. Curtiss, C. F., "Kinetic Theory of Nonspherical Molecules," *J. Chem. Phys.*, **24**, 225–41, 1956.

97. Curtiss, C. F. and Muckenfuss, C., "Kinetic Theory of Non-spherical Molecules. II," *J. Chem. Phys.*, **26**, 1619; *Ibid*, **36**, 1957.

98. Muckenfuss, C. and Curtiss, C. F., "Kinetic Theory of Non-spherical Molecules. III," *J. Chem. Phys.*, **29**, 1257–72, 1958.

99. Livingston, P. M. and Curtiss, C. F., "Kinetic Theory of Non-spherical Molecules. IV. Angular Momentum Transport Coefficient," *J. Chem. Phys.*, **31**, 1643–5, 1959.

100. Curtiss, C. F. and Dahler, J. S., "Kinetic Theory of Non-spherical Molecules. V," *J. Chem. Phys.*, **38**, 2352–62, 1963.

101. Sandler, S. I. and Dahler, J. S., "Transport Properties of Polyatomic Fluids. II. A Dilute Gas of Spherocylinders," *J. Chem. Phys.*, **44**, 1229–37, 1966.

102. Kagan, Yu. and Afana'sev, A. M., "On the Kinetic Theory of Gases with Rotational Degrees of Freedom," *Zh. Eksp. Teor. Fiz.*, **41**, 1536–45, 1961; English translation: *Sov. Phys.—JETP*, **14**, 1096–101, 1962.

103. Dahler, J. S. and Sather, N. F., "Kinetic Theory of Loaded Spheres. I," *J. Chem. Phys.*, **38**, 2363–82, 1963.

104. Sandler, S. I. and Dahler, J. S., "Kinetic Theory of Loaded Spheres. II," *J. Chem. Phys.*, **43**, 1750–9, 1965.

105. Pidduck, F. B., "The Kinetic Theory of a Special Type of Rigid Molecules," *Proc. Roy. Soc.* (*London*), **A101**, 101–12, 1922.

106. Condiff, D. W., Lu, W.-K., and Dahler, J. S., "Transport Properties of Polyatomic Fluids, a Dilute Gas of Perfectly Rough Spheres," *J. Chem. Phys.*, **42**, 3445–75, 1965.

107. McLaughlin, I. L. and Dahler, J. S., "Transport Properties of Polyatomic Fluids. III. The Transport–Relaxation Equations for a Dilute Gas of Rough Spheres," *J. Chem. Phys.*, **44**, 4453–9, 1966.

108. Waldmann, L., "Kinetische Theorie des Lorentz-Gases ans Rotierenden Molekillen," *Z. Naturforsch*, **18a**, 1033–48, 1963.

109. Dahler, J. S., "Introductory Comments on the Theory of Transport in Polyatomic Fluids," in *Proc. International Seminar on the Transport Properties of Gases*, Brown University, Providence, R. I., 85–96, 1964.

110. Pople, J. A., "The Statistical Mechanics of Assemblies of Axially Symmetric Molecules. I. General Theory," *Proc. Roy. Soc.* (*London*), **221A**, 498–507, 1954.

111. Pople, J. A., "The Statistical Mechanics of Assemblies of Axially Symmetric Molecules. II. Second Virial Coefficients," *Proc. Roy. Soc.* (*London*), **221A**, 508–16, 1954.

112. Castle, B. J., Jansen, L., and Dawson, J. M., "On the Second Virial Coefficients for Assemblies of Nonspherical Molecules," *J. Chem. Phys.*, **24**, 1078–83, 1956.

113. Saksena, M. P. and Saxena, S. C., "Second Virial Coefficient of Non-Polar Non-Spherical Molecules," *Phys. Letters*, **18**, 120–2, 1965.

114. McCourt, F. R. and Snider, R. F., "Thermal Conductivity of a Gas with Rotational States," *J. Chem. Phys.*, **41**, 3185–94, 1964.

115. McCourt, F. R. and Snider, R. F., "Transport Properties of Gases with Rotational States. II," *J. Chem. Phys.*, **43**, 2276–83, 1965.

116. Kagan, Yu. and Maksimov, L. A., "Transport Phenomena in a Paramagnetic Gas," *Zh. Eksp. Theor. Fiz.*, **41**, 842–52, 1961; English translation: *Sov. Phys.—JETP*, **14**, 604–10, 1962.

117. Dahler, J. S., "Transport Phenomena in a Fluid Composed of Diatomic Molecules," *J. Chem. Phys.*, **30**, 1447–75, 1959.

118. Brout, R., "Rotational Energy Transfer in Diatomic Molecules," *J. Chem. Phys.*, **22**, 1189–90, 1954.

119. Belov, V. A. and Dubner, V. M., "Angular Distribution and Transport Cross Sections," *Teplofiz. Vys. Temp.*, **4**, 872–7, 1966; English translation: *High Temp.*, **4**, 806–7, 1966.

120. O'Toole, J. T. and Dahler, J. S., "Molecular Friction in Dilute Gases," *J. Chem. Phys.*, **33**, 1496–1504, 1960.

121. Sather, N. F. and Dahler, J. S., "Molecular Friction in Dilute Gases. II. Thermal Relaxation of Translational and Rotational Degrees of Freedom," *J. Chem. Phys.*, **35**, 2029–37, 1961.

122. Sather, N. F. and Dahler, J. S., "Molecular Friction in Dilute Gases. III. Rotational Relaxation in Polyatomic Fluids," *J. Chem. Phys.*, **37**, p. 1947, 1962.

123. Bjerre, A., "Kinetic Theory of Nonspherical Molecules," *J. Chem. Phys.*, **48**, 3540–4, 1968.

124. Morse, T. F., "Kinetic Model for Gases with Internal Degrees of Freedom," *Phys. Fluids*, **7**, 159–69, 1964.

125. Brau, C., "Kinetic Theory of Polyatomic Gases: Models for the Collision Processes," *Phys. Fluids*, **10**, 48–55, 1967.

126. Gioumousis, G. and Curtiss, C. F., "Molecular Collisions. I. Formal Theory and the Pauli Principle," *J. Chem. Phys.*, **29**, 996–1001, 1958.

127. Gioumousis, G. and Curtiss, C. F., "Molecular Collisions. II. Diatomic Molecules," *J. Math. Phys.*, **2**, 96–104, 1961.

128. Gioumousis, G., "Molecular Collisions. III. Symmetric Top Molecules," *J. Math. Phys.*, **2**, 723–7, 1961.

129. Gioumousis, G. and Curtiss, C. F., "Molecular Collisions. IV. Nearly Spherical Rigid Body Approximation," *J. Math. Phys.*, **3**, 1059–72, 1962.

130. Curtiss, C. F. and Hardisson, A., "Molecular Collisions. V. Nearly Spherical Potentials," *J. Chem. Phys.*, **46**, 2618–33, 1967.

131. Curtiss, C. F., "Molecular Collisions. VI. Diagrammatic Methods," *J. Chem. Phys.*, **48**, 1725–31, 1968.

132. Biolsi, L. and Curtiss, C. F., "Molecular Collisions. VII. Nuclear Spin and Statistics Effects," *J. Chem. Phys.*, **48**, 4508–16, 1968.

133. Curtiss, C. F., "Molecular Collisions. VIII.," *J. Chem. Phys.*, **49**, 1952–7, 1968.

134. Curtiss, C. F. and Bernstein, R. B., "Molecular Collisions. IX. Restricted Distorted-Wave Approximation for Rotational Excitation and Scattering of Diatomic Molecules," *J. Chem. Phys.*, **50**, 1168–76, 1969.

135. Fenstermaker, R. W., Curtiss, C. F., and Bernstein, R. B., "Molecular Collisions. X. Restricted-Distorted-Wave-Born and First-Order Sudden Approximations for Rotational Excitation of Diatomic Molecules," *J. Chem. Phys.*, **51**, 2439–48, 1969.

136. Wang, C. C. S. and Uhlenbeck, G. E., "On the Transport Phenomena in Rarefied Gases," University of Michigan, Ann Arbor, Mich., Report No. CM-443, Feb. 20, 1948.

137. Wang, C. C. S. and Uhlenbeck, G. E., "Transport Phenomena in Polyatomic Gases," University of Michigan, Ann Arbor, Mich., Report No. CM-681, July 10, 1951.

138. Wang, C. C. S., Uhlenbeck, G. E., and de Boer, J., "The Heat Conductivity and Viscosity of Polyatomic Gases," Part C, Vol. II of *Studies in Statistical Mechanics* (de Boer, J. and Uhlenbeck, G. E., Editors), North-Holland Publishing Co., Amsterdam, 243–68, 1964.

139. Finkelstein, L. and Harris, S., "Kernel of the Linearized Wang Chang–Uhlenbeck Collision Operator," *Phys. Fluids*, **9**, 8–11, 1966.

140. Finkelstein, L., "Structure of Boltzmann Collision Operator," *Phys. Fluids*, **8**, 431–6, 1965.

141. Hanson, F. B. and Morse, T. F., "Kinetic Models for a Gas with Internal Structure," *Phys. Fluids*, **10**, 345–53, 1967.

142. Taxman, N., "Classical Theory of Transport Phenomena in Dilute Polyatomic Gases," *Phys. Rev.*, **110**, 1235–9, 1958.

143. Mason, E. A. and Monchick, L., "Heat Conductivity of Polyatomic and Polar Gases," *J. Chem. Phys.*, **36**, 1622–39, 1962.

144. Monchick, L., Yun, K. S., and Mason, E. A., "Relaxation Effects in the Transport Properties of a Gas of Rough Spheres," *J. Chem. Phys.*, **38**, 1282–7, 1963.

145. Monchick, L., Yun, K. S., and Mason, E. A., "Formal Kinetic Theory of Transport Phenomena in Polyatomic Gas Mixtures," *J. Chem. Phys.*, **39**, 654–69, 1963.

146. Monchick, L., Pereira, A. N. G., and Mason, E. A., "Heat Conductivity of Polyatomic and Polar Gases and Gas Mixtures," *J. Chem. Phys.*, **42**, 3241–56, 1965.

147. Monchick, L., Munn, R. J., and Mason, E. A., "Thermal Diffusion in Polyatomic Gases: A Generalized Stefan-Maxwell Diffusion Equation," *J. Chem. Phys.*, **45**, 3051–8, 1966.

148. Monchick, L., Sandler, S. I., and Mason, E. A., "Thermal Diffusion in Polyatomic Gases: Non-Spherical Interactions," *J. Chem. Phys.*, **49**, 1178–84, 1968.

149. Sandler, S. I. and Dahler, J. S., "Kinetic Theory of Loaded Spheres. IV. Thermal Diffusion in a Dilute-Gas Mixture of D_2 and HT," *J. Chem. Phys.*, **47**, 2621–30, 1967.

150. Sandler, S. I. and Mason, E. A., "Thermal Diffusion in a Loaded Sphere-Smooth Sphere Mixture: A Model for ⁴He–HT and ³He–HD," *J. Chem. Phys.*, **47**, 4653–8, 1967.

151. Alievskii, M. Ya and Zhdanov, V. M., "Transport and Relaxation Phenomena in Polyatomic Gas Mixtures," *Soviet Phys.—JETP*, **28**, 116–21, 1969.

152. Zhdanov, V. M., "The Kinetic Theory of a Polyatomic Gas," *Soviet Phys.—JETP*, **26** 1187–91, 1968.

153. Grad, H., "Note on N-Dimensional Hermite Polynomials," *Commun. Pure Appl. Math.*, **2**, 325–30, 1949.

154. Zhdanov, V. M., Kagan, Yu., and Sazykin, A., "Effect of Viscous Transfer of Momentum on Diffusion in a Gas Mixture," *Sov. Phys.—JETP*, **15**, 596–602, 1962.

155. Waldmann, L. and Trübenbacher, E., "Formale Kinetische Theorie von Gasgemischen aus Arregbaren Molekülen," *Z. Naturforsc.*, **17a**, 363–76, 1962.

156. Buckingham, A. D. and Pople, J. A., "The Statistical Mechanics of Imperfect Polar Gases, Part I. Second Virial Coefficients," *Trans. Faraday Soc.*, **51**, 1173–9, 1955.

157. Buckingham, A. D. and Pople, J. A., "The Statistical Mechanics of Imperfect Polar Gases, Part 2. Dielectric Polarization," *Trans. Faraday Soc.*, **51**, 1179–83, 1955.

158. Saxena, S. C. and Joshi, K. M., "Second Virial Coefficient of Polar Gases," *Phys. Fluids*, **5**, 1217–22, 1962.

159. Krieger, F. J., "The Viscosity of Polar Gases," Rand Corporation, Santa Monica, California, Research Memorandum RM-646, 20 pp., 1951.

160. Stockmayer, W. H., "Second Virial Coefficient of Polar Gases," *J. Chem. Phys.*, **9**, 398–402, 1941.

161. Stockmayer, W. H., "Second Virial Coefficients of Polar Gas Mixtures," *J. Chem. Phys.*, **9**, 863–70, 1941.

162. Joshi, K. M. and Saxena, S. C., "Viscosity of Polar Gases," *Physica*, **27**(3), 329–36; *Ibid.* **27**(12), p. 1101, 1961.

163. Liley, P. E., "Collision Integrals for the Viscosity of Polar Gases," *J. Chem. Eng. Data*, **5**, 307–8, 1960.

164. Itean, E. C., Glueck, A. R., and Svehla, R. A., "Collision Integrals for a Modified Stockmayer Potential," NASA Technical Note D-481, 29 pp., 1961.

165. Monchick, L. and Mason, E. A., "Transport Properties of Polar Gases," *J. Chem. Phys.*, **35**, 1676–97, 1961.

166. Hornig, J. F. and Hirschfelder, J. O., "Concept of Inter-molecular Forces in Collisions," *Phys. Rev.*, **103**, 908–17, 1956.

167. Mason, E. A., Vanderslice, J. T., and Yos, J. M., "Transport Properties of High-Temperature Multicomponent Gas Mixtures," *Phys. Fluids*, **2**, 688–94, 1959.

168. Mason, E. A. and Monchick, L., "Transport Properties of Polar-Gas Mixtures," *J. Chem. Phys.*, **36**, 2746–57, 1962.

169. Mott, N. F. and Massey, H. S., *The Theory of Atomic Collisions*, Clarendon Press, Oxford, 388 pp., 1949.

170. Bernstein, R. B., "Quantum Effects in Elastic Molecular Scattering," in *Molecular Beams*, Vol. X of *Advances in Chemical Phys.* (Ross, J., Editor), Interscience Publishers, New York, 75–134, 1966.

171. Uehling, E. A. and Uhlenbeck, G. E., "Transport Phenomena in Einstein–Bose and Fermi–Dirac Gases," *Phys. Rev.*, **43**, 552–61, 1933.

172. Buckingham, R. A. and Gal, E., "Applications of Quantum Theory to the Viscosity of Dilute Gases," in *Advances in Atomic and Molecular Physics*, Vol. 4 (Bates, D. R. and Estermann, Editors), Academic Press, New York, 37–61, 1968.

173. de Boer, J., "Transport-Properties of Gaseous Helium at Low Temperatures," Chapter 18 of Vol. I in *Progress in Low Temperature Physics* (Gorter, C. J., Editor), North-Holland Publishing Co., Amsterdam, 381–406, 1955.

174. Mori, H., Oppenheim, I., and Ross, J., "Some Topics in Quantum Statistics: The Wigner Function and Transport Theory," Part C of *Studies in Statistical Mechanics*, Vol. I (de Boer, J. and Uhlenbeck, G. E., Editors), North-Holland Publishing Co., Amsterdam, 217–98, 1962.

175. Waldman, L., "The Boltzmann Equation–for Gases from Spin Particles," *A. Naturforsch.*, **13a**, 609–20, 1958.

176. Snider, R. F., "Quantum-Mechanical Modified Boltzmann Equation for Degenerate Internal States," *J. Chem. Phys.*, **32**, 1051–60, 1960.

177. Snider, R. F., "Perturbation Variation Methods for a Quantum Boltzmann Equation," *J. Math. Phys.*, **5**, 1580–7, 1964.

178. Hoffman, D. K., Mueller, J. J., and Curtiss, C. F., "Quantum-Mechanical Boltzmann Equation," *J. Chem. Phys.*, **43**, 2878–84, 1965.

179. Hoffman, D. K., "On a Derivation of a Quantum-Mechanical Linearized Boltzmann Equation," *J. Chem. Phys.*, **44**, 2644–51, 1966.

180. Prigogine, I. and Résibois, P., "On the Approach to Equilibrium of a Quantum Gas," *Physica*, **24**, 705–816, 1958.

181. Prigogine, I. and Ono, S., "On the Transport Equation in Quantum Gases," *Physica*, **25**, 171–8, 1959.

182. Prigogine, I. and Balescu, R., "Irreversible Processes in Gases. I. The Diagram Technique," *Physica*, **25**, 281–301, 1959; "II. The Equation of Evolution," *Physica*, **25**, 302–23, 1959.

183. Mueller, J. J. and Curtiss, C. F., "Quantum-Mechanical Kinetic Theory of Loaded Spheres," *J. Chem. Phys.*, **46**, 283–302, 1967.

184. Mueller, J. J. and Curtiss, C. F., "Quantum-Mechanical Kinetic Theory of Loaded Spheres. II. The Classical Limit," *J. Chem. Phys.*, **46**, 1252–64, 1967.

185. de Boer, J. and Bird, R. B., "Quantum Corrections to Transport Properties at High Temperatures," *Phys. Rev.*, **83**, 1259–60, 1951.

186. de Boer, J. and Bird, R. B., "Quantum Corrections to the Transport Coefficients of Gases at High Temperatures," *Physica*, **20**, 185–98, 1954.

187. Saxena, S. C., Kelley, J. G., and Watson, W. W., "Temperature Dependence of the Thermal Diffusion Factor for Helium, Neon and Argon," *Phys. Fluids*, **4**, 1216–25, 1961.

188. Choi, S. and Ross, J., "Quantum Corrections for Transport Coefficients," *J. Chem. Phys.*, **33**, 1324–31, 1960.

189. Buckingham, R. A. and Corner, J., "Tables of Second Virial and Low-Pressure Joule–Thomson Coefficients for Intermolecular Potentials with Exponential Repulsion," *Proc. Roy. Soc.*, **A189**, 118–29, 1948.

190. Imam-Rahajoe, S., Curtiss, C. F., and Bernstein, R. B., "Numerical Evaluation of Quantum Effects on Transport Cross Sections," *J. Chem. Phys.*, **42**, 530–6, 1965.

191. Munn, R. J., Smith, F. J., Mason, E. A., and Monchick, L., "Transport Collision Integrals for Quantum Gases Obeying a 12-6 Potential," *J. Chem. Phys.*, **42**, 537–9, 1965.

192. Curtiss, C. F. and Power, R. S., "An Expansion of Binary Collision Phase Shifts in Powers of h," *J. Chem. Phys.*, **40**, 2145–50, 1964.

193. Wood, H. T. and Curtiss, C. F., "Quantum Corrections to the Transport Cross Sections," *J. Chem. Phys.*, **41**, 1167–73, 1964.

194. Munn, R. J., Mason, E. A., and Smith, F. J., "Some Aspects of the Quantal and Semiclassical Calculation of Phase Shifts and Cross Sections for Molecular Scattering and Transport," *J. Chem. Phys.*, **41**, 3978–88, 1964; Erratum Ibid., **43**, 2158, 1965.

195. Smith, F. J., Mason, E. A., and Vanderslice, J. T., "Higher-Order Stationary-Phase Approximations in Semiclassical Scattering," *J. Chem. Phys.*, **42**, 3257–64, 1965.

196. Burnstein, R. B., Curtiss, C. F., Imam-Rahajoe, S., and Wood, H. T., "Numerical Evaluation of Barrier Penetration and Resonance Effects on Phase Shifts," *J. Chem. Phys.*, **44**, 4072–81, 1966.

197. Massey, H. S. W. and Mohr, C. B. O., "Free Paths and Transport Phenomena in Gases and the Quantum Theory of Collisions. I. The Rigid Sphere Model," *Proc. Roy. Soc. (London)*, **A141**, 434–53, 1933.

198. Massey, H. S. W. and Mohr, C. B. O., "Free Paths and Transport Phenomena in Gases and the Quantum Theory of Collisions. II. The Determination of the Laws of Force Between Atoms and Molecules," *Proc. Roy. Soc. (London)*, **A144**, 188–205, 1934.

199. Massey, H. S. W. and Buckingham, R. A., "The Low-Temperature Properties of Gaseous Helium," *Proc. Roy. Soc. (London)*, **A168**, 378–89, 1938.

200. Massey, H. S. W. and Buckingham, R. A., "The Low Temperature Properties of Gaseous Helium, Errata," *Proc. Roy. Soc. (London)*, **A169**, 205, 1938.

201. Slater, J. C. and Kirkwood, J. G., "The van der Waals Forces in Gases," *Phys. Rev.*, **37**, 682–97, 1931.

202. Buckingham, R. A., Hamilton, J., and Massey, H. S. W., "The Low-Temperature Properties of Gaseous Helium. II," *Proc. Roy. Soc. (London)*, **A179**, 103–22, 1941.

203. de Boer, J., "Transport Phenomena of Gaseous He at Very Low Temperatures," *Physica*, **10**, 348–56, 1943.

204. Keller, W. E., "Calculation of the Viscosity of Gaseous He^3 and He^4 at Low Temperatures," *Phys. Rev.*, **105**, 41–5, 1957.

205. Monchick, L., Mason, E. A., Munn, R. J., and Smith, F. J., "Transport Properties of Gaseous He^3 and He^4," *Phys. Rev.*, **139**, A1076–82, 1965.

206. Larsen, S. Y., Witte, K., and Kilpatrick, J. E., "On the Quantum-Mechanical Pair-Correlation Function of He^4 at Low Temperatures," *J. Chem. Phys.*, **44**, 213–20, 1966.

207. de Boer, J. and Michels, A., "Quantum-Mechanical Calculation of the Second-Virial Coefficient of Helium at Low Temperatures," *Physica*, **6**, 409–20, 1939.

208. Mason, E. A. and Rice, W. E., "The Intermolecular Potentials of Helium and Hydrogen," *J. Chem. Phys.*, **22**, 522–35, 1954.

209. Buckingham, R. A. and Temperley, H. N. V., "The Viscosity of Liquid He^3," *Phys. Rev.*, **78**, 482, 1950.

210. de Boer, J. and Cohen, E. G. D., "The Viscosity of Gaseous He^3 at Low Temperatures," *Physica*, **17**, 993–1000, 1951.

211. Buckingham, R. A. and Seriven, R. A., "Diffusion in Gaseous Helium at Low Temperatures," *Proc. Phys. Soc. (London)*, **65A**, 376–7, 1952.

212. Cohen, E. G. D., Offerhaus, M. J., and de Boer, J., "The Transport Properties and Equation of State of Gaseous Mixtures of the Helium Isotopes," *Physica*, **20**, 501–15, 1954.

213. Halpern, O. and Buckingham, R. A., "Symmetry Effects in Gas Kinetics. I. The Helium Isotopes," *Phys. Rev.*, **98**, 1626–31, 1955.

214. Larsen, D. M., "Binary Mixtures of Dilute Bose Gases with Repulsive Interactions at Low Temperatures," *Ann. Phys.*, **24**, 89–101, 1963.

215. Cohen, E. G. D., Offerhaus, M. J., van Leeuwen, J. M. J., Roos, B. W., and de Boer, J., "The Transport Properties and the Equation of State of Gaseous Para- and Ortho-Hydrogen and Their Mixtures Below 40 K," *Physica*, **22**, 791–815, 1956.

216. Buckingham, R. A., Davies, A. R., and Gilles, D. C., "Symmetry Effects in Gas Kinetics. II. Ortho- and Para-hydrogen," *Proc. Phys. Soc. (London)*, **71**, 457–69, 1958.

217. Takayanagi, K. and Ohno, K., "Collisions Between Non-Spherical Molecules. I. Molecular Collisions in Hydrogen Gas at Lower Temperatures," *Prog. Theor. Phys. (Kyoto)*, **13**, 243–59, 1955.

218. Niblett, P. D. and Takayanagi, K., "The Calculation of Some Properties of Hydrogen Gas at Low Temperatures," *Proc. Roy. Soc. (London)*, **A250**, 224–47, 1959.

219. Waldmann, L., "The Basic Kinetic Equations for Para-Ortho-Hydrogen Mixtures," *Physica*, **30**, 17–37, 1964.

220. Hartland, A. and Lipsicas, M., "Quantum Symmetry Effects in Hydrogen Gas," *Phys. Letters*, **3**, 212–3, 1963.

221. Diller, D. E. and Mason, E. A., "Low-Temperature Transport Properties of Gaseous H_2, D_2, and HD," *J. Chem. Phys.*, **44**, 2604–9, 1966.

222. Buckingham, R. A. and Fox, J. W., "The Coefficient of Viscosity of Atomic Hydrogen from 25 to 300 K," *Proc. Roy. Soc. (London)*, **A267**, 102–18, 1962.

223. Buckingham, R. A., Fox, J. W., and Gal, E., "The Coefficients of Viscosity and Thermal Conductivity of Atomic Hydrogen from 1 to 400 K," *Proc. Roy. Soc. (London)*, **A284**, 237–51, 1965.

224. Browning, R. and Fox, J. W., "The Coefficient of Viscosity of Atomic Hydrogen and the Coefficient of Mutual Diffusion for Atomic and Molecular Hydrogen," *Proc. Roy. Soc. (London)*, **A278**, 274–86, 1964.

225. Konowalow, D. D., Hirschfelder, J. O., and Linder, B., "Low-Temperature, Low-Pressure Transport Coefficients for Gaseous Oxygen and Sulfur Atoms," *J. Chem. Phys.*, **31**, 1575–9, 1959.

226. Knuth, E. L. and Fisher, S. S., "Low-Temperature Viscosity Cross Sections Measured on a Supersonic Argon Beam," *J. Chem. Phys.*, **48**, 1674–84, 1968.

227. Spitzer, L., *Physics of Fully Ionized Gases*, Interscience Publishers, New York, 1962.

228. Ahtye, W. F., "A Critical Evaluation of Methods for Calculating Transport Coefficients of a Partially Ionized Gas," in *Proc. Heat Transfer and Fluid Mechanics Institute* (Giedt, W. H. and Levy, S., Editors), Stanford University Press, 211–25, 1964.

229. Ahtye, W. F., "A Critical Evaluation of Methods for Calculating Transport Coefficients of Partially and Fully Ionized Gases," NASA TN D-2611, 110 pp., 1965.

230. Brokaw, R. S., "Transport Properties of High Temperature Gases," NASA TM X-52315, 15 pp. and 12 figures, 1967.

231. Amdur, I. and Mason, E. A., "Properties of Gases at Very High Temperatures," *Phys. Fluids*, **1**, 370–83, 1958.

232. Kamnev, A. B. and Leonas, V. B., "Kinetic Coefficients for Inert Gases at High Temperatures," *Teplofiz. Vys. Temp.*, **4**, 288–9, 1966.

233. Balyaev, Y. N. and Leonas, B. V., "Kinetic Coefficients of Molecular Oxygen and Nitrogen," *Teplofiz. Vys. Temp.*, **4**, 732–3, 1966; English translation: *High Temp.*, **4**, 686, 1966.

234. Amdur, I. and Ross, J., "On the Calculation of Properties of Gases at Elevated Temperatures," *Combust. Flame*, **2**, 412–20, 1958.

235. Amdur, I., "An Experimental Approach to the Determination of Gaseous Transport Properties at Very High Temperatures," in *Proceedings of the Conference on Physical Chemistry in Aerodynamics and Space Flight* (Greenberg, M., Editor), Pergamon Press, New York, Vol. 3, 228–35, 1961.

236. Amdur, I., "High Temperature Transport Properties of Gases; Limitations of Current Calculating Methods in the Light of Recent Experimental Data," *Am. Inst. Chem. Eng. J.*, **8**, 521–6, 1962.

237. Brokaw, R. S., "Energy Transport in High Temperature and Reacting Gases," in *Proceedings of the Conference on Physical Chemistry in Aerodynamics and Space Flight* (Greenberg, M., Editor), Pergamon Press, New York, Vol. 3, 238–52, 1961.

238. Yos, J. M., "Transport Properties of Nitrogen, Hydrogen, Oxygen, and Air to 30,000 K," AVCO Technical Memorandum RAD-TM-63-7, 65 pp., 1963.

239. Yos, J. M., "Revised Transport Properties for High Temperature Air and its Contents," AVCO Technical Release, 50 pp., 28 Nov. 1967.

240. Spitzer, L. and Harm, R., "Transport Phenomena in a Completely Ionized Gas," *Phys. Rev.*, **89**, 977–81, 1953.

241. Vanderslice, J. T., Weissman, S., Mason, E. A., and Fallon, R. J., "High-Temperature Transport Properties of Dissociating Hydrogen," *Phys. Fluids*, **5**, 155–64, 1962.

242. Grier, N. T., "Calculation of Transport Properties and Heat-Transfer Parameters of Dissociating Hydrogen," NASA TN D-1406, 64 pp., 1962.

243. Yun, K. S., Weissman, S., and Mason, E. A., "High-Temperature Transport Properties of Dissociating Nitrogen and Dissociating Oxygen," *Phys. Fluids*, **5**, 672–8, 1962.

244. Bade, W. L., Mason, E. A., and Yun, K. S., "Transport Properties of Dissociated Air," *J. Am. Rocket Soc.*, **31**, 1151–3, 1961.

245. Yun, K. S. and Mason, E. A., "Collision Integrals for the Transport Properties of Dissociating Air at High Temperatures," *Phys. Fluids*, **5**, 380–6, 1962.

246. Krupenie, P. H., Mason, E. A., and Vanderslice, J. T., "Interaction Energies and Transport Coefficients of Li + H and O + H Gas Mixtures at High Temperatures," *J. Chem. Phys.*, **39**, 2399–408, 1963.

247. Belov, V. A. and Klyuchnikov, N. I., "Collision Integrals for the LiH System Viscosity of an LiH Mixture," *Teplofiz. Vys. Temp.*, **3**, 645–8, 1965; English translation: *High Temp.*, **3**, 594–7, 1965.

248. Davies, R. H., Mason, E. A., and Munn, R. J., "High-Temperature Transport Properties of Alkali Meta Vapors," *Phys. Fluids*, **8**, 444–52, 1965.

249. Tchen, C. M., "Kinetic Equations for Fully Ionized Plasmas," Chapter IV, in *Kinetic Processes in Gases and Plasmas* (Hochstim, A. R., Editor), Academic Press, New York, 101–14, 1969.

250. Lewis, M. B., "The Boltzmann and Fokker-Planck Equations," Chapter V, in *Kinetic Processes in Gases and Plasmas* (Hochstim, A. R., Editor), Academic Press, New York, 115–39, 1969.

251. Hochstim, A. R. and Massel, G. A., "Calculations of Transport Coefficients in Ionized Gases," Chapter VI, in *Kinetic*

Processes in Gases and Plasmas (Hochstim, A. R., Editor), Academic Press, New York, 141–255, 1969.

252. Devoto, R. S., "Transport Properties of Ionized Monatomic Gases," *Phys. Fluids*, 9, 1230–40, 1966.

253. Devoto, R. S., "Third Approximation to the Viscosity of Multicomponent Mixtures," *Phys. Fluids*, 10, 2704–6, 1967.

254. Devoto, R. S., "Transport Coefficients of Partially Ionized Argon," *Phys. Fluids*, 10, 354–64, 1967.

255. Li, C. P. and Devoto, R. S., "Fifth and Sixth Approximations to the Electron Transport Coefficients," *Phys. Fluids*, 11, 448–50, 1968.

256. Devoto, R. S., "Simplified Expressions for the Transport Properties of Ionized Monatomic Gases," *Phys. Fluids*, 10, 2105–12, 1967.

257. Mason, E. A. and Sherman, M. P., "Effect of Resonant Charge Exchange on Heat Conduction in Plasmas," *Phys. Fluids*, 9, 1989–91, 1966.

258. Chmieleski, R. M. and Ferziger, J. H., "Transport Properties of a Non-equilibrium Partially Ionized Gas," *Phys. Fluids*, 10, 364–71, 1967.

259. Sandler, S. I. and Mason, E. A., "Transport Properties of Almost-Lorentzian Mixtures," *Phys. Fluids*, 12, 71–7, 1969.

260. Meador, W. E., "A Semiempirical Collision Model for Plasmas," *NASA TR R-310*, 32 pp., 1969.

261. Devoto, R. S. and Li, C. P., "Transport Coefficients of Partially Ionized Helium," *J. Plasma Phys.*, 2, 17–32, 1968.

262. Kulik, P. P., Panevin, I. G., and Khvesyuk, V. I., "Theoretical Calculation of the Viscosity, Thermal Conductivity and Prandtl Number for Argon in the Presence of Ionization," *Teplofiz. Vys. Temp.*, 1, 56–63, 1963; English translation: *High Temp.*, 1, 45–51, 1963.

263. Devoto, R. S., "Transport Coefficients of Partially Ionized Krypton and Xenon," *AIAA J.*, 7, 199–204, 1969.

264. Devoto, R. S., "Transport Coefficients of Partially Ionized Hydrogen," *J. Plasma Phys.*, 2, 617–31, 1968.

265. Devoto, R. S., "Comments on Transport Properties of Hydrogen," *AIAA J.*, 4, 1149–50, 1966.

266. Grier, N. T., "Calculation of Transport Properties of Ionizing Atomic Hydrogen," *NASA TN D-3186*, 85 pp., 1966.

267. Mason, E. A., Munn, R. J., and Smith, F. J., "Transport Coefficients of Ionized Gases," *Phys. Fluids*, 10, 1827–32, 1967.

268. Smith, F. J., Mason, E. A., and Munn, R. J., "Classical Collision Integrals for the Repulsive Screened Coulomb Potential," *Phys. Fluids*, 8, 1907–8, 1965.

269. Beshinske, R. J. and Curtiss, C. F., "A Statistical Derivation of the Hydrodynamic Equations of Change for a System of Ionized Molecules. I. General Equations of Change and the Maxwell Equations," *J. Statistical Phys.*, 1, 163–74, 1969.

270. Dalgarno, A. and Smith, F. J., "The Viscosity and Thermal Conductivity of Atomic Hydrogen," *Proc. Roy. Soc. (London)*, A267, 417–23, 1962.

271. Dalgarno, A., "Transport Properties of Atomic Hydrogen," in *Proc. Conference on Phys. Chem. in Aerodynamics and Space Flight* (Greenberg, M., Editor), Pergamon Press, New York, Vol. 3, 236–7, 1961.

272. Everhart, E., Stone, G., and Carbone, R. J., "Classical Calculation of Differential Cross Section for Scattering from a Coulomb Potential with Exponential Screening," *Phys. Rev.*, 99, 1287–90, 1955.

273. Lane, G. H. and Everhart, E., "Calculations of Total Cross Sections for Scattering from Coulomb Potentials with Exponential Screening," *Phys. Rev.*, 117, 920–4, 1960.

274. Curtiss, C. F., "Transport Phenomena in Gases," *Ann Rev. Phys. Chem.*, 18, 125–34, 1967.

275. Cohen, E. G. D., "Transport Phenomena in Dense Gases," in *Proceedings of the International Seminar on the Transport Properties of Gases*, Brown University, Providence, R. I., 125–42, 1964.

276. Cohen, E. G. D., "Kinetic Theory of Dense Gases," in *Lectures in Theoretical Physics*, Vol. IX C, *Kinetic Theory* (Brittin, W. E., Barut, A. O., and Guenin, M., Editors), Gordon and Breach, Science Publishers, Inc., New York, 791 pp., 1967.

277. Cohen, E. G. D., "The Kinetic Theory of Dense Gases," in *Fundamental Problems in Statistical Mechanics II* (Cohen, E. G. D., Editor), North-Holland Publishing Co., Amsterdam, 228–75, 1968.

278. Ernst, M. H., Haines, L. K., and Dorfman, J. R., "Theory of Transport Coefficients for Moderately Dense Gases," *Rev. Mod. Phys.*, 41, 296–316, 1969.

279. Ernst, M. H., "Transport Coefficients from Time Correlation Functions," in *Lectures in Theoretical Physics*, Vol. IX C, *Kinetic Theory* (Brittin, W. E., Barut, A. O., and Guenin, M., Editors), Gordon and Breach, Science Publishers, Inc., New York, 791 pp., 1967.

280. Cohen, E. G. D., "On the Connection Between Various Derivations of the Boltzmann Equation," *Physica*, 27, 163–84, 1961.

281. Sengers, J. V. and Cohen, E. G. D., "Statistical Mechanical Derivation of the Generalized Boltzmann Equation for a Fluid Consisting of Rigid Spherical Molecules," *Physica*, 27, 230–44, 1961.

282. Cohen, E. G. D., "On the Generalization of the Boltzmann Equation to General Order in the Density," *Physica*, 28, 1025–44, 1962.

283. Cohen, E. G. D., "Cluster Expansions and the Hierarchy I. Nonequilibrium Distribution Functions," *Physica*, 28, 1045–59, 1962.

284. Cohen, E. G. D., "Cluster Expansions and the Hierarchy II. Equilibrium Distribution Functions," *Physica*, 28, 1060–73, 1962.

285. Cohen, E. G. D., "On the Kinetic Theory of Dense Gases," *J. Math. Phys.*, 4, 183–9, 1963.

286. Piccirelli, R. A., "Some Properties of the Long-Time Values of the Probability Densities for Moderately Dense Gases," *J. Math. Phys.*, 7, 922–34, 1966.

287. García-Colin, L. S., "A Theory of the Hydrodynamical State for Dense Gases," in *Lectures in Theoretical Physics*, Vol. IX C, *Kinetic Theory* (Brittin, W. E., Barut, A. O., and Guenin, M., Editors), Gordon and Breach, Science Publishers, Inc., New York, 791 pp., 1967.

288. García-Colin, L. S. and Green, M. S., "Definition of Temperature in the Kinetic Theory of Dense Gases," *Phys. Rev.*, 150, 153–8, 1966.

289. Ernst, M. H., "Transport Coefficients and Temperature Definition," *Physica*, 32, 252–72, 1966.

290. Dymond, J. H. and Alder, B. J., "Van der Waals Theory of Transport in Dense Fluids," *J. Chem. Phys.*, 45, 2061–8, 1966.

291. Longuet-Higgins, H. C. and Pople, J. A., "Transport Properties of a Dense Fluid of Hard Spheres," *J. Chem. Phys.*, 25, 884–9, 1956.

292. Longuet-Higgins, H. C. and Valleau, J. P., "Transport of Energy and Momentum in a Dense Fluid of Hard Spheres," *Faraday Soc. Discuss.*, **22**, 47–53, 1956.

293. Valleau, J. P., "Transport of Energy and Momentum in a Dense Fluid of Rough Spheres," *Mol. Phys.*, **1**, 63–7, 1958.

294. Longuet-Higgins, H. C. and Valleau, J. P., "Transport Coefficients of Dense Fluids of Molecules Interacting According to a Square Well Potential," *Mol. Phys.*, **1**, 284–94, 1958.

295. Valleau, J. P., "Transport in Dense Square-Well Fluid Mixtures," *J. Chem. Phys.*, **44**, 2626–32, 1966.

296. Naghizadeh, J., "Transport in a Two Component Square-Well Fluid," *J. Chem. Phys.*, **39**, 3406–11, 1963.

297. McLaughlin, I. L. and Davis, H. T., "Kinetic Theory of Dense Fluid Mixtures. I. Square-Well Model," *J. Chem. Phys.*, **45**, 2020–31, 1966.

298. McCoy, B. J., Sandler, S. I., and Dahler, J. S., "Transport Properties of Polyatomic Fluids. IV. The Kinetic Theory of a Dense Gas of Perfectly Rough Spheres," *J. Chem. Phys.*, **45**, 3485–512, 1966.

299. Sandler, S. I. and Dahler, J. S., "Kinetic Theory of Loaded Spheres. III. Transport Coefficients for the Dense Gas," *J. Chem. Phys.*, **46**, 3520–31, 1967.

300. Sather, N. F. and Dahler, J. S., "Approximate Theory of Viscosity and Thermal Conductivity in Dense Polyatomic Fluids," *Phys. Fluids*, **5**, 754–68, 1962.

301. O'Toole, J. T. and Dahler, J. S., "On the Kinetic Theory of a Fluid Composed of Rigid Spheres," *J. Chem. Phys.*, **32**, 1097–106, 1960.

302. Livingston, P. M. and Curtiss, C. F., "Kinetic Theory of Moderately Dense, Rigid-Sphere Gases," *Phys. Fluids*, **4**, 816–32, 1961.

303. Ono, S. and Shizume, T., "Statistical Mechanics of Transport Phenomena in Gases at Moderate Densities," *J. Phys. Soc. (Japan)*, **18**, 29–54, 1963.

304. Snider, R. F. and Curtiss, C. F., "Kinetic Theory of Moderately Dense Gases," *Phys. Fluids*, **1**, 122–38, 1958.

305. Irving, J. H. and Kirkwood, J. G., "The Statistical Mechanical Theory of Transport Processes. IV. The Equation of Hydrodynamics," *J. Chem. Phys.*, **18**, 817–29, 1950.

306. Reference withdrawn.

307. Snider, R. F. and McCourt, F. R., "Kinetic Theory of Moderately Dense Gases: Inverse Power Potentials," *Phys. Fluids*, **6**, 1020–5, 1963.

308. Curtiss, C. F., McElroy, M. B., and Hoffman, D. K., "The Transport Properties of a Moderately Dense Lennard-Jones Gas," *Int. J. Eng. Sci.*, **3**, 269–83, 1965.

309. García-Colin, L. S. and Flores, A., "On the Transport Coefficients of Moderately Dense Gases," *Physica*, **32**, 289–303, 1966.

310. García-Colin, L. S. and Flores, A., "The Generalization of Choh-Uhlenbeck's Method in the Kinetic Theory of Dense Gases," *J. Math. Phys.*, **7**, 254–9, 1966.

311. Reference withdrawn.

312. Stogryn, D. E. and Hirschfelder, J. O., "Initial Pressure Dependence of Thermal Conductivity and Viscosity," *J. Chem. Phys.*, **31**, 1545–54, 1959.

313. Hill, T. L., "Molecular Cluster in Imperfect Gases," *J. Chem. Phys.*, **23**, 617–22, 1955.

314. Hill, T. L., *Statistical Mechanics*, Chapter 5, McGraw-Hill Book Co., Inc., New York, 432 pp., 1956.

315. Stogryn, D. E. and Hirschfelder, J. O., "Contribution of Bound, Metastable and Free Molecules to the Second Virial Coefficient and Some Properties of Double Molecules," *J. Chem. Phys.*, **31**, 1531–45, 1959.

316. Barua, A. K. and Das Gupta, A., "Pressure Dependence of the Viscosity of Superheater Steam," *Trans. Faraday Soc.*, **59**(490), 2243–7, 1963.

317. Das Gupta, A. and Barua, A. K., "Calculation of the Viscosity of Ammonia at Elevated Pressures," *J. Chem. Phys.*, **42**, 2849–51, 1965.

318. Pal, A. K. and Barua, A. K., "Effect of Cluster Formation on the Viscosity of Dense Gases," *Indian J. Phys.*, **41**(5), 323–6, 1967.

319. Singh, Y., Deb, S. K., and Barua, A. K., "Dimerization and the Initial Pressure Dependence of the Viscosity of Polar Gases," *J. Chem. Phys.*, **46**, 4036–40, 1967.

320. Pal, A. K. and Barua, A. K., "Viscosity of Some Quadrupolar Gases and Vapors," *J. Chem. Phys.*, **48**, 872–4, 1968.

321. Singh, Y. and Bhattacharyya, P. K., "Thermal Conductivity and Viscosity of Moderately Dense Quadrupolar Gases," *J. Phys. B (Proc. Phys. Soc.)*, **1**(2), 922–8, 1968.

322. Singh, Y. and Das Gupta, A., "Formations of Dimers in Quadrupolar Gases," *J. Phys. B (Proc. Phys. Soc.)*, **1**(2), 914–21, 1968.

323. Singh, Y. and Manna, A., "Thermal Conductivity and Viscosity of Moderately Dense Dipolar Gases," *J. Phys. B (Atom. Mol. Phys.)*, **2**(2), 294–302, 1969.

324. Singh, Y., Deb, S. K., and Barua, A. K., "Dimerization and the Initial Pressure Dependence of the Viscosity of Polar Gases," *J. Chem. Phys.*, **46**(10), 4036–40, 1967.

325. Kim, S. K. and Ross, J., "Viscosity of Moderately Dense Gases," *J. Chem. Phys.*, **42**, 263–71, 1965.

326. Hollinger, H. B. and Curtiss, C. F., "Kinetic Theory of Dense Gases," *J. Chem. Phys.*, **33**, 1386–1402, 1960.

327. Hollinger, H. B., "Molecular Chaos and the Boltzmann Equation," *J. Chem. Phys.*, **36**, 3208–20, 1962.

328. Hoffman, D. K. and Curtiss, C. F., "Kinetic Theory of Dense Gases. III. The Generalized Enskog Equation," *Phys. Fluids*, **7**, 1887–97, 1964.

329. Hoffman, D. K. and Curtiss, C. F., "Kinetic Theory of Dense Gases. IV. Transport Virial Coefficients," *Phys. Fluids*, **8**, 667–82, 1965.

330. Hoffman, D. K. and Curtiss, C. F., "Kinetic Theory of Dense Gases. V. Evaluation of the Second Transport Virial Coefficients," *Phys. Fluids*, **8**, 890–5, 1965.

331. Bennett, D. E. and Curtiss, C. F., "Density Effects on the Transport Coefficients of Gaseous Mixtures," *J. Chem. Phys.*, **51**, 2811–25, 1969.

332. Sengers, J. V., "Density Expansion of the Viscosity of a Moderately Dense Gas," *Phys. Rev. Letters*, **15**, 515–7, 1965.

333. Sengers, J. V., "Triple Collision Contribution to the Transport Coefficients of a Rigid Sphere Gas," *Phys. Fluids*, **9**, 1333–47, 1966.

334. Sengers, J. V., "Divergence in the Density Expansion of the Transport Coefficients of a Two-Dimensional Gas," *Phys. Fluids*, **9**, 1685–96, 1966.

335. Sengers, J. V., "Triple Collision Effects in the Thermal Conductivity and Viscosity of Moderately Dense Gases," *AEDC-TR-69-68*, 156 pp., 1969.

336. Sengers, J. V., "Triple Collision Contributions to the Transport Coefficients of Gases," in *Lectures in Theoretical Physics*, Vol. IX C, *Kinetic Theory* (Brittin, W. E., Barut, A. O., and Guenin, M., Editors), Gordon and Breach, Science Publishers, Inc., New York, 791 pp., 1967.

337. Dorfman, J. R. and Cohen, E. G. D., "On the Density Expansion of the Pair Distribution Function for a Dense Gas Not in Equilibrium," *Phys. Letters*, **16**, 124–5, 1965.

338. Dorfman, J. R. and Cohen, E. G. D., "Difficulties in the Kinetic Theory of Dense Gases," *J. Math. Phys.*, **8**, 282–97, 1967.

339. Dorfman, J. R., "The Binary Collision Expansion Method in Kinetic Theory," in *Lectures in Theoretical Physics*, Vol. IX C, *Kinetic Theory* (Brittin, W. E., Barut, A. O., and Guenin, M., Editors), Gordon and Breach, Science Publishers, Inc., New York, 791 pp., 1967.

340. Dorfman, J. R., "Transport Coefficients for Dense Gases," in *Dynamics of Fluids and Plasmas*, Academic Press, Inc., New York, 199–212, 1966.

341. Stecki, J., "On the Divergence of Ternary Scattering Operator in Two Dimensions," *Phys. Letters*, **19**, 123–4, 1965.

342. Andrews, F. C., "On the Solution of the BBGKY Equations for a Dense Classical Gas," *J. Math. Phys.*, **6**, 1496–1505, 1965.

343. Andrews, F. C., "On the Validity of the Density Expansion Solution of the BBGKY Equations," *Phys. Letters*, **21**, 170–1, 1966.

344. Fujita, S., "On the Nonpower Density Expansion of Transport Coefficients," *Proc. Natl. Acad. Sci. (USA)*, **56**, 794–800, 1966.

345. Fujita, S., "Does a Logarithmic Term Exist in the Density Expansion of a Transport Coefficient," *Phys. Letters*, **24A**, 235–6, 1967.

346. Sengers, J. V., "Thermal Conductivity and Viscosity of Simple Fluids," *Int. J. Heat Mass Transfer*, **8**, 1103–16, 1965.

347. Sengers, J. V., "Transport Properties of Compressed Gases," in *Recent Advances in Engineering Science* (Eringen, A. C., Editor), Gordon and Breach, Science Publishers, Inc., New York, Vol. 3, 153–96, 1968.

348. Hanley, H. J. M., McCarty, R. D., and Sengers, J. V., "Density Dependence of Experimental Transport Coefficients of Gases," *J. Chem. Phys.*, **50**, 857–70, 1969.

349. Iman-Rahajoe, S. and Curtiss, C. F., "Collisional Transfer Contributions in the Quantum Theory of Transport Co-efficients," *J. Chem. Phys.*, **47**, 5269–89, 1967.

350. Grossmann, S., "Occupation Number Representation with Localized One-Particle Functions (Macroscopic Description of Quantum Gases I)," *Physica*, **29**, 1373–92, 1963.

351. Grossmann, S., "Macroscopic Time Evolution and In-homogeneous Master-equation (Macroscopic Description of Quantum Gases)," *Physica*, **30**, 779–807, 1964.

352. Grossmann, S., "On Transport Theory in Real Gases," *Nuovo Cimento*, **37**, 698–713, 1965.

353. Grossmann, S., "Transport Coefficients in Moderately Dense Gases," *Z. Naturforsch.*, **20a**, 861–9, 1965.

354. Baerwinkel, K. and Grossmann, S., "On the Derivation of the Boltzmann-Landau Equation from the Quantum Mechanical Hierarchy," *Z. Phys.*, **198**, 277–87, 1967.

355. Fujita, S., "Generalized Boltzmann Equation for a Quantum Gas Obeying Classical Statistics," *J. Math. Phys.*, **7**, 1004–8, 1965.

356. Fujita, S., "Connected-Diagram Expansion of Transport Coefficients. II. Quantum Gas Obeying Boltzmann Statistics," *Proc. Natl. Acad. Sci. (USA)*, **56**, 16–21, 1966.

357. Morita, T., "Derivation of the Generalized Boltzmann Equation in Quantum Statistical Mechanics," *J. Math. Phys.*, **7**, 1039–45, 1966.

358. Helfand, E., "The Correlation Function Method," in *Proceedings of the International Seminar on the Transport Properties of Gases*, Brown University, Providence, R. I., 143–67, 1964.

359. Kawasaki, K. and Oppenheim, I., "Triple Collision Operators in the Transport Theory of Dense Gases," *Phys. Letters*, **11**, 124–6, 1964.

360. Kawasaki, K. and Oppenheim, I., "Correlation-Function Method for the Transport Coefficients of Dense Gases. I. First Density Correction to the Shear Viscosity," *Phys. Rev.*, **136**, A1519–34, 1964.

361. Kawasaki, K. and Oppenheim, I., "Correlation-Function Method for the Transport Coefficients of Dense Gases. II. First Density Correction to the Shear Viscosity for Systems with Attractive Forces," *Phys. Rev.*, **139**, 649–63, 1965.

362. Kawasaki, K. and Oppenheim, I., "Logarithmic Term in the Density Expansion of Transport Coefficients," *Phys. Rev.*, **139**, 1763–8, 1965.

363. Frisch, H. L. and Berne, B., "High-Temperature Expansion of Thermal Transport Coefficients," *J. Chem. Phys.*, **43**, 250–6, 1965.

364. Storer, R. G. and Frisch, H. L., "Transport Coefficients for Systems with Steep Intermolecular Potentials," *J. Chem. Phys.*, **43**, 4539–40, 1965.

365. Prigogine, I., "Transport Processes, Correlation Functions, and Reciprocity Relations in Dense Media," in *Liquids: Structure. Properties Solid Interactions* (Hughel, T. J., Editor), Elsevier Publishing Co., Amsterdam, 384 pp., 1965.

366. Cohen, E. G. D., Dorfman, J. R., and Ernst, M. H. J. J., "Transport Coefficients from Correlation Functions and Distribution Functions," *Phys. Letters*, **12**, 319–20, 1964.

367. Ernst, M. H., Dorfman, J. R., and Cohen, E. G. D., "Transport Coefficients in Dense Gases. I. The Dilute and Moderately Dense Gas," *Physica*, **31**, 493–521, 1965.

368. Ernst, M. H., "Formal Theory of Transport Coefficients to General Order in the Density," *Physica*, **32**(2), 209–43, 1966.

369. Ernst, M. H., "Hard Sphere Transport Coefficients from Time Correlation Functions," *Physica*, **32**(2), 273–88, 1966.

370. Certain, P. R. and Bruch, L. W., "Intermolecular Forces," *MTP International Review of Science, Physical Chemistry Series One*. Vol. 1 of *Theoretical Chemistry* (Buckingham, A. D., Consultant Editor; Brown, W. B., Vol. Editor), Butterworth and Co., Publishers, 113–65, 1972.

371. Zwanzig, R., "Method for Funding the Density Expansion of Transport Coefficients of Gases," *Phys. Rev.*, **129**, 486–94, 1963.

372. Zwanzig, R., "Elementary Derivation of Time-Correlation Formulas for Transport Coefficients," *J. Chem. Phys.*, **40**, 2527–33, 1964.

373. Weinstock, J., "Cluster Formulation of the Exact Equation for the Evolution of a Classical Many-Body System," *Phys. Rev.*, **132**, 454–69, 1963.

374. Weinstock, J., "Generalized Master Equation for Quantum-Mechanical Systems to All Orders in Density," *Phys. Rev.*, **136**, A879–88, 1964.

375. Weinstock, J., "Nonanalyticity of Transport Coefficients and the Complete Density Expansion of Momentum Correlation Functions," *Phys. Rev.*, **140**, A460–5, 1965.

376. Weinstock, J., "Divergence in the Density Expansion of Quantum-Mechanical Transport Coefficients," *Phys. Rev. Letters*, **17**, 130–2, 1966.

377. Weinstock, J., "Density Expansion of Quantum Mechanical Transport Coefficients," in *Lectures in Theoretical Physics*, Vol. IX C, *Kinetic Theory* (Brittin, W. E., Barut, A. O., and Guenin, M., Editors), Gordon and Breach, Science Publishers, Inc., New York, 791 pp., 1967.

378. Williams, R. H. and Weinstock, J., "Failure of the Weak Coupling Model in the Transport Theory of Dense Real Gases," *Phys. Rev.*, **169**, 196–9, 1968.

379. Goldman, R., "Higher Order Behavior in the Boltzmann Expansion of the Bogoliubov–Born–Green–Kirkwood–Yvon Hierarchy," *Phys. Rev. Letters*, **17**, 910–2, 1966.

380. García-Colin, L. S. and Flores, A., "Note on the Transport Coefficients of a Moderately Dense Gas," *Physica*, **32**, 444–9, 1966.

381. Chao, F and García-Colin, L. S., "Density Expansions of the Transport Coefficients for a Moderately Dense Gas," *Phys. Fluids*, **9**, 382–9, 1966.

382. Stecki, J. and Taylor, H. S., "On the Areas of Equivalence of the Bogoliubov Theory and the Prigogine Theory of Irreversible Processes in Classical Gases," *Rev. Mod. Phys.*, **37**, 762–73, 1965.

383. Prigogine, I. and Resibois, P., "On the Kinetics of the Approach to Equilibrium," *Physica*, **27**, 629–46, 1961.

384. Beenakker, J. J. M., "The Influence of Electric and Magnetic Fields on the Transport Properties of Polyatomic Dilute Gases," *Festkörperprobleme VIII*, 276 pp., 1968.

385. Resibois, P., "Structure of the Three-Particle Scattering Operator in Classical Gases," *J. Math. Phys.*, **4**, 166–73, 1963.

386. Resibois, P., "On the Asymptotic Form of the Transport Equation in Dense Homogeneous Gases," *Phys. Letters*, **9**, 139–41, 1964.

387. Reisbois, P., "On the Connection Between the Kinetic Approach and the Correlation-Function Method for Thermal Transport Coefficients," *J. Chem. Phys.*, **41**, 2979–92, 1964.

388. Brocas, J. and Resibois, P., "On the Equivalence Between the Master Equation and the Functional Approaches to the Generalized Transport Equation," *Physica*, **32**, 1050–64, 1966.

389. Nicolis, G. and Severne, G., "Nonstationary Contributions to the Bulk Viscosity and Other Transport Coefficients," *J. Chem. Phys.*, **44**, 1477–86, 1966.

390. Senftleben, H., "Influence of a Magnetic Field on the Thermal Conductivity of a Paramagnetic Gas," *Phys. Z.*, **31**, 961–3, 1930.

391. Trautz, M. and Fröschel, E., "Note on the Influence of a Magnetic Field on the Viscosity of O_2," *Phys. Z.*, **33**, 947, 1932.

392. Engelhardt, H. and Sach, H., "The Influence of a Magnetic Field on the Viscosity of O_2," *Phys. Z.*, **33**, 724–7, 1932.

393. Senftleben, H. and Gladisch, H., "The Influence of Magnetic Fields on the Viscosity of Gases," *Ann. Phys.*, **30**, 713–27, 1937.

394. Senftleben, H. and Gladisch, H., "The Effect of Magnetic Fields on the Internal Viscosity of Gases (Investigation of Nictric Acid)," *Ann. Phys.*, **33**, 471–6, 1938.

395. Senftleben, H. and Pietzner, J., "The Effect of Magnetic Fields on the Heat Conduction of (Paramagnetic) Gases," *Ann. Phys.*, **16**, 907–29, 1933.

396. Senftleben, H. and Pietzner, J., "The Influence of Magnetic Fields on the Thermal Conductivity of Gases, Part I," *Ann. Phys.*, **27**, 108–16, 1936.

397. Senftleben, H. and Pietzner, J., "The Influence of Magnetic Fields on the Thermal Conductivity of Gases, Part III," *Ann. Phys.*, **27**, 117–22, 1936.

398. Senftleben, H. and Pietzner, J., "The Effect of Magnetic Fields on the Thermal Conductivity of Gases. IV. Mixtures of Oxygen with Diamagnetic Gases," *Ann. Phys.*, **30**, 541–54, 1937.

399. Reiger, E., "The Influence of Magnetic Fields on the Thermal Conductivity of Gases (Temperature Dependence)," *Ann. Phys.*, **31**, 453–72, 1938.

400. Torwegge, H., "Action of Magnetic Fields on the Thermal Conductivity Power of NO and NO_2," *Ann. Phys.*, **33**, 459–70, 1938.

401. Gorter, C. J., "The Interpretation of the Senftleben Effect," *Naturwissenschaften*, **26**, p. 140, 1938.

402. Zernike, F. and Van Lier, C., "Theory of the Senftleben Effect," *Physica*, **6**, 961–71, 1939.

403. Beenakker, J. J. M., Scoles, G., Knaap, H. F. P., and Jonkman, R. M., "The Influence of a Magnetic Field on the Transport Properties of Diatomic Molecules in the Gaseous State," *Phys. Letters*, **2**, 5–6, 1962.

404. Korving, J., Hulsman, H., Knaap, H. F. P., and Beenakker, J. J. M., "The Influence of a Magnetic Field on the Viscosity of CH_4 and CF_4 (Rough Spherical Molecules)," *Phys. Letters*, **17**, 33–4, 1965.

405. Beenakker, J. J. M., Hulsman, H., Knaap, H. F. P., Korving, J., and Scoles, G., "The Influence of a Magnetic Field on the Viscosity and Other Transport Properties of Gaseous Diatomic Molecules," from *Advances in Thermophysical Properties at Extreme Temperatures and Pressures* (Gratch, S., Editor), ASME Symp., Purdue University, Lafayette, Ind., 216–20, 1965.

406. Korving, J., Hulsman, H., Scoles, G., Knaap, H. F. P., and Beenakker, J. J. M., "The Influence of a Magnetic Field on the Transport Properties of Gases of Polyatomic Molecules, Part I. Viscosity," *Physica*, **36**, 177–97, 1967.

407. Gorelik, L. L. and Sinitsyn, V. V., "Influence of a Magnetic Field on the Thermal Conductivity of Gases with Nonspherical Molecules," *Zh. Eksp. Teor. Phys. (USSR)*, **46**, 401–2, 1964; English translation: *Soviet Phys.—JETP*, **19**, 272–3, 1964.

408. Gorelik, L. L., Redkoborodyi, Yu. N., and Sinitsyn, V. V., "Influence of a Magnetic Field on the Thermal Conductivity of Gases with Nonspherical Molecules," *Zh. Eksp. Teor. Phys. (USSR)*, **48**, 761–5, 1965; English translation: *Soviet Phys.—JETP*, **21**, 503–5, 1965.

409. Waldmann, L., "Dilute Monatomic Gases, Accuracy and Limits of Applicability of Transport Equation," in *Proceedings of the International Seminar on the Transport Properties of Gases*, Brown University, Providence, R. I., 59–84, 1964.

410. Kagan, Yu. and Maksimov, L. A., "Kinetic Theory of Gases Taking into Account Rotational Degrees of Freedom in an External Field," *Zh. Eksp. Teor. Phys. (USSR)*, **51**, 1893–908, 1966; English translation: *Soviet Phys.—JETP*, **24**, 1272–81, 1967.

411. McCourt, F. R. and Snider, R. F., "Thermal Conductivity of a Gas of Rotating Diamagnetic Molecules in an Applied Magnetic Field," *J. Chem. Phys.*, **46**, 2387–98, 1967.

412. Knaap, H. F. P. and Beenakker, J. J. M., "Heat Conductivity and Viscosity of a Gas of Nonspherical Molecules in a Magnetic Field," *Physica*, **33**, 643–70, 1967.

413. Tip, A., "The Influence of Angular Momentum Anisotropy on the Heat Conductivity of Dilute Diatomic Gases," *Physica*, **37**, 82–96, 1967.

414. Levi, A. C. and McCourt, F. R., "Odd Terms in Angular Momentum and Transport Properties of Polyatomic Gases in a Field," *Physica*, **38**, 415–37, 1968.

415. Tip, A., Levi, A. C., and McCourt, F. R., "Magnetic Dispersion Relations in the Senftleben–Beenakker Effect," *Physica*, **40**, 435–45, 1968.

416. Tip, A., "Some Aspects of the Influence of a Magnetic Field on Transport Phenomena in Dilute Gases," Ph.D. Thesis, Leiden, 86 pp., 1969.

417. Hooyman, G. J., Mazur, P., and de Groot, S. R., "Coefficients of Viscosity for a Fluid in a Magnetic Field or in a Rotating System," *Physica*, **21**, 355–9, 1955.

418. Korvig, J., Hulsman, H., Knaap, H. F. P., and Beenakker, J. J. M., "Transverse Momentum Transport in Viscous Flow of Diatomic Gases in a Magnetic Field," *Phys. Letters*, **21**, 5–7, 1966.

419. Kikoin, I. K., Balashov, K. I., Lazarev, S. D., and Neushtadt, R. E., "On the Influence of a Magnetic Field on Viscous Gas Flow," *Phys. Letters*, **24A**, 165–6, 1967.

420. Kikoin, I. K., Balashov, K. I., Lazarev, S. D., and Neushtadt, P. E., "Viscous Flow of Gases in Strong Magnetic Fields," *Phys. Letters*, **26A**, 650–1, 1968.

421. Tip, A., "On the Magnetic Field Dependence of the Transport Properties of the Hydrogen Isotopic Molecules H_2, D_2 and HD," *Phys. Letters*, **25A**, 409–10, 1967.

422. Korvig, J., Knaap, H. F. P., Gordon, R. G., and Beenakker, J. J. M., "The Influence of a Magnetic Field on the Transport Properties of Polyatomic Gases; A Comparison of Theory with Experiments," *Phys. Letters*, **24A**, 755–6, 1967.

423. Korvig, J., Honeywell, W. I., Bose, T. K., and Beenakker, J. J. M., "The Influence of a Magnetic Field on the Transport Properties of Gases of Polyatomic Molecules. Part II, Thermal Conductivity," *Physica*, **36**, 198–214, 1967.

424. Gorelik, L. L. and Sinitsyn, V. V., "On the Influence of a Magnetic Field on the Thermal Conductivity of Gases," *Physica*, **41**, 486–8, 1969.

425. Levi, A. C., McCourt, F. R., and Hajdu, J., "Burnett Coefficients in a Magnetic Field. I. General Formulation for a Polyatomic Gas," *Physica*, **42**, 347–62, 1969.

426. Levi, A. C., McCourt, F. R., and Beenakker, J. J. M., "Burnett Coefficients in a Magnetic Field. II. The Linear Effects," *Physica*, **42**, 363–87, 1969.

427. McCourt, F. R., Knaap, H. F. P., and Moraal, H., "The Senftleben–Beenakker Effects for a Gas of Rough Spherical Molecules. I. The Thermal Conductivity," *Physica*, **43**, 485–512, 1969.

428. Gorelik, L. L., Nikolaevskii, V. G., and Sinitsyn, V. V., "Transverse Heat Transfer in a Molecular-Thermal Stream Produced in a Gas of Nonspherical Molecules in the Presence of a Magnetic Field," *JETP Letters*, **4**, 307–10, 1966.

429. Tip, A., "On the Senftleben–Beenakker Effect in Mixtures. I. The Magnetic Field Dependence of the Shear Viscosity Tensor in Mixtures of Diamagnetic Gases," *Physica*, **37**, 411–22, 1967.

430. Tip, A., de Vries, A. E., and Los, J., "Thermal Diffusion and the Senftleben Effect," *Physica*, **32**, 1429–36, 1966.

431. Vugts, H. F., Tip, A., and Los, J., "The Senftleben Effect on Diffusion," *Physica*, **38**, 579–86, 1968.

432. Senftleben, H., "The Influence of Electrical Fields on the Transport Phenomena in Gases," *Ann. Phys.*, VII, **15**(5–6), 273–7, 1965.

433. Amme, R. C., "Viscoelectric Effect in Gases," *Phys. Fluids*, **7**, 1387–8, 1964.

434. Borman, V. D., Gorelik, L. L., Nikolaev, B. I., and Sinitsyn, V. V., "Influence of Alternating Electric Field on Transport Phenomena in Polar Gases," *JETP Letters*, **5**, 85–7, 1967.

435. Borman, V. D., Nikolaev, B. I., and Nikolaev, N. I., "Transport Phenomena in a Mixture of Monatomic and Polar Gases," *Z. Eksp. Teor. Fiz. (USSR)*, **51**, 579–85, 1966; English translation: *Soviet Phys.—JETP*, **24**, 387–91, 1967.

436. Gallinaro, G., Meneghetti, G., and Scoles, G., "Viscoelectric Effect in Polar Polyatomic Gases," *Phys. Letters*, **24A**, 451–2, 1967.

437. Levi, A. C., McCourt, F. R., and Tip, A., "Electric Field Senftleben–Beenakker Effects," *Physica*, **39**, 165–204, 1968.

438. Green, M. S. and Sengers, J. V., Editors, "Critical Phenomena," Proc. Conf. held in Washington, D.C., April 1965, National Bureau of Standards Miscellaneous Publications 273, 242 pp., 1966.

439. Sengers, J. V., "Behavior of Viscosity and Thermal Conductivity of Fluids Near the Critical Point," Critical Phenomena 165–78, NBS Publ. 273, 1966.

440. Sengers, J. V. and Sengers, A. L., "The Critical Region," *Chem. Eng. News*, **46**, 104–18, 1968.

441. Deutch, J. M. and Zwanzig, R., "Anomalous Specific Heat and Viscosity of Binary van der Waals Mixtures," *J. Chem. Phys.*, **46**, 1612–20, 1967.

442. Fixman, M., "Comments on Transport Coefficients in the Gas Critical Region," *J. Chem. Phys.*, **48**, 4329–30, 1968.

443. Teague, R. K. and Pings, C. J., "Refractive Index and the Lorentz-Lorenz Function for Gaseous and Liquid Argon, Including a Study of the Coexistence Curve Near the Critical State," *J. Chem. Phys.*, **48**, 4973–84, 1968.

444. Cercignani, C. and Sernagiotto, F., "Cylindrical Poiseuille Flow of a Rarefied Gas," *Phys. Fluids*, **9**, 40–4, 1966.

445. Reid, R. C. and Sherwood, T. K., "Viscosity," in *The Properties of Gases and Liquids: Their Estimation and Correlation*, McGraw-Hill Book Co., New York, Chapter 9, 395–455, 1966.

446. Westenberg, A. A., "A Critical Survey of the Major Methods for Measuring and Calculating Dilute Gas Transport Properties," in *Advances in Heat Transfer*, Academic Press, Inc., New York, Vol. 3, 253–302, 1966.

447. Hilsenrath, J. and Touloukian, Y. S., "The Viscosity, Thermal Conductivity, and Prandtl Number for Air, O_2, N_2, NO, H_2, CO, CO_2, H_2O, He and Ar," *Trans. ASME*, **76**, 967–85, 1954.

448. Hilsenrath, J., Beckett, C. W., Benedict, W. S., Fano, L., Hoge, H. J., Masi, J. F., Nuttall, R. L., Touloukian, Y. S., and Woolley, H. W., "Tables of Thermal Properties of Gases," from *Tables of Thermodynamic and Transport Properties of Air, Argon, Carbon Dioxide, Carbon Monoxide, Hydrogen, Nitrogen, Oxygen, and Steam*, NBS Circular 564, Pergamon Press, Oxford, 478 pp., 1960.

449. Svehla, R. A., "Estimated Viscosities and Thermal Conductivities of Gases at High Temperatures," NASA TR R-132, 120 pp., 1962.

450. Svehla, R. A., "Thermodynamic and Transport Properties for the Hydrogen-Oxygen System," NASA SP-3011, 419 pp., 1964.

451. Simon, H. A., Liu, C. S., and Hartnett, J. P., "Properties of Hydrogen: Carbon Dioxide, and Carbon Dioxide: Nitrogen Mixtures," NASA CR-387, 133 pp., 1966.

452. Gonzalez, M., Eakin, B. E., Lee, A. L., *Viscosity of Natural Gases*, American Petroleum Institute Publication (Associated with Research Project 65), 109 pp., 1970.

453. Lee, A. L., *Viscosity of Light Hydrocarbons*, American Petroleum Institute, New York, 128 pp., 1965.

454. ASTM Viscosity Index Calculated from Kinematic Viscosity, *ASTM Data Series DS 39a* (Formerly STP 168), American Society for Testing and Materials, 1916 Race St., Philadelphia, Pa., 964 pp., 1965.

455. Margenau, H. and Kestner, N. R., *Theory of Intermolecular Forces*, Pergamon Press, New York, 360 pp., 1969.

456. Hirschfelder, J. O., Editor, *Intermolecular Forces*, Vol. XII of *Advances in Chemical Phys.*, Interscience Publishers, 643 pp., 1967.

457. Margenau, H., "Van der Waals Forces," *Rev. Mod. Phys.*, **11**, 1–35, 1939.

458. Fitts, D. D., "Statistical Mechanics: A Study of Intermolecular Forces," *Ann. Rev. Phys. Chem.*, **17**, 59–82, 1966.

459. Pauly, H. and Toennies, J. P., "The Study of Intermolecular Potentials with Molecular Beams at Thermal Energies," in *Advances in Atomic and Molecular Physics* (Bates, D. R. and Estermann, I., Editors), Academic Press, New York, Vol. 1, 408 pp., 1965.

460. Lichten, W., "Resonant Charge Exchange in Atomic Collisions," in *Advances in Chemical Physics* (Prigogine, I., Editor), Interscience Publishers, Vol. XIII, 398 pp., 1967.

461. Buckingham, R. A., "The Present Status of Intermolecular Potentials for Calculations of Transport Properties," in *Proceedings of the Conference on Physical Chemistry in Aerodynamics and Space Flight, Planetary and Space Science*, Pergamon Press, New York, Vol. 3, 205–16, 1961.

462. Dalgarno, A., "Intermolecular Potentials for Ionic Systems," in *Proceedings of the Conference on Physical Chemistry in Aerodynamics and Space Flight, Planetary and Space Science*, Pergamon Press, New York, Vol. 3, 217–20, 1961.

463. Walker, R. E., Monchick, L., Westenberg, A. A., and Favin, S., "High Temperature Gaseous Diffusion Experiments and Intermolecular Potential Energy Functions," in *Proceedings of the Conference on Physical Chemistry in Aerodynamics and Space Flight, Planetary and Space Science*, Pergamon Press, New York, Vol. 3, 221–7, 1961.

464. Treanor, C. E. and Skinner, G. T., "Molecular Interactions at High Temperatures," in *Proceedings of the Conference on Physical Chemistry in Aerodynamics and Space Flight, Planetary and Space Science*, Pergamon Press, New York, Vol. 3, 253–70, 1961.

465. Whalley, E., "Zero-Point Energy: A Contribution to Intermolecular Forces," *Trans. Faraday Soc.*, **54**, 1613–21, 1958.

466. Dahler, J. S. and Hirschfelder, J. O., "Long-Range Intermolecular Forces," *J. Chem. Phys.*, **25**, 986–1005, 1956.

467. Meath, W. J. and Hirschfelder, J. O., "Long-Range (Retarded) Intermolecular Forces," *J. Chem. Phys.*, **44**, 3210–5, 1966.

468. Chang, T. Y., "Long-Range Interatomic Forces," *Mol. Phys.*, **13**, 487–8, 1967.

469. Wilson, J. N., "On the London Potential Between Pairs of Rare-Gas Atoms," *J. Chem. Phys.*, **43**, 2564–5, 1965.

470. Chang, T. Y., "Moderately Long-Range Interatomic Forces," *Rev. Mod. Phys.*, **39**, 911–42, 1967.

471. Meath, W. J. and Hirschfelder, J. O., "Relativistic Intermolecular Forces, Moderately Long Range," *J. Chem. Phys.*, **44**, 3197–3209, 1966.

472. Cottrell, T. L., "Intermolecular Repulsive Forces," *Faraday Soc., Discuss.*, **22**, 10–16, 1956.

473. Barua, A. K. and Chatterjee, S., "Repulsive Energy Between Hydrogen and Helium Atoms," *Mol. Phys.*, **7**, 433–8, 1964.

474. Brown, W. B., "Interatomic Forces at Very Short Range," *Faraday Soc. Discuss.*, **40**, 140–9, 1965.

475. Hirschfelder, J. O., "Perturbation Theory for Exchange Forces. I," *Chem. Phys. Letters*, **1**, 325–9, 1967.

476. Hirschfelder, J. O., "Perturbation Theory for Exchange Forces. II," *Chem. Phys. Letters*, **1**, 363–8, 1967.

477. Jansen, L. and Slawsky, Z. I., "Deviations from Additivity of the Intermolecular Field at High Densities," *J. Chem. Phys.*, **22**, 1701–4, 1954.

478. Sherwood, A. E. and Prausnitz, J. M., "Intermolecular Potential Functions and the Second and Third Virial Coefficients," *J. Chem. Phys.*, **41**, 429–37, 1964.

479. Sherwood, A. E., de Rocco, A. G., and Mason, E. A., "Nonadditivity of Intermolecular Forces: Effects on the Third Virial Coefficient," *J. Chem. Phys.*, **44**, 2984–94, 1966.

480. Hamann, S. D. and Lambert, J. A., "The Behavior of Fluids of Quasi-Spherical Molecules. I. Gases at Low Densities," *Aust. J. Chem.*, **7**, 1–17, 1954.

481. Hamann, S. D. and Lambert, J. A., "The Behavior of Fluids of Quasi-Spherical Molecules. II. High Density Gases and Liquids," *Aust. J. Chem.*, **7**, 18–27, 1954.

482. Bennett, L. A. and Vines, R. G., "The Molecular Complexity of Polar Organic Vapors," *Aust. J. Chem.*, **8**, 451–4, 1955.

483. Lichten, W., "Resonant Charge Exchange in Atomic Collisions," in *Advances in Chemical Phys.* (Prigogine, I., Editor), Interscience Publishers, New York, Vol. 13, 398 pp., 1967.

484. Hasted, J. B., "Recent Measurements on Charge Transfer," in *Advances in Atomic and Molecular Phys.* (Bates, D. R. and Estermann, I., Editors), Academic Press, New York, 465 pp., 1968.

485. Amdur, I., "Intermolecular Potentials from Scattering Experiments: Results, Applications, and Limitations," in *Progress in Int. Res. on Thermodynamic and Transport Properties*, Second Symp. on Thermophysical Properties, ASME, New York, 369–77, 1962.

486. Amdur, I. and Jordan, J. E., "Elastic Scattering of High-Energy Beams: Repulsive Forces," in *Molecular Beams*, Vol. 10 of *Advance in Chemical Physics* (Ross, J., Editor), Interscience Publishers, New York, 29–73, 1966.

487. Hirschfelder, J. O., "Determination of Intermolecular Forces," *J. Chem. Phys.*, **43**, S199–S201, 1965.

488. Woolley, H. W., "Empirical Intermolecular Potential for Inert Gas Atoms," *J. Chem. Phys.*, **32**, 405–9, 1960.

489. Saxena, S. C. and Mathur, B. P., "Thermal Diffusion in Binary Gas Mixtures and Intermolecular Forces," *Rev. Mod. Phys.*, **37**, 316–25, 1965.

490. Saxena, S. C. and Mathur, B. P., "Thermal Diffusion in Isotopic Gas Mixtures and Intermolecular Forces," *Rev. Mod. Phys.*, **38**, 380–90, 1966.

491. Axilrod, B. M., "A Survey of Some Empirical and Semi-Empirical Interatomic and Intermolecular Potentials," NBS Tech. Note 246, 52 pp., 1966.

492. Mann, J. B., "Collision Integrals and Transport Properties for Gases Obeying an Exponential Repulsive Potential: Application to Hydrogen and Helium," Los Alamos Scientific Lab, Rept. LA-2383, 85 pp., 1960.

493. Monchick, L., "Collision Integrals for the Exponential Repulsive Potential," *Phys. Fluids*, **2**, 695–700, 1959.

494. Kihara, T., Taylor, M. H., and Hirschfelder, J. O., "Transport Properties for Gases Assuming Inverse Power Intermolecular Potentials," *Phys. Fluids*, **3**, 715–20, 1960.

495. Munn, R. J., Mason, E. A., and Smith, F. J., "Collision Integrals for the Exponential Attractive Potential," *Phys. Fluids*, **8**, 1103–5, 1965.

496. Brokaw, R. S., "Estimated Collision Integrals for the Exponential Attractive Potential," *Phys. Fluids*, **4**, 944–6, 1961.

497. Hirschfelder, J. O., Bird, R. B., and Spotz, E. L., "The Transport Properties for Nonpolar Gases," *J. Chem. Phys.*, **16**, 968–81, 1948.

498. Hirschfelder, J. O., Bird, R. B., and Spotz, E. L., "The Transport Properties for Nonpolar Gases," *J. Chem. Phys.*, **17**, 1343–4, 1949.

499. Liley, P. E., "Collision Integrals for the Lennard-Jones (6–12) Potential," Purdue University, TPRC Rept. 15, 15 pp., 1963.

500. Smith, F. L., Mason, E. A., and Munn, R. J., "Transport Collision Integrals for Gases Obeying 9-6 and 28-7 Potentials," *J. Chem. Phys.*, **42**, 1334–9, 1965.

501. Klein, M. and Smith, F. J., "Tables of Collision Integrals for the (m, 6) Potential Function for 10 Values of m," *J. Res. Natl. Bur. Std.—A. Phys. and Chem.*, **72A**, 359–423, 1968.

502. Mason, E. A., "Transport Properties of Gases Obeying a Modified Buckingham (Exp-six) Potential," *J. Chem. Phys.*, **22**, 169–86, 1954.

503. Mason, E. A., "Higher Approximations for the Transport Properties of Binary Gas Mixtures. II. Applications," *J. Chem. Phys.*, **27**, 782–90, 1957.

504. Smith, F. J. and Munn, R. J., "Automatic Calculation of the Transport Collision Integrals with Tables for the Morse Potential," *J. Chem. Phys.*, **41**, 3560–8, 1964.

505. Samoilov, E. V. and Tsitelauri, N. N., "Collision Integrals for the Morse Potential," *Teplofiz. Vys. Temp.*, **2**, 565–72, 1964.

506. Smith, F. J., Munn, R. J., and Mason, E. A., "Transport Properties of Quadrupolar Gases," *J. Chem. Phys.*, **46**, 317–21, 1967.

507. Barker, J. A., Fock, W., and Smith, F., "Calculation of Gas Transport Properties and the Interaction of Argon Atoms," *Phys. Fluids*, **7**, 897–903, 1964.

508. Guggenheim, E. A. and McGlashan, M. L., "Interaction Between Argon Atoms," *Proc. Roy. Soc. (London)*, **A255**, 456–76, 1960.

509. Pollara, L. Z. and Funke, P. T., "Note on a New Potential Function," *J. Chem. Phys.*, **31**, 855–6, 1959.

510. Saxena, S. C. and Joshi, K. M., "Second Virial and Zero Pressure Joule-Thomson Coefficients of Nonpolar Quasi-Spherical Molecules," *Indian J. Phys.*, **36**, 422–30, 1962.

511. Saxena, S. C. and Joshi, K. M., "Second Virial Coefficient of Polar Gases," *Phys. Fluids*, **5**, 1217–22, 1962.

512. Saxena, S. C., Joshi, K. M., and Ramaswamy, S., "Zero Pressure Joule-Thomson Coefficient of Polar Gases," *Indian J. Pure Appl. Phys.*, **1**, 420–6, 1963.

513. Saksena, M. P. and Saxena, S. C., "Equilibrium Properties of Gases and Gaseous Mixtures," *Nat. Inst. Sci. (India)*, **32A**, 177–95, 1966.

514. Saxena, S. C. and Saksena, M. P., "Certain Equilibrium Properties of Gases and Gas Mixtures on Steeper Lennard-Jones and Stockmayer Type Potentials," *Def. Sci. J.*, **17**, 79–94, 1967.

515. Saksena, M. P., Nain, V. P. S., and Saxena, S. C., "Second Virial and Zero-Pressure Joule-Thomson Coefficients of Polar and Nonpolar Gases and Gas Mixtures," *Indian J. Phys.*, **41**, 123–33, 1967.

516. Varshni, V. P., "Intermolecular Potential Function for Helium," *J. Chem. Phys.*, **45**, 3894–5, 1966.

517. Dymond, J. H., Rigby, M., and Smith, E. B., "Intermolecular Potential-Energy Function for Simple Molecules," *J. Chem. Phys.*, **42**, 2801–6, 1965.

518. Nain, V. P. S. and Saxena, S. C., "On the Appropriateness of Dymond Rigby and Smith Intermolecular Potential," *Chem. Phys. Letters*, **1**, 46–7, 1967.

519. Feinberg, M. J. and deRocco, G., "Intermolecular Forces: The Triangle Well and Some Comparisons with the Square Well and Lennard-Jones," *J. Chem. Phys.*, **41**, 3439–50, 1964.

520. de Rocco, A. G. and Hoover, W. G., "Second Virial Coefficient for the Spherical Shell Potential," *J. Chem. Phys.*, **36**(4), 916–26, 1962.

521. de Rocco, A. G., Spurling, T. H., and Storvick, T. S., "Intermolecular Forces in Globular Molecules. II. Multipolar Gases with a Spherical-Shell Central Potential," *J. Chem. Phys.*, **46**, 599–602, 1967.

522. Spurling, T. H. and de Rocco, A. G., "Intermolecular Forces in Globular Molecules. III. A Comparison of the Spherical Shell and Kihara Models," *Phys. Fluids*, **10**, 231–2, 1967.

523. Storvick, T. S., Spurling, T. H., and de Rocco, A. G., "Intermolecular Forces in Globular Molecules. IV. Additive Third Virial Coefficients and Quadrupolar Corrections," *J. Chem. Phys.*, **46**, 1498–1506, 1967.

524. McKinley, M. D. and Reed, T. M., "Intermolecular Potential-Energy Functions for Pairs of Simple Polyatomic Molecules," *J. Chem. Phys.*, **42**, 3891–9, 1965.

525. Lawley, K. P. and Smith, E. B., "Contribution of Off-Centre Dipoles to the Second Virial Coefficients of Polar Gases," *Trans. Faraday Soc.*, **59**, 301–8, 1963.

526. Dymond, J. H. and Smith, E. B., "Off-Center Dipole Model and the Second Virial Coefficients of Polar Gases," *Trans. Faraday Soc.*, **60**, 1378–85, 1964.

527. Spurling, T. H. and Mason, E. A., "On the Off-Center Dipole Model for Polar Gases," *J. Chem. Phys.*, **46**, 404–5, 1967.

528. Carra, S. and Konowalow, D. D., "An Improved Intermolecular Potential Function," *Nuovo Cimento*, **34**, 205–14, 1964.

529. Nain, V. P. S. and Saxena, S. C., "Second Virial Coefficient of Nonpolar Gases and Gas Mixtures and Buckingham-Carra-Konowalow Potential," *Indian J. Phys.*, **41**, 199–208, 1967.

530. Nain, V. P. S. and Saksena, M. P., "The Modified-BCK Potential for Nonpolar Molecules," *Chem. Phys. Letters*, **1**, 125–6, 1967.

531. Konowalow, D. D., "Comment on the Modified Buckingham-Carra-Konowalow Potential for Nonpolar Molecules," *Chem. Phys. Letters*, **2**, 179–81, 1968.

532. Mason, E. A. and Monchick, L., "Methods for the Determination of Intermolecular Forces," *Adv. Chem. Phys.*, **12**, 329–87, 1967.

533. Bernstein, R. B. and Muckerman, J. T., "Determination of Intermolecular Forces Via Low-Energy Molecular Beam Scattering," *Adv. Chem. Phys.*, **12**, 389–486, 1967.

534. Birnbaum, G., "Microwave Pressure Broadening and Its Application to Intermolecular Forces," *Adv. Chem. Phys.*, **12**, 487–548, 1967.

535. Bloom, M. and Oppenheim, I., "Intermolecular Forces Determined by Nuclear Magnetic Resonance," *Adv. Chem. Phys.*, **12**, 549–99, 1967.

536. Flynn, L. W. and Thodos, G., "Lennard-Jones Force Constants from Viscosity Data: Their Relationship to Critical Properties," *Am. Inst. Chem. Eng. J.*, **8**, 362–5, 1962.

537. Stiel, L. I. and Thodos, G., "Lennard-Jones Force Constants Predicted from Critical Properties," *J. Chem. Eng. Data*, **7**, 234–6, 1962.

538. Saksena, M. P. and Saxena, S. C., "On Possible Correlation Between Potential Parameters and Critical or Boiling Point Constants," *Indian J. Pure Appl. Phys.*, **4**, 86–7, 1966.

539. Konowalow, D. D. and Guberman, S. L., "Estimation of Morse Potential Parameters from the Critical Constants and the Acentric Factor," *Ind. Eng. Chem. Fundam.*, **7**, 622–5, 1968.

540. Reed, T. M., and McKinley, M. D., "Estimation of Lennard-Jones Potential Energy Parameters from Liquid Densities," *J. Chem. Eng. Data*, **9**, 553–6, 1964.

541. Barker, J. A. and Leonard, P. J., "Long-Range Interaction Forces Between Inert Gas Atoms," *Phys. Letters*, **13**, 127–8, 1964.

542. Munn, R. J., "On the Calculation of the Dispersion-Forces Coefficient Directly from Experimental Transport Data," *J. Chem. Phys.*, **42**, 3032–3, 1965.

543. Mason, E. A. and Vanderslice, J. T., "High Energy Elastic Scattering of Atoms, Molecules and Ions," in *Atomic and Molecular Process* (Bates, D. R., Editor), Academic Press, New York, 663–95, 1962.

544. Kamnev, A. B. and Leonas, V. B., "Experimental Determination of the Repulsion Potential and the Kinetic Properties of Noble Gases at High Temperatures," *Teplofiz. Vys. Temp.*, **3**, 744–6, 1965.

545. Zumino, B. and Keller, J. B., "Determination of Intermolecular Potentials from Thermodynamic Data and the Law of Corresponding States," *J. Chem. Phys.*, **30**, 1351–3, 1959.

546. Munn, R. J., "Interaction Potential of the Inert Gases. I," *J. Chem. Phys.*, **40**(5), 1439–46, 1964.

547. Munn, R. J. and Smith, F. J., "Interaction Potential of the Inert Gases, II," *J. Chem. Phys.*, **43**, 3998–4002, 1965.

548. Klein, M., "Determination of Intermolecular Potential Functions from Macroscopic Measurements," *J. Res. Natl. Bur. Std.*, **70A**, 259–69, 1966.

549. Hanley, H. J. M. and Klein, M., "On the Selection of the Intermolecular Potential Function: Application of Statistical Mechanical Theory to Experiment," NBS Tech. Note 360, 82 pp., 1967.

550. Hanley, H. J. M. and Klein, M., "Selection of the Intermolecular Potential Function: III. From the Isotopic Thermal Diffusion Factor," *J. Chem. Phys.*, **50**, 4765–70, 1969.

551. Klein, M. and Hanley, H. J. M., "Selection of the Intermolecular Potential. Part 2—From Data of State and Transport Properties Taken in Pairs," *Trans. Faraday Soc.*, **64**, 2927–38, 1968.

552. Muller, C. R. and Brackett, J. W., "Quantum Calculation of the Sensitivity of Diffusion, Viscosity, and Scattering Experiments to the Intermolecular Potential," *J. Chem. Phys.*, **40**, 654–61, 1964.

553. Bird, R. B., Hirschfelder, J. O., and Curtiss, C. F., "Theoretical Calculation of the Equation of State and Transport Properties of Gases and Liquids," *Trans. Am. Soc. Mech. Eng.*, 1011–38, 1954.

554. Srivastava, B. N. and Madan, M. P., "The Temperature Dependence of Viscosity of Nonpolar Gases," *Proc. Natl. Acad. Sci. (India)*, **21A**, 254–60, 1952.

555. Hawksworth, W. A., "A Shorter Method of Calculating Lennard-Jones (12–6) Potential Parameters from Gas Viscosity Data," *J. Chem. Phys.*, **35**, 1534, 1961.

556. Whalley, E. and Schneider, W. G., "The Lennard-Jones 12:6 Potential and the Viscosity of Gases," *J. Chem. Phys.*, **20**, 657–61, 1952.

557. Robinson, J. D. and Ferron, J. R., "Direct Determination of Intermolecular Potentials from Transport Data," Preprint 33A of *Am. Inst. Chem. Eng.*, *Symp. on Transport Properties*, *Part II*, Sixty-First Annual Meeting, Los Angeles, Calif., 39 pp., 5 Tables and 2 Figures, 1968.

558. Mason, E. A. and Rice, W. E., "The Intermolecular Potentials for Some Simple Nonpolar Molecules," *J. Chem. Phys.*, **22**, 843–51, 1954.

559. Hanley, H. J. M., "The Viscosity and Thermal Conductivity Coefficients of Dilute Argon Between 100 and 2000 K," NBS Tech. Note No. 333, 23 pp., 1966.

560. Hanley, H. J. M., "Comparison of the Lennard-Jones, Exp-6, and Kihara Potential Functions from Viscosity Data of Dilute Argon," *J. Chem. Phys.*, **44**, 4219–22, 1966.

561. Hanley, H. J. M. and Childs, G. E., "The Viscosity and Thermal Conductivity Coefficients of Dilute Neon, Krypton, and Xenon," NBS Tech. Note No. 352, 24 pp., 1967.

562. de Rocco, A. G. and Halford, J. O., "Intermolecular Potentials of Argon, Methane and Ethane," *J. Chem. Phys.*, **28**, 1152–4, 1958.

563. Milligan, J. H. and Liley, P. E., "Lennard-Jones Potential Parameter Variation as Determined from Viscosity Data for Twelve Gases," Paper No. 64-HT-20, 8 pp., 1964.

564. Saran, A., "Potential Parameters for Like and Unlike Interactions on Morse Potential Model," *Indian J. Phys.*, **37**, 491–9, 1963.

565. Pal, A. K., "Intermolecular Forces and Viscosity of Some Polar Organic Vapors," *Indian J. Phys.*, **41**, 823–7, 1967.

566. Chakraborti, P. K., "Gas Properties at High Temperatures on the Exponential Model," *Indian J. Phys.*, **35**, 354–60, 1961.

567. Saxena, S. C., "Thermal Conductivity and Force Between Like Molecules," *Indian J. Phys.*, **29**, 587–602, 1955.

568. Srivastava, K. P., "Force Constants for Like Molecules on Exp-Six Model From Thermal Conductivity," *Indian J. Phys.*, **31**, 404–14, 1957.

569. Srivastava, B. N. and Madan, M. P., "Intermolecular Force and Coefficient of Self-Diffusion," *Phil. Mag.*, **43**, 968–75, 1952.

570. Amdur, I. and Schatzki, T. F., "Diffusion Coefficients of the Systems Xe–Xe and Ar–Xe," *J. Chem. Phys.*, **27**, 1049–54, 1957.

571. Vugts, H. F., Boerboom, A. J. H., and Los, J., "Measurements of Relative Diffusion Coefficients of Argon," *Physica*, **44**, 219–26, 1969.

572. Srivastava, B. N. and Madan, M. P., "Intermolecular Force Constants from Thermal Diffusion and Other Properties of Gases," *J. Chem. Phys.*, **21**, 807–15, 1953.

573. Saxena, S. C. and Srivastava, B. N., "Second Approximation to the Thermal Diffusion Factor on the Lennard-Jones 12–6 Model," *J. Chem. Phys.*, **23**, 1571–4, 1955.

574. Madan, M. P., "Potential Parameters for Krypton," *J. Chem. Phys.*, **27**, 113–5, 1957.

575. Saxena, S. C., Kelley, J. G., and Watson, W. W., "Temperature Dependence of the Thermal Diffusion Factor for Helium, Neon, and Argon," *Phys. Fluids*, **4**, 1216–25, 1961.

576. Yntema, J. L. and Schneider, W. G., "On the Intermolecular Potentials of Helium," *J. Chem. Phys.*, **16**, 646–50, 1950.

577. Whalley, E. and Schneider, W. G., "Intermolecular Potentials of Argon, Krypton, and Xenon," *J. Chem. Phys.*, **23**, 1644–50, 1955.

578. Schamp, H. W., Mason, E. A., Richardson, A. C. B., and Altman, A., "Compressibility and Intermolecular Forces in Gases: Methane," *Phys. Fluids*, **1**, 329–37, 1958.

579. Schamp, H. W., Mason, E. A., and Su, K., "Compressibility and Intermolecular Forces in Gases. II. Nitrous Oxide," *Phys. Fluids*, **5**, 769–75, 1962.

580. Barua, A. K., "Intermolecular Potential of Helium," *Indian J. Phys.*, **34**, 76–84, 1960.

581. Barua, A. K., "Force Parameters for Some Nonpolar Molecules on the Exp 6-8 Model," *J. Chem. Phys.*, **31**, 957–60, 1959.

582. Srivastava, I. B., "Intermolecular Potential and Properties of Argon," *Indian J. Phys.*, **34**, 539–48, 1960.

583. Srivastava, I. B. and Barua, A. K., "Intermolecular Potentials of H_2 and D_2," *Indian. J. Phys.*, **35**, 320–2, 1961.

584. Barua, A. K. and Saran, A., "The Difference in the Intermolecular Potential of H_2 and D_2," *Physica*, **29**, 1393–6, 1963.

585. Mason, E. A., Amdur, I., and Oppenheim, I., "Differences in the Spherical Intermolecular Potentials of Hydrogen and Deuterium," *J. Chem. Phys.*, **43**, 4458–63, 1965.

586. Gambhir, R. S. and Saxena, S. C., "Zero-Pressure Joule-Thomson Coefficient for a Few Nonpolar Gases on the Morse Potential," *Indian J. Phys.*, **37**, 540–2, 1963.

587. Ahlert, R. C. and Vogl, W., "Lennard-Jones Parameters for Methane," *Am. Inst. Chem. Eng. J.*, **12**, 1025–6, 1966.

588. Saxena, S. C., "Zero-Pressure Joule-Thomson Coefficient and Exponential-Six Intermolecular Potential," *Chem. Phys. Letters*, **4**, 81–3, 1969.

589. Saksena, M. P., Gandhi, J. M., and Nain, V. P. S., "Determination of Force Constants for the Spherically Symmetric Potential Functions," *Chem. Phys. Letters*, **1**, 424–6, 1967.

590. Whalley, E., "The Difference in the Intermolecular Forces of H_2O and D_2O," *Trans. Faraday Soc.*, **53**, 1578–85, 1957.

591. Whalley, E., "Intermolecular Forces and Crystal Properties of Methane," *Phys. Fluids*, **2**, 335–6, 1959.

592. Whalley, E. and Falk, M., "Difference of Intermolecular Potentials of CH_3OH and CH_3OD," *J. Chem. Phys.*, **34**, 1569–71, 1961.

593. Saran, A. and Barua, A. K., "Intermolecular Potentials for Inert Gas Atoms," *Canadian J. Phys.*, **42**, 2026–9, 1964.

594. Brown, J. S., "Interatomic Potential Parameters of Solid Neon and Argon," *Proc. Phys. Soc.*, **89**, 987–92, 1966.

595. Mikolaj, P. G. and Pings, C. J., "Direct Determination of the Intermolecular Potential Function for Argon from X-Ray Scattering Data," *Phys. Rev. Letters*, **16**, 4–6, 1966.

596. Axilrod, B. M., "Comments on the Rosen Interaction Potential of Two Helium Atoms," *J. Chem. Phys.*, **38**, 275–7, 1963.

597. Nesbet, R. K., "Interatomic Potentials for HeNe, HeAr, and NeAr," *J. Chem. Phys.*, **48**, 1419–20, 1968.

598. Beck, D. E., "Interatomic Potentials for Helium and Molecules of Helium Isotopes," *J. Chem. Phys.*, **50**, 541–2, 1969.

599. Fender, B. E. F., "Potential Parameters of Krypton," *J. Chem. Phys.*, **35**, 2243–5, 1961.

600. Bahethi, O. P. and Saxena, S. C., "Intermolecular Potentials for Krypton," *Indian J. Phys.*, **3**, 12–15, 1964.

601. Barua, A. K. and Chakraborti, P. K., "Krypton–Krypton Molecular Interaction," *Physica*, **27**, 753–62, 1961.

602. Chakraborti, P. K., "Potential Energy Curve for the Interaction of Two Xenon Atoms," *Physica*, **29**, 227–33, 1963.

603. Chakraborti, P. K., "Intermolecular Potential of Radon," *J. Chem. Phys.*, **44**, 3137–8, 1966.

604. Srivastava, B. N. and Saxena, S. C., "Generalized Relations for the Thermal Diffusion Factor of Inert Gas Mixtures with One Invariable Constituent," *Physica*, **22**, 253–62, 1956.

605. Konowalow, D. D., Taylor, M. H., and Hirschfelder, J. O., "Second Virial Coefficient for the Morse Potential," *Phys. Fluids*, **4**, 622–8, 1961.

606. Konowalow, D. D. and Hirschfelder, J. O., "Intermolecular Potential Functions for Nonpolar Molecules," *Phys. Fluids*, **4**, 629–36, 1961.

607. Bahethi, O. P. and Saxena, S. C., "Morse Potential Parameters for Hydrogen," *Indian J. Pure Appl. Phys.*, **2**, 267–9, 1964.

608. Bahethi, O. P. and Saxena, S. C., "Morse Potential Parameters for Helium," *Phys. Fluids*, **6**, 1774–5, 1963.

609. Konowalow, D. D. and Carra, S., "Determination and Assessment of Morse Potential Functions for Some Nonpolar Gases," *Phys. Fluids*, **8**, 1585–9, 1965.

610. Konowalow, D. D. and Carra, S., "Central Potential for Polyatomic Molecules. I. A Survey of Morse Potential Determined Separately from Viscosity and Second Virial Coefficient," *Nuovo Cimento*, **44**, 133–8, 1966.

611. Konowalow, D. D., "Central Potentials for Nonpolar Polyatomic Molecules," *Phys. Fluids*, **9**, 23–7, 1966.

612. Konowalow, D. D., "Relationship Between Pitzer's Acentric Factor and the Morse Intermolecular Potential Function," *J. Chem. Phys.*, **46**, 818–9, 1967.

613. Saxena, S. C. and Bahethi, O. P., "Transport Properties of Some Simple Nonpolar Gases on the Morse Potential," *Mol. Phys.*, **7**, 183–9, 1963.

614. Hirschfelder, J. O., Bird, R. B., and Spotz, E. L., "The Transport Properties of Gases and Gaseous Mixtures. II," *Chem. Rev.*, **44**, 205–31, 1949.

615. Mason, E. A., "Forces Between Unlike Molecules and the Properties of Gaseous Mixtures," *J. Chem. Phys.*, **23**, 49–56, 1955.

616. Srivastava, B. N. and Srivastava, K. P., "Combination Rules for Potential Parameters of Unlike Molecules on Exp-Six Model," *J. Chem. Phys.*, **24**, 1275–6, 1956.

617. Saxena, S. C. and Gambhir, R. S., "Second Virial Coefficient of Gases and Gaseous Mixtures on the Morse Potential," *Mol. Phys.*, **6**, 577–83, 1963.

618. Srivastava, K. P., "Unlike Molecular Interactions and Properties of Gas Mixtures," *J. Chem. Phys.*, **28**, 543–9, 1958.

619. Bahethi, O. P., Gambhir, R. S., and Saxena, S. C., "Properties of Gases and Gas Mixtures with a Morse Potential," *Z. Naturforsch.*, **19a**, 1478–85, 1964.

620. Srivastava, I. B., "Determination of Unlike Interactions from Binary Viscosity," *Indian J. Phys.*, **35**, 86–91, 1961.

621. Saxena, S. C. and Gandhi, J. M., "Thermal Conductivity of Multicomponent Mixtures of Inert Gases," *Rev. Mod. Phys.*, **35**, 1022–32, 1963.

622. Gambhir, R. S. and Saxena, S. C., "Thermal Conductivity of Binary and Ternary Mixtures of Krypton, Argon, and Helium," *Mol. Phys.*, **11**, 233–41, 1966.

623. Gandhi, J. M. and Saxena, S. C., "Thermal Conductivity of Binary and Ternary Mixtures of Helium, Neon and Xenon," *Mol. Phys.*, **12**, 57–68, 1967.

624. Mathur, S., Tondon, P. K., and Saxena, S. C., "Thermal Conductivity of Binary, Ternary and Quaternary Mixtures of Rare Gases," *Mol. Phys.*, **12**, 569–79, 1967.

625. Gambhir, R. S. and Saxena, S. C., "Thermal Conductivity of the Gas Mixtures: $Ar-D_2$, $Kr-D_2$ and $Ar-Kr-D_2$," *Physica*, **32**, 2037–43, 1966.

626. Gandhi, J. M. and Saxena, S. C., "Thermal Conductivities of the Gas Mixtures D_2-He, D_2-Ne, and $D_2-He-Ne$," *Brit. J. Appl. Phys.*, **18**, 807–12, 1967.

627. Mathur, S., Tondon, P. K., and Saxena, S. C., "Thermal Conductivity of the Gas Mixtures: D_2-Xe, $D_2-Ne-Kr$, $D_2-Ne-Ar$, and $D_2-Ar-Kr-Xe$," *J. Phys. Soc. Japan*, **25**, 530–5, 1968.

628. Saxena, S. C. and Gupta, G. P., "Thermal Conductivity of Binary, Ternary, and Quaternary Mixtures of Polyatomic Gases," in *Proceedings of the Seventh Conference on Thermal Conductivity*, NBS Special Publ. 302, 605–13, 1968.

629. Amdur, I., Ross, J., and Mason, E. A., "Intermolecular Potentials for the Systems CO_2-CO_2 and CO_2-N_2O," *J. Chem. Phys.*, **20**, 1620–3, 1952.

630. Amdur, I. and Shuler, L. M., "Diffusion Coefficients of the Systems $CO-CO$ and $CO-N_2$," *J. Chem. Phys.*, **38**, 188–92, 1963.

631. Amdur, I. and Beatty, J. W., "Diffusion Coefficients of Hydrogen Isotopes," *J. Chem. Phys.*, **42**, 3361–4, 1965.

632. Amdur, I. and Malinauskas, A. P., "Diffusion Coefficients of the Systems $He-T_2$ and $He-TH$," *J. Chem. Phys.*, **42**, 3355–60, 1965.

633. Mason, E. A., Annis, B. K., and Islam, M., "Diffusion Coefficients of T_2-H_2 and T_2-D_2: The Nonequivalence of the H_2 and D_2 Cross Sections," *J. Chem. Phys.*, **42**, 3364–6, 1965.

634. Srivastava, K. P., "Mutual Diffusion of Binary Mixtures of Helium, Argon and Xenon at Different Temperatures," *Physica*, **25**, 571–8, 1959.

635. Srivastava, K. P. and Barua, A. K., "The Temperature Dependence of Interdiffusion Coefficient for Some Pairs of Rare Gases," *Indian J. Phys.*, **23**, 229–40, 1959.

636. Paul, R. and Srivastava, I. B., "Mutual Diffusion of the Gas Pairs H_2-Ne, H_2-Ar, and H_2-Xe at Different Temperatures," *J. Chem. Phys.*, **35**, 1621–4, 1961.

637. Srivastava, B. N. and Srivastava, I. B., "Studies on Mutual Diffusion of Polar–Nonpolar Gas Mixtures," *J. Chem. Phys.*, **36**, 2616–20, 1962.

638. Srivastava, I. B., "Mutual Diffusion of Binary Mixtures of Ammonia with He, Ne and Xe," *Indian J. Phys.*, **36**, 193–9, 1962.

639. Walker, R. E. and Westenberg, A. A., "Molecular Diffusion Studies in Gases at High Temperature. II. Interpretation of Results on the $He-N_2$ and CO_2-N_2 Systems," *J. Chem. Phys.*, **29**, 1147–53, 1958.

640. Walker, R. E. and Westenberg, A. A., "Molecular Diffusion Studies in Gases at High Temperature. III. Results and Interpretation of the He–Ar System," *J. Chem. Phys.*, **31**, 519–22, 1959.

641. Walker, R. E. and Westenberg, A. A., "Molecular Diffusion Studies in Gases at High Temperature. IV. Results and Interpretation of the CO_2-O_2, CH_4-O_2, H_2-O_2, $CO-O_2$, and H_2O-O_2," *J. Chem. Phys.*, **32**, 436–42, 1960.

642. Westenberg, A. A. and Frazier, G., "Molecular Diffusion Studies in Gases at High Temperatures. V. Results for the H_2-Ar System," *J. Chem. Phys.*, **36**, 3499–500, 1962.

643. Saxena, S. C. and Mathur, B. P., "Central Molecular Potentials, Combination Rules and Properties of Gases and Gas Mixtures," *Chem. Phys. Letters*, **1**, 224–6, 1967.

644. Mathur, B. P. and Saxena, S. C., "Measurement of the Concentration Diffusion Coefficient for He–Ar and Ne–Kr by a Two-Bulb Method," *Appl. Sci. Res.*, **18**, 325–35, 1968.

645. Srivastava, B. N. and Madan, M. P., "Thermal Diffusion of Gas Mixtures and Forces Between Unlike Molecules," *Proc. Phys. Soc. (London)*, **66A**, 277–87, 1953.

646. Saxena, S. C., "Thermal Diffusion of Gas Mixtures and Determination of Force Constants," *Indian J. Phys.*, **29**, 131–40, 1955.

647. Saxena, S. C., "Higher Approximations to Diffusion Coefficients and Determination of Force Constants," *Indian J. Phys.*, **29**, 453–60, 1955.

648. Srivastava, B. N., "Comments. Determination of Potential Parameters from Thermal Diffusion," *Phys. Fluids*, **4**, 526, 1961.

649. Madan, M. P., "Transport Properties of Some Gas Mixtures," *Proc. Natl. Inst. Sci. (India)*, **19**, 713–9, 1953.

650. Saxena, S. C., "Transport Coefficients and Force Between Unlike Molecules," *Indian J. Phys.*, **31**, 146–55, 1957.

651. Srivastava, B. N. and Srivastava, K. P., "Force Constants for Unlike Molecules on Exp-Six Model from Thermal Diffusion," *Physica*, **23**, 103–17, 1957.

652. Srivastava, K. P., "Intermolecular Potentials for Unlike Interaction on Exp-Six Model," *J. Chem. Phys.*, **26**, 579–81, 1957.

653. Mathur, B. P. and Saxena, S. C., "Composition Dependence of the Thermal Diffusion Factor in Binary Gas Mixtures," *Z. Naturforsch.*, **22a**, 164–9, 1967.

654. Mathur, B. P., Nain, V. P. S., and Saxena, S. C., "A Note on the Composition Dependence of the Thermal Diffusion Factor of Ar–He System," *Z. Naturforsch.*, **22a**, 840, 1967.

655. Nain, V. P. S. and Saxena, S. C., "Composition Dependence of the Thermal Diffusion Factor of Binary Gas Systems," *J. Chem. Phys.*, **51**, 1541–5, 1969.

656. Mathur, B. P., Joshi, R. K., and Saxena, S. C., "Thermal Diffusion Factors from the Measurements on a Trennschaukel: Ar–He and Kr–Ne," *J. Chem. Phys.*, **46**, 4601–3, 1967.

657. Saxena, V. K., Nain, V. P. S., and Saxena, S. C., "Thermal-Diffusion Factors from the Measurements on a Trennschaukel: Ne–Ar and Ne–Xe," *J. Chem. Phys.*, **48**, 3681–5, 1968.

658. Taylor, W. L., Weissman, S., Haubach, W. J., and Pickett, P. T., "Thermal-Diffusion Factors for the Neon–Xenon System," *J. Chem. Phys.*, **50**, 4886–98, 1969.

659. Weissman, S., Saxena, S. C., and Mason, E. A., "Intermolecular Forces from Diffusion and Thermal Diffusion Measurements," *Phys. Fluids*, **3**, 510–8, 1960.

660. Weissman, S., Saxena, S. C., and Mason, E. A., "Diffusion and Thermal Diffusion in $Ne-CO_2$," *Phys. Fluids*, **4**, 643–8, 1961.

661. Mason, E. A., Islam, M., and Weissman, S., "Thermal Diffusion and Diffusion in Hydrogen–Krypton Mixtures," *Phys. Fluids*, **7**, 1011–22, 1964.

662. McQuarrie, D. A. and Hirschfelder, J. O., "Intermediate-Range Intermolecular Forces in H_2^+," *J. Chem. Phys.*, **47**, 1775–80, 1967.

663. Kim, H. and Hirschfelder, J. O., "Energy of Interaction Between Two Hydrogen Atoms by the Gaussian-Type Functions," *J. Chem. Phys.*, **47**, 1005–8, 1967.

664. Certain, P. R., Hirschfelder, J. O., Kolos, W., and Wolniewicz, L., "Exchange and Coulomb Energy of H_2 Determined by Various Perturbation Methods," *J. Chem. Phys.*, **49**, 24–34, 1968.

665. Mason, E. A., Ross, J., and Schatz, P. N., "Energy of Interaction Between a Hydrogen Atom and a Helium Atom," *J. Chem. Phys.*, **25**, 626–9, 1956.

666. Ross, J. and Mason, E. A., "The Energy of Interaction of He^+ and H^-," *Astrophys. J.*, **124**, 485–7, 1956.

667. Mason, E. A. and Hirschfelder, J. O., "Short-Range Intermolecular Forces, I," *J. Chem. Phys.*, **26**, 173–82, 1957.

668. Mason, E. A. and Hirschfelder, J. O., "Short-Range Intermolecular Forces. II. H_2–H_2 and H_2–H," *J. Chem. Phys.*, **26**, 756–66, 1957.

669. Mason, E. A. and Vanderslice, J. T., "Delta-Function Model for Short-Range Intermolecular Forces. I. Rare Gases," *J. Chem. Phys.*, **28**, 432–8, 1958.

670. Vanderslice, J. T. and Mason, E. A., "Interaction Energies for the H–H_2 and H_2–H_2 System," *J. Chem. Phys.*, **33**, 492–4, 1960.

671. Vanderslice, J. T. and Mason, E. A., "Quantum Mechanical Calculations of Short-Range Intermolecular Forces," *Rev. Mod. Phys.*, **32**, 417–21, 1960.

672. Fallon, R. J., Mason, E. A., and Vanderslice, J. T., "Energies of Various Interactions Between Hydrogen and Helium Atoms and Ions," *Astrophys. J.*, **131**, 12–14, 1960.

673. Mason, E. A. and Vanderslice, J. T., "Interaction Energies and Scattering Cross-Sections of Hydrogen Ions in Helium," *J. Chem. Phys.*, **27**, 917–27, 1957.

674. Mason, E. A. and Vanderslice, J. T., "Scattering Cross Sections and Interaction Energies of Low-Velocity He^+ Ions in Helium," *Phys. Rev.*, **108**, 293–4, 1957.

675. Mason, E. A. and Vanderslice, J. T., "Interaction Energy and Scattering Cross Sections of H^- Ions in Helium," *J. Chem. Phys.*, **28**, 253–7, 1958.

676. Mason, E. A. and Vanderslice, J. T., "Interactions of H^- Ions and H Atoms with Ne, Ar, and H_2," *J. Chem. Phys.*, **28**, 1070–4, 1958.

677. Mason, E. A., Schamp, H. W., and Vanderslice, J. T., "Interaction Energy and Mobility of Li^+ Ions in Helium," *Phys. Rev.*, **112**, 445–8, 1958.

678. Mason, E. A. and Vanderslice, J. T., "Mobility of Hydrogen Ions (H^+, H_2^+, H_3^+) in Hydrogen," *Phys. Rev.*, **114**, 497–502, 1959.

679. Mason, E. A. and Vanderslice, J. T., "Determination of the Binding Energy of He_2^+ from Ion Scattering Data," *J. Chem. Phys.*, **29**, 361–5, 1958.

680. Mason, E. A. and Vanderslice, J. T., "Binding Energy of Ne_2^+ from Ion Scattering Data," *J. Chem. Phys.*, **30**, 599–600, 1959.

681. Cloney, R. D., Mason, E. A., and Vanderslice, J. T., "Binding Energy of Ar_2^+ from Ion Scattering Data," *J. Chem. Phys.*, **36**, 1103–4, 1962.

682. Vanderslice, J. T., Mason, E. A., Maisch, W. G., and Lippincott, E. R., "Ground-State of Hydrogen by the Rydberg–Klein–Rees Method," *J. Mol. Spectroscopy*, **3**, 17–29, 1959; *Errata*: **5**, 83, 1960.

683. Vanderslice, J. T., Mason, E. A., and Lippincott, E. R., "Interactions Between Ground-State Nitrogen Atoms and Molecules. The N–N, N–N_2, and N_2–N_2 Interactions," *J. Chem. Phys.*, **30**, 129–36, 1959.

684. Vanderslice, J. T., Mason, E. A., and Maisch, W. G., "Interactions Between Oxygen and Nitrogen: O–N, O–N_2, and O_2–N_2," *J. Chem. Phys.*, **31**, 738–46, 1959.

685. Vanderslice, J. T., Mason, E. A., and Maisch, W. G., "Interactions Between Ground-State Oxygen Atoms and Molecules: O–O and O_2–O_2," *J. Chem. Phys.*, **32**, 515–24, 1960.

686. Fallon, R. J., Vanderslice, J. T., and Mason, E. A., "Potential Energy Curves of Hydrogen Fluoride," *J. Chem. Phys.*, **32**, 698–700, 1960.

687. Fallon, R. J., Vanderslice, J. T., and Mason, E. A., "Potential Energy Curves for Lithium Hydride," *J. Chem. Phys.*, **32**, 1453–5, 1960; Erratum: "Potential Energy Curves for HF and LiH," *J. Chem. Phys.*, **33**, 944, 1960.

688. Tobias, I., Fallon, R. J., and Vanderslice, J. T., "Potential Energy Curves for CO," *J. Chem. Phys.*, **33**, 1638–40, 1960.

689. Vanderslice, J. T., Mason, E. A., Maisch, W. G., and Lippincott, E. R., "Potential Curves for N_2, NO, and O_2," *J. Chem. Phys.*, **33**, 614–5, 1960.

690. Konowalow, D. D. and Hirschfelder, J. O., "More Potential Parameters for O–O, N–N, and N–O Interactions," *Phys. Fluids*, **4**, 637–42, 1961.

691. Tobias, I. and Vanderslice, J. T., "Potential Energy Curves for the $X' \sum_g^+$ and $B' \sum_u^+$ States of Hydrogen," *J. Chem. Phys.*, **35**, 1852–5, 1961.

692. Vanderslice, J. T., "Modification of the Rydberg–Klein–Rees Method for Obtaining Potential Curves for Doublet States Intermediate Between Hund's Cases (a) and (b)," *J. Chem. Phys.*, **37**, 384–8, 1962.

693. Krupenie, P. H., Mason, E. A., and Vanderslice, J. T., "Interaction Energies and Transport Coefficients of Li + H and O + H Gas Mixtures at High Temperatures," *J. Chem. Phys.*, **39**, 2399–2408, 1963.

694. Weissman, S., Vanderslice, J. T., and Battino, R., "On the Recalculation of the Potential Curves for the Ground States of I_2 and H_2," *J. Chem. Phys.*, **39**, 2226–8, 1963.

695. Knof, H., Mason, E. A., and Vanderslice, J. T., "Interaction Energies, Charge Exchange Cross Sections, and Diffusion Cross Sections for N^+– N and O^+– O Collisions," *J. Chem. Phys.*, **40**, 3548–53, 1964.

696. Krupenie, P. H. and Weissman, S., "Potential-Energy Curves for CO and CO^+," *J. Chem. Phys.*, **43**, 1529–34, 1965.

697. Benesch, W., Vanderslice, J. T., Tilford, S. G., and Wilkinson, P. G., "Potential Curves for the Observed States of N_2 Below 11 eV," *Astrophys. J.*, **142**, 1227–40, 1965.

698. Benesch, W., Vanderslice, J. T., Tilford, S. G., and Wilkinson, P. G., "Franck–Condon Factors for Observed Transitions in N_2 Above 6 eV," *Astrophys. J.*, **143**, 236–52, 1966.

699. Benesch, W., Vanderslice, J. T., Tilford, S. G., and Wilkinson, P. G., "Franck–Condon Factors for Permitted Transitions in N_2," *Astrophys. J.*, **144**, 408–18, 1966.

700. Stiel, L. I. and Thodos, G., "The Normal Boiling Points and Critical Constants of Saturated Aliphatic Hydrocarbons," *Am. Inst. Chem. Eng. J.*, **8**, 527–9, 1962.

701. Thodos, G., "Critical Constants of the Naphthenic Hydrocarbons," *Am. Inst. Chem. Eng. J.*, **2**, 508–13, 1956.

702. Thodos, G., "Critical Constants of the Aromatic Hydrocarbons," *Am. Inst. Chem. Eng. J.*, **3**, 428–31, 1957.

703. Thodos, G., "Critical Constants of Unsaturated Aliphatic Hydrocarbons," *Am. Inst. Chem. Eng. J.*, **1**, 165–8, 1955.

704. Thodos, G., "Critical Constants of Saturated Aliphatic Hydrocarbons," *Am. Inst. Chem. Eng. J.*, **1**, 168–73, 1955.

705. Forman, J. C. and Thodos, G., "Critical Temperatures and Pressures of Hydrocarbons," *Am. Inst. Chem. Eng. J.*, **4**, 356–61, 1958.

706. Forman, J. C. and Thodos, G., "Critical Temperatures and Pressures of Organic Compounds," *Am. Inst. Chem. Eng. J.*, **6**, 206–9, 1960.

707. Ekiner, O. and Thodos, G., "The Critical Temperatures and Critical Pressures of Binary Mixtures of Aliphatic Hydrocarbons," *J. Appl. Chem.*, **15**, 393–7, 1965.

708. Ekiner, O. and Thodos, G., "Critical Temperatures and Pressures of the Ethane–*n*-Heptane System," *Canadian J. Chem. Eng.*, **43**(4), 205–8, 1965.

709. Ekiner, O. and Thodos, G., "Critical Temperatures and Critical Pressures of the Ethane–*n*–Pentane System," *J. Chem. Eng. Data*, **11**, 154–5, 1966.

710. Grieves, R. B. and Thodos, G., "The Critical Temperatures and Critical Pressures of Binary Systems: Hydrocarbons of All Types and Hydrogen," *Am. Inst. Chem. Eng. J.*, **6**, 561–6, 1960.

711. Grieves, R. B. and Thodos, G., "The Critical Temperatures and Critical Pressures of Binary Mixtures of the Fixed Gases and Aliphatic Hydrocarbons," *Soc. Pet. Eng. J.*, 194–202, 1962.

712. Grieves, R. B. and Thodos, G., "The Critical Temperatures of Multicomponent Hydrocarbon Systems," *Am. Inst. Chem. Eng. J.*, **8**, 550–3, 1962.

713. Grieves, R. B. and Thodos, G., "The Critical Pressures of Multicomponent Hydrocarbon Mixtures and the Critical Densities of Binary Hydrocarbon Mixtures," *Am. Inst. Chem. Eng. J.*, **9**, 25–30, 1963.

714. Grieves, R. B. and Thodos, G., "The Critical Temperatures of Ternary Hydrocarbon Systems," *Ind. Eng. Chem. Fundam.*, **1**, 45–8, 1962.

715. Mehra, V. S. and Thodos, G., "The Methane–Propane–*n*-Pentane System, Critical Temperatures and Pressures of Ternary Systems from Limited Data," *J. Chem. Eng. Data*, **7**, 497–9, 1962.

716. Cota, H. M. and Thodos, G., "Critical Temperatures and Critical Pressures of Hydrocarbon Mixtures, Methane–Ethane–*n*-Butane System," *J. Chem. Eng. Data*, **7**, 62–5, 1962.

717. Forman, J. C. and Thodos, G., "Experimental Determination of Critical Temperatures and Pressures of Mixtures: The Methane–Ethane–*n*-Butane System," *Am. Inst. Chem. Eng. J.*, **8**, 209–13, 1962.

718. Ekiner, O. and Thodos, G., "Critical Temperatures and Critical Pressures of the Ethane–*n*-Pentane–*n*-Heptane System," *J. Chem. Eng. Data*, **11**, 457–60, 1966.

719. Grieves, R. B. and Thodos, G., "Critical Temperatures and Pressures of Ternary Hydrocarbon Mixtures: The Ethane–Propane–*n*-Butane System," *J. Appl. Chem.*, **13**, 466–70, 1963.

720. Mehra, V. S. and Thodos, G., "Critical Temperatures and Critical Pressures for the Ethane–*n*-Butane–*n*-Pentane System," *J. Appl. Chem.*, **14**, 265–8, 1964.

721. Ekiner, O. and Thodos, G., "Interaction Model for Critical Temperatures of Multicomponent Mixtures of Methane-Free Alphatic Hydrocarbons," *Am. Inst. Chem. Eng. J.*, **11**, 897–900, 1965.

722. Ekiner, O. and Thodos, G., "Critical Temperatures of Methane-Aliphatic Hydrocarbon Mixtures," *Ind. Eng. Chem. Fundam.*, **6**, 222–4, 1967.

723. Ekiner, O. and Thodos, G., "Interaction Model for Critical Pressures of Multicomponent Methane-Free Aliphatic Hydrocarbon Mixtures," *Chem. Eng. Sci.*, **21**, 353–60, 1966.

724. Rastogi, R. P. and Girdhar, H. L., "Molecular Interaction in Saturated Hydrocarbons," *J. Chem. Phys.*, **36**, 998–1000, 1962.

725. Gunn, R. D., Chuch, P. L., and Prausnitz, J. M., "Predictions of Thermodynamic Properties of Dense Gas Mixtures Containing One or More of the Quantum Gases," *Am. Inst. Chem. Eng. J.*, 937–41, 1966.

726. Gambill, W. R., "Predict Critical Temperature," *Chem. Eng.*, **66**, 181–4, 1959.

727. Gambill, W. R., "How to Predict Critical Pressure," *Chem. Eng.*, **66**, 157–60, 1959.

728. Gambill, W. R., "How to Predict PVT Relations," *Chem. Eng.*, **66**, 195–202, 1959.

729. Keyes, F. G., "A Summary of Viscosity and Heat-Conduction Data for He, Ar, H_2, O_2, N_2, CO, CO_2, H_2O and Air," *Trans. Am. Soc. Mech. Engrs.*, **73**, 589–96, 1951.

730. Gambill, W. R., "Estimate Low-Pressure Gas Viscosity," *Chem. Eng.*, **65**, 169–72, 1958.

731. Westenberg, A. A., "Present Status of Information on Transport Properties Applicable to Combustion Research," *Combust. Flame*, **1**(3), 346–59, 1957.

732. Sutton, J. R., "A Method of Calculating the Viscosities of Polar Gases," from *Progress in International Research on Thermodynamic and Transport Properties* (Masi, J. F. and Tsai, D. H., Editors), Academic Press, New York, 266–70, 1962.

733. Klimov, V. L., "Approximated Equations for Collision Integrals $\Omega^{(1,s)*}$," *Teplofiz. Vys. Temp.*, **3**, 807–8, 1965; English translation: *High Temp.*, **3**, 747–8, 1965.

734. Kim, S. K. and Ross, J., "On the Determination of Potential Parameters from Transport Coefficients," *J. Chem. Phys.*, **46**, 818, 1967.

735. Brokaw, R. S., "Predicting Transport Properties of Dilute Gases," *Ind. Eng. Chem. Process Des. Dev.*, **8**, 240–53, 1969.

736. Bromley, L. A. and Wilke, C. R., "Viscosity Behavior of Gases," *Ind. Eng. Chem.*, **43**, 1641–8, 1951.

737. Holmes, J. T. and Baerns, M. G., "Predicting Physical Properties of Gases and Gas Mixtures," *Chem. Eng.*, **72**, 103–8, 1965.

738. Weintraub, M. and Corey, P. E., "High-Temperature Viscosity of Gases Estimated Quickly," *Chem. Eng.*, **74**(22), 204, 1967.

739. Brokaw, R. S., "Alignment Charts for Transport Properties Viscosity, Thermal Conductivity, and Diffusion Coefficients for Nonpolar Gases and Gas Mixtures at Low Density," NASA TR R-81, 23 pp., 1961.

740. Brokaw, R. S., "Recent Advances Concerning the Transport Properties of Dilute Gases," *Int. J. Eng. Sci.*, **3**(3), 251–67, 1965.

741. Licht, W. and Stechert, D. G., "The Variation of the Viscosity of Gases and Vapors with Temperature," *J. Phys. Chem.*, **48**, 23–47, 1944.

742. Rogers, J. D., Zeigler, K., and McWilliams, P., "Hydrogen Transport Property Correlations," *J. Chem. Eng. Data*, **7**, 179–82, 1962.

743. Fiore, A. W., "Viscosity of Air," *J. Spacecr. Rockets*, **3**(5), 756–8, 1966.

744. Bertram, M. H., "Comment on Viscosity of Air," *J. Spacecr. Rockets*, **4**(2), 287, 1967.

745. Fiore, A. W., "Reply by Author to M. H. Bertram's Comment," *J. Spacecr. Rockets*, **4**(2), 288, 1967.

746. Kestin, J. and Wang, H. E., "On the Correlation of Experimental Viscosity Data," *Physica*, **24**, 604–8, 1958.

747. Smith, A. S. and Brown, G. G., "Correlating Fluid Viscosity," *Ind. Eng. Chem.*, **35**, 705–11, 1943.

748. Whalley, E., "The Viscosity of Gases and the Theory of Corresponding States," *Can. J. Chem.*, **32**, 485–91, 1954.

749. Othmer, D. F. and Josefowitz, S., "Correlating Viscosities of Gases with Temperature and Pressure," *Ind. Eng. Chem.*, **38**, 111–6, 1946.

750. Gambill, W. R., "Hot T and P Change Gas Viscosity," *Chem. Eng.*, **65**(21), 157–62, 1958.

751. Bruges, E. A., Latto, B., and Ray, A. K., "New Correlations and Tables of the Coefficient of Viscosity of Water and Steam up to 1000 Bar and 1000 C," *Int. J. Heat Mass Transfer*, **9**, 465–80, 1966.

752. Lee, A. L., Starling, K. E., Dolan, J. P., and Ellington, R. T., "Viscosity Correlation for Light Hydrocarbon Systems," *Am. Inst. Chem. Eng. J.*, **10**, 694–7, 1964.

753. Lee, A. L. and Ellington, R. T., "Viscosity of *n*-Pentane," *J. Chem. Eng. Data*, **10**, 101–4, 1965.

754. Gonzalez, M. H. and Lee, A. L., "Graphical Viscosity Correlation for Hydrocarbons," *Am. Inst. Chem. Eng. J.*, **14**, 242–4, 1968.

755. Gegg, D. G. and Purchas, D. B., "Estimation of Viscosity of Gases," *Br. Chem. Eng.*, **10**, 850–1, 1965.

756. Shimotake, H. and Thodos, G., "Viscosity: Reduced-State Correlation for the Inert Gases," *Am. Inst. Chem. Eng. J.*, **4**, 257–62, 1958.

757. Trappeniers, N. J., Botzen, A., Ten Seldam, C. A., Van den Berg, H. R., and Van Oosten, J., "Corresponding States for the Viscosity of Noble Gases up to High Densities," *Physica*, **31**, 1681–91, 1965.

758. Brebach, W. J. and Thodos, G., "Viscosity-Reduced State Correlation for Diatomic Gases," *Ind. Eng. Chem.*, **50**, 1095–100, 1958.

759. Stiel, L. I. and Thodos, G., "Viscosity of Hydrogen in the Gaseous and Liquid States for Temperatures up to 5000," *Ind. Eng. Chem. Fundam.*, **2**, 233–7, 1963.

760. Rosenbaum, B. M. and Thodos, G., "Viscosity Correlation for Para-Hydrogen in the Gaseous and Liquid States," *J. Spacecr. Rockets*, **4**, 122–4, 1967.

761. Lo, H. Y., Carroll, D. L., and Stiel, L. I., "Viscosity of Gaseous Air at Moderate and High Pressures," *J. Chem. Eng. Data*, **11**, 540–4, 1966.

762. Kennedy, J. T. and Thodos, G., "The Transport Properties of Carbon Dioxide," *Am. Inst. Chem. Eng. J.*, **7**, 625–31, 1961.

763. Groenier, W. S. and Thodos, G., "Viscosity and Thermal Conductivity of Ammonia in the Gaseous and Liquid States," *J. Chem. Eng. Data*, **6**, 240–4, 1961.

764. Theiss, R. V. and Thodos, G., "Viscosity and Thermal Conductivity of Water: Gaseous and Liquid States," *J. Chem. Eng. Data*, **8**, 390–5, 1963.

765. Stiel, L. I. and Thodos, G., "The Viscosity of Nonpolar Gases at Normal Pressures," *Am. Inst. Chem. Eng. J.*, **7**, 611–5, 1961.

766. Mathur, G. P. and Thodos, G., "The Viscosity of Dissociated and Undissociated Gases for Temperatures up to 10,000 K," *Am. Inst. Chem. Eng. J.*, **9**, 596–600, 1963.

767. Stiel, L. I. and Thodos, G., "The Viscosity of Polar Gases at Normal Pressures," *Am. Inst. Chem. Eng. J.*, **8**, 229–32, 1962.

768. Starling, K. E. and Ellington, R. T., "Viscosity Correlations for Nonpolar Dense Fluids," *Am. Inst. Chem. Eng. J.*, **10**, 11–5, 1964.

769. Lennert, D. A. and Thodos, G., "Application of the Enskog Relationships for Prediction of the Transport Properties of Simple Substances," *Ind. Eng. Chem. Fundam.*, **4**, 139–41, 1965.

770. Elzinga, D. J. and Thodos, G., "The Transport Properties of p-Hydrogen from the Enskog Theory," *Cryogenics*, **6**(4), 216–21, 1966.

771. Jossi, J. A., Stiel, L. I., and Thodos, G., "The Viscosity of Pure Substances in the Dense Gaseous and Liquid Phases," *Am. Inst. Chem. Eng. J.*, **8**, 59–63, 1962.

772. Stiel, L. I. and Thodos, G., "The Viscosity of Polar Substances in the Dense Gaseous and Liquid Regions," *Am. Inst. Chem. Eng. J.*, **10**, 275–7, 1964.

773. Simon, H. A., Liu, C. S., and Hartnett, J. P., "Properties of Hydrogen–Nitrogen and Hydrogen–Carbon Dioxide Mixtures," *Int. J. Heat Mass Transfer*, **8**(8), 1176–8, 1965.

774. Rogers, J. D., Zeigler, R. K., and McWilliams, P., "Hydrogen Transport Property Correlations Part II," Los Alamos Scientific Laboratory Report LA-2719, 40 pp., 1962.

775. Childs, G. E. and Hanley, H. J. M. "The Viscosity and Thermal Conductivity Coefficients of Dilute Nitrogen and Oxygen," NBS Tech. Note 350, 27 pp., 1966.

776. Brush, S. G. and Lawrence, J. D., "Transport Coefficients for the Square Well Potential Model," UCRL-7376, 25 pp., 1963.

777. Kessel'man, P. M. and Chernyshev, S. K., "Thermal Properties of Some Hydrocarbons at High Temperatures," *Teplofiz. Vys. Temp.*, **3**, 700–7, 1965; English translation: *High Temp.*, **3**, 651–7, 1965.

778. Partington, J., *An Advanced Treatise on Physical Chemistry*, Longmans, Green and Co., London, Vol. I, 943 pp., 1949.

779. Enskog, D., "Kinetic Theory of Processes in Moderately Low Pressure Gases," Inaugural Dissertation, Uppsala, Sweden, 1917. As quoted in Ref. 669.

780. Gambell, W. R., "To Get Viscosity for a Gas Mixture," *Chem. Eng.*, **65**(23), 157–60, 1958.

781. Buddenberg, J. W. and Wilke, C. R., "Calculation of Gas Mixture Viscosities," *Ind. Eng. Chem.*, **41**, 1345–7, 1949.

782. Wilke, C. R., "A Viscosity Equation for Gas Mixtures," *J. Chem. Phys.*, **18**, 517–9, 1950.

783. Saxena, S. C. and Narayanan, T. K. S., "Multicomponent Viscosities of Gaseous Mixtures at High Temperatures," *Ind. Eng. Chem. Fundam.*, **1**, 191–5, 1962.

784. Mathur, S. and Saxena, S. C., "A Quick and Approximate Method for Estimating the Viscosity of Multicomponent Gas Mixtures," *Indian J. Pure Appl. Phys.*, **3**, 138–40, 1965.

785. Mathur, S. and Saxena, S. C., "Viscosity of Polar Gas Mixtures: Wilkes' Method," *Appl. Sci. Res.*, **15A**, 404–10, 1965.

786. Mathur, S. and Saxena, S. C., "Viscosity of Polar–Nonpolar Gas Mixtures: Empirical Method," *Indian J. Phys.*, **39**, 278–82, 1965.

787. Herning, F. and Zipperer, L., "Calculation of the Viscosities of Technical Gas Mixtures from the Viscosity of the Individual Gases," *Gas Wasserfach*, **79**, 49–54, 69–73, 1936.

788. Tondon, P. K. and Saxena, S. C., "Calculation of Viscosities of Mixtures Containing Polar Gases," *Indian J. Pure Appl. Phys.*, **6**, 475–8, 1968.

789. Dean, D. E. and Stiel, L. I., "The Viscosity of Nonpolar Gas Mixtures at Moderate and High Pressures," *Am. Inst. Chem. Eng. J.*, **11**, 526–32, 1965.

790. Strunk, M. R., Custead, W. G., and Stevenson, G. L., "The Prediction of the Viscosity of Nonpolar Binary Gaseous Mixtures at Atmospheric Pressure," *Am. Inst. Chem. Eng. J.*, **10**, 483–6, 1964.

791. Strunk, M. R. and Fehsenfeld, G. D., "The Prediction of the Viscosity of Multicomponent, Nonpolar Gaseous Mixtures at Atmospheric Pressure," *Am. Inst. Chem. Eng. J.*, **11**, 389–90, 1965. (Tabular material has been deposited with the American Documentation Institute, Photoduplication Service, Library of Congress, Washington 25, D.C., as ADI Document 8254, 12 pp.)

792. Ulybin, S. A., "Temperature Dependence of the Viscosity of Rarefied Gas Mixtures," *Teplofiz. Vys. Temp.*, **2**, 583–7, 1964; English translation: *High Temp.*, **2**, 526–30, 1964.

793. Saxena, S. C., "Comments on the Ulybin et al. Method of Calculating Thermal Conductivities of Mixtures of Chemically Non-Reacting Gases at Ordinary Pressures," *Mol. Phys.*, **18**, 123–7, 1970.

794. Cowling, T. G., "Appendix, The Theoretical Basis of Wassiljewa's Equation," *Proc. Roy. Soc. (London)*, **A263**, 186–7, 1961.

795. Cowling, T. G., Gray, P., and Wright, P. G., "The Physical Significance of Formulae for the Thermal Conductivity and Viscosity of Gaseous Mixtures," *Proc. Roy. Soc. (London)*, **A276**, 69–82, 1963.

796. Francis, W. E., "Viscosity Equations for Gas Mixtures," *Trans. Faraday Soc.*, **54**, 1492–7, 1958.

797. Brokaw, R. S., "Approximate Formulas for the Viscosity and Thermal Conductivity of Gas Mixtures," *J. Chem. Phys.*, **29**, 391–7, 1958.

798. Brokaw, R. S., "Approximate Formulas for the Viscosity and Thermal Conductivity of Gas Mixtures. II," *J. Chem. Phys.*, **42**, 1140–6, 1965.

799. Hansen, C. F., "Interpretation of Linear Approximations for the Viscosity of Gas Mixtures," *Phys. Fluids*, **4**, 926–7, 1961.

800. Wright, P. G. and Gray, P., "Collisional Interference Between Unlike Molecules Transporting Momentum or Energy in Gases," *Trans. Faraday Soc.*, **58**, 1–16, 1962.

801. Burnett, D., "Viscosity and Thermal Conductivity of Gas Mixtures. Accuracy of Some Empirical Formulas," *J. Chem. Phys.*, **42**, 2533–40, 1965.

802. Yos, J. M., "Approximate Equations for the Viscosity and Translational Thermal Conductivity of Gas Mixtures," AVCO Missiles Space and Electronics Group Rept., Wilmington, Mass., 56 pp., 1967.

803. Saxena, S. C. and Gambhir, R. S., "Semi-Empirical Formulae for the Viscosity and Translational Thermal Conductivity of Gas Mixtures," *Proc. Phys. Soc.*, **81**, 788–9, 1963.

804. Saxena, S. C. and Gambhir, R. S., "A Semi-Empirical Formula for the Viscosity of Multicomponent Gas Mixtures," *Indian J. Pure Appl. Phys.*, **1**, 208–15, 1963.

805. Mathur, S. and Saxena, S. C., "A Semi-Empirical Formula for the Viscosity of Polar Gas Mixtures," *Br. J. Appl. Phys.*, **16**, 389–94, 1965.

806. Gambhir, R. S. and Saxena, S. C., "Translational Thermal Conductivity and Viscosity of Multicomponent Gas Mixtures," *Trans. Faraday Soc.*, **60**, 38–44, 1964.

807. Saksena, M. P. and Saxena, S. C., "Viscosity of Multicomponent Gas Mixtures," *Proc. Natl. Inst. Sci. (India)*, **31A**, 18–25, 1965.

808. Mathur, S. and Saxena, S. C., "Viscosity of Multicomponent Gas Mixtures of Polar Gases," *Appl. Sci. Res.*, **15**, 203–15, 1965.

809. Brokaw, R. S., Svehla, R. A., and Baker, C. E., "Transport Properties of Dilute Gas Mixtures," NASA TN D-2580, 15 pp., 1965.

810. Saxena, S. C. and Gambhir, R. S., "Viscosity and Translational Thermal Conductivity of Gas Mixtures," *Br. J. Appl. Phys.*, **14**, 436–38, 1963.

811. Gandhi, J. M. and Saxena, S. C., "An Approximate Method for the Simultaneous Prediction of Thermal Conductivity and Viscosity of Gas Mixtures," *Indian J. Pure Appl. Phys.*, **2**, 83–5, 1964.

812. Mason, E. A. and Saxena, S. C., "Approximate Formula for the Thermal Conductivity of Gas Mixtures," *Phys. Fluids*, **1**, 361–9, 1958.

813. Tondon, P. K. and Saxena, S. C., "Modification of Brokaw's Method for Calculating Viscosity of Mixtures of Gases," *Ind. Eng. Chem. Fundam.*, **7**, 314, 1968.

814. Brokaw, R. S., "Viscosity of Gas Mixtures," NASA TN D-4496, 25 pp., 1968.

815. Gupta, G. P. and Saxena, S. C., "Calculation of Viscosity and Diffusion Coefficients of Nonpolar Gas Mixtures at Ordinary Pressures," *Am. Inst. Chem. Eng. J.*, **14**, 519–20, 1968. (See also document No. 9883 with the American Documentation Institute, Photoduplication Service, Library of Congress, Washington 25, D.C.)

816. Saxena, S. C. and Agrawal, J. P., "Interrelation of Thermal Conductivity and Viscosity of Binary Gas Mixtures," *Proc. Phys. Soc.*, **80**, 313–5, 1962.

817. Saxena, S. C. and Tondon, P. K., "Thermal Conductivity of Multicomponent Mixtures of Rare Gases," in *Proceedings of the Fourth Symposium on Thermophysical Properties* (Moszynski, J. R., Editor), The American Society of Mechanical Engineers, New York, 398–404, 1968.

818. Saxena, S. C. and Gupta, G. P., "Experimental Data and Prediction Procedures for Thermal Conductivity of Multicomponent Mixtures of Nonpolar Gases," *J. Chem. Eng. Data*, **15**(1), 98–107, 1970.

819. Gupta, S. C., "Transport Coefficients of Binary Gas Mixtures," *Physica*, **35**, 395–404, 1967.

820. Gupta, G. P. and Saxena, S. C., "Prediction of Thermal Conductivity of Pure Gases and Mixtures," *Supp. Def. Sci. J.*, **17**, 21–34, 1967.

821. Gandhi, J. M. and Saxena, S. C., "Correlation Between Thermal Conductivity and Diffusion of Gases and Gas Mixtures of Monatomic Gases," *Proc. Phys. Soc.*, **87**, 273–9, 1966.

822. Mathur, S. and Saxena, S. C., "Relations Between Thermal Conductivity and Diffusion Coefficients of Pure and Mixed Polyatomic Gases," *Proc. Phys. Soc.*, **89**, 753–64, 1966.

823. Nain, V. P. S. and Saxena, S. C., "Measurement of the Concentration Diffusion Coefficient for Ne–Ar, Ne–Xe, Ne–H_2, Xe–H_2, H_2–N_2, and H_2–O_2 Gas Systems," *Appl. Sci. Res.* (in press).

824. Malinauskas, A. P. and Silverman, M. D., "Gaseous Diffusion in Neon-Noble Gas Systems," *J. Chem. Phys.*, **50**, 3263–70, 1969.

825. Wright, P. G., "A Method of Obtaining Sutherland-Wassiljewa Coefficients," in *Proceedings Leeds Philosophical and Literary Soc., Scientific Section*, Vol. IX, Pt. VIII, 215–21, 1964.

826. Huck, R. J. and Thornton, E., "Sutherland-Wassiljewa Coefficients for the Viscosity of Binary Rare Gas Mixtures," *Proc. Phys. Soc.*, **92**, 244–52, 1967.

827. O'Neal, C. and Brokaw, R. S., "Relation Between Thermal Conductivity and Viscosity for Some Nonpolar Gases," *Phys. Fluids*, **5**, 567–74, 1962.

828. Saxena, V. K. and Saxena, S. C., "Thermal Conductivity of Krypton and Xenon in the Temperature Range 350–1500 K," *J. Chem. Phys.*, **51**, 3361–8, 1969.

829. Saxena, S. C., Gupta, G. P., and Saxena, V. K., "Measurement of the Thermal Conductivity of Nitrogen (350 to 1500 K) by the Column Method," in *Proceedings of the Eighth Conference on Thermal Conductivity* (Ho, C. Y. and Taylor, R. E., Editors), Plenum Press, New York, 125–39, 1969.

830. Saxena, S. C. and Gupta, G. P., "The Column Method of Measuring Thermal Conductivity of Gases: Results on Carbon Monoxide and Oxygen in the Temperature Range 350 to 1500 K," AIAA 4th Thermophysics Conf., Paper No. 69–603, 8 pp., 1969.

831. Dunstan, A. E. and Thole, F. B., *The Viscosity of Liquids*, Longmans, Green and Co., London, 91 pp., 1914.

832. Hatschek, E., *The Viscosity of Liquids*, D. Van Nostrand Co., New York, 239 pp., 1928.

833. Barr, G., *A Monograph of Viscometry*, Oxford University Press, London, 318 pp., 1931.

834. Van Wazer, J. R., Lyons, J. W., Kim, K. Y., and Colwell, R. E., *Viscosity and Flow Measurement: A Laboratory Handbook of Rheology*, Interscience Publishers, New York, 406 pp., 1963.

835. Kestin, J., "Direct Determination of the Viscosity of Gases at High Pressures and Temperatures," in *Proc. Second Biennial Gas Dynamics Symp. on Transport Properties in Gases* (Cambel, A. B. and Fenn, J. B., Editors), NorthWestern University Press, Evanston, Ill., 182 pp., 1958.

836. Hagan, G., "The Movement of Water in Narrow Cylindrical Tubes," *Ann. Phys.*, **46**, 423–42, 1839.

837. Poiseville, J. L. M., Mém. Savants É'trangers, **9**, p. 433, 1846; *Compt. Rend.*, **11**, 961, p. 1041, 1840; **12**, 112, 1841; **15**, 1167, 1842.

838. Fryer, G. M., "A Theory of Gas Flow Through Capillary Tubes," *Proc. Roy. Soc. (London)*, **A293**, 329–41, 1966.

839. Shimotake, H. and Thodos, G., "The Viscosity of Ammonia: Experimental Measurements for the Dense Gaseous Phase and a Reduced State Correlation for the Gaseous and Liquid Regions," *Am. Inst. Chem. Eng. J.*, **9**, 68–72, 1963.

840. Flynn, G. P., Hanks, R. V., Lemaire, N. A., and Ross, J., "Viscosity of Nitrogen, Helium, Neon, and Argon from −78.5 to 100 C Below 200 Atmospheres," *J. Chem. Phys.*, **38**, 154–62, 1963.

841. Giddings, J. G., Kao, J. T. F., and Kobayashi, R., "Development of a High-Pressure Capillary-Tube Viscometer and Its Application to Methane, Propane, and Their Mixtures in the Gaseous and Liquid Regions," *J. Chem. Phys.*, **45**(2), 578–86, 1966.

842. Carr, N. L., Parent, J. D., and Peck, R. E., "Viscosity of Gases and Gas Mixtures at High Pressures," *Chem. Eng. Prog. Symp. Ser.*, **51**(16), 91–9, 1955.

843. Graham, T., "On the Motion of Gases," *Phil. Trans.*, **136**, 573–631, 1846; **139**, 349–91, 1849.

844. Edwards, R. S., "On the Effect of Temperature on the Viscosity of Air," *Proc. Roy. Soc. (London)*, **A117**, 245–57, 1927.

845. Williams, F. A., "The Effect of Temperature on the Viscosity of Air," *Proc. Roy. Soc. (London)*, **A110**, 141–67, 1926.

846. Rankine, A. O., "The Effect of Temperature on the Viscosity of Air," *Proc. Roy. Soc. (London)*, **A111**, 219–23, 1926.

847. Kenney, M. J., Sarjant, R. J., and Thring, M. W., "The Viscosity of Mixtures of Gases at High Temperatures," *Br. J. Appl. Phys.*, **7**, 324–9, 1956.

848. Bonilla, C. F., Brooks, R. D., and Walker, P. L., "The Viscosity of Steam and of Nitrogen at Atmospheric Pressure and High Temperatures," in *Proceedings of the General Discussion on Heat Transfer*, The Institution of Mechanical Engineers, London, 167–73, 1951.

849. White, C. M., "Streamline Flow Through Curved Pipes," *Proc. Roy. Soc. (London)*, **A123**, 645–63, 1929.

850. Bonilla, C. F., Wang, S. J., and Weiner, H., "The Viscosity of Steam, Heavy-Water Vapor, and Argon at Atmospheric Pressure up to High Temperatures," *Trans. Am. Soc. Mech. Eng.*, **78**, 1285–9, 1956.

851. McCoubrey, J. C. and Singh, N. M., "Intermolecular Forces in Quasi-Spherical Molecules," *Trans. Faraday Soc.*, **53**, 877–83, 1957.

852. McCoubrey, J. C. and Singh, N. M., "The Vapor Phase Viscosities of the Pentanes," *J. Phys. Chem.*, **67**, 517–8, 1963.

853. Salzberg, H. W., "A Simple Gas Viscosity Experiment," *J. Chem. Educ.*, **42**, 663, 1965.

854. Trautz, M. and Weizel, W., "Determination of the Viscosity of Sulfur Dioxide and its Mixtures with Hydrogen," *Ann. Phys.*, **78**, 305–69, 1925.

855. Rankine, A. O., "On a Method of Determining the Viscosity of Gases, Especially Those Available only in Small Quantities," *Proc. Roy. Soc. (London)*, **83A**, 265–76, 1910.

856. Rankine, A. O., "On the Viscosities of the Gases of the Argon Group," *Proc. Roy. Soc. (London)*, **83A**, 516–25, 1910.

857. Rankine, A. O., "Viscosity of Gases of the Argon Group," *Proc. Roy. Soc. (London)*, **84A**, 181–92, 1910.

858. Rankine, A. O., "A Simple Viscometer for Gases," *J. Sci. Instrum.*, **1**, 105–11, 1924.

859. Rankine, A. O. and Smith, C. J., "On the Viscosity and Molecular Dimensions of Gaseous Ammonia, Phosphine, and Arsine," *Phil. Mag.*, **43**, 603–14, 1921.

860. Rankine, A. O., "The Viscosity and Molecular Dimensions of Gaseous Cyanogen," *Proc. Roy. Soc. (London)*, **99A**, 331–6, 1921.

861. Rankine, A. O. and Smith, C. J., "On the Viscosities and Molecular Dimensions of Methane, Sulphuretted Hydrogen and Cyanogen," *Phil. Mag.*, **42**, 615–20, 1921.

862. Comings, E. W. and Egly, R. S., "Viscosity of Ethylene and of Carbon Dioxide under Pressure," *Ind. Eng. Chem.*, **33**, 1224–9, 1941.

863. Baron, J. D., Roof, J. G., and Wells, F. W., "Viscosity of Nitrogen, Methane, Ethane, and Propane at Elevated Temperature and Pressure," *J. Chem. Eng. Data*, **4**, 283–8, 1959.

Hagen, G. O.
Poiseville

864. Heath, H. R., "The Viscosity of Gas Mixtures," *Proc. Phys. Soc.* (*London*), **66B**, 362–7, 1953.

865. Thornton, E., "Viscosity and Thermal Conductivity of Binary Gas Mixtures: Xenon–Krypton, Xenon–Argon, Xenon–Neon, and Xenon–Helium," *Proc. Phys. Soc.* (*London*), **76**, 104–12, 1960.

866. Thornton, E., "Viscosity and Thermal Conductivity of Binary Gas Mixtures: Krypton–Argon, Krypton–Neon, and Krypton–Helium," *Proc. Phys. Soc.* (*London*), **77**, 1166–9, 1961.

867. Thornton, E. and Baker, W. A. D., "Viscosity and Thermal Conductivity of Binary Gas Mixtures: Argon–Neon, Argon–Helium, and Neon–Helium," *Proc. Phys. Soc.* (*London*), **80**, 1171–5, 1962.

868. Raw, C. J. G. and Ellis, C. P., "High-Temperature Gas Viscosities. I. Nitrous Oxide and Oxygen," *J. Chem. Phys.*, **28**, 1198–1200, 1958.

869. Ellis, C. P. and Raw, C. J. G., "High-Temperature Gas Viscosities. II. Nitrogen, Nitric Oxide, Boron Trifluoride, Silicon Tetrafluoride, and Sulfur Hexafluoride," *J. Chem. Phys.*, **30**, 574–6, 1959.

870. Hawksworth, W. A., Nourse, H. H. E., and Raw, C. J. G., "High-Temperature Gas Viscosities. III. NO–N$_2$O Mixtures," *J. Chem. Phys.*, **37**, 918–9, 1962.

871. Raw, C. J. G. and Tang, H., "Viscosity and Diffusion Coefficients of Gaseous Sulfur Hexafluoride-Carbon Tetrafluoride Mixtures," *J. Chem. Phys.*, **39**, 2616–8, 1963.

872. Burch, L. G. and Raw, C. J. G., "Transport Properties of Polar-Gas Mixtures. I. Viscosities of Ammonia-Methylamine Mixtures," *J. Chem. Phys.*, **47**, 2798–2801, 1967.

873. Chang, K. C., Hesse, R. J., and Raw, C. J. G., "Transport Properties of Polar Gas Mixtures SO$_2$ + SO$_2$F$_2$ Mixtures," *Trans. Faraday Soc.*, **66**, 590–6, 1970.

874. Rigby, M. and Smith, E. B., "Viscosities of Inert Gases," *Trans. Faraday Soc.*, **62**, 54–8, 1966.

875. Clarke, A. G. and Smith, E. B., "Low-Temperature Viscosities of Argon, Krypton, and Xenon," *J. Chem. Phys.*, **48**, 3988–91, 1968.

876. Clarke, A. G. and Smith, E. B., "Low-Temperature Viscosities and Intermolecular Forces of Simple Gases," *J. Chem. Phys.*, **51**, 4156–61, 1969.

877. Dawe, R. A. and Smith, E. B., "Viscosity of Argon at High Temperatures," *Science*, **163**, 675–6, 1969.

878. Timrot, D. L., "Determination of the Viscosity of Steam and Water at High Temperatures and Pressures," *J. Phys.* (*USSR*), **2**, 419–35, 1940.

879. Makavetskas, R. A., Popov, V. N., and Tsederberg, N. V., "Experimental Study of the Viscosity of Helium and Nitrogen," *Teplofiz. Vys. Temp.*, **1**(2), 191–7, 1963.

880. Makavetskas, R. A., Popov, V. N., and Tsederberg, N. V., "An Experimental Investigation of the Viscosity of Mixtures of Nitrogen and Helium," *Teplofiz. Vys. Temp.*, **1**(3), 348–55, 1963.

881. Vasilesco, V., "Experimental Research on the Viscosity of Gases at High Temperatures," *Ann. Phys.*, **20**, 137–76, 1945.

882. Lazarre, F. and Vodar, B., "Determination of the Viscosity of Nitrogen Compressed, Up to 3000 Kg cm^2," *Compt. Rend.*, **242**, 468, 1956.

883. Lazarre, F. and Vodar, B., "Measurement of the Viscosity of Compressed Nitrogen up to 3000 Atmospheres," in *Conference on Thermodynamic and Transport Properties of Fluids*, London, 159–62, 1957.

884. Luker, J. A. and Johnson, C. A., "Viscosity of Helium, Oxygen, Helium–Oxygen, Helium–Steam, and Oxygen–Steam Mixtures at High Temperatures and Pressures," *J. Chem. Eng. Data*, **4**, 176–82, 1959.

885. Andreev, I. I., Tsederberg, V. N., and Popov, V. N., "Experimental Investigation of the Viscosity of Argon," *Teploenergetika*, **13**(8), 78–81, 1966.

886. Rivkin, S. L. and Levin, A. Ya., "Experimental Study of the Viscosity of Water and Steam," *Teploenergetika*, **13**(4), 79–83, 1966.

887. Lee, D. I. and Bonilla, C. F., "The Viscosity of the Alkali Metal Vapors," *Nuc. Eng. Des.*, **7**, 445–69, 1968.

888. Barua, A. K., Afzal, M., Flynn, G. P., and Ross, J., "Viscosity of Hydrogen, Deuterium, Methane, and Carbon Monoxide from −50 to 150 C Below 200 Atmospheres," *J. Chem. Phys.*, **41**, 374–8, 1964.

889. Gracki, J. A., Flynn, G. P., and Ross, J., "Viscosity of Nitrogen, Helium, Hydrogen, and Argon from −100 to 24 C up to 150–250 Atmospheres," *J. Chem. Phys.*, **51**, 3856–63, 1969.

890. Kao, J. T. F. and Kobayashi, R., "Viscosity of Helium and Nitrogen and Their Mixtures at Low Temperatures and Elevated Pressures," *J. Chem. Phys.*, **47**, 2836–49, 1967.

891. Michels, A. and Gibson, R. O., "The Measurement of the Viscosity of Gases at High Pressures—The Viscosity of Nitrogen to 1000 Atmospheres," *Proc. Roy. Soc.* (*London*), **A134**, 288–307, 1931.

892. Michels, A., Schipper, A. C. J., and Rintoul, W. H., "The Viscosity of Hydrogen and Deuterium at Pressures up to 2000 Atmospheres," *Physica*, **19**, 1011–28, 1953.

893. Michels, A., Botzen, A., and Schuurman, W., "The Viscosity of Argon at Pressures up to 2000 Atmospheres," *Physica*, **20**, 1141–8, 1954.

894. Michels, A., Botzen, A., and Schuurman, W., "The Viscosity of Carbon Dioxide Between 0 and 75 C and at Pressures up to 2000 Atmospheres," *Physica*, **23**, 95–102, 1957.

895. Trappeniers, N. J., Botzen, A., Van den Berg, H. R., and Van Oosten, J., "The Viscosity of Neon Between 25 C and 75 C at Pressures up to 1800 Atmospheres. Corresponding States for the Viscosity of the Noble Gases up to High Densities," *Physica*, **30**, 985–6, 1964.

896. Trappeniers, N. J., Botzen, A., Van Oosten, J., and Van den Berg, H. R., "The Viscosity of Krypton Between 25 and 75 C and at Pressures up to 2000 Atmospheres," *Physica*, **31**, 945–52, 1965.

897. Bond, W. N., "The Viscosity of Air," *Proc. Phys. Soc.*, **49**, 205–13, 1937.

898. Rigden, P. J., "The Viscosity of Air, Oxygen, and Nitrogen," *Phil. Mag.*, **25**, 961–81, 1938.

899. Thacker, R. and Rowlinson, J. S., "The Physical Properties of Some Polar Solutions, Part 2. The Viscosities of the Mixed Vapours," *Trans. Faraday Soc.*, **50**, 1158–63, 1954.

900. Chakraborti, P. K. and Gray, P., "Viscosities of Gaseous Mixtures Containing Polar Gases: Mixtures with One Polar Constituent," *Trans. Faraday Soc.*, **61**, 2422–34, 1965.

901. Chakraborti, P. K. and Gray, P., "Viscosities of Gaseous Mixtures Containing Polar Gases: More than One Polar Constituent," *Trans. Faraday Soc.*, **62**, 1769–75, 1966.

902. Lambert, J. D., Cotton, K. J., Pailthorpe, M. W., Robinson, A. M., Scrivins, J., Vale, W. R. F., and Young, R. M., "Transport Properties of Gaseous Hydrocarbons," *Proc. Roy. Soc.* (*London*), **A231**, 280–90, 1955.

903. Shimotake, H. and Thodos, G., "Viscosity of Sulfur Dioxide at 200 C for Pressures up to 3500 PSI," *J. Chem. Eng. Data*, **8**, 88–90, 1968.

904. Reynes, E. G. and Thodos, G., "The Viscosity of Argon, Krypton, and Xenon in the Dense Gaseous Region," *Physica*, **30**, 1529–42, 1964.

905. DeWitt, K. J. and Thodos, G., "Viscosities of Binary Mixtures in the Dense Gaseous State: The Methane–Tetrafluoromethane System," *Physica*, **32**, 1459–72, 1966.

906. DeWitt, K. J. and Thodos, G., "Viscosities of Binary Mixtures in the Dense Gaseous State: The Methane–Carbon Dioxide System," *Can. J. Chem. Eng.*, **44**(3), 148–51, 1966.

907. Reynes, E. G. and Thodos, G., "Viscosity of Helium, Neon, and Nitrogen in the Dense Gaseous Region," *J. Chem. Eng. Data*, **11**, 137–40, 1966.

908. Eakin, B. E. and Ellington, R. T., "Improved High Pressure Capillary Tube Viscometer," *Petroleum Trans. AIME*, **216**, 85–91, 1959.

909. Starling, K. E., Eakin, B. E., and Ellington, R. T., "Liquid, Gas, and Dense-Fluid Viscosity of Propane," *Am. Inst. Chem. Eng. J.*, **6**, 438–42, 1960.

910. Eakin, B. E., Starling, K. E., Dolan, J. P., and Ellington, R. T., "Liquid, Gas, and Dense Fluid Viscosity of Ethane," *J. Chem. Eng. Data*, **7**, 33–6, 1962.

911. Dolan, J. P., Starling, K. E., Lee, A. L., Eakin, B. E., and Ellington, R. T., "Liquid, Gas, and Dense Fluid Viscosity of *n*-Butane," *J. Chem. Eng. Data*, **8**, 396–9, 1963.

912. Dolan, J. P., Ellington, R. T., and Lee, A. L., "Viscosity of Methane–*n*-Butane Mixtures," *J. Chem. Eng. Data*, **9**, 484–7, 1964.

913. Gonzalez, M. H. and Lee, A. L., "Viscosity of Isobutane," *J. Chem. Eng. Data*, **11**, 357–9, 1966.

914. Lee, A. L., Gonzalez, M. H., and Eakin, B. E., "Viscosity of Methane–*n*-Decane Mixtures," *J. Chem. Eng. Data*, **11**, 281–7, 1966.

915. Gonzalez, M. H., Bukacek, R. F., and Lee, A. L., "Viscosity of Methane," *Soc. Pet. Eng. J.*, **7**(1), 75–9, 1967.

916. Gonzalez, M. H. and Lee, A. L., "Viscosity of 2,2-Dimethylpropane," *J. Chem. Eng. Data*, **13**, 66–9, 1968.

917. Hanley, H. J. M. and Childs, G. E., "Discrepancies Between Viscosity Data for Simple Gases," *Science*, **159**, 1114–7, 1968.

918. Maxwell, J. D., "On the Viscosity or Internal Friction of Air and Other Gases," *Phil. Trans. Roy. Soc. (London)*, **156**, 249–59, 1866.

919. Craven, P. M. and Lambert, J. D., "The Viscosities of Organic Vapours," *Proc. Roy. Soc. (London)*, **A205**, 439–49, 1951.

920. Van Itterbeek, A. and Claes, A., "Viscosity of Gaseous Oxygen at Low Temperatures. Dependence on the Pressure," *Physica*, **3**, 275–81, 1936.

921. Van Itterbeek, A. and Claes, A., "Measurements on the Viscosity of Hydrogen and Deuterium Gas Between 293 K and 14 K," *Physica*, **5**(10), 938–44, 1938.

922. Van Itterbeek, A. and Keesom, W. H., "Measurements on the Viscosity of Helium Gas Between 293 and 1.6 K," *Physica*, **5**, 257–69, 1938.

923. Van Itterbeek, A. and Van Paemel, O., "Measurement on the Velocity of Sound as a Function of Pressure in Oxygen Gas at Liquid Oxygen Temperatures. Calculation of the Sound Virial Coefficient and the Specific Heat," *Physica*, **5**(7), 593–604, 1938.

924. Van Itterbeek, A. and Van Paemel, O., "Measurements of the Viscosity of Neon, Hydrogen, Deuterium, and Helium as a Function of the Temperature Between Room Temperature and Liquid-Hydrogen Temperatures," *Physica*, **7**, 265–72, 1940.

925. Keesom, W. H. and Macwood, G. E., "The Viscosity of Liquid Helium," *Physica*, **5**, 737–44, 1938.

926. Keesom, W. H. and Macwood, G. E., "The Viscosity of Hydrogen Vapor," *Physica*, **5**, 749–52, 1938.

927. Macwood, G. E., "The Theory of the Measurement of Viscosity and Slip of Fluids by the Oscillating Disk Method. I," *Physica*, **5**, 374–84, 1938.

928. Macwood, G. E., "The Theory of the Measurement of Viscosity and Slip of Fluids by the Oscillating Disk Method. II," *Physica*, **5**, 763–8, 1938.

929. Van Itterbeek, A. and Keesom, W. H., "Measurement of the Viscosity of Oxygen Gas at Liquid-Oxygen Temperatures," *Physica*, **2**, 97–103, 1935.

930. Van Itterbeek, A. and Van Paemel, O., "Measurements of the Viscosity of Argon Gas at Room Temperature and Between 90 and 55 K," *Physica*, **5**, 1009–12, 1938.

931. Van Itterbeek, A., Van Paemel, O., and Van Lierde, J., "Measurements on the Viscosity of Gas Mixtures," *Physica*, **13**, 88–96, 1947.

932. Rietveld, A. O., Van Itterbeek, A., and Van Den Berg, G. J., "Measurement on the Viscosity of Mixtures of Helium and Argon," *Physica*, **19**, 517–24, 1953.

933. Rietveld, A. O. and Van Itterbeek, A., "Measurements on the Viscosity of Ne–Ar Mixtures Between 300 and 70 K," *Physica*, **22**, 785–90, 1956.

934. Rietveld, A. O. and Van Itterbeek, A., "Viscosity of Mixtures of H_2 and HD Between 300 and 14 K," *Physica*, **23**, 838–42, 1957.

935. Coremans, J. M. J., Van Itterbeek, A., Beenakker, J. J. M., Knaap, H. F. P., and Zandbergen, P., "The Viscosity of Gaseous He, Ne, H_2, and D_2 Below 80 K," *Physica*, **24**, 557–76, 1958.

936. Coremans, J. M. J., Van Itterbeek, A., Beenakker, J. J. M., Knaap, H. F. P., and Zandbergen, P., "The Viscosity of Gaseous HD Below 80 K," *Physica*, **24**, 1102–4, 1958.

937. Rietveld, A. O., Van Itterbeek, A., and Velds, C. A., "Viscosity of Binary Mixtures of Hydrogen Isotopes and Mixtures of Helium and Neon," *Physica*, **25**, 205–16, 1959.

938. Sutherland, B. P. and Maass, O., "Measurement of the Viscosity of Gases over a Large Temperature Range," *Can. J. Res.*, **6**, 428–43, 1932.

939. Mason, S. G. and Maass, O., "Measurement of Viscosity in the Critical Region. Ethylene," *Can. J. Res.*, **18B**, 128–37, 1940.

940. Johnston, H. L. and McCloskey, K. E., "Viscosities of Several Common Gases Between 90 K and Room Temperature," *J. Phys. Chem.*, **44**, 1038–58, 1940.

941. Johnston, H. L. and Grilly, E. R., "Viscosities of Carbon Monoxide, Helium, Neon, and Argon Between 80 and 300 K. Coefficients of Viscosity," *J. Phys. Chem.*, **46**, 948–63, 1942.

942. Kestin, J. and Pilarezyk, K., "Measurement of the Viscosity of Five Gases at Elevated Pressures by the Oscillating Disk Method," *Trans. ASME*, **76**, 987–99, 1954.

943. Kestin, J. and Wang, H. E., "Corrections for the Oscillating Disk Viscometer," *J. Appl. Mechanics Trans. ASME*, **79**, 197–206, 1957.

944. Kestin, J. and Wang, H. E., "The Viscosity of Five Gases: A Re-Evaluation," *Trans. ASME*, **80**, 11–7, 1958.

945. Kestin, J., Leidenfrost, W., and Liu, C. Y., "On Relative Measurements of the Viscosity of Gases by the Oscillating Disk Method," *Z. Angew. Math. Phys. (ZAMP)*, **10**, 558–64, 1959.

946. Kestin, J. and Leidenfrost, W., "The Viscosity of Helium," *Physica*, **25**, 537–55, 1959.

947. Kestin, J. and Leidenfrost, W., "The Effect of Moderate Pressures on the Viscosity of Five Gases," from *Thermodynamics and Transport Properties of Gases and Liquids* (Touloukian, Y. S., Editor), *ASME Symposium*, McGraw-Hill, 321–38, 1959.

948. Kestin, J. and Moszynski, J. R., "Instruments for the Measurement of the Viscosity of Steam and Compressed Water," *Trans. ASME*, **80**, 1009–14, 1958.

949. Mariens, P. and Van Paemel, O., "Theory and Experimental Verification of the Oscillating Disk Method for Viscosity Measurements in Fluids," *Appl. Sci. Res.*, **A5**(5), 411–24, 1955.

950. Dash, J. G. and Taylor, R. D., "Hydrodynamics of Oscillating Disks in Viscous Fluids: Density and Viscosity of Normal Fluid in Pure He⁴ from 1.2 K to the Lambda Point," *Phys. Rev.*, **105**(1), 7–24, 1957.

951. Newell, G. F., "Theory of Oscillation Type Viscometers. V. Disk Oscillating Between Fixed Plates," *Z. Angew. Math. Phys. (ZAMP)*, **10**(2), 160–74, 1959.

952. Kestin, J. and Leidenfrost, W., "An Absolute Determination of the Viscosity of Eleven Gases over a Range of Pressures," *Physica*, **25**, 1033–62, 1959.

953. Kestin, J. and Leidenfrost, W., "The Effect of Pressure on the Viscosity of N_2–CO_2 Mixtures," *Physica*, **25**, 525–36, 1959.

954. Iwasaki, H. and Kestin, J., "The Viscosity of Argon–Helium Mixtures," *Physica*, **29**, 1345–72, 1963.

955. Iwasaki, H., Kestin, J., and Nagashima, A., "Viscosity of Argon–Ammonia Mixtures," *J. Chem. Phys.*, **40**, 2988–95, 1964.

956. Kestin, J. and Nagashima, A., "Viscosity of Neon–Helium and Neon–Argon Mixtures at 20 and 30 C," *J. Chem. Phys.*, **40**, 3648–54, 1964.

957. Kestin, J. and Nagashima, A., "Viscosity of the Isotopes of Hydrogen and their Intermolecular Force Potentials," *Phys. Fluids*, **7**, 730–4, 1964.

958. Breetveld, J. D., Di Pippo, R., and Kestin, J., "Viscosity and Binary Diffusion Coefficient on Neon–Carbon Dioxide Mixtures at 20 and 30 C," *J. Chem. Phys.*, **45**, 124–6, 1966; Comment, *Ibid*, **46**, 1541, 1967.

959. Kestin, J., Kobayashi, Y., and Wood, R. T., "The Viscosity of Four Binary Gaseous Mixtures at 20 and 30 C," *Physica*, **32**, 1065–89, 1966.

960. Di Pippo, R., Kestin, J., and Oguchi, K., "Viscosity of Three Binary Gaseous Mixtures," *J. Chem. Phys.*, **46**, 4758–64, 1967.

961. Kestin, J. and Yata, J., "Viscosity and Diffusion Coefficient of Six Binary Mixtures," *J. Chem. Phys.*, **49**, 4780–91, 1968.

962. Di Pippo, R., Kestin, J., and Whitelaw, J. H., "A High-Temperature Oscillating Disk Viscometer," *Physica*, **32**, 2064–80, 1966.

963. Clifton, D. G., "Measurement of the Viscosity of Krypton," *J. Chem. Phys.*, **38**, 1123–31, 1963.

964. Pal, A. K. and Barua, A. K., "Viscosity of Hydrogen–Nitrogen and Hydrogen–Ammonia Gas Mixtures," *J. Chem. Phys.*, **47**, 216–8, 1967.

965. Kestin, J. and Whitelaw, J. H., "A Relative Determination of the Viscosity of Several Gases by the Oscillating Disk Method," *Physica*, **29**(4), 335–56, 1963.

966. Pal, A. K. and Barua, A. K., "Viscosity and Intermolecular Potentials of Hydrogen Sulphide," *Trans. Faraday Soc.*, **63**, 341–6, 1967.

967. Pal, A. K., "Intermolecular Forces and Viscosity of Some Polar Organic Vapours," *Indian J. Phys.*, **41**, 823–7, 1967.

968. Pal, A. K. and Barua, A. K., "Viscosity of Polar–Nonpolar Gas Mixtures," *Indian J. Phys.*, **41**, 713–8, 1967.

969. Pal, A. K. and Barua, A. K., "Intermolecular Potentials and Viscosities of Some Polar Organic Vapours," *Br. J. Appl. Phys. (J. Phys. D)*, **1**, 71–6, 1968.

970. Gururaja, G. J., Tirunarayanan, M. A., and Ramachandran, R., "Dynamic Viscosity of Gas Mixtures," *J. Chem. Eng. Data*, **12**(4), 562–7, 1967.

971. Gilchrist, L., "An Absolute Determination of the Viscosity of Air," *Phys. Rev.*, **1**, 124–40, 1913.

972. Harrington, E. L., "A Redetermination of the Absolute Value of the Coefficient of Viscosity of Air," *Phys. Rev.*, **8**, 738–51, 1916.

973. Yen, K. L., "An Absolute Determination of the Coefficients of Viscosity of Hydrogen, Nitrogen, and Oxygen," *Phil. Mag.*, **38**, 582–97, 1919.

974. Van Dyke, K. S., "The Coefficients of Viscosity and of Slip of Air and of Carbon Dioxide by the Rotating Cylinder Method," *Phys. Rev.*, **21**, 250–65, 1923.

975. Millikan, R. A., "Coefficients of Slip in Gases and the Law of Reflection of Molecules from the Surfaces of Solids and Liquids," *Phys. Rev.*, **B21**, 217–38, 1923.

976. Stacy, L. J., "A Determination by the Constant Deflection Method of the Value of the Coefficient of Slip for Rough and for Smooth Surfaces in Air," *Phys. Rev.*, **21**, 239–49, 1923.

977. States, M. N., "The Coefficient of Viscosity of Helium and the Coefficients of Slip of Helium and Oxygen by the Constant Deflection Method," *Phys. Rev.*, **21**, 662–71, 1923.

978. Blankenstein, E., "Coefficients of Slip and Momentum Transfer in Hydrogen, Helium, Air and Oxygen," *Phys. Rev.*, **22**, 582–9, 1923.

979. Day, R. K., "Variation of the Vapor Viscosities of Normal and Isopentane with Pressure by the Rotating Cylinder Method," *Phys. Rev.*, **40**, 281–90, 1932.

980. Houston, W. V., "The Viscosity of Air," *Phys. Rev.*, **52**, 751–7, 1937.

981. Kellstrom, G., "A New Determination of the Viscosity of Air by the Rotating Cylinder Method," *Phil. Mag.*, **23**, 313–38, 1937.

982. Reamer, H. H., Cokelet, G., and Sage, B. H., "Viscosity of Fluids at High Pressures, Rotating Cylinder Viscometer and the Viscosity of *n*-Pentane," *Anal. Chem.*, **31**, 1422–8, 1959.

983. Carmichael, L. T. and Sage, B. H., "Viscosity of Ethane at High Pressures," *J. Chem. Eng. Data*, **8**, 94–8, 1963.

984. Carmichael, L. T., Reamer, H. H., and Sage, B. H., "Viscosity of Ammonia at High Pressures," *J. Chem. Eng. Data*, **8**, 400–4, 1963.

985. Carmichael, L. T. and Sage, B. H., "Viscosity and Thermal Conductivity of Nitrogen–*n*-Heptane and Nitrogen–*n*-Octane Mixtures," *Am. Inst. Chem. Eng. J.*, **12**, 559–62, 1966.

986. Carmichael, L. T., Berry, V., and Sage, B. H., "Viscosity of a Mixture of Methane and *n*-Butane," *J. Chem. Eng. Data*, **12**, 44–7, 1967.

987. Ishida, Y., "Determination of Viscosities and of the Stokes-Millikan Law Constant by the Oil-Drop Method," *Phys. Rev.*, **21**, 550–63, 1923.

988. Hawkins, G. A., Solberg, H. L., and Potter, A. A., "The Viscosity of Water and Superheated Steam," *Trans. ASME*, **57**(7), 395–400, 1935.

989. Hubbard, R. M. and Brown, G. G., "The Rolling Ball Viscometer," *Ind. Eng. Chem., Anal. Educ.*, **15**, 212–8, 1943.

990. Bicher, L. B. and Katz, D. L., "Viscosities of the Methane-Propane System," *Ind. Eng. Chem.*, **35**, 754–61, 1943.

991. Swift, G. W., Christy, J. A., Heckes, A. A., and Kurata, F., "Determining Viscosity of Liquefied Gaseous Hydrocarbons at Low Temperatures and High Pressures," *Chem. Eng. Prog.*, **54**, 47–50, 1958.

992. Swift, G. W., Lohrenz, J., and Kurata, F., "Liquid Viscosities Above the Normal Boiling Point for Methane, Ethane, Propane and *n*-Butane," *Am. Inst. Chem. Eng. J.*, **6**, 415–9, 1960.

993. Huang, E. T. S., Swift, G. W., and Kurata, F., "Viscosities of Methane and Propane at Low Temperatures and High Pressures," *Am. Inst. Chem. Eng. J.*, **12**, 932–6, 1966.

994. Huang, E. T. S., Swift, G. W., and Kurata, F., "Viscosities and Densities of Methane-Propane Mixtures at Low Temperatures and High Pressures," *Am. Inst. Chem. Eng. J.*, **13**, 846–50, 1967.

995. Herzfeld, K. F. and Litovitz, T. A., *Absorption and Dispersion of Ultrasonic Waves*, Academic Press, Inc., New York, 535 pp., 1959.

996. Carnevale, E. H., Carey, C. A., and Larsen, G. S., "Experimental Determination of the Transport Properties of Gases," Panametrics Technical Report AFML-TR-65-141, 57 pp., August 1965.

997. Carnevale, E. H., Wolnik, S., Larson, G., Carey, C., and Wares, G. W., "Simultaneous Ultrasonic and Line Reversal Temperature Determination in a Shock Tube," *Phys. Fluids*, **10**, 1459–67, 1967.

998. Carnevale, E. H., Lynnworth, L. C., and Larson, G. S., "Ultrasonic Determination of Transport Properties of Monatomic Gases at High Temperatures," *J. Chem. Phys.*, **46**, 3040–7, 1967.

999. Carnevale, E. H., Larson, G., Lynnworth, L. C., Carey, C., Panaro, M., and Marshall, T., "Experimental Determination of Transport Properties of High Temperature Gases," NASA CR-789, 67 + A44, June 1967.

1000. Carnevale, E. H., Carey, C., Marshall, T., and Uva, S., "Experimental Determination of Gas Properties at High Temperatures and/or Pressures," Panametrics Rept. AEDC-TR-68-105, 107 pp., June 1968.

1001. Carey, C., Carnevale, E. H., Uva, S., and Marshall, T., "Experimental Determination of Gas Properties at High Temperatures and/or Pressures," Panametrics Rept. AEDC-TR-69-78, 51 pp., March 1969.

1002. Ahtye, W. F., "A Critical Evaluation of the Use of Ultrasonic Absorption for Determining High-Temperature Gas Properties," NASA TN D-4433, 66 pp., March 1968.

1003. Madigosky, W. M., "Density Dependence of the Bulk Viscosity in Argon," *J. Chem. Phys.*, **46**, 4441–4, 1967.

1004. Carey, C. A., Carnevale, E. H., and Marshall, T., "Experimental Determination of the Transport Properties of Gases, Past II. Heat Transfer and Ultrasonic Measurements," Panametrics Rept. AFML-TR-65-141, Pt. II, 96 pp., September 1966.

1005. Hartunian, R. A. and Marrone, P. V., "Viscosity of Dissociated Gases from Shock-Tube Heat-Transfer Measurements," *Phys. Fluids*, **4**, 535–43, 1961.

1006. Emmons, H. W., "Arc Measurement of High-Temperature Gas Transport Properties," *Phys. Fluids*, **10**, 1125–36, 1967.

1007. Schreiber, P. W., Schumaker, K. H., and Benedetto, K. R., "Experimental Determination of Plasma Transport Properties," in *Proceedings of the Eighth Conference on Thermal Conductivity* (Ho, C. Y. and Taylor, R. E., Editors), Plenum Press, New York, 249–63, 1969.

1008. Frenkel, J., "Kinetic Theory of Liquids," Dover Publications, Inc., New York, 488 pp., 1955.

1009. Green, H. S., *The Molecular Theory of Fluids*, North-Holland Publishing Co., Amsterdam, 264 pp., 1952.

1010. Rice, S. A. and Gray, P., *The Statistical Mechanics of Simple Liquids. An Introduction to the Theory of Equilibrium and Non-Equilibrium Phenomena*, Interscience Publishers, New York, 582 pp., 1965.

1011. Kirkwood, J. G., *Theory of Liquids* (Alder, B. J., Editor), Gordon and Breach, Science Publishers, New York, 140 pp., 1968.

1012. Rice, S. A., "The Kinetic Theory of Dense Fluids," Colloquium Lectures in Pure and Applied Science, No. 9, Mobil Oil Corp. Research Dept. Field Research Lab., Dallas, Texas, 308 pp., 1964.

1013. Kimball, G. E., "The Liquid State," Chapter III of *A Treatise on Physical Chemistry* (Taylor, H. S. and Glasstone, S., Editors), D. Van Nostrand Co., Inc., New York, Vol. II of 3rd Edition, 701 pp., 1951.

1014. Levelt, J. M. H. and Cohen, E. G. D., "A Critical Study of Some Theories of the Liquid State Including a Comparison with Experiment," Part B in *Studies in Statistical Mechanics*, North-Holland Publishing Co., Amsterdam, 249 pp., 1962.

1015. Brush, S. G., "Theories of Liquid Viscosity," University of California, Lawrence Radiation Lab., Livermore, Calif., Rept. No. UCRL-6400, 106 pp., 1961.

1016. Partington, J. R., "An Advanced Treatise on Physical Chemistry," Vol. II of *The Properties of Liquids*, Longmans, Green and Co., New York, 448 pp., 1951.

1017. Hildebrand, J. H., "Models and Molecules—Seventh Spiers Memorial Lecture," *Faraday Soc. Discus.*, **15**, 9–23, 1953.

1018. Andrade, E. N. da C., "A Theory of the Viscosity of Liquids —Part I," *Phil. Mag.*, **17**, 497–511, 1934.

1019. Andrade, E. N. da C., "A Theory of the Viscosity of Liquids —Part II," *Phil. Mag.*, **17**, 698–732, 1934.

1020. Eyring, H., "Viscosity, Plasticity, and Diffusion as Examples of Absolute Reaction Rates," *J. Chem. Phys.*, **4**, 283–91, 1936.

1021. Glasstone, S., Laidler, K. J., and Eyring, H., *The Theory of Rate Processes*, McGraw-Hill, New York, 611 pp., 1941.

1022. Ewell, R. H. and Eyring, H., "Theory of the Viscosity of Liquids as a Function of Temperature and Pressure," *J. Chem. Phys.*, **5**, 726–36, 1937.

1023. Eyring, H. and Hirschfelder, J. O., "The Theory of the Liquid State," *J. Phys. Chem.*, **41**, 249–57, 1937.

1024. Hirschfelder, J. O., Stevenson, D., and Eyring, H., "A Theory of Liquid Structure," *J. Chem. Phys.*, **5**, 896–912, 1937.

1025. Walter, J. and Eyring, H., "A Partition Function for Normal Liquids," *J. Chem. Phys.*, **9**, 393–7, 1941.

1026. Eyring, H., Ree, T., and Hirai, N., "Significant Structures in the Liquid State. I," *Proc. Natl. Acad. Sci.*, **44**, 683–8, 1958.

1027. Fuller, E. J., Ree, T., and Eyring, H., "Significant Structures in Liquids. II," *Proc. Natl. Acad. Sci.*, **45**, 1594–9, 1959.

1028. Carlson, C. M., Eyring, H., and Ree, T., "Significant Structures in Liquids. III," *Proc. Natl. Acad. Sci.*, **46**, 333–6, 1960.

1029. Thomson, T. R., Eyring, H., and Ree, T., "Significant Structures in Liquids. IV. Liquid Chlorine," *Proc. Natl. Acad. Sci.*, **46**, 336–43, 1960.

1030. Ree, F. H., Ree, T., and Eyring, H., "Relaxation Theory of Transport Problems in Condensed Systems," *Ind. Eng. Chem.*, **50**, 1036–40, 1958.

1031. Carlson, C. M., Eyring, H., and Ree, T., "Significant Structures in Liquids. V. Thermodynamic and Transport Properties of Molten Metals," *Proc. Natl. Acad. Sci.*, **46**, 649–59, 1960.

1032. Eyring, H. and Ree, T., "Significant Liquid Structures. VI. The Vacancy Theory of Liquids," *Proc. Natl. Acad. Sci.*, **47**, 526–37, 1961.

1033. Ree, T. S., Ree, T., and Eyring, H., "Significant Liquid Structure Theory. IX. Properties of Dense Gases and Liquids," *Proc. Natl. Acad. Sci.*, **48**, 501–17, 1962.

1034. Lu, W-C., Ree, T., Gerrard, V. G., and Eyring, H., "Significant Structure Theory Applied to Molten Salts," *J. Chem. Phys.*, **49**, 797–804, 1968.

1035. Lennard-Jones, J. E. and Devonshire, A. F., "Critical Phenomena in Gases—I," *Proc. Roy. Soc. (London)*, **163A**, 53–70, 1937.

1036. Lennard-Jones, J. E. and Devonshire, A. F., "Critical Phenomena in Gases. II. Vapour Pressures and Boiling Points," *Proc. Roy. Soc. (London)*, **165A**, 1–11, 1938.

1037. Pople, J. A., "Molecular Association in Liquids. III. A Theory of Cohesion of Polar Liquids," *Proc. Roy. Soc. (London)*, **215A**, 67–83, 1952.

1038. Furth, R., "On the Theory of the Liquid State. III. The Hole Theory of the Viscous Flow of Liquids," *Proc. Camb. Phil. Soc.*, **37**, 281–90, 1941.

1039. Furth, R., "On the Theory of the Liquid State. I. The Statistical Treatment of the Thermodynamics of Liquids by the Theory of Holes," *Proc. Camb. Phil. Soc.*, **37**, 252–75, 1941.

1040. Eisenschitz, R., "The Effect of Temperature on the Thermal Conductivity and Viscosity of Liquids," *Proc. Phys. Soc. (London)*, **59**, 1030–6, 1947.

1041. Wentorf, R. H., Buehler, R. J., Hirschfelder, J. O., and Curtiss, C. F., "Lennard-Jones and Devonshire Equation of State of Compressed Gases and Liquids," *J. Chem. Phys.*, **18**, 1484–500, 1950.

1042. Kirkwood, J. G., "Critique of the Free Volume Theory of the Liquid State," *J. Chem. Phys.*, **18**, 380–2, 1950.

1043. Rowlinson, J. S. and Curtiss, C. F., "Lattice Theories of the Liquid State," *J. Chem. Phys.*, **19**, 1519–29, 1951.

1044. Buehler, R. J., Wentorf, R. H., Hirschfelder, J. O., and Curtiss, C. F., "The Free Volume for Rigid Sphere Molecules," *J. Chem. Phys.*, **19**, 61–71, 1951.

1045. Dahler, J. S., Hirschfelder, J. O., and Thacher, H. C., "Improved Free-Volume Theory of Liquids. I," *J. Chem. Phys.*, **25**, 249–60, 1956.

1046. Dahler, J. S. and Hirschfelder, J. O., "Improved Free-Volume Theory of Liquids. II," *J. Chem. Phys.*, **32**, 330–49, 1960.

1047. Chung, H. S. and Dahler, J. S., "Improved Free Volume Theory of Liquids. III. Approximate Theory of Molecular Correlations in Liquids," *J. Chem. Phys.*, **37**, 1620–30, 1962.

1048. De Boer, J., "Cell-Cluster Theory for the Liquid State. I," *Physica*, **20**, 655–64, 1954.

1049. Cohen, E. G. D., De Boer, J., and Salsburg, Z. W., "A Cell-Cluster Theory for the Liquid State. II," *Physica*, **21**, 137–47, 1955.

1050. Dahler, J. S. and Cohen, E. G. D., "Cell-Cluster Theory for the Liquid State. VI. Binary Liquid Solutions and Hole Theory," *Physica*, **26**, 81–102, 1960.

1051. Collins, F. C. and Raffel, H., "Approximate Treatment of the Viscosity of Idealized Liquids. I. The Collisional Contribution," *J. Chem. Phys.*, **22**, 1728–33, 1956.

1052. Mayer, J. E. and Montroll, E., "Molecular Distribution," *J. Chem. Phys.*, **9**, 2–16, 1941.

1053. Mayer, J. E., "Integral Equations Between Distribution Functions of Molecules," *J. Chem. Phys.*, **15**, 187–201, 1947.

1054. Born, M. and Green, H. S., "A General Kinetic Theory of Liquids. I. The Molecular Distribution Functions," *Proc. Roy. Soc. (London)*, **A188**, 10–8, 1946.

1055. Born, M. and Green, H. S., "A General Kinetic Theory of Liquids. III. Dynamical Properties," *Proc. Roy. Soc. (London)*, **A190**, 455–74, 1947.

1056. Green, H. S., "A General Kinetic Theory of Liquids. II. Equilibrium Properties," *Proc. Roy. Soc. (London)*, **A189**, 103–16, 1947.

1057. Kirkwood, J. G., Buff, F. P., and Green, M. S., "The Statistical Mechanical Theory of Transport Processes. III. The Coefficients of Shear and Bulk Viscosity of Liquids," *J. Chem. Phys.*, **17**, 988–94, 1949.

1058. Kirkwood, J. G., "Statistical Mechanics of Fluid Mixtures," *J. Chem. Phys.*, **3**, 300–13, 1935.

1059. Kirkwood, J. G. and Salsburg, Z. W., "The Statistical Mechanical Theory of Molecular Distribution Functions in Liquids," *Faraday Soc. Discuss.*, **15**, 28–34, 1953.

1060. Kirkwood, J. G., "Molecular Distribution in Liquids," *J. Chem. Phys.*, **7**, 919–25, 1939.

1061. Kirkwood, J. G. and Boggs, E. M., "The Radial Distribution Function in Liquids," *J. Chem. Phys.*, **10**, 394–402, 1942.

1062. Kirkwood, J. G., Maun, E. K., and Alder, B. J., "Radial Distribution Function and the Equation of State of a Fluid Composed of Rigid Spherical Molecules," *J. Chem. Phys.*, **18**, 1040–7, 1950.

1063. Kirkwood, J. G., Lewinson, V. A., and Alder, B. J., "Radial Distribution Functions and the Equation of State of Fluids Composed of Molecules Interacting According to the Lennard-Jones Potential," *J. Chem. Phys.*, **20**, 929–38, 1952.

1064. De Boer, J., "Theories of the Liquid State," *Proc. Roy. Soc. (London)*, **A215**, 4–29, 1952.

1065. Eisenschitz, R., "Transport Processes in Liquids," *Proc. Roy. Soc. (London)*, **A215**, 29–36, 1952.

1066. Andrade, E. N. da C., "Viscosity of Liquids," *Proc. Roy. Soc. (London)*, **A215**, 36–43, 1952.

1067. Collins, F. C. and Navidi, M. H., "The Calculation of the Free Volumes of Liquids from Measurements of Sonic Velocity," *J. Chem. Phys.*, **22**, 1254–5, 1954.

1068. Eisenschitz, R., "The Steady Non-Uniform State for a Liquid," *Proc. Phys. Soc.*, **A62**, 41–9, 1949.

1069. Rice, S. A. and Allnatt, A. R., "On the Kinetic Theory of Dense Fluids. VI. Singlet Distribution Function for Rigid Spheres with an Attractive Potential," *J. Chem. Phys.*, **34**, 2144–55, 1961.

1070. Allnatt, A. R. and Rice, S. A., "On the Kinetic Theory of Dense Fluids. VII. The Doublet Distribution Function for Rigid Spheres with an Attractive Potential," *J. Chem. Phys.*, **34**, 2156–65, 1961.

1071. Hiroike, K., Gray, P., and Rice, S. A., "On the Kinetic Theory of Dense Fluids. XIX. Comments on and a Re-derivation of the Kinetic Equations," *J. Chem. Phys.*, **42**, 3134–43, 1965.

1072. Lowry, B. A., Rice, S. A., and Gray, P., "On the Kinetic Theory of Dense Fluids. XVII. The Shear Viscosity," *J. Chem. Phys.*, **40**, 3673–83, 1964.

1073. Wei, C. C. and Davis, H. T., "Kinetic Theory of Dense Fluid Mixtures. III. The Doublet Distribution Functions of the Rice–Allnatt Model," *J. Chem. Phys.*, **46**, 3456–67, 1967.

1074. Wei, C. C. and Davis, H. T., "Kinetic Theory of Dense Fluid Mixtures. II. Solution to the Singlet Distribution Functions for the Rice–Allnatt Model," *J. Chem. Phys.*, **45**, 2533–44, 1966.

1075. de Boer, J., "Quantum Properties of the Condensed State," in *Proc. Intl. Conf. Theor., Phys.*, Kyoto and Tokyo, 507–30, 1953.

1076. Dahler, J. S., "Calculation of the Radial Distribution Function from the Cell Theory of Liquids," *J. Chem. Phys.*, **29**, 1082–5, 1958.

1077. Davis, H. T., Rice, S. A., and Sengers, J. V., "On the Kinetic Theory of Dense Fluids. IX. The Fluid of Rigid Spheres with a Square-Well Attraction," *J. Chem. Phys.*, **35**, 2210–33, 1961.

1078. Davis, H. T. and Luks, K. D., "Transport Properties of a Dense Fluid of Molecules Interacting with a Square-Well Potential," *J. Phys. Chem.*, **69**, 869–80, 1965.

1079. Luks, K. D., Miller, M. A., and Davis, H. T., "Transport Properties of a Dense Fluid of Molecules Interacting with a Square-Well Potential: Part II," *Am. Inst. Chem. Eng. J.*, **12**, 1079–86, 1966.

1080. Kadanoff, L. P. and Martin, P. C., "Hydrodynamic Equations and Correlation Functions," *Ann. Phys.*, **24**, 419–69, 1963.

1081. Forster, D., Martin, P. C., and Yip, S., "Moment Method Approximation for the Viscosity of Simple Liquids: Application to Argon," *Phys. Rev.*, **170**, 160–3, 1968.

1082. Zwanzig, R. W., Kirkwood, J. G., Stripp, K. F., and Oppenheim, I., "The Statistical Mechanical Theory of Transport Processes. VI. A Calculation of the Coefficients of Shear and Bulk Viscosity of Liquids," *J. Chem. Phys.*, **21**, 2050–5, 1953.

1083. Levelt, J. M. H. and Hurst, R. P., "Quantum Mechanical Cell Model of the Liquid State. I," *J. Chem. Phys.*, **32**, 96–104, 1960.

1084. Alder, B. J. and Wainwright, T. E., "Studies in Molecular Dynamics. I. General Method," *J. Chem. Phys.*, **31**, 459–66, 1959.

1085. Bueche, F., "Viscosity of Entangled Polymers, Theory of Variation with Shear Rate," *J. Chem. Phys.*, **48**, 4781–4, 1968.

1086. Sharp, P. and Bloomfield, V. A., "Intrinsic Viscosity of Wormlike Chains with Excluded-Volume Effects," *J. Chem. Phys.*, **48**, 2149–55, 1968.

1087. Ishihara, A., "Viscosity of Rodlike Molecules in Solution," *J. Chem. Phys.*, **49**, 257–60, 1968.

1088. Ullman, R., "Intrinsic Viscosity of Wormlike Polymer Chains," *J. Chem. Phys.*, **49**, 5486–97, 1968.

1089. Imai, S., "Intrinsic Viscosity of Polyelectrolytes," *J. Chem. Phys.*, **50**, 2107–15, 1969.

1090. Helfand, E. and Rice, S. A., "Principle of Corresponding States for Transport Properties," *J. Chem. Phys.*, **32**, 1642–4, 1960.

1091. Rogers, J. D. and Brickwedde, F. G., "Comparison of Saturated-Liquid Viscosities of Low Molecular Substances According to the Quantum Principle of Corresponding States," *Physica*, **32**, 1001–18, 1966.

1092. Boon, J. P. and Thomaes, G., "The Viscosity of Liquefied Gases," *Physica*, **29**, 208–14, 1963.

1093. Boon, J. P., Legros, J. C., and Thomaes, G., "On the Principle of Corresponding States for the Viscosity of Simple Liquids," *Physica*, **33**, 547–57, 1967.

1094. Boon, J. P. and Thomaes, G., "The Fluidity of Binary Mixtures," *Physica*, **28**, 1074–6, 1962.

1095. Boon, J. P. and Thomaes, G., "The Fluidity of Argon–Methane and Krypton–Methane Mixtures," *Physica*, **29**, 123–8, 1963.

1096. Fontaine-Limbourg, M. C., Legros, J. C., Boon, J. P., and Thomaes, G., "The Fluidity of Argon–Oxygen and Methane–Deuteromethane Mixtures," *Physica*, **31**, 396–400, 1965.

1097. Holleman, Th. and Hijmans, J., "A Principle of Corresponding States for the Thermodynamic Excess Functions of Binary Mixtures of Chain Molecules," *Physica*, **28**, 604–16, 1962.

1098. Gambill, W. R., "How to Calculate Liquid Viscosity Without Experimental Data," *Chem. Eng.*, **66**(1), 127–30, 1959.

1099. Gambill, W. R., "How P and T Change Liquid Viscosity," *Chem. Eng.*, **66**(3), 123–6, 1959.

1100. Lennert, D. A. and Thodos, G., "Thermal Pressure Applied to the Prediction of Viscosity of Simple Substances in the Dense Gaseous and Liquid Regions," *Am. Inst. Chem. Eng. J.*, **11**, 155–8, 1965.

1101. Dolan, J. P., Starling, K. E., Lee, A. L., Eakin, B. E., and Ellington, R. T., "Liquid, Gas and Dense Fluid Viscosity of *n*-Butane," *J. Chem. Eng. Data*, **8**, 396–9, 1963.

1102. Lee, A. L. and Ellington, R. T., "Viscosity of *n*-Decane in the Liquid Phase," *J. Chem. Eng. Data*, **10**, 346–8, 1965.

1103. Othmer, D. F. and Conwell, J. W., "Correlating Viscosity and Vapor Pressure of Liquids," *Ind. Eng. Chem.*, **37**, 1112–5, 1945.

1104. Othmer, D. F. and Silvis, S. J., "Correlating Viscosities," *Ind. Eng. Chem.*, **42**, 527–8, 1950.

1105. Thomas, L. H., "The Dependence of the Viscosities of Liquids on Reduced Temperature, and a Relation of Viscosity, Density, and Chemical Constitution," *J. Chem. Soc.*, **Part II**, 573–9, 1946.

1106. Auluck, F. C., De, S. C., and Kothari, D. S., "The Hole Theory of Liquid State," *Proc. Natl. Inst. Sci.*, **10**(4), 397–405, 1944.

1107. Das, T. R., Ibrahim, S. H., and Kuloor, N. R., "Correlations for Determining Normal Boiling Point and Kinematic Viscosity of Organic Liquids," *Indian. J. Tech.*, **7**, 131–8, 1969.

1108. Gambill, W. R., "How to Estimate Mixture Viscosities," *Chem. Eng.*, **66**(5), 151–2, 1959.

1109. Katti, P. K. and Chaudhri, M. M., "Viscosities of Binary Mixtures of Benzyl Acetate with Dioxane, Aniline and *m*-Cresol," *J. Chem. Eng. Data*, **9**, 442–3, 1964.

1110. Katti, P. K. and Prakash, O., "Viscosities of Binary Mixtures of Carbon Tetrachloride with Methanol and Isopropyl Alcohol," *J. Chem. Eng. Data*, **11**, 46–7, 1966.

1111. Katti, P. K., Chaudhri, M. M., and Prakash, O., "Viscosities of Binary Mixtures Involving Benzene, Carbon Tetrachloride, and Cyclohexane," *J. Chem. Eng. Data*, **11**, 593–4, 1966.

1112. Katti, P. K. and Prakash, O., "Boiling Points and Viscosities of Binary Mixtures of Ethanol and Carbon Tetrachloride," *Indian Chem. Engineer (Trans.)*, **8**, 69–72, 1966.

1113. Heric, E. L., "On the Viscosity of Ternary Mixtures," *J. Chem. Eng. Data*, **11**, 66–8, 1966.

1114. Kalidas, R. and Laddha, G. S., "Viscosity of Ternary Liquid Mixtures," *J. Chem. Eng. Data*, **9**, 142–5, 1964.

1115. Huang, E. T. S., Swift, G. W., and Kurata, F., "Viscosities and Densities of Methane–Propane Mixtures at Low Temperatures and High Pressures," *Am. Inst. Chem. Eng. J.*, **13**, 846–50, 1967.

1116. Reynolds, O., "On the Theory of Lubrication and its Application to Mr. Beauchamp Tower's Experiments, Including an Experimental Determination of the Viscosity of Olive Oil," *Phil. Trans.*, **177**, 157–234, 1886.

1117. Lipkin, M. R., Davison, J. A., and Kurtz, S. S., "Viscosity of Propane, Butane, and Isobutane," *Ind. Eng. Chem.*, **34**, 976–8, 1942.

1118. Boon, J. P. and Thomaes, G., "The Viscosity of Liquid Deuteromethane," *Physica*, **28**, 1197–8, 1962.

1119. Legros, J. C. and Thomaes, G., "The Viscosity of Liquid Xenon," *Physica*, **31**, 703–5, 1965.

1120. Denny, V. E. and Ferenbaugh, R., "Properties of Superheated Liquids: Viscosity of Carbon Tetrachloride," *J. Chem. Eng. Data*, **12**, 397–8, 1967.

1121. Mullin, J. W. and Osman, M. M., "Diffusivity, Density, Viscosity, and Refractive Index of Nickel Ammonium Sulfate Aqueous Solutions," *J. Chem. Eng. Data*, **12**, 516–7, 1967.

1122. Swindells, J. F., Coe, J. R., and Godfrey, T. B., 'Absolute Viscosity of Water at 20 C," *J. Res. Natl. Bur. Stand.*, **48**, 1–31, 1952.

1123. Van Itterbeek, A., Zink, H., and van Paemel, O., "Viscosity Measurements in Liquefied Gases," *Cryogenics*, **2**(4), 210–1, 1962.

1124. Van Itterbeek, A., Zink, H., and Hellemans, J., "Viscosity of Liquefied Gases at Pressures Above One Atmosphere," *Physica*, **32**, 489–93, 1966.

1125. Van Itterbeek, A., Hellemans, J., Zink, H., and Van Cauteren, M., "Viscosity of Liquefied Gases at Pressures Between 1 and 100 Atmosphere," *Physica*, **32**, 2171–2, 1966.

1126. Hubbard, R. M. and Brown, G. G., "Viscosity of *n*-Pentane," *Ind. Eng. Chem.*, **35**, 1276–80, 1943.

1127. Chacon-Tribin, H., Loftus, J., and Salterfield, C. N., "Viscosity of the Vandium Pentoxide-Potassium Sulfate Eutectic," *J. Chem. Eng. Data*, **11**, 44–5, 1966.

1128. Riebling, E. F., "Improved Counterbalanced Sphere Viscometer for Use to 1750 C," *Rev. Sci. Instrum.*, **34**, 568–72, 1963.

1129. Moynihan, C. T. and Cantor, S., "Viscosity and its Temperature Dependence in Molten BeF_2," *J. Chem. Phys.*, **48**, 115–9, 1968.

1130. Cantor, S., Ward, W. T., and Moynihan, C. T., "Viscosity and Density in Molten BeF_2–LiF Solutions," *J. Chem. Phys.*, **50**, 2874–9, 1969.

1131. Cottingham, D. M., "Simple Viscometer for Use with Low Melting Point Metals," *Br. J. Appl. Phys.*, **12**, 625–8, 1961.

1132. Welber, B., "Damping of a Torsionally Oscillating Cylinder in Liquid Helium at Various Temperatures and Densities," *Phys. Rev.*, **119**, 1816–22, 1960.

1133. Welber, B. and Qumby, S. L., "Measurement of the Product of Viscosity and Density of Liquid Helium with a Torsional Crystal," *Phys. Rev.*, **107**(3), 645–6, 1957.

1134. Webeler, R. W. H. and Hammer, D. C., "Viscosity × Normal Density of Liquid Helium in a Temperature Interval about the Lambda Point," *Phys. Letters*, **15**, 233–4, 1965.

1135. Webeler, R. W. H. and Hammer, D. C., "Viscosity Coefficients and the Phonon Density Temperature Dependence in Liquid ^{4}He," *Phys. Letters*, **19**, 533–4, 1965.

1136. Webeler, R. W. H. and Hammer, D. C., "Viscosity Coefficients for Liquid Helium-3 in the Interval 0.36 to 2.6 K," *Phys. Letters*, **21**, 403–4, 1966.

1137. De Bock, A., Grevendonk, W., and Awouters, H., "Pressure Dependence of the Viscosity of Liquid Argon and Liquid Oxygen, Measured by Means of a Torsionally Vibrating Quartz Crystal," *Physica*, **34**, 49–52, 1967.

1138. De Bock, A., Grevendonk, W., and Herreman, W., "Shear Viscosity of Liquid Argon," *Physica*, **37**, 227–32, 1967.

1139. Solov'ev, A. N. and Kaplun, A. B., "The Vibration Method of Measuring the Viscosity of Liquids," *Teplofiz. Vys. Temp.*, **3**, 139–48, 1965.

1140. Krutin, V. N. and Smirnitskii, I. B., "Measurement of the Viscosity of Newtonian Fluids by Means of Vibratory Probes," *Sov. Phys.-Acoustics*, **12**, 42–5, 1966.

1141. Andrade, E. N. da C. and Dodd, C., "The Effect of an Electric Field on the Viscosity of Liquids," *Proc. Roy. Soc. (London)*, **A187**, 296–337, 1946.

1142. Andrade, E. N. da C. and Dodd, C., "The Effect of an Electric Field on the Viscosity of Liquids. II," *Proc. Roy. Soc. (London)*, **A204**, 449–64, 1951.

1143. Kincaid, J. F., Eyring, H., and Stearn, A. E., "The Theory of Absolute Reaction Rates and its Application to Viscosity and Diffusion in the Liquid State," *Chem. Rev.*, **28**, 301–65, 1941.

1144. Schrieber, P. W., Hunter, A. M., and Benedetto, K. R., "Argon Plasma Viscosity Measurements," AIAA Third Fluid and Plasma Dynamics Conf., Los Angeles, Calif., AIAA Paper No. 70–775, 9 pp., June 29–July 1, 1970.

1145. Dedit, A., Galperin, B., Vermesse, J., and Vodar, B., "Enregistrement, En Fonction du Temps, Des Déplacements D'une Colonne De Mercure Placée A L'intérieur D'une Enceinte Hautes Pressions. Application A La Mesure Du Coefficient de Viscosite' Des Gaz Sous Hautes Pressions," *J. Phys. Appliq.*, **26**, 189A–193A, 1965.

1146. Kao, J. T. F., Ruska, W., and Kobayashi, R., "Theory and Design of an Absolute Viscometer for Low Temperature-High Pressure Applications," *Rev. Sci. Instrum.*, **39**, 824–34, 1968.

1147. Masiá, A. P., Paniego, A. R., and Pinto, J. M. G., "Fuerzas Intermoleculares a Partir de Medidas de Viscosidad en Fase Vapor," *An. de Fis. Quim.*, **LXIII-B**, 1093–1102, 1967.

1148. Peña, M. D. and Esteban, F., "Viscosidad de Vapores Organicos," *An. Fis. Quim.*, **62A**, 337–46, 1966.

1149. Peña, M. D. and Esteban, F., "Viscosity of Quasi-Spherical Molecules in Vapor Phase," *An. Fis. Quim.*, **62A**, 347–57, 1966.

1150. Stefanov, B. I., Timrot, D. L., Totskii, E. E., and Chu, Wen-hao, "Viscosity and Thermal Conductivity of the

Vapors of Sodium and Potassium," *Teplofiz. Vys. Temp.*, **4**, 141–2, 1966.

1151. Dawe, R. A. and Smith, E. B., "Viscosities of the Inert Gases at High Temperatures," *J. Chem. Phys.*, **52**, 693–703, 1970.

1152. Dawe, R. A., Maitland, G. C., Rigby, M., and Smith, E. B., "High Temperature Viscosities and Intermolecular Forces of Quasi-Spherical Molecules," *Trans. Faraday Soc.*, **66**, 1955–65, 1970.

1153. Comings, E. W. and Egly, R. S., "Viscosity of Gases and Vapors at High Pressures," *Ind. Eng. Chem.*, **32**, 714–8, 1940.

1154. Meyer, G. R. and Thodos, G., "Viscosity and Thermal Conductivity of Sulfur Dioxide in the Gaseous and Liquid States," *J. Chem. Eng. Data*, **7**, 532–6, 1962.

1155. Flynn, L. W. and Thodos, G., "The Viscosity of Hydrocarbon Gases at Normal Pressures," *J. Chem. Eng. Data*, **6**, 457–9, 1961.

1156. Belov, V. A., "Viscosity of Partially Ionized Hydrogen," *Teplofiz. Vys. Temp.*, **5**, 37–43, 1967.

1157. Agaev, N. A. and Yusibova, A. D., "Viscosity of Heavy Water at High Pressures," *At. Energ.*, **23**, 149–51, 1967.

1158. Kessel'man, P. M. and Litvinov, A. S., "Calculation of Viscosity of Gas Mixtures at Atmospheric Pressure," *Inzh.-Fiz. Zh.*, **10**, 385–92, 1966.

1159. Lefrancois, B., "Viscosité des Gaz Sous Haute Pression Corps Purs," *Chem. Ind. Génie Chim.*, **98**, 1377–80, 1967.

1160. Barbe, C., "Calcul Automatique des Paramétres de Transport des Melanges de Gaz," *Entropie*, **20**, 49–55, 1968.

1161. Aksarailian, A. and Cerceau, O., "Cálculo Teórico de la Viscosidad de Metano y del Cloruro de Metilo," *Acta Cient. Venez.*, **16**, 54–7, 1965.

1162. Singh, Y. and Das Gupta, A., "Transport and Equilibrium Properties of Polar Gases," *J. Chem. Phys.*, **52**, 3064–7, 1970.

1163. Singh, Y. and Das Gupta, A., "Transport Properties of Polar–Quadrupolar Gas Mixtures," *J. Chem. Phys.*, **52**, 3055–63, 1970.

1164. Fenstermaker, R. W., Curtiss, C. F., and Bernstein, R. B., "Molecular Collisions. X. Restricted–Distorted-Wave-Born and First-Order Sudden Approximations for Rotational Excitation of Diatomic Molecules," *J. Chem. Phys.*, **51**, 2439–48, 1969.

1165. Curtiss, C. F., "Molecular Collisions. XI," *J. Chem. Phys.*, **52**, 1078–81, 1970.

1166. Curtiss, C. F., "Molecular Collisions. XII. Generalized Phase Shifts," *J. Chem. Phys.*, **52**, 4832–41, 1970.

1167. Biolsi, L., "Molecular Collisions. XIII. Nuclear Spin and Statistics Effects for Nearly Spherical Potentials," *J. Chem. Phys.*, **53**, 165–77, 1970.

1168. Pattengill, M. D., Curtiss, C. F., and Bernstein, R. B., "Molecular Collisions. XIV. First Order Approximation of the Generalized Phase Shift Treatment of Rotational Excitation: Atom-Rigid Rotor," *J. Chem. Phys.*, **54**, 2197–207, 1971.

1169. Pattengill, M. D., Curtiss, C. F., and Bernstein, R. B., "Molecular Collisions. XV. Classical Limit of the Generalized Phase Shift Treatment of Rotational Excitation: Atom-Rigid Rotor," *J. Chem. Phys.*, **55**, 3682–93, 1971.

1170. Pattengill, M. D., LaBudde, R. A., Bernstein, R. B., and Curtiss, C. F., "Molecular Collisions. XVI. Comparison of GPS with Classical Trajectory Calculations of Rotational Inelasticity for the Ar–N_2 System," *J. Chem. Phys.*, **55**, 5517–22, 1971.

1171. Curtiss, C. F., "Transport Properties of a Gas of Diatomic Molecules," *J. Chem. Phys.*, **54**, 872–7, 1971.

1172. Tip, A., "Transport Equations for Dilute Gases with Internal Degrees of Freedom. II. The Generalized Master Equation Approach," *Physica*, **53**, 183–92, 1971.

1173. Stevens, G. A., "Transport Properties of Methane," *Physica*. **46**, 539–49, 1968.

1174. Sengers, J. V., "Triple Collision Effects in the Transport Properties for a Gas of Hard Spheres," in *Kinetic Equations* (Liboff, R. L. and Rostoker, N., Editors), Gordon and Breach, Science Publishers, Inc., New York, 137–93, 1971.

1175. Kestin, J., Paykoc, E., and Sengers, J. V., "Viscosity of Helium, Argon and Nitrogen as a Function of Density," Arnold Engrg. Development Center Rept. No. AEDC-TR-71-190, 38 pp., 1971.

1176. Sengers, J. V., "Transport Properties of Gases and Binary Liquids Near the Critical Point," NASA CR-2112, 67 pp., 1972.

1177. Sengers, J. V., "Transport Processes Near the Critical Point of Gases and Binary Liquids in the Hydrodynamic Regime," *Ber. Bunsenges. Phys. Chem. (Z. Elektrochem.)*, **76**, 234–49, 1972.

1178. Hunter, L. W. and Curtiss, C. F., "Molecular Collisions. XVII. Formal Theory of Rotational and Vibrational Excitation in Collisions of Polyatomic Molecules," *J. Chem. Phys.*, **58**, 3884–96, 1973.

1179. Hunter, L. W. and Curtiss, C. F., "Molecular Collisions. XVIII. Restricted Distorted Wave Approximation to Rotational and Vibrational Excitation of Polyatomic Molecules," *J. Chem. Phys.*, **58**, 3897–3902, 1973.

1180. Hulsman, H. and Burgmans, A. L. J., "The Five Shear Viscosity Coefficients of a Polyatomic Gas in a Magnetic Field," *Phys. Letters*, **29A**, 629–30, 1969.

1181. Moraal, H., McCourt, F. R., and Knaap, H. F. P., "The Senftleben-Beenakker Effects for a Gas of Rough Spherical Molecules. II. The Viscosity Scheme," *Physica*, **45**, 455–68, 1969.

1182. Korving, J., "Viscosity of Ammonia in High Magnetic Fields," *Physica*, **46**, 455–68, 1970.

1183. Tommasini, F., Levi, A. C., Scoles, G., de Groot, J. J., van den Broeke, J. W., van den Meijdenberg, C. J. N., and Beenakker, J. J. M., "Viscosity and Thermal Conductivity of Polar Gases in an Electric Field," *Physica*, **49**, 299–341, 1970.

1184. Hulsman, H., van Waasdijk, E. J., Burgmans, A. L. J., Knaap, H. F. P., and Beenakker, J. J. M., "Transverse Momentum Transport in Polyatomic Gases under the Influence of a Magnetic Field," *Physica*, **50**, 53–76, 1970.

1185. Hulsman, H. and Knaap, H. F. P., "Experimental Arrangements for Measuring the Five Independent Shear-Viscosity Coefficients in a Polyatomic Gas in a Magnetic Field," *Physica*, **50**, 565–72, 1970.

1186. Beenakker, J. J. M. and McCourt, F. R., "Magnetic and Electric Effects on Transport Properties," *Ann. Rev. Phys. Chem.*, **21**, 47–72, 1970.

1187. Mo, K. C., Gubbins, K. E., and Dufty, J. W., "Perturbation Theory for Dense Fluid Transport Properties," in *Proceedings of the Sixth Symposium on Thermophysical Properties*, *Am. Soc. Mech. Eng.*, 158–67, 1973.

1188. Tham, M. K. and Gubbins, K. E., "Kinetic Theory of Multicomponent Dense Fluid Mixtures of Rigid Spheres," *J. Chem. Phys.*, **55**, 268–79, 1971.

1189. Wakeham, W. A., Kestin, J., Mason, E. A., and Sandler, S. I., "Viscosity and Thermal Conductivity of Moderately Dense Gas Mixtures," *J. Chem. Phys.*, **57**, 295–301, 1972.

1190. Tham, M. J. and Gubbins, K. E., "Correspondence Principle for Transport Properties of Dense Fluids," *Ind. Eng. Chem. Fundam.*, **8**, 791–5, 1969.

1191. Tham, M. J. and Gubbins, K. E., "Correspondence Principle for Transport Properties of Dense Fluids. Nonpolar Polyatomic Fluids," *Ind. Eng. Chem. Fundam.*, **9**, 63–70, 1970.

1192. Hahn, H-S., Mason, E. A., Miller, E. J., and Sandler, S. I., "Dynamic Shielding Effects in Partially Ionized Gases," *J. Plasma Phys.*, **7**, 285–92, 1972.

1193. Curtiss, C. F., "Transport Properties of a Gas of Diatomic Molecules. II," *J. Chem. Phys.*, **55**, 947–9, 1971.

1194. Pal, A. K. and Bhattacharyya, "Viscosity of Binary Polar-Gas Mixtures," *J. Chem. Phys.*, **51**, 828–31, 1969.

1195. Brokaw, R. S., "Viscosity of Binary Polar-Gas Mixtures," *J. Chem. Phys.*, **52**, 2796–7, 1970.

1196. Hogervorst, W., "Transport and Equilibrium Properties of Simple Gases and Forces Between Like and Unlike Atoms," *Physica*, **51**, 77–89, 1971.

1197. Kong, C. L., "Combining Rules for Intermolecular Potential Parameters. I. Rules for the Dymond-Alder Potential," *J. Chem. Phys.*, **59**, 1953–8, 1973.

1198. Kong, C. L., "Combining Rules for Intermolecular Potential Parameters. II. Rules for the Lennard-Jones (12–6) Potential and the Morse Potential," *J. Chem. Phys.*, **59**, 2464–7, 1973.

1199. Alvarez-Rizzatti, M. and Mason, E. A., "Estimation of Dipole–Quadrupole Dispersion Energies," *J. Chem. Phys.*, **59**, 518–22, 1973.

1200. Sutherland, W., "The Viscosity of Gases and Molecular Force," *Phil. Mag.*, **36**, 507–31, 1893.

1201. Hattikudur, U. R. and Thodos, G., "Equations for the Collision Integrals $\Omega^{(1,1)*}$ and $\Omega^{(2,2)*}$," *J. Chem. Phys.*, **52**, 4313, 1970.

1202. Neufeld, P. D., Janzen, A. R., and Aziz, R. A., "Empirical Equations to Calculate 16 of the Transport Collision Integrals $\Omega^{(l,g)*}$ for the Lennard-Jones (12–6) Potential," *J. Chem. Phys.*, **57**, 1100–2, 1972.

1203. Dymond, J. H., "Corresponding States: A Universal Reduced Potential Energy Function for Spherical Molecules," *J. Chem. Phys.*, **54**, 3675–81, 1971.

1204. Kestin, J., Ro, S. T., and Wakeham, W., "An Extended Law of Corresponding States for the Equilibrium and Transport Properties of the Noble Gases," *Physica*, **58**, 165–211, 1972.

1205. Neufeld, P. D. and Aziz, R. A., "Test of Three New Corresponding States Potentials for Ne, Ar, Kr and Xe with Application to Thermal Diffusion," *J. Chem. Phys.*, **59**, 2234–43, 1973.

1206. Dymond, J. H., Rigby, M., and Smith, E. B., "Intermolecular Potential Energy Function for Simple Molecules," *J. Chem. Phys.*, **42**, 2801–6, 1965.

1207. Dymond, J. H. and Alder, B. J., "Pair Potential for Argon," *J. Chem. Phys.*, **51**, 309–20, 1969.

1208. Guevara, F. A., McInteer, B. B., and Wageman, W. E., "High-Temperature Viscosity Ratios for Hydrogen, Helium, Argon, and Nitrogen," *Phys. Fluids*, **12**, 2493–505, 1969.

1209. Goldblatt, M., Guevara, F. A., and McInteer, B. B., "High Temperature Viscosity Ratios for Krypton," *Phys. Fluids*, **13**, 2873–4, 1970.

1210. Guevara, F. A. and Stensland, G., "High Temperature Viscosity Ratios for Neon," *Phys. Fluids*, **14**, 746–8, 1971.

1211. Goldblatt, M. and Wageman, W. E., "High Temperature Viscosity Ratios for Xenon," *Phys. Fluids*, **14**, 1024–5, 1971

1212. Kestin, J., Wakeham, W., and Watanabe, K., "Viscosity, Thermal Conductivity and Diffusion Coefficient of Ar–Ne and Ar–Kr Gaseous Mixtures in the Temperature Range 25–700 C," *J. Chem. Phys.*, **53**, 3773–80, 1970.

1213. Kestin, J., Ro, S. T., and Wakeham, W. A., "Viscosity of the Binary Gaseous Mixture Neon–Krypton," *J. Chem. Phys.*, **56**, 4086–91, 1972.

1214. Kestin, J., Ro, S. T., and Wakeham, W. A., "Viscosity of the Noble Gases in the Temperature Range 25–700 C," *J. Chem. Phys.*, **56**, 4119–24, 1972.

1215. Kestin, J., Ro, S. T., and Wakeham, W. A., "Viscosity of the Binary Gases Mixture Helium–Nitrogen," *J. Chem. Phys.*, **56**, 4036–42, 1972.

1216. McAllister, R. A., "The Viscosity of Liquid Mixtures," *Am. Inst. Chem. Eng. J.*, **6**, 427–31, 1960.

1217. Saxena, S. C., "A Semi-Empirical Formula for the Viscosity of Liquid Mixtures," *Chem. Phys. Letters*, **19**, 32–4, 1973.

1218. Saxena, S. C., "Viscosity of Multicomponent Mixtures of Gases," in *Proceedings of the A.S.M.E. 6th Symposium on Thermophysical Properties*, 100–10, August 6–8, 1973.

Numerical Data

Data Presentation and Related General Information

1. SCOPE OF COVERAGE

Presented in this volume are 1803 sets of viscosity data on 59 pure fluids and 129 systems of fluid mixtures. These substances were selected based on consideration of scientific and technological interest and needs.

Viscosity is strongly and intricately dependent on the shape and structure of the molecules. Consequently, different varieties and complexities of molecules and their different combinations in the mixtures have been selected. It is hoped that such an investigation of the viscosity of different categories of fluid molecules and their combinations will help in elucidating the various ways in which the viscosity of fluids and fluid mixtures can vary with changes in such variables as temperature, density (or pressure), and mixture composition.

The pure fluids include 13 elements, 10 inorganic compounds, and 36 organic compounds, and were originally selected to match parallel programs for thermal conductivity and for specific heat, the tables resulting from which have been published in Volumes 3 and 6, respectively. The data on pure fluids have been critically evaluated, analyzed, and synthesized, and "recommended reference values" are presented for the saturated liquid, saturated vapor, and gaseous states, with the available experimental data given in the departure plots.

The fluid mixtures selected include 99 binary systems, 8 ternary systems, 3 quaternary systems, and 19 multicomponent systems. These are further divided into monatomic–monatomic, monatomic–nonpolar polyatomic, monatomic–polar polyatomic, nonpolar polyatomic–nonpolar polyatomic, nonpolar polyatomic–polar polyatomic, and polar polyatomic–polar polyatomic systems. The data on fluid mixtures have been smoothed graphically and the smoothed values as well as the experimental data are presented as a function of composition, density, or temperature in both graphical and tabular forms. Those experimental data originally reported in the research document as a function of pressure have been converted to functions of density. The experimental data for binary mixtures with composition dependence have been fitted with equations of the Sutherland type, and the Sutherland coefficients have been calculated and are presented in this volume.

2. PRESENTATION OF DATA

The viscosity data and information for each pure fluid are presented separately for three physical states: saturated liquid, saturated vapor, and gaseous. For each physical state, the material presented consists of a discussion, a tabulation of the recommended viscosity values, and a departure plot.

In the discussion, the available experimental data and information are reviewed and assessed, the considerations involved in arriving at the recommendation of the viscosity values are discussed, the theoretical or empirical equation used in curve fitting is given, and the estimated accuracy of the recommended values is stated. Recommended values are presented in tabular form, accompanied by indications of phase transition temperatures where these fall within the range of the tabulation. A departure plot, or plots, showing the concordance between the various experimental and/or theoretical values and the recommended values is given if sufficient experimental data are available.

In preparing the departure plots the following definition is used:

Percent departure

$$= \frac{\text{Experimental data} - \text{Recommended value}}{\text{Recommended value}} \times 100$$

By the above definition, departures are positive if the experimental data are greater than the recommended values and vice versa. Extrapolation of the values beyond the limits of the table is not recommended. If, however, this must be done, the departure plots should be examined to obtain an indication of the probable trend in the values in regions not yet experimentally studied.

The viscosity data and information for each system of fluid mixtures are presented separately for three different dependences: composition, density, and temperature. Those data originally reported as a function of pressure have been converted to be as a function of density. A consistent numbering system for tables and figures is adopted. Thus, a table numbered as 60-G(C)E, for example, lists the experimental (E) viscosity data as a function of composition (C) for gaseous (G) argon–helium (60) mixtures. The viscosity variation is shown in terms of the mole fraction of the heavier component in the mixture. A table numbered as 60-G(D)E deals with the experimental data as a function of the density (D) of the gaseous argon–helium mixtures. Similarly a table numbered as 60-G(T)E reports experimental data as a function of temperature (T). In each case the remaining variables are specified while reporting a given set of data. Also the data of different workers on a given system for the same dependence are grouped together in the same table and listed in the order of increasing temperatures. If all the experimental viscosity data on a given system for the same dependence are not easily accommodated in one figure, these are distributed in a set of figures identically numbered.

The graphically smoothed viscosity values at equally spaced twenty-one entries of the mole fraction of the heavier component in the gaseous binary system and at the temperature of measurement are reported in a table numbered as G(C)S. These tables giving the composition (C) dependence of viscosity are also included for each system along with the above-mentioned 3 sets of tables. Similarly the smoothed values for round density and temperature are reported in tables numbered as G(D)S and G(T)S, respectively. In these different categories of data, whenever a liquid system is involved instead of a gaseous system the first letter G is replaced by L. In an analogous manner the letter V is used to signify the vapor state.

The experimental data for ternary, quaternary, and multicomponent systems are also grouped together in the light of their molecular structure, but are not further processed like those for binary systems except in a few cases which are either pure air or mixtures of air and other fluids. Treating air as a pure component the data on systems air–carbon dioxide, air–methane, air–ammonia, air–hydrogen chloride, and air–hydrogen sulfide have also been smoothed.

It is hoped that a better understanding of the viscosity of binary systems will help in predicting the viscosity of systems containing more than two components, for it is impossible in practice to measure the viscosity of mixtures with all the possible combinations of components. The data reported here for complex systems will serve to check the various predictive schemes either already developed or to be developed.

3. SYMBOLS AND ABBREVIATIONS USED IN THE FIGURES AND TABLES

Most abbreviations and symbols used are those generally accepted in engineering and scientific practice and convention.

In this volume the word "data" is reserved for an experimentally determined quantity, while quantities determined by calculation or estimation are referred to as values.

The notations "n.m.p.," "n.b.p.," and "c.p." refer to normal melting point, normal boiling point, and critical point, respectively. Numbers in square brackets in the discussion and those signified by the notation "Reference" on the departure plot correspond to the *References to Data Sources* listed at the end of this *Numerical Data* section.

In the departure plots, curve numbers are surrounded either by circles or squares, the latter being used to indicate a single data point. Solid lines are used in the plot to connect experimental data points and dotted lines indicate calculated or correlated values. When the percent departure for any of the data points falls outside the range of the departure plot, the numerical value of the departure is correctly given at the data point with a vertical arrow pointing up or down from the data point to the given value to indicate the fact that the value is beyond the range of the plot.

In the tables and figures for systems of mixtures, the term "mole fraction" is used to denote the ratio of the number of molecules of one kind present in a given mixture to the total number of molecules. Thus, in an argon–helium mixture when the stated mole fraction of argon is 0.20, it implies that in the mixture argon is 20% by the number of molecules, and hence that 1/5 of the total volume is argon. The mole fraction of a given component will often vary between the extreme limits 0 and 1 referring to its complete absence and presence, respectively.

4. CONVENTION FOR BIBLIOGRAPHIC CITATION

For the following types of documents the bibliographic information is cited in the sequences given below.

Journal Article

    a. Author(s)—The names and initials of all authors are given. The last name is written first, followed by initials.

    b. Title of the article—The title of a journal article is enclosed in quotation marks.

    c. Name of the Journal—The abbreviated name of the journal is given as used in *Chemical Abstracts*.

    d. Series, volume, and issue number—If the series is designated by a letter, no comma is used between the letter for series and the numeral for volume, and they are both in bold-face type. In case series is also designated by a numeral, a comma is used between the numeral for series and the numeral for volume, and only the numeral denoting volume is boldfaced. No comma is used between the numerals denoting volume and issue number. The numeral for issue number is enclosed in parentheses.

    e. Pages—The inclusive page numbers of the article.

    f. Year—The year of publication.

Report

    a. Author(s).

    b. Title of report—The title of a report is enclosed in quotation marks.

    c. Name of the sponsoring agency and report number.

    d. Part.

    e. Pages.

    f. Year.

    g. ASTIA's AD number—This is enclosed in square brackets whenever available.

Book

    a. Author(s).

    b. Title—The title of a book is underlined.

    c. Volume.

    d. Edition.

    e. Publisher.

    f. Location of the publisher.

    g. Pages.

    h. Year.

5. NAME, FORMULA, MOLECULAR WEIGHT, TRANSITION TEMPERATURES, AND PHYSICAL CONSTANTS OF ELEMENTS AND COMPOUNDS

The table given here contains information on the molecular weight, transition temperatures, and physical constants of the elements and compounds included in this volume and of a few selected compounds in addition. This information is very useful in data correlation and synthesis. The molecular weights are based on the values given in the article entitled "Atomic Weights of the Elements 1971," published in *Pure and Applied Chemistry*, Vol. 30, Nos. 3–4, 639–49, 1972, by the International Union of Pure and Applied Chemistry. The electric dipole moments are quoted from the compilation of Nelson, Like, and Maryott, National Standard Reference Data Series—National Bureau of Standards, NSRDS-NBS 10, 49 pp., 1967.

6. CONVERSION FACTORS FOR UNITS OF VISCOSITY

The conversion factors for units of viscosity given in the table are based upon the following defined values and conversion factors given in NBS Special Publication 330, 1972:

Standard acceleration of free fall = $980.665 \text{ cm s}^{-2}$

$$1 \text{ in} = 2.54 \text{ cm}$$

$$1 \text{ lb} = 453.59237 \text{ g}$$

Name, Formula, Molecular Weight, Transition Temperatures, and Physical Constants of Elements and Compounds

Name	Formula	Molecular Weight	Density (25 C), g cm^{-3}	Melting (or Triple) Point, K	Normal Boiling Point, K	Critical Temp., K	C_p (25°C), cal g^{-1}K^{-1}	C_v (25°C), cal g^{-1}K^{-1}	Dipole Moment, Debyes
Acetone	C_3H_6O	58.080	0.933 (l)†	178	329	508	0.528 (l)		2.88
Acetylene	C_2H_2	26.038	1.077 -3**	179	189	309	0.407	0.329	0
Air		28.966	1.184 -3	60	79b, 82d	133	0.240	0.172	
Ammonia	NH_3	17.030	0.601 -3	195	240	405	0.515	0.387	1.47
Argon	Ar	39.948	1.634 -3	84	88	151	0.125	0.075	0
Benzene	C_6H_6	78.113	0.876 (l)	279	353	563	0.415 (l)	--	0
Boron Trifluoride	BF_3	67.805		146	172	261	--	--	0
Bromine	Br_2	159.808		266	332	584	0.113	--	0
i-Butane	$i\text{-}C_4H_{10}$	58.123		114	262	408	0.404		0.132
n-Butane	$n\text{-}C_4H_{10}$	58.123	2.491 -3	137	273	426	0.409	0.358	≤0.05
Carbon Dioxide	CO_2	44.010	1.811 -3	216(5 atm)	195	304	0.203	0.158	0
Carbon Monoxide	CO	28.010	1.145 -3	68	81	134	0.249	0.177	0.112
Carbon Tetrachloride	CCl_4	153.823	1.589 (l)	250	350	556	0.204 (l)	--	0
Chlorine	Cl_2	70.906	2.944 -3	172	239	417	0.114*	0.084	0
Chloroform	$CHCl_3$	119.378	1.469 (l)	210	334	536	0.228 (l)	--	1.01
n-Decane	$C_{10}H_{22}$	142.284	0.728 (l)	243	447	619	0.527 (l)	--	
Deuterium	D_2	4.028	0.165 -3	19(.16 atm)	24	38	1.731*	1.241	0
Diethylamine	$C_4H_{11}N$	73.138	0.711 (l)	233	329	496	0.516 (l)	--	1.11
Ethane	C_2H_6	30.069	1.243 -3	90	185	305	0.422	0.335	0
Ethyl Alcohol	C_2H_6O	46.069	0.789 (l)	159±3	351	516	0.580 (l)	--	1.69
Ethyl Ether	$C_4H_{10}O$	74.123	0.716 (l)	157(α), 150(β)	308	467	0.559 (l)	--	1.15
Ethylene	C_2H_4	28.054	1.155 -3	104	170	283	0.374	0.297	0
Ethylene Glycol	$C_2H_6O_2$	62.068	1.100 (l)	258	471		0.575 (l)	--	2.28
Fluorine	F_2	37.997	1.553 -3	54	85	144	0.197*	0.152	0
Freon 11	CCl_3F	137.368	5.840 -3	162	297	471	0.136*	0.125	0.45
Freon 12	CCl_2F_2	120.914	5.045 -3	116	243	385	0.146	0.128	0.51
Freon 13	$CClF_3$	104.459	4.388 -3	91	191	302	0.153*	0.138	0.50
Freon 21	$CHCl_2F$	102.923	4.284 -3	138	282	451	0.141*	0.119	1.29
Freon 22	$CHClF_2$	86.469	3.588 -3	113	233	369	0.151	0.133	1.42
Freon 113	$C_2Cl_3F_3$	187.376	1.564 (l)	238	321	487	0.225 (l)		
Freon 114	$C_2Cl_2F_4$	170.922	7.012 -3	179	276	419	0.170	0.157	0.5
Glycerol	$C_3H_8O_3$	92.095	1.263 (l)	291	563		0.567 (l)		
Helium	He	4.003	0.164 -3		4	5.4	1.240*	0.748	0
n-Heptane	C_7H_{16}	100.203	0.681 (l)	183	371	540	0.536 (l)	--	
n-Hexane	C_6H_{14}	86.177	0.657 (l)	178	342	508	0.543 (l)	--	
Hydrogen	H_2	2.016	0.082 -3	14	20	33	3.420	2.438	0
Hydrogen Chloride	HCl	36.461	1.502 -3	160±2	188	325	0.191*	0.140	1.08
Hydrogen Iodide	HI	127.912		223	238	423	0.054*		
Hydrogen Sulfide	H_2S	34.076	1.409 -3	190	213	374	0.240*	0.157	0.97
Iodine	I_2	253.809	4.93 (s)	387	458	785	0.052 (s)	--	0
Krypton	Kr	83.80	3.429 -3	116	120	210	0.059*	0.035	0
Methane	CH_4	16.043	0.657 -3	90	112	190	0.533	0.409	0
Methyl Alcohol	CH_4O	32.042	0.789 (l)	175	338	513	0.602 (l)		1.70
Methyl Chloride	CH_3Cl	50.488		175	249	416	0.193		
Methyl Formate	$C_2H_4O_2$	60.052	0.974 (l)	174	305	487	0.516	--	
Neon	Ne	20.179	0.824 -3	25	27	44	0.246*	0.150	0
Nitric Oxide	NO	30.000	1.228 -3	111	121	180	0.238	0.167	0.153
Nitrogen	N_2	28.013	1.146 -3	63	78	126	0.249	0.178	0
Nitrogen Peroxide	NO_2	46.006	1.44 (l)	263	295	431	0.369 (l)		0.316
Nitrous Oxide	N_2O	44.013		176±7	184	310	0.209*	0.170	0.167
n-Nonane	C_9H_{20}	128.257	0.714 (l)	220	424	594	0.529 (l)	--	
n-Octane	C_8H_{18}	114.230	0.701 (l)	216	399	569	0.530 (l)	--	
Oxygen	O_2	31.999	1.310 -3	55	90	155	0.220	0.157	0
n-Pentane	C_5H_{12}	72.150	0.621 (l)	144	309	470	0.561 (l)		
Cyclopropane	C_3H_6	42.080	0.720 (l)	146	240			--	
Propane	C_3H_8	44.096	1.854 -3	86	231	369	0.400	0.350	0.084
Propylene	C_3H_6	42.080	0.514 (l)	88	226	365	0.370	0.320	
Radon	Rn	222		202	211	377			0
Sulfur Dioxide	SO_2	64.059	2.679 -3	198	263	430	0.149*	0.081	1.63
Toluene	C_7H_8	92.140	1.028 (l)	178	384	594	0.410 (l)	--	0.36
Tritium	T_2	6.032		21	26	44			0
Water	H_2O	18.015	0.997 (l)	273	373	647	0.998 (l)	--	1.85
Xenon	Xe	131.30	5.397 -3	16J	165	290	0.0378*	0.0227*	0

* For ideal gas state.

** The notation –3 signifies 10^{-3}, so that 1.077 –3 means 1.077 x 10^{-3}, etc.

† (l) and (s) designate liquid and solid state, respectively.

Conversion Factors for Units of Viscosity

MULTIPLY by appropriate factor to OBTAIN →	$N\,s\,m^{-2}$ $(kg\,s^{-1}\,m^{-1})$	$Pa\,s$ $(kg\,s^{-1}\,m^{-1})$	Poise $(dyne\,s\,cm^{-2})$ $(g\,s^{-1}\,cm^{-1})$	centipoise	micropoise	$lb_f\,s\,ft^{-2}$	poundal s ft^{-2} $(lb_m\,s^{-1}\,ft^{-1})$	$lb_m\,hr^{-1}\,ft^{-1}$	$slug\,hr^{-1}\,ft^{-1}$
$N\,s\,m^{-2}$ $(kg\,s^{-1}\,m^{-1})$	1	1	10	1×10^3	1×10^7	2.08854×10^{-2}	0.671969	2.41909×10^3	75.1876
$Pa\,s$ $(kg\,s^{-1}\,m^{-1})$	1	1	10	1×10^3	1×10^7	2.08854×10^{-2}	0.671969	2.41909×10^3	75.1876
Poise $(dyne\,s\,cm^{-2})$ $(g\,s^{-1}\,cm^{-1})$	0.1	0.1	1	1×10^2	1×10^6	2.08854×10^{-3}	6.71969×10^{-2}	2.41909×10^2	7.51876
centipoise	1×10^{-3}	1×10^{-3}	1×10^{-2}	1	1×10^4	2.08854×10^{-5}	6.71969×10^{-4}	2.41909	7.51876×10^{-2}
micropoise	1×10^{-7}	1×10^{-7}	1×10^{-6}	1×10^{-4}	1	2.08854×10^{-9}	6.71969×10^{-6}	2.41909×10^{-4}	7.51876×10^{-6}
$lb_f\,s\,ft^{-2}$	47.8803	47.8803	4.78803×10^2	4.78803×10^4	4.78803×10^8	1	32.1740	1.15827×10^5	3.60000×10^3
poundal s ft^{-2} $(lb_m\,s^{-1}\,ft^{-1})$	1.48816	1.48816	14.8816	1.48816×10^3	1.48816×10^7	3.10810×10^{-2}	1	3.60000×10^3	1.11891×10^2
$lb_m\,hr^{-1}\,ft^{-1}$	4.13379×10^{-4}	4.13379×10^{-4}	4.13379×10^{-3}	0.413379	4.13379×10^3	8.63360×10^{-6}	2.77778×10^{-4}	1	3.10810×10^{-2}
$slug\,hr^{-1}\,ft^{-1}$	1.33001×10^{-2}	1.33001×10^{-2}	0.133001	13.3001	1.33001×10^5	2.77778×10^{-4}	8.93724×10^{-3}	32.1740	1

Numerical Data on Viscosity

1. ELEMENTS

TABLE 1-L(T). VISCOSITY OF LIQUID ARGON

DISCUSSION

SATURATED LIQUID

A search of the literature has revealed seven sets of experimental data [19, 20, 43, 44, 189, 246, 268], covering a temperature range from the melting point to the critical point. The various sets are not mutually consistent. The data of Zhdanova [268] covers the wider range of temperature but are lower than the other data.

The correlation was made by adjusting an equation

$$\log \mu = A + B/T$$

to the data from 125 K down to the melting point. Above that range the curve was smoothed graphically. Values computed by the method of Jossi et al. [100] for the saturated liquid near the critical point are not in good agreement with the recommended values, but served to estimate the critical viscosity.

The accuracy is of about 2% between the melting point and around the boiling point, but above, to the critical point, there is a need for more accurate data, the accuracy being not better than ± 10%.

RECOMMENDED VALUES

[Temperature, T, K; Viscosity, μ, 10^{-3} N s m^{-2}]

SATURATED LIQUID

T	μ
85	0.2813
90	0.2396
95	0.2075
100	0.1823
105	0.1622
110	0.1458
115	0.1323
120	0.1210
125	0.1115
130	0.1010
135	0.0890
140	0.0750
145	0.0603
150	0.0447
151*	0.0279

* Crit. Temp.

3

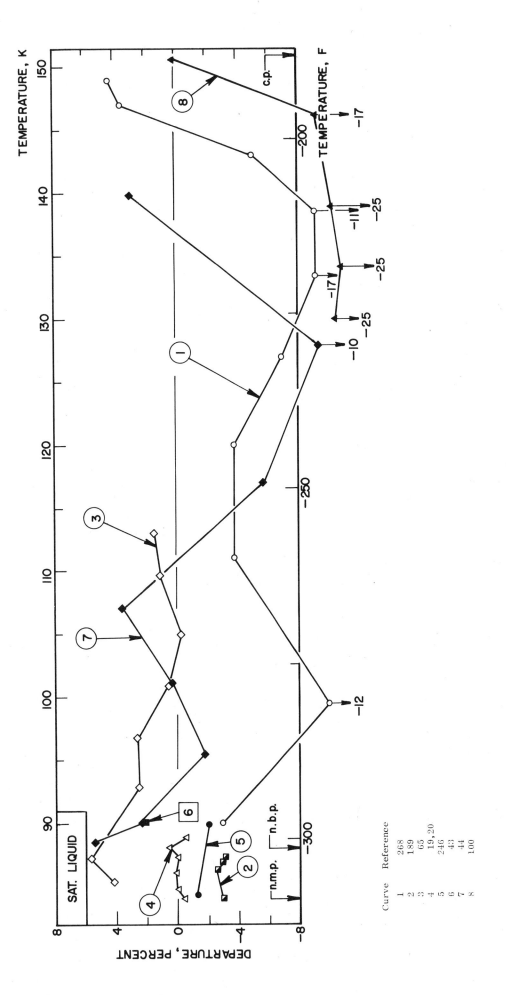

FIGURE 1-L(T). DEPARTURE PLOT FOR VISCOSITY OF LIQUID ARGON

Curve	Reference
1	268
2	189
3	65
4	19, 20
5	246
6	43
7	44
8	100

TABLE 1-V(T). VISCOSITY OF ARGON VAPOR

RECOMMENDED VALUES

[Temperature, T, K; Viscosity, μ, 10^{-3} N s m^{-2}]

SATURATED VAPOR

T	μ
85	0.00720
90	0.00765
95	0.00810
100	0.00855
105	0.00903
110	0.00955
115	0.01010
120	0.01070
125	0.0114
130	0.0122
135	0.0132
140	0.0145
145	0.0162
146	0.0166
147	0.0172
148	0.0180
149	0.0193
151*	0.0279

* Crit. Temp.

DISCUSSION

SATURATED VAPOR

Recommended values for the viscosity of the saturated vapor were computed by means of the correlation technique devised by Jossi et al. [100] using the recommended value of the 1 atm gas, and the density values given by Din [49].

Their accuracy is of about ± 5%.

TABLE 1-G(T). VISCOSITY OF GASEOUS ARGON

RECOMMENDED VALUES

[Temperature, T, K; Viscosity, μ, 10^{-6} N s m^{-2}]

GAS

T	μ	T	μ	T	μ
60	5.34	450	31.16	850	48.3
70	6.09	460	31.67	860	48.6
80	6.83	470	32.17	870	49.0
90	7.57	480	32.67	880	49.4
100	8.34	490	33.16	890	49.7
110	9.11	500	33.65	900	50.1
120	9.91	510	34.1	910	50.4
130	10.70	520	34.6	920	50.8
140	11.49	530	35.1	930	51.1
150	12.27	540	35.6	940	51.5
160	13.04	550	36.0	950	51.8
170	13.80	560	36.5	960	52.2
180	14.55	570	36.9	970	52.5
190	15.29	580	37.4	980	52.8
200	16.01	590	37.8	990	53.2
210	16.73	600	38.3	1000	53.5
220	17.44	610	38.7	1050	55.2
230	18.13	620	39.1	1100	56.8
240	18.82	630	39.6	1150	58.3
250	19.49	640	40.0	1200	59.9
260	20.16	650	40.4	1250	61.3
270	20.81	660	40.9	1300	62.8
280	21.45	670	41.3	1350	64.2
290	22.09	680	41.7	1400	65.6
300	22.72	690	42.1	1450	67.0
310	23.33	700	42.5	1500	68.4
320	23.94	710	42.9	1550	69.7
330	24.54	720	43.3	1600	71.0
340	25.13	730	43.7	1650	72.3
350	25.72	740	44.1	1700	73.5
360	26.29	750	44.5	1750	74.8
370	26.86	760	44.9	1800	76.0
380	27.42	770	45.3	1850	77.2
390	27.97	780	45.7	1900	78.4
400	28.52	790	46.0	1950	79.5
410	29.06	800	46.4	2000	80.7
420	29.59	810	46.8	2100	82.9
430	30.12	820	47.2	2200	85.1
440	30.64	830	47.5		
		840	47.9		

DISCUSSION

GAS

There are 30 sets of experimental data available for the viscosity of Argon, covering an overall range of temperature from 58 to 1868 K. Experimental data below normal temperature are those of Johnston and Grilly [17], Schmitt [193], Van Paemel [248, 253] Rietveld [180, 181],Filippova [62] and Flynn [64]. Experimental data extending to high temperature are those of Trautz and Zink [233, 234], Bonilla et al. [18], Rigby and Smith[182] and [254, 255] Vasilesco. Other experimental results are in immediate temperature range above or around normal temperature.

To analyze the data, use was made of the theoretical expression for viscosity:

$$\mu = 266.93 \; \frac{M \sqrt{T}}{\sigma^2 \Omega(T^*)} \; f_\mu$$

The group $\sigma^2\Omega(T^*)/f_\mu$ was computed from the experimental data and plotted as a function of 1/T. A smooth curve was drawn through the values obtained, and a table generated. Recommended values were calculated from the above formula using the value of y_{calc} interpolated from the table.

Recommended values are thought to be accurate to within two percent. Kestin and Whitelaw [117] and Di Pippo [52] values, on one side, and Vasilesco values, on the other, are diverging from the recommended curve. This discrepancy has been already pointed out by Hanley and Childs [85].

By assuming that $\Omega(T^*)/f_\mu$ is unity at the Boyle temperature* one obtains the value of $\sigma = 3.431$ for the collision diameter, which is quite in agreement with values found for typical interaction potentials.

Curves 31 and 32 are correlations given by other authors [3, 121].

* $T_B = 407.79$ K.

6

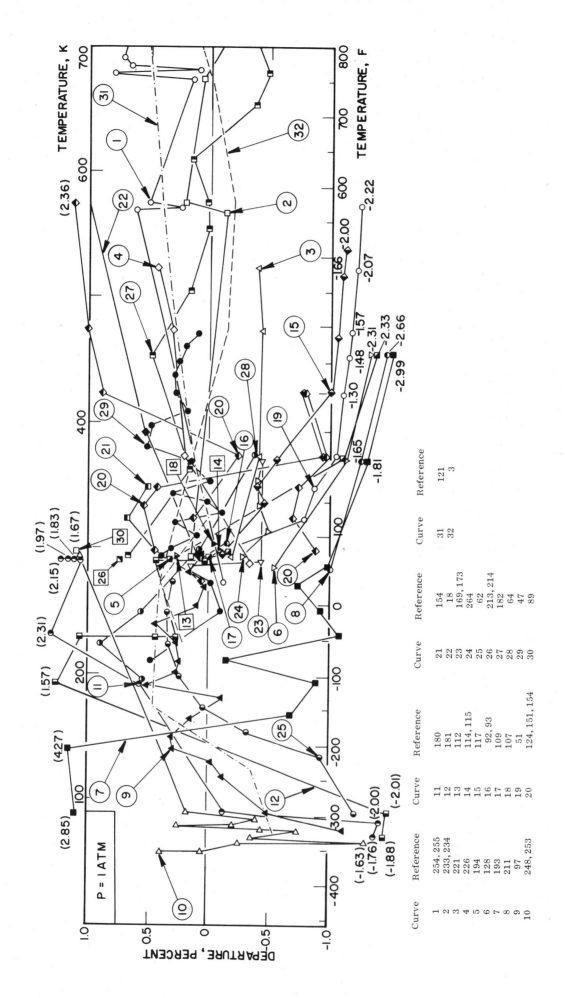

FIGURE 1-G(T). DEPARTURE PLOT FOR VISCOSITY OF GASEOUS ARGON

7

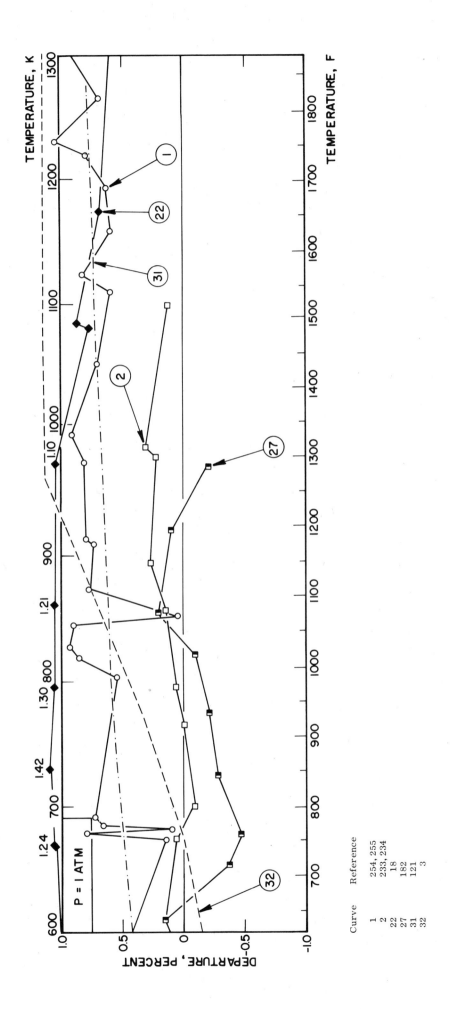

FIGURE 1-G(T).   DEPARTURE PLOT FOR VISCOSITY OF GASEOUS ARGON (continued)

Curve	Reference
1	254,255
2	233,234
22	18
27	182
31	121
32	3

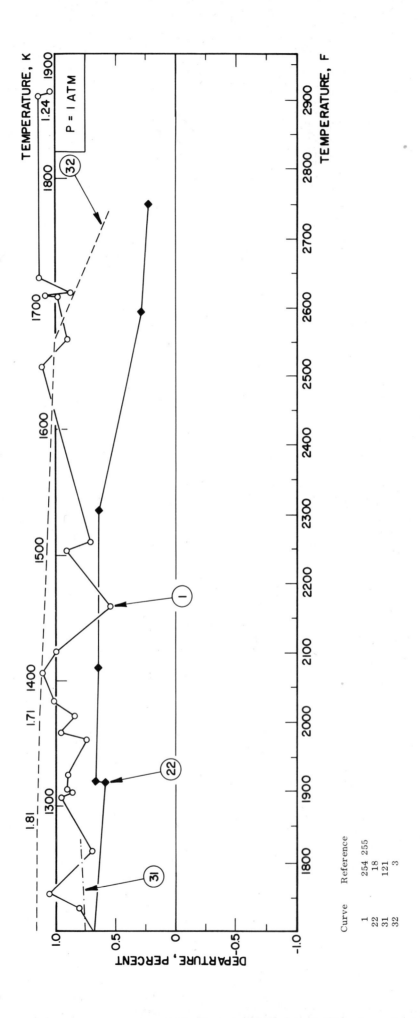

FIGURE 1-G(T). DEPARTURE PLOT FOR VISCOSITY OF GASEOUS ARGON (continued)

Curve	Reference
1	254 255
22	18
31	121
32	3

TABLE 2-G(T).   VISCOSITY OF GASEOUS BROMINE

DISCUSSION

GAS

Two sets of experimental data were retrieved from the literature, the results of Rankine [173] and from Braune et al. [21] which are in good agreement.

With the aid of the theoretical relation $\mu = K\sqrt{T}/(\sigma^2\Omega)$, $\sigma^2\Omega$ was computed from the experimental data, and adjusted to a quadratic equation which was used to generate the recommended values of viscosity. The accuracy is about ±2 percent.

RECOMMENDED VALUES

[Temperature, T, K; Viscosity, μ, N s m^{-2} · 10^{-6}]

GAS

T	η	T	η
		500	25.1
		510	25.6
		520	26.0
280	14.6	530	26.5
290	15.1	540	27.0
300	15.5	550	27.5
310	16.0	560	28.0
320	16.4	570	28.4
330	16.9	580	28.9
340	17.4	590	29.4
350	17.9	600	29.9
360	18.3	610	30.3
370	18.8	620	30.8
380	19.3	630	31.3
390	19.8	640	31.8
400	20.3	650	32.2
410	20.7	660	32.7
420	21.2	670	33.2
430	21.7	680	33.6
440	22.2	690	34.1
450	22.7	700	34.6
460	23.1	710	35.0
470	23.6	720	35.5
480	24.1	730	36.0
490	24.6	740	36.4
		750	36.9
		760	37.4
		770	37.8
		780	38.3
		790	38.7
		800	39.2

10

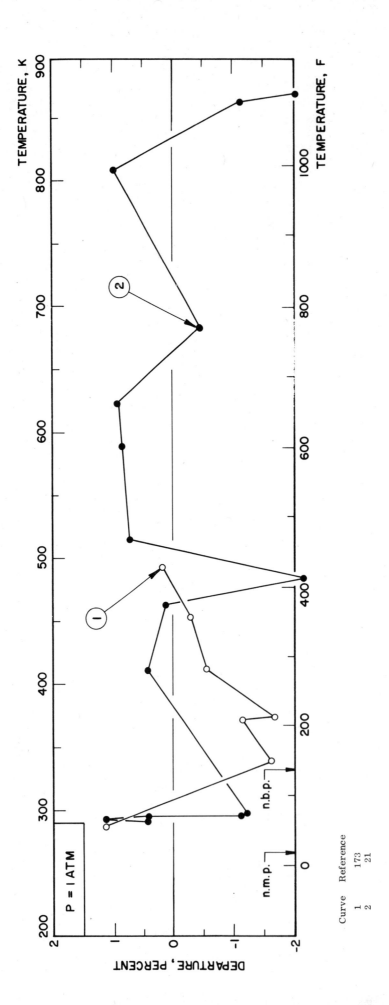

FIGURE 2-G(T). DEPARTURE PLOT FOR VISCOSITY OF GASEOUS BROMINE

TABLE 3-G(T). VISCOSITY OF GASEOUS CHLORINE

DISCUSSION

GAS

Eight sets of experimental data were found in the literature [225, 236, 257, 22, 295, 296, 297, 264] and some computed values were given by Andrussow [3]. They cover a range going from 280 K to 772 K. Andrussow gives values to 1273 K.

Use was made of the theoretical relation $\mu = K\sqrt{T}/(\sigma^2\Omega)$ to get $\sigma^2\Omega$. The latter was plotted as a function of 1/T and a quadratic equation was found to represent the data. From the adjusted curve of $\sigma^2\Omega$, the recommended values of viscosity were computed. The accuracy is thought to be ±2 percent.

RECOMMENDED VALUES

[Temperature, T, K; Viscosity, u, N s m$^{-2} \cdot 10^{-6}$]

GAS

T	μ	T	μ
270	12.36	550	24.0
280	12.81	560	24.3
290	13.27	570	24.7
300	13.71	580	25.1
310	14.16	590	25.4
320	14.60	600	25.8
330	15.04	610	26.2
340	15.48	620	26.5
350	15.92	630	26.9
360	16.35	640	27.2
370	16.78	650	27.6
380	17.20	660	27.9
390	17.62	670	28.3
400	18.04	680	28.6
410	18.45	690	29.0
420	18.87	700	29.3
430	19.28	710	29.6
440	19.68	720	30.0
450	20.09	730	30.3
460	20.49	740	30.6
470	20.88	750	31.0
480	21.28	760	31.3
490	21.67	770	31.6
500	22.06	780	32.0
510	22.4	790	32.3
520	22.8	800	32.6
530	23.2		
540	25.6		

12

FIGURE 3-G(T). DEPARTURE PLOT FOR VISCOSITY OF GASEOUS CHLORINE

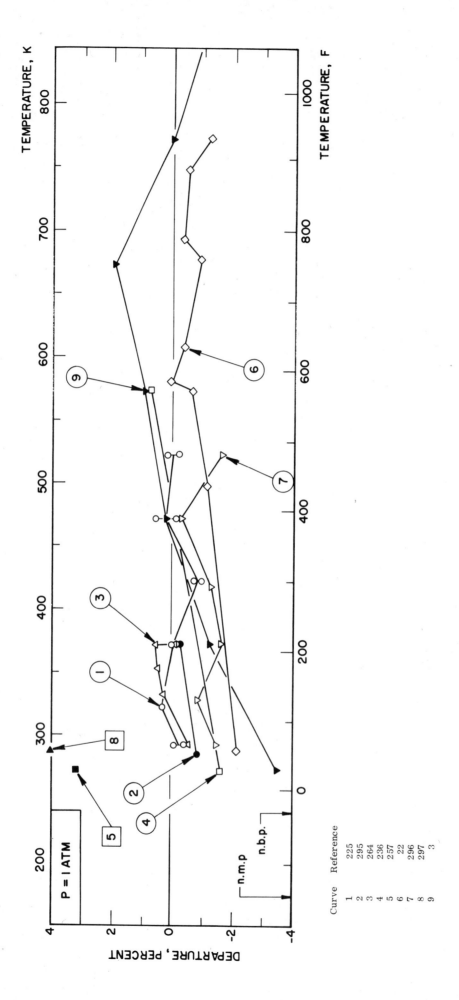

TABLE 4-G(T). VISCOSITY OF GASEOUS DEUTERIUM

DISCUSSION

GAS

Twelve sets of experimental data were found in the literature [6, 11, 39, 109, 111, 155, 245, 247, 253, 279, 298, 299]. They cover a range from 12 K to 423K. Semi-theoretical values were computed by Andrussow [3] from 223 K to 1273 K. Only experimental values were taken into consideration in generation of the recommended values.

The correlation was made by using the theoretical relation $\mu = K\sqrt{T}/(\sigma^2\Omega)$. The group $\sigma^2\Omega$ was obtained from the experimental data and plotted as a function of $1/T$ and a smooth curve drawn through the points. The accuracy is estimated to be ± 1 percent around room temperature, but is only ± 5 percent at the lowest temperature.

RECOMMENDED VALUES

[Temperature, T, K; Viscosity, μ, 10^{-6} N sec m^{-2}]

GAS

T	μ	T	μ
12	0.774	200	9.55
13	0.849	210	9.88
14	0.922	220	10.22
15	0.995	230	10.54
		240	10.87
16	1.068	250	11.20
17	1.141	260	11.51
18	1.213	270	11.82
19	1.285	280	12.14
20	1.357	290	12.43
25	1.715	300	12.74
30	2.054	310	13.03
35	2.382	320	13.32
40	2.70	330	13.60
45	3.01	340	13.88
50	3.30	350	14.16
60	3.86	360	14.45
70	4.39	370	14.73
80	4.88	380	15.01
90	5.35	390	15.27
100	5.79	400	15.54
110	6.21	410	15.80
120	6.62	420	16.06
130	7.01	430	16.32
140	7.39	440	16.58
150	7.77	450	16.84
160	8.14	460	17.09
170	8.49	470	17.34
180	8.85	480	17.58
190	9.20	490	17.82
		500	18.05

14

FIGURE 4-G(T).   DEPARTURE PLOT FOR VISCOSITY OF GASEOUS DEUTERIUM

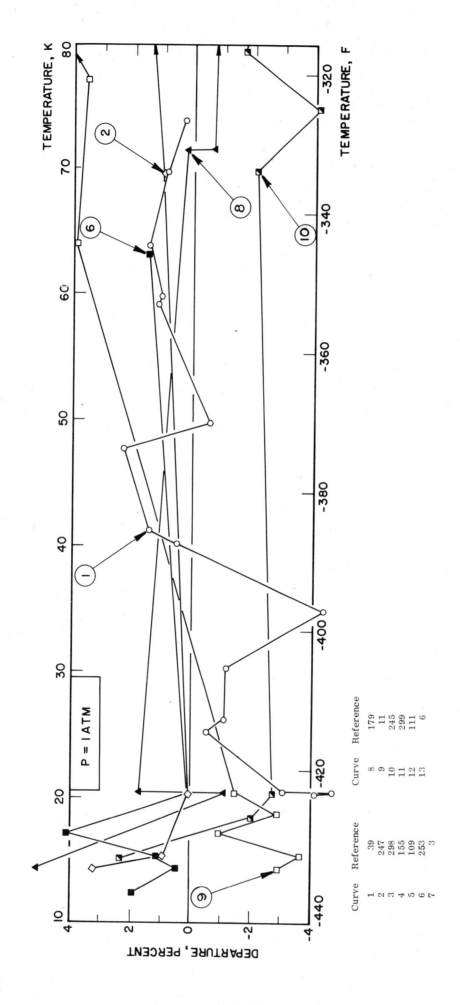

15

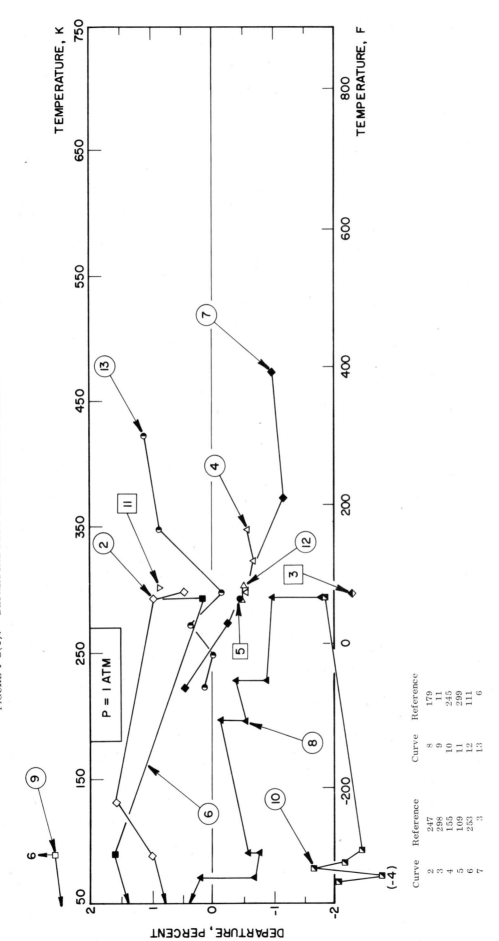

FIGURE 4-G(T). DEPARTURE PLOT FOR VISCOSITY OF GASEOUS DEUTERIUM (continued)

DISCUSSION

GAS

Two sets of data were found in the literature, the results of Franck and Stober [304] and those of Kanda [101]. The temperature range covered by the first author is larger than the temperature range covered by the second, but the disagreement in considerable. Although both measurements were made by the oscillating disk method, Franck thinks that his gas was of greater purity.

The recommended values are based on Franck's data which were used to obtain $\sigma^2\Omega$ from the theoretical relation $\mu = K\sqrt{T}/(\sigma^2\Omega_{22})$. The values of $\sigma^2\Omega$ were plotted versus $1/T$ and fitted to a quadratic equation. From the adjusted equation, the recommended values of the viscosity were computed.

TABLE 5-G(T).   VISCOSITY OF GASEOUS FLUORINE

RECCMMENDED VALUES

[Temperature, T, K; Viscosity, μ, 10^{-6} N sec m^{-2}]

GAS

T	μ	T	μ
90	7.66	300	23.6
100	8.56	310	24.3
110	9.45	320	24.9
120	10.33	330	25.5
130	11.19	340	26.1
140	12.03	350	26.7
150	12.86	360	27.3
160	13.7	370	27.9
170	14.5	380	28.4
180	15.3	390	29.0
190	16.0	400	29.5
200	16.8	410	30.1
210	17.5	420	30.6
220	18.2	430	31.2
230	19.0	440	31.7
240	19.7	450	32.2
250	20.3	460	32.7
260	21.0	470	33.3
270	21.7	480	33.8
280	22.4	490	34.3
290	23.0	500	34.8

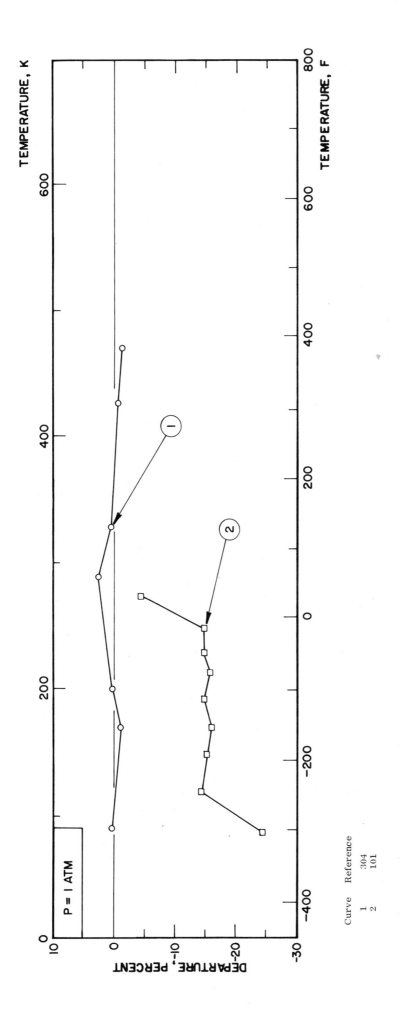

FIGURE 5-G(T).   DEPARTURE PLOT FOR VISCOSITY OF GASEOUS FLUORINE

DISCUSSION

GAS

Thirty-nine sets of experimental data were found in the literature covering an overall temperature range from 1.25 K to 2344 K. At temperatures higher than normal, the results of Trautz [221, 224, 232, 233], Kestin [107, 108, 109, 110, 112, 114, 115, 117], Makavetskas [148], Guevarra [812] and di Pippo [51] are in good agreement (within about 2%). At temperatures lower than normal, the results of Johnston [97], Van Itterbeek [179, 181, 247] and Van Paemel [253] were in fair agreement. Below 4 K the results of Van Itterbeek [251] and those of Becker [11] disagree by about 20%.

To make the correlation the expression:

$$\mu = \frac{K\sqrt{T}}{\sigma^2 \Omega(T^*)}$$

was computed and plotted as a function of 1/T. A curve was drawn through the points and smoothed. From this smoothed curve recommended values of the viscosity were generated.

The accuracy is thought to be about ±1% at temperatures higher than normal, about ±3% down to 20 K, but can be about ±10% below 20 K.

TABLE 6-G(T).   VISCOSITY OF GASEOUS HELIUM

RECOMMENDED VALUES

[Temperature, T, K; Viscosity, μ, N s m^{-2} · 10^{-6}]

GAS

T	μ	T	μ	T	μ	T	μ
1.25	0.364	150	12.34	500	28.27	850	40.44
1.50	0.422	160	12.87	510	28.65	860	40.77
2.0	0.545	170	13.40	520	29.03	870	41.08
2.5	0.675	180	13.93	530	29.41	880	41.39
3.0	0.804	190	14.45	540	29.78	890	41.70
3.5	0.927	200	14.97	550	30.16	900	42.01
4.0	1.045	210	15.48	560	30.53	910	42.32
4.5	1.159	220	16.00	570	30.90	920	42.63
5.0	1.268	230	16.51	580	31.27	930	42.94
6.0	1.474	240	17.00	590	31.63	940	43.24
7.0	1.666	250	17.50	600	31.99	950	43.35
8	1.848	260	17.99	610	32.35	960	43.85
9	2.020	270	18.48	620	32.71	970	44.14
10	2.183	280	18.95	630	33.07	980	44.44
15	2.896	290	19.42	640	33.42	990	44.74
20	3.502	300	19.89	650	33.77	1000	45.04
25	4.046	310	20.35	660	34.11	1050	46.5
30	4.553	320	20.81	670	34.46	1100	48.0
35	5.030	330	21.26	680	34.81	1150	49.4
40	5.479	340	21.70	690	35.15	1200	50.8
45	5.902	350	22.14	700	35.49	1250	52.1
50	6.304	360	22.58	710	35.84	1300	53.5
60	7.057	370	23.01	720	36.18	1350	54.8
70	7.758	380	23.44	730	36.52	1400	56.1
80	8.414	390	23.86	740	36.85	1450	57.3
90	9.038	400	24.28	750	37.19	1500	58.6
100	9.631	410	24.69	760	37.52	1550	59.8
110	10.20	420	25.10	770	37.85	1600	61.0
120	10.75	430	25.51	780	38.19	1650	62.2
130	11.29	440	25.92	790	38.51	1700	63.4
140	11.81	450	26.32	800	38.84	1750	64.5
		460	26.71	810	39.16	1800	65.7
		470	27.10	820	39.49	1850	66.8
		480	27.50	830	39.81	1900	67.9
		490	27.88	840	40.13	1950	69.0
						2000	70.0
						2100	72.2
						2200	74.2
						2300	76.3
						2400	78.3
						2500	80.2

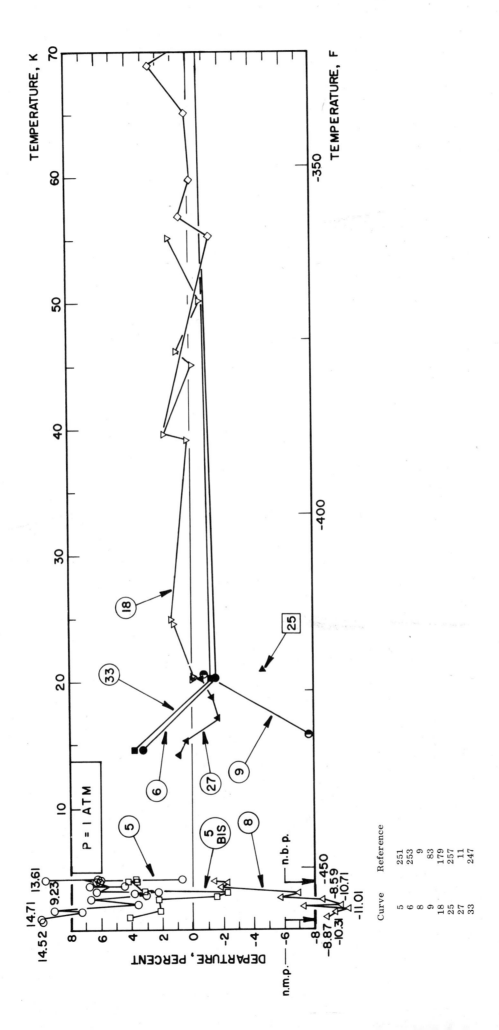

FIGURE 6-G(T).   DEPARTURE PLOT FOR VISCOSITY OF GASEOUS HELIUM

20

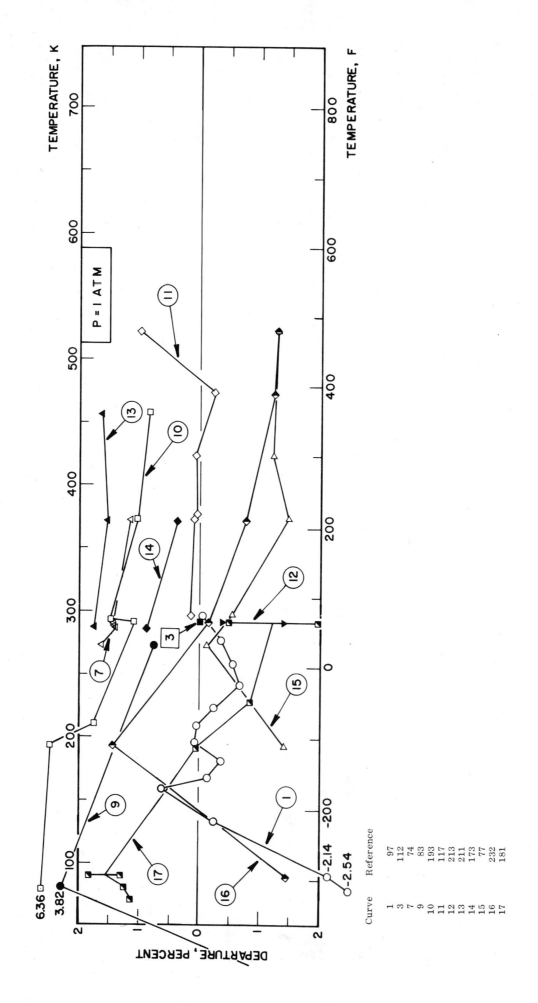

FIGURE 6-G(T).   DEPARTURE PLOT FOR VISCOSITY OF GASEOUS HELIUM   (continued)

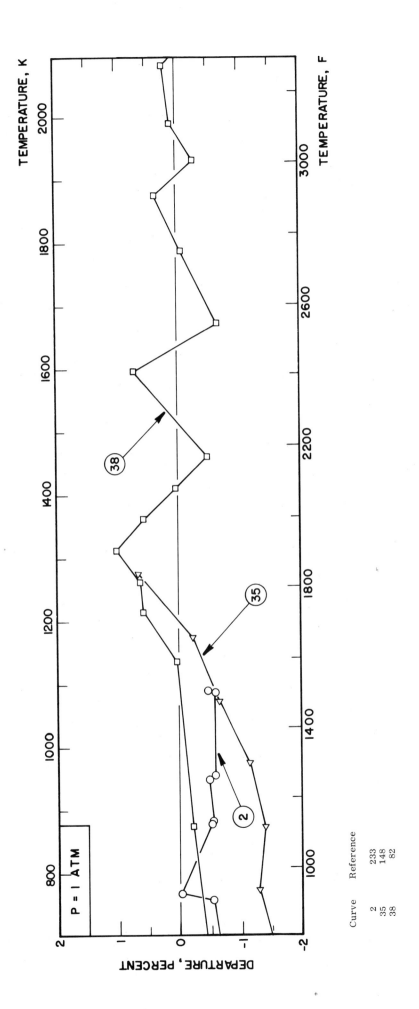

FIGURE 6-G(T).   DEPARTURE PLOT FOR VISCOSITY OF GASEOUS HELIUM   (continued)

Curve	Reference
2	233
35	148
38	82

FIGURE 6-G(T). DEPARTURE PLOT FOR VISCOSITY OF GASEOUS HELIUM (continued)

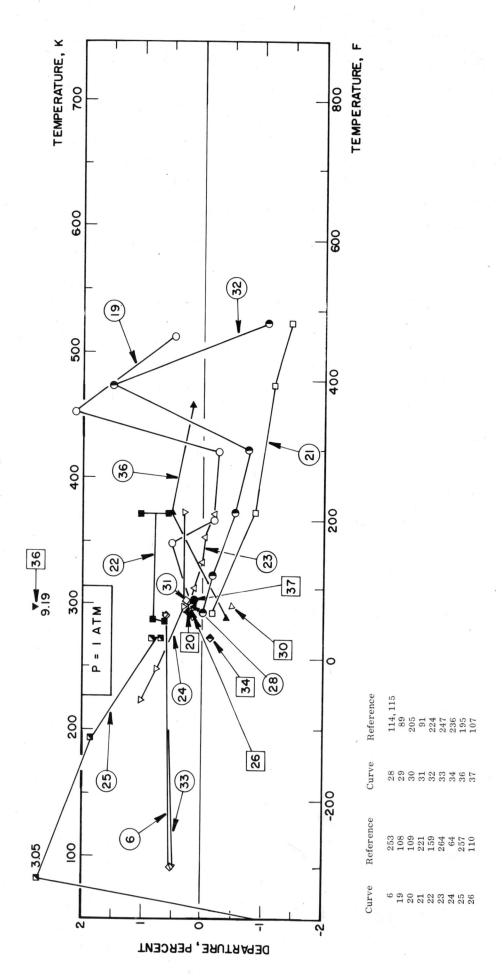

Curve	Reference	Curve	Reference
6	253	28	114, 115
19	108	29	89
20	109	30	205
21	221	31	91
22	159	32	224
23	264	33	247
24	64	34	236
25	257	36	195
26	110	37	107

FIGURE 6-G(T). DEPARTURE PLOT FOR VISCOSITY OF GASEOUS HELIUM (continued)

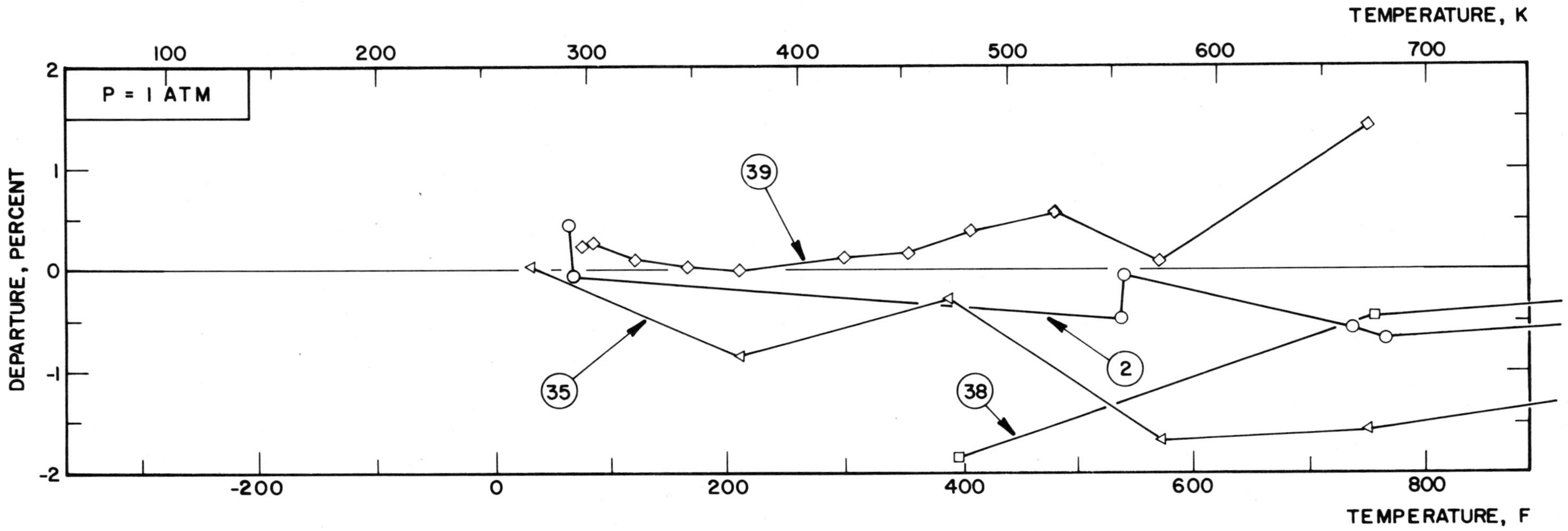

Curve	Reference
2	233
35	148
38	82
39	51

P = 1 ATM

DEPARTURE, PERCENT

TEMPERATURE, K

TEMPERATURE, F

TABLE 7-L(T). VISCOSITY OF LIQUID HYDROGEN

DISCUSSION

SATURATED LIQUID

Seven sets of experimental data were found in the literature [48 , 96 , 187, 239, 253, 256, 283]. They cover the range from 14 to 32 K, although only Diller [48 , 283] gives values above the normal boiling point to about the critical temperature.

The results below the normal boiling point were least square fitted to an equation

$$\log \mu = A + B/T$$

while the results above the normal boiling point were smoothed graphically.

The accuracy is thought to be about ± 5%.

RECOMMENDED VALUES

[Temperature, T, K; Viscosity, μ, N s m^{-2} · 10^{-3}]

SATURATED LIQUID

T	μ
14	0.0259
15	0.0225
16	0.0200
17	0.0179
18	0.0163
19	0.0149
20	0.0139
21	0.0129
22	0.0120
23	0.0112
24	0.0103
25	0.00955
26	0.00881
27	0.00822
28	0.00759
29	0.00699
30	0.00640
31	0.00585
32	0.00485
33	0.00380
33*	0.00364

* crit. temp.

FIGURE 7-L(T).   DEPARTURE PLOT FOR VISCOSITY OF LIQUID HYDROGEN

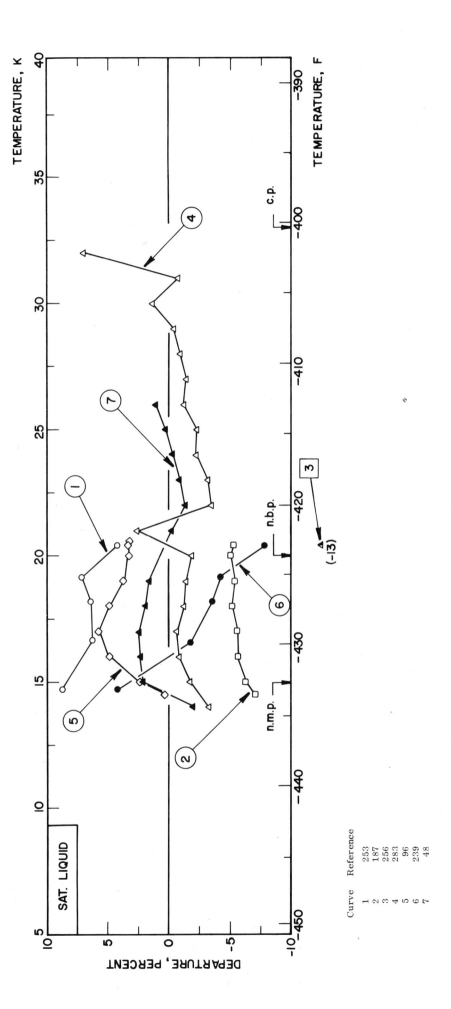

TABLE 7-V(T). VISCOSITY OF HYDROGEN VAPOR

DISCUSSION

SATURATED VAPOR

Recommended values for the viscosity of the saturated vapor have been generated using the excess viscosity concept as outlined by Jossi, Stiel and Thodos [100]. From a reduced excess viscosity curve versus reduced temperature, the excess viscosity was obtained and added to the recommended values for the 1 atm gas.

The accuracy is thought to be about ± 5%.

RECOMMENDED VALUES

[Temperature, T, K, Viscosity, μ, 10^{-3} N s m^{-2}]

SATURATED VAPOR

T	μ
20	0.00109
21	0.00116
22	0.00121
23	0.00128
24	0.00134
25	0.00140
26	0.00146
27	0.00152
28	0.00161
29	0.00174
30	0.00183
31	0.00201
32	0.00227
33	0.00279
33*	0.00364

* Crit. Temp.

TABLE 7-G(T). VISCOSITY OF GASEOUS HYDROGEN

RECOMMENDED VALUES

[Temperature, T, K; Viscosity, μ, N s m^{-2} · 10^{-6}]

GAS

T	μ	T	μ	T	μ	T	μ
10	0.50	250	7.90	550	13.6	850	18.5
15	0.80	260	8.11	560	13.8	860	18.6
20	1.09	270	8.32	570	14.0	870	18.8
25	1.36	280	8.53	580	14.1	880	18.9
		290	8.73	590	14.3	890	19.1
30	1.61	300	8.94	600	14.5	900	19.2
35	1.86	310	9.14	610	14.7	910	19.4
40	2.09	320	9.35	620	14.8	920	19.5
45	2.31	330	9.54	630	15.0	930	19.7
50	2.52	340	9.74	640	15.2	940	19.8
55	2.71	350	9.94	650	15.3	950	20.0
60	2.91	360	10.14	660	15.5	960	20.1
70	3.27	370	10.33	670	15.6	970	20.3
80	3.60	380	10.52	680	15.8	980	20.4
90	3.92	390	10.72	690	16.0	990	20.5
100	4.21	400	10.91	700	16.1	1000	20.7
110	4.49	410	11.10	710	16.3	1050	21.4
120	4.77	420	11.28	720	16.5	1100	22.2
130	5.04	430	11.47	730	16.6	1150	22.9
140	5.31	440	11.66	740	16.8	1200	23.6
150	5.57	450	11.84	750	16.9	1250	24.3
160	5.82	460	12.02	760	17.1	1300	25.0
170	6.07	470	12.21	770	17.3	1350	25.6
180	6.31	480	12.39	780	17.4	1400	26.3
190	6.55	490	12.57	790	17.6	1450	27.0
200	6.78	500	12.74	800	17.7	1500	27.6
210	7.01	510	12.9	810	17.8	1550	28.2
220	7.24	520	13.1	820	18.0	1600	28.9
230	7.46	530	13.3	830	18.2	1650	29.5
240	7.68	540	13.5	840	18.3	1700	30.1
						1750	30.7
						1800	31.3
						1850	31.9
						1900	32.5
						1950	33.1
						2000	33.6

DISCUSSION

GAS

Fifty two sets of experimental data were found in the literature. Trautz and his school have produced 13 sets of data covering a temperature range from 195 K to 1100 K [7, 206, 220, 221, 222, 226, 227, 228, 229, 230, 231, 233, 234]. Only the results of Guevara [82] cover a higher temperature range (from 296 to 2334 K). In the low temperature range there are results of Johnston (down to 90 K) [98] , Van Itterbeek's school [178, 179, 244, 247, 250, 253], and of Coremans [39] down to 20 K.

To correlate the data, use was made of the theoretical expression

$$\mu = \frac{K \cdot \sqrt{T}}{\sigma^2 \Omega(T^*)}$$

to compute $\sigma^2 \Omega$ which was plotted as a function of 1/T and smoothed. Recommended values were computed from the smoothed curve which was forced to be close to the accurate values at normal temperature from Kestin [98, 111, 114–115, 117] and Majumdar [146].

The accuracy is thought to be ± 2% in the range 90 K to 1100 K and about ±4% outside this range.

28

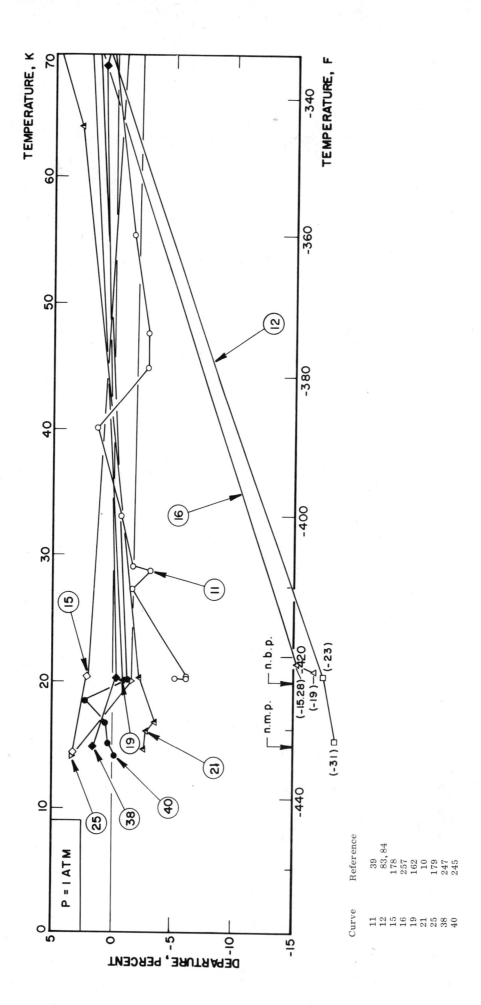

FIGURE 7-G(T).   DEPARTURE PLOT FOR VISCOSITY OF GASEOUS HYDROGEN

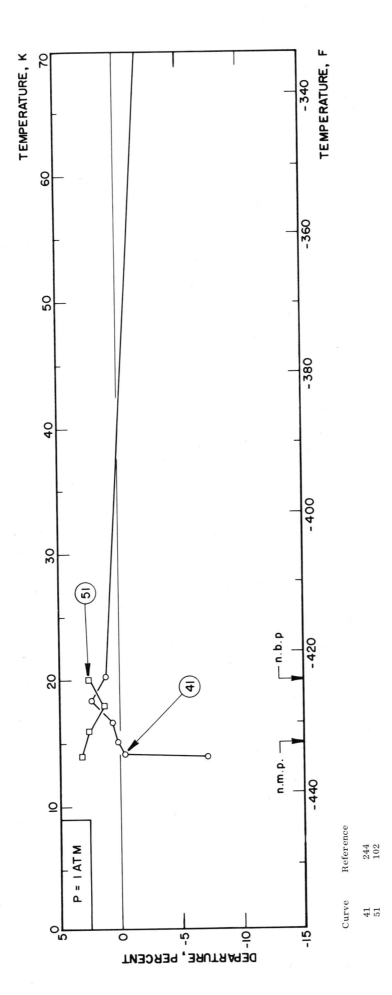

FIGURE 7-G(T). DEPARTURE PLOT FOR VISCOSITY OF GASEOUS HYDROGEN (continued)

FIGURE 7-G(T).   DEPARTURE PLOT FOR VISCOSITY OF GASEOUS HYDROGEN   (continued)

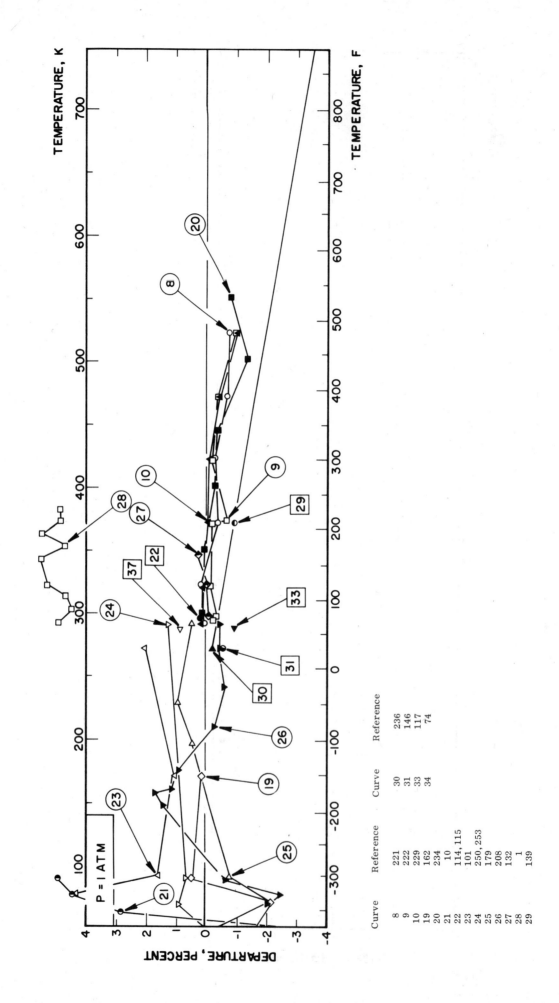

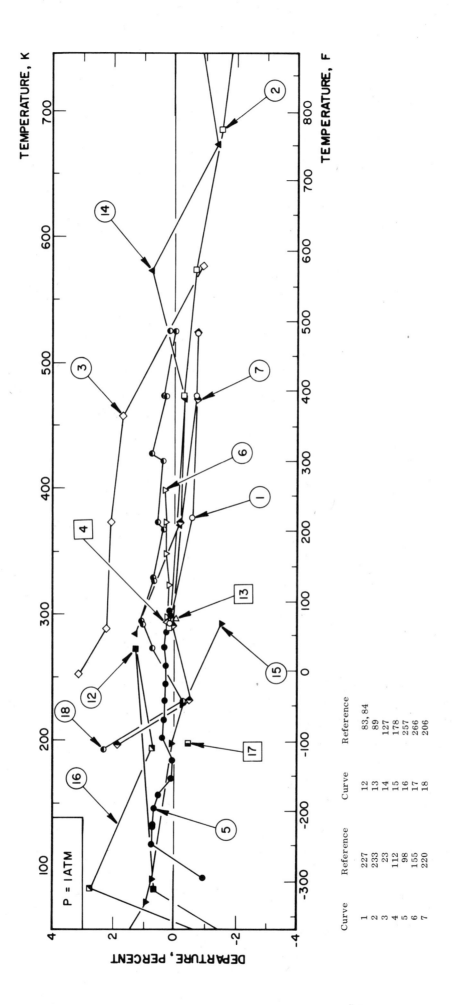

FIGURE 7-G(T).   DEPARTURE PLOT FOR VISCOSITY OF GASEOUS HYDROGEN   (continued)

Curve	Reference		Curve	Reference
1	227		12	83, 84
2	233		13	89
3	23		14	127
4	112		15	178
5	98		16	257
6	155		17	266
7	220		18	206

32

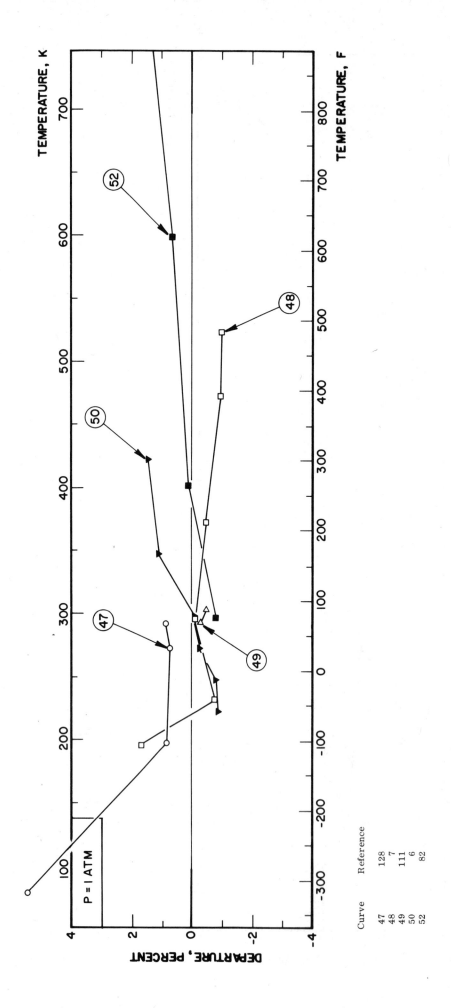

FIGURE 7-G(T).   DEPARTURE PLOT FOR VISCOSITY OF GASEOUS HYDROGEN (continued)

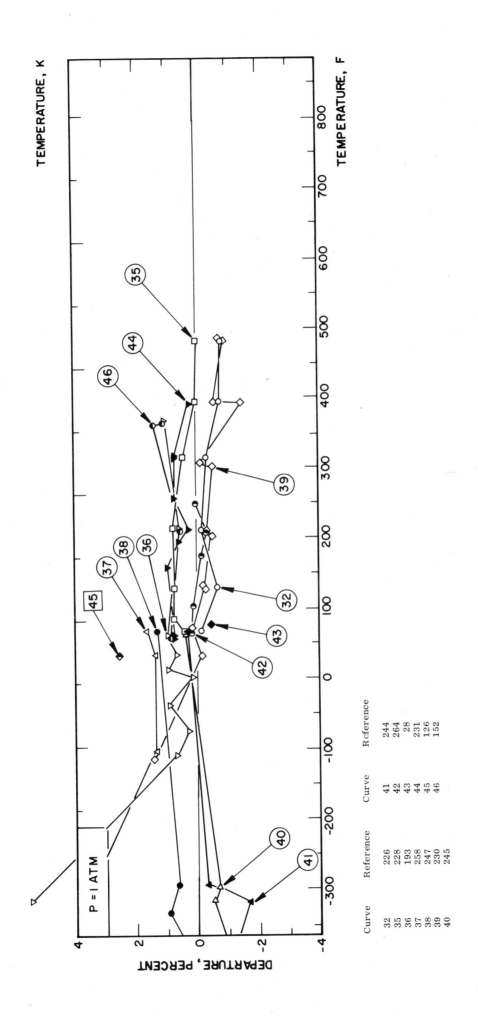

FIGURE 7-G(T). DEPARTURE PLOT FOR VISCOSITY OF GASEOUS HYDROGEN (continued)

Curve	Reference	Curve	Reference
32	226	41	244
35	228	42	264
36	193	43	28
37	258	44	231
38	247	45	126
39	230	46	152
40	245		

34

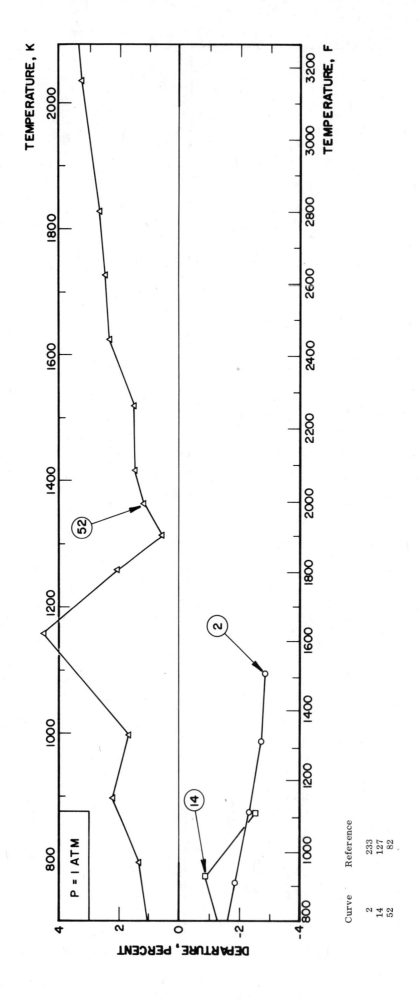

FIGURE 7-G(T). DEPARTURE PLOT FOR VISCOSITY OF GASEOUS HYDROGEN (continued)

Curve	Reference
2	233
14	127
52	82

TABLE 8-G(T).   VISCOSITY OF GASEOUS IODINE

DISCUSSION

GAS

Two sets of experimental data were found in the literature. Those of Rankine [311] covering a range from 396 K to 520 K and those of Braune and Linke [22] covering a range from 379 K to 795 K. They are in good agreement.

The data were fitted to the equation $\mu = K\sqrt{T}/(\sigma^2 \Omega_{22})$. The group $\sigma^2 \Omega$ was calculated from the data and fitted to a quadratic equation in 1/T, from which adjusted $\sigma^2 \Omega$ were derived to generate recommended values of viscosity. The accuracy is thought to be better than ±1 percent.

RECOMMENDED VALUES
[Temperature, T, K; Viscosity, μ, 10^{-6} N s m^{-2}]

GAS

T	μ	T	μ
370	17.28	550	25.1
380	17.72	560	25.5
390	18.16	570	26.0
400	18.60	580	26.4
410	19.04	590	26.8
420	19.48	600	27.2
430	19.92	610	27.6
440	20.55	620	28.1
450	20.79	630	28.5
460	21.23	640	28.9
470	21.66	650	29.3
480	22.09	660	29.7
490	22.53	670	30.1
500	22.96	680	30.5
510	23.4	690	30.9
520	23.8	700	31.3
530	24.3		
540	24.7		

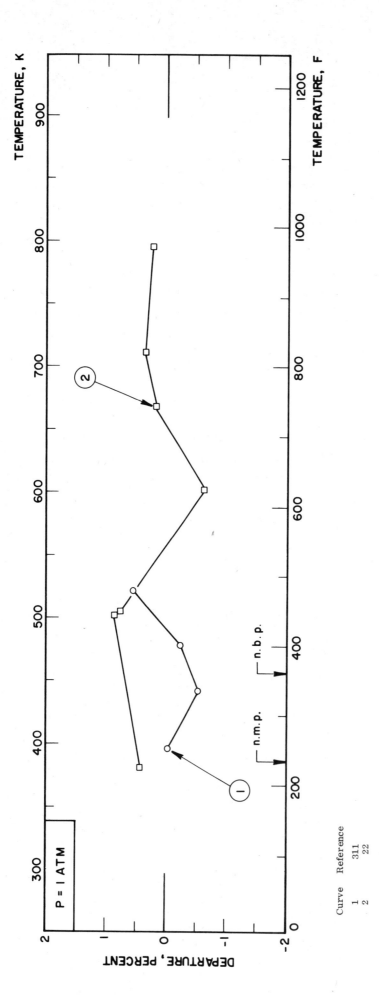

FIGURE 8-G(T).   DEPARTURE PLOT FOR VISCOSITY OF GASEOUS IODINE

TABLE 9-G(T). VISCOSITY OF GASEOUS KRYPTON

RECOMMENDED VALUES

[Temperature, T, K; Viscosity, μ, N.m^{-1}.sec^{-1}.10^{-6}]

GAS

T	μ	T	μ	T	μ
100	9.29	500	39.14	900	60.1
110	10.08	510	39.8	910	60.5
120	10.87	520	40.4	920	61.0
130	11.68	530	41.0	930	61.4
140	12.51	540	41.6	940	61.9
150	13.35	550	42.1	950	62.3
160	14.19	560	42.7	960	62.8
170	15.03	570	43.3	970	63.2
180	15.87	580	43.9	980	63.6
190	16.72	590	44.4	990	64.1
200	17.55	600	45.0	1000	64.5
210	18.39	610	45.6	1010	64.9
220	19.21	620	46.1	1020	65.4
230	20.03	630	46.7	1030	65.8
240	20.84	640	47.2	1040	66.2
250	21.64	650	47.7	1050	66.6
260	22.44	660	48.3	1060	67.0
270	23.22	670	48.8	1070	67.5
280	24.00	680	49.4	1080	67.8
290	24.77	690	49.9	1090	68.3
300	25.53	700	50.4	1100	68.7
310	26.28	710	50.9	1110	69.1
320	27.03	720	51.4	1120	69.5
330	27.76	730	51.9	1130	69.9
340	28.49	740	52.4	1140	70.3
350	29.21	750	52.9	1150	70.7
360	29.92	760	53.4	1160	71.1
370	30.62	770	53.9	1170	71.5
380	31.32	780	54.4	1180	71.9
390	32.01	790	54.9	1190	72.3
400	32.69	800	55.4	1200	72.7
410	33.37	810	55.9	1210	73.0
420	34.03	820	56.4	1220	73.4
430	34.70	830	56.8	1230	73.8
440	35.35	840	57.3	1240	74.2
450	36.00	850	57.8	1250	74.6
460	36.64	860	58.2	1260	75.0
470	37.27	870	58.7	1270	75.3
480	37.90	880	59.2	1280	75.7
490	38.53	890	59.6	1290	76.1

DISCUSSION

GAS

Experimental data for viscosity of krypton reported in the literature are those of Clifton [312], Trautz [232], Kestin [109], Rankine [169], Nasini [159], Uchiyama [236], Trappeniers [313], and Rigby and Smith [182]. Data given by Carvalho [45] seems to come from other authors, while values given by Andrussow [3] are computed values.

Among the data covering a wide range of temperature, those of Clifton were very scattered, but those of Rigby and Smith were found reliable.

To analyze the data, use was made of the theoretical expression for viscosity:

$$\mu = 266.93 \frac{\sqrt{MT}}{\sigma^2 \Omega(T^*)} f_\mu$$

The group $[\sigma^2 \Omega(T^*)/f_\mu]$ was computed from the experimental data, and plotted as a function of 1/T, and a smooth curve drawn. The curve obtained has been compared with the similar curve obtained for argon, assuming that reduced viscosity values are the same at the Boyle temperature. A table was then generated, from which recommended values were computed. These should be accurate to well within 2.5 percent below 1150 K and five percent for all higher temperatures tabulated.

TABLE 9-G(T). VISCOSITY OF GASEOUS KRYPTON (continued)

RECOMMENDED VALUES

[Temperature, T, K; Viscosity, μ, N.m^{-1}.sec^{-1}.10^{-6}]

GAS

T	μ
1300	76.5
1310	76.8
1320	77.2
1330	77.6
1340	77.9
1350	78.3
1360	78.7
1370	79.0
1380	79.4
1390	79.7
1400	80.1
1410	80.5
1420	80.8
1430	81.2
1440	81.5
1450	81.9
1460	82.2
1470	82.6
1480	82.9
1490	83.3
1500	83.6

FIGURE 9-G(T). DEPARTURE PLOT FOR VISCOSITY OF GASEOUS KRYPTON

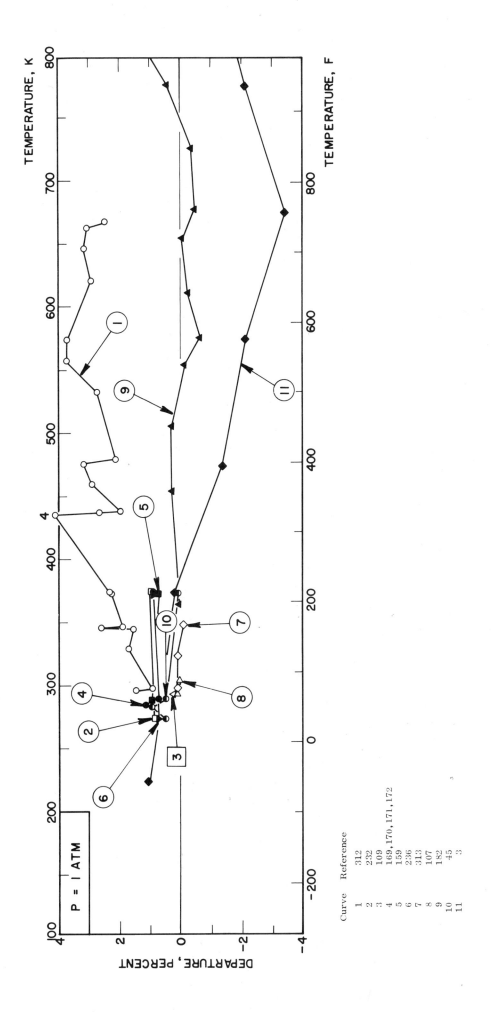

Curve	Reference
1	312
2	232
3	109
4	169,170,171,172
5	159
6	236
7	313
8	107
9	182
10	45
11	3

40

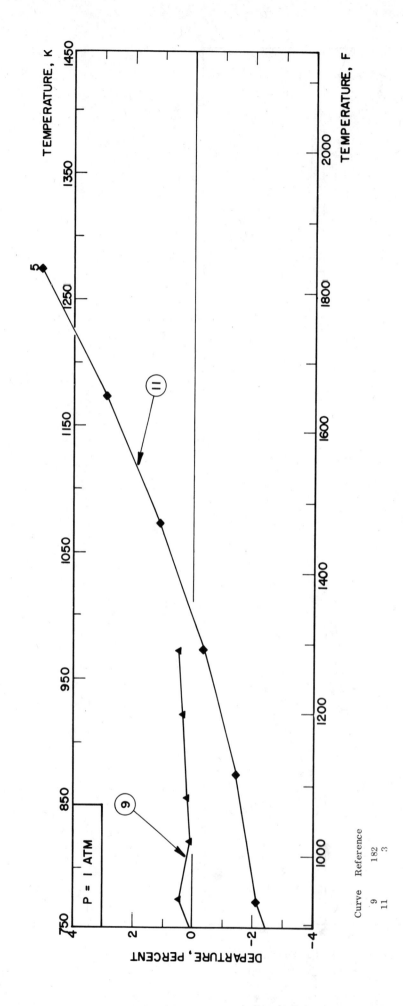

FIGURE 9-G(T). DEPARTURE PLOT FOR VISCOSITY OF GASEOUS KRYPTON (continued)

TABLE 10-L(T).   VISCOSITY OF LIQUID NEON

DISCUSSION

SATURATED LIQUID

Two sets of experimental data were found in the literature, by Forster
[65] and by Huth [282]. They were discussed by Bewilogua [15] who states that
the accuracy expected is about ±10%.

Below 29 K, the Forster data were fitted to an equation:

$\log \mu = A + B/T$

Above 29 K, the recommended curve was obtained from a graphical extrapolation
of the equation fit to join the estimated value at the critical point.

RECOMMENDED VALUES

[Temperature, T, K; Viscosity, μ, N s m^{-2} · 10^3]

SATURATED LIQUID

T	μ
25	0.151
26	0.139
27	0.127
28	0.116
29	0.105
30	0.098
31	0.091
32	0.084
33	0.078
34	0.072
35	0.0668
36	0.0619
37	0.0562
38	0.0517
39	0.0473
40	0.0427
41	0.0387
42	0.0343
43	0.0309
44	0.0269
44*	0.0167

* Crit. Temp.

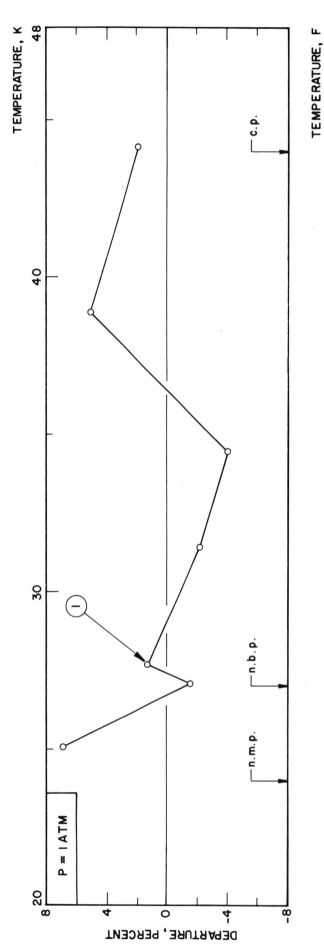

FIGURE 10-L(T).   DEPARTURE PLOT FOR VISCOSITY OF LIQUID NEON

TABLE 10-V(T).   VISCOSITY OF NEON VAPOR

DISCUSSION

SATURATED VAPOR

Recommended values for the viscosity of the saturated vapor have been generated using the excess viscosity concept as outlined by Jossi, Stiel and Thodos [100].

From a curve of the reduced excess viscosity versus reduced temperature, the excess viscosity was obtained and added to the recommended values of the 1 atm gas.

The accuracy is thought to be about ± 5%.

RECOMMENDED VALUES

[Temperature, T, K, Viscosity, μ, 10^{-3} N s m^{-2}]

SATURATED VAPOR

T	μ
27	0.00463
28	0.00485
29	0.00504
30	0.00524
31	0.00543
32	0.00564
33	0.00583
34	0.00604
35	0.00622
36	0.00644
37	0.00671
38	0.00703
39	0.00739
40	0.00781
41	0.00836
42	0.00913
43	0.0102
44	0.0121
44*	0.0167

*Crit. Temp.

TABLE 10-G(T).   VISCOSITY OF GASEOUS NEON

RECOMMENDED VALUES

[Temperature, T, K; Viscosity, μ, N s m⁻² · 10⁻⁶]

GAS

T	μ	T	μ	T	μ
20	3.38	350	35.19	700	55.2
25	4.28	360	35.85	710	55.7
30	5.15	370	36.52	720	56.2
35	5.97	380	37.16	730	56.7
40	6.75	390	37.82	740	57.2
45	7.52	400	38.45	750	57.6
50	8.24	410	39.09	760	58.1
60	9.63	420	39.70	770	58.6
70	10.91	430	40.33	780	59.1
80	12.11	440	40.94	790	59.6
90	13.27	450	41.53	800	60.0
100	14.36	460	42.15	810	60.5
110	15.42	470	42.73	820	61.0
120	16.45	480	43.33	830	61.4
130	17.47	490	43.91	840	61.9
140	18.46	500	44.48	850	62.3
150	19.43	510	45.1	860	62.8
160	20.36	520	45.6	870	63.2
170	21.28	530	46.2	880	63.7
180	22.18	540	46.8	890	64.1
190	23.04	550	47.3	900	64.6
200	23.91	560	47.9	910	65.0
210	24.76	570	48.4	920	65.5
220	25.58	580	48.9	930	65.9
230	26.40	590	49.5	940	66.3
240	27.19	600	50.0	950	66.8
250	27.97	610	50.6	960	67.2
260	28.74	620	51.1	970	67.6
270	29.50	630	51.6	980	68.0
280	30.25	640	52.1	990	68.5
290	30.99	650	52.7	1000	68.9
300	31.71	660	53.2	1050	71.0
310	32.44	670	53.7	1100	73.0
320	33.14	680	54.2	1150	75.0
330	33.84	690	54.7	1200	76.9
340	34.52				

DISCUSSION

GAS

Twenty-two sets of experimental data were found in the literature [28, 39, 51, 54, 64, 97, 109, 117, 146, 170, 179, 180, 213, 218, 222, 224, 232, 233, 236, 247, 253, 264]. They cover a temperature range from 16 K to 1100 K.
Above normal temperature, in the high temperature range there are sets by Trautz [222, 224, 232, 233], Edwards [54], Kestin [109, 117] and di Pippo [51] which are in good agreement. Below normal temperature, three sets, of the Van Itterbeek school[179-80,247,253],of Coremans [39] and of Johnston [97], are in fair agreement.

To correlate the data, use was made of the theoretical expression

$$\mu = \frac{K\sqrt{T}}{\sigma^2 \Omega(T^*)}$$

from which the group $\sigma^2\Omega$ was computed, plotted as a function of 1/T and smoothed. Recommended values were generated from the smoothed curve.

The accuracy is thought to be ±2%, although a higher figure is expected at low temperature.

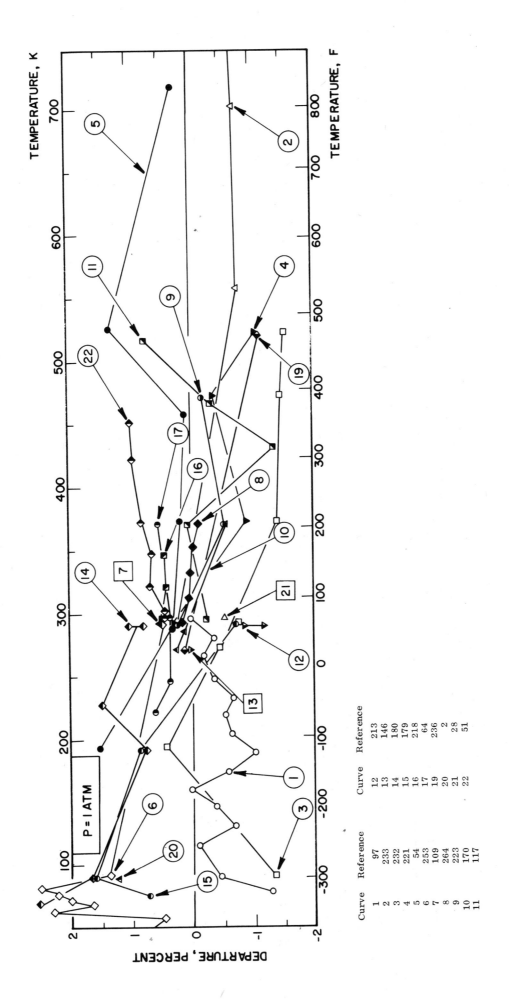

FIGURE 10-G(T). DEPARTURE PLOT FOR VISCOSITY OF GASEOUS NEON

Curve	Reference		Curve	Reference
1	97		12	213
2	233		13	146
3	232		14	180
4	221		15	179
5	54		16	218
6	253		17	64
7	109		19	236
8	264		20	2
9	223		21	28
10	170		22	51
11	117			

46

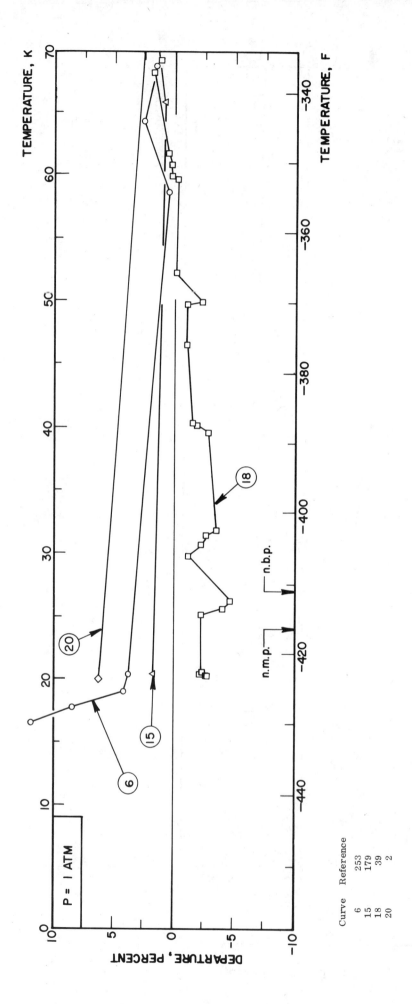

FIGURE 10-G(T). DEPARTURE PLOT FOR VISCOSITY OF GASEOUS NEON (continued)

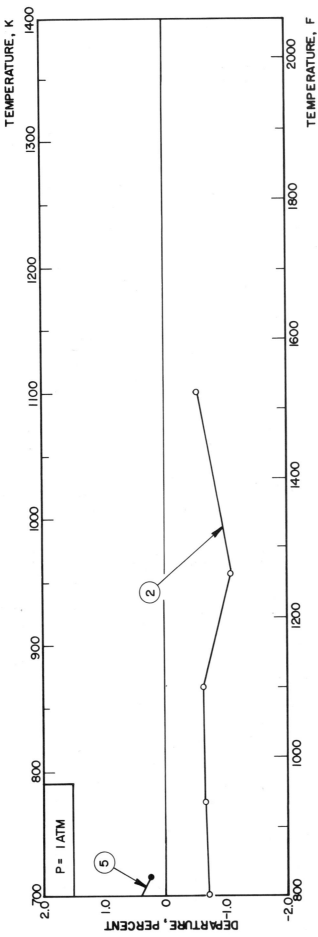

FIGURE 10-G(T). DEPARTURE PLOT FOR VISCOSITY OF GASEOUS NEON (continued)

TABLE 11-L(T). VISCOSITY OF LIQUID NITROGEN

RECOMMENDED VALUES

[Temperature, T, K; Viscosity, μ, 10^{-3} N s m^{-2}]

SATURATED LIQUID

T	μ
60	0.360
65	0.274
70	0.217
75	0.1768
80	0.1480
85	0.1266
90	0.1101
95	0.0972
100	0.0869
105	0.0785
110	0.0708
115	0.0599
120	0.0484
125	0.0316
126*	0.0191

* Crit. Temp.

DISCUSSION

SATURATED LIQUID

Six sets of experimental data were found in the literature. The data of Rudenko [188] and of Forster [65] covers the temperature range from the boiling point to the vicinity of the critical point. Other data by Rudenko [189] and Van Itterbeek [240, 241] and Boon [20] are below or about the boiling point. The various sets are not mutually consistant, and it is difficult to assess their reliability. They were adjusted by least squares to an equation

$$\log \mu = A + B/T$$

below 105 K and smoothed graphically between 105 K and the critical point.

Values for the saturated liquid computed with the correlating technique of Jossi et al. [100] are not in good agreement with the recommended values. The accuracy of the correlation must be about ±10% and there is a need for reliable experimental data.

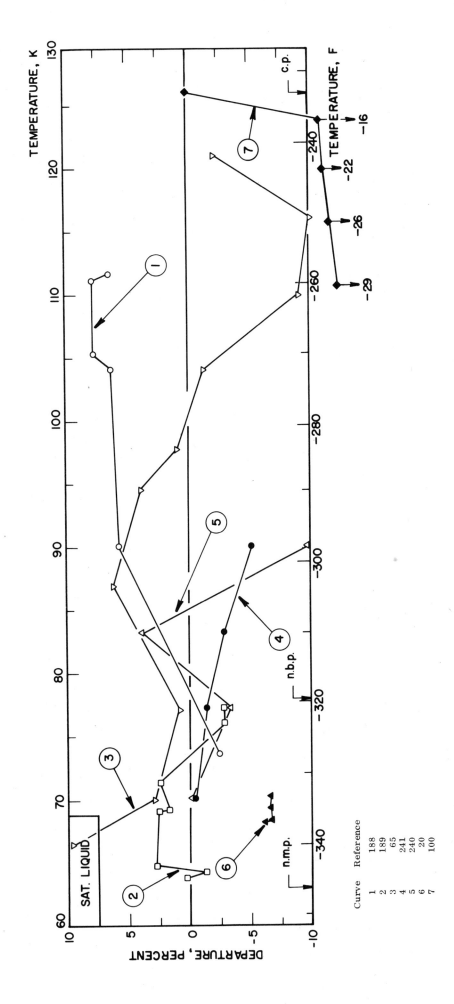

FIGURE 11-L(T).   DEPARTURE PLOT FOR VISCOSITY OF LIQUID NITROGEN

50

TABLE 11-V(T). VISCOSITY OF NITROGEN VAPOR

DISCUSSION

SATURATED VAPOR

The recommended values for the viscosity of the saturated vapor of nitrogen were computed by means of the correlation technique of Jossi, Stiel and Thodos [100], using density values given by Din [49] and the recommended values for the 1 atm gas.

They are thought reliable to ±2% at low temperature with a probable error of ±5% in the vicinity of the critical point.

RECOMMENDED VALUES

[Temperature, T, E; Viscosity, μ, 10^{-3} N s m^{-2}]

SATURATED VAPOR

T	μ
80	.00560
85	.00596
90	.00636
95	.00680
100	.00728
105	.00782
110	.00842
115	.00925
120	.01068
121	.0110
122	.0115
123	.0120
124	.0129
125	.0144
126*	.0191

* Crit. Temp.

TABLE 11-G(T).   VISCOSITY OF GASEOUS NITROGEN

RECOMMENDED VALUES

[Temperature, T, K; Viscosity, μ, 10^{-6} N s m^{-2}]

GAS

T	μ	T	μ	T	μ
80	5.59	450	24.08	850	36.55
90	6.22	460	24.45	860	36.82
100	6.87	470	24.82	870	37.08
110	7.52	480	25.18	880	37.34
120	8.15	490	25.54	890	37.60
130	8.78	500	25.90	900	37.86
140	9.40	510	26.25	910	38.12
150	10.00	520	26.60	920	38.37
160	10.59	530	26.95	930	38.63
170	11.18	540	27.29	940	38.88
180	11.75	550	27.63	950	39.12
190	12.31	560	27.96	960	39.38
200	12.86	570	28.30	970	39.63
210	13.40	580	28.63	980	39.87
220	13.93	590	28.95	990	40.12
230	14.45	600	29.27	1000	40.36
240	14.96	610	29.59	1010	40.6
250	15.46	620	29.91	1020	40.8
260	15.96	630	30.23	1030	41.1
270	16.45	640	30.54	1040	41.3
280	16.92	650	30.85	1050	41.6
290	17.40	660	31.15	1060	41.8
300	17.86	670	31.46	1070	42.0
310	18.32	680	31.76	1080	42.3
320	18.77	690	32.06	1090	42.5
330	19.21	700	32.35	1100	42.7
340	19.65	710	32.65	1110	43.0
350	20.08	720	32.94	1120	43.2
360	20.50	730	33.23	1130	43.4
370	20.92	740	33.52	1140	43.6
380	21.33	750	33.80	1150	43.9
390	21.74	760	34.09	1160	44.1
400	22.14	770	34.37	1170	44.3
410	22.54	780	34.65	1180	44.5
420	22.93	790	34.93	1190	44.8
430	23.32	800	35.20	1200	45.0
440	23.70	810	35.48	1210	45.2
		820	35.75	1220	45.4
		830	36.02	1230	45.6
		840	36.29	1240	45.8

DISCUSSION

GAS

There are 34 sets of experimental data available for the viscosity of nitrogen [5, 18, 50, 59, 64, 67, 91, 95, 98, 99, 109, 110, 112, 115, 117, 138, 146, 148, 151, 152, 175, 183, 201, 220, 222, 227, 233, 236, 349, 254, 255, 257, 264, 266, 326]. The overall temperature range covered, is from 78 to 2500 K. In general there is a good agreement between the various investigators, although the high temperature results from Di Pippo and Kestin [50], Kestin and Whitelaw [117] on one hand, and those of Vasilesco [254-255], and Bonilla [18] on the other lie on opposite sides of the curve, indicating some systematic deviation, a fact already pointed out by Hanley and Childs [85]. A semi-theoretical evaluation was made by Andrussow [3], and is in good agreement.

To correlate the data, use was made of the theoretical expression:

$$\mu = 266.93 \, \frac{M\sqrt{T}}{\sigma^2 \Omega(T^*)} \, f_\mu$$

The group $\sigma^2 \Omega(T^*)/f_\mu$ was computed from the experimental data and plotted as a function of 1/T. The curve obtained was smoothed, using the similar curve for argon as a guide, to make the comparison reduction factors used were the ratio of collision diameter and the ratio of Boyle temperature. A table was generated in order to compute recommended values.

The accuracy is thought to be about ± 2 percent.

TABLE 11-G(T).   VISCOSITY OF GASEOUS NITROGEN   (continued)

RECOMMENDED VALUES

[Temperature, T, K; Viscosity, μ, 10^{-6} N s m^{-2}]

GAS

T	μ
1250	46.1
1260	46.3
1270	46.5
1280	46.7
1290	46.9
1300	47.1
1310	47.3
1320	47.5
1330	47.8
1340	48.0
1350	48.2
1360	48.4
1370	48.6
1380	48.8
1390	49.0
1400	49.2
1410	49.4
1420	49.6
1430	49.8
1440	50.0
1450	50.2
1460	50.4
1470	50.6
1480	50.8
1490	51.0
1500	51.2
1550	52.1
1600	53.1
1650	54.0
1700	54.9
1750	55.8
1800	56.7
1850	57.6
1900	58.5
1950	59.3
2000	60.1
2100	61.8
2200	63.4

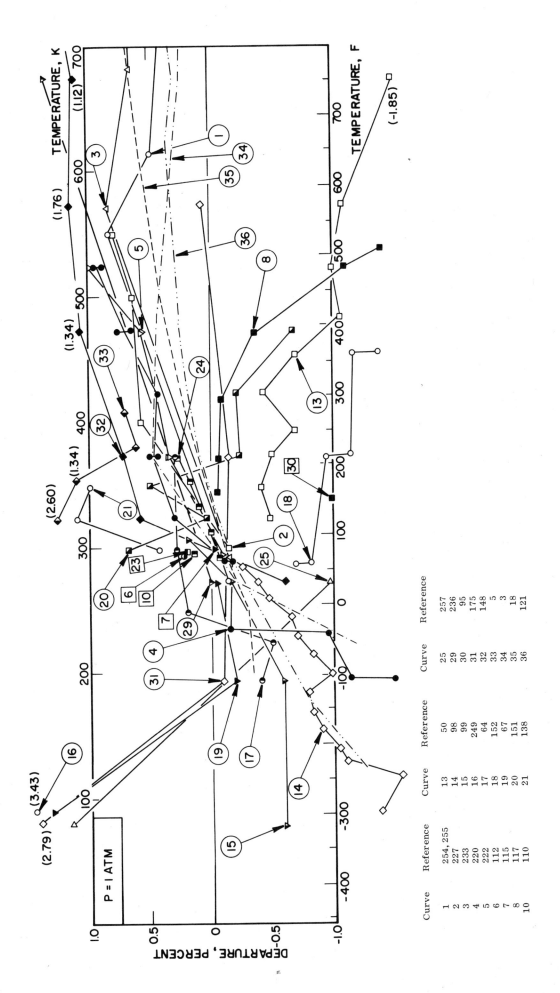

FIGURE 11-G(T).   DEPARTURE PLOT FOR VISCOSITY OF GASEOUS NITROGEN

54

FIGURE 11-G(T). DEPARTURE PLOT FOR VISCOSITY OF GASEOUS NITROGEN (continued)

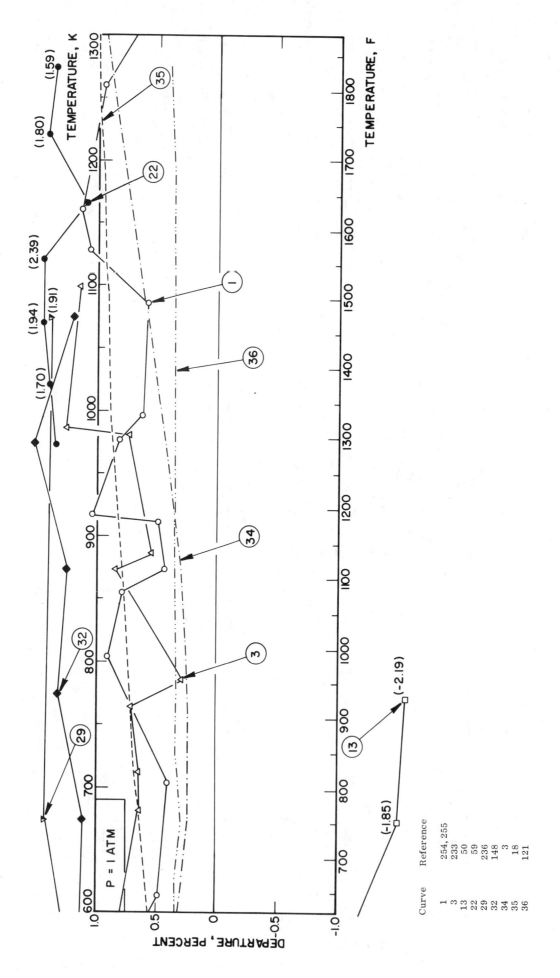

Curve	Reference
1	254, 255
3	233
13	50
22	59
29	236
32	148
34	3
35	18
36	121

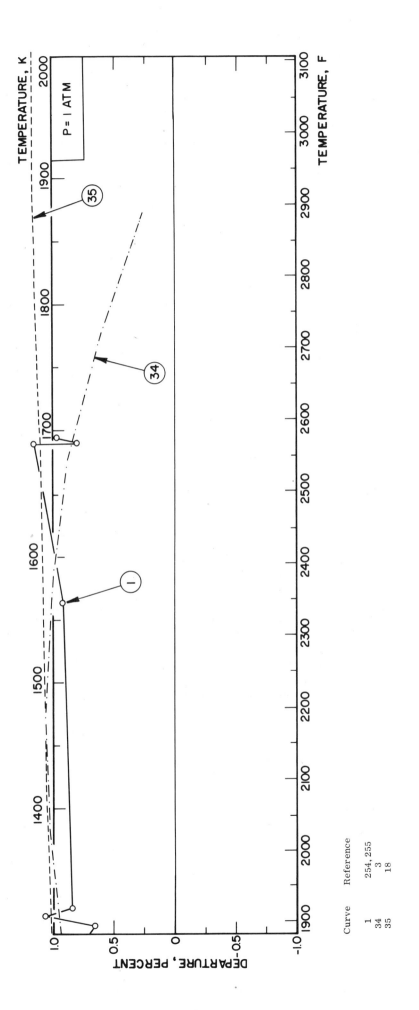

FIGURE 11-G(T).   DEPARTURE PLOT FOR VISCOSITY OF GASEOUS NITROGEN (continued)

Curve	Reference
1	254, 255
34	3
35	18

TABLE 12-L(T). VISCOSITY OF LIQUID OXYGEN

DISCUSSION

SATURATED LIQUID

There are eight sets of experimental data in the literature [19, 43, 69, 188, 189, 240, 241, 253]. All data, except those of Rudenko [188] are below the boiling point, and are in good agreement, although the results of Galkov [69] seem less accurate and higher.

All the data below 125 K were fitted by least squares to an equation

$$\log \mu = A + B/T$$

The recommended values below 125 K were joined smoothly to the value of the viscosity at the critical point, by a hand drawn curve. They are in disagreement with the experimental values of Rudenko [188].

The accuracy is estimated as about ±3% below 125 K and drops to ±15% from 125 K to the critical temperature.

RECOMMENDED VALUES

[Temperature, T, K; Viscosity, μ, 10^{-3} N s m^{-2}]

SATURATED LIQUID

T	μ
55	0.804
60	0.593
65	0.459
70	0.368
75	0.304
80	0.257
85	0.222
90	0.195
95	0.173
100	0.1560
105	0.1418
110	0.1300
115	0.1201
120	0.1117
125	0.1040
130	0.0960
135	0.0875
140	0.0780
145	0.0665
150	0.0510
154*	0.0259

* Crit. Temp.

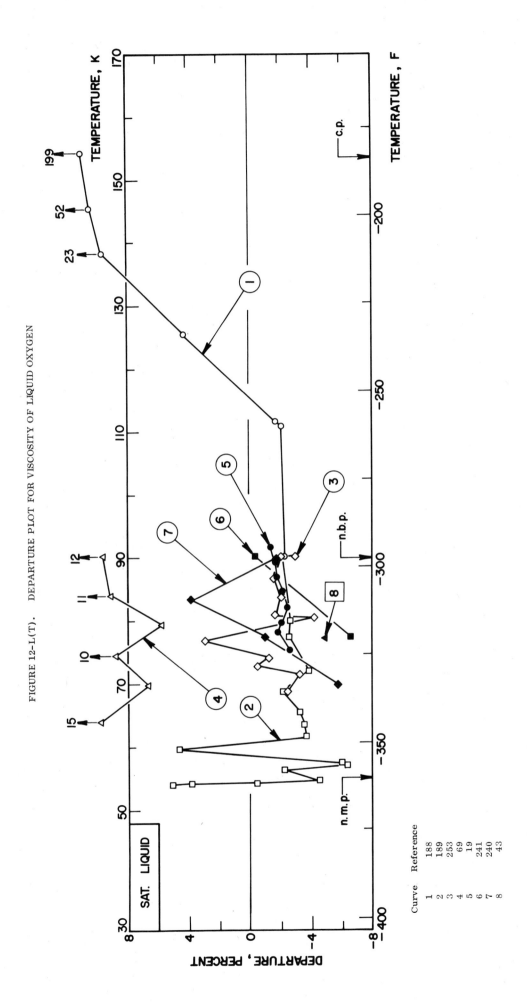

FIGURE 12-L(T). DEPARTURE PLOT FOR VISCOSITY OF LIQUID OXYGEN

57

TABLE 12-V(T).   VISCOSITY OF OXYGEN VAPOR

DISCUSSION

RECOMMENDED VALUES

[Temperature, T, K; Viscosity, μ, 10^{-3} N s m^{-2}]

SATURATED VAPOR

Recommended values of the viscosity of the saturated vapor were computed by the technique of Jossi et al. [100] using the recommended value of the viscosity of the dilute gas and a generalized correlation of the excess viscosity versus the reduced temperature which was established with the values of several gases using also the correlation of Jossi et al. In the complete absence of any experimental data, no accuracy estimation is made.

SATURATED VAPOR

T	μ
80	.00627
90	.00698
100	.00772
110	.00860
120	.00949
130	.01057
140	.01231
150	.0161
154*	.0259

* Crit. Temp.

TABLE 12-G(T). VISCOSITY OF GASEOUS OXYGEN

DISCUSSION

GAS

There are 20 sets of experimental data available for the viscosity of Oxygen, covering an overall range of temperature from 72 to 2500 K. They are those reported by Trautz and al [222, 227, 232, 233, Johnston [98], Van Itterbeek [243-249], Kestin [109-110], Makita [150], Uchiyama [236], Volker [258], Markowski [152], Raw and Ellis [176], Majumdar [146], Yen [266], Rigden [183], Vogel [257], Wobser and Muller [264] and Bonilla [18]. A previous correlation made by Keyes [121], and a semi-theoretical evaluation by Andrussow [3] both are in good agreement with the present correlation.

At low temperature, the data of Johnston appears smoother than those of Van Itterbeek while Volker's data diverge greatly (about 20%). At room temperature, there is good agreement between Kestin, Rigden, Majumdar and Yen. At high temperature, there are discrepancies between the data of Trautz, Bonilla and of Raw and Ellis. The Johnston data were given more weight at low temperatures.

To correlate the data, use was made of the expression:

$$\mu = 266.93 \, \frac{\sqrt{MT}}{\sigma^2 \Omega(T^*)} \, f_\mu$$

The group $\sigma^2 \Omega(T^*)/f_\mu$ was computed from the experimental data, and plotted as a function of $1/T$. To help smoothing, the curve obtained has been compared with the similar curve obtained for Argon, the resulting curve was chosen so as to match the results of Majumdar, Rigden and Kestin, at room temperature, and a table was generated which was used for computing recommended values.

The accuracy is of the order of ±2 percent but is better than one percent around room temperature.

RECOMMENDED VALUES

[Temperature, T, K; Viscosity, μ, 10^{-6} N s m^{-2}]

GAS

T	μ	T	μ	T	μ
80	6.27	450	28.28	850	43.8
90	6.98	460	28.74	860	44.1
100	7.68	470	29.20	870	44.4
110	8.39	480	29.65	880	44.7
120	9.12	490	30.10	890	45.1
130	9.85	500	30.54	900	45.4
140	10.56	510	31.0	910	45.7
150	11.27	520	31.4	920	46.0
160	11.96	530	31.8	930	46.3
170	12.65	540	32.3	940	46.7
180	13.33	550	32.7	950	47.0
190	13.99	560	33.1	960	47.3
200	14.65	570	33.5	970	47.6
210	15.29	580	33.9	980	47.9
220	15.93	590	34.3	990	48.2
230	16.55	600	34.7	1000	48.5
240	17.17	610	35.2	1050	50.0
250	17.77	620	35.5	1100	51.4
260	18.37	630	35.9	1150	52.9
270	18.96	640	36.3	1200	54.2
280	19.54	650	36.7	1250	55.6
290	20.11	660	37.1	1300	56.9
300	20.67	670	37.4	1350	58.2
310	21.23	680	37.8	1400	59.5
320	21.77	690	38.2	1450	60.7
330	22.31	700	38.5	1500	61.9
340	22.84	710	38.9	1550	63.1
350	23.37	720	39.3	1600	64.3
360	23.89	730	39.6	1650	65.5
370	24.40	740	40.0	1700	66.6
380	24.91	750	40.3	1750	67.7
390	25.41	760	40.7	1800	68.8
400	25.89	770	41.1	1850	69.9
410	26.39	780	41.4	1900	71.0
420	26.87	790	41.7	1950	72.1
430	27.35	800	42.1	2000	73.1
440	27.82	810	42.4		
		820	42.8		
		830	43.1		
		840	43.4		

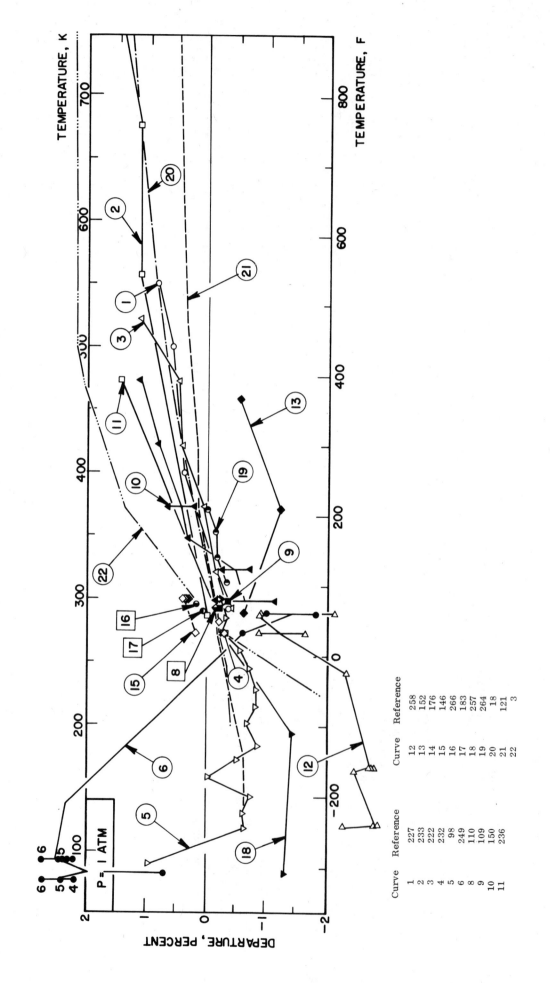

FIGURE 12-G(T). DEPARTURE PLOT FOR VISCOSITY OF GASEOUS OXYGEN

61

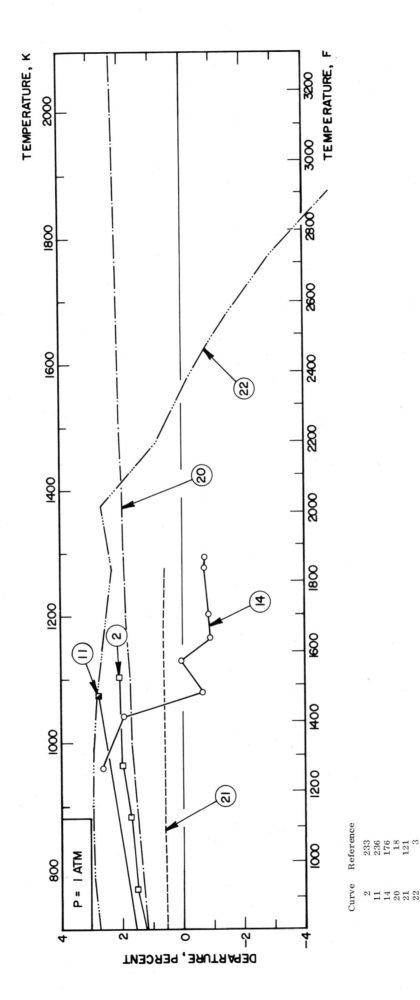

FIGURE 12-G(T). DEPARTURE PLOT FOR VISCOSITY OF GASEOUS OXYGEN (continued)

DISCUSSION

TABLE 13-G(T).   VISCOSITY OF GASEOUS XENON

RECOMMENDED VALUES

[Temperature, T, K; Viscosity, μ, N.m⁻¹.sec⁻¹.10⁻⁶]

GAS

Experimental data for the viscosity of xenon reported in the literature were those of Trautz[232,373],Rankine[170],Nasini[159], Kestin[107], Uchiyama [236], and Rigby and Smith[183]. The latter, and those of Trautz, covers the widest range of temperature, and are in good agreement.

The analysis was made with the help of the theoretical expression for viscosity:

$$\mu = 266.93 \frac{\sqrt{MT}}{\sigma^2 \Omega(T^*)} f_\mu$$

The group $[\sigma^2 \Omega(T^*)/f_\mu]$ was computed from the experimental data and plotted as a function of $1/T$. The curve obtained has been compared with the similar curve obtained for argon, the scaling being made using the ratio of Boyle temperature, and the ratio of the collision diameter estimated from the data. Good agreement was found, and a table was then generated, from which recommended values were computed. The recommended values should be accurate to within one percent below 1000 K and to less than five percent for all higher temperatures tabulated.

T	μ	T	μ	T	μ
120	10.09	500	36.85	900	58.6
130	10.81	510	37.5	910	59.0
140	11.52	520	38.1	920	59.5
150	12.23	530	38.7	930	60.0
160	12.95	540	39.3	940	60.4
170	13.65	550	39.9	950	60.9
180	14.37	560	40.5	960	61.4
190	15.10	570	41.1	970	61.8
200	15.85	580	41.7	980	62.3
210	16.59	590	42.3	990	62.7
220	17.35	600	42.9	1000	63.2
230	18.10	610	43.4	1010	63.6
240	18.84	620	44.0	1020	64.1
250	19.60	630	44.6	1030	64.5
260	20.35	640	45.1	1040	65.0
270	21.09	650	45.7	1050	65.4
280	21.83	660	46.3	1060	65.8
290	22.57	670	46.8	1070	66.3
300	23.31	680	47.4	1080	66.7
310	24.03	690	47.9	1090	67.1
320	24.75	700	48.4	1100	67.6
330	25.48	710	49.0	1110	68.0
340	26.19	720	49.5	1120	68.4
350	26.90	730	50.0	1130	68.8
360	27.60	740	50.6	1140	69.3
370	28.29	750	51.1	1150	69.7
380	28.98	760	51.6	1160	70.1
390	29.67	770	52.1	1170	70.5
400	30.35	780	52.6	1180	70.9
410	31.02	790	53.2	1190	71.3
420	31.69	800	53.7	1200	71.7
430	32.36	810	54.2	1210	72.2
440	33.02	820	54.7	1220	72.6
450	33.67	830	55.2	1230	73.0
460	34.31	840	55.6	1240	73.4
470	34.95	850	56.1	1250	73.8
480	35.59	860	56.6	1260	74.2
490	36.23	870	57.1	1270	74.6
		880	57.6	1280	75.0
		890	58.1	1290	75.4

TABLE 13-G(T).   VISCOSITY OF GASEOUS XENON (continued)

RECOMMENDED VALUES

[Temperature, T, K; Viscosity, μ, N.m^{-1}.sec^{-1},10^{-6}]

GAS

T	μ
1300	75.7
1310	76.1
1320	76.5
1330	76.9
1340	77.3
1350	77.7
1360	78.1
1370	78.4
1380	78.8
1390	79.2
1400	79.6
1410	80.0
1420	80.3
1430	80.7
1440	81.1
1450	81.4
1460	81.8
1470	82.2
1480	82.5
1490	82.9
1500	83.3

64

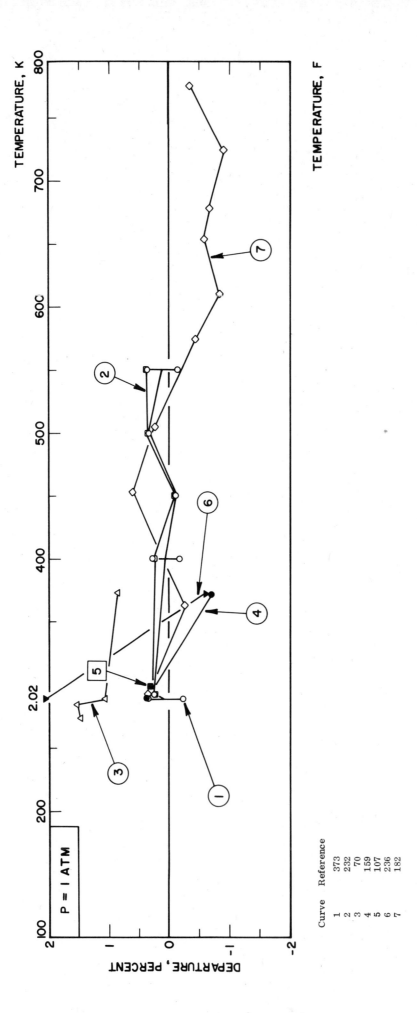

FIGURE 13-G(T). DEPARTURE PLOT FOR VISCOSITY OF GASEOUS XENON

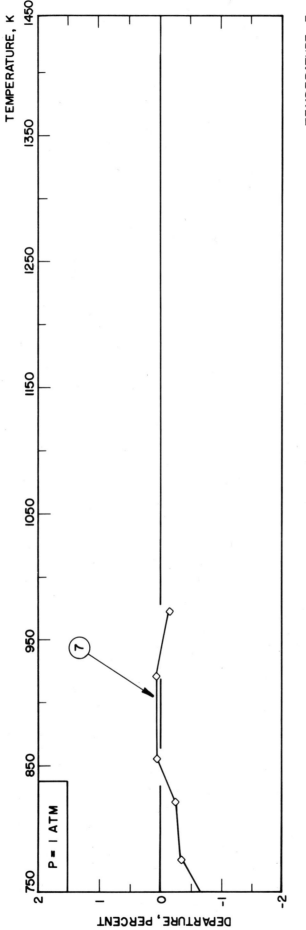

FIGURE 13-G(T). DEPARTURE PLOT FOR VISCOSITY OF GASEOUS XENON (continued)

Curve   Reference
  7   182

65

2. INORGANIC COMPOUNDS

TABLE 14-L(T). VISCOSITY OF LIQUID AMMONIA

RECOMMENDED VALUES

[Temperature, T, K; Viscosity, μ, 10^{-3} N s m^{-2}]

SATURATED LIQUID

T	μ
240	0.285
250	0.246
260	0.215
270	0.190
280	0.169
290	0.152
300	0.1370
310	0.1247
320	0.1141
330	0.1050
340	0.0971
350	0.0885
360	0.0795
370	0.0702
380	0.0607
390	0.0507
400	0.0395
405*	0.0249

* Crit. Temp.

DISCUSSION

SATURATED LIQUID

Five sets of experimental data were found in the literature. These are the values given by Stakelbeck [202], Pleskov [167] and Carmichael [29,31], Pinevich [168]. The more recent set of the latter disagree with the older. Generally the sets are not mutually consistent, and there is a need for more accurate data, in the whole range of temperature.

An equation of the type:

$$\log \mu = A + B/T + \delta$$

was used. The residual δ was smoothed graphically, and a table generated.

There is no means to evaluate the accuracy of this correlation due to the big discrepancies in observed values.

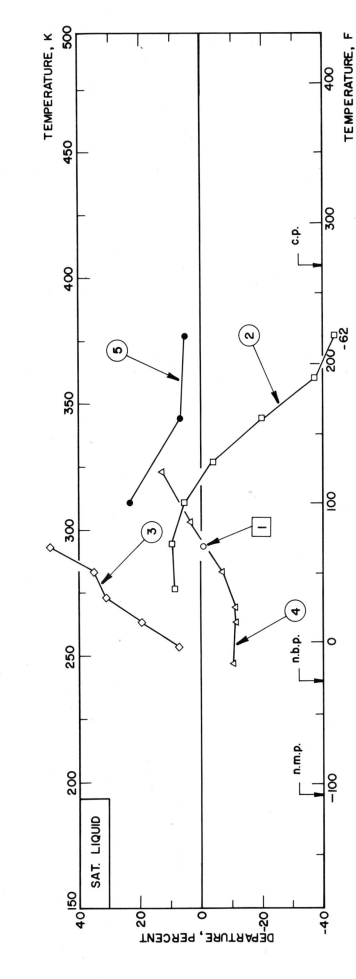

FIGURE 14-L(T).   DEPARTURE PLOT FOR VISCOSITY OF LIQUID AMMONIA

Curve	Reference
1	168
2	31
3	202
4	167
5	29

TABLE 14-V(T). VISCOSITY OF AMMONIA VAPOR

DISCUSSION

SATURATED VAPOR

Recommended values for the viscosity of the saturated vapor of ammonia were computed with the correlation equation given by Jossi, Stiel and Thodos [100]. This equation gives excess viscosity as a function of the reduced density. The values for the density of the saturated vapor used, were those given by Din [49].

The recommended values for the 1 atm gas together with the excess viscosity gave the recommended values.

The accuracy is thought to be ± 3%.

RECOMMENDED VALUES

[Temperature, T, K; Viscosity, μ, 10^{-3} N s m^{-2}]

SATURATED VAPOR

T	μ
240	0.00925
250	0.00959
260	0.00994
270	0.01030
280	0.01067
290	0.01105
300	0.01145
310	0.01186
320	0.01229
330	0.01274
340	0.01322
350	0.01375
360	0.01435
370	0.01506
380	0.01594
390	0.01715
400	0.0195
405*	0.0249

* Crit. Temp.

TABLE 14-G(T).   VISCOSITY OF GASEOUS AMMONIA

DISCUSSION

GAS

The literature revealed 16 sets of data covering a wide temperature range from 196 K to about 1,000 K, although between 196 K and room temperature there is a lack of data. The high temperature values are not very consistent, except below 500 K, where the values of Trautz [222] show a good internal consistency.

The correlation was made by using the theoretical relation:

$$\mu = \frac{K\sqrt{T}}{\sigma^2 \Omega} \qquad (1)$$

The group $\sigma^2\Omega$ was obtained from the experimental data and plotted as a function of $1/T$. A polynomial was least square fitted to the values of $\sigma^2\Omega$. Values obtained from this polynomial were used in eq. 1 to give the recommended values. Previous correlations by Keyes [122] and Vukalovich [259], shows increasing divergence at high and low temperature.

The accuracy is about ±2% below 500 K but may reach ±5% at higher temperature.

RECOMMENDED VALUES

[Temperature, T, K; Viscosity, μ, 10^{-6} N s m^{-2}]

GAS

T	μ	T	μ
200	6.89	600	21.4
210	7.21	610	21.7
220	7.53	620	22.1
230	7.86	630	22.5
240	8.19	640	22.9
250	8.53	650	23.2
260	8.87	660	23.6
270	9.21	670	24.0
280	9.56	680	24.3
290	9.91	690	24.7
300	10.27	700	25.1
310	10.62	710	25.5
320	10.98	720	25.8
330	11.34	730	26.2
340	11.70	740	26.6
350	12.06	750	26.9
360	12.43	760	27.3
370	12.80	770	27.7
380	13.16	780	28.0
390	13.53	790	28.4
400	13.90	800	28.8
410	14.27	810	29.1
420	14.64	820	29.5
430	15.01	830	29.8
440	15.38	840	30.2
450	15.76	850	30.6
460	16.13	860	30.9
470	16.50	870	31.3
480	16.88	880	31.6
490	17.25	890	32.0
500	17.63	900	32.4
510	18.0	910	32.7
520	18.4	920	33.1
530	18.8	930	33.4
540	19.1	940	33.8
550	19.5	950	34.1
560	19.9	960	34.5
570	20.3	970	34.8
580	20.6	980	35.2
590	21.0	990	35.5
		1000	35.9

72

FIGURE 14-G(T).   DEPARTURE PLOT FOR VISCOSITY OF GASEOUS AMMONIA

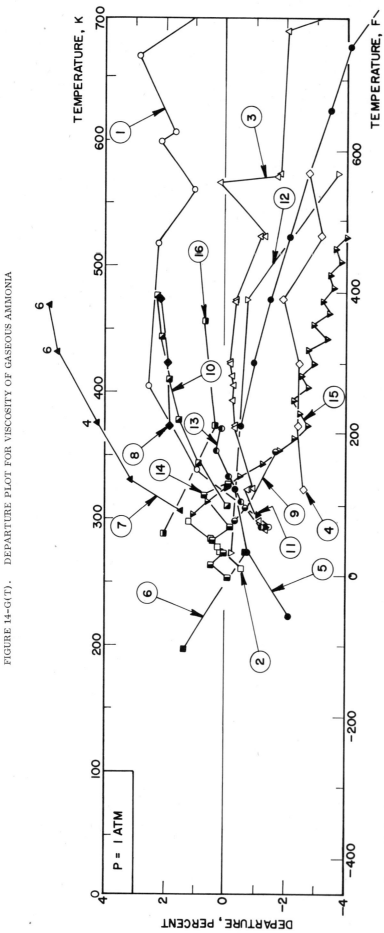

Curve	Reference		Curve	Reference
1	22		10	29
2	237		11	92
3	222		12	236
4	150		13	264
5	122		14	202
6	257		15	259
7	163		16	56
8	199			
9	35			

73

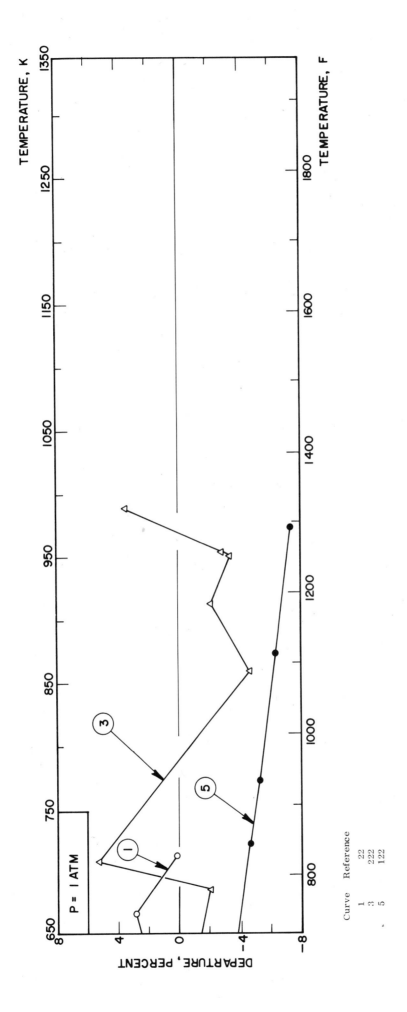

FIGURE 14-G(T). DEPARTURE PLOT FOR VISCOSITY OF GASEOUS AMMONIA (continued)

TABLE 15-G(T). VISCOSITY OF GASEOUS BORON TRIFLUORIDE

DISCUSSION

GAS

Six sets of experimental data were found in the literature [59, 288, 289, 145, 290, 291]. They cover a range of temperature from ~190 K to 973 K, however, the gas decomposes above 700 K. Good agreement exist among the authors.

To correlate the data, use was made of the theoretical relation

$$\mu = K \sqrt{T}/(\sigma^2 \Omega).$$

The group $\sigma^2 \Omega$ was computed from the data, plotted as a function of $1/T$, and it was found that a quadratic equation could fit the data. From the computed $\sigma^2 \Omega$, recommended values of the viscosity were calculated. The accuracy is of the order of ± 1 percent or better.

RECOMMENDED VALUES

[Temperature, T, K; Viscosity, μ, N s m^{-2} · 10^{-6}]

GAS

T	η	T	η
		450	23.96
		460	24.39
		470	24.82
		480	25.25
190	11.66	490	25.68
200	12.15	500	26.10
210	12.64	510	26.5
220	13.14	520	26.9
230	13.63	530	27.4
240	14.13	540	27.8
250	14.62	550	28.2
260	15.11	560	28.6
270	15.60	570	29.0
280	16.09	580	29.4
290	16.57	590	29.8
300	17.06	600	30.2
310	17.54	610	30.6
320	18.02	620	31.0
330	18.49	630	31.4
340	18.96	640	31.8
350	19.43	650	32.1
360	19.90	660	32.5
370	20.36	670	32.9
380	20.82	680	33.3
390	21.28	690	33.6
400	21.73	700	34.0
410	22.18		
420	22.63		
430	23.07		
440	23.52		

75

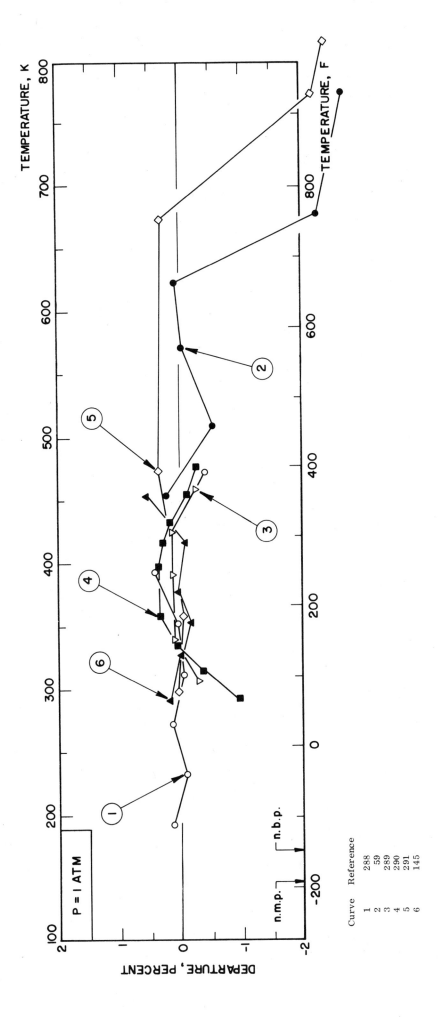

FIGURE 15-G(T). DEPARTURE PLOT FOR VISCOSITY OF GASEOUS BORON TRIFLUORIDE

TABLE 16-G(T). VISCOSITY OF GASEOUS HYDROGEN CHLORIDE

DISCUSSION

GAS

Four sets of experimental data were found in the literature [206, 228, 236, 309]. They are in good agreement. They cover a temperature range from 273 K to 524 K.

Use was made of the theoretical relation $\mu = K\sqrt{T}/[\sigma^2 \Omega_{22}(T^*)]$ to obtain the values of $\sigma^2 \Omega$ from the experimental data. These were plotted as a function of $1/T$ and adjusted to a quadratic equation. From this equation, recommended values of the viscosity were computed. The accuracy is of the order of ±1 percent.

RECOMMENDED VALUES

[Temperature, T, K; Viscosity, μ, 10^{-6} N s m^{-2}]

GAS

T	μ	T	μ
250	12.08	450	21.99
260	12.60	460	22.45
270	13.12	470	22.91
280	13.64	480	23.37
290	14.16	490	23.82
300	14.67	500	24.27
310	15.18	510	24.7
320	15.69	520	25.2
330	16.19	530	25.6
340	16.69	540	26.0
350	17.19	550	26.5
360	17.68	560	26.9
370	18.17	570	27.4
380	18.66	580	27.8
390	19.15	590	28.2
400	19.63	600	28.8
410	20.11	610	29.0
420	20.58	620	29.5
430	21.05	630	29.9
440	21.52	640	30.3
		650	30.6

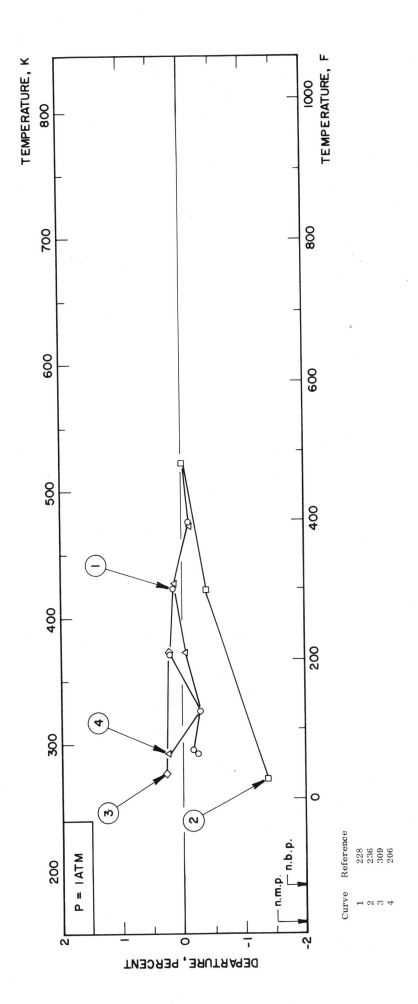

FIGURE 16-G(T). DEPARTURE PLOT FOR VISCOSITY OF GASEOUS HYDROGEN CHLORIDE

DISCUSSION

TABLE 17-G(T).   VISCOSITY OF GASEOUS HYDROGEN IODIDE

GAS

Three sets of experimental data were found in the literature. Two are from Trautz's group [225, 296] and the other from Harle [309]. One point of the latter author is not in good agreement with the two other sets. The temperature range goes from 293 K to 525 K. Use was made of the theoretical relation $\mu = K\sqrt{T}/(\sigma^2\Omega)$ to obtain the values of $\sigma^2\Omega$ from the experimental data. These were plotted as a function of 1/T and adjusted to a quadratic equation. From this equation, recommended values of the viscosity were computed. The accuracy is of the order of ±1 percent.

RECOMMENDED VALUES

[Temperature, T, K; Viscosity, μ, 10^{-6} N s m^{-2}]

GAS

T	μ	T	μ
250	15.88	450	28.09
260	16.50	460	28.68
270	17.12	470	29.27
280	17.75	480	29.85
290	18.37	490	30.43
300	18.99	500	31.01
310	19.61	510	31.6
320	20.23	520	32.2
330	20.84	530	32.7
340	21.46	540	33.3
350	22.07	550	33.9
360	22.69	560	34.4
370	23.29	570	35.0
380	23.90	580	35.6
390	24.51	590	36.1
400	25.11	600	36.7
410	25.71	610	37.2
420	26.31	620	37.8
430	26.91	630	38.3
440	27.50	640	38.8
		650	39.3

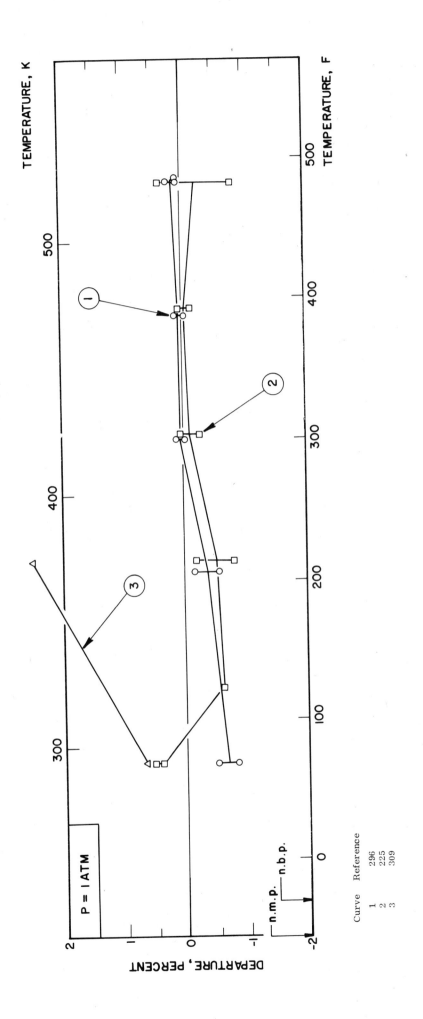

FIGURE 17-G(T).   DEPARTURE PLOT FOR VISCOSITY OF GASEOUS HYDROGEN IODIDE

TABLE 18-G(T).   VISCOSITY OF GASEOUS HYDROGEN SULFIDE

RECOMMENDED VALUES

[Temperature T, K; Viscosity, μ, 10^{-6} N s m^{-2}]

GAS

T	μ	T	μ
		400	16.89
		410	17.30
		420	17.70
270	11.32	430	18.18
280	11.76	440	18.50
290	12.21		
300	12.65	450	18.90
310	13.09	460	19.29
320	13.52	470	19.68
330	13.95	480	20.07
340	14.38	490	20.46
350	14.81	500	20.85
360	15.23		
370	15.65		
380	16.06		
390	16.48		

DISCUSSION

GAS

Four sets of data were found in the literature [80, 174, 236, 310]. Their agreement is not outstanding.

To correlate the data, the theoretical relation $\mu = K\sqrt{T}/(\sigma^2 \Omega_{22})$ was used. From the data, $\sigma^2 \Omega$ was computed and plotted as a function of 1/T. A linear equation was found to fit the data, and from this equation and the above relation, recommended values were generated. The accuracy is about ± 2.5 percent.

81

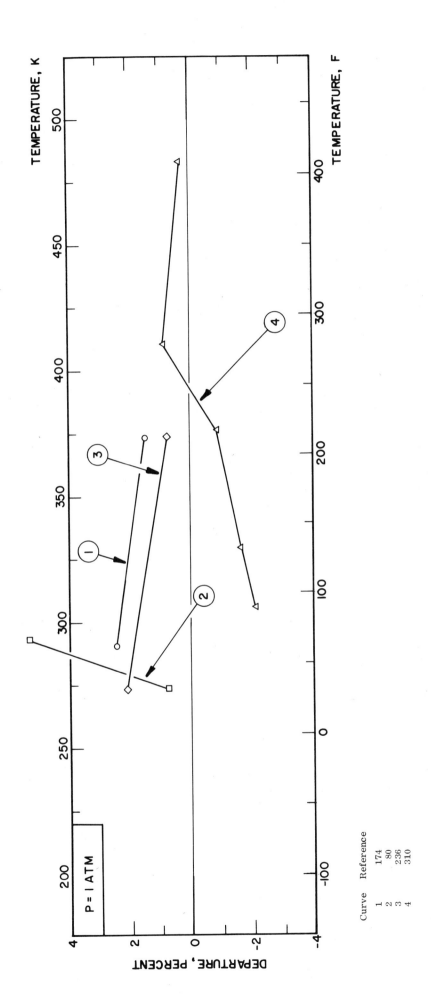

FIGURE 18-G(T).   DEPARTURE PLOT FOR VISCOSITY OF GASEOUS HYDROGEN SULFIDE

Curve	Reference
1	174
2	80
3	236
4	310

TABLE 19-G(T).   VISCOSITY OF GASEOUS NITRIC OXIDE

DISCUSSION

GAS

Ten sets of experimental data on the viscosity of nitric oxide were found in the literature [59, 80, 98, 121, 126, 236, 257, 264, 315, 316]. They fall in a small temperature range, except those of Ellis and Raw [59] which go up to 1356 K.

To analyze the data, use was made of the theoretical expression for viscosity:

$$\mu = 266.93 \frac{\sqrt{MT}}{\sigma^2 \Omega(T^*)} \frac{f_\mu}{\mu}$$

The group $[\sigma^2 \Omega(T^*)/f_\mu] \mu = y_{obs}$ was computed from the experimental data and plotted as a function of 1/T. The curve obtained was smoothed and a table generated. Recommended values were calculated by introducing, into the above formula, the value of y_{calc} interpolated from the table. The recommended values are thought to be accurate to two percent below 1250 K and five percent for all higher temperatures tabulated.

RECOMMENDED VALUES

[Temperature, T, K; Viscosity, μ, N.m^{-1}.sec^{-1}.10^{-6}]

GAS

T	μ	T	μ	T	μ
110	7.81	450	26.33	800	39.0
120	8.49	460	26.76	810	39.3
130	9.16	470	27.18	820	39.7
140	9.83	480	27.59	830	40.0
150	10.49	490	28.00	840	40.3
160	11.14	500	28.41	850	40.6
170	11.78	510	28.8	860	40.9
180	12.42	520	29.2	870	41.2
190	13.04	530	29.6	880	41.5
200	13.65	540	30.0	890	41.8
210	14.26	550	30.4	900	42.1
220	14.85	560	30.8	910	42.4
230	15.44	570	31.1	920	42.7
240	16.02	580	31.5	930	42.9
250	16.58	590	31.9	940	43.2
260	17.14	600	32.3	950	43.5
270	17.69	610	32.6	960	43.8
280	18.23	620	33.0	970	44.1
290	18.76	630	33.3	980	44.4
300	19.29	640	33.7	990	44.6
310	19.80	650	34.1	1000	44.9
320	20.31	660	34.4	1050	46.3
330	20.81	670	34.7	1100	47.6
340	21.31	680	35.1	1150	48.9
350	21.80	690	35.4	1200	50.2
360	22.28	700	35.8	1250	51.4
370	22.75	710	36.1	1300	52.7
380	23.22	720	36.4	1350	53.9
390	23.68	730	36.8	1400	55.0
400	24.14	740	37.1	1450	56.2
410	24.59	750	37.4	1500	57.3
420	25.03	760	37.8		
430	25.47	770	38.1		
440	25.90	780	38.4		
		790	38.7		

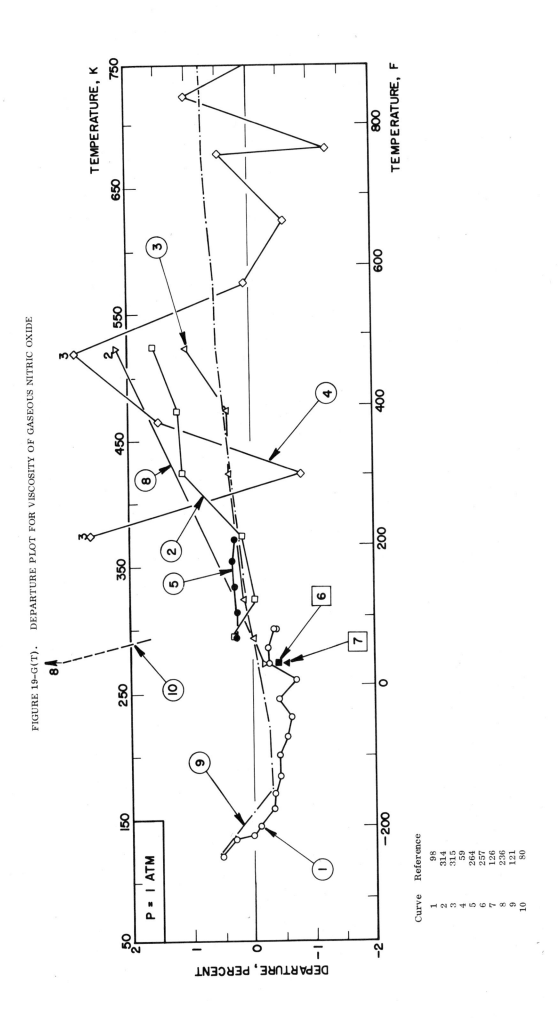

FIGURE 19-G(T).   DEPARTURE PLOT FOR VISCOSITY OF GASEOUS NITRIC OXIDE

84

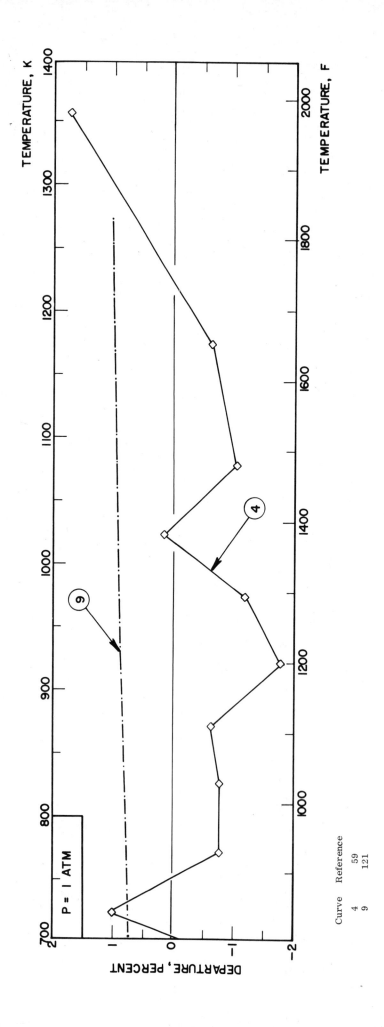

FIGURE 19-G(T). DEPARTURE PLOT FOR VISCOSITY OF GASEOUS NITRIC OXIDE (continued)

TABLE 20-G(T).   VISCOSITY OF GASEOUS NITROGEN PEROXIDE

DISCUSSION

GAS

Three sets of experimental data were found in the literature; the results of Petker et al. [372], Beer [312] and Timrot et al. [318].   They relate to the equilibrium mixture $N_2O_4 \rightleftharpoons 2\ NO_2$.   Although the procedure was not theoretically well founded, use was made of the relation $\mu = K\sqrt{T}/(\sigma^2\,\Omega_{22})$ which allowed an easy interpolation on a graph of $\sigma^2\,\Omega$ versus $1/T$.   The data of Beer appeared of poor consistency, while the data of Petker and of Timrot were in agreement, particularly at high temperature.

The recommended values are thought to be accurate to ±3 percent at low temperature and better at high temperature.

RECOMMENDED VALUES

[Temperature, T, K; Viscosity, μ, 10^{-6} N s m^{-2}]

GAS

T	μ
300	13.01
310	13.87
320	14.76
330	15.70
340	16.67
350	17.69
360	18.75
370	19.65
380	20.20
390	20.76
400	21.31
410	21.86
420	22.40
430	22.95
440	23.49
450	24.02

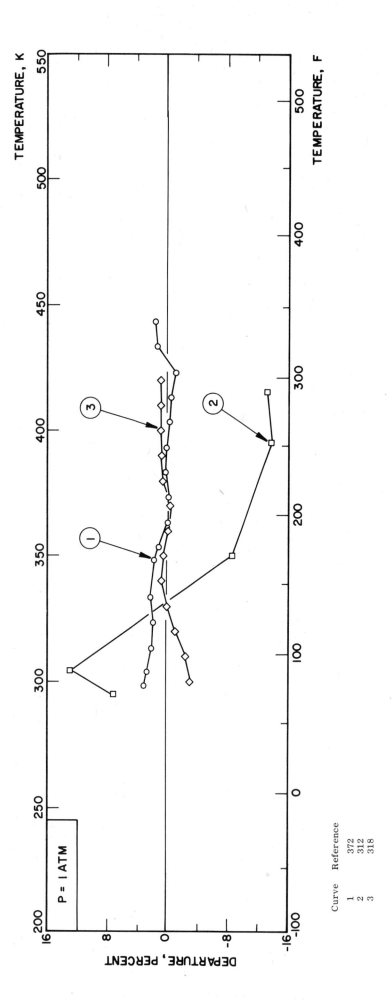

FIGURE 20–G(T).   DEPARTURE PLOT FOR VISCOSITY OF GASEOUS NITROGEN PEROXIDE

TABLE 21-G(T). VISCOSITY OF GASEOUS NITROUS OXIDE

RECOMMENDED VALUES

[Temperature, T, K; Viscosity, μ, N.m^{-1}.sec^{-1}.10^{-6}]

GAS

T	μ	T	μ	T	μ
180	9.01	500	23.56	850	35.6
190	9.52	510	23.9	860	35.9
200	10.03	520	24.3	870	36.2
210	10.54	530	24.7	880	36.5
220	11.05	540	25.1	890	36.8
230	11.56	550	25.5	900	37.1
240	12.06	560	25.9	910	37.4
250	12.55	570	26.2	920	37.7
260	13.04	580	26.6	930	38.0
270	13.53	590	27.0	940	38.3
280	14.02	600	27.3	950	38.6
290	14.49	610	27.7	960	38.9
300	14.97	620	28.0	970	39.1
310	15.44	630	28.4	980	39.4
320	15.90	640	28.7	990	39.7
330	16.36	650	29.1	1000	40.0
340	16.82	660	29.4	1050	41.4
350	17.27	670	29.8	1100	42.7
360	17.72	680	30.1	1150	44.0
370	18.16	690	30.5	1200	45.3
380	18.60	700	30.8	1250	46.6
390	19.03	710	31.1	1300	47.8
400	19.46	720	31.5	1350	49.0
410	19.89	730	81.8	1400	50.2
420	20.31	740	32.1	1450	51.4
430	20.73	750	32.5	1500	52.5
440	21.14	760	32.8		
450	21.56	770	33.1		
460	21.96	780	33.4		
470	22.37	790	33.7		
480	22.77	800	34.1		
490	23.16	810	34.4		
		820	34.7		
		830	35.0		
		840	35.3		

DISCUSSION

GAS

Eight sets of experimental data for the viscosity of nitrous oxide were found in the literature [35, 63, 98, 176, 201, 225, 236, 257], the range of temperature covered is from 185 up to 1296 K, although above 800 K, the Raw and Ellis [176] data must pertain to a mixture of nitrous oxide and its dissociation products.

To correlate the data, use was made of the following theoretical expression:

$$\mu = 266.93 \frac{\sqrt{MT}}{\sigma^2 \, \Omega(T^*)} f_\mu$$

The group $[\sigma^2 \Omega(T^*)/f_\mu]$ was computed from the experimental data, plotted as a function of $1/T$, and a table generated from the smooth curve obtained. Recommended values were computed from the theoretical formula, with the values of $[\sigma^2 \Omega(T^*)/f_\mu]$ interpolated from the table.

Good agreement is found with Keyes [121] correlation, and a fair agreement with Andrussow [3] semi-theoretical approach, although our curve and theirs were both extrapolated in the dissociation temperature range. Below 850 K, the recommended values are thought to be accurate to within two percent. The accuracy at higher temperatures is more difficult to assess, as it depends on the degree of dissociation.

88

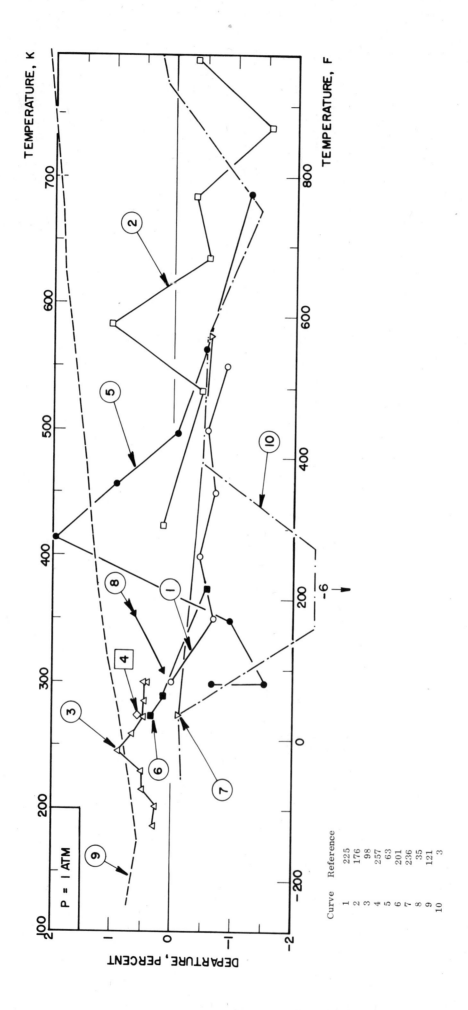

FIGURE 21-G(T). DEPARTURE PLOT FOR VISCOSITY OF GASEOUS NITROUS OXIDE

89

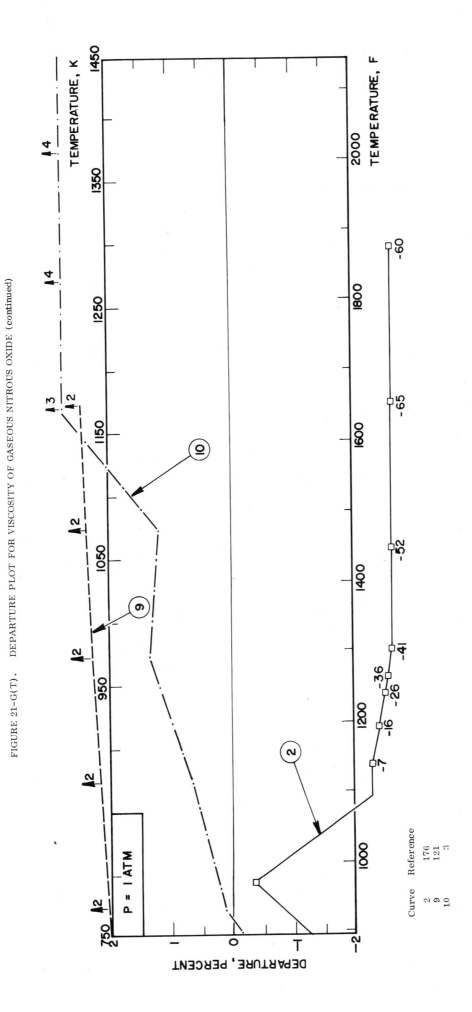

FIGURE 21-G(T). DEPARTURE PLOT FOR VISCOSITY OF GASEOUS NITROUS OXIDE (continued)

90

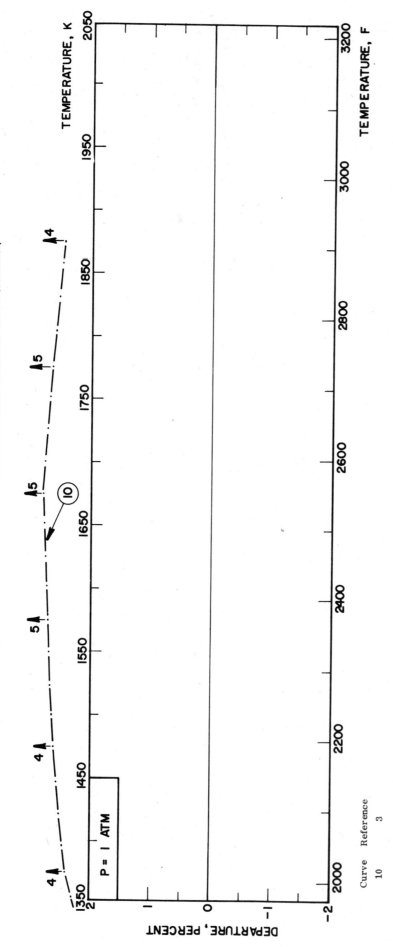

FIGURE 21-G(T). DEPARTURE PLOT FOR VISCOSITY OF GASEOUS NITROUS OXIDE (continued)

TABLE 22-G(T). VISCOSITY OF GASEOUS SULFUR DIOXIDE

DISCUSSION

GAS

Twelve sets of experimental data were found in the literature [35, 202, 206, 231, 233, 236, 257, 264, 296, 310, 321, 322]. They show large discrepancies, evan at ambient temperature, and at low temperature.

The correlation was made by using the theoretical relation
$\mu = K \sqrt{T}/[\sigma^2 \Omega_{22}(T^*)]$. From the data $\sigma^2 \Omega$ was computed and plotted as a function of $1/T$. A smooth curve was drawn through the experimental points, and fitted to a quadratic equation, from which recommended values were generated. Values computed by Andrussow [3] with the aid of a semi-theoretical relation are in fair agreement, except at low temperature. The accuracy must be of the order of ±3 percent.

RECOMMENDED VALUES

[Temperature, T, K; Viscosity, μ, 10^{-6} N s m^{-2}]

GAS

T	μ	T	μ	T	μ
200	8.62	500	21.3	800	32.1
210	9.06	510	21.7	810	32.4
220	9.51	520	22.1	820	32.7
230	9.96	530	22.5	830	33.1
240	10.40	540	22.8	840	33.4
250	10.84	550	23.2	850	33.7
260	11.28	560	23.6	860	34.0
270	11.72	570	24.0	870	34.4
280	12.16	580	24.4	880	34.7
290	12.60	590	24.7	890	35.0
300	13.04	600	25.1	900	35.3
310	13.47	610	25.5	910	35.6
320	13.90	620	25.8	920	35.9
330	14.33	630	26.2	930	36.3
340	14.76	640	26.6	940	36.6
350	15.18	650	26.6	950	36.9
360	15.61	660	27.3	960	37.2
370	16.03	670	27.6	970	37.5
380	16.45	680	28.0	980	37.8
390	16.86	690	28.3	990	38.1
400	17.28	700	28.7	1000	38.4
410	17.69	710	29.1	1050	39.9
420	18.10	720	29.4	1100	41.3
430	18.51	730	29.7	1150	42.8
440	18.91	740	30.1	1200	44.2
450	19.32	750	30.4	1250	45.4
460	19.72	760	30.7		
470	20.12	770	31.1		
480	20.51	780	31.4		
490	20.91	790	31.8		

92

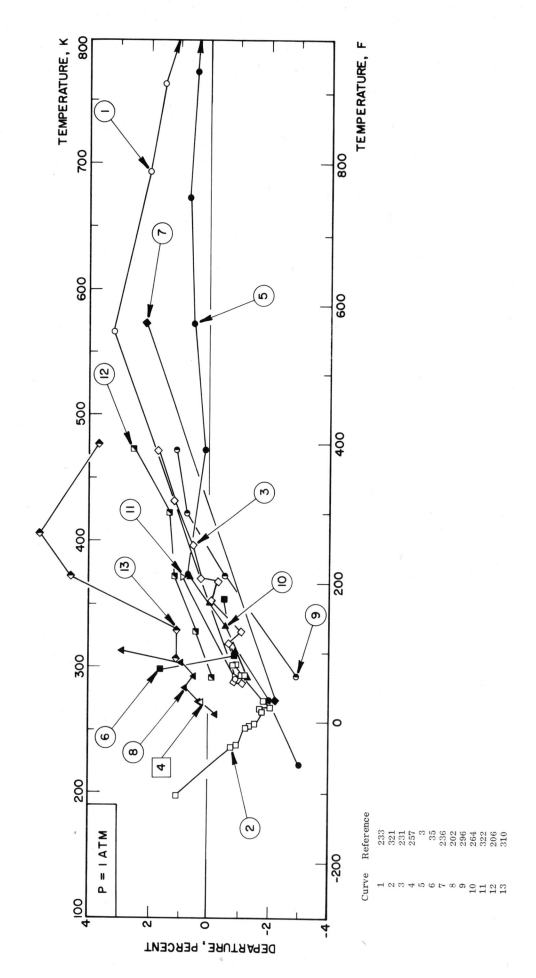

FIGURE 22–G(T).   DEPARTURE PLOT FOR VISCOSITY OF GASEOUS SULFUR DIOXIDE

FIGURE 22-G(T). DEPARTURE PLOT FOR VISCOSITY OF GASEOUS SULFUR DIOXIDE (continued)

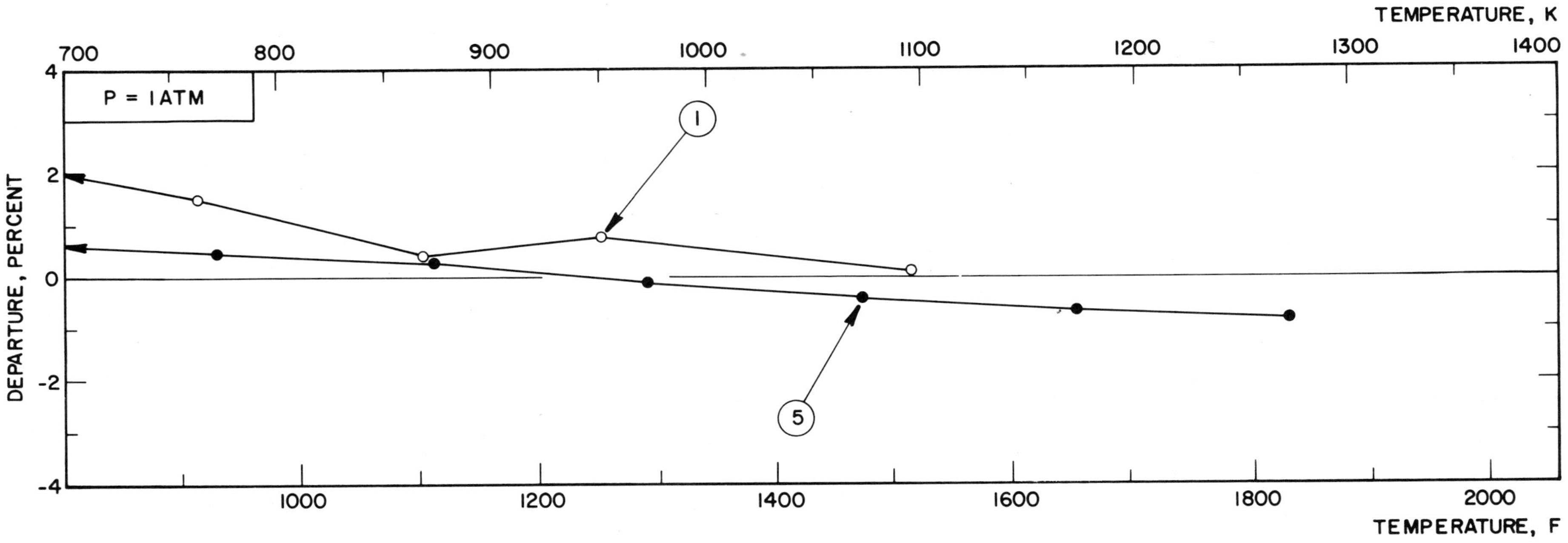

Curve	Reference
1	233
5	3

TABLE 23-L(T).   VISCOSITY OF LIQUID WATER

DISCUSSION

SATURATED LIQUID

A recommended equation was released following the work of the Sixth International Conference on Steam (1964 Skeleton Tables):

$$\mu(\text{Nsm}^{-2}) = 2.414 \times 10^{-2} \times 10^{\,247.8/T - 140}$$

An excellent review of the subject was done by Kestin [118]. The viscosity of water at atmospheric pressure and at a temperature of 20 C was measured accurately by Swindells [284] and by Roscoe [285] so that accordingly a reasonable value is

$$\mu_{293.15} = (1.002 \pm 0.001)\ 10^{-3}\ \text{N sec m}^{-2}$$

The first equation above covers the range 273.15 K to 573.15 K and was adopted to generate the present recommended values.

At the 7th International Conference on Steam several papers were presented, based on a unique equation to represent the whole p, T, μ domain [156, 185, 210]. The correlated values fall within the tolerances of the 1964 Skeleton Tables, and so do a correlation by Bruges [27] which is extended to the critical point.

Tanishita's [210] values were used to generate recommended values between 573.15 K and the critical point.

The accuracy stated in the 1964 Skeleton Table is ±2.5%.

RECOMMENDED VALUES

[Temperature, T, K; Viscosity, μ, 10^{-3} N s m^{-2}]

SATURATED LIQUID

T	μ	T	μ
273.15	1.753	450	0.1514
280	1.422	460	0.1430
290	1.083	470	0.1355
		480	0.1288
300	0.823	490	0.1228
310	0.672	500	0.1174
320	0.560	510	0.1125
330	0.476	520	0.1081
340	0.411	530	0.1041
350	0.360	540	0.1003
360	0.319	550	0.0969
370	0.285	560	0.0937
380	0.258	570	0.0902
390	0.234	580	0.0865
400	0.2149	590	0.0827
410	0.1983	600	0.0788
420	0.1840	610	0.0750
430	0.1716	620	0.0711
440	0.1608	630	0.0663
		640	0.0587
		647*	0.0416

* Crit. Temp.

TABLE 23-V(T).   VISCOSITY OF WATER VAPOR

DISCUSSION

SATURATED VAPOR

The 6th International Conference on Steam agreed on an equation representing the excess viscosity from 1 bar pressure to saturation pressure in the range 373.15 K to 573.15 K. The subject has been discussed at length by Kestin [118] in his presentation of the 1964 International Skeleton Table.

The equation is

$$(\mu - \mu_1) = (5.90 \, t - 1858) \qquad \text{(t in deg} \quad) \qquad (1)$$

where

$$\mu_1 = (80.4 + 0.407 \, t) \ 10^{-7} \ \text{N sec m}^{-2} \qquad (2)$$

Equation (1) is largely based on determination at Brown University [113, 116] as primary references. Equation (2) is the same as that used to generate the 1 atm. gas recommended values. The tolerance stated is ±1%.

At the 7th International Conference on Steam, two papers were presented [156, 210] based on a unique equation for representing the whole p, T, μ domain, and a paper by Bruges [27] which takes also into account new results by Ray [286]. Their correlated values fall close to the tolerance of the International Skeleton Table (1964). The recommended values of this work was interpolated from the values of the latter, in the range 373.15 K to 575.15 K, but, above this temperature Tanishita's values were used to generate the recommended values.

RECOMMENDED VALUES

[Temperature, T, K; Viscosity, μ, 10^{-6} N s m^{-2}]

SATURATED VAPOR

T	μ
373.15	12.03
380	12.29
390	12.68
400	13.05
410	13.43
420	13.79
430	14.15
440	14.50
450	14.86
460	15.20
470	15.54
480	15.89
490	16.23
500	16.59
510	16.95
520	17.34
530	17.73
540	18.14
550	18.61
560	19.10
570	19.63
580	20.32
590	21.23
600	22.23
610	23.52
620	25.23
630	27.60
640	31.17
647*	41.6

* Crit. Temp.

DISCUSSION

GAS

The Sixth International Conference on the Properties of Steam charged a panel with the task of producing new tables on transport properties. The result was the recommendation of the equation:

$$\mu = (80.4 + 0.407\ t)\ 10^{-7}\ \text{N sec m}^{-2} \qquad (t\ \text{in C}) \qquad (1)$$

which served for the representation of the viscosity of superheated steam in the range 100-700 C, in the International Skeleton table (1964).

This equation is based on Shifrin's [197] results as a primary reference. An excellent discussion on the subject, can be found in a paper by Kestin[118]. The tolerances are ±1% in the range 373-573 K and ±3% in the range 573-973 K.

Several papers presented at the 7th International Conference (Tokyo, 1968) were dealing with the subject. Three of these are based on a unique equation for the representation in the whole p, T, μ domain, instead of 4 equations representing four separate domains (Tanishita [210], Miyabe [156] Rivkin [185]). Another paper, by Bruges [27] which is an extension of a previous work [26] uses several equations characteristic of different domains, and includes the experimental results of Latto [135].

Based on the same primary sources of references the values obtained in the different correlation fall well within the tolerances given by the International Skeleton table (1964). Therefore the recommended values were generated from the above equation (1).

In view of the wide acceptance of our basic equation and the numerous detailed discussion in the technical literature coupled with pressing requirements of time, no departure plot appears.

TABLE 23-G(T).   VISCOSITY OF GASEOUS WATER

GAS

RECOMMENDED VALUES

[Temperature, T, K; Viscosity, μ, 10^{-6} N s m^{-2}]

T	μ	T	μ
		650	23.38
		660	23.78
		670	24.19
280	8.32	680	24.60
290	8.73	690	25.01
300	9.13	700	25.41
310	9.54	710	25.82
320	9.95	720	26.23
330	10.35	730	26.63
340	10.76	740	27.04
350	11.17	750	27.45
360	11.57	760	27.85
370	11.98	770	28.25
380	12.39	780	28.67
390	12.80	790	29.08
400	13.20	800	29.48
410	13.61	810	29.89
420	14.02	820	30.30
430	14.42	830	30.70
440	14.83	840	31.11
450	15.24	850	31.52
460	15.64	860	31.92
470	16.05	870	32.33
480	16.46	880	32.74
490	16.87	890	33.15
500	17.27	900	33.55
510	17.68	910	33.96
520	18.09	920	34.37
530	18.49	930	34.77
540	18.90	940	35.18
550	19.31	950	35.59
560	19.71	960	35.99
570	20.12	970	36.40
580	20.53	980	36.81
590	20.94	990	37.22
600	21.34	1000	37.62
610	21.75		
620	22.16		
630	22.56		
640	22.97		

3. ORGANIC COMPOUNDS

TABLE 24-G(T). VISCOSITY OF GASEOUS ACETONE

GAS

DISCUSSION

Five sets of experimental data were found in the literature [41, 215, 236, 175, 257]. They are in good agreement, except for the results of Uchiyama [236].

Use was made of the theoretical relation $\mu = K\sqrt{T}/(\sigma^2\Omega)$ to obtain the values of $\sigma^2\Omega$ from the experimental data. These were plotted as a function of $1/T$ and adjusted to a quadratic equation. From this equation. recommended values of the viscosity were computed. The accuracy is of the order of ± 1.5 percent.

RECOMMENDED VALUES

[Temperature, T, K; Viscosity, μ, N s m$^{-2} \cdot 10^6$]

GAS

T	η
250	6.78
260	6.96
270	7.15
280	7.35
290	7.56
300	7.77
310	7.99
320	8.21
330	8.44
340	8.67
350	8.90
360	9.14
370	9.38
380	9.63
390	9.88
400	10.13
410	10.38
420	10.64
430	10.89
440	11.15
450	11.42
460	11.68
470	11.95
480	12.21
490	12.48
500	12.75
510	13.0
520	13.3
530	13.6
540	13.9
550	14.1
560	14.4
570	14.7
580	15.0
590	15.3
600	15.5
610	15.8
620	16.1
630	16.4
640	16.7
650	16.9

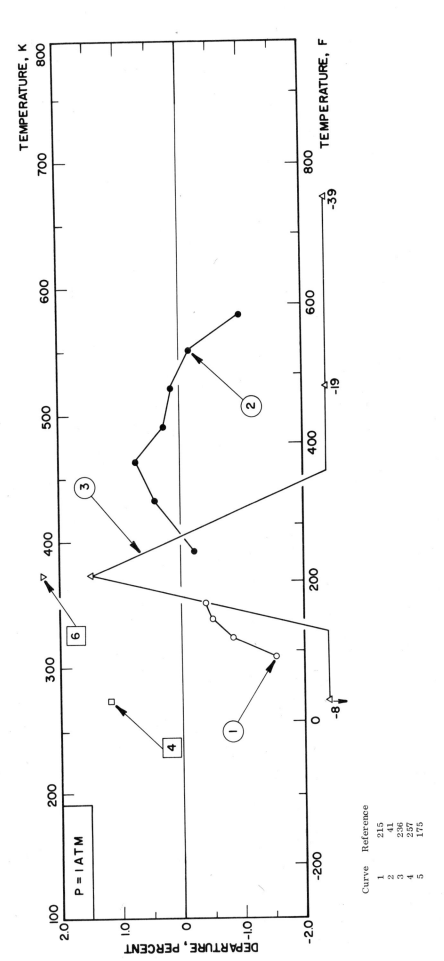

FIGURE 24-G(T).   DEPARTURE PLOT FOR VISCOSITY OF GASEOUS ACETONE

TABLE 25-G(T). VISCOSITY OF GASEOUS ACETYLENE

DISCUSSION

Experimental data for the viscosity of acetylene found in the literature are those of Kiyama et al. [124], Uchiyama [236], Wobser et al. [264], Vogel [257], Adzumi [1], and Titani [373]. The temperature range covered by the investigators is very narrow: from 273 to 523 K.

The analysis was performed using the theoretical relation:

$$\mu = 266.93 \, \frac{\sqrt{MT}}{\sigma^2 \, \Omega(T^*)} \, f_\mu$$

from which the group $\dfrac{\sigma^2 \Omega(T^*)}{f_\mu} = y_{obs}$ was computed. The values obtained were plotted as a function of 1/T, and the curve compared with a similar curve for methane, using appropriate reduction factors. From the smooth curve drawn, a table was generated and recommended values computed.

The fit of the data is within 2 percent except for one value of Uchiyama [236].

RECOMMENDED VALUES

[Temperature, T, K; Viscosity, μ, N.m^{-1}.sec^{-1}.10^{-6}]

GAS

T	μ
270	9.35
280	9.67
290	10.00
300	10.33
310	10.66
320	10.98
330	11.30
340	11.61
350	11.93
360	12.25
370	12.56
380	12.87
390	13.18
400	13.48
410	13.78
420	14.09
430	14.39
440	14.68
450	14.98
460	15.27
470	15.55
480	15.84
490	16.12
500	16.41
510	16.69
520	16.97
530	17.24
540	17.52
550	17.79
560	18.06
570	18.33
580	18.59
590	18.86
600	19.12

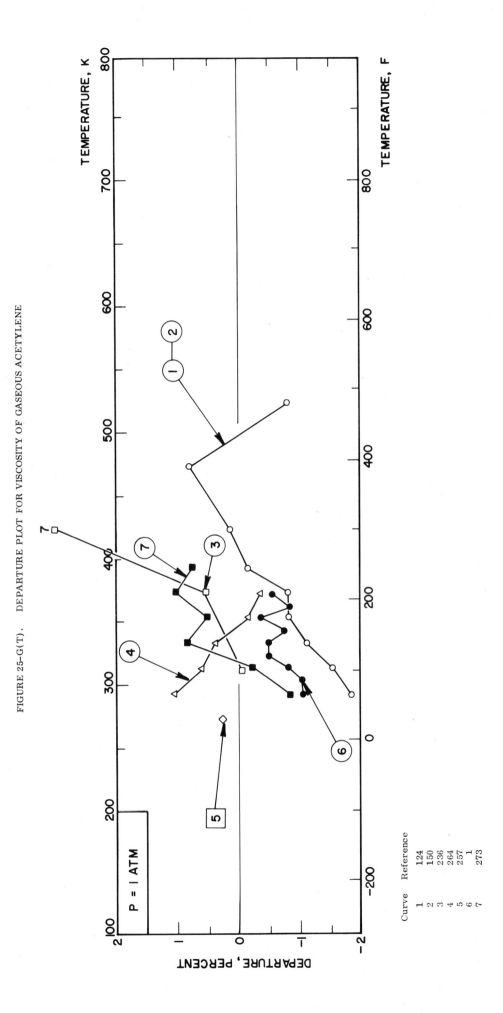

FIGURE 25-G(T). DEPARTURE PLOT FOR VISCOSITY OF GASEOUS ACETYLENE

TABLE 26-G(T).   VISCOSITY OF GASEOUS BENZENE

DISCUSSION

GAS

Seven sets of experimental data were found in the literature. Most of them are reliable [41, 215, 154, 236, 257, 287, 175]. High temperature data of Uchiyama [236] were not taken into account.

The correlation was made by using the theoretical relation
$\mu = K\sqrt{T}/(\sigma^2\Omega)$. From the data, the group $\sigma^2\Omega$ was computed and plotted as a function of $1/T$. A second degree polynomial was fitted to the data, and recommended values were generated using the polynomial. The accuracy is thought to be about ±1 percent over the whole range.

RECOMMENDED VALUES

[Temperature, T, K; Viscosity, 10^{-6} N sec m^{-2}]

GAS

T	μ	T	μ
270	6.90	450	11.43
280	7.15	460	11.68
290	7.40	470	11.93
300	7.65	480	12.18
310	7.90	490	12.42
320	8.16	500	12.67
330	8.41	510	12.9
340	8.66	520	13.2
350	8.92	530	13.4
360	9.17	540	13.7
370	9.42	550	13.9
380	9.67	560	14.1
390	9.93	570	14.4
400	10.18	580	14.6
410	10.43	590	14.9
420	10.68	600	15.1
430	10.93	610	15.3
440	11.18	620	15.6
		630	15.8
		640	16.0
		650	16.3

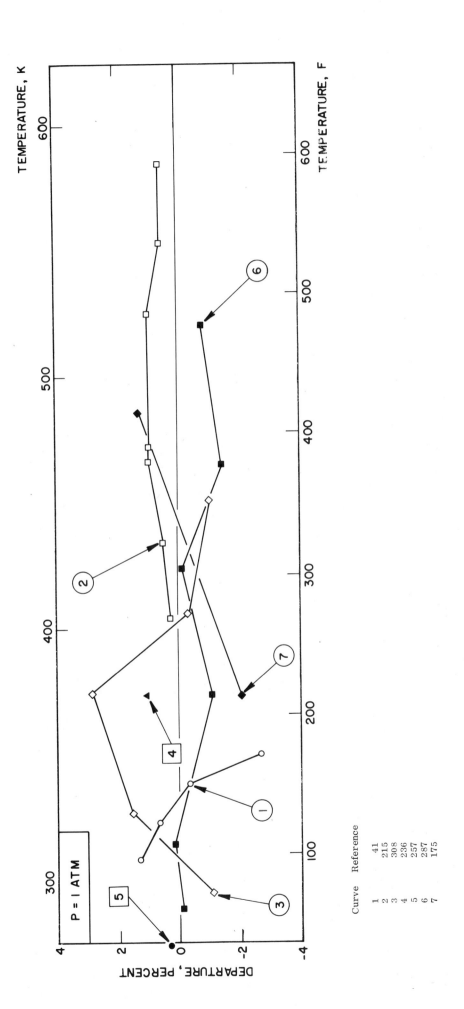

FIGURE 26-G(T). DEPARTURE PLOT FOR VISCOSITY OF GASEOUS BENZENE

TABLE 27-L(T). VISCOSITY OF LIQUID BROMOTRIFLUOROMETHANE

RECOMMENDED VALUES

[Temperature, T, K, Viscosity, μ, 10^{-3} N s m^{-2}]

SATURATED LIQUID

T	μ
170	0.936
180	0.746
190	0.609
200	0.507
210	0.430
220	0.370
230	0.322
240	0.284
250	0.253
260	0.228
270	0.206
280	0.188
290	0.173
300	0.1594
310	0.1445
320	0.1260
330	0.0985
340	0.0460
340*	0.0346

DISCUSSION

SATURATED LIQUID

Two sets of experimental data were found in the literature, those of Gordon [79] and of Lilios [140] and a single value at 30°C in a commercial note [275]. The latter is not reliable, but the two sets show good consistency. They were least square fitted to an equation:

$$\log \mu = A + B/T$$

from which the recommended values were generated.

The accuracy is thought to be ±3%.

* Crit. Temp.

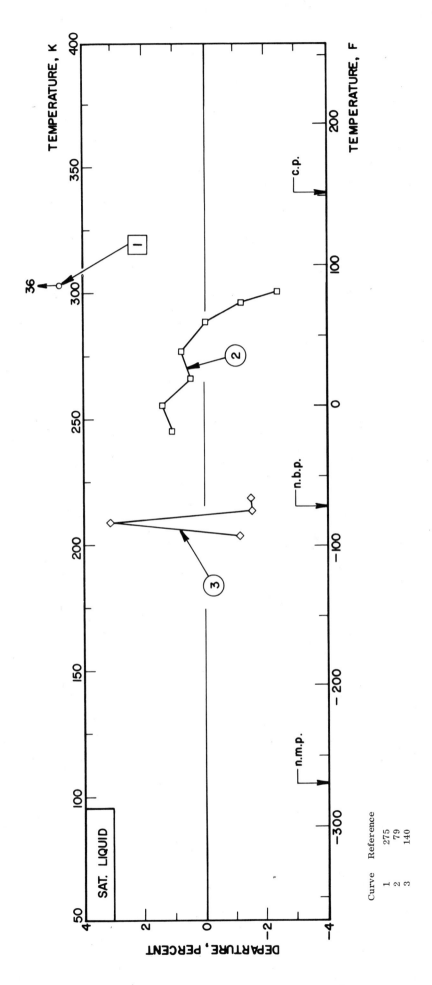

FIGURE 27-L(T).   DEPARTURE PLOT FOR VISCOSITY OF LIQUID BROMOTRIFLUOROMETHANE

TABLE 27-V(T). VISCOSITY OF BROMOTRIFLUOROMETHANE VAPOR

DISCUSSION

SATURATED VAPOR

Recommended values for the viscosity of the saturated vapor were estimated with the method outlined by Stiel and Thodos[207] using the excess viscosity concept. The density values were taken from a manufacturer's technical note.

The accuracy is thought to be ±5%.

RECOMMENDED VALUES

[Temperature, T, K, Viscosity, μ, 10^{-3} N s m^{-2}]

SATURATED VAPOR

T	μ
210	0.01119
220	0.01171
230	0.01233
240	0.01287
250	0.01345
260	0.01410
270	0.01480
280	0.01555
290	0.01635
300	0.0172
310	0.0183
320	0.0197
330	0.0220
340	0.0300
340*	0.0346

* Crit. Temp.

TABLE 27-G(T). VISCOSITY OF GASEOUS BROMOTRIFLUOROMETHANE

RECOMMENDED VALUES

[Temperature, T, K; Viscosity, μ, 10^{-6} N s m^{-2}]

GAS

T	μ
230	12.21
240	12.72
250	13.22
260	13.73
270	14.22
280	14.71
290	15.20
300	15.78
310	16.15
320	16.62
330	17.09
340	17.55
350	18.01
360	18.46
370	18.91
380	19.35
390	19.79
400	20.22
410	20.65
420	21.08
430	21.50
440	21.92
450	22.33
460	22.74
470	23.14
480	23.55
490	23.95
500	24.34

DISCUSSION

GAS

Three sets of experimental data were found in the literature [177, 235, 262]. They cover the temperature range from 230 K to 423 K, with no overlap from one set to another. They were least square fitted to a quadratic equation:

$$\sigma^2 \Omega = \frac{K \sqrt{T}}{\mu} = f(T)$$

from which recommended values were generated.

The accuracy is of the order of $\pm 2\%$.

108

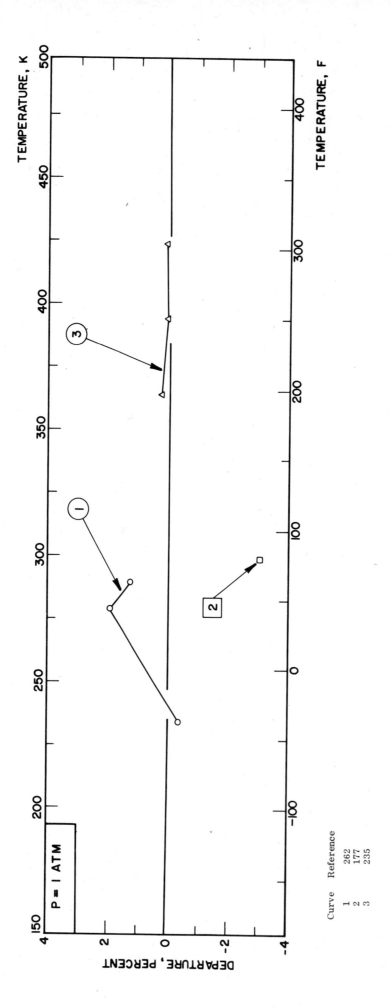

FIGURE 27-G(T).   DEPARTURE PLOT FOR VISCOSITY OF GASEOUS BROMOTRIFLUOROMETHANE

TABLE 28-L(T).   VISCOSITY OF LIQUID i-BUTANE

DISCUSSION

SATURATED LIQUID

There are two sets of experimental data for the viscosity of liquid iso-butane. The data of Lipkin [142] covers the range of temperature from 210 K to 311 K, while there are two points of Gonzalez [78] at 311 and 344 K. They are quite diverging. The data of Lipkin were fitted to an equation

$$\log \mu = A + B/T$$

from 210 to 320 K. Above this temperature to the critical temperature there is a lack of data. The recommended value at 320 K was joined graphically to the value of the viscosity at the critical temperature computed by the method of Jossi, Stiel and Thodos [100].

The accuracy is considered to be about ±3% below 320 K and ±10% or even a larger uncertainty at higher temperatures.

RECOMMENDED VALUES

[Temperature, T, K; Viscosity, μ, 10^{-3} N s m^{-2}]

SATURATED LIQUID

T	μ
190	0.748
200	0.606
210	0.500
220	0.421
230	0.359
240	0.311
250	0.272
260	0.240
270	0.214
280	0.193
290	0.175
300	0.1593
310	0.1462
320	0.1349
330	0.1250
340	0.1155
350	0.1055
360	0.0950
370	0.0850
380	0.0745
390	0.0640
400	0.0510
408*	0.0233

* Crit. Temp.

110

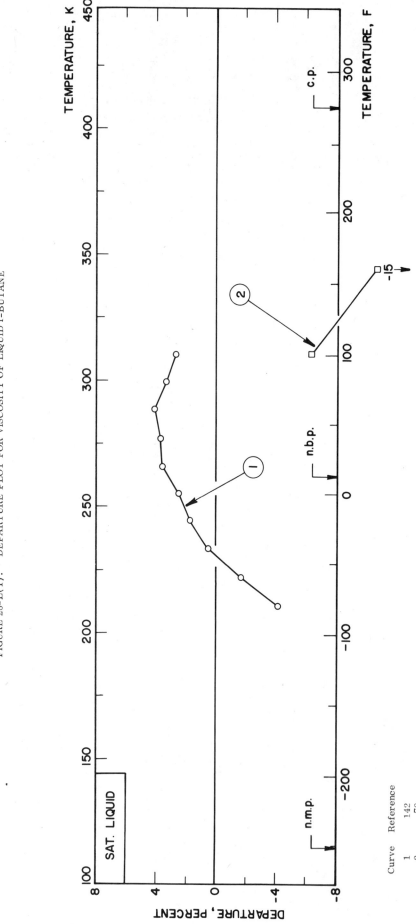

FIGURE 28-L(T).   DEPARTURE PLOT FOR VISCOSITY OF LIQUID i-BUTANE

DISCUSSION

TABLE 28-V(T). VISCOSITY OF i-BUTANE VAPOR

RECOMMENDED VALUES

[Temperature, T, K; Viscosity, μ, 10^{-3} N s m^{-2}]

SATURATED VAPOR

SATURATED VAPOR

Recommended values of the viscosity of the saturated vapor were computed by the correlation technique of Jossi, Stiel and Thodos [100] using the recommended values of viscosity of the dilute gas and a generalized correlation of the excess viscosity versus reduced temperature from other gases using also the correlation of Jossi, Stiel and Thodos [100].

The accuracy is of about 2% except where approaching the critical temperature, where it may reach about ±5%.

T	μ
270	.00699
280	.00727
290	.00757
300	.00786
310	.00816
320	.00845
330	.00876
340	.00917
350	.00966
360	.01025
370	.01092
380	.01190
390	.01326
400	.0154
408*	.0233

* Crit. Temp.

TABLE 28-G(T).   VISCOSITY OF GASEOUS *i*-BUTANE

DISCUSSION

GAS

There are four sets [133, 191, 216, 236] and a single value [89] of experimental data for the viscosity of gaseous i-butane. The temperature range covered is very narrow, going only from 293 to 407 K. There are large discrepancies between the values reported.

The analysis was made by computing the group $\sigma^2 \Omega(T^*)/f \mu = y_{obs}$ from the theoretical relation:

$$\mu = 266.93 \frac{\sqrt{MT}}{\sigma^2 \Omega(T^*)} f \mu$$

and plotting the values obtained as a function of 1/T. From the curve obtained, a table was generated after graphical smoothing. Recommended values were interpolated from the table. The accuracy is about ±5%.

RECOMMENDED VALUES

[Temperature, T, K; Viscosity, μ, 10^{-6} N s m^{-2}]

GAS

T	μ
270	6.88
280	7.12
290	7.36
300	7.60
310	7.84
320	8.08
330	8.32
340	8.57
350	8.81
360	9.05
370	9.29
380	9.52
390	9.76
400	10.00
410	10.24
420	10.47
430	10.71
440	10.94
450	11.17
460	11.40
470	11.63
480	11.86
490	12.09
500	12.31
510	12.53
520	12.75

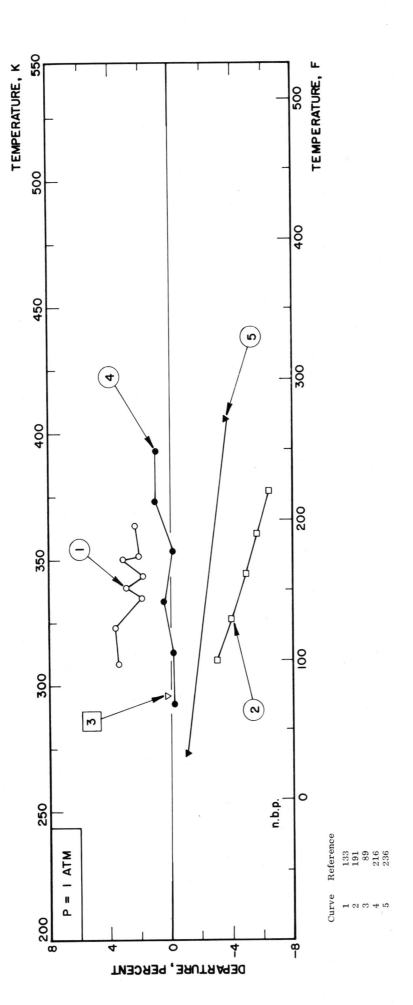

FIGURE 28-G(T). DEPARTURE PLOT FOR VISCOSITY OF GASEOUS i-BUTANE

TABLE 29-L(T). VISCOSITY OF LIQUID n-BUTANE

RECOMMENDED VALUES

[Temperature, T, K; Viscosity, μ, 10^{-3} N s m^{-2}]

SATURATED LIQUID

T	μ
180	0.688
190	0.571
200	0.482
210	0.414
220	0.361
230	0.318
240	0.283
250	0.255
260	0.231
270	0.211
280	0.194
290	0.177
300	0.1613
310	0.1463
320	0.1325
330	0.1193
340	0.1092
350	0.0984
360	0.0881
370	0.0784
380	0.0694
390	0.0612
400	0.0548
410	0.0470
420	0.0360
426*	0.0240

* Crit. Temp.

DISCUSSION

SATURATED LIQUID

The literature revealed four sets of experimental data, the results of Lipkin [142], Swift [209], Krueger [129], and of Carmichael and Sage [32]. The latter are at two temperatures, one of which is close to the critical temperature. One of the values of Swift is at 373 K. There is thus a lack of data between these two temperatures, the other data being between 183 and 373 K. The agreement between the sets is good.

The correlation was made by using an equation

$$\log \mu = A + B/T + \delta$$

which was least square fitted below 280 K with $\delta = 0$, the residual δ being smoothed from 280 K to 400 K. Above this temperature a hand drawn curve was used to join the recommended value to the viscosity at the critical point computed by means of the technique of Jossi, Stiel and Thodos [200].

The accuracy of the recommended value is estimated as ±3% throughout.

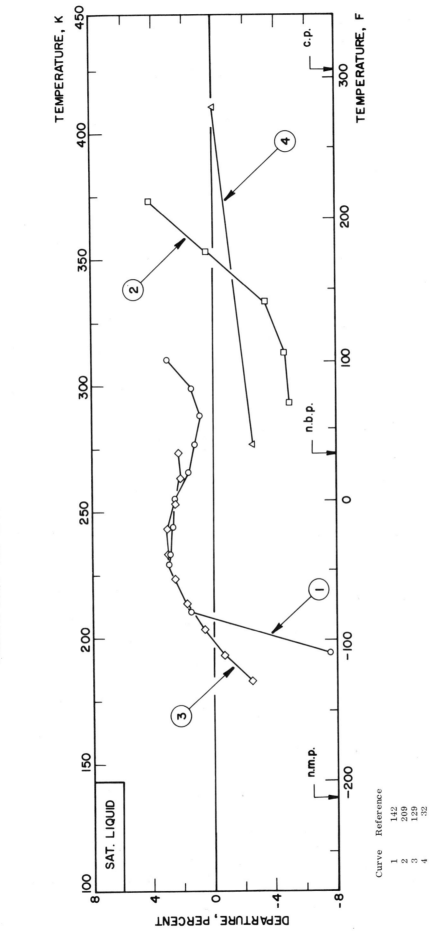

FIGURE 29-L(T).   DEPARTURE PLOT FOR VISCOSITY OF LIQUID n-BUTANE

TABLE 29-V(T). VISCOSITY OF n-BUTANE VAPOR.

DISCUSSION

SATURATED VAPOR

Recommended values for the viscosity of the saturated vapor were generated by means of the correlating technique of Jossi, Stiel and Thodos [100]. Data for the excess viscosity of several gases were plotted as a function of reduced temperature. From the curve obtained, the excess viscosity for n-butane was determined and used together with the recommended values of the 1 atm gas, to generate the recommended curve.

The accuracy is estimated as ±3%.

RECOMMENDED VALUES

[Temperature, T, K; Viscosity, μ, 10^{-3} N s m^{-2}]

SATURATED VAPOR	
T	μ
270	.00692
280	.00719
290	.00747
300	.00776
310	.00805
320	.00836
330	.00864
340	.00894
350	.00930
360	.00975
370	.01029
380	.01092
390	.01166
400	.0128
410	.0143
420	.0168
426*	.0239

* Crit. Temp.

TABLE 29-G(T). VISCOSITY OF GASEOUS n-BUTANE

RECOMMENDED VALUES

[Temperature, T, K; Viscosity, μ, 10^{-6} N s m^{-2}]

GAS

T	μ
270	6.86
280	7.09
290	7.33
300	7.57
310	7.81
320	8.05
330	8.29
340	8.53
350	8.77
360	9.01
370	9.25
380	9.49
390	9.72
400	9.96
410	10.20
420	10.43
430	10.67
440	10.90
450	11.13
460	11.36
470	11.59
480	11.82
490	12.05
500	12.27
510	12.50
520	12.72

DISCUSSION

GAS

There are eight sets of experimental data available for the viscosity of n-butane [32, 52, 130, 133, 191, 216, 236, 264]. They cover an overall temperature range going from 273 to 510 K. Agreement between authors is not outstanding and more weight was given to the experimental data of Wobser and Muller [264] whose work with other gases was found reliable.

The correlation was made by computing the group $\sigma^2 \Omega(T^*)/f \mu = y_{obs}$ from the theoretical expression

$$\mu = 266.93 \frac{\sqrt{M} \; T}{\sigma^2 \Omega(T^*)} f_\mu$$

and plotting the values obtained as a function of 1/T. From the curve obtained, a table was generated after graphical smoothing. Recommended values, interpolated from the table, are thought to be accurate to within three percent.

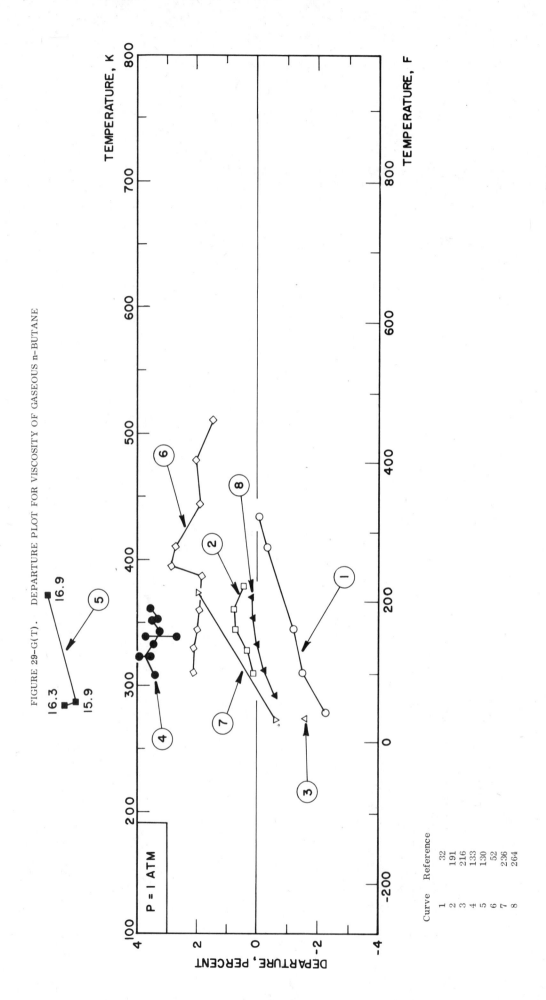

FIGURE 29–G(T). DEPARTURE PLOT FOR VISCOSITY OF GASEOUS n–BUTANE

Curve	Reference
1	32
2	191
3	216
4	133
5	130
6	52
7	236
8	264

TABLE 30-L(T).   VISCOSITY OF LIQUID CARBON DIOXIDE

DISCUSSION

SATURATED LIQUID

Three sets of experimental data were found in the literature, from Novikov[161], Stakelbeck[202], and Warburg[260]. They cover a narrow range from 255 K to the critical point and were assumed to be of equal reliability in deriving the recommended values from them.

Values computed by the correlation technique of Jossi, Stiel and Thodos [100] for the saturated liquid near the critical point are in fair agreement with the recommended values. The accuracy of this correlation is thought to be about ±3%.

RECOMMENDED VALUES

[Temperature, T, K; Viscosity, μ, 10^{-3} N s m^{-2}]

SATURATED LIQUID

T	μ
255	0.1222
260	0.1146
265	0.1077
270	0.1016
275	0.0960
280	0.0908
285	0.0861
290	0.0790
295	0.0704
300	0.0596
304*	0.0316

* Crit. Temp.

120

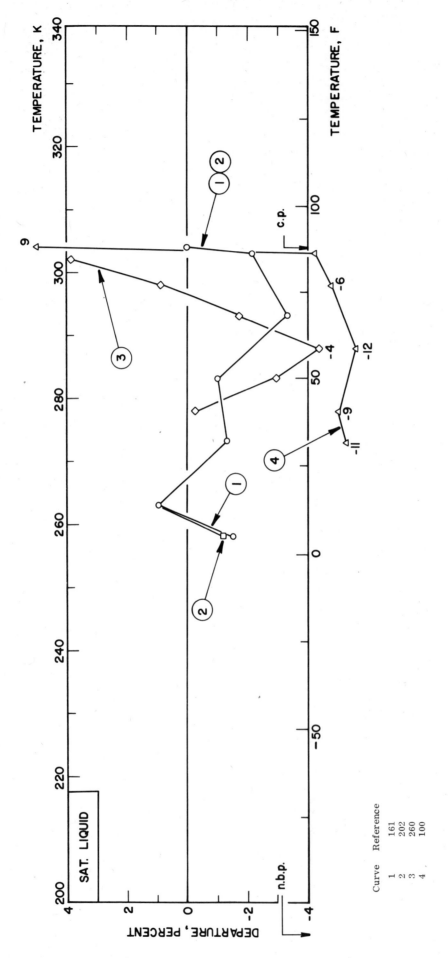

FIGURE 30-L(T). DEPARTURE PLOT FOR VISCOSITY OF LIQUID CARBON DIOXIDE

Curve	Reference
1	161
2	202
3	260
4	100

TABLE 30-V(T).   VISCOSITY OF CARBON DIOXIDE VAPOR

DISCUSSION

SATURATED VAPOR

Values for the viscosity of the saturated vapor were computed by the method of Jossi, Stiel and Thodos [100] using the recommended values for the 1 atm gas, and the density values of Din [49]. They are thought to be reliable to about ±5%, but may be in large error (±10%) in the vicinity of the critical point.

RECOMMENDED VALUES

[Temperature, T, K; Viscosity, μ, 10^{-3} N s m^{-2}]

SATURATED VAPOR

T	μ
216.56	0.01103
220	0.01129
230	0.01201
240	0.01273
250	0.01347
260	0.01428
270	0.01523
280	0.01649
290	0.01874
300	0.02282
304*	0.03160

* Crit. Temp.

TABLE 30-G(T). VISCOSITY OF GASEOUS CARBON DIOXIDE

GAS

DISCUSSION

Experimental data for the viscosity of carbon dioxide reported in the literature are twenty eight in number. They cover a temperature range from 175 to 1686 K.

To correlate the data, use was made of the theoretical expression

$$\mu = 266.93 \; \frac{\sqrt{M}\ T}{\sigma^2 \Omega(T^*)} \; f_\mu$$

The group $\sigma^2\Omega(T^*)/f_\mu$ was computed from the experimental data and plotted as a function of $1/T$. A table was generated from the smooth curve obtained. Recommended values were computed from the theoretical formula, with the values of $\sigma^2\Omega(T^*)/f_\mu$ interpolated from the table. Previous correlations made by Keyes [121] and a semi-theoretical evaluation made by Andrussow [3] were in good agreement with the present work.

At high temperature, the results of Di Pippo [51] are generally higher than the recommended curve, while experimental values of Vasilesco [254-255], Kompaneetz [127] and Trautz [219, 225, 233] are lower, indicating systematic divergence. The accuracy is about ±2%, but may reach ±5% at high temperature.

RECOMMENDED VALUES

GAS

[Temperature, T, K; Viscosity, μ, 10^{-6} N s m^{-2}]

T	μ	T	μ	T	μ
170	8.79	550	25.4	950	38.3
180	9.26	560	25.7	960	38.6
190	9.74	570	26.1	970	38.9
200	10.22	580	26.5	980	39.1
210	10.71	590	26.8	990	39.4
220	11.19	600	27.2	1000	39.7
230	11.68	610	27.6	1050	41.1
240	12.16	620	27.9	1100	42.4
250	12.63	630	28.3	1150	43.7
260	13.12	640	28.6	1200	44.9
270	13.59	650	29.0	1250	46.2
280	14.06	660	29.3	1300	47.4
290	14.53	670	29.6	1350	48.6
300	14.99	680	30.0	1400	49.7
310	15.45	690	30.3	1450	50.9
320	15.91	700	30.6	1500	52.0
330	16.36	710	31.0	1550	53.1
340	16.81	720	31.3	1600	54.2
350	17.26	730	31.6	1650	55.2
360	17.70	740	32.0	1700	56.3
370	18.14	750	32.3	1750	57.3
380	18.57	760	32.6	1800	58.3
390	19.00	770	32.9	1850	59.3
400	19.42	780	33.2	1900	60.3
410	19.85	790	33.5	1950	61.2
420	20.26	800	33.9	2000	62.2
430	20.68	810	34.2		
440	21.09	820	34.5		
450	21.50	830	34.8		
460	21.90	840	35.1		
470	22.30	850	35.4		
480	22.69	860	35.7		
490	23.09	870	36.0		
500	23.48	880	36.3		
510	23.9	890	36.6		
520	24.3	900	36.9		
530	24.6	910	37.1		
540	25.0	920	37.4		
		930	37.7		
		940	38.0		

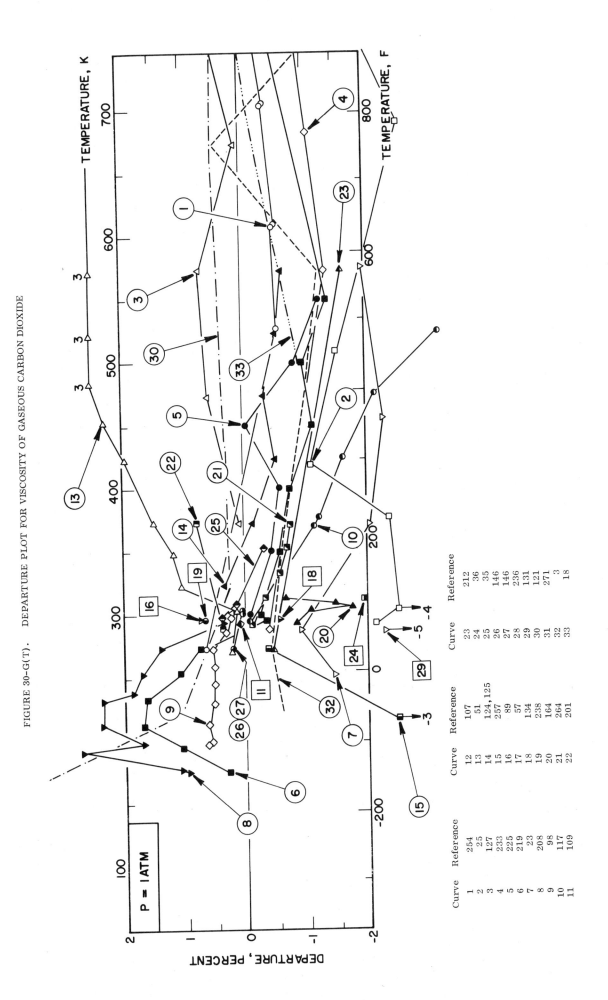

FIGURE 30-G(T). DEPARTURE PLOT FOR VISCOSITY OF GASEOUS CARBON DIOXIDE

123

124

FIGURE 30-G(T).   DEPARTURE PLOT FOR VISCOSITY OF GASEOUS CARBON DIOXIDE (continued)

DISCUSSION

GAS

There are eight sets of experimental data available for the viscosity of carbon monoxide. They are those reported by [6, 97, 139, 201, 227, 257, 264, 270], covering the temperature range from 80 to 550 K. In general a good agreement was found between the data of the various investigators.

In the analysis use was made of the theoretical expression:

$$\mu = 266.93 \frac{\sqrt{MT}}{\sigma^2 \, \Omega(T^*)} f_\mu$$

The group $[\sigma^2 \, \Omega(T^*)/f_\mu = y_{obs}$ was computed from the experimental data and plotted as a function of $1/T$. A curve was drawn through the points, using for guidance similar curves for argon and nitrogen, which were reduced to the same units making use of the ratios of the collision diameters evaluated from the data, and the ratios of the Boyle temperatures. From the resulting curve, a table was generated which was used to compute recommended values.

A previous correlation was made by Keyes [121] and a semi-theoretical evaluation made by Andrussow [3], both are in good agreement with the present work. The recommended values are thought to be accurate to about one percent below 500 K, two percent from 500 to 1000 K, and within five percent for all higher temperatures tabulated.

TABLE 31-G(T).   VISCOSITY OF GASEOUS CARBON MONOXIDE

RECOMMENDED VALUES

[Temperature, T, K; Viscosity, μ, N.m^{-1}.sec^{-1}.10^{-6}]

GAS

T	μ	T	μ	T	μ
80		450	24.10	850	36.8
90		460	24.48	860	37.0
80	5.40	470	24.85	870	37.3
90	6.06	480	25.22	880	37.6
		490	25.59	890	37.8
100	6.70	500	25.95	900	38.1
110	7.34	510	26.3	910	38.3
120	7.98	520	26.7	920	38.6
130	8.61	530	27.0	930	38.9
140	9.23	540	27.4	940	39.1
150	9.84	550	27.7	950	39.4
160	10.44	560	28.0	960	39.6
170	11.03	570	28.4	970	39.9
180	11.61	580	28.7	980	40.1
190	12.18	590	29.0	990	40.4
200	12.74	600	29.4	1000	40.6
210	13.29	610	29.7	1010	40.8
220	13.82	620	30.0	1020	41.1
230	14.35	630	30.3	1030	41.3
240	14.87	640	30.7	1040	41.6
250	15.38	650	31.0	1050	41.8
260	15.88	660	31.3	1060	42.1
270	16.38	670	31.6	1070	42.3
280	16.87	680	31.9	1080	42.5
290	17.34	690	32.2	1090	42.8
300	17.81	700	32.5	1100	43.0
310	18.27	710	32.8	1110	43.2
320	18.73	720	33.1	1120	43.5
330	19.18	730	33.4	1130	43.7
340	19.62	740	33.7	1140	43.9
350	20.05	750	34.0	1150	44.2
360	20.48	760	34.3	1160	44.4
370	20.91	770	34.5	1170	44.6
380	21.32	780	34.8	1180	44.9
390	21.74	790	35.1	1190	45.1
400	22.14	800	35.4	1200	45.3
410	22.54	810	35.7	1210	45.5
420	22.94	820	35.9	1220	45.7
430	23.33	830	36.2	1230	46.0
440	23.72	840	36.5	1240	46.2

TABLE 31-G(T). VISCOSITY OF GASEOUS CARBON MONOXIDE (continued)

RECOMMENDED VALUES

[Temperature, T, K; Viscosity, μ, N.m^{-1}.sec^{-1}.10^{-6}]

GAS

T	μ
1250	46.4
1260	46.6
1270	46.8
2180	47.1
1290	47.3
1300	47.5
1310	47.7
1320	47.9
1330	48.1
1340	48.3
1350	48.5
1360	48.7
1370	49.0
1380	49.2
1390	49.4
1400	49.6
1410	49.8
1420	50.0
1430	50.2
1440	50.4
1450	50.6
1460	50.8
1470	51.0
1480	51.2
1490	51.4
1500	51.6

127

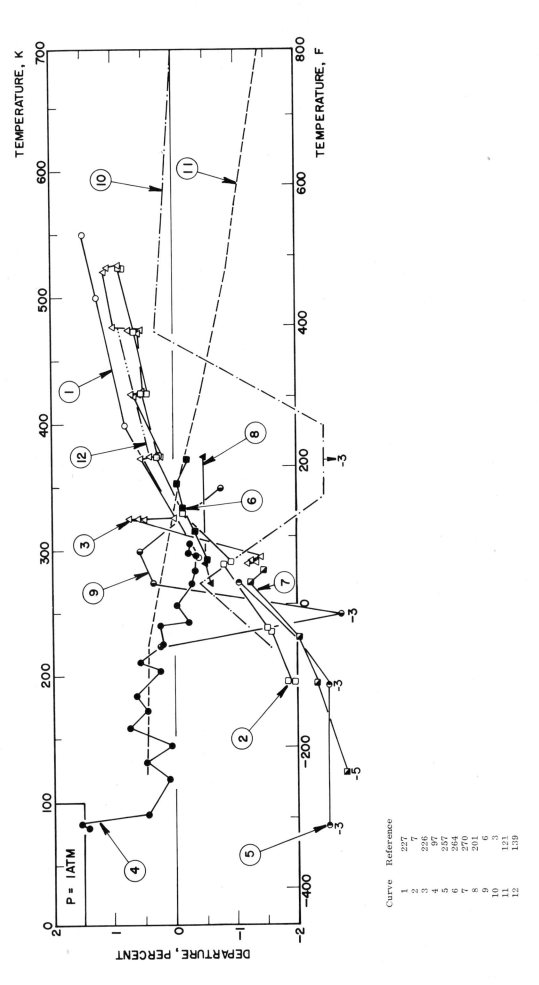

FIGURE 31-G(T). DEPARTURE PLOT FOR VISCOSITY OF GASEOUS CARBON MONOXIDE

128

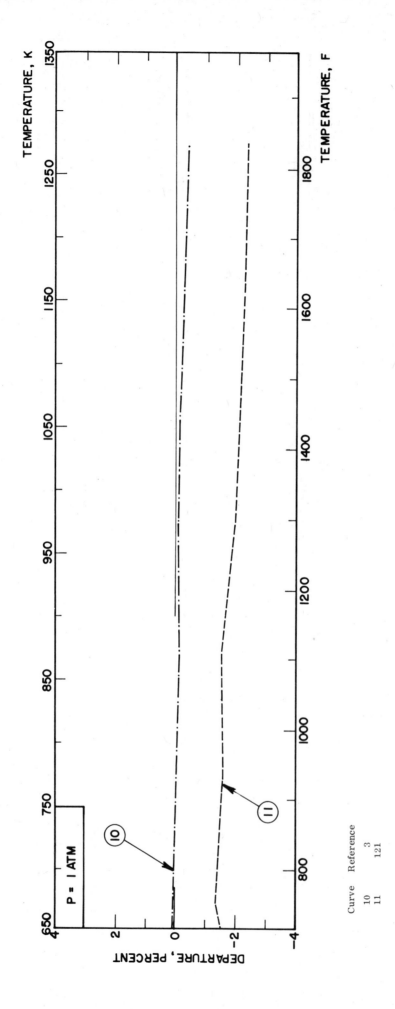

FIGURE 31-G(T). DEPARTURE PLOT FOR VISCOSITY OF GASEOUS CARBON MONOXIDE (continued)

TABLE 32-G(T). VISCOSITY OF GASEOUS CARBON TETRACHLORIDE

DISCUSSION

GAS

Five sets of experimental data were found in the literature [22,215,292, 293,294]. They are in reasonable agreement.

The experimental data were used to calculate $\sigma^2\Omega$ in the relation $\mu = K, \widehat{T}/(\sigma^2\Omega)$. The values of $\sigma^2\Omega$ were then plotted on a graph as a function of 1/T. It was found that these values could be represented by a quadratic equation, which was used to generate the recommended values of the viscosity. The accuracy is of the order of ±3 percent.

RECOMMENDED VALUES

[Temperature, T, K; Viscosity, μ, N s m^{-2} · 10^{-6}]

GAS

T	μ	T	μ
280	9.30	550	17.5
290	9.64	560	17.7
300	9.97	570	18.0
310	10.30	580	18.3
320	10.63	590	18.5
330	10.96	600	18.8
340	11.28	610	19.1
350	11.60	620	19.3
360	11.92	630	19.6
370	12.23	640	19.8
380	12.54	650	20.1
390	12.85	660	20.3
400	13.16	670	20.6
410	13.46	680	20.8
420	13.76	690	21.1
430	14.06	700	21.3
440	14.36	710	21.6
450	14.65	720	21.8
460	14.94	730	22.1
470	15.23	740	22.3
480	15.52	750	22.5
490	15.81	760	22.8
500	16.09	770	23.0
510	16.4	780	23.2
520	16.7	790	23.5
530	16.9	800	23.7
540	17.2		

130

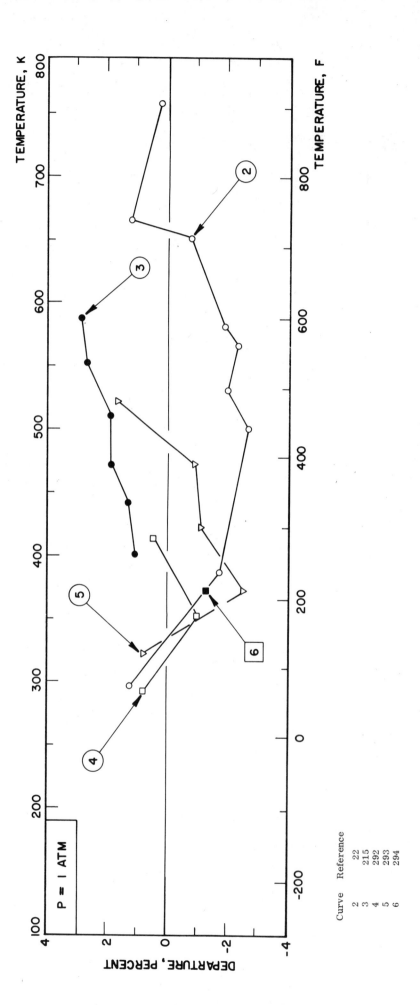

FIGURE 32-G(T). DEPARTURE PLOT FOR VISCOSITY OF GASEOUS CARBON TETRACHLORIDE

TABLE 33-G(T). VISCOSITY OF GASEOUS CARBON TETRAFLUORIDE

DISCUSSION

GAS

Five sets of experimental data were found in the literature [144, 145, 235, 262, 265]. They cover a temperature range from 230 K to 460 K. From the data, the values of:

$$\sigma^2 \Omega = \frac{K \sqrt{T}}{\mu}$$

were computed and fitted to a quadratic equation in 1/T. Recommended values in the range 230 to 500 K were generated from this adjusted equation.

The agreement of the data of the various authors is generally good, and the accuracy is thought to be ±2% in the whole temperature range.

RECOMMENDED VALUES

[Temperature, T, K; Viscosity, μ, N s m^{-2} · 10^{-6}]

GAS	
T	μ
230	13.86
240	14.41
250	14.94
260	15.47
270	15.99
280	16.50
290	17.01
300	17.50
310	17.99
320	18.48
330	18.96
340	19.43
350	19.89
360	20.35
370	20.80
380	21.25
390	21.69
400	22.13
410	22.56
420	22.99
430	23.41
440	23.83
450	24.24
460	24.65
470	25.05
480	25.45
490	25.85
500	26.24

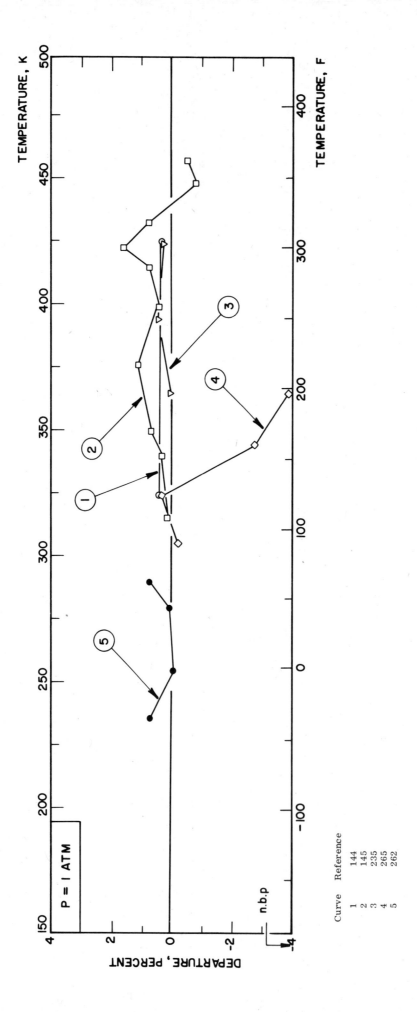

FIGURE 33-G(T). DEPARTURE PLOT FOR VISCOSITY OF GASEOUS CARBON TETRAFLUORIDE

TABLE 34-L(T). VISCOSITY OF LIQUID CHLORODIFLUOROMETHANE

DISCUSSION

SATURATED LIQUID

Six sets of experimental data were found in the literature [14, 58, 79, 123, 166, 371, 184] and two sets of values given manufacturer's technical notes [275, 276].

They were fitted to an equation

$$\log \mu = A + B/T$$

in the range of temperature going from 168 K to 320 K. Above this temperature, a curve was drawn to join the estimated value of the viscosity at the critical temperature. The experimental data shows increasing divergence as the temperature increases.

The accuracy is of about ± 15%.

RECOMMENDED VALUES

[Temperature, T, K, Viscosity, μ, 10^{-3} N s m^{-2}]

SATURATED LIQUID

T	μ
170	0.770
180	0.647
190	0.554
200	0.481
210	0.424
220	0.378
230	0.340
240	0.309
250	0.2824
260	0.2602
270	0.2412
280	0.2248
290	0.2105
300	0.1980
310	0.1870
320	0.1772
330	0.1670
340	0.1500
350	0.1320
360	0.1050
369*	0.0305

* Crit. Temp.

134

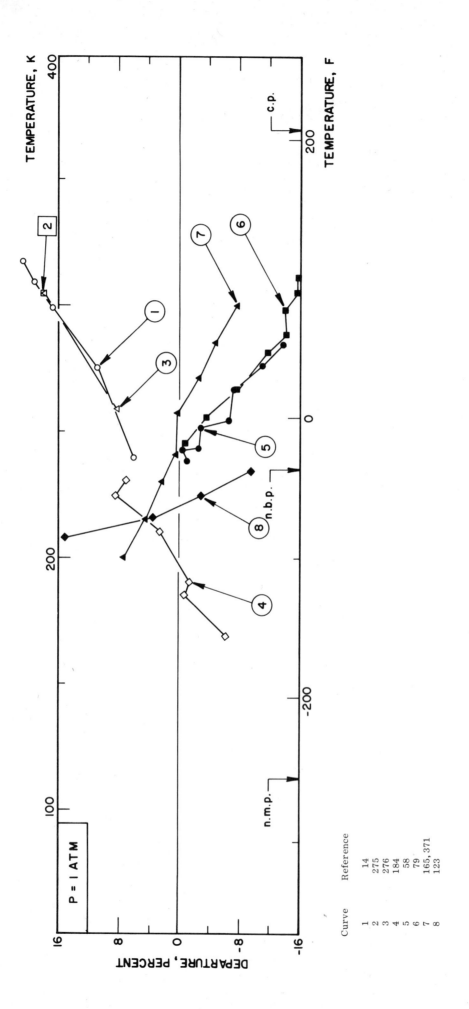

FIGURE 34-L(T). DEPARTURE PLOT FOR VISCOSITY OF LIQUID CHLORODIFLUOROMETHANE

TABLE 34-V(T).   VISCOSITY OF CHLORODIFLUOROMETHANE VAPOR

DISCUSSION

RECOMMENDED VALUES

[Temperature, T, K, Viscosity, μ, 10^{-3} N s m^{-2}]

SATURATED VAPOR

Recommended values for the viscosity of the saturated vapor were estimated by the method of Stiel and Thodos [207] using the recommended values for the 1 atm gas and the density values given in a manufacturer's technical note.

The accuracy is thought to be ±5% although this figure may rise to about ±10% around the critical temperature.

SATURATED VAPOR

T	μ
230	0.01000
240	0.01043
250	0.01087
260	0.01132
270	0.01180
280	0.01233
290	0.01290
300	0.01350
310	0.01415
320	0.01485
330	0.01560
340	0.01640
350	0.0177
360	0.0199
369*	0.0305

* Crit. Temp.

TABLE 34-G(T).   VISCOSITY OF GASEOUS CHLORODIFLUOROMETHANE

DISCUSSION

GAS

Eight sets of experimental data were found in the literature [14, 40, 136-7, 149, 150, 235, 262, 265].   Six sets were used to compute the values of

$$\sigma^2 \Omega = K \sqrt{\frac{T}{\mu}}$$

which were fitted to a quadratic equation in 1/T, from which recommended values were computed.

The data of Coughlin [40] and Latto [136-7] which became available later were found to fit well with the calculated values.

The accuracy is of about ±2%.

RECOMMENDED VALUES

[Temperature, T, K; Viscosity, μ, 10^{-6} N s m^{-2}]

GAS

T	μ
250	10.86
260	11.29
270	11.72
280	12.14
290	12.57
300	12.99
310	13.41
320	13.82
330	14.24
340	14.65
350	15.06
360	15.47
370	15.88
380	16.28
390	16.73
400	17.08
410	17.48
420	17.87
430	18.26
440	18.65
450	19.04
460	19.42
470	19.80
480	20.18
490	20.56
500	20.93

137

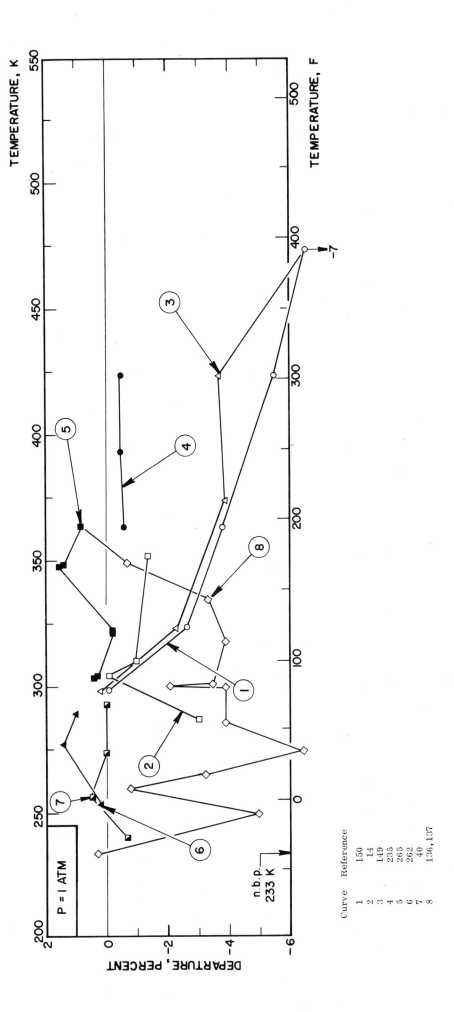

FIGURE 34-G(T). DEPARTURE PLOT FOR VISCOSITY OF GASEOUS CHLORODIFLUOROMETHANE

Curve	Reference
1	150
2	14
3	149
4	235
5	265
6	262
7	40
8	136,137

TABLE 35-G(T). VISCOSITY OF GASEOUS CHLOROFORM

RECOMMENDED VALUES

[Temperature, T, K; Viscosity, μ N s m^{-2} · 10^{-3}]

DISCUSSION

GAS

Four sets of experimental data were found in the literature [22, 215, 236, 257]. They are in reasonable agreement.

Use was made of the theoretical relation $\mu = K\sqrt{T}/(\sigma^2\Omega)$ to obtain values of $\sigma^2\Omega$ from the experimental data. These were plotted as a function of $1/T$ and adjusted to a quadratic equation. From this equation, recommended values of the viscosity were computed. The accuracy is of the order of ± 2.5 percent.

GAS

T	μ
250	8.60
260	8.95
270	9.29
280	9.63
290	9.98
300	10.32
310	10.66
320	11.00
330	11.34
340	11.68
350	12.02
360	12.36
370	12.69
380	13.02
390	13.36
400	13.69
410	14.02
420	14.35
430	14.68
440	15.01
450	15.33
460	15.66
470	15.98
480	16.30
490	16.62
500	16.94
510	17.3
520	17.6
530	17.9
540	18.2
550	18.5
560	18.8
570	19.1
580	19.4
590	19.7
600	20.0
610	20.3
620	20.6
630	20.9
640	21.2
650	21.6

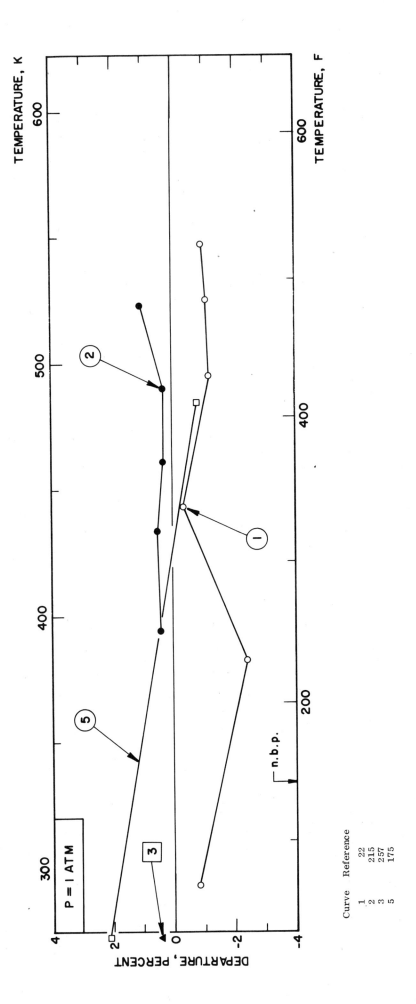

FIGURE 35-G(T).   DEPARTURE PLOT FOR VISCOSITY OF GASEOUS CHLOROFORM

TABLE 36-L(T).   VISCOSITY OF LIQUID CHLOROPENTAFLUOROETHANE

RECOMMENDED VALUES

[Temperature, T, K, Viscosity, μ, 10^{-3} N s m^{-2}]

DISCUSSION

SATURATED LIQUID

Three sets of experimental data were found in the literature, those of Lilios[140], Gordon [79] and Phillips [371].

They were least square fitted to an equation:

$\log \mu = A + B/T$

from which the recommended values were generated, in the range 190 K to 310 K. Above the latter temperature to the critical, recommended values were read from a curve joining the estimated value at the critical temperature.

The accuracy is thought to be ±5%.

SATURATED LIQUID

T	μ
190	0.9610
200	0.7738
210	0.6361
220	0.5323
230	0.4524
240	0.3897
250	0.3398
260	0.2994
270	0.2663
280	0.2388
290	0.2158
300	0.1930
310	0.1710
320	0.1485
330	0.1235
340	0.0945
350	0.053
353*	0.029

* Crit. Temp.

141

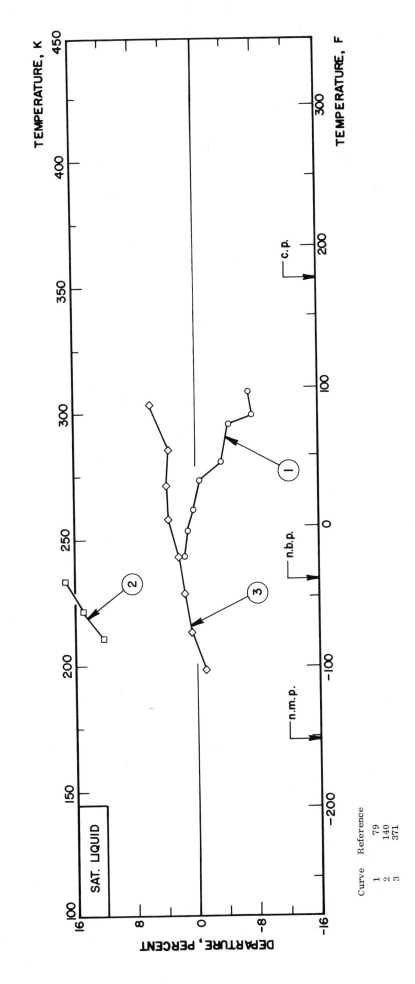

FIGURE 36-L(T).   DEPARTURE PLOT FOR VISCOSITY OF LIQUID CHLOROPENTAFLUOROETHANE

TABLE 36-V(T). VISCOSITY OF CHLOROPENTAFLUOROETHANE VAPOR

DISCUSSION

SATURATED VAPOR

Recommended values for the saturated vapor was computed by the method of Stiel and Thodos [207] which makes use of the excess viscosity concept. Reduced excess viscosity versus reduced temperature were obtained for several refrigerants. Values read from that curve were used together with the recommended values for the 1 atm gas to get the present values.

The accuracy is thought to be ±5% near the boiling point, but may reach more than ±10% when approaching the critical point.

RECOMMENDED VALUES

[Temperature, T, K, Viscosity, μ, 10^{-3} N s m^{-2}]

SATURATED VAPOR

T	μ
230	0.01006
240	0.01044
250	0.01092
260	0.01141
270	0.01189
280	0.01235
290	0.01298
300	0.0136
310	0.0143
320	0.0151
330	0.0167
340	0.0176
350	0.0211
353*	0.0287

* Crit. Temp.

TABLE 36-G(T).   VISCOSITY OF GASEOUS CHLOROPENTAFLUOROETHANE

RECOMMENDED VALUES

[Temperature, T, K; Viscosity, μ, 10^{-6} N s m^{-2}]

GAS

T	μ
250	10.83
260	11.22
270	11.61
280	12.00
290	12.38
300	12.76
310	13.15
320	13.53
330	13.90
340	14.28
350	14.66
360	15.03
370	15.40
380	15.77
390	16.14
400	16.50
410	16.86
420	17.22
430	17.58
440	17.93
450	18.29
460	18.64
470	18.99
480	19.34
490	19.68
500	20.02

DISCUSSION

GAS

Five sets of experimental data were found in the literature, covering a range of temperature from about 250 K to 473 K [40 , 143 , 235 , 262 , 265].

From the data, the values of

$$\sigma^2 \Omega = \frac{K \sqrt{T}}{\mu}$$

were computed, and adjusted to a quadratic equation in 1/T, from which in turn, recommended values were generated. The agreement from set to set is generally good, and the accuracy is thought to be ±2%.

144

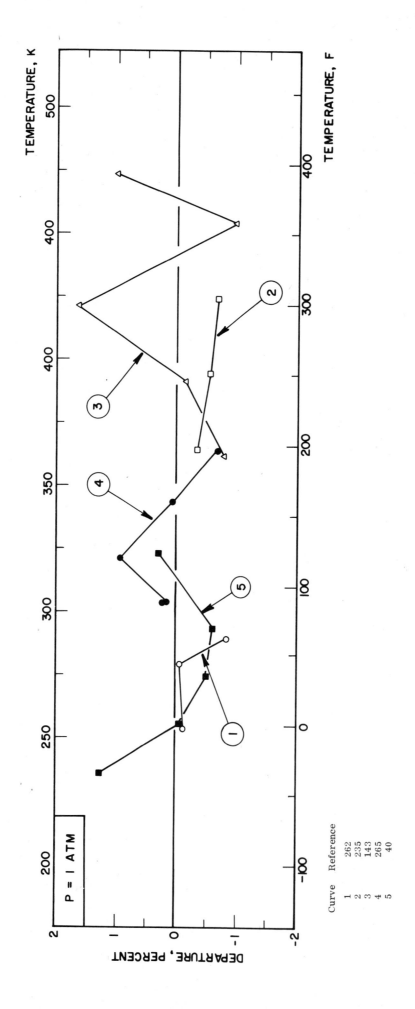

FIGURE 36—G(T).   DEPARTURE PLOT FOR VISCOSITY OF GASEOUS CHLOROPENTAFLUOROETHANE

TABLE 37-L(T). VISCOSITY OF LIQUID CHLOROTRIFLUOROMETHANE

DISCUSSION

SATURATED LIQUID

Two sets of data were found in the literature, [79, 371] and one point value was found in a commercial technical note [276]. The two sets of data are in disagreement, one being lower and the other higher than the recommended curve which was generated by a least square adjustment of:

$$\log \mu = A + B/T$$

and which covers the temperature range of 200 K to 270 K. Above the latter temperature, a curve was drawn to joint the estimated viscosity at the critical temperature.

The accuracy of the recommended values are thought to be ±7%, in the range 200 to 270 K but may reach ±15% at higher temperatures.

RECOMMENDED VALUES

[Temperature, T, K, Viscosity, μ, 10^{-3} N s m^{-2}]

SATURATED LIQUID

T	μ
170	0.459
180	0.383
190	0.326
200	0.282
210	0.248
220	0.220
230	0.197
240	0.179
250	0.1629
260	0.1497
270	0.1335
280	0.1140
290	0.0870
300	0.052
302*	0.029

* Crit. temp.

146

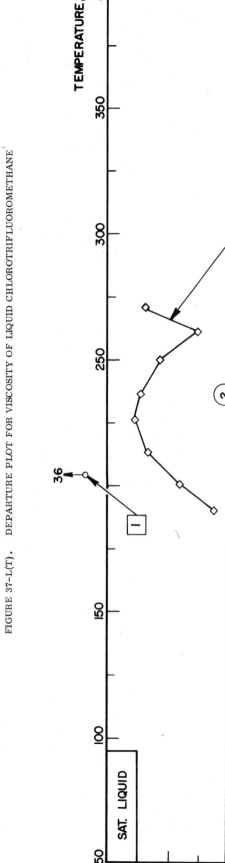

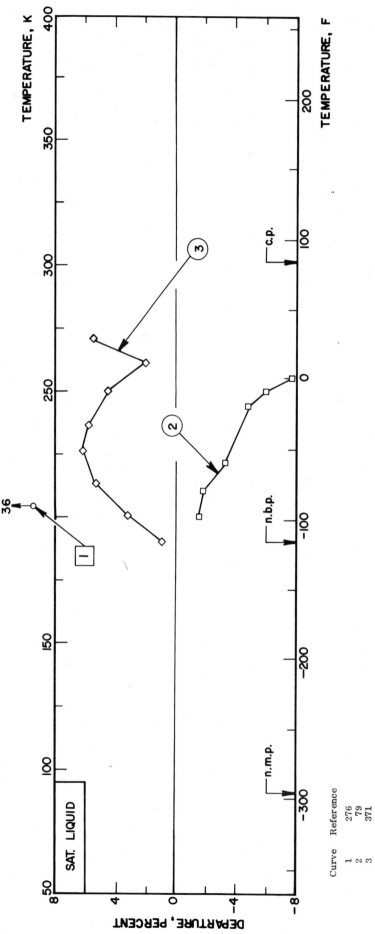

FIGURE 37-L(T). DEPARTURE PLOT FOR VISCOSITY OF LIQUID CHLOROTRIFLUOROMETHANE

TABLE 37-V(T). VISCOSITY OF CHLOROTRIFLUOROMETHANE VAPOR

RECOMMENDED VALUES

[Temperature, T, K, Viscosity, μ, 10^{-3} N s m^{-2}]

DISCUSSION

SATURATED VAPOR

Recommended values for the viscosity of the saturated vapor were estimated with the density values of a manufacturer's technical note [276] and with the recommended values for the 1 atm gas, using the correlating equation of Stiel and Thodos [207].

The accuracy is thought to be ±5 percent.

SATURATED VAPOR

T	μ
190	0.00974
200	0.01027
210	0.01080
220	0.01133
230	0.01188
240	0.01247
250	0.01315
260	0.01380
270	0.01475
280	0.01580
290	0.01740
300	0.0224
302*	0.0289

* Crit. Temp.

TABLE 37-G(T).   VISCOSITY OF GASEOUS CHLOROTRIFLUOROMETHANE

DISCUSSION

GAS

Six sets of experimental data were found in the literature [137, 177, 235, 261, 262, 265]. They cover the range 220 K to 423 K. The correlation was made by computing

$$\sigma^2 \Omega = K \frac{\sqrt{T}}{\mu}$$

which was least square fitted to a quadratic equation in 1/T, with all the data, except those of Latto [137] which became available later, and were found to fit reasonably well.

The accuracy is thought to be ± 2% in the whole range.

RECOMMENDED VALUES

[Temperature, T, K; Viscosity, μ, 10^{-6} N s m^{-2}]

GAS

T	μ
230	11.57
240	11.97
250	12.37
260	12.79
270	13.21
280	13.63
290	14.06
300	14.49
310	14.92
320	15.36
330	15.80
340	16.24
350	16.68
360	17.13
370	17.57
380	18.02
390	18.46
400	18.91
410	19.36
420	19.81
430	20.26
440	20.71
450	21.15
460	21.60
470	22.05
480	22.50
490	22.95
500	23.39

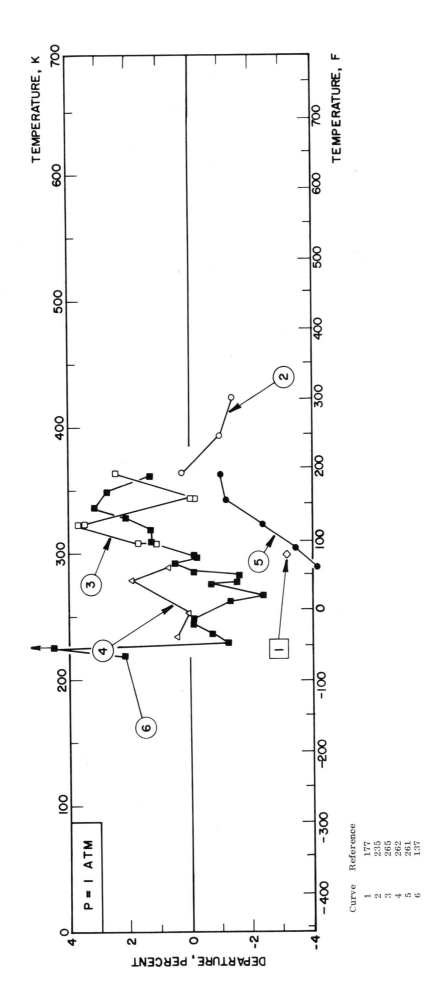

FIGURE 37-G(T). DEPARTURE PLOT FOR VISCOSITY OF GASEOUS CHLOROTRIFLUOROMETHANE

TABLE 38-L(T).   VISCOSITY OF LIQUID DICHLORODIFLUOROMETHANE

DISCUSSION

SATURATED LIQUID

Nine sets of data were found in the literature [14, 58, 79, 123, 140, 275, 276, 277, 371]. They cover a temperature range going from about 200 K to 340 K. The results of Benning [14] covers about this whole range but are much higher, particularly at high temperature. The more recent results of Phillips [371] in the range 200 to 310 K seems more consistent and also those of Eisele [58] in the same range. They were adjusted by least square to an equation:

$$\log \mu = A + B/T$$

Above 340 K, a curve was drawn to join the recommended value to the estimated value at the critical temperature. Recommended values were read from this curve and smoothed.

The accuracy of the recommended values are thought to be about ±5%, although around the critical point the figure may reach ±10%.

RECOMMENDED VALUES

[Temperature, T, K, Viscosity, μ, 10^{-3} N s m^{-2}]

SATURATED LIQUID

T	μ
170	1.210
180	0.969
190	0.794
200	0.664
210	0.565
220	0.488
230	0.426
240	0.377
250	0.337
260	0.303
270	0.275
280	0.252
290	0.231
300	0.2139
310	0.1989
320	0.1857
330	0.1741
340	0.1600
350	0.1445
360	0.1275
370	0.1055
380	0.0750
385*	0.0310

* Crit. temp.

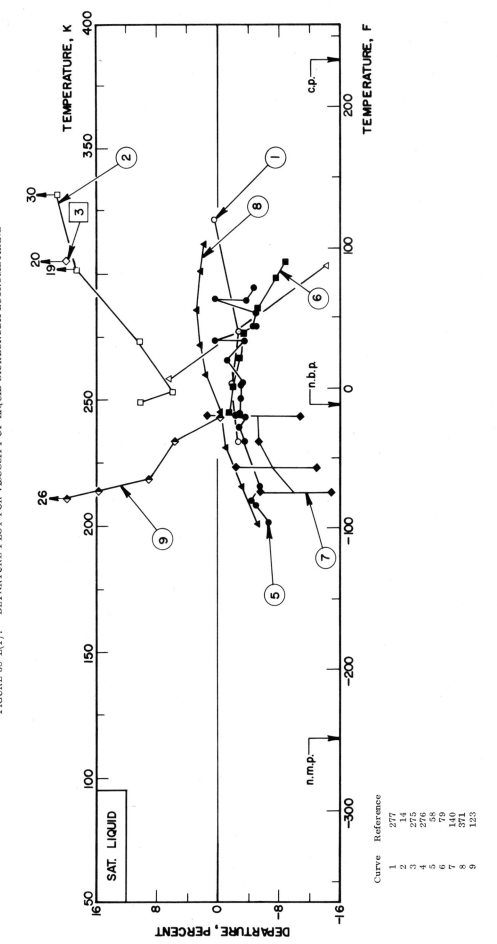

FIGURE 38-L(T).   DEPARTURE PLOT FOR VISCOSITY OF LIQUID DICHLORODIFLUOROMETHANE

TABLE 38-V(T). VISCOSITY OF DICHLORODIFLUOROMETHANE VAPOR

RECOMMENDED VALUES

[Temperature, T, K, Viscosity, μ, 10^{-3} N s m^{-2}]

SATURATED VAPOR

T	μ
240	0.01016
250	0.01058
260	0.01102
270	0.01148
280	0.01196
290	0.01246
300	0.01300
310	0.01356
320	0.01415
330	0.01480
340	0.01550
350	0.01640
360	0.01746
370	0.01900
380	0.02220
385*	0.03102

* Crit. Temp.

DISCUSSION

SATURATED VAPOR

Recommended values for the viscosity of the saturated vapor were estimated with the correlating equation of Stiel and Thodos [207] with the density values of a manufacturer's technical note [277] and our recommended values for the 1 atm gas.

The accuracy is thought to be ±5% although a higher discrepancy may occur around the critical point.

TABLE 38-G(T). VISCOSITY OF GASEOUS DICHLORODIFLUOROMETHANE

GAS

DISCUSSION

Ten sets of experimental data were found in the literature [14, 28, 137, 149, 150, 235, 261, 262, 265, 267]. They cover a temperature range from 250 K to 473 K. The smoothed values of Makita [149, 150] at temperatures higher than 373 K diverge increasingly with temperature. They were not taken into account in the adjustment of the equation:

$$\sigma^2 \Omega = \frac{K \sqrt{T}}{\mu} = f(1/T)$$

which was least square fitted, and from which recommended values were generated.

Data by Žalandik [267] and Latto [137] were only available later and were found to fit very well with the recommended values. Likewise for values from a DuPont report [277].

The accuracy of this correlation is thought to be ±3%.

RECOMMENDED VALUES

[Temperature, T, K; Viscosity, μ, N s m^{-2} · 10^{-6}]

GAS

T	μ
250	10.57
260	10.98
270	11.40
280	11.80
290	12.21
300	12.60
310	13.00
320	13.39
330	13.78
340	14.17
350	14.54
360	14.92
370	15.29
380	15.66
390	16.03
400	16.39
410	16.74
420	17.10
430	17.45
440	17.80
450	18.14
460	18.49
470	18.83
480	19.16
490	19.49
500	19.83

153

154

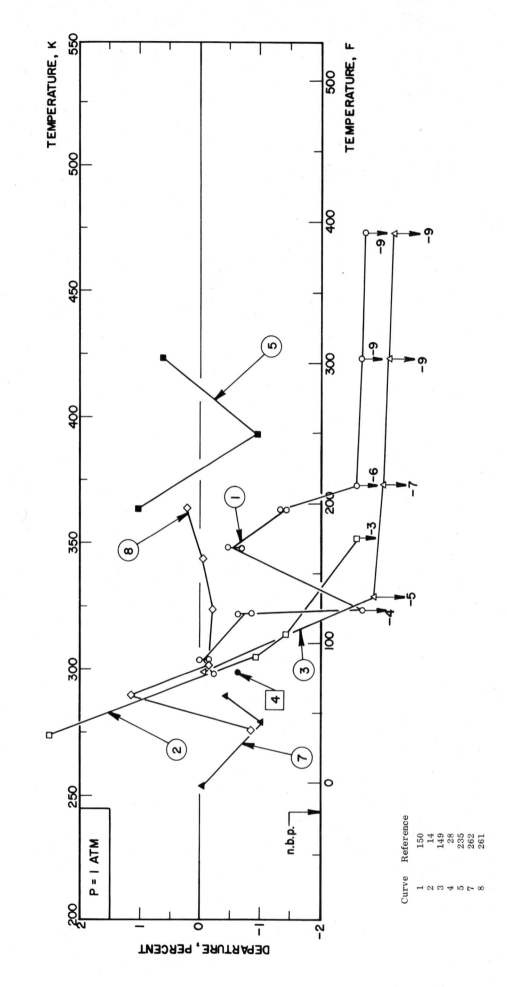

FIGURE 38-G(T). DEPARTURE PLOT FOR VISCOSITY OF GASEOUS DICHLORODIFLUOROMETHANE

TABLE 39-L(T). VISCOSITY OF LIQUID DICHLOROFLUOROMETHANE

DISCUSSION

RECOMMENDED VALUES

[Temperature, T, K, Viscosity, μ, 10^{-3} N s m^{-2}]

SATURATED LIQUID

Three sets of experimental data were found in the literature. These are the results of Benning [14], Phillips [166] and Kinser [123]. A single value was also found in a manufacturers technical note [275]. The results of Phillips cover the whole range of 200 K to 350 K and were given more weight in the adjustment of

$$\log \mu = A + B/T$$

This adjustment curve was used to generate recommended values in the range 170 to 390 K. Above the latter temperature, to the critical a curve was drawn graphically to join smoothly the estimated viscosity value at the critical point.

The recommended values are thought to be reliable to ±5% below 390 K. Between 390 K and the critical temperature they should be considered tentative.

SATURATED LIQUID	
T	μ
170	2.019
180	1.590
190	1.283
200	1.059
210	0.889
220	0.759
230	0.657
240	0.575
250	0.509
260	0.455
270	0.410
280	0.372
290	0.340
300	0.313
310	0.289
320	0.269
330	0.251
340	0.235
350	0.2207
360	0.2082
370	0.1971
380	0.1871
390	0.1770
400	0.1655
410	0.1525
420	0.1370
430	0.1190
440	0.0970
450	0.055
451*	0.032

* Crit. Temp.

156

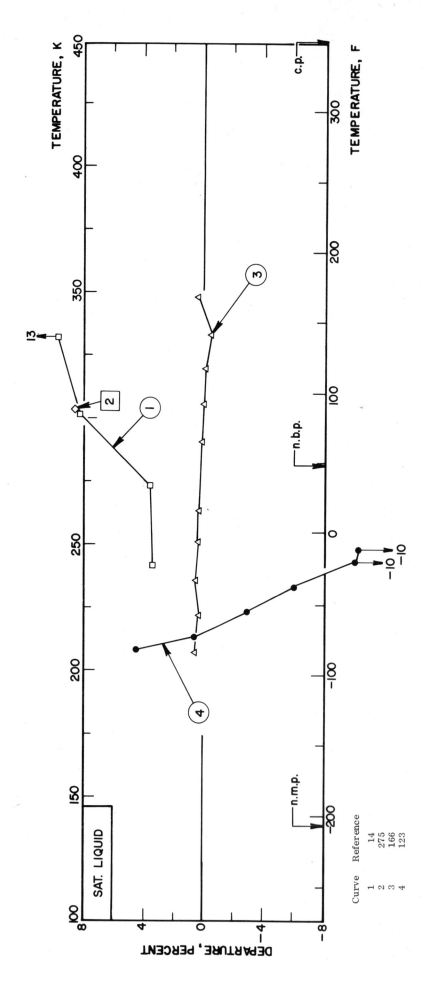

FIGURE 39-L(T). DEPARTURE PLOT FOR VISCOSITY OF LIQUID DICHLOROFLUOROMETHANE

TABLE 39-V(T). VISCOSITY OF DICHLOROFLUOROMETHANE VAPOR

RECOMMENDED VALUES

[Temperature, T, K, Viscosity, μ, 10^{-3} N s m^{-2}]

DISCUSSION

SATURATED VAPOR

Recommended values for the viscosity of the saturated vapor were generated by means of the method of Stiel and Thodos [207]which made use of the excess
viscosity concept. Reduced excess viscosities as a function of reduced temperature were gotten from a curve generated from data on other refrigerants.

The accuracy is thought to be ±5%.

SATURATED VAPOR

T	μ
280	0.01089
290	0.01128
300	0.01168
310	0.01208
320	0.01247
330	0.01291
340	0.01334
350	0.01380
360	0.01429
370	0.01482
380	0.01537
390	0.01593
400	0.01664
410	0.01743
420	0.01823
430	0.01945
440	0.02131
450	0.02584
451*	0.03227

* Crit. Temp.

TABLE 39-G(T). VISCOSITY OF GASEOUS DICHLOROFLUOROMETHANE

DISCUSSION

GAS

Seven sets of experimental data were found in the literature [14, 143, 149, 150, 235, 261, 265]. The results of Makita [149, 150] shows an increasing divergence from other results as the temperature increase, while the results of Benning [14] are systematically lower. They were not used in the generation of the recommended values, which was done by fitting a quadratic equation to the expression

$$\sigma^2 \Omega = \frac{K\sqrt{T}}{\mu}.$$

derived from the experimental viscosities.

The accuracy is thought to be ±2%.

RECOMMENDED VALUES

[Temperature, T, K; Viscosity, μ, 10^{-6} N s m^{-2}]

GAS

T	μ
280	10.89
290	11.26
300	11.63
310	11.99
320	12.34
330	12.73
340	13.09
350	13.45
360	13.82
370	14.18
380	14.54
390	14.90
400	15.26
410	15.62
420	15.97
430	16.33
440	16.68
450	17.03
460	17.38
470	17.73
480	18.08
490	18.43
500	18.77

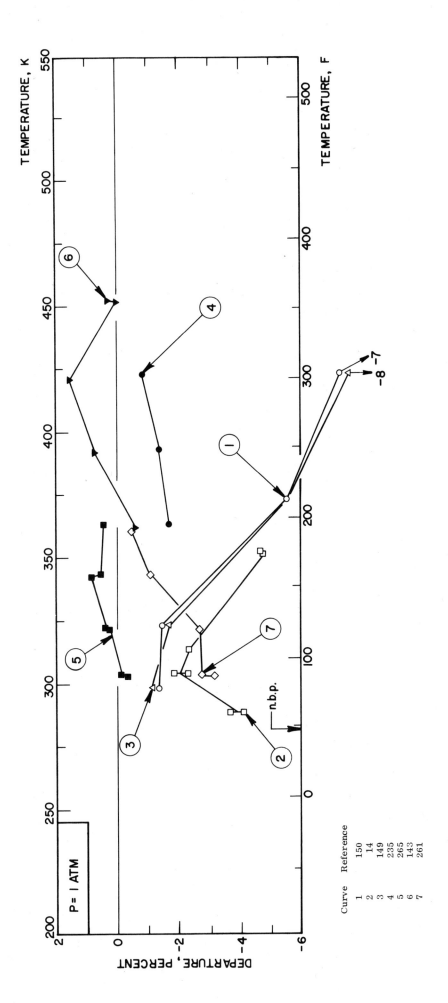

FIGURE 39–G(T).   DEPARTURE PLOT FOR VISCOSITY OF GASEOUS DICHLOROFLUOROMETHANE

TABLE 40-L(T).   VISCOSITY OF LIQUID DICHLOROTETRAFLUOROETHANE

DISCUSSION

SATURATED LIQUID

Two sets of experimental data were found in the literature, and two sets of values were found in manufacturer's technical notes [275, 276]. The experimental results are those of Phillips [371] and those of Kinser [123] covering temperature ranges of 200-330 K and 200-270 K respectively. They were least square fitted to an equation

$$\log \mu = A + B/T$$

from which recommended values were generated in the temperature range 170-360 K. Above this temperature a curve was drawn to join smoothly the estimated viscosity value at the critical temperature.

The accuracy is thought to be about ±5%.

RECOMMENDED VALUES

[Temperature, T, K, Viscosity, μ, 10^{-3} N s m^{-2}]

SATURATED LIQUID

T	μ
170	4.422
180	3.177
190	2.363
200	1.811
210	1.423
220	1.143
230	0.936
240	0.779
250	0.658
260	0.564
270	0.488
280	0.427
290	0.377
300	0.335
310	0.301
320	0.272
330	0.247
340	0.226
350	0.2072
360	0.1912
370	0.1772
380	0.1607
390	0.1410
400	0.117
410	0.087
419*	0.031

* Crit. Temp.

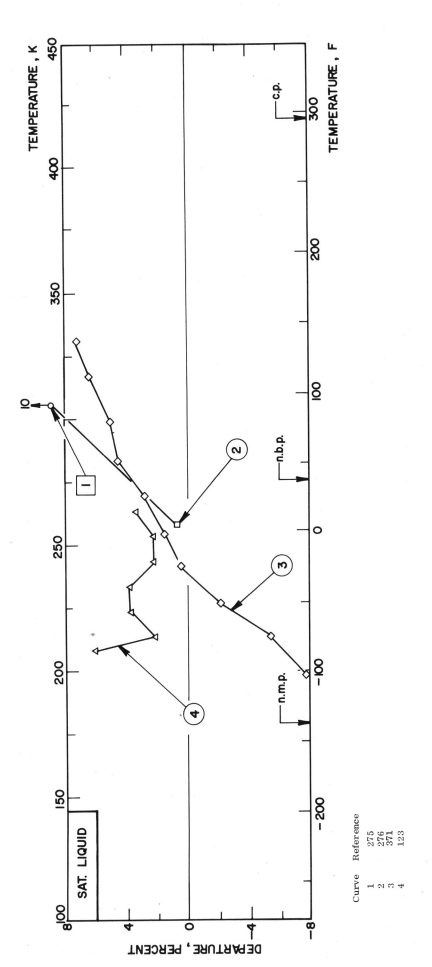

FIGURE 40-L(T). DEPARTURE PLOT FOR VISCOSITY OF LIQUID DICHLOROTETRAFLUOROETHANE

TABLE 40-V(T).   VISCOSITY OF DICHLOROTETRAFLUOROETHANE VAPOR

RECOMMENDED VALUES

[Temperature, T, K, Viscosity, μ, 10^{-3} N s m^{-2}]

SATURATED VAPOR

T	μ
270	0.01063
280	0.01096
290	0.01133
300	0.01172
310	0.01213
320	0.01256
330	0.01293
340	0.01342
350	0.01396
360	0.01455
370	0.01520
380	0.01590
390	0.01685
400	0.0182
410	0.0203
419*	0.0311

* Crit. Temp.

DISCUSSION

SATURATED VAPOR

Recommended values for the viscosity of the saturated vapor we obtained with the method of Stiel and Thodos [207] which makes use of the excess viscosity concept as a function of density. Saturated vapor density were taken from a manufacturer's technical note [279].

The accuracy is thought to be ±5% although the figure may rise to ±10% when approaching the critical temperature.

TABLE 40-G(T). VISCOSITY OF GASEOUS DICHLOROTETRAFLUOROETHANE

RECOMMENDED VALUES

[Temperature, T, K; Viscosity, μ, 10^{-6} N s m^{-2}]

GAS

T	μ
230	9.44
240	9.72
250	10.02
260	10.32
270	10.63
280	10.95
290	11.27
300	11.59
310	11.92
320	12.25
330	12.59
340	12.92
350	13.26
360	13.60
370	13.94
380	14.28
390	14.62
400	14.97
410	15.31
420	15.66
430	16.00
440	16.34
450	16.69
460	17.04
470	17.38
480	17.73
490	18.07
500	18.42

DISCUSSION

GAS

Five sets of experimental data were found in the literature [12 , 143 , 235 , 261 , 265], covering an overall temperature range, from 233 to 473 K. The agreement between the sets is generally good.

From the experimental values of viscosity we computed

$$\sigma^2 \Omega = K \frac{\sqrt{T}}{\mu}$$

which we fitted to a quadratic equation in 1/T. From this equation the recommended values of viscosity were computed, reversing the procedure.

The accuracy is thought to be ±2% throughout.

164

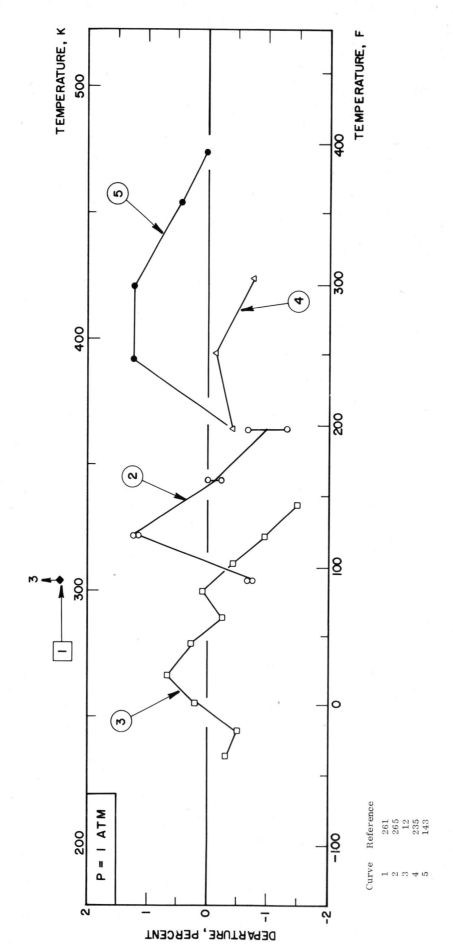

FIGURE 40-G(T). DEPARTURE PLOT FOR VISCOSITY OF GASEOUS DICHLOROTETRAFLUOROETHANE

TABLE 41-L(T).   VISCOSITY OF LIQUID 1,1-DIFLUOROETHANE

DISCUSSION

SATURATED LIQUID

Only one set of experimental data by Phillips was found in the literature[371].
An equation of the type:

$$\log \mu = A + B/T$$

was least square fitted to the data, and recommended values were generated
from the equation.

There is no means to assess the accuracy.

RECOMMENDED VALUES

[Temperature, T, K; Viscosity, μ, N s m^{-2} · 10^{-3}]

SATURATED LIQUID

T	μ
200	0.593
205	0.541
210	0.497
215	0.457
220	0.423
225	0.392
230	0.365
235	0.341
240	0.319
245	0.300
250	0.282
255	0.266
260	0.251
265	0.238
270	0.226
275	0.2145
280	0.2041
285	0.1946
290	0.1857
295	0.1774
300	0.1698
305	0.1626
310	0.1559
315	0.1496
320	0.1436

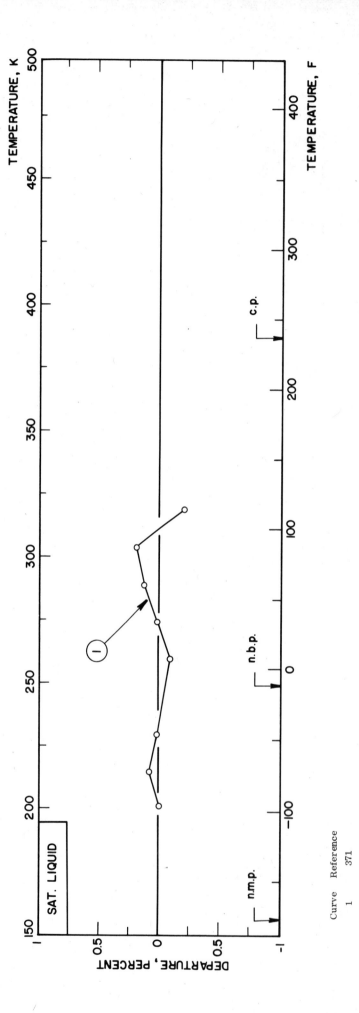

FIGURE 41-L(T). DEPARTURE PLOT FOR VISCOSITY OF LIQUID 1,1-DIFLUOROETHANE

Curve   Reference

1   371

TABLE 42-L(T). VISCOSITY OF LIQUID ETHANE

DISCUSSION

SATURATED LIQUID

Four sets of experimental data were found in the literature. The experimental data of Swift [209] cover the higher temperature range from 193 K to the critical point. The data from Galkov and Gerf [69] and those of Gerf and Galkov [70] cover the low temperature range from 100 K to 170 K, while the data of Di Geronimo [46] are in an intermediate range.

A least square fit of an equation of the type

$$\log \mu = A + B/T$$

was computed with all the data below 250 K, while the data from 250 K to the critical point were smoothed graphically. All data seems to be of equal reliability. Values for the saturated liquid were computed with the correlation equation of Jossi, Stiel and Thodos [100] used with orthobaric density data of Din [49]. These are in fair agreement with the experimental data, but diverges with decreasing temperature. The overall accuracy of this correlation is estimated to ±3%.

RECOMMENDED VALUES

[Temperature, T, K; Viscosity, μ, 10^{-3} N s m^{-2}]

SATURATED LIQUID

T	μ
100	0.882
110	0.635
120	0.482
130	0.383
140	0.314
150	0.2639
160	0.2270
170	0.1987
180	0.1766
190	0.1567
200	0.1392
210	0.1242
220	0.1107
230	0.0994
240	0.0888
250	0.0794
260	0.0708
270	0.0616
280	0.0540
290	0.0460
300	0.0361
305*	0.0217

* Crit. Temp.

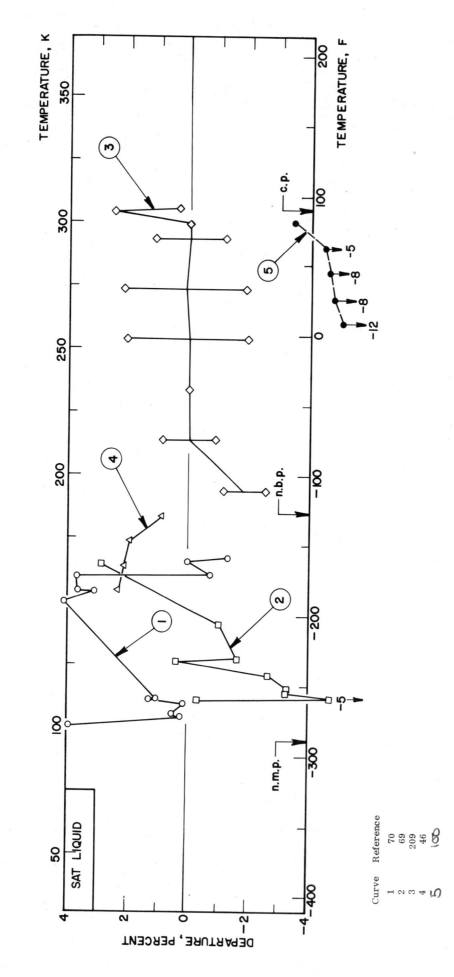

FIGURE 42-L(T). DEPARTURE PLOT FOR VISCOSITY OF LIQUID ETHANE

TABLE 42-V(T).   VISCOSITY OF ETHANE VAPOR

DISCUSSION

SATURATED VAPOR

   The recommended values for the viscosity of ethane in the saturated
vapor state were computed by the correlation technique of Jossi, Stiel and
Thodos [100] using the recommended value for the 1 atm gas, and the density
data given by Din [49] . They are reliable to about ±5%.

RECOMMENDED VALUES

[Temperature, T, K; Viscosity, μ, 10^{-3} N s m^{-2}]

SATURATED VAPOR

T	μ
190	0.00619
200	0.00653
210	0.00689
220	0.00725
230	0.00766
240	0.00808
250	0.00856
260	0.00908
270	0.00970
280	0.01046
290	0.01174
300	0.01433
305*	0.02166

* Crit. Temp.

TABLE 42-G(T). VISCOSITY OF GASEOUS ETHANE

DISCUSSION

GAS

Eleven sets of experimental data on ethane were found in the literature, covering a temperature range from 194 to 523 K, [1, 30, 41, 47, 53, 133, 146, 200, 229, 257, 269] . The agreement is generally good.

The analysis of data was made with the help of the theoretical expression for viscosity:

$$\mu = 266.93 \frac{\sqrt{MT}}{\sigma^2 \Omega(T^*)} f_\mu$$

From the experimental data, the group $\sigma^2 \Omega(T^*)/f_\mu = \dfrac{\sqrt{MT}}{\mu} = y_{obs}$ was computed and plotted as a function of $1/T$. The curve was smoothed and a table was generated, from which recommended values were computed.

Most of the results lie within ± 2 percent of the calculated values.

RECOMMENDED VALUES

[Temperature, T, K; Viscosity, μ, 10^{-6} N s m^{-2}]

GAS

T	μ	T	μ	T	μ
190	6.13	450	13.57	750	20.4
		460	13.82	760	20.6
200	6.43	470	14.08	770	20.8
210	6.74	480	14.33	780	21.0
220	7.04	490	14.58	790	21.2
230	7.35	500	14.82	800	21.4
240	7.65	510	15.1	810	21.6
		520	15.3	820	21.8
250	7.96	530	15.6	830	22.0
260	8.26	540	15.8	840	22.2
270	8.56	550	16.0	850	22.4
280	8.86	560	16.3	860	22.6
290	9.15	570	16.5	870	22.8
		580	16.7	880	23.0
300	9.45	590	17.0	890	23.1
310	9.74	600	17.2	900	23.3
320	10.03	610	17.4	910	23.5
330	10.32	620	17.6	920	23.7
340	10.60	630	17.9	930	23.9
350	10.88	640	18.1	940	24.1
360	11.16	650	18.3	950	24.2
370	11.44	660	18.5	960	24.4
380	11.71	670	18.7	970	24.6
390	11.98	680	18.9	980	24.8
400	12.25	690	19.2	990	25.0
410	12.52	700	19.4	1000	25.1
420	12.78	710	19.6		
430	13.05	720	19.8		
440	13.31	730	20.0		
		740	20.2		

171

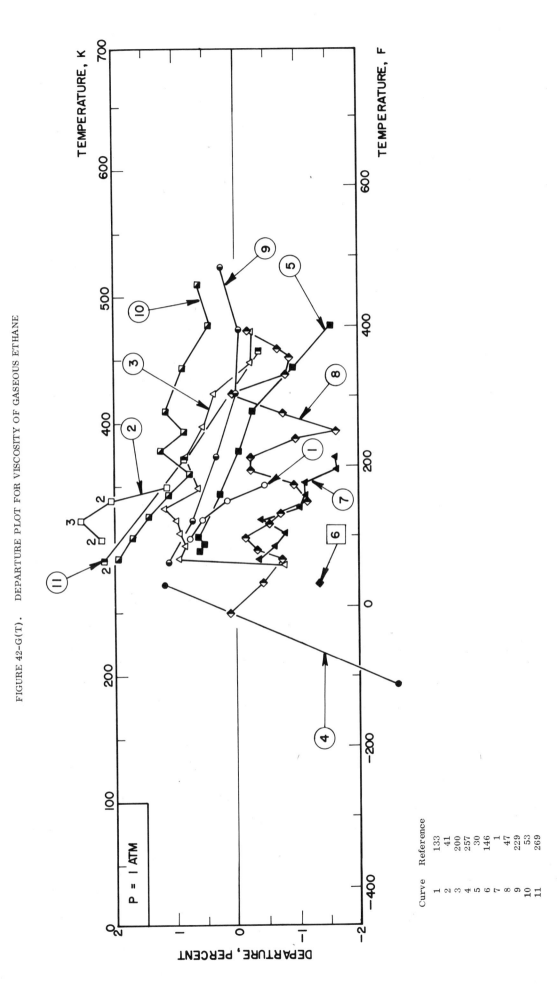

FIGURE 42-G(T). DEPARTURE PLOT FOR VISCOSITY OF GASEOUS ETHANE

Curve	Reference
1	133
2	41
3	200
4	257
5	30
6	146
7	1
8	47
9	229
10	53
11	269

P = 1 ATM

DISCUSSION

GAS

Six sets of experimental data were found in the literature [175, 215, 236, 257, 300, 301]. They are in the range from 373 K to about 600 K except for a single value of 273 K from Vogel [257]. Two points above 500 K from Khalilov [300], and three points from Uchiyama [236] show large divergence and were discarded.

The adjustment was made by using the relation $\mu = K \cdot \overline{T}/(\sigma^2 \Omega)$ to get $\sigma^2 \Omega$ which was plotted as a function of 1/T. A quadratic equation was fitted to these data, and was used to generate the recommended values. The accuracy is thought to be ±2 percent.

TABLE 43-G(T). VISCOSITY OF GASEOUS ETHYL ALCOHOL

RECOMMENDED VALUES

[Temperature, T, K; Viscosity, μ, 10^{-6} N sec m^{-2}]

GAS

T	μ	T	μ
270	8.14	450	13.10
280	8.43	460	13.37
290	8.71	470	13.63
		480	13.89
		490	14.15
300	9.00	500	14.41
310	9.28	510	14.67
320	9.56	520	14.93
330	9.84	530	15.18
340	10.11	540	15.43
350	10.39	550	15.69
360	10.67	560	15.93
370	10.95	570	16.18
380	11.22	580	16.44
390	11.49	590	16.68
400	11.77	600	16.93
410	12.04		
420	12.30		
430	12.57		
440	12.84		

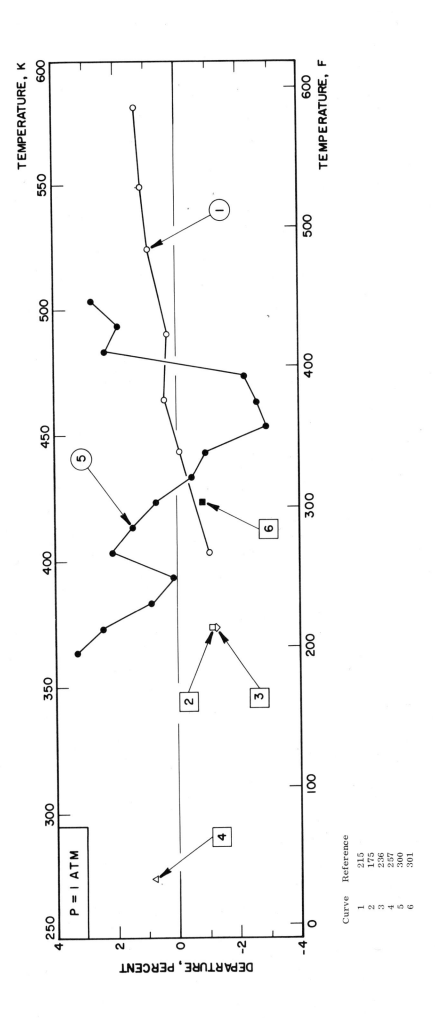

FIGURE 43-G(T). DEPARTURE PLOT FOR VISCOSITY OF GASEOUS ETHYL ALCOHOL

TABLE 44-L(T). VISCOSITY OF LIQUID ETHYLENE

RECOMMENDED VALUES

[Temperature, T, K; Viscosity, μ, 10^{-3} N s m^{-2}]

SATURATED LIQUID

T	μ
100	0.801
110	0.563
120	0.420
130	0.328
140	0.265
150	0.220
160	0.187
170	0.162
180	0.143
190	0.128
200	0.1153
210	0.1052
220	0.0967
230	0.0890
240	0.0805
250	0.0710
260	0.0600
270	0.0465
280	0.0295
283*	0.0219

* Crit. Temp.

DISCUSSION

SATURATED LIQUID

Three sets of experimental data were found in the literature. The data of Gerf [70] covers the lower temperature range below 169 K, while those of Rudenko [188] covers the temperature range between 169 K and the critical temperature. The data of Mason and Maass [153] are around the critical point.

The data of Gerf and Galkov, fitted to an equation

$$\log \mu = A + B/T,$$

shows a good consistency. The data of Rudenko shows big discrepancies in the vicinity of the critical point.

The recommended values of the viscosity below 220 K were joined graphically to smoothly intersect the value of the viscosity at the critical point calculated by the method of Jossi, Stiel and Thodos [100] which was found to be in good agreement with the experimental values of Mason et al. [153].

The accuracy is thought to be of about ±3% below 220 K and of about ±6% to the critical point.

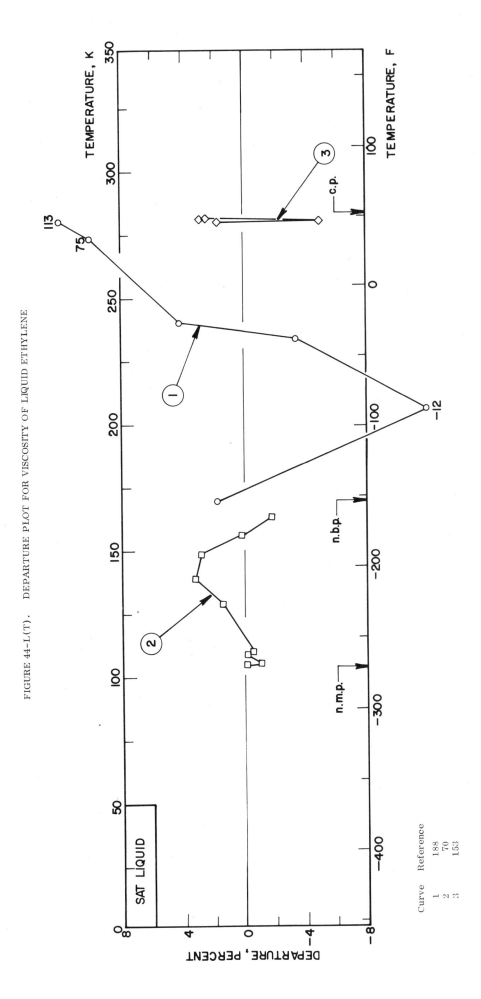

FIGURE 44-L(T).   DEPARTURE PLOT FOR VISCOSITY OF LIQUID ETHYLENE

TABLE 44-V(T).   VISCOSITY OF ETHYLENE VAPOR

RECOMMENDED VALUES

[Temperature, T, K; Viscosity, μ, 10^{-3} N s m^{-2}]

SATURATED VAPOR

T	μ
190	0.00685
200	0.00725
210	0.00766
220	0.00807
230	0.00857
240	0.00912
250	0.0099
260	0.0109
270	0.0126
280	0.0160
283*	0.0219

* Crit. Temp.

DISCUSSION

SATURATED VAPOR

Recommended values of the viscosity of the saturated vapor were computed by the correlation technique of Jossi, Stiel and Thodos [100] using the recommended values of viscosity of the dilute gas and a generalized correlation of the excess viscosity versus the reduced temperature from other gases using also the correlation technique of Jossi, Stiel and Thodos. No accuracy estimate is possible due to the complete absence of any experimental data.

TABLE 44-G(T). VISCOSITY OF GASEOUS ETHYLENE

GAS

DISCUSSION

There are twelve sets of experimental data available for the viscosity of ethylene [6, 23, 27-38, 133, 160, 206, 219, 222, 227, 237, 257, 270]. They cover a temperature range from 190 K to 1270 K. The agreement between the data of the various authors is generally good, except at low temperature where serious divergence exist.

To perform the analysis use was made of the theoretical relation

$$\mu = 266.93 \frac{\sqrt{MT}}{\sigma^2 \Omega(T^*)} f_\mu$$

from which the group $\sigma^2 \Omega(T^*)/f_\mu$ was computed. The values obtained were plotted as a function of $1/T$ and a smooth curve drawn using, as a guide, the similar curve obtained for methane, brought to the same scale with adequate reduction factors.

The recommended values are thought to be accurate within $\pm 2\%$ in the range from 250 to about 600 K.

RECOMMENDED VALUES

[Temperature, T, K; Viscosity, μ, 10^{-6} N s m^{-2}]

GAS

T	μ	T	μ	T	μ
190	6.74	550	17.6	900	25.5
200	7.08	560	17.8	910	25.7
210	7.42	570	18.1	920	25.9
220	7.76	580	18.3	930	26.1
230	8.09	590	18.6	940	26.3
240	8.43	600	18.8	950	26.5
250	8.76	610	19.1	960	26.7
260	9.09	620	19.3	970	26.9
270	9.42	630	19.6	980	27.1
280	9.75	640	19.8	990	27.3
290	10.07	650	20.0	1000	27.5
300	10.39	660	20.3	1050	28.4
310	10.71	670	20.5	1100	29.3
320	11.03	680	20.7	1150	30.2
330	11.34	690	21.0	1200	31.1
340	11.65	700	21.2	1250	31.9
350	11.96	710	21.4	1300	32.8
360	12.27	720	21.7	1350	33.6
370	12.57	730	21.9	1400	34.4
380	12.87	740	22.1	1450	35.2
390	13.16	750	22.3	1500	36.0
400	13.46	760	22.6		
410	13.75	770	22.8		
420	14.04	780	23.0		
430	14.32	790	23.2		
440	14.61	800	23.4		
450	14.89	810	23.6		
460	15.17	820	23.8		
470	15.45	830	24.1		
480	15.72	840	24.3		
490	15.99	850	24.5		
500	16.26	860	24.7		
510	16.5	870	24.9		
520	16.8	880	25.1		
530	17.1	890	25.3		
540	17.3				

178

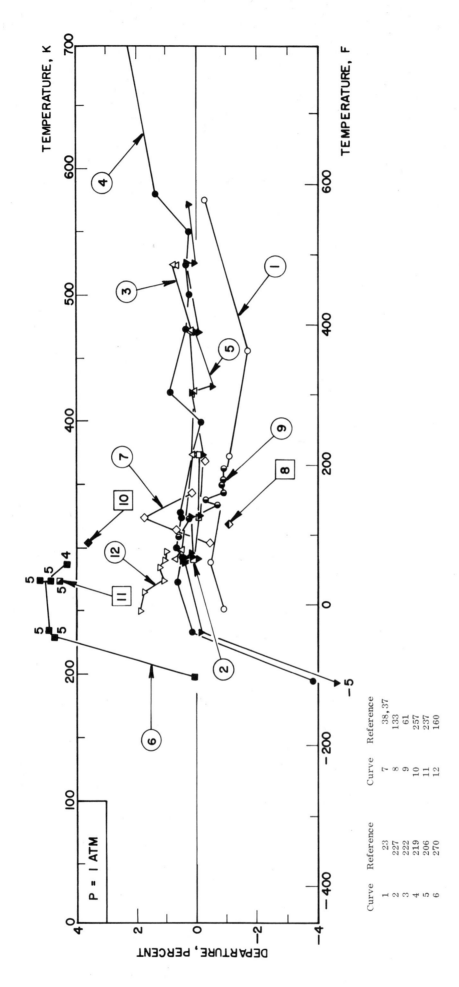

FIGURE 44-G(T).   DEPARTURE PLOT FOR VISCOSITY OF GASEOUS ETHYLENE

FIGURE 44-G(T). DEPARTURE PLOT FOR VISCOSITY OF GASEOUS ETHYLENE (continued)

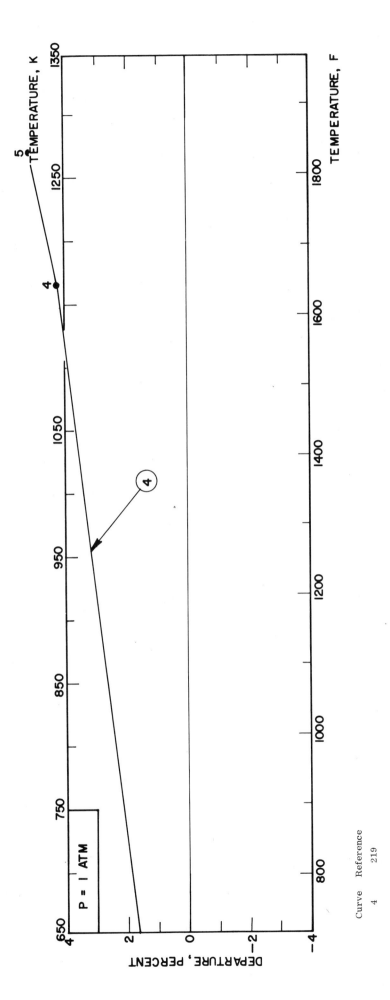

Curve   Reference
  4   219

TABLE 45-G(T). VISCOSITY OF GASEOUS ETHYL ETHER

DISCUSSION

GAS

Seven sets of experimental data were found in the literature [41, 175, 215, 236, 257, 302, 303], covering a temperature range going from 273 K to 582 K. The agreement is generally good except for single values from Uchiyama [236] and Pedersen [303].

Use was made of the theoretical relation $\mu = K_{\mu} \overline{T}/(\sigma^2 \Omega)$ to obtain the values of $\sigma^2 \Omega$ from the experimental data. These were plotted as a function of 1/T and adjusted to a quadratic equation. From this equation, recommended values of the viscosity were computed. The accuracy is of the order of ±2 percent.

RECOMMENDED VALUES

[Temperature, T, K; Viscosity, μ, 10^{-6} N sec m^{-2}]

GAS

T	μ
250	6.28
260	6.55
270	6.82
280	7.08
290	7.35
300	7.61
310	7.87
320	8.13
330	8.39
340	8.65
350	8.90
360	9.15
370	9.40
380	9.65
390	9.90
400	10.14
410	10.39
420	10.63
430	10.87
440	11.11
450	11.34
460	11.58
470	11.81
480	12.04
490	12.27
500	12.50
510	12.7
520	13.0
530	13.2
540	13.4
550	13.6
560	13.8
570	14.1
580	14.3
590	14.5
600	14.7
610	14.9
620	15.1
630	15.3
640	15.5
650	15.7

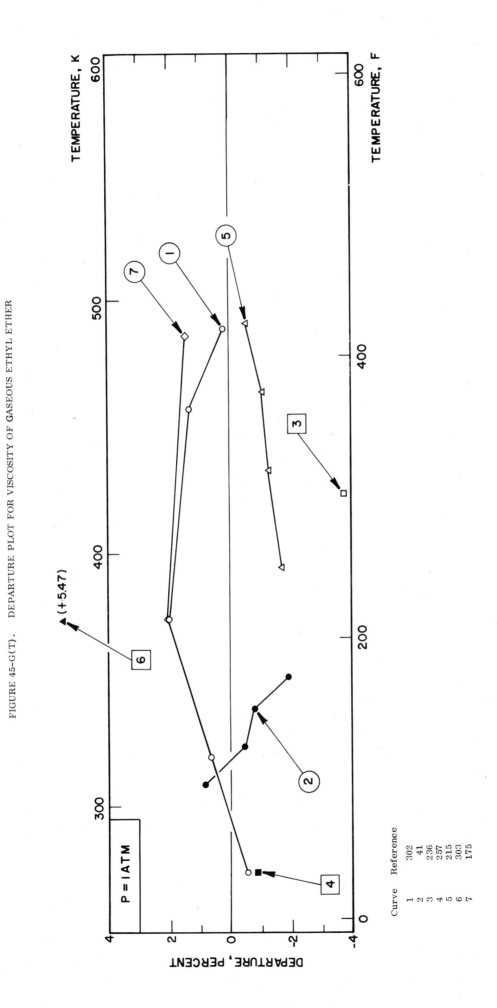

FIGURE 45-G(T). DEPARTURE PLOT FOR VISCOSITY OF GASEOUS ETHYL ETHER

TABLE 46-G(T). VISCOSITY OF GASEOUS n-HEPTANE

RECOMMENDED VALUES

[Temperature, T, K; Viscosity, μ, N.m^{-1}.sec^{-1}.10^{-6}]

GAS	
T	μ
270	5.54
280	5.72
290	5.90
300	6.08
310	6.26
320	6.44
330	6.62
340	6.81
350	6.99
360	7.18
370	7.37
380	7.55
390	7.74
400	7.93
410	8.12
420	8.31
430	8.50
440	8.68
450	8.87
460	9.06
470	9.25
480	9.44
490	9.62
500	9.81
510	10.00
520	10.19
530	10.37
540	10.56
550	10.74
560	10.92
570	11.11
580	11.29

DISCUSSION

GAS

There are six sets of experimental data available in the literature for the viscosity of n-heptane [133,236,300,305,306,307]. They cover a temperature range from normal temperature to 548 K.

The data were analyzed by using the theoretical relation

$$\mu = 266.93 \frac{MT}{\sigma^2 \Omega(T^*)} f_\mu$$

from which the group $[\sigma^2 \Omega(T^*)/f_\mu = y_{obs}$ was computed. The values obtained were plotted as a function of 1/T and a smooth curve drawn, using as a guide the similar curve obtained for methane brought to the same scale by appropriate reduction factors.

Considerable discrepancies exists between various investigators, and more weight was given to the results of Agaev [306] and Carmichael [307]. The data of Khalilov are widely scattered and seem to be accurate to three percent between 300 and 580 K. The recommended values should be accurate to three percent between 300 and 580 K and to within five percent for all other tabulated temperatures.

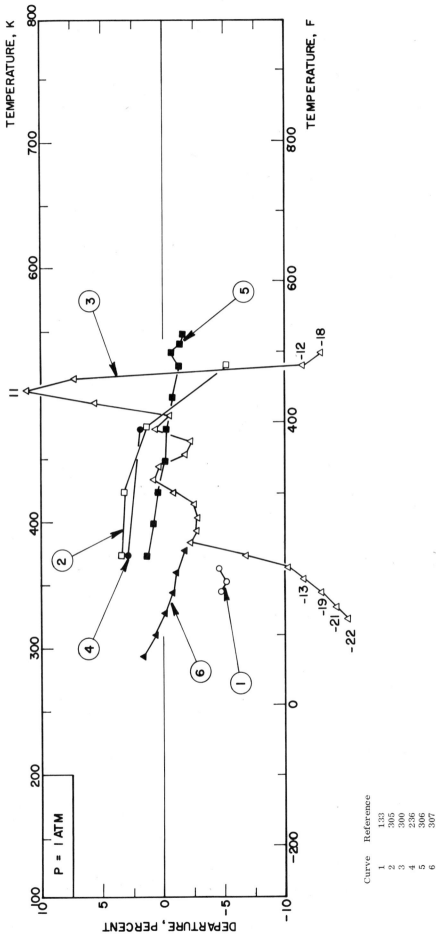

FIGURE 46-G(T).   DEPARTURE PLOT FOR VISCOSITY OF GASEOUS n–HEPTANE

Curve	Reference
1	133
2	305
3	300
4	236
5	306
6	307

DISCUSSION

GAS

Seven sets of experimental data on the viscosity of n-hexane were found in the literature [41, 133, 215, 236, 300, 306, 308]. They cover a temperature range, going from 273 to 873 K.

The analysis of the data was performed using the theoretical relation: $\eta = 266.93 \ \frac{\sqrt{MT} \ f_\mu}{\sigma^2 \Omega(T^*)}$ from which the group $[\sigma^2 \Omega(T^*)]/f_\mu = y_{obs}$ was computed. The values obtained were plotted as a function of $1/T$, and a smooth curve drawn, using as a guide a similar curve obtained for methane, brought to the same scale by appropriate reduction factors.

There are considerable discrepancies between various investigators, the data of Khalilov [300] seeming particularly incorrect.

TABLE 47-G(T).   VISCOSITY OF GASEOUS n-HEXANE

RECOMMENDED VALUES

[Temperature, T, K; Viscosity, μ, 10^{-6} N s m^{-2}]

GAS

T	μ		T	μ
270	6.07		600	12.80
280	6.27		610	13.00
290	6.46		620	13.20
			630	13.38
			640	13.57
300	6.66		650	13.77
310	6.87		660	13.96
320	7.07		670	14.15
330	7.27		680	14.34
340	7.48		690	14.53
350	7.69		700	14.71
360	7.90		710	14.90
370	8.11		720	15.09
380	8.31		730	15.27
390	8.52		740	15.46
400	8.73		750	15.64
410	8.94		760	15.82
420	9.15		770	16.00
430	9.36		780	16.18
440	9.56		790	16.36
450	9.77		800	16.54
460	9.98		810	16.72
470	10.18		820	16.89
480	10.39		830	17.07
490	10.60		840	17.25
500	10.80		850	17.42
510	11.00		860	17.60
520	11.21		870	17.77
530	11.41		880	17.94
540	11.61		890	18.11
550	11.81		900	18.29
560	12.01			
570	12.21			
580	12.41			
590	12.60			

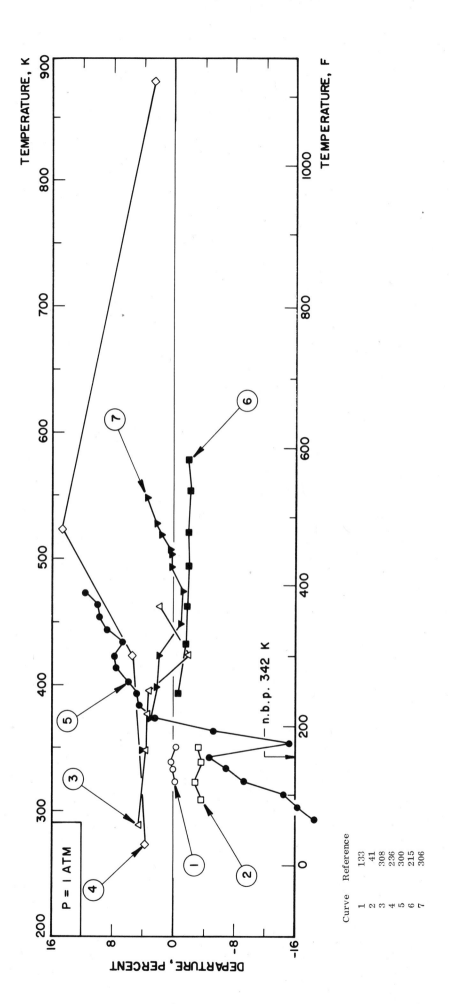

FIGURE 47–G(T).   DEPARTURE PLOT FOR VISCOSITY OF GASEOUS n–HEXANE

TABLE 48-L(T). VISCOSITY OF LIQUID METHANE

DISCUSSION

SATURATED LIQUID

There are six sets of experimental data in the literature, covering an overall range of temperature from 88 K to about the critical temperature. They were adjusted by least squares to the equation

$$\log \mu = A + B/T$$

in the range of temperature from 88 K to 150 K and smoothed graphically from 150 to the critical temperature.

The experimental data of Boon and Thomaes [20] in the range 90 K to 114 K appear the most reliable and are in good agreement with those of Di Geronimo [46]. The experimental data of Rudenko[188] and Swift et al. [209] show big discrepancies. Computed values by the method of Jossi, Stiel and Thodos [100] for the saturated liquid near the critical point are in fair agreement with the recommended values.

The reliability is thought to be of the order of ±10% for temperatures above 150 K, of ±5% down to 120 K and ±2% in the lower temperature range.

RECOMMENDED VALUES

[Temperature, T, K, Viscosity, μ, 10^{-3} N s m^{-2}]

SATURATED LIQUID

T	μ
90	0.2048
100	0.1525
110	0.1226
120	0.0980
130	0.0826
140	0.0714
150	0.0629
160	0.0545
170	0.0449
180	0.0349
190	0.0214
190*	0.0165

* Crit. Temp.

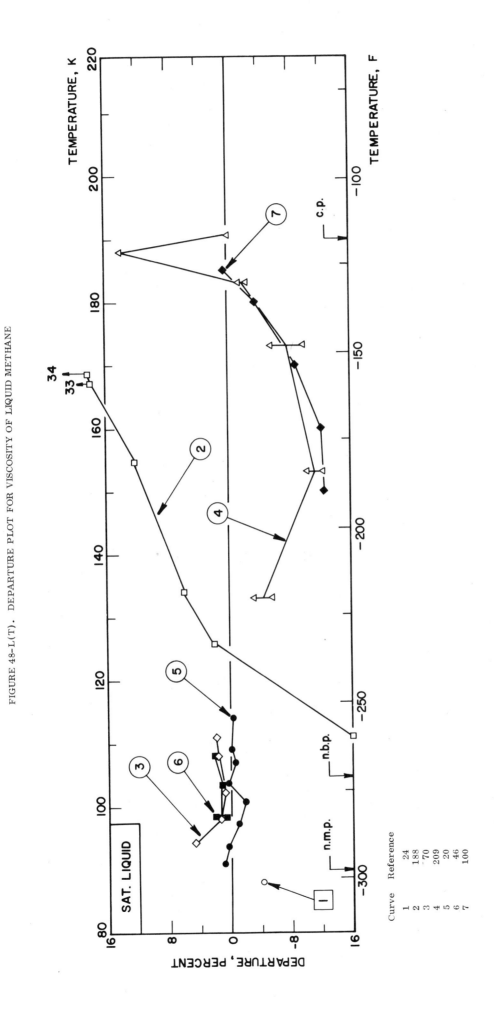

FIGURE 48-L(T). DEPARTURE PLOT FOR VISCOSITY OF LIQUID METHANE

TABLE 48-V(T).   VISCOSITY OF METHANE VAPOR

DISCUSSION

SATURATED VAPOR

The recommended values for the viscosity of methane in the saturated vapor state were computed by the correlation technique of Jossi, Stiel and Thodos [100] using the recommended value for the 1 atm gas and the density values given by Din [49].

They are reliable to about ±5%.

RECOMMENDED VALUES

[Temperature, T, K. Viscosity, μ, 10^{-3} N s m^{-2}]

SATURATED VAPOR

T	μ
90	.00372
100	.00407
110	.00442
120	.00478
130	.00511
140	.00553
150	.00606
160	.00678
170	.00771
180	.00906
185	.01008
186	.01038
188	.01124
190	.01312
190*	.01645

* Crit. Temp.

TABLE 48-G(T).   VISCOSITY OF GASEOUS METHANE

RECOMMENDED VALUES

[Temperature, T, K; Viscosity, μ, 10^{-6} N s m^{-2}]

GAS

T	μ	T	μ	T	μ
70	3.00	400	14.24	750	22.8
80	3.36	410	14.53	760	23.0
90	3.72	420	14.81	770	23.2
100	4.06	430	15.09	780	23.4
110	4.42	440	15.37	790	23.6
120	4.78	450	15.64	800	23.8
130	5.11	460	15.91	810	24.0
140	5.53	470	16.18	820	24.2
150	5.90	480	16.44	830	24.4
160	6.28	490	16.70	840	24.6
170	6.65	500	16.96	850	24.8
180	7.02	510	17.2	860	25.0
190	7.39	520	17.5	870	25.2
200	7.76	530	17.7	880	25.4
210	8.12	540	18.0	890	25.6
220	8.48	550	18.2	900	25.7
230	8.84	560	18.5	910	25.9
240	9.19	570	18.7	920	26.1
250	9.53	580	19.0	930	26.3
260	9.87	590	19.2	940	26.5
270	10.21	600	19.4	950	26.7
280	10.54	610	19.7	960	26.9
290	10.87	620	19.9	970	27.0
300	11.20	630	20.1	980	27.2
310	11.52	640	20.4	990	27.4
320	11.84	650	20.6	1000	27.6
330	12.15	660	20.8		
340	12.46	670	21.0		
350	12.77	680	21.3		
360	13.07	690	21.5		
370	13.37	700	21.7		
380	13.66	710	21.9		
390	13.95	720	22.1		
		730	22.3		
		740	22.5		

GAS

DISCUSSION

There are 27 sets of experimental data available for the viscosity of methane, covering an overall temperature range from 78 to 772 K contained in 28 references [1, 5, 6, 16, 33, 34, 35, 37, 47, 72, 88, 89, 94, 93, 110, 132, 133, 139, 146, 174, 190, 196, 229, 233, 236, 242, 257, 264]. Agreement between investigators is generally good at moderate and high temperatures, but, Uchiyama [236] results at 873 and 1073 K seems to be in serious error (above 30%). At low temperature, more weight was given to Johnston [98] data.

To analyze the data, use was made of the theoretical expression for viscosity:

$$\mu = 266.93 \frac{\sqrt{MT}}{\sigma^2 \Omega(T^*)} f_\mu$$

The group $[\sigma^2 \Omega(T^*)/f_\mu] = y_{obs}$, was computed from the experimental data and plotted as a function of 1/T. The curve obtained has been compared with the similar curve obtained for argon, using reduction factors which were the ratio of Boyle temperature and the ratio of collision diameter evaluated from the data at the Boyle temperature.

Good agreement exists with the correlation given by Keyes [121] in the region where experimental data are available, but a divergence exists at high temperature, and at 73 K. The recommended values should be accurate to within two percent between 100 and 750 K and within five percent for all other temperatures tabulated.

The accuracy is about ±2% throughout.

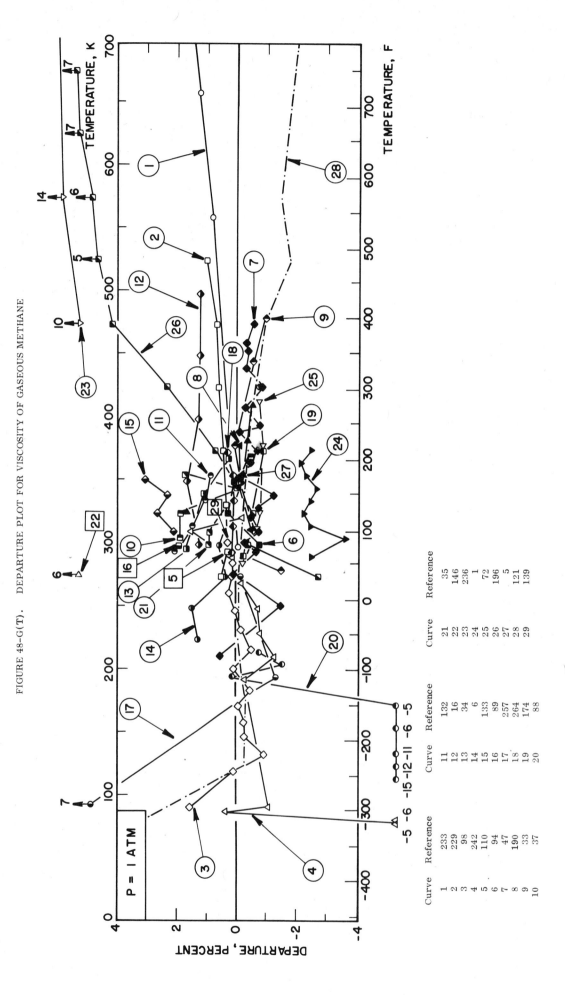

FIGURE 48-G(T). DEPARTURE PLOT FOR VISCOSITY OF GASEOUS METHANE

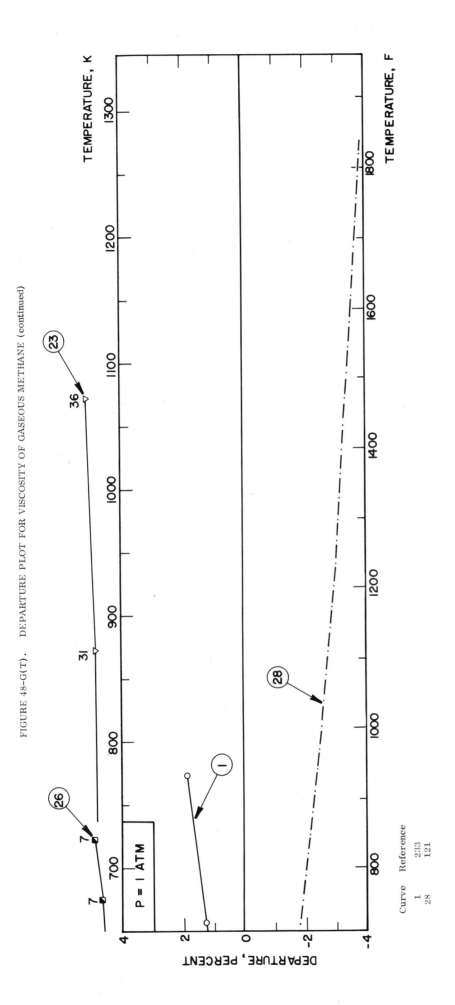

FIGURE 48-G(T).   DEPARTURE PLOT FOR VISCOSITY OF GASEOUS METHANE (continued)

TABLE 49-G(T). VISCOSITY OF GASEOUS METHYL ALCOHOL

DISCUSSION

GAS

Five sets of experimental data were found in the literature. These are the results of Craven [41], Titani [215], Reid [301], Uchiyama [236] and Khalilov [300]. The last two sources listed do not seem very reliable.

The experimental data of the first three authors were used to obtain $\sigma^2\Omega$ in the relation $\mu = K\sqrt{T}/(\sigma^2\Omega)$. The values of $\sigma^2\Omega$ were then plotted on a graph as a function of $1/T$. It was found that these values could be represented by a quadratic equation, which was used to generate the recommended values of the viscosity. The accuracy is of the order of ± 1 percent.

RECOMMENDED VALUES

[Temperature, T, K; Viscosity, μ, 10^{-6} N s m^{-2}]

GAS

T	μ	T	μ
250	8.29	450	14.81
260	8.60	460	15.14
270	8.91	470	15.47
280	9.23	480	15.81
290	9.55	490	16.14
300	9.87	500	16.47
310	10.19	510	16.8
320	10.51	520	17.1
330	10.84	530	17.5
340	11.17	540	17.8
350	11.49	550	18.1
360	11.82	560	18.5
370	12.15	570	18.8
380	12.48	580	19.1
390	12.81	590	19.4
400	13.15	600	19.8
410	13.48	610	20.1
420	13.80	620	20.4
430	14.14	630	20.8
440	14.47	640	21.1
		650	21.3

193

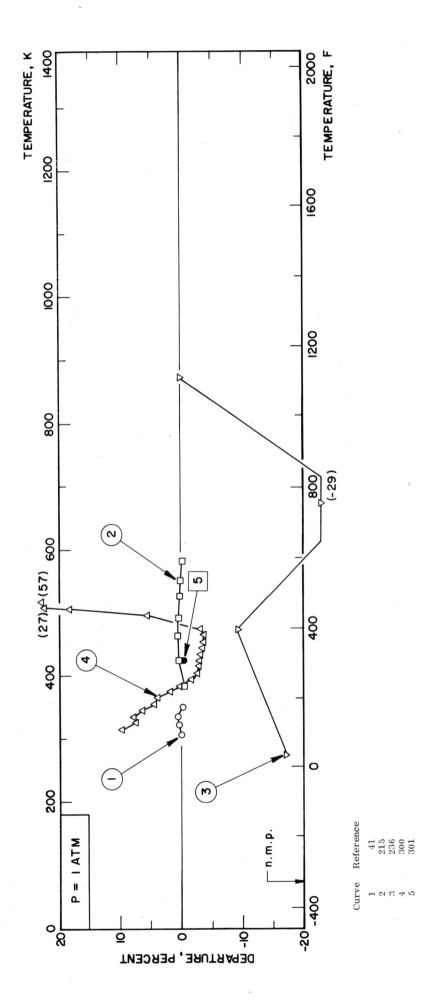

FIGURE 49-G(T).   DEPARTURE PLOT FOR VISCOSITY OF GASEOUS METHYL ALCOHOL

TABLE 50-L(T). VISCOSITY OF LIQUID METHYL CHLORIDE

DISCUSSION

SATURATED LIQUID

Two sets of experimental data were found in the literature [14, 202] and one set of data was also found in a manufacturer's technical note [272].

The experimental data were least square fitted to an equation

$$\log \mu = A + B/T$$

from which recommended values were generated.

The experimental data of Stakelbeck [202] are higher than the recommended values, while the data of Benning [14] are lower. The manufacturer's data seems an extrapolation above 320 K of the latter. A curve was drawn to join the recommended value at 320 K to the estimated value at the critical temperature.

Below 320 K the accuracy is thought to be about ±7%, but may be about 15% above 320 K.

RECOMMENDED VALUES

[Temperature, T, K; Viscosity, μ, N s m⁻² · 10⁻³]

SATURATED LIQUID

T	μ
230	0.357
240	0.335
250	0.316
260	0.300
270	0.285
280	0.272
290	0.261
300	0.251
310	0.242
320	0.233
330	0.223
340	0.214
350	0.205
360	0.195
370	0.184
380	0.172
390	0.158
400	0.139
410	0.107
416*	0.028

* Crit. Temp.

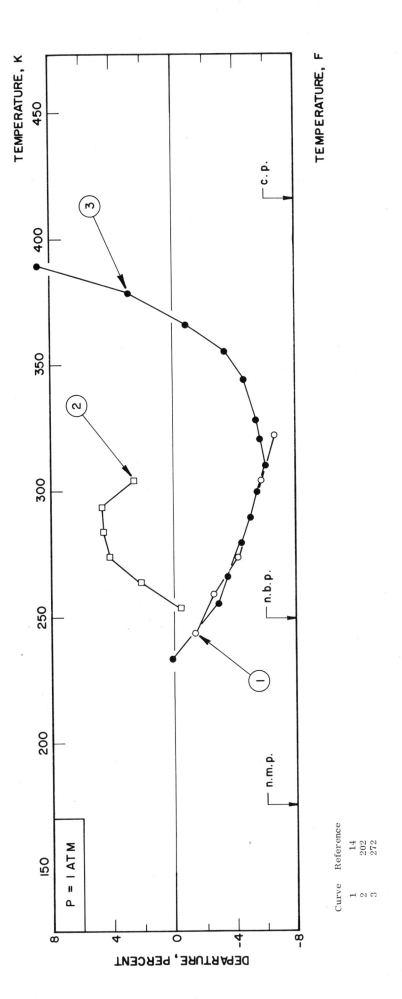

FIGURE 50-L(T). DEPARTURE PLOT FOR VISCOSITY OF LIQUID METHYL CHLORIDE

TABLE 50-V(T).   VISCOSITY OF METHYL CHLORIDE VAPOR

DISCUSSION

SATURATED VAPOR

Recommended values for the viscosity of the saturated vapor was done by means of the method of Stiel and Thodos [100] which make use of the excess viscosity concept.

Reduced excess viscosity as a function of reduced temperature from a curve built from data on other refrigerants were used to generate the excess viscosity which was added to the recommended values for the 1 atm gas to generate the present recommended values.

The accuracy is about ±3 percent.

RECOMMENDED VALUES

[Temperature, T, K, Viscosity, μ, 10^{-3} N s m^{-2}]

SATURATED VAPOR

T	μ
240	0.00895
250	0.00929
260	0.00963
270	0.00998
280	0.01035
290	0.01071
300	0.01113
310	0.01153
320	0.01195
330	0.01240
340	0.01291
350	0.01341
360	0.01397
370	0.01462
380	0.01538
390	0.01625
400	0.0175
410	0.0196
416*	0.0278

* Crit. Temp.

TABLE 50-G(T). VISCOSITY OF GASEOUS METHYL CHLORIDE

DISCUSSION

GAS

Nine sets of experimental data were found in the literature [14, 22, 23, 42, 202, 236, 257, 272, 273]. There is a good consistency among the different sets, except the value of Uchiyama[236]. The range of temperature covered goes from 250 to 600 K. Computed values of

$$\sigma^2 \Omega = \frac{K \sqrt{T}}{\mu}$$

were least square fitted to a quadratic equation in 1/T, from which recommended values were generated.

The accuracy of the correlation is of about ± 2%.

RECOMMENDED VALUES

[Temperature, T, K, Viscosity, μ, 10^{-6} N s m^{-2}]

GAS

T	μ
250	9.29
260	9.63
270	9.96
280	10.31
290	10.65
300	11.00
310	11.34
320	11.69
330	12.04
340	12.38
350	12.73
360	13.08
370	13.43
380	13.78
390	14.12
400	14.47
410	14.82
420	15.16
430	15.51
440	15.86
450	16.20
460	16.55
470	16.89
480	17.23
490	17.57
500	17.92
510	18.3
520	18.6
530	18.9
540	19.3
550	19.6
560	19.9
570	20.3
580	20.6
590	21.0
600	21.3
610	21.6
620	21.9
630	22.3
640	22.6
650	22.9
660	23.3

198

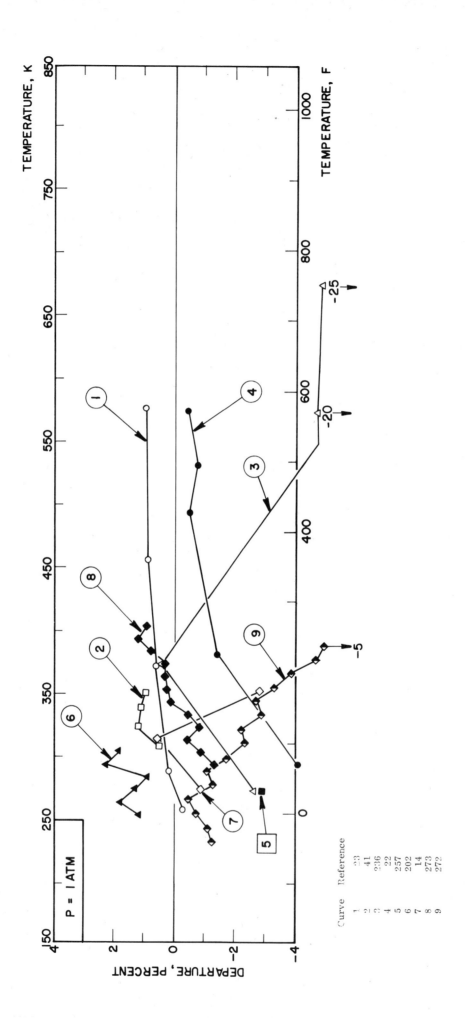

FIGURE 50-G(T). DEPARTURE PLOT FOR VISCOSITY OF GASEOUS METHYL CHLORIDE

TABLE 51-L(T). VISCOSITY OF LIQUID OCTAFLUOROCYCLOBUTANE

DISCUSSION

SATURATED LIQUID

Two sets of experimental data were found in the literature, those of Gordon [79] and Lilios [140]. They cover a range of temperature from 243 K to 300 K. The agreement is generally good. The data were least square fitted to an equation

$$\log \mu = A + B/T$$

from which recommended values were generated in the range 240 to 310 K. Above this temperature, a curve was drawn to join smoothly the estimated viscosity value at the critical temperature.

The accuracy is thought to be ±5%, although the figure may be higher when approaching the critical point.

RECOMMENDED VALUES

[Temperature, T. K: Viscosity, u, N s m^{-2} · 10^{-3}]

SATURATED LIQUID

T	u
240	0.955
245	0.867
250	0.790
255	0.722
260	0.663
265	0.610
270	0.564
275	0.522
280	0.485
285	0.452
290	0.421
295	0.394
300	0.370
305	0.348
310	0.327
315	0.308
320	0.289
325	0.272
330	0.257
335	0.242
340	0.228
345	0.215
350	0.203
355	0.190
360	0.175
365	0.159
370	0.142
375	0.124
380	0.103
385	0.081
388*	0.030

* Crit. Temp.

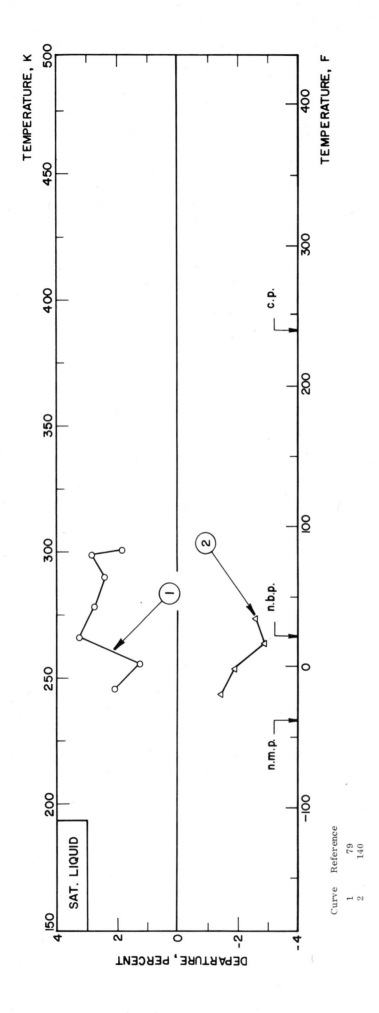

FIGURE 51-L(T).   DEPARTURE PLOT FOR VISCOSITY OF LIQUID OCTAFLUOROCYCLOBUTANE

TABLE 51-V(T). VISCOSITY OF OCTAFLUOROCYCLOBUTANE VAPOR

DISCUSSION

SATURATED VAPOR

Recommended values for the viscosity of the saturated vapor were estimated using the correlation equation of Stiel and Thodos [100], with the density values taken from a manufacturer's technical note [280], and with our recommended values for the 1 atm gas.

The accuracy is thought to be ±5% although this figure may rise to ±10% around the critical temperature.

RECOMMENDED VALUES

[Temperature, T, K, Viscosity, μ, 10^{-3} N s m^{-2}]

SATURATED VAPOR

T	μ
260	0.01046
270	0.01083
280	0.01121
290	0.01164
300	0.01209
310	0.01257
320	0.01310
330	0.01370
340	0.01435
350	0.0151
360	0.0160
370	0.0172
380	0.0195
388*	0.0302

* Crit. Temp.

TABLE 51-G(T). VISCOSITY OF GASEOUS OCTAFLUOROCYCLOBUTANE

DISCUSSION

GAS

Three sets of experimental data were found in the literature [235, 262, 265]. They cover a range from 290 K to 423 K.

The correlation was made on computed values of

$$\sigma^2 \Omega = K \frac{\sqrt{T}}{\mu}$$

which were least square fitted to a linear equation in 1/T. From this equation, recommended values were computed.

The accuracy is thought to be ±2%.

RECOMMENDED VALUES

[Temperature, T, K; Viscosity, μ, N s m^{-2} · 10^{-6}]

GAS

T	μ
270	10.83
280	11.20
290	11.56
300	11.93
310	12.28
320	12.64
330	12.99
340	13.33
350	13.67
360	14.00
370	14.34
380	14.67
390	14.99
400	15.32
410	15.64
420	15.95
430	16.27
440	16.50

FIGURE 51-G(T). DEPARTURE PLOT FOR VISCOSITY OF GASEOUS OCTAFLUOROCYCLOBUTANE

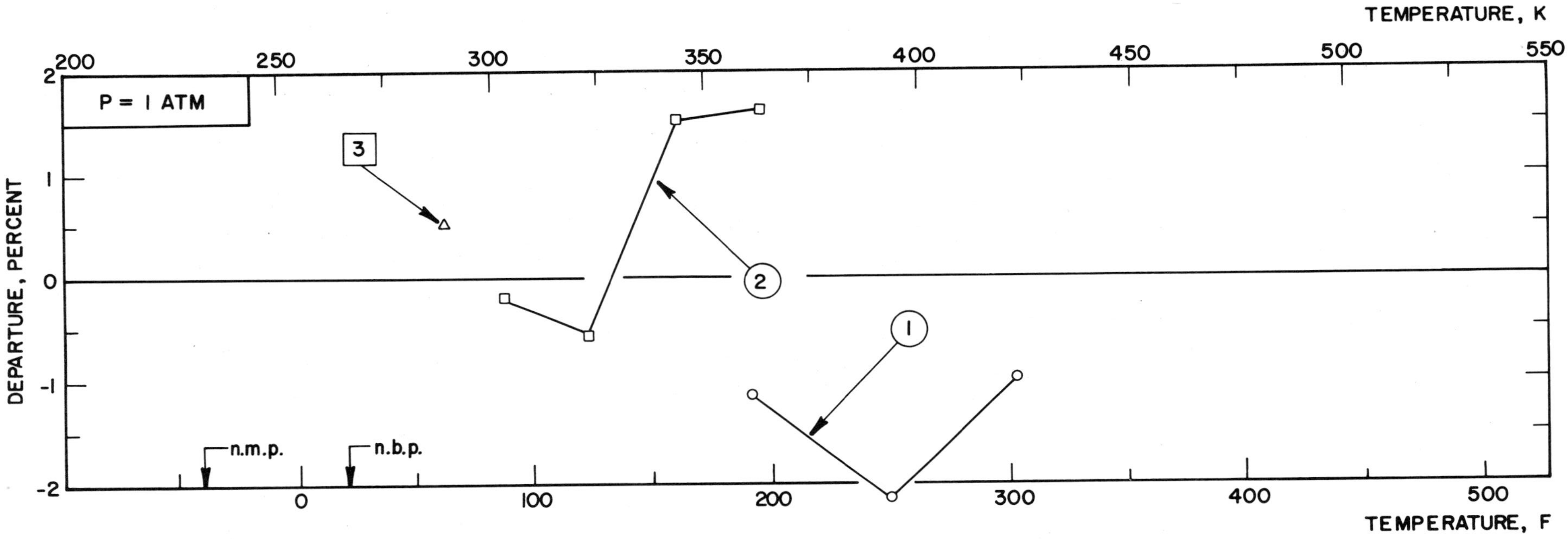

Curve	Reference
1	235
2	265
3	262

TABLE 52-G(T).   VISCOSITY OF GASEOUS n-OCTANE

DISCUSSION

GAS

Five sets of experimental data were found in the literature, those of Lambert [133], McCoubrey [308], Uchiyama [236], Agaev [306], and Carmichael [307]. The data of Lambert are systematically higher and only one point of Uchiyama fits, with the results of the three other authors. The points from Agaev and those from McCoubrey overlaps in a small region, but the curve drawn through the data of these two authors do not indicate the same trend as the results of Carmichael at lower temperature.

The sets of Agaev, McCoubrey and Carmichael were used to generate recommended values by fitting them to the theoretical relation $\mu = K\sqrt{T}/(\sigma^2 \Omega_{22})$, values of $\sigma^2 \Omega$ being computed from the data and adjusted to a quadratic equation in $1/_5$, which was then used to generate the recommended values. The accuracy is poor and could well be ±5 percent.

RECOMMENDED VALUES

[Temperature, T, K; Viscosity, μ, 10^{-6} N s m^{-2}]

GAS

T	μ	T	μ
300	5.64	500	9.2
310	5.82	510	9.4
320	5.99	520	9.5
330	6.17	530	9.7
340	6.34	540	9.9
350	6.52	550	10.1
360	6.69	560	10.2
370	6.87	570	10.4
380	7.05	580	10.6
390	7.22	590	10.7
400	7.40	600	10.9
410	7.58	610	11.1
420	7.76	620	11.3
430	7.93	630	11.4
440	8.11	640	11.6
450	8.29	650	11.8
460	8.47		
470	8.64		
480	8.82		
490	9.00		

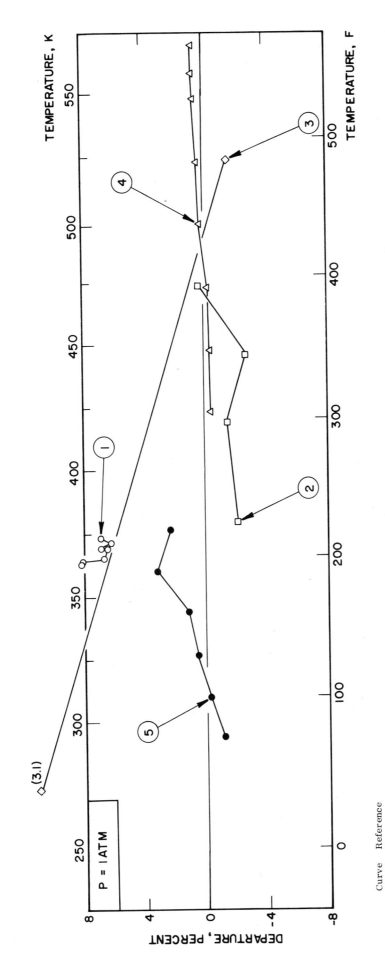

FIGURE 52-G(T).   DEPARTURE PLOT FOR VISCOSITY OF GASEOUS n-OCTANE

Curve	Reference
1	133
2	308
3	236
4	306
5	307

TABLE 53-G(T). VISCOSITY OF GASEOUS n-PENTANE

RECOMMENDED VALUES

[Temperature, T, K; Viscosity, μ, N.m^{-1}.sec^{-1}.10^{-6}]

GAS

T	μ
270	6.38
280	6.59
290	6.81
300	7.02
310	7.24
320	7.46
330	7.68
340	7.90
350	8.12
360	8.35
370	8.56
380	8.78
390	9.01
400	9.23
410	9.45
420	9.67
430	9.88
440	10.10
450	10.32
460	10.54
470	10.76
480	10.97
490	11.19
500	11.40
510	11.61
520	11.82
530	12.04
540	12.25
550	12.46

DISCUSSION

GAS

There are 8 sets of experimental data available for the viscosity of n-pentane covering a range of temperature from 273 to 579 K [133, 215, 236, 294, 300, 319, 320, 374].

The analysis was performed using the theoretical relation:

$$\mu = 266.93 \frac{\sqrt{MT}}{\sigma^2 \, \Omega(T^*)} f_\mu$$

from which the group $[\sigma^2 \Omega(T^*)/f_\mu] = y_{obs}$ was computed. The values obtained were plotted as a function of 1/T, and the curve obtained was compared with a similar curve for methane, using appropriate reduction factors.

In the table generation, more weight was given to the high temperature data of Agaev [374], and those of Titani [215], and at lower temperature to those of McCoubrey [320]. Errors of up to five percent below 350 K and up to ten percent for higher temperatures appear reasonable estimates.

FIGURE 53-G(T). DEPARTURE PLOT FOR VISCOSITY OF GASEOUS n-PENTANE

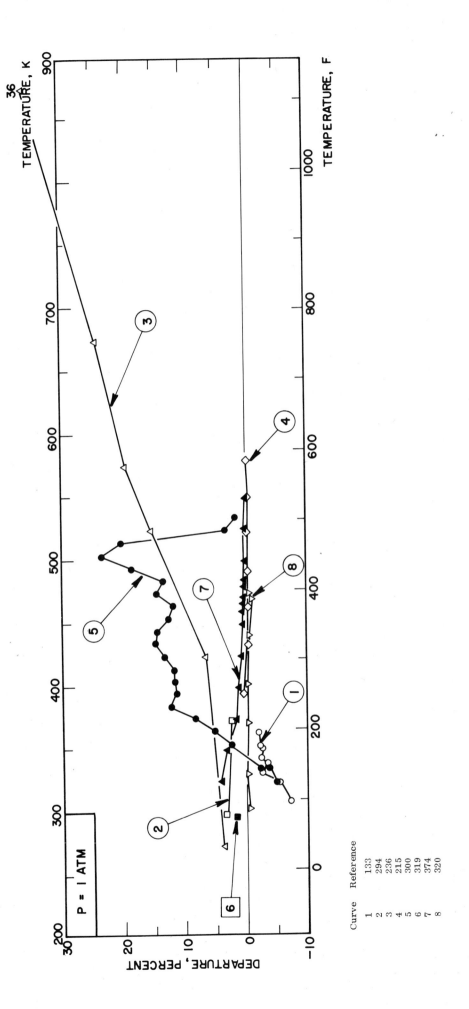

Curve	Reference
1	133
2	294
3	236
4	215
5	300
6	319
7	374
8	320

TABLE 54-L(T). VISCOSITY OF LIQUID PROPANE

DISCUSSION

SATURATED LIQUID

Six sets of experimental data were found in the literature. The data reported by Gnapp [75] covers the widest temperature range and were given more weight in the correlation. The data of Swift [209] are in the range from 240 K to about the critical temperature. The data of Lipkin [142] are intermediate while those of Gerf and Galkov [69, 70] and of Kruger [129] covers a lower temperature range.

The correlating technique was to use an equation:

$$\log \mu = A + B/T + \delta$$

to represent the experimental values, where the residual δ was smoothed graphically and used to generate a table from which recommended values were computed.

The accuracy should be about ±3% except near the critical temperature where larger discrepancies are expected. Values computed through the correlating technique of Jossi, Stiel and Thodos [100] are in fair agreement in the vicinity of the critical point, but diverges gradually with decreasing temperature.

RECOMMENDED VALUES

[Temperature, T, K; Viscosity, μ, 10^{-3} N s m^{-2}]

SATURATED LIQUID

T	μ
80	19.16
90	7.521
100	3.793
110	2.258
120	1.501
130	1.081
140	0.8234
150	0.6425
160	0.5376
170	0.4525
180	0.3887
190	0.3391
200	0.2996
210	0.2651
220	0.2360
230	0.2111
240	0.1895
250	0.1708
260	0.1549
270	0.1408
280	0.1282
290	0.1168
300	0.1055
310	0.0940
320	0.0829
330	0.0724
340	0.0617
350	0.0512
360	0.0410
369*	0.0214

* Crit. Temp.

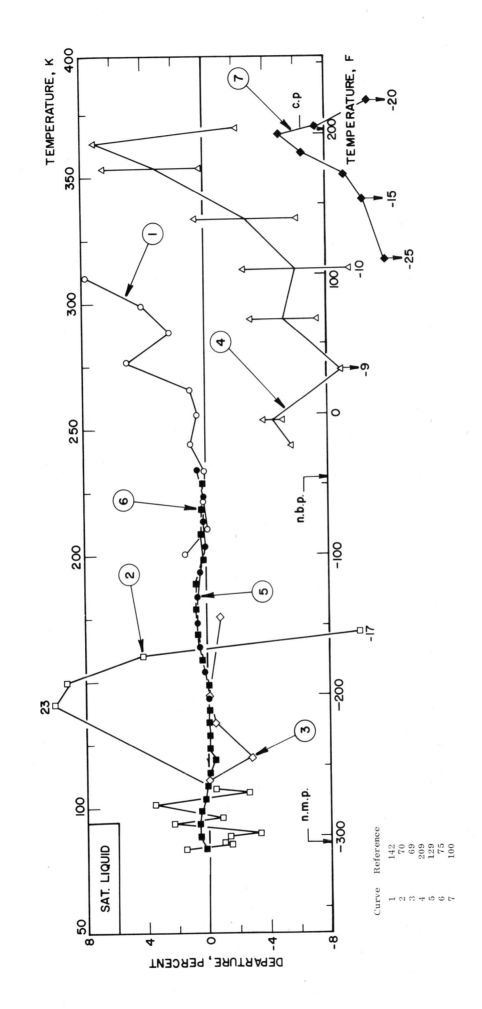

F[GURE 54-L(T). DEPARTURE PLOT FOR VISCOSITY OF LIQUID PROPANE

TABLE 54-V(T).   VISCOSITY OF PROPANE VAPOR

RECOMMENDED VALUES

[Temperature, T, K; Viscosity, μ, 10^{-3} N s m^{-2}]

SATURATED VAPOR

T	μ
280	0.00798
290	0.00838
300	0.00880
310	0.00925
320	0.00973
330	0.01025
340	0.01100
350	0.01208
360	0.01490
369*	0.02140

* Crit. Temp.

DISCUSSION

SATURATED VAPOR

Recommended values for the viscosity of the saturated vapor were computed through the correlating technique of Jossi, Stiel and Thodos [100] using the recommended values of the 1 atm gas and with the density values reported by Din [49].

The accuracy of the correlation is of about ±5% with larger deviation expected near the critical point.

TABLE 54-G(T). VISCOSITY OF GASEOUS PROPANE

DISCUSSION

GAS

There are 15 sets of experimental data available for the viscosity of propane [1, 5, 30, 38, 71, 72, 126, 133, 192, 200, 203-204, 225, 229, 236, 264]. However they cover a narrow range of temperature (from 273 to 548 K) with the exception of some data of Uchiyama [236] going up to 1073 K. These may well be in serious error (about 40%) and were not considered in the analysis.

The analysis of data was made using the theoretical relation for viscosity:

$$\mu = 266.93 \frac{\sqrt{M\,T}}{\sigma^2 \Omega(T^*)} f\mu$$

to compute the group $\sigma^2 \Omega(T^*)/f\mu$ from the experimental data.

The values obtained were then plotted as a function of $1/T$. A graphical smoothing was operated, and from the table generated recommended values were computed.

More weight was given to some authors, mainly: Wobser [264] and Trautz [225, 229] whose work with other gases is well known.

The accuracy is within two percent.

RECOMMENDED VALUES

[Temperature, T, K; Viscosity, μ, 10^{-6} N s m^{-2}]

GAS	
T	μ
270	7.47
280	7.73
290	7.99
300	8.25
310	8.52
320	8.78
330	9.04
340	9.30
350	9.56
360	9.82
370	10.08
380	10.34
390	10.59
400	10.85
410	11.10
420	11.35
430	11.60
440	11.84
450	12.09
460	12.34
470	12.58
480	12.82
490	13.06
500	13.30
510	13.54
520	13.77
530	14.01
450	14.24
550	14.47
560	14.70
570	14.93
580	15.15
590	15.38
600	15.60

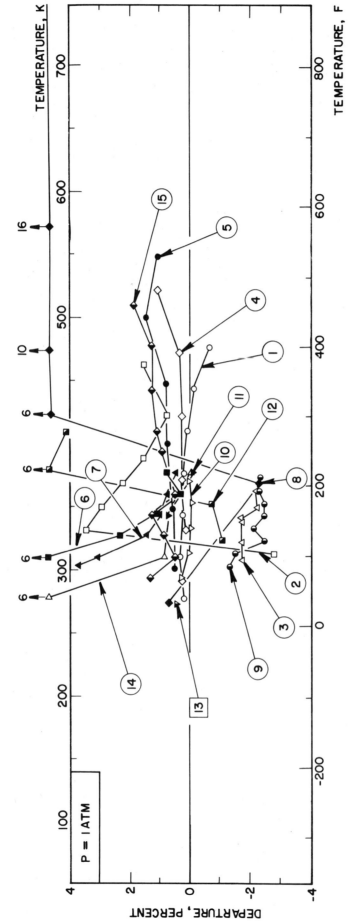

FIGURE 54-G(T). DEPARTURE PLOT FOR VISCOSITY OF GASEOUS PROPANE

Curve	Reference
1	30
2	200
3	133
4	229
5	225
6	192
7	38
8	236

Curve	Reference
9	1
10	264
11	71
12	5
13	126
14	72
15	203, 204

P = 1 ATM

TABLE 55-L(T).   VISCOSITY OF LIQUID PROPYLENE

DISCUSSION

SATURATED LIQUID

Four sets of experimental data were found in the literature [69, 70, 75, 160] covering a range of temperature from 88 K to 270 K. There is thus a lack of experimental data from the latter temperature to the critical. The agreement between the sets is generally good.

An equation of the type:

$$\log \mu = A + B/T + \delta$$

was used to generate the recommended values. The residual δ was smoothed graphically from 88 K to the critical temperature.

The accuracy is thought to be $\pm 3\%$.

RECOMMENDED VALUES

[Temperature, T, K; Viscosity, μ, 10^{-3} N s m^{-2}]

SATURATED LIQUID

T	μ
90	12.25
100	4.523
110	2.327
120	1.425
130	0.975
140	0.723
150	0.568
160	0.462
170	0.384
180	0.326
190	0.282
200	0.2466
210	0.2178
220	0.1932
230	0.1724
240	0.1548
250	0.1399
260	0.1260
270	0.1130
280	0.1009
290	0.0896
300	0.0789
310	0.0686
320	0.0582
330	0.0485
340	0.0404
350	0.0330
360	0.0271
365*	0.0239

* Crit. Temp.

214

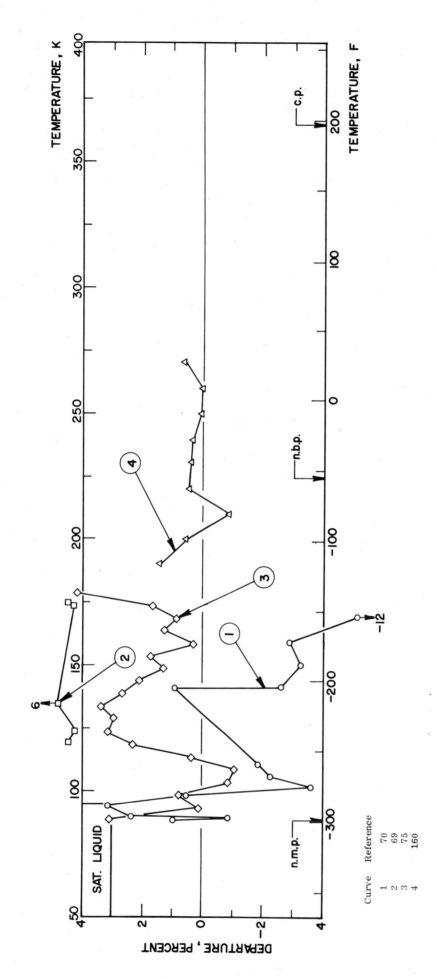

FIGURE 55–L(T). DEPARTURE PLOT FOR VISCOSITY OF LIQUID PROPYLENE

TABLE 55-V(T). VISCOSITY OF PROPYLENE VAPOR

DISCUSSION

SATURATED VAPOR

Recommended values for the viscosity of the saturated vapor were generated using the correlation technique of Jossi, Stiel and Thodos [100].

A generalized curve of the excess viscosity versus reduced temperature for several gases was drawn. The excess viscosity for propylene, read from that curve, and the recommended value of the gas at 1 atm, were used to generate the recommended value for the viscosity of the saturated vapor was obtained.

The accuracy is thought to be ±5%.

RECOMMENDED VALUES

[Temperature, T, K; Viscosity, μ, 10^{-3} N s m^{-2}]

SATURATED VAPOR

T	μ
210	.00604
220	.00637
230	.00672
240	.00707
250	.00742
260	.00779
270	.00815
280	.00852
290	.00888
300	.00930
310	.00986
320	.01050
330	.01127
340	.01233
350	.01400
360	.01669
365*	.02394

* Crit. Temp.

TABLE 55-G(T). VISCOSITY OF GASEOUS PROPYLENE

DISCUSSION

GAS

Two sets of experimental values were found in the literature. These are smoothed values given by Neduzhii [160] and one value at 0°C by Titani [216].

The theoretical expression

$$\mu = \frac{K\sqrt{T}}{\sigma^2 \Omega} \qquad (1)$$

was used to obtain values of $\sigma^2 \Omega$ as a function of $1/T$. These values were fitted by least square to the equation:

$$\sigma^2 \Omega = A + B/T \qquad (2)$$

which represents the data well in the experimental range. The results were used in equation (1) to obtain the recommended values.

The accuracy is thought to be of about ± 2%. Above 320 K the table values are extrapolated.

RECOMMENDED VALUES

[Temperature, T, K; Viscosity, μ, 10^{-6} N s m^{-2}]

GAS

T	μ
210	6.04
220	6.36
230	6.67
240	6.98
250	7.28
260	7.59
270	7.89
280	8.19
290	8.49
300	8.78
310	9.08
320	9.36
330	9.64
340	9.93
350	10.22
360	10.50

217

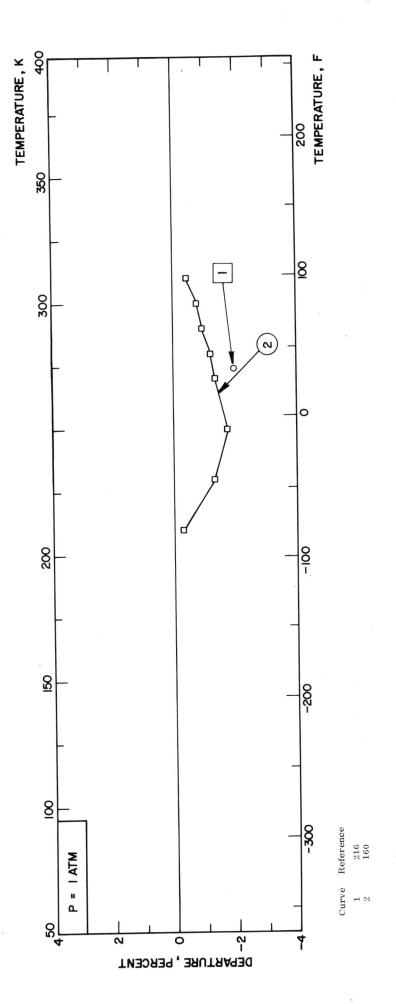

FIGURE 55-G(T). DEPARTURE PLOT FOR VISCOSITY OF GASEOUS PROPYLENE

DISCUSSION

TABLE 56-G(T). VISCOSITY OF GASEOUS TOLUENE

RECOMMENDED VALUES

[Temperature, T, K; Viscosity, μ, 10^{-6} N s m^{-2}]

GAS

Only one set of experimental data, by Nasini et al. [287] was found in the literature. Using the theoretical relation $\mu = K \cdot \sqrt{T}/[\sigma^2 \Omega_{22}(T^*)]$, $\sigma^2 \Omega$ was computed from the data and adjusted to a linear equation in $1/T$, from which the recommended values were generated.

There is no means to assess the accuracy of the recommended data.

GAS

T	μ
330	7.82
340	8.06
350	8.30
360	8.54
370	8.78
380	9.02
390	9.25
400	9.48
410	9.71
420	9.94
430	10.17
440	10.40
450	10.62
460	10.89
470	11.06
480	11.28
490	11.50
500	11.7
510	11.9
520	12.2
530	12.4
540	12.6
550	12.8

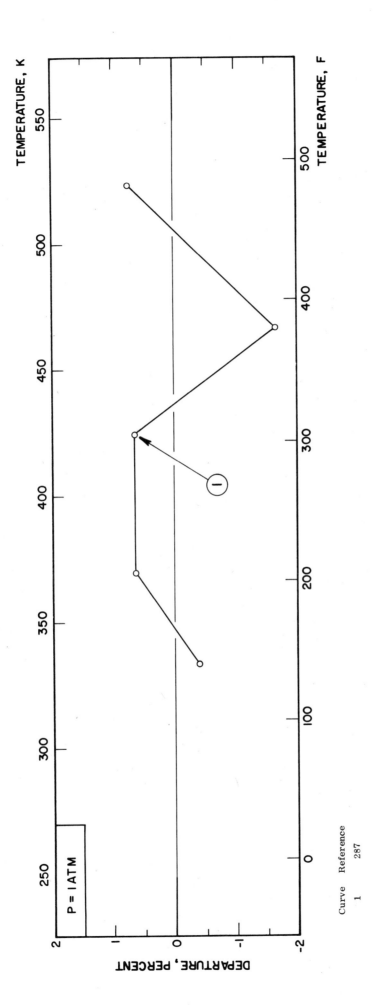

FIGURE 56-G(T). DEPARTURE PLOT FOR VISCOSITY OF GASEOUS TOLUENE

TABLE 57-L(T). VISCOSITY OF LIQUID TRICHLOROFLUOROMETHANE

DISCUSSION

SATURATED LIQUID

Nine sets of data were found in the literature [14, 58, 79, 123, 371, 184, 274, 275, 276], covering a temperature range from about 200 K to about 350 K. They are of equal reliability. They were fitted to an equation

$$\log \mu = A + B/T$$

in the range 170 to 390 K. From the latter temperature to the critical (471 K) a curve was drawn graphically to join the value of the viscosity at the critical temperature, estimated by the method of Stiel and Thodos [207].

The recommended curve is thought to be reliable to ±5% below 390 K and may be reliable to about ±10% from 390 K to the critical point.

RECOMMENDED VALUES

[Temperature, T, K, Viscosity, μ 10^{-3} N s m^{-2}]

SATURATED LIQUID	
T	μ
170	3.514
180	2.670
190	2.088
200	1.674
210	1.370
220	1.142
230	0.968
240	0.831
250	0.722
260	0.635
270	0.563
280	0.504
290	0.454
300	0.413
310	0.377
320	0.346
330	0.320
340	0.297
350	0.2764
360	0.2586
370	0.2428
380	0.2287
390	0.2161
400	0.2025
410	0.1865
420	0.1690
430	0.1505
440	0.1305
450	0.108
460	0.084
470	0.057
471*	0.033

* Crit. Temp.

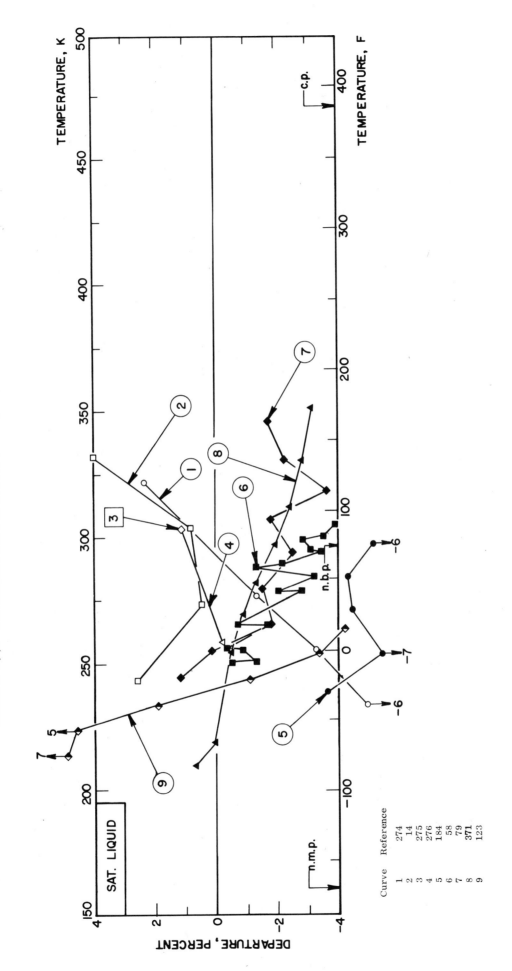

FIGURE 57-L(T).   DEPARTURE PLOT FOR VISCOSITY OF LIQUID TRICHLOROFLUOROMETHANE

Curve	Reference
1	274
2	14
3	275
4	276
5	184
6	58
7	79
8	371
9	123

TABLE 57-V(T).   VISCOSITY OF TRICHLOROFLUOROMETHANE VAPOR

DISCUSSION

RECOMMENDED VALUES

[Temperature, T, K, Viscosity, μ, 10^{-3} N s m^{-2}]

SATURATED VAPOR

SATURATED VAPOR

T	μ
290	0.01063
300	0.01100
310	0.01137
320	0.01174
330	0.01212
340	0.01250
350	0.01290
360	0.01332
370	0.01377
380	0.01422
390	0.01470
400	0.01520
410	0.01575
420	0.01645
430	0.01730
440	0.01830
450	0.0197
460	0.0218
470	0.0278
471*	0.0333

Recommended values for the viscosity of the saturated vapor were estimated with the correlating equations of Stiel and Thodos [207]. Their accuracy is thought to be ±5% although this figure may rise to ±10% near to the critical point.

* Crit. temp.

TABLE 57-G(T). VISCOSITY OF GASEOUS TRICHLOROFLUOROMETHANE

RECOMMENDED VALUES

[Temperature, T, K, Viscosity, μ, 10^{-6} N s m^{-2}]

GAS

T	μ
230	8.87
240	9.14
250	9.42
260	9.71
270	10.01
280	10.32
290	10.63
300	10.95
310	11.27
320	11.59
330	11.92
340	12.25
350	12.58
360	12.92
370	13.26
380	13.60
390	13.94
400	14.29
410	14.63
420	14.98
430	15.32
440	15.67
450	16.02
460	16.38
470	16.73
480	17.08
490	17.43
500	17.78

DISCUSSION

GAS

There are four sets of data for R 11, those of Tsui [238], McCullum [143] and of Benning[12,14]. Data are also reported in a manufacturer's bulletin [274].

The correlation was made by computing

$$(\sigma^2 \Omega) = \frac{K\sqrt{T}}{\mu}$$

and adjusting the values obtained to a quadratic equation in 1/T. Recommended values and calculated values of the viscosity were computed through the equation obtained. The accuracy of the correlation is thought to be ±2%.

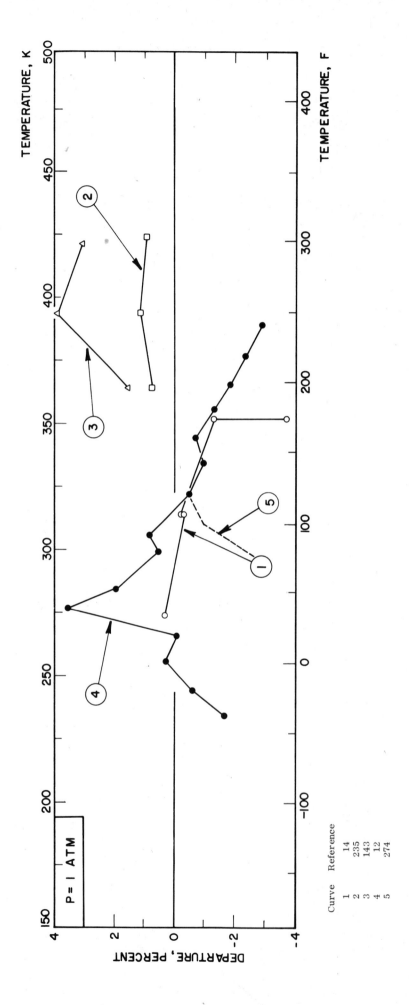

FIGURE 57-G(T).   DEPARTURE PLOT FOR VISCOSITY OF GASEOUS TRICHLOROFLUOROMETHANE

TABLE 58-L(T). VISCOSITY OF LIQUID TRICHLOROTRIFLUOROETHANE

RECOMMENDED VALUES

[Temperature, T, K, Viscosity, μ, 10^{-3} N s m^{-2}]

SATURATED LIQUID

T	μ
230	2.206
240	1.779
250	1.461
260	1.217
270	1.028
280	0.879
290	0.760
300	0.663
310	0.584
320	0.518
330	0.463
340	0.417
350	0.377
360	0.343
370	0.314
380	0.289
390	0.266
400	0.246
410	0.228
420	0.210
430	0.193
440	0.175
450	0.155
460	0.133
470	0.108
480	0.077
487*	0.030

* Crit. Temp.

DISCUSSION

SATURATED LIQUID

Four sets of experimental data [13, 14, 123, 140] and two sets of values from manufacturer's technical notes[275, 276] were found in the literature. There is a good consistency between sets except for the data of Kinser [123] at low temperature.

An adjustment was made to an equation

$$\log \mu = A + B/T$$

and recommended values generated from this equation in the range 230 K to 390 K. Above this temperature, a curve was drawn to join smoothly the value of the viscosity at the critical point estimated by the method of Stiel and Thodos [207].

The accuracy is about ±5% below 390 K, but may be poorer above.

226

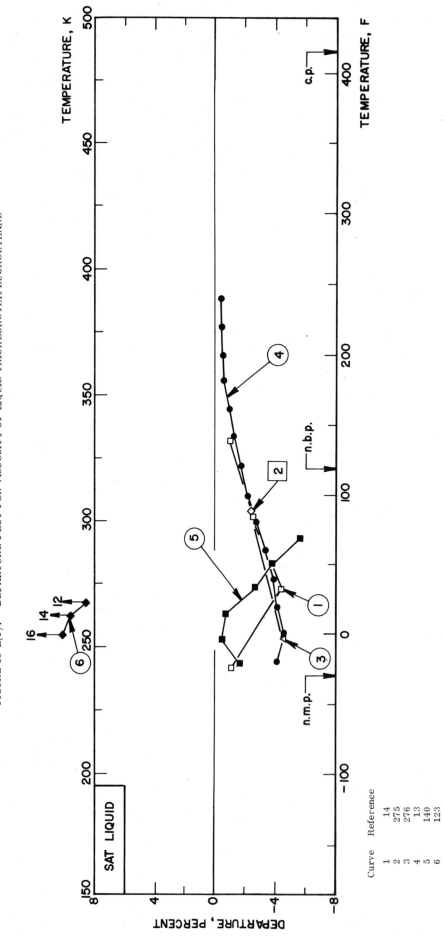

FIGURE 58-L(T).   DEPARTURE PLOT FOR VISCOSITY OF LIQUID TRICHLOROTRIFLUOROETHANE

TABLE 58-V(T). VISCOSITY OF TRICHLOROTRIFLUOROETHANE VAPOR

DISCUSSION

RECOMMENDED VALUES

[Temperature, T, K, Viscosity, μ, 10^{-3} N s m^{-2}]

SATURATED VAPOR

Recommended values for the viscosity of R 113 were obtained using the excess viscosity concept. The method used was the method of Stiel and Thodos [207]. The excess viscosity was taken from a curve of reduced excess viscosity versus reduced temperature constructed from the values of density and of viscosity from other refrigerants.

The accuracy is thought to be about ± 5% although the figure may be higher around the critical temperature.

SATURATED VAPOR

T	μ
320	0.01081
330	0.01109
340	0.01136
350	0.01164
360	0.01189
370	0.01218
380	0.01248
390	0.01282
400	0.01318
410	0.01357
420	0.01405
430	0.01446
440	0.01504
450	0.0157
460	0.0166
470	0.0178
480	0.0200
487*	0.0296

* Crit. Temp.

TABLE 58-G(T). VISCOSITY OF GASEOUS TRICHLOROTRIFLUOROETHANE

DISCUSSION

GAS

One set of experimental data can be found in the literature, by Benning and Markwood [14]. Other informations are from manufacturer's technical notes [275, 276, 13]. The latter [13] is stated as due to Benning and McHarness. They were used to obtain the values of

$$\sigma^2 \Omega = \frac{K \sqrt{T}}{\mu}$$

which were fitted to a quadratic equation in 1/T. From the adjusted equation recommended values of viscosity were generated.

The accuracy is thought to be of ±2%. New data from other sources are needed.

RECOMMENDED VALUES

[Temperature, T, K; Viscosity, μ, 10^{-6} N s m^{-2}]

GAS

T	μ
230	8.61
240	8.87
250	9.13
260	9.38
270	9.63
280	9.88
290	10.12
300	10.35
310	10.59
320	10.81
330	11.04
340	11.26
350	11.49
360	11.69
370	11.90
380	12.12
390	12.32
400	12.52

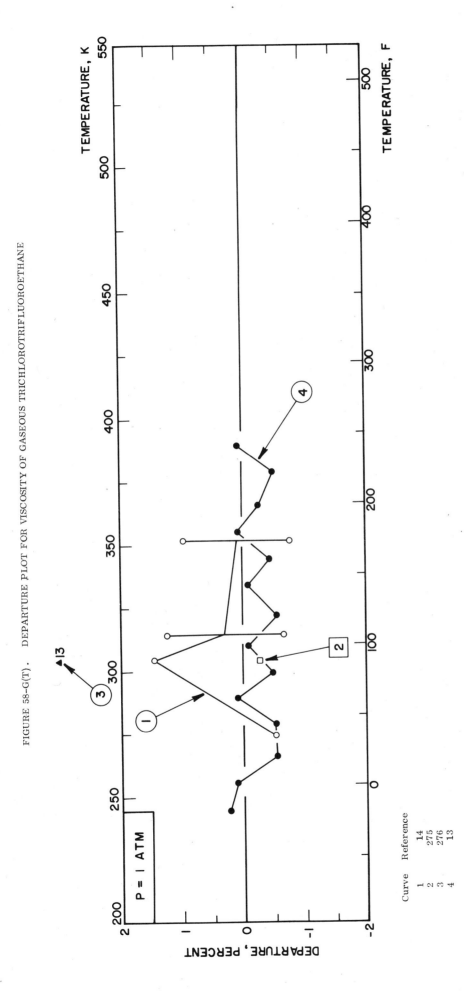

FIGURE 58-G(T). DEPARTURE PLOT FOR VISCOSITY OF GASEOUS TRICHLOROTRIFLUOROETHANE

TABLE 59-L(T). VISCOSITY OF LIQUID TRIFLUOROMETHANE

DISCUSSION

SATURATED LIQUID

Only one set of experimental data by Phillips [166] was found in the literature, covering a temperature range from 190 to 260 K. The author gives also an equation of the type:

$$\log \mu = A + B/T + C/T^2 + D/T^3$$

which was adopted in the range 200–260 K, and from which recommended values were generated. Above the latter temperature the curve was joined smoothly to the estimated viscosity at the critical temperature.

Deviations of the experimental data from the curve are low, but no means exist to assess the accuracy of the data.

Above 260 K the accuracy may be very poor, and the values must be considered as tentative values.

RECOMMENDED VALUES

[Temperature, T, K; Viscosity, μ, N s m^{-2} · 10^{-3}]

SATURATED LIQUID

T	μ
170	0.425
175	0.392
180	0.363
185	0.338
190	0.316
195	0.296
200	0.278
205	0.262
210	0.247
215	0.233
220	0.220
225	0.208
230	0.197
235	0.186
240	0.176
245	0.166
250	0.157
255	0.148
260	0.139
265	0.130
270	0.121
275	0.109
280	0.098
285	0.083
290	0.062
293*	0.029

* Crit. Temp.

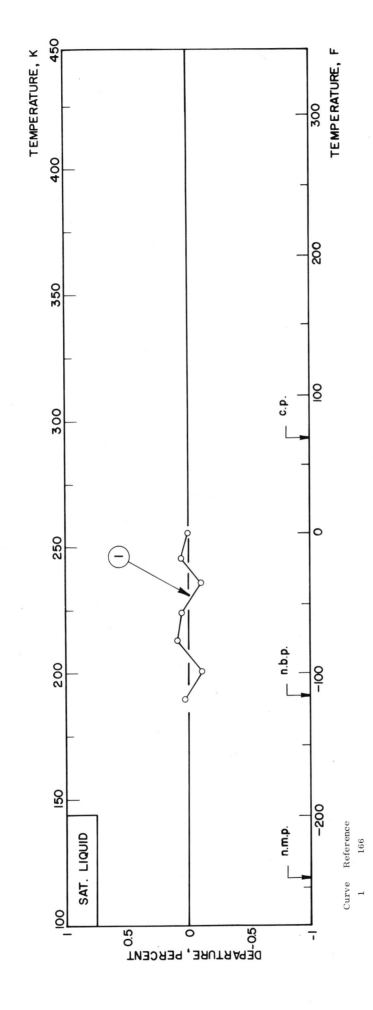

FIGURE 59-L(T).   DEPARTURE PLOT FOR VISCOSITY OF LIQUID TRIFLUOROMETHANE

TABLE 59-V(T). VISCOSITY OF TRIFLUOROMETHANE VAPOR

RECOMMENDED VALUES

[Temperature, T, K, Viscosity, μ, 10^{-3} N s m^{-2}]

SATURATED VAPOR

T	μ
200	0.00979
210	0.01034
220	0.01102
230	0.01169
240	0.01243
250	0.0132
260	0.0142
270	0.0154
280	0.0170
290	0.0213
293*	0.0288

* Crit. Temp.

DISCUSSION

SATURATED VAPOR

Recommended values for the viscosity of the saturated vapor were computed with the method of Stiel and Thodos [207] which makes use of the excess viscosity concept. A graph of reduced excess viscosity versus reduced temperature, was constructed for several refrigerants. From this graph, the excess viscosity was read and used with the recommended values for the 1 atm gas, to generate the present values.

The accuracy is poor and should be about ±10% close to the boiling point, but may reach ±20% when approaching the critical point.

TABLE 59-G(T). VISCOSITY OF GASEOUS TRIFLUOROMETHANE

DISCUSSION

GAS

Five sets of experimental data were found in the literature [40, 143, 235, 262, 265]. They cover a temperature range from 230 to 470 K.

Computed values of

$$\sigma^2 \Omega = \frac{K\sqrt{T}}{\mu}$$

were least square fitted to a quadratic equation in 1/T from which recommended values were generated.

The accuracy of the correlation is about ± 2%.

RECOMMENDED VALUES

[Temperature, T, K, Viscosity, μ, 10^{-6} N s m^{-2}]

GAS

T	μ
230	11.31
240	11.84
250	12.36
260	12.88
270	13.40
280	13.90
290	14.40
300	14.90
310	15.39
320	15.88
330	16.36
340	16.84
350	17.31
360	17.77
370	18.23
380	18.69
390	19.14
400	19.59
410	20.03
420	20.47
430	20.90
440	21.33
450	21.76
460	22.18
470	22.59
480	23.00
490	23.41
500	23.82

234

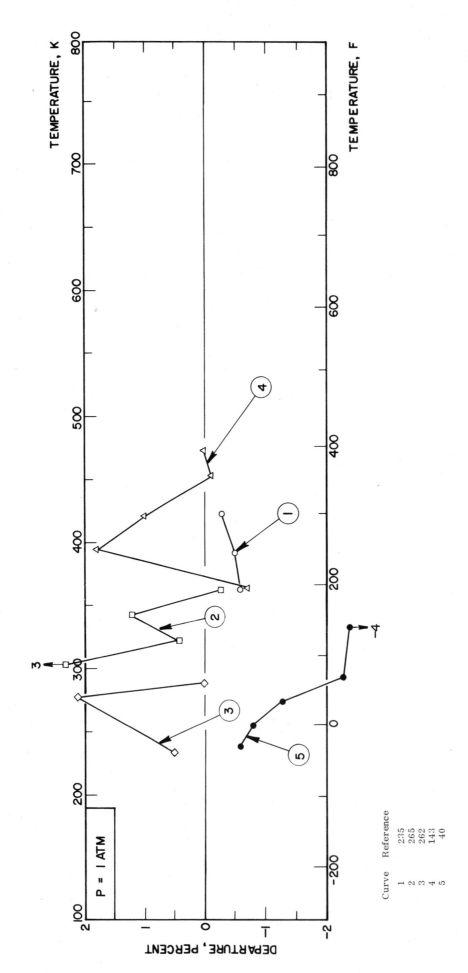

FIGURE 59-G(T). DEPARTURE PLOT FOR VISCOSITY OF GASEOUS TRIFLUOROMETHANE

4. BINARY SYSTEMS

BINARY SYSTEMS

The viscosity data (expressed in N s m^{-2}) for ninety-nine binary systems are presented in Figures and Tables 60 through 158. Each Figure and Table includes data on a single binary system and it is further divided into as many as three different sections to accommodate data with composition, density, and temperature dependences. Those data originally reported in the research document as a function of pressure have been converted to be as a function of density.

In graphical smoothing of the data for a binary system giving the composition dependence at a particular temperature, the two end points, referring to the two pure components, were regarded as correct, and then, consistent with the accuracy of the data, a smooth curve was drawn through the experimental points. This approach, which was adopted in almost all cases, has many implications. The reliability of the viscosity data for pure fluids is generally better than that for the mixtures obtained on the same apparatus. This is because in principle a better theoretical mechanistic formulation of the viscometer is accomplished for pure fluids. Also in relative measurements, viscometers are calibrated at the end points with pure fluids and consequently these are most reliable of all the reported data points. A reconsideration of the data of a particular worker will then be necessary in case his data on pure fluids is significantly different from the most probable values. A greater reliance can be placed in such cases on the relative changes in viscosity with the variable parameter than on the absolute values.

A close look at the viscosity data of the binary systems as displayed in various figures reveals that no general common trends in the variation of viscosity with temperature, composition, and density exist. It appears that the viscosity of a binary gaseous system always increases with temperature for a given composition and density of the mixture. On the other hand the viscosity of several of the liquid systems examined such as sodium chlorate - sodium nitrate, iron - carbon, lead - tin, carbon tetrachloride - octamethylcyclotetrasiloxane, n-decane - methane, ethane - ethylene, and ethylene - methane exhibit the opposite trend, viz. the viscosity decreases with increasing temperature.

The variation of viscosity with composition is rather complex. Some systems such as argon - krypton, helium - neon, argon - ammonia, liquid benzene - octamethylcyclotetrasiloxane, carbon monoxide - hydrogen, carbon monoxide - oxygen, liquid carbon tetrachloride - octamethylcyclotetrasiloxane, ethylene - oxygen, hydrogen - nitric oxide, etc. exhibit a monotonic increase in the viscosity with increasing proportion of the heavier component in the mixture. Similarly, for many systems such as argon - neon, neon - krypton, krypton - xenon, neon - xenon, argon - sulfur dioxide, liquid benzene - n-hexane, carbon dioxide - nitrogen, carbon dioxide - oxygen, carbon dioxide - propane, carbon monoxide - ethylene, ethylene - nitrogen, methane - propane, nitrous oxide - propane, carbon dioxide - sulfur dioxide, their viscosity is found to systematically decrease with the increasing proportion of the heavier component in the mixture. For many other systems such as argon - helium, argon - xenon, helium - krypton, helium - xenon, ethane - hydrogen, ethylene - hydrogen, hydrogen - propane, carbon dioxide - hydrogen chloride, hydrogen - ammonia, hydrogen - ethyl ether, hydrogen - sulfur dioxide, methane - ammonia, methane - sulfur dioxide, carbon tetrachloride - methanol, etc. the viscosity exhibits a maximum at a certain value of the mole fraction of the heavier component in the mixture. In the liquid carbon tetrachloride - isopropyl alcohol and benzene - cyclohexane systems, a minimum is observed in the viscosity versus mole fraction of the heavier component. Thus, examples of all possible variations have been encountered while treating the data on binary systems.

The dependence of viscosity on density is also likewise complicated. For most of the systems such as argon - neon, helium - krypton, argon - hydrogen, argon - nitrogen, helium - carbon dioxide, helium - nitrogen, krypton - carbon dioxide, n-butane - methane, carbon dioxide - methane, carbon dioxide - nitrogen, carbon tetrafluoride - methane, methane - nitrogen, methane - propane, the viscosity is found to increase with density. Of all the systems examined here only the viscosity of helium - hydrogen system is found to decrease with density and this dependence is feeble.

It may be noted that even for mixtures of nonpolar and spherically symmetric rare gas molecules the viscosity variation is not systematic and does not fall in one characteristic category. This stresses the need for a careful study of the predictive procedures and thorough analysis of the available data on viscosity of fluid mixtures.

The experimental data for ternary, quaternary, and multicomponent systems are presented in Tables 159 through 188. These data are not further processed like binary systems except in a few cases which are either pure air or its mixtures with other substances.

TABLE 60-G(C)E. EXPERIMENTAL VISCOSITY DATA AS A FUNCTION OF COMPOSITION FOR GASEOUS ARGON-HELIUM MIXTURES

Cur. No.	Fig. No.	Ref. No.	Author(s)	Temp. (K)	Pressure (atm)	Mole Fraction of Ar	Viscosity (N s m^{-2} x 10^{-6})	Remarks
1	60-G(C)	165	Rietveld, A.O., Van Itterbeek, A., and Van den Berg, G.J.	72.0		1.000	6.35	Ar: purity not specified, He: hydrogen free; oscillating disk method, relative measurements; mixture composition corrected for thermal diffusion effect; precision about 1.0%; L_1 = 0.365%, L_2 = 0.598%, L_3 = 1.709%.
						0.828	6.79	
						0.657	7.21	
						0.557	7.52	
						0.538	7.57	
						0.4585	7.78	
						0.391	8.01	
						0.357	8.08	
						0.258	8.45	
						0.159	8.34	
						0.000	7.98	
2	60-G(C)	165	Rietveld, A.O., et al.	81.1		1.000	7.05	Same remarks as for curve 1 except L_1 = 0.507%, L_2 = 0.713%, L_3 = 1.873%.
						0.828	7.37	
						0.657	7.97	
						0.557	8.28	
						0.538	8.28	
						0.4585	8.55	
						0.391	8.72	
						0.357	8.83	
						0.258	9.19	
						0.159	9.02	
						0.000	8.59	
3	60-G(C)	165	Rietveld, A.O., et al.	90.2		1.000	7.60	Same remarks as for curve 1 except L_1 = 0.411%, L_2 = 0.713%, L_3 = 1.908%.
						0.828	8.28	
						0.657	8.61	
						0.557	8.89	
						0.538	8.95	
						0.391	9.35	
						0.357	9.48	
						0.258	9.69	
						0.159	9.71	
						0.000	9.10	
						0.000	9.15	
4	60-G(C)	165	Rietveld, A.O., et al.	192.5		1.000	15.38	Same remarks as for curve 1 except L_1 = 0.305%, L_2 = 0.411%, L_3 = 0.829%.
						0.887	15.74	
						0.8055	15.96	
						0.801	15.94	
						0.711	16.13	
						0.622	16.25	
						0.494	16.62	
						0.465	16.58	
						0.411	16.81	
						0.303	16.88	
						0.200	16.64	
						0.1055	16.07	
						0.000	14.71	
						0.000	14.48	
5	60-G(C)	165	Rietveld, A.O., et al.	229.5		1.000	17.68	Same remarks as for curve 1 except L_1 = 0.054%, L_2 = 0.093%, L_3 = 0.218%.
						1.000		
						0.8865	18.08	
						0.805	18.33	
						0.800	18.38	
						0.710	18.54	
						0.621	18.70	
						0.464	18.96	
						0.409	19.06	
						0.301	19.17	
						0.199	18.74	
						0.105	17.99	
						0.000	16.42	
						0.000	16.27	
6	60-G(C)	211	Tanzler, P.	288.2		100.00	22.20	Ar: prepared by method of Ramsay and Teavers, He: spectroscopically analyzed for purity, prepared by heating Mondzite sand to glowing; capillary transpiration method; L_1 = 0.200%, L_2 = 0.396%, L_3 = 1.178%.
						95.074	22.31	
						90.93	22.43	
						85.715	22.53	
						80.744	22.66	
						77.055	22.64	
						68.458	22.66	
						61.193	23.03	
						53.374	22.99	
						29.147	22.80	
						19.215	22.26	
						0.000	19.66	

TABLE 60-G(C)E. EXPERIMENTAL VISCOSITY DATA AS A FUNCTION OF COMPOSITION FOR GASEOUS ARGON-HELIUM MIXTURES (continued)

Cur. No.	Fig. No.	Ref. No.	Author(s)	Temp. (K)	Pressure (mm Hg)	Mole Fraction of Ar	Viscosity ($N \ s \ m^{-2} \times 10^{-6}$)	Remarks
7	60-G(C)	165	Rietveld, A.O., Van Iterbeek, A., and Van den Berg, G.J.	291.1		1.000	21.85, 21.68	Same remarks as for curve 1 except $L_1 = 0.138\%$, $L_2 = 0.186\%$, $L_3 = 0.444\%$.
						0.828	22.28	
						0.657	22.70	
						0.557	23.06	
						0.538	23.02	
						0.4585	22.93	
						0.391	22.96	
						0.357	22.75	
						0.258	22.46	
						0.159	21.76	
						0.000	19.35	
8	60-G(C)	165	Rietveld, A.O., et al.	291.1		1.000	21.72	Same remarks as for curve 1 except $L_1 = 0.249\%$, $L_2 = 0.297\%$, $L_3 = 0.514\%$.
						1.000	21.75	
						0.8865	22.23	
						0.805	22.43	
						0.800	22.40	
						0.710	22.81	
						0.621	22.94	
						0.464	23.11	
						0.409	23.04	
						0.301	22.90	
						0.199	22.29	
						0.105	21.01	
						0.000	19.14	
						0.000	19.11	
9	60-G(C)	213	Thornton, E. and Baker, W.A.D.	291.2	700	1.000	22.0	Ar: 99.8 pure, He: spectroscopically pure; modified Rankine viscometer, relative measurements; uncertainties: mixture composition $\pm 0.3\%$, viscosity $\pm 1.0\%$; $L_1 = 0.196\%$, $L_2 = 0.249\%$, $L_3 = 0.548\%$.
						0.914	22.2	
						0.844	22.4	
						0.782	22.5	
						0.720	22.7	
						0.645	22.7	
						0.574	22.8	
						0.520	22.9	
						0.438	22.8	
						0.299	22.7	
						0.208	22.2	
						0.061	20.5	
						0.000	19.4	
10	60-G(C)	223	Trautz, M. and Kipphan, K.F.	293		1.0000	22.11	Gas purity: He $<1\%$ Ne, Ar $<0.5\%$ N_2; method of Trautz and Weizel, calibrated with air; $L_1 = 0.000\%$, $L_2 = 0.000\%$, $L_3 = 0.000\%$.
						0.6180	22.91	
						0.5094	22.96	
						0.0000	19.73	
11	60-G(C)	223	Trautz, M. and Kipphan, K.F.	373		1.0000	26.84	Same remarks as for curve 10 except $L_1 = 0.144\%$, $L_2 = 0.211\%$, $L_3 = 0.394\%$.
						0.6180	27.45	
						0.5094	27.50	
						0.0000	23.20	
12	60-G(C)	211	Tanzler, P.	373.2		100.00	27.56	Same remarks as for curve 6 except $L_1 = 0.165\%$, $L_2 = 0.230\%$, $L_3 = 0.429\%$.
						95.074	27.56	
						90.930	27.70	
						85.715	27.83	
						80.744	27.91	
						77.055	27.84	
						68.458	27.90	
						61.193	28.06	
						53.374	27.88	
						20.147	27.53	
						19.215	26.64	
						0.000	23.55	
13	60-G(C)	211	Tanzler, P.	456.2		100.00	32.27	Same remarks as for curve 6 except $L_1 = 0.124\%$, $L_2 = 0.204\%$, $L_3 = 0.526\%$.
						95.074	32.17	
						90.930	32.32	
						85.715	32.48	
						80.744	32.52	
						68.458	32.50	
						61.193	32.44	
						19.215	30.42	
						0.000	26.91	
14	60-G(C)	223	Trautz, M. and Kipphan, K.F.	473		1.0000	32.08	Same remarks as for curve 10 except $L_1 = 0.000\%$, $L_2 = 0.000\%$, $L_3 = 0.000\%$.
						0.6180	32.50	
						0.0000	27.15	
15	60-G(C)	223	Trautz, M. and Kipphan, K.F.	523		1.0000	34.48	Same remarks as for curve 10 except $L_1 = 0.000\%$, $L_2 = 0.000\%$, $L_3 = 0.000\%$.
						0.6180	34.88	
						0.0000	29.03	

TABLE 60-G(C)S. SMOOTHED VISCOSITY VALUES AS A FUNCTION OF COMPOSITION FOR GASEOUS ARGON-HELIUM MIXTURES

Mole Fraction of Ar	72.0 K [Ref. 165]	81.1 K [Ref. 165]	90.2 K [Ref. 165]	192.5 K [Ref. 165]	229.5 K [Ref. 165]	288.2 K [Ref. 211]	291.1 K [Ref. 165]	291.1 K [Ref. 165]
0.00	7.98	8.59	9.12	14.48	16.35	19.66	19.35	19.11
0.05	8.23	8.90	9.40	15.45	17.57	20.60	20.31	20.10
0.10	8.34	9.02	9.60	16.02	17.94	21.36	21.10	20.92
0.15	8.37	9.07	9.70	16.39	18.41	21.92	21.68	21.72
0.20	8.37	9.07	9.74	16.64	18.75	22.33	22.12	22.28
0.25	8.32	9.03	9.70	16.79	19.00	22.63	22.44	22.64
0.30	8.24	8.96	9.60	16.88	19.17	22.84	22.67	22.87
0.35	8.13	8.86	9.47	16.89	19.15	22.98	22.83	23.00
0.40	8.00	8.74	9.32	16.84	19.09	23.06	22.93	23.07
0.45	7.85	8.60	9.17	16.71	19.01	23.08	22.97	23.07
0.50	7.70	8.44	9.03	16.57	18.93	23.10	22.98	23.06
0.55	7.54	8.27	8.90	16.44	18.85	23.08	22.95	23.03
0.60	7.40	8.09	8.76	16.31	18.76	23.04	22.89	22.97
0.65	7.24	7.92	8.63	16.19	18.67	22.98	22.78	22.91
0.70	7.09	7.76	8.49	16.07	18.57	22.90	22.69	22.81
0.75	6.95	7.61	8.35	15.97	18.46	22.80	22.56	22.67
0.80	6.82	7.48	8.21	15.86	18.34	22.69	22.41	22.50
0.85	6.70	7.36	8.07	15.76	18.20	22.58	22.24	22.33
0.90	6.57	7.25	7.93	15.67	18.04	22.46	22.06	22.13
0.95	6.50	7.15	7.80	15.57	17.87	22.34	21.87	21.93
1.00	6.35	7.05	7.68	15.47	17.68	22.20	21.68	21.73

Mole Fraction of Ar	291.3 K [Ref. 213]	293.0 K [Ref. 223]	373.0 K [Ref. 223]	373.2 K [Ref. 211]	456.2 K [Ref. 211]	473.0 K [Ref. 223]	523.0 K [Ref. 223]
0.00	19.40	19.73	23.20	23.55	26.91	27.15	29.03
0.05	20.31	22.63	23.80	24.42	28.26	27.76	29.82
0.10	21.14	21.31	24.43	25.26	29.22	28.39	30.63
0.15	21.74	21.81	25.02	26.08	29.94	28.99	31.38
0.20	22.15	22.19	25.56	26.76	30.52	29.52	32.07
0.25	22.41	22.46	26.04	27.25	30.98	30.00	32.66
0.30	22.60	22.66	26.43	27.58	31.34	31.40	33.17
0.35	22.72	22.80	26.76	27.80	31.65	31.82	33.60
0.40	22.80	22.89	27.04	27.92	31.90	32.16	33.98
0.45	22.85	22.94	27.24	27.98	32.08	31.44	34.29
0.50	22.87	22.97	27.38	28.00	32.24	31.68	34.53
0.55	22.86	22.96	27.46	27.99	32.35	31.87	34.72
0.60	22.82	22.93	27.48	27.98	32.43	32.01	34.84
0.65	22.78	22.88	27.49	27.95	32.48	32.12	34.92
0.70	22.70	22.82	27.46	27.91	32.50	32.19	34.96
0.75	22.61	22.74	27.42	27.86	32.50	32.22	34.96
0.80	22.50	22.64	27.34	27.81	32.48	32.23	34.92
0.85	22.37	22.53	27.24	27.75	32.45	32.22	34.82
0.90	22.21	22.40	27.13	27.69	32.40	32.18	34.75
0.95	22.05	22.26	27.00	27.63	32.35	32.13	34.62
1.00	21.88	22.11	26.85	27.56	32.28	32.08	34.48

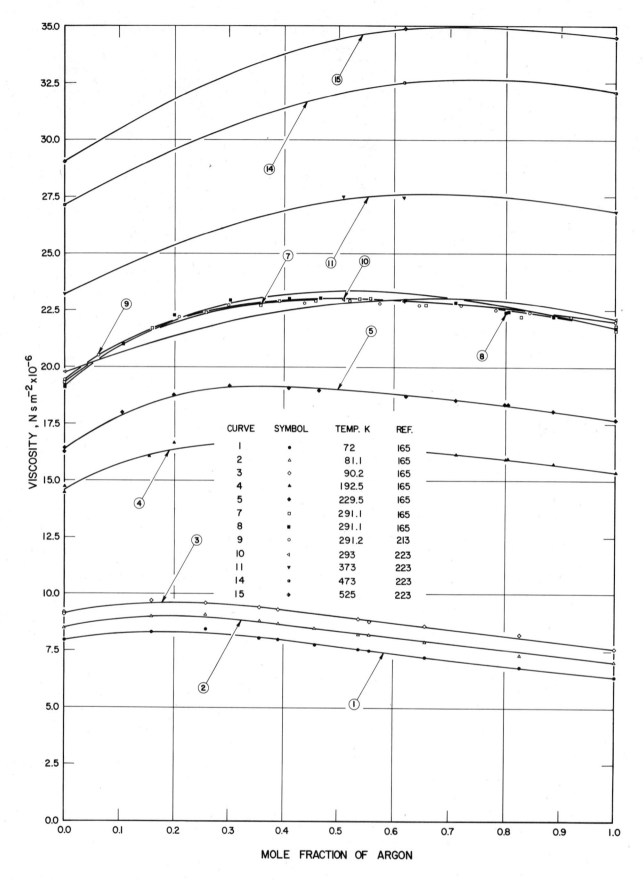

CURVE	SYMBOL	TEMP. K	REF.
1	•	72	165
2	△	81.1	165
3	◇	90.2	165
4	▲	192.5	165
5	◆	229.5	165
7	□	291.1	165
8	■	291.1	165
9	○	291.2	213
10	◁	293	223
11	▼	373	223
14	◉	473	223
15	◈	525	223

VISCOSITY, $N\,s\,m^{-2} \times 10^{-6}$

MOLE FRACTION OF ARGON

FIGURE 60-G(C). VISCOSITY DATA AS A FUNCTION OF COMPOSITION
FOR GASEOUS ARGON – HELIUM MIXTURES

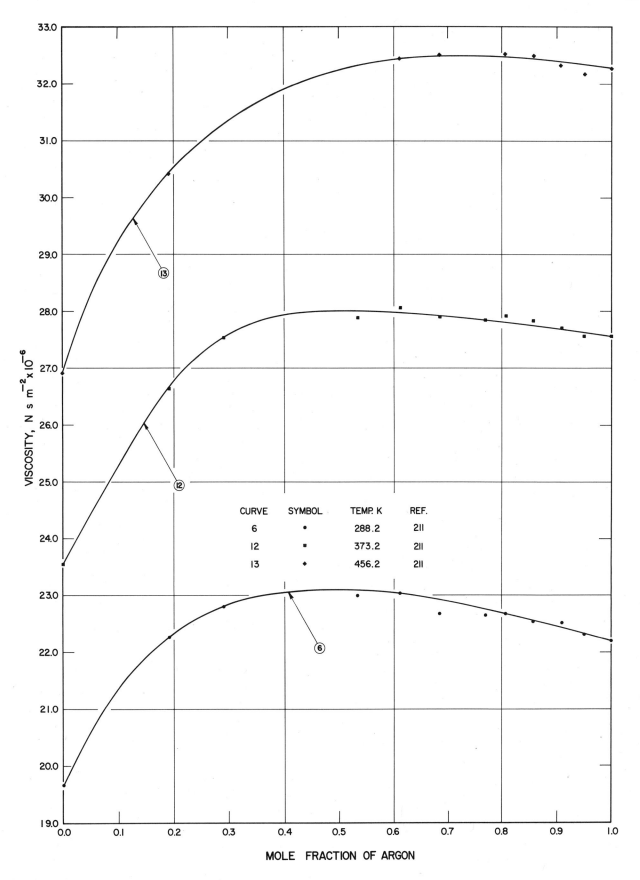

CURVE | SYMBOL | TEMP. K | REF.
6 | • | 288.2 | 211
12 | ■ | 373.2 | 211
13 | ♦ | 456.2 | 211

MOLE FRACTION OF ARGON

FIGURE 60-G(C). VISCOSITY DATA AS A FUNCTION OF COMPOSITION
FOR GASEOUS ARGON-HELIUM MIXTURES (continued)

TABLE 60-G(D)E.   EXPERIMENTAL VISCOSITY DATA AS A FUNCTION OF DENSITY FOR GASEOUS ARGON-HELIUM MIXTURES

Cur. No.	Fig. No.	Ref. No.	Author(s)	Mole Fraction of Ar	Temp. (K)	Density (g cm^{-3})	Viscosity (N s m^{-2} x10^{-6})	Remarks
1	60-G(D)	91	Iwasaki, H. and Kestin, J.	1.0000	293.2	0.001684	22.275	Ar: 99.997 pure, He: 99.99 pure; oscillating disk viscometer; accuracy of absolute measurements of pure fluids and of relative measurements of mixtures with respect to pure fluids is 0.1 to 0.2%.
						0.009403	22.362	
						0.017944	22.462	
						0.034916	22.681	
						0.052123	22.954	
						0.069120	23.221	
						0.088147	23.572	
2	60-G(D)	91	Iwasaki, H. and Kestin, J.	0.801	293.2	0.001352	22.707	Same remarks as for curve 1.
						0.008325	22.778	
						0.001349	22.711	
						0.008308	22.775	
						0.015220	22.859	
						0.022387	22.932	
						0.029405	23.025	
						0.036642	23.115	
						0.043813	23.202	
						0.051011	23.302	
						0.058095	23.409	
						0.065387	23.520	
						0.071995	23.615	
3	60-G(D)	91	Iwasaki, H. and Kestin, J.	0.629	293.2	0.001108	23.095	Same remarks as for curve 1.
						0.006404	23.150	
						0.012295	23.192	
						0.016262	23.220	
						0.023522	23.296	
						0.029335	23.371	
						0.033182	23.391	
						0.038620	23.441	
						0.046485	23.524	
						0.052188	23.600	
						0.057799	23.656	
4	60-G(D)	91	Iwasaki, H. and Kestin, J.	0.366	293.2	0.000725	23.161	Same remarks as for curve 1.
						0.004363	23.181	
						0.007996	23.205	
						0.011684	23.234	
						0.015261	23.253	
						0.018872	23.281	
						0.022740	23.296	
						0.026175	23.322	
						0.029571	23.356	
						0.033003	23.382	
						0.036717	23.411	
5	60-G(D)	91	Iwasaki, H. and Kestin, J.	0.193	293.2	0.000460	22.528	Same remarks as for curve 1.
						0.002774	22.527	
						0.005055	22.540	
						0.007343	22.539	
						0.009701	22.549	
						0.012014	22.551	
						0.014246	22.570	
						0.016539	22.573	
						0.018868	22.587	
						0.021145	22.593	
						0.023312	22.603	
6	60-G(D)	91	Iwasaki, H. and Kestin, J.	0.137	293.2	0.000371	22.027	Same remarks as for curve 1.
						0.002263	22.033	
						0.004143	22.040	
						0.006065	22.038	
						0.007845	22.042	
						0.009744	22.053	
						0.011640	22.046	
						0.013460	22.056	
						0.015299	22.067	
						0.017150	22.063	
						0.018879	22.074	

TABLE 60-G(D)E.   EXPERIMENTAL VISCOSITY DATA AS A FUNCTION OF DENSITY FOR GASEOUS ARGON-HELIUM MIXTURES (continued)

Cur. No.	Fig. No.	Ref. No.	Author(s)	Mole Fraction of Ar	Temp. (K)	Density (g cm^{-3})	Viscosity (N s m^{-2} x 10^{-6})	Remarks
7	60-G(D)	91	Iwasaki, H. and Kestin, J.	0.058	293.2	0.000254 0.001539 0.002809 0.004060 0.005348 0.006641 0.007849 0.009136 0.010400 0.011418 0.012825	20.902 20.913 20.901 20.902 20.901 20.899 20.904 20.900 20.898 20.897 20.902	Same remarks as for curve 1.
8	60-G(D)	91	Iwasaki, H. and Kestin, J.	0.000	293.2	0.000169 0.003565 0.005790 0.008477	19.604 19.597 19.586 19.577	Same remarks as for curve 1.
9	60-G(D)	91	Iwasaki, H. and Kestin, J.	1.000	303.2	0.001611 0.009849 0.01808 0.03495 0.05235 0.06893 0.08567	22.944 23.048 23.136 23.356 23.628 23.902 24.206	Same remarks as for curve 1.
10	60-G(D)	91	Iwasaki, H. and Kestin, J.	0.789	303.2	0.001304 0.007714 0.021341 0.035070 0.048620 0.062392 0.068892	23.396 23.454 23.605 23.769 23.950 24.162 24.260	Same remarks as for curve 1.
11	60-G(D)	91	Iwasaki, H. and Kestin, J.	0.577	303.2	0.000994 0.006075 0.016226 0.026487 0.036742 0.046956 0.052386	23.748 23.796 23.883 23.983 24.088 24.221 24.281	Same remarks as for curve 1.
12	60-G(D)	91	Iwasaki, H. and Kestin, J.	0.390	303.2	0.000728 0.004428 0.011719 0.019209 0.026597 0.033959 0.037575	23.811 23.843 23.884 23.931 23.994 24.059 24.086	Same remarks as for curve 1.
13	60-G(D)	91	Iwasaki, H. and Kestin, J.	0.214	303.2	0.000469 0.002883 0.007591 0.012415 0.017168 0.021879 0.024140	23.239 23.244 23.258 23.269 23.293 23.317 23.327	Same remarks as for curve 1.
14	60-G(D)	91	Iwasaki, H. and Kestin, J.	0.125	303.2	0.000346 0.002143 0.005505 0.008974 0.012404 0.015807 0.017505	22.430 22.445 22.445 22.454 22.459 22.468 22.463	Same remarks as for curve 1.
15	60-G(D)	91	Iwasaki, H. and Kestin, J.	0.061	303.2	0.000248 0.001515 0.002744 0.006545 0.010269 0.012732	21.481 21.495 21.487 21.477 21.488 21.485	Same remarks as for curve 1.
16	60-G(D)	91	Iwasaki, H. and Kestin, J.	0.000	303.2	0.000162 0.001830 0.003444 0.005886 0.008275	20.094 20.094 20.080 20.076 20.071	Same remarks as for curve 1.

TABLE 60-G(D)S.　SMOOTHED VISCOSITY VALUES AS A FUNCTION OF DENSITY FOR GASEOUS ARGON-HELIUM MIXTURES

Density (g cm⁻³)	Mole Fraction of Argon							
	0.000 (293.2 K) [Ref. 91]	0.058 (293.2 K) [Ref. 91]	0.137 (293.2 K) [Ref. 91]	0.193 (293.2 K) [Ref. 91]	0.366 (293.2 K) [Ref. 91]	0.629 (293.2 K) [Ref. 91]	0.801 (293.2 K) [Ref. 91]	1.000 (293.2 K) [Ref. 91]
0.010	19.575	20.900	22.050	22.549	23.220	23.170	22.800	22.350
0.020		20.912	22.075	22.590	23.287	23.240	22.915	22.471
0.030				22.630	23.360	23.326	23.032	22.608
0.040					23.449	23.437	23.151	22.760
0.050						23.575	23.280	22.921
0.060						23.720	23.434	23.079
0.070							23.588	23.235
0.080								23.405
0.090								23.610

Density (g cm⁻³)	Mole Fraction of Argon							
	0.000 (293.2 K) [Ref. 91]	0.061 (293.2 K) [Ref. 91]	0.125 (293.2 K) [Ref. 91]	0.214 (293.2 K) [Ref. 91]	0.390 (293.2 K) [Ref. 91]	0.577 (293.2 K) [Ref. 91]	0.789 (293.2 K) [Ref. 91]	1.000 (293.2 K) [Ref. 91]
0.010	20.068	21.485	22.447	23.260	23.870	23.830	23.480	23.040
0.020		21.490	22.470	23.302	23.940	23.918	23.585	23.160
0.030				23.372	24.105	24.015	23.700	23.287
0.040					24.110	24.130	23.827	23.430
0.050						23.260	23.970	23.582
0.060							24.120	23.760
0.070							24.272	23.946
0.080								24.120

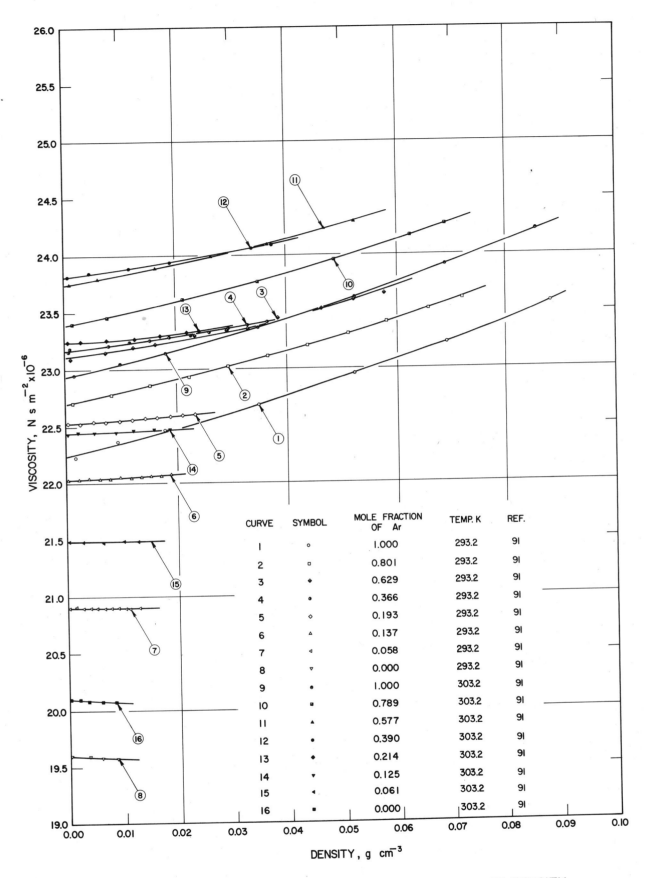

FIGURE 60-G(D). VISCOSITY DATA AS A FUNCTION OF DENSITY
FOR GASEOUS ARGON-HELIUM MIXTURES

TABLE 60-G(T)E.   EXPERIMENTAL VISCOSITY DATA AS A FUNCTION OF TEMPERATURE FOR GASEOUS ARGON-HELIUM MIXTURES

Cur. No.	Fig. No.	Ref. No.	Author(s)	Mole Fraction of Ar	Pressure (atm)	Temp. (K)	Viscosity (N s m^{-2} x 10^{-6})	Remarks
1	60-G(T)	211	Tanzler, P.	1.00000	74.85	285.2	22.00	Ar: prepared by method of Ramsey and Teavers, He: spectroscopically analyzed for purity; prepared by heating Mondzite sane to glowing; capillary transpiration method.
					74.84	372.8	27.46	
					74.62	456.2	32.31	
2	60-G(T)	211	Tanzler, P.	0.95074	74.87	285.8	22.19	Same remarks as for curve 1.
					75.10	373.0	27.45	
					74.52	455.9	32.18	
3	60-G(T)	211	Tanzler, P.	0.9093	75.10	284.5	22.17	Same remarks as for curve 1.
					75.06	372.8	27.68	
					75.00	456.3	32.44	
4	60-G(T)	211	Tanzler, P.	0.85715	75.17	286.9	22.44	Same remarks as for curve 1.
					75.81	373.1	27.84	
					76.19	457.5	32.54	
5	60-G(T)	211	Tanzler, P.	0.80744	74.95	292.9	22.94	Same remarks as for curve 1.
					75.36	372.8	27.90	
					75.20	456.3	32.50	
6	60-G(T)	211	Tanzler, P.	0.77055	75.76	293.7	23.01	Same remarks as for curve 1.
					75.65	373.0	27.85	
7	60-G(T)	211	Tanzler, P.	0.68458	75.19	295.3	23.16	Same remarks as for curve 1.
					75.03	372.7	27.27	
					75.61	456.3	32.53	
8	60-G(T)	211	Tanzler, P.	0.61193	75.35	294.8	23.41	Same remarks as for curve 1.
					75.33	372.6	28.07	
					75.75	456.8	32.44	
9	60-G(T)	211	Tanzler, P.	0.53374	75.31	294.1	23.34	Same remarks as for curve 1.
					75.29	372.7	27.85	
10	60-G(T)	211	Tanzler, P.	0.29174	76.11	292.3	23.03	Same remarks as for curve 1.
					76.05	373.1	27.52	
11	60-G(T)	211	Tanzler, P.	0.19215	75.49	292.1	22.46	Same remarks as for curve 1.
					75.78	373.0	26.58	
					75.67	456.2	30.39	
12	60-G(T)	211	Tanzler, P.	0.00000	75.50	288.5	19.69	Same remarks as for curve 1.
					75.00	372.8	23.48	
					75.66	457.8	26.99	

TABLE 60-G(T)S.　　SMOOTHED VISCOSITY VALUES AS A FUNCTION OF TEMPERATURE FOR GASEOUS ARGON-HELIUM MIXTURES

Temp. (K)	Mole Fraction of Argon					
	0.0000 [Ref. 211]	0.1922 [Ref. 211]	0.2917 [Ref. 211]	0.5337 [Ref. 211]	0.6119 [Ref. 211]	0.6846 [Ref. 211]
275						22.12
287.5	19.66				22.96	
290			22.92	23.10		
300	20.24	22.92	23.46	23.67	23.74	23.40
310			24.02	24.24		
312.5	20.82	23.59				
320			24.58	24.81		
325	21.39	24.24	24.86	25.10	25.28	24.72
330			25.14	25.39		
337.5	21.95	24.87				
340			25.40	25.97		
350	22.50	25.48	26.24	26.53	26.78	26.00
360			26.80			
362.5		26.08				
370			27.35	27.68		
375	23.60	26.68	27.63	27.96	28.18	27.40
380			27.90	28.26		
400	24.66	27.86			29.54	28.92
425	25.70	29.00			30.86	30.50
450	26.70	30.12			32.10	32.10
462.5	27.19					

Temp. (K)	Mole Fraction of Argon					
	0.7706 [Ref. 211]	0.8074 [Ref. 211]	0.8572 [Ref. 211]	0.9093 [Ref. 211]	0.9507 [Ref. 211]	1.0000 [Ref. 211]
275		21.83		21.56	21.50	21.35
287.5			22.45			
290	22.78					
300	23.38	23.40	23.30	23.15	23.07	22.96
310	24.00					
312.5						
320	24.60					
325	24.90	24.98	24.90	24.71	24.60	24.55
330	25.21					
337.5						
340	25.82					
350	26.42	26.53	26.46	26.28	26.10	26.11
360	27.04					
362.5						
370	27.65					
375	27.94	28.04	27.96	27.82	27.56	27.61
380	28.26					
400		29.47	29.43	29.30	29.00	29.09
425		30.84	30.80	30.70	30.44	30.54
450		32.16	32.13	32.08	31.84	31.97
462.5						

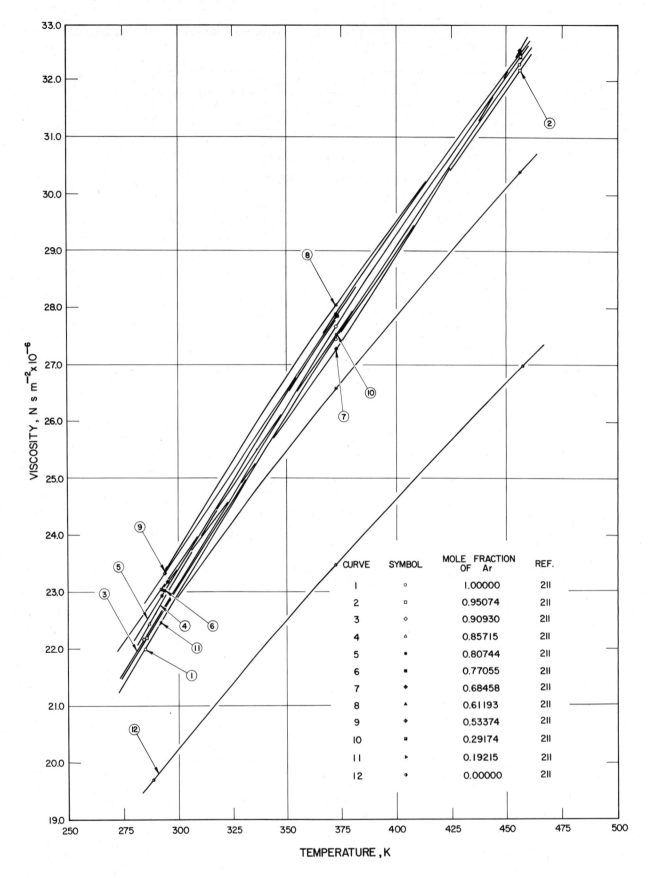

CURVE	SYMBOL	MOLE FRACTION OF Ar	REF.
1	○	1.00000	211
2	□	0.95074	211
3	◇	0.90930	211
4	△	0.85715	211
5	●	0.80744	211
6	■	0.77055	211
7	◆	0.68458	211
8	▲	0.61193	211
9	◆	0.53374	211
10	■	0.29174	211
11	▶	0.19215	211
12	●	0.00000	211

VISCOSITY, N s m^{-2} x 10^{-6}

TEMPERATURE , K

FIGURE 60-G(T). VISCOSITY DATA AS A FUNCTION OF TEMPERATURE
FOR GASEOUS ARGON-HELIUM MIXTURES

TABLE 61–G(C)E.   EXPERIMENTAL VISCOSITY DATA AS A FUNCTION OF COMPOSITION FOR GASEOUS ARGON-KRYPTON MIXTURES

Cur. No.	Fig. No.	Ref. No.	Author(s)	Temp. (K)	Pressure mm Hg	Mole Fraction of Kr	Viscosity (N s m^{-2} x 10^{-6})	Remarks
1	61–G(C)	278	Thornton, E.	291.2	700	1.000	24.8	Kr: 99-100 pure, balance Xe; Ar:
						0.865	24.5	99.8 pure; modified Rankine visco-
						0.777	24.5	meter, relative measurements;
						0.673	24.2	uncertainties: mixture composition
						0.546	23.9	±0.3%, viscosity ±1.0%; L$_1$ =
						0.443	23.6	0.095%, L$_2$ = 0.167%, L$_3$ = 0.410%.
						0.330	23.3	
						0.228	23.0	
						0.109	22.6	
						0.000	22.1	

TABLE 61–G(C)S.   SMOOTHED VISCOSITY VALUES AS A FUNCTION OF COMPOSITION FOR GASEOUS ARGON-KRYPTON MIXTURES

Mole Fraction of Kr	291.2 K [Ref. 278]
0.00	22.10
0.05	22.34
0.10	22.56
0.15	22.75
0.20	22.91
0.25	23.62
0.30	23.21
0.35	23.35
0.40	23.48
0.45	23.62
0.50	23.75
0.55	23.88
0.60	24.01
0.65	24.12
0.70	24.24
0.75	24.34
0.80	24.44
0.85	24.54
0.90	24.63
0.95	24.72
1.00	24.80

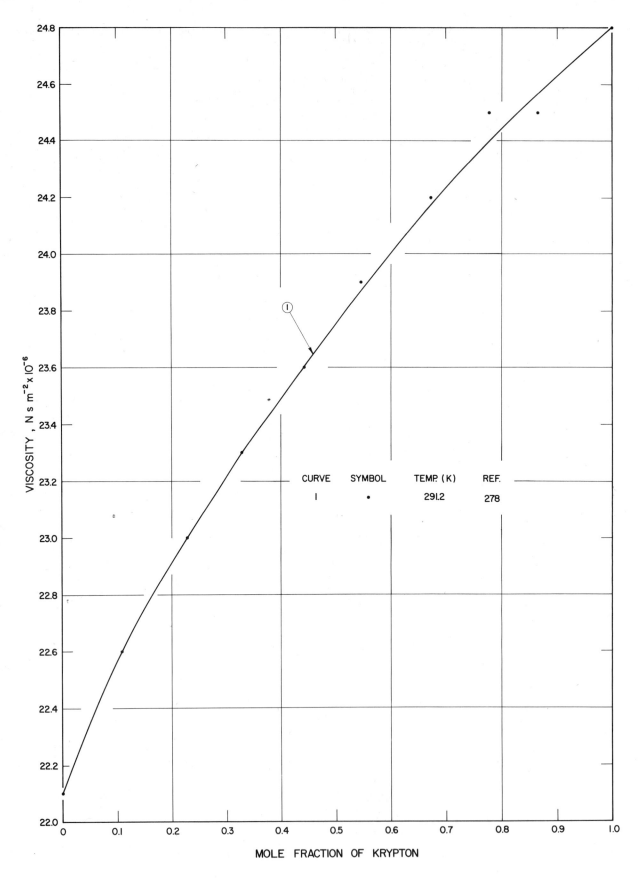

FIGURE 61-G(C). VISCOSITY DATA AS A FUNCTION OF COMPOSITION
FOR GASEOUS ARGON—KRYPTON MIXTURES

TABLE 62-G(C)E. EXPERIMENTAL VISCOSITY DATA AS A FUNCTION OF COMPOSITION FOR GASEOUS ARGON-NEON MIXTURES

Cur. No.	Fig. No.	Ref. No.	Author(s)	Temp. (K)	Pressure (mm Hg)	Mole Fraction of Ar	Viscosity ($N \ s \ m^{-2} \times 10^{-6}$)	Remarks
1	62-G(C)	180	Rietveld, A.O. and Van Itterbeek, A.	72.3		1.0000	6.38	Gas purities are not specified; oscillating disk viscometer, relative measurements; precision about 1%; $L_1 = 0.027\%$, $L_2 = 0.071\%$, $L_3 = 0.189\%$.
						0.8300	7.00	
						0.6707	7.70	
						0.5011	8.52	
						0.3231	9.52	
						0.1613	10.56	
						0.0000	11.72	
2	62-G(C)	180	Rietveld, A.O. and Van Itterbeek, A.	90.3		1.0000	7.75	Same remarks as for curve 1 except $L_1 = 0.195\%$, $L_2 = 0.317\%$, $L_3 = 0.733\%$.
						0.8390	8.51	
						0.6713	9.22	
						0.4828	10.18	
						0.3265	11.16	
						0.1634	12.37	
						0.0000	13.52	
3	62-G(C)	180	Rietveld, A.O. and Van Itterbeek, A.	193.4		1.0000	15.29	Same remarks as for curve 1 except $L_1 = 0.424\%$, $L_2 = 0.638\%$, $L_3 = 1.461\%$.
						0.8298	16.38	
						0.6690	17.32	
						0.5024	18.75	
						0.3292	20.27	
						0.1698	21.79	
						0.0000	23.52	
4	62-G(C)	180	Rietveld, A.O. and Van Itterbeek, A.	229.0		1.0000	18.03	Same remarks as for curve 1 except $L_1 = 0.196\%$, $L_2 = 0.276\%$, $L_3 = 0.624\%$.
						1.0000	18.00	
						1.0000	17.88	
						0.8320	19.24	
						0.6507	20.63	
						0.5017	21.82	
						0.4308	22.58	
						0.3348	23.39	
						0.1654	25.00	
						0.0000	26.70	
5	62-G(C)	180	Rietveld, A.O. and Van Itterbeek, A.	291.1		1.0000	22.15	Same remarks as for curve 1 except $L_1 = 0.163\%$, $L_2 = 0.224\%$, $L_3 = 0.407\%$.
						1.0000	22.06	
						0.8323	23.39	
						0.6757	24.69	
						0.4970	26.36	
						0.3227	27.93	
						0.1693	29.61	
						0.0000	31.29	
						0.0000	31.40	
6	62-G(C)	213	Thornton, E. and Baker, W.A.D.	291.2	700.0	1.000	22.0	Ar: impurities not exceeding 0.2%; He: spectroscopically pure; modified Rankine viscometer, relative measurements; uncertainties: mixture composition ±0.3%, viscosity ±1.0%; $L_1 = 0.460\%$, $L_2 = 0.599$, $L_3 = 1.238\%$
						0.900	22.8	
						0.803	23.6	
						0.726	23.9	
						0.638	24.7	
						0.541	25.5	
						0.436	26.7	
						0.328	27.8	
						0.221	28.5	
						0.157	29.0	
						0.000	30.7	
7	62-G(C)	221	Trautz, M. and Binkele, H.E.	293		0.0000	30.92	Ar: Linde Co., commercial grade, 99.8-99.5 purity, Ne: Linde Co., commercial grade, 99.0-99.5 purity; capillary method, r = 0.2019 mm; accuracy <±0.4%; $L_1 = 0.000\%$, $L_2 = 0.000\%$, $L_3 = 0.000\%$.
						0.2680	28.08	
						0.6091	25.04	
						0.7420	24.01	
						1.0000	22.13	
8	62-G(C)	221	Trautz, M. and Binkele, H.E.	373		0.0000	36.23	Same remarks as for curve 7 except $L_1 = 0.000\%$, $L_2 = 0.000\%$, $L_3 = 0.000\%$.
						0.2680	33.13	
						0.6091	29.90	
						0.7420	28.85	
						1.0000	26.93	
9	62-G(C)	221	Trautz, M. and Binkele, H.E.	473		0.0000	42.20	Same remarks as for curve 7 except $L_1 = 0.000\%$, $L_2 = 0.000\%$, $L_3 = 0.000\%$.
						0.2680	38.90	
						0.6091	35.29	
						0.7420	34.13	
						1.0000	32.22	
10	62-G(C)	221	Trautz, M. and Binkele, H.E.	523		0.0000	45.01	Same remarks as for curve 7 except $L_1 = 0.107\%$, $L_2 = 0.170\%$, $L_3 = 0.273\%$.
						0.2680	41.50	
						0.6091	37.93	
						0.7420	36.58	
						1.0000	34.60	

TABLE 62-G(C)S. SMOOTHED VISCOSITY VALUES AS A FUNCTION OF COMPOSITION FOR GASEOUS ARGON-NEON MIXTURES

Mole Fraction of Ar	72.3 K [Ref. 180]	90.3 K [Ref. 180]	193.4 K [Ref. 180]	229.0 K [Ref. 180]	291.1 K [Ref. 180]
0.00	11.72	13.52	23.52	26.70	31.38
0.05	11.37	13.14	22.95	26.16	30.08
0.10	11.01	12.76	22.42	25.64	30.24
0.15	10.65	12.38	21.90	25.13	29.70
0.20	10.31	12.01	21.40	24.63	29.16
0.25	9.98	11.66	20.90	24.14	28.66
0.30	9.66	11.32	20.44	23.65	28.16
0.35	9.35	11.00	19.98	23.18	27.68
0.40	9.06	10.68	19.54	22.72	27.20
0.45	8.80	10.38	19.10	22.28	26.73
0.50	8.53	10.10	18.68	21.85	26.26
0.55	8.28	9.83	18.27	21.43	25.80
0.60	7.03	9.57	17.86	21.03	25.34
0.65	7.80	9.32	17.28	20.62	24.90
0.70	7.56	9.10	17.12	20.22	24.48
0.75	7.34	8.87	16.79	19.82	24.06
0.80	7.12	8.65	16.47	19.42	23.76
0.85	6.92	8.44	16.16	19.04	23.26
0.90	6.74	8.21	15.86	18.67	22.87
0.95	6.56	8.00	15.56	18.34	22.48
1.00	6.38	7.75	15.29	18.02	22.11

Mole Fraction of Ar	291.2 K [Ref. 213]	293 K [Ref. 221]	373 K [Ref. 221]	473 K [Ref. 221]	523 K [Ref. 221]
0.00	30.89	30.92	36.23	42.20	45.01
0.05	30.31	30.38	35.65	41.57	44.32
0.10	29.75	29.82	35.08	40.95	43.65
0.15	29.21	29.30	34.45	40.32	43.00
0.20	28.70	28.78	33.90	39.72	42.37
0.25	28.22	28.29	33.36	39.14	41.72
0.30	27.82	27.78	32.81	38.55	41.13
0.35	27.25	27.30	32.30	37.99	40.55
0.40	26.79	26.82	31.80	37.44	40.00
0.45	26.34	26.38	31.30	36.90	39.47
0.50	25.90	25.92	30.85	36.40	38.92
0.55	25.46	25.52	30.40	35.89	38.41
0.60	25.03	25.10	29.99	35.40	37.91
0.65	24.61	24.72	29.60	34.92	37.43
0.70	24.20	24.35	29.20	34.50	37.00
0.75	23.81	23.94	28.82	34.07	36.60
0.80	23.43	23.60	28.48	33.67	36.19
0.85	23.06	23.22	28.08	33.30	35.79
0.90	22.71	22.88	27.68	32.92	35.40
0.95	22.36	22.50	27.30	32.58	35.00
1.00	22.00	22.13	26.93	32.22	34.60

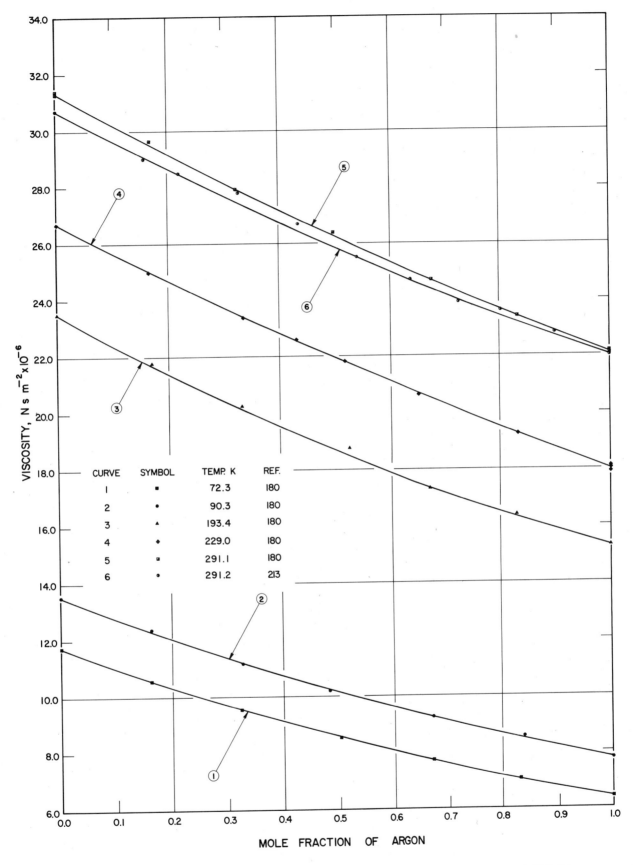

CURVE	SYMBOL	TEMP. K	REF.
1	■	72.3	180
2	•	90.3	180
3	▲	193.4	180
4	◆	229.0	180
5	▫	291.1	180
6	•	291.2	213

MOLE FRACTION OF ARGON

VISCOSITY, N s m^{-2} x 10^{-6}

FIGURE 62-G(C). VISCOSITY DATA AS A FUNCTION OF COMPOSITION
FOR GASEOUS ARGON – NEON MIXTURES

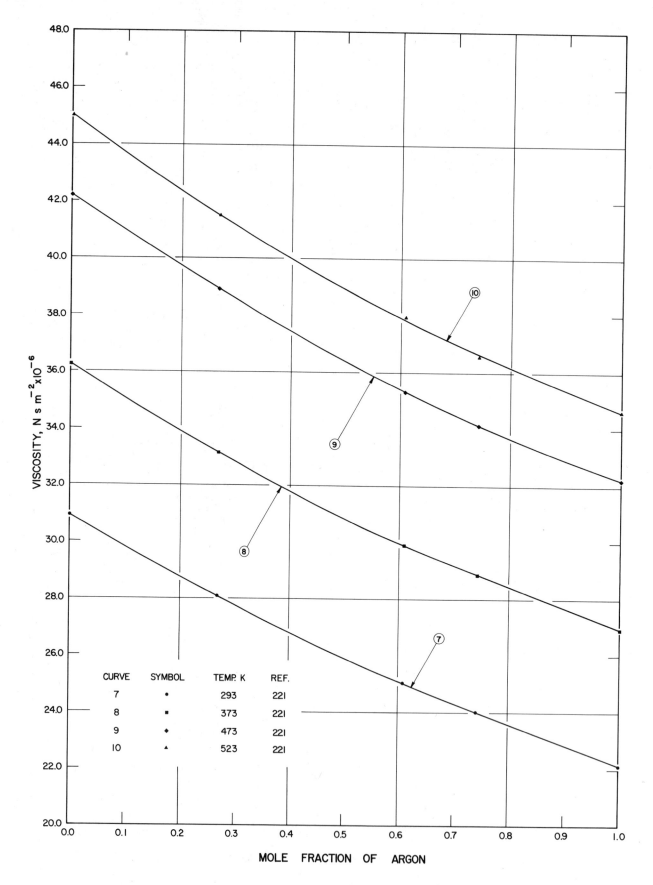

FIGURE 62-G(C).   VISCOSITY DATA AS A FUNCTION OF COMPOSITION
FOR GASEOUS ARGON-NEON MIXTURES (continued)

TABLE 62-G(D)E. EXPERIMENTAL VISCOSITY DATA AS A FUNCTION OF DENSITY FOR GASEOUS ARGON-NEON MIXTURES

Cur. No.	Fig. No.	Ref. No.	Author(s)	Mole Fraction of Ar	Temp. (K)	Density (g cm^{-3})	Viscosity (N s m^{-2} x 10^{-6})	Remarks
1	62–G(D)	323	Kestin, J. and Nagashima, A.	0.000	293.2	0.04037	31.597	Ar: 99.997 pure, Ne: 99.991 pure; oscillating disk viscometer; accuracy ±0.1%, ratios of viscosity values ±0.04%.
						0.03294	31.572	
						0.02495	31.536	
						0.01640	31.497	
						0.008471	31.476	
						0.0008526	31.412	
						0.03744	31.608	
						0.02910	31.577	
						0.02068	31.543	
						0.01166	31.499	
						0.004197	31.473	
						0.0008502	31.450	
2	62–G(D)	323	Kestin, J. and Nagashima, A.	0.402	293.2	0.04133	27.527	Same remarks as for curve 1.
						0.03342	27.434	
						0.02517	27.356	
						0.01723	27.296	
						0.009201	27.215	
						0.001192	27.163	
3	62–G(D)	323	Kestin, J. and Nagashima, A.	0.668	293.2	0.04901	25.314	Same remarks as for curve 1.
						0.04192	25.222	
						0.03513	25.129	
						0.02806	25.055	
						0.02092	24.985	
						0.01408	24.893	
						0.006994	24.847	
						0.001435	24.790	
4	62–G(D)	323	Kestin, J. and Nagashima, A.	0.901	293.2	0.04980	23.610	Same remarks as for curve 1.
						0.04006	23.458	
						0.03213	23.351	
						0.02392	23.251	
						0.01569	23.155	
						0.008007	23.062	
						0.001622	23.003	
5	62–G(D)	323	Kestin, J. and Nagashima, A.	1.000	293.2	0.08723	23.608	Same remarks as for curve 1.
						0.06875	23.258	
						0.06010	23.123	
						0.05103	22.970	
						0.04260	22.836	
						0.03399	22.711	
						0.02519	22.587	
						0.01675	22.488	
						0.008399	22.384	
						0.001684	22.300	
6	62–G(D)	323	Kestin, J. and Nagashima, A.	0.000	303.2	0.03733	32.364	Same remarks as for curve 1.
						0.03156	32.346	
						0.02263	32.300	
						0.01638	32.298	
						0.008003	32.246	
						0.0008327	32.213	
7	62–G(D)	323	Kestin, J. and Nagashima, A.	0.402	303.2	0.03872	28.203	Same remarks as for curve 1
						0.03192	28.094	
						0.02433	28.030	
						0.01612	27.974	
						0.008851	27.923	
						0.001146	27.856	
8	62–G(D)	323	Kestin, J. and Nagashima, A.	0.668	303.2	0.04547	25.966	Same remarks as for curve 1.
						0.03845	25.865	
						0.02914	25.775	
						0.02022	25.671	
						0.01034	25.581	
						0.001378	25.475	
9	62–G(D)	323	Kestin, J. and Nagashima, A.	0.901	303.2	0.04407	24.201	Same remarks as for curve 1.
						0.03887	24.125	
						0.03102	24.016	
						0.02310	23.904	
						0.01542	23.818	
						0.007688	23.738	
						0.001569	23.666	
10	62–G(D)	323	Kestin, J. and Nagashima, A.	1.000	303.2	0.08430	24.236	Same remarks as for curve 1.
						0.06913	23.960	
						0.05766	23.725	
						0.04953	23.597	
						0.04086	23.479	
						0.03268	23.353	
						0.02421	23.243	
						0.01635	23.142	
						0.008095	23.044	
						0.001650	22.980	

TABLE 62-G(D)S.   SMOOTHED VISCOSITY VALUES AS A FUNCTION OF DENSITY FOR GASEOUS ARGON-NEON MIXTURES

Density (g cm⁻³)	Mole Fraction of Argon				
	0.000 (293.2 K) [Ref. 323]	0.402 (293.2 K) [Ref. 323]	0.668 (293.2 K) [Ref. 323]	0.901 (293.2 K) [Ref. 323]	1.000 (293.2 K) [Ref. 323]
0.0000	31.50			23.00	
0.0025	31.49	27.174	24.790	23.02	
0.0050	31.48	27.190	24.820	23.04	22.338
0.0100	31.48	27.223	24.875	23.08	22.400
0.0150	31.49	27.260	24.930	23.14	22.460
0.0200	31.50	27.302	24.980	23.20	22.525
0.0250	31.52	27.349	25.040	23.26	22.588
0.0300	31.54	27.401	25.098	23.33	22.655
0.0350	31.56	27.560	25.153	23.39	22.725
0.0400	31.60	27.515	25.210	23.46	22.795
0.0450		27.576	25.265	23.54	22.870
0.0500			25.320	23.62	22.946
0.0550					23.025
0.0600					23.105
0.0650					23.190
0.0700					23.280
0.0750					23.370
0.0800					23.465
0.0900					23.665
0.0950					23.770
0.1000					23.875

Density (g cm⁻³)	Mole Fraction of Argon				
	0.000 (303.2 K) [Ref. 323]	0.402 (303.2 K) [Ref. 323]	0.668 (303.2 K) [Ref. 323]	0.901 (303.2 K) [Ref. 323]	1.000 (303.2 K) [Ref. 323]
0.0025	32.230	27.174			
0.0050	32.239	27.189	25.512	23.705	
0.0100	32.258	27.223	25.560	23.758	23.070
0.0150	32.278	27.260	25.615	23.810	23.130
0.0200	32.297	27.302	25.667	23.870	23.188
0.0250	32.317	27.349	25.716	23.930	23.250
0.0300	32.336	27.400	25.770	23.995	23.315
0.0350	32.356	27.456	25.824	24.067	23.383
0.0400	32.374	27.515	25.885	24.140	23.455
0.0450	32.393	27.517	25.958	24.213	23.526
0.0500			26.035	24.295	23.601
0.0550					23.680
0.0600					23.762
0.0650					23.850
0.0700					23.945
0.0750					24.040
0.0800					24.142
0.0850					24.250
0.0900					24.365

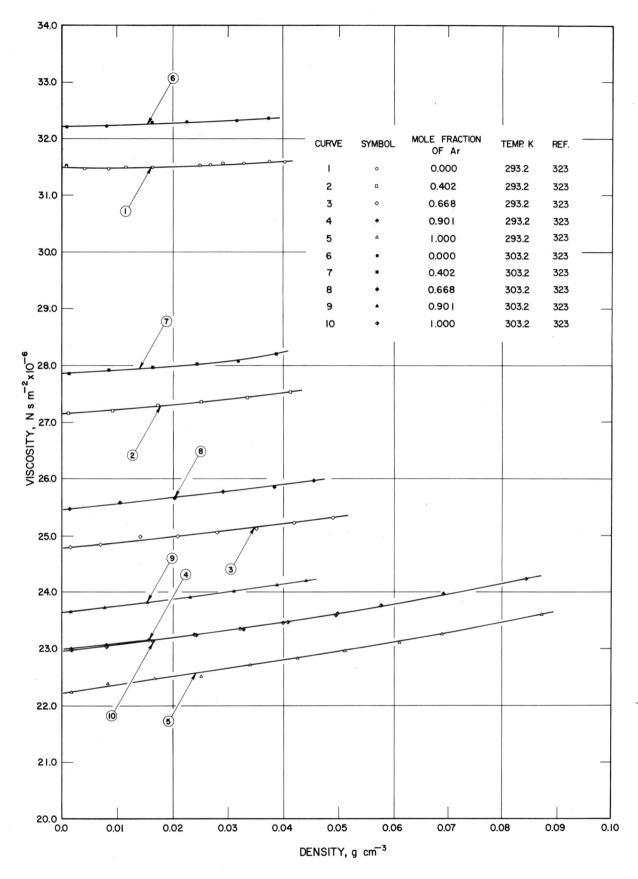

CURVE	SYMBOL	MOLE FRACTION OF Ar	TEMP. K	REF.
1	○	0.000	293.2	323
2	□	0.402	293.2	323
3	◇	0.668	293.2	323
4	◆	0.901	293.2	323
5	△	1.000	293.2	323
6	●	0.000	303.2	323
7	■	0.402	303.2	323
8	◆	0.668	303.2	323
9	▲	0.901	303.2	323
10	◆	1.000	303.2	323

FIGURE 62 – G (D). VISCOSITY DATA AS A FUNCTION OF DENSITY
FOR GASEOUS ARGON – NEON MIXTURES

TABLE 63-G(C)E.   EXPERIMENTAL VISCOSITY DATA AS A FUNCTION OF COMPOSITION FOR GASEOUS
ARGON-XENON MIXTURES

Cur. No.	Fig. No.	Ref. No.	Author(s)	Temp. (K)	Pressure mm Hg	Mole Fraction of Xe	Viscosity (N s m^{-2} x 10^{-6})	Remarks
1	63-G(C)	324	Thornton, E.	291.2	700	1.000	22.5	Xe: 99-100 pure, balance Kr,
						0.905	22.6	Ar: 99.8 pure; modified Rankine
						0.792	22.8	viscometer, relative measurements;
						0.701	22.9	uncertainties: mixture composition
						0.598	22.9	±0.3%, viscosity ±1.0%; L_1 =
						0.498	23.0	0.153%, L_2 = 0.189%, L_3 = 0.444%
						0.405	22.9	
						0.300	22.9	
						0.213	22.8	
						0.109	22.4	
						0.000	22.1	

TABLE 63-G(C)S.   SMOOTHED VISCOSITY VALUES AS A FUNCTION OF COMPOSITION FOR GASEOUS ARGON-XENON MIXTURES

Mole Fraction of Xe	291.2 K [Ref. 324]
0.00	22.10
0.05	22.29
0.10	22.47
0.15	22.62
0.20	22.74
0.25	22.82
0.30	22.88
0.35	22.91
0.40	22.94
0.45	22.95
0.50	22.96
0.55	22.95
0.60	22.94
0.65	22.91
0.70	22.87
0.75	22.82
0.80	22.77
0.85	22.71
0.90	22.64
0.95	22.57
1.00	22.50

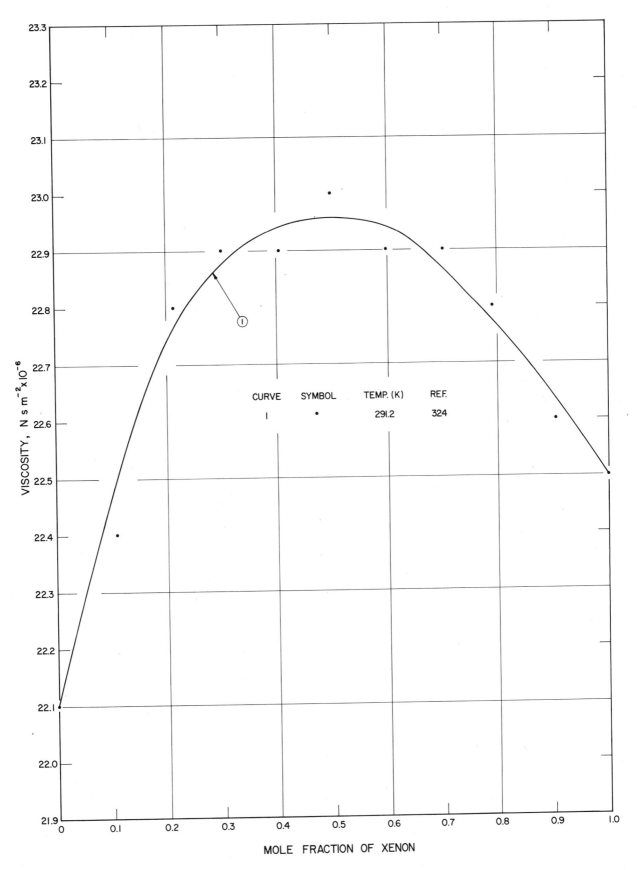

FIGURE 63-G(C). VISCOSITY DATA AS A FUNCTION OF COMPOSITION
FOR GASEOUS ARGON-XENON MIXTURES

TABLE 64-G(C)E.   EXPERIMENTAL VISCOSITY DATA AS A FUNCTION OF COMPOSITION FOR GASEOUS HELIUM-KRYPTON MIXTURES

Cur. No.	Fig. No.	Ref. No.	Author(s)	Temp. (K)	Pressure (mm Hg)	Mole Fraction of Kr	Viscosity (N s m^{-2} x 10^{-6})	Remarks
1	64-G(C)	325	Nasini, A.G. and Rossi, C.	283.2		0.0000	19.52	He and Kr: commercial grade; capillary method; precision ±0.2-0.3%; L_1 = 0.054%, L_2 = 0.098%, L_3 = 0.285%.
						0.1021	23.35	
						0.2046	24.97	
						0.3086	25.61	
						0.4995	25.54	
						0.7098	25.16	
						0.8100	24.93	
						0.8845	24.75	
						0.9454	24.64	
						1.0000	24.41	
2	64-G(C)	278	Thornton, E.	291.2	700	1.000	24.8	Kr: 99-100 pure, balance Xe, He: spectroscopically pure; modified Rankine viscometer, relative measurements; uncertainties: mixture composition ±0.3%, viscosity ±1.0%; L_1 = 0.206%, L_2 = 0.294%, L_3 = 0.548%.
						0.891	25.2	
						0.797	25.4	
						0.698	25.9	
						0.600	26.0	
						0.439	26.3	
						0.353	26.4	
						0.272	26.2	
						0.151	24.9	
						0.069	22.9	
						0.000	19.4	
3	64-G(C)	325	Nasini, A.G. and Rossi, C.	373.2		0.0000	23.35	Same remarks as for curve 1 except L_1 = 0.056%, L_2 = 0.077%, L_3 = 0.195%.
						0.1021	27.85	
						0.2046	30.01	
						0.3086	31.09	
						0.4995	31.42	
						0.7098	31.27	
						0.8100	31.15	
						0.8845	30.96	
						0.9454	30.76	
						1.0000	30.68	

TABLE 64-G(C)S.   SMOOTHED VISCOSITY VALUES AS A FUNCTION OF COMPOSITION FOR GASEOUS HELIUM-KRYPTON MIXTURES

Mole Fraction of Kr	283.2 K [Ref. 325]	291.2 K [Ref. 278]	373.2 K [Ref. 325]
0.00	19.52	19.40	23.35
0.05	21.80	22.15	25.81
0.10	23.30	23.94	27.73
0.15	24.29	24.94	28.97
0.20	24.90	25.61	29.95
0.25	25.28	26.06	30.64
0.30	25.56	26.30	31.03
0.35	25.68	26.40	31.22
0.40	25.66	26.44	31.33
0.45	25.62	26.42	31.40
0.50	25.56	26.36	31.41
0.55	25.46	26.26	31.40
0.60	25.36	26.14	31.37
0.65	25.26	26.00	31.33
0.70	25.16	25.84	31.28
0.75	25.06	25.70	31.22
0.80	24.95	25.53	31.16
0.85	24.82	25.36	31.06
0.90	24.70	25.18	30.94
0.95	24.56	25.00	30.80
1.00	24.42	24.82	30.66

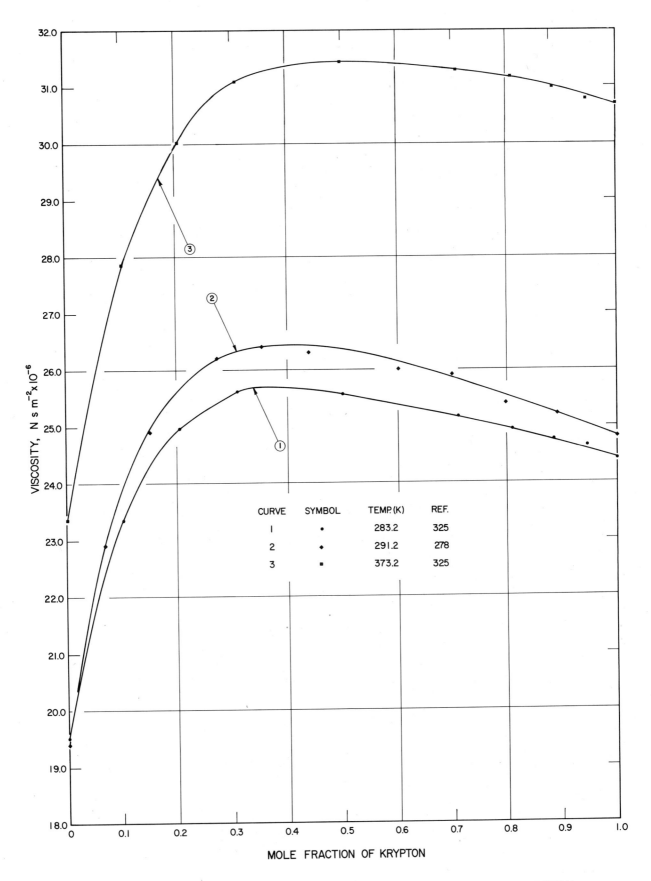

FIGURE 64-G(C). VISCOSITY DATA AS A FUNCTION OF COMPOSITION
FOR GASEOUS HELIUM-KRYPTON MIXTURES

TABLE 64-G(D)E.　　EXPERIMENTAL VISCOSITY DATA AS A FUNCTION OF DENSITY FOR GASEOUS HELIUM-KRYPTON MIXTURES

Cur. No.	Fig. No.	Ref. No.	Author(s)	Mole Fraction of Kr	Temp. (K)	Density (g cm^{-3})	Viscosity (N s m^{-2} x 10^{-6})	Remarks
1	64-G(D)	326	Kestin, J., Kobayashi, Y., and Wood, R.T.	1.000	293.2	0.09126	26.013	Kr: 99.99 pure, He: 99.995 pure; oscillating disk viscometer; uncertainties: mixture composition ±0.002%, viscosity ±0.10%, viscosity ratios ±0.04%.
						0.08253	25.901	
						0.07000	25.737	
						0.06342	25.662	
						0.05413	25.530	
						0.04137	25.402	
						0.03566	25.310	
						0.02653	25.203	
						0.01765	25.121	
						0.01045	25.057	
						0.004016	24.996	
2	64-G(D)	326	Kestin, J., et al.	0.6737	293.2	0.05892	26.478	Same remarks as for curve 1.
						0.03775	26.254	
						0.01205	26.021	
						0.002437	25.931	
3	64-G(D)	326	Kestin, J., et al.	0.4924	293.2	0.04513	26.752	Same remarks as for curve 1.
						0.02738	26.604	
						0.009019	26.449	
						0.001825	26.378	
4	64-G(D)	326	Kestin, J., et al.	0.3881	293.2	0.03800	26.752	Same remarks as for curve 1.
						0.02139	26.620	
						0.007274	26.533	
						0.001474	26.462	
5	64-G(D)	326	Kestin, J., et al.	0.3239	293.2	0.03164	26.633	Same remarks as for curve 1.
						0.01822	26.530	
						0.006256	26.439	
						0.001265	26.391	
6	64-G(D)	326	Kestin, J., et al.	0.2823	293.2	0.02762	26.470	Same remarks as for curve 1.
						0.01647	26.408	
						0.005528	26.335	
						0.001118	26.285	
7	64-G(D)	326	Kestin, J., et al.	0.1909	293.2	0.01974	25.736	Same remarks as for curve 1.
						0.01191	25.699	
						0.003994	25.671	
						0.0008098	25.638	
8	64-G(D)	326	Kestin, J., et al.	0.1415	293.2	0.01591	24.991	Same remarks as for curve 1.
						0.009441	24.964	
						0.003191	24.947	
						0.0006443	24.909	
9	64-G(D)	326	Kestin, J., et al.	0.1068	293.2	0.01106	24.223	Same remarks as for curve 1.
						0.007778	24.235	
						0.002602	24.202	
						0.0005277	24.180	
10	64-G(D)	326	Kestin, J., et al.	1.0000	303.2	0.08791	26.782	Same remarks as for curve 1.
						0.06985	26.531	
						0.05972	26.392	
						0.04487	26.208	
						0.03436	26.073	
						0.01710	25.909	
						0.01026	25.820	
						0.003421	25.759	
11	64-G(D)	326	Kestin, J., et al.	0.6737	303.2	0.04816	27.156	Same remarks as for curve 1.
						0.02561	26.946	
						0.01166	26.802	
						0.002356	26.716	
12[*]	64-G(D)	326	Kestin, J., et al.	0.4924	303.2	0.04357	27.487	Same remarks as for curve 1.
						0.02640	27.336	
						0.008717	27.210	
						0.001765	27.151	
13	64-G(D)	326	Kestin, J., et al.	0.3881	303.2	0.02961	27.425	Same remarks as for curve 1.
						0.02431	27.385	
						0.02019	27.344	
						0.007034	27.266	
						0.001425	27.205	
14	64-G(D)	326	Kestin, J., et al.	0.3239	303.2	0.03025	27.317	Same remarks as for curve 1.
						0.01817	27.249	
						0.006007	27.168	
						0.001224	27.129	
15	64-G(D)	326	Kestin, J., et al.	0.2823	303.2	0.02320	27.164	Same remarks as for curve 1.
						0.01606	27.119	
						0.005346	27.044	
						0.001097	27.011	

TABLE 64-G(D)E.   EXPERIMENTAL VISCOSITY DATA AS A FUNCTION OF DENSITY FOR GASEOUS
HELIUM-KRYPTON MIXTURES (continued)

Cur. No.	Fig. No.	Ref. No.	Author(s)	Mole Fraction of Kr	Temp. (K)	Density (g cm^{-3})	Viscosity (N s m^{-2}x10^{-6})	Remarks
16	64-G(D)	326	Kestin, J., et al.	0.1909	303.2	0.01944	26.393	Same remarks as for curve 1.
						0.01153	26.355	
						0.003862	26.315	
						0.0007836	26.282	
17	64-G(D)	326	Kestin, J., et al.	0.1415	303.2	0.01547	25.625	Same remarks as for curve 1.
						0.009458	25.606	
						0.003044	25.663	
						0.0006230	25.532	
18	64-G(D)	326	Kestin, J., et al.	0.1068	303.2	0.01202	24.831	Same remarks as for curve 1.
						0.007522	24.824	
						0.002516	24.809	
						0.0005102	24.771	
19*	64-G(D)	326	Kestin, J., et al.	0.0000	303.2	0.003927	20.074	Same remarks as for curve 1.
						0.003691	20.075	
						0.003198	20.072	
						0.002767	20.077	
						0.002378	20.069	
						0.002023	20.073	
						0.001203	20.077	
						0.0007979	20.071	
						0.0004884	20.070	
						0.0001656	20.068	

TABLE 64-G(D)S. SMOOTHED VISCOSITY VALUES AS A FUNCTION OF DENSITY FOR GASEOUS HELIUM-KRYPTON MIXTURES

Density (g cm⁻³)	Mole Fraction of Krypton								
	0.1068 (293.2 K) [Ref. 326]	0.1415 (293.2 K) [Ref. 326]	0.1909 (293.2 K) [Ref. 326]	0.2823 (293.2 K) [Ref. 326]	0.3239 (293.2 K) [Ref. 326]	0.3881 (293.2 K) [Ref. 326]	0.4924 (293.2 K) [Ref. 326]	0.6737 (293.2 K) [Ref. 326]	1.0000 (293.2 K) [Ref. 326]
0.00125	24.188	24.920	25.644						
0.00250	24.205	24.940	25.663	26.297	26.402	26.476	26.387		
0.00375	24.215	24.958	25.678						
0.00500	24.225	24.970	25.690	26.320	26.427	26.510	26.418	25.975	25.005
0.00625	24.232	24.980	25.700						
0.00750	24.235	24.988	25.708	26.341	26.452	26.539	26.449		
0.00875	24.235		25.712						
0.01000	24.232	24.995	25.716	26.362	26.475	26.570	26.479	26.014	25.054
0.01125	24.224								
0.01250		25.000	25.720	26.382		26.595	26.508		
0.01500		25.000	25.724	26.400	26.520	26.618	26.536	26.050	25.098
0.01625		25.000							
0.01750			25.724	26.416			26.565		
0.02000			25.700	26.432	26.561	26.659	26.593	26.092	25.142
0.02250				26.446			26.618		
0.02500				26.459	26.599	26.692	26.642	26.134	25.190
0.02750				26.472			26.664		
0.03000					26.631	26.718	26.682	26.178	25.241
0.03500					26.659	26.740	26.712	26.274	25.295
0.03750					26.674	26.750			
0.04000						26.760	26.735	26.274	25.350
0.04500							26.752	26.322	25.408
0.05000								26.380	25.468
0.05500								26.435	25.524
0.06000								26.490	25.590
0.06500									25.652
0.07000									25.719
0.07500									25.785
0.08000									25.855
0.08500									25.921
0.09000									25.994
0.09500									26.066
0.10000									26.140

TABLE 64-G(D)S.   SMOOTHED VISCOSITY VALUES AS A FUNCTION OF DENSITY FOR GASEOUS
HELIUM-KRYPTON MIXTURES (continued)

Density (g cm⁻³)	Mole Fraction of Krypton								
	0.0000 (303.2 K) [Ref. 326]	0.1068 (303.2 K) [Ref. 326]	0.1415 (303.2 K) [Ref. 326]	0.1909 (303.2 K) [Ref. 326]	0.2823 (303.2 K) [Ref. 326]	0.3239 (303.2 K) [Ref. 326]	0.3881 (303.2 K) [Ref. 326]	0.6737 (303.2 K) [Ref. 326]	1.0000 (303.2 K) [Ref. 326]
0.00050	20.070								
0.00100	20.072								
0.00125		24.783	25.565	26.292					
0.00150	20.073								
0.00200	20.073								
0.00250	20.074	24.810	25.585	26.320	27.023	27.138	27.215	26.718	
0.00300	20.074								
0.00350	20.075								
0.00375		24.820	25.601	26.340					
0.00400	20.075								
0.00450	20.076								
0.00500	20.076	24.828	25.612	26.352	27.044	27.160	27.239	26.749	
0.00625		24.831		26.362					
0.00750		24.831	25.628	26.369	27.064	27.179	27.262		
0.00875		24.833							
0.01000		24.833	25.636	26.370	27.084	27.198	27.284	26.808	25.048
0.01125		24.835							
0.01250		24.835	25.638	26.372		27.215			
0.01500			25.634	26.378	27.118	27.320	27.306	26.874	
0.01625			25.632						
0.01750				26.378					
0.02000				26.378	27.148	27.262	27.359	26.914	25.142
0.02500					27.174	27.292	27.389	26.962	
0.03000						27.316	27.415	27.008	25.240
0.03250							27.428		
0.03500						27.380		27.060	
0.04000								27.092	25.350
0.04500								27.131	
0.05000								27.168	25.468
0.06000									25.590
0.07000									25.719
0.08000									25.855
0.09000									25.995
0.10000									26.140

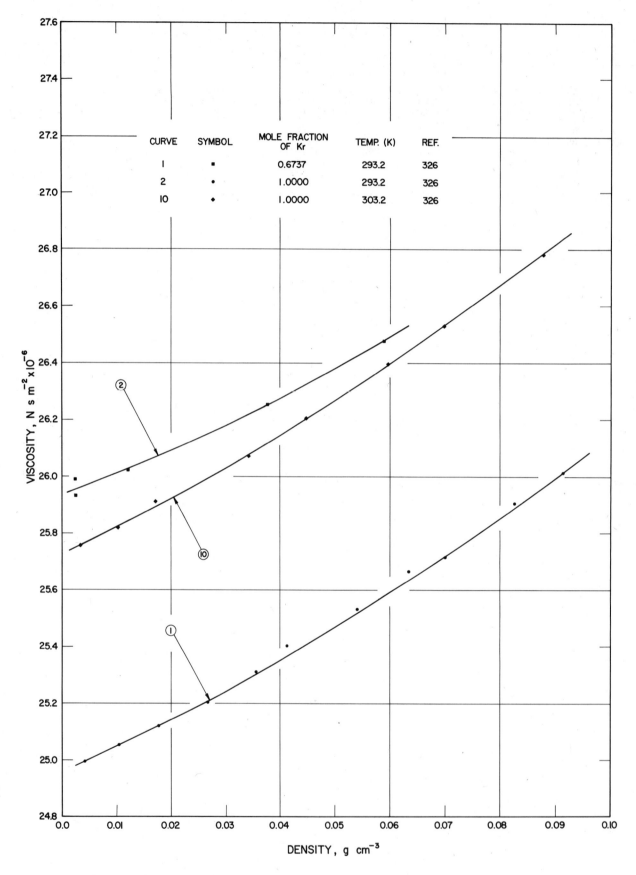

CURVE	SYMBOL	MOLE FRACTION OF Kr	TEMP. (K)	REF.
1	■	0.6737	293.2	326
2	•	1.0000	293.2	326
10	◆	1.0000	303.2	326

FIGURE 64-G(D).   VISCOSITY DATA AS A FUNCTION OF DENSITY
FOR GASEOUS HELIUM - KRYPTON MIXTURES

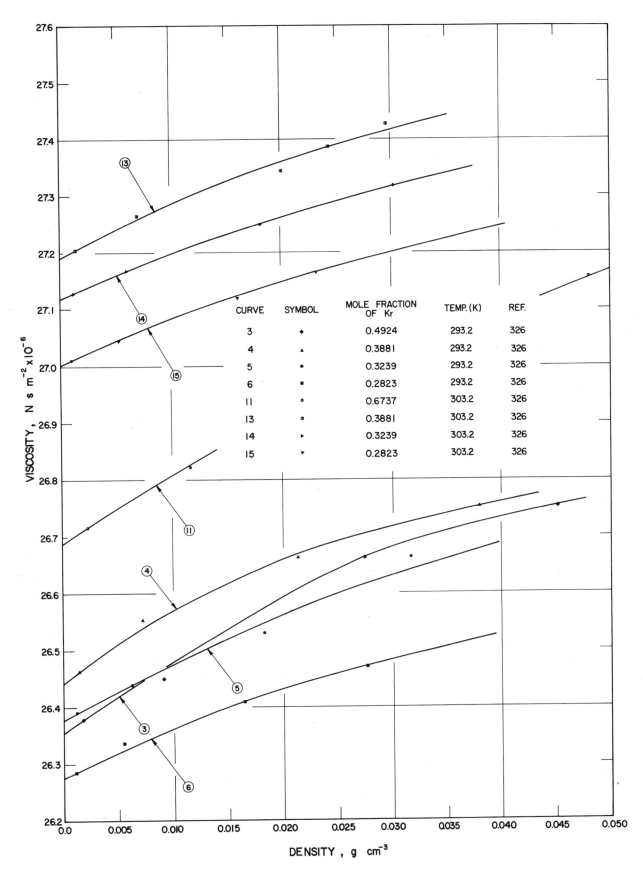

CURVE	SYMBOL	MOLE FRACTION OF Kr	TEMP. (K)	REF.
3	♦	0.4924	293.2	326
4	▲	0.3881	293.2	326
5	●	0.3239	293.2	326
6	■	0.2823	293.2	326
11	◦	0.6737	303.2	326
13	▫	0.3881	303.2	326
14	▸	0.3239	303.2	326
15	▼	0.2823	303.2	326

FIGURE 64-G(D). VISCOSITY DATA AS A FUNCTION OF DENSITY
FOR GASEOUS HELIUM-KRYPTON MIXTURES (continued)

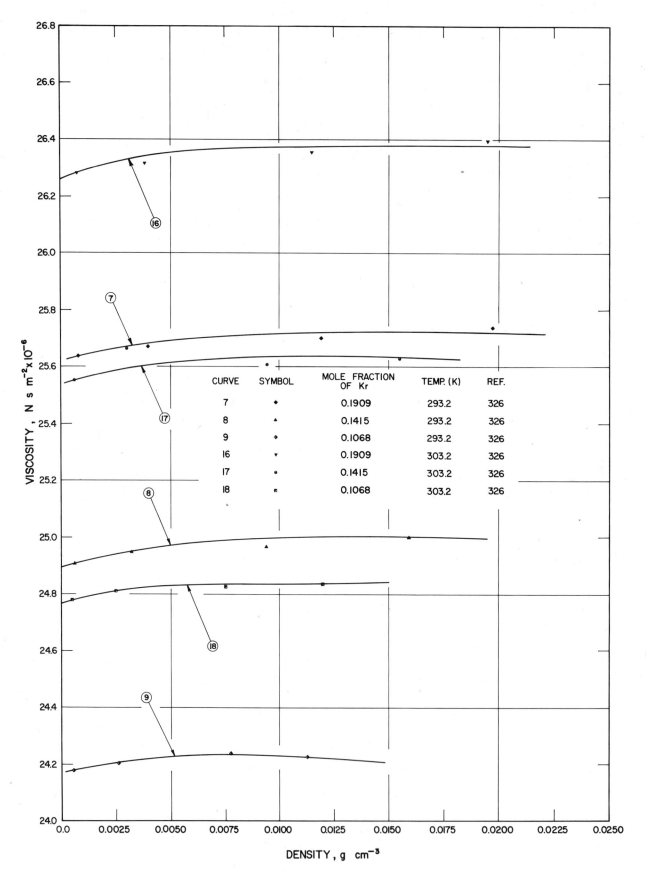

CURVE	SYMBOL	MOLE FRACTION OF Kr	TEMP. (K)	REF.
7	◆	0.1909	293.2	326
8	▲	0.1415	293.2	326
9	◈	0.1068	293.2	326
16	▼	0.1909	303.2	326
17	○	0.1415	303.2	326
18	▫	0.1068	303.2	326

FIGURE 64 – G (D).   VISCOSITY DATA AS A FUNCTION OF DENSITY
FOR GASEOUS HELIUM-KRYPTON MIXTURES (continued)

TABLE 65-G(C)E.　EXPERIMENTAL VISCOSITY DATA AS A FUNCTION OF COMPOSITION FOR GASEOUS HELIUM-NEON MIXTURES

Cur. No.	Fig. No.	Ref. No.	Author(s)	Temp. (K)	Pressure (mm Hg)	Mole Fraction of Ne	Viscosity $(N\ s\ m^{-2} \times 10^{-6})$	Remarks
1	65-G(C)	179	Rietveld, A.O., Van Itterbeek, A., and Velds, C.A.	20.4	40.0 19.0 13.0 9.0 7.0	0.000 0.256 0.492 0.720 1.000	3.50 3.67 3.69 3.61 3.51	He and Ne: purities not specified; oscillating disk viscometer, relative measurements; uncertainties: 2-3%, more at low temperatures; $L_1 = 1.041\%$, $L_2 = 1.649\%$, $L_3 = 2.770\%$.
2	65-G(C)	179	Rietveld, A.O., et al.	65.8	58.0 36.0 26.0 21.0 17.0	0.000 0.258 0.509 0.761 1.000	7.45 9.15 9.96 10.32 10.45	Same remarks as for curve 1 except $L_1 = 0.808\%$, $L_2 = 1.400\%$, $L_3 = 2.925\%$.
3	65-G(C)	179	Rietveld, A.O., et al.	90.2	40.0 25.0 18.0 15.0 12.0	0.000 0.251 0.491 0.755 1.000	9.12 11.35 12.51 13.19 13.50	Same remarks as for curve 1 except $L_1 = 0.394\%$, $L_2 = 0.608\%$, $L_3 = 1.107\%$.
4	65-G(C)	179	Rietveld, A.O., et al.	194.0	57.0 37.0 27.0 21.0 18.0	0.000 0.244 0.482 0.759 1.000	14.93 18.82 21.10 22.73 23.60	Same remarks as for curve 1 except $L_1 = 0.002\%$, $L_2 = 0.005\%$, $L_3 = 0.011\%$.
5	65-G(C)	325	Nasini, A.G. and Rossi, C.	284.2		0.0000 0.0340 0.2861 0.4995 0.6804 0.7850 0.9091 0.9461 0.9900	19.29 20.00 24.20 26.60 27.80 28.45 29.17 29.31 29.50	Ne: commercial grade, 99% pure; He: commercial grade; capillary method; precision ±0.2-0.3%; $L_1 = 0.134\%$, $L_2 = 0.286\%$, $L_3 = 0.679\%$.
6	65-G(C)	213	Thornton, E. and Baker, W.A.D.	291.2	700	1.000 0.894 0.783 0.655 0.565 0.393 0.250 0.158 0.000	30.8 30.7 29.9 29.2 28.1 26.4 24.4 22.8 19.2	Ne and He: spectroscopically pure; modified Rankine viscometer, relative measurements; uncertainties: mixture composition ±0.3%, viscosity ±1.0%; $L_1 = 0.237\%$, $L_2 = 0.437\%$, $L_3 = 1.091\%$.
7	65-G(C)	221	Trautz, M. and Binkele, H.E.	293.0		1.0000 0.7341 0.4376 0.2379 0.0000	30.92 29.71 27.02 24.29 19.41	He and Ne: Linde Co., commercial grade, 99.0-99.5 purity; capillary method, r = 0.2019 mm; $L_1 = 0.220\%$, $L_2 = 0.395\%$, $L_3 = 0.844\%$.
8	65-G(C)	179	Rietveld, A.O., Van Itterbeek, A., and Velds, C.A.	293.1	58.0 38.0 28.0 22.0 19.0	0.000 0.262 0.498 0.752 1.000	19.61 24.76 27.72 29.73 30.97	Same remarks as for curve 1 except $L_1 = 0.041\%$, $L_2 = 0.064\%$, $L_3 = 0.115\%$.
9	65-G(C)	221	Trautz, M. and Binkele, H.E.	373.0		1.0000 0.7341 0.4376 0.2379 0.0000	36.23 34.79 31.71 28.46 22.81	Same remarks as for curve 7 except $L_1 = 0.061\%$, $L_2 = 0.135\%$, $L_3 = 0.303\%$.
10	65-G(C)	325	Nasini, A.G. and Rossi, C	373.2		0.0000 0.0340 0.2861 0.4995 0.6804 0.7850 0.9091 0.9461 0.9900	23.35 24.18 29.21 32.00 33.62 34.31 35.25 33.35 35.49	Same remarks as for curve 5 except $L_1 = 0.104$, $L_2 = 0.162\%$, $L_3 = 0.313\%$.
11	65-G(C)	221	Trautz, M. and Binkele, H.E.	473.0		1.0000 0.7341 0.4376 0.2379 0.0000	42.20 40.56 37.02 33.27 26.72	Same remarks as for curve 7 except $L_1 = 0.105\%$, $L_2 = 0.181\%$, $L_3 = 0.377\%$.

TABLE 65-G(C)E.   EXPERIMENTAL VISCOSITY DATA AS A FUNCTION OF COMPOSITION FOR GASEOUS HELIUM-NEON MIXTURES (continued)

Cur. No.	Fig. No.	Ref. No.	Author(s)	Temp. (K)	Pressure (mm Hg)	Mole Fraction of Ne	Viscosity (N s m^{-2}x10^{-6})	Remarks
12	65-G(C)	221	Trautz, M. and Binkele, H. E.	523.0		1.0000	45.01	Same remarks as for curve 7 except $L_1 = 0.049\%$, $L_2 = 0.098\%$, $L_3 = 0.197\%$.
						0.7341	43.10	
						0.2379	35.55	
						0.0000	28.53	

TABLE 65-G(C)S.   SMOOTHED VISCOSITY VALUES AS A FUNCTION OF COMPOSITION FOR GASEOUS HELIUM-NEON MIXTURES

Mole Fraction	20.4 K [Ref. 179]	65.8 K [Ref. 179]	90.2 K [Ref. 179]	194.0 K [Ref. 179]	284.2 K [Ref. 325]	291.2 K [Ref. 213]
0.00	3.50	7.45	9.12	14.93	19.29	19.20
0.05	3.60	7.72	9.72	15.88	20.32	20.40
0.10	3.65	8.04	10.28	16.77	21.28	21.56
0.15	3.68	8.30	10.70	17.58	22.08	22.60
0.20	3.69	8.60	11.10	18.30	22.95	23.52
0.25	3.70	8.82	11.44	18.95	23.56	24.40
0.30	3.70	9.08	11.76	19.55	24.40	25.18
0.35	3.70	9.25	12.00	20.02	25.01	25.90
0.40	3.70	9.47	12.26	20.50	25.59	26.60
0.45	3.70	9.67	12.50	20.90	26.10	27.20
0.50	3.70	9.80	12.68	21.30	26.60	27.78
0.55	3.70	9.92	12.80	21.60	27.03	28.30
0.60	3.70	10.08	12.98	21.91	27.42	28.72
0.65	3.70	10.19	13.08	22.21	27.75	29.12
0.70	3.70	10.22	13.10	22.50	28.11	29.48
0.75	3.70	10.30	13.23	22.75	28.40	29.72
0.80	3.68	10.38	13.35	22.98	28.68	30.00
0.85	3.67	10.42	13.40	23.15	29.40	30.29
0.90	3.64	10.48	13.40	23.30	29.15	30.50
0.95	3.62	10.50	13.50	23.50	29.35	30.70
1.00	3.61	10.50	13.51	23.60	29.50	30.80

Mole Fraction	293.0 K [Ref. 221]	293.1 K [Ref. 179]	373.0 K [Ref. 221]	373.2 K [Ref. 325]	473.0 K [Ref. 221]	523.0 K [Ref. 221]
0.00	19.41	19.61	22.81	23.35	26.72	25.53
0.05	20.61	20.89	24.20	24.59	28.32	30.29
0.10	21.72	21.98	25.43	25.75	29.80	31.81
0.15	22.73	22.76	26.60	26.80	31.17	33.30
0.20	23.70	23.81	27.70	27.76	32.42	34.58
0.25	24.55	24.60	28.71	28.65	33.62	25.92
0.30	25.32	25.34	29.65	29.43	34.70	37.08
0.35	26.09	26.03	30.50	30.18	35.67	38.10
0.40	26.78	26.66	31.28	30.85	26.55	39.00
0.45	27.40	27.24	31.99	31.45	37.35	39.80
0.50	27.92	27.78	32.60	32.00	38.10	40.51
0.55	28.41	28.27	33.18	32.51	38.78	41.20
0.60	28.89	28.72	33.70	33.00	39.38	41.70
0.65	29.28	29.12	34.13	33.45	39.88	42.32
0.70	29.60	29.49	34.55	33.82	40.30	42.80
0.75	29.90	29.82	34.90	34.18	40.70	43.26
0.80	30.15	30.10	35.25	34.50	41.10	43.58
0.85	30.40	30.36	35.58	34.80	41.40	44.08
0.90	30.60	30.58	35.86	35.09	41.58	44.40
0.95	30.80	30.78	36.10	35.30	41.93	44.71
1.00	31.00	30.97	36.34	35.49	42.20	45.01

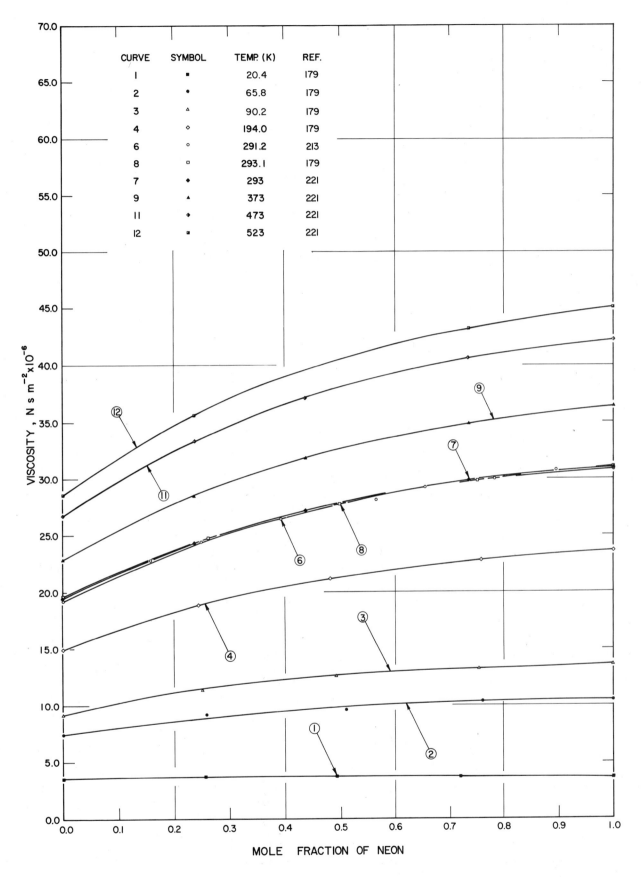

CURVE	SYMBOL	TEMP. (K)	REF.
1	■	20.4	179
2	●	65.8	179
3	△	90.2	179
4	◇	194.0	179
6	○	291.2	213
8	□	293.1	179
7	◆	293	221
9	▲	373	221
11	◆	473	221
12	■	523	221

VISCOSITY, $N \, s \, m^{-2} \times 10^{-6}$

MOLE FRACTION OF NEON

FIGURE 65-G(C). VISCOSITY DATA AS A FUNCTION OF COMPOSITION
FOR GASEOUS HELIUM-NEON MIXTURES

TABLE 65-G(D)E.   EXPERIMENTAL VISCOSITY DATA AS A FUNCTION OF DENSITY FOR GASEOUS
HELIUM-NEON MIXTURES

Cur. No.	Fig. No.	Ref. No.	Author(s)	Mole Fraction of Ne	Temp. (K)	Density (g cm^{-3})	Viscosity (N s m^{-2} x 10^{-6})	Remarks
1	65-G(D)	323	Kestin, J. and Nagashima, A.	0.741	293.2	0.02545	30.036	Ne: 99.991 pure, He: 99.989 pure; oscillating disk viscometer; accuracy ±0.1%, ratios of viscosity values ±0.4%.
						0.02308	30.043	
						0.01984	30.016	
						0.01627	29.995	
						0.01325	29.994	
						0.009941	29.968	
						0.006707	29.973	
						0.003337	29.959	
						0.000687	29.948	
2	65-G(D)	323	Kestin, J. and Nagashima, A.	0.567	293.2	0.01986	28.628	Same remarks as for curve 1.
						0.01842	28.619	
						0.01625	28.608	
						0.01354	28.611	
						0.01081	28.589	
						0.008046	28.593	
						0.005266	28.578	
						0.002779	28.571	
						0.000562	28.546	
3	65-G(D)	323	Kestin, J. and Nagashima, A.	0.350	293.2	0.01253	26.204	Same remarks as for curve 1.
						0.009980	26.215	
						0.006385	26.224	
						0.003217	26.225	
						0.0004179	26.182	
4	65-G(D)	323	Kestin, J. and Nagashima, A.	0.154	293.2	0.008369	23.048	Same remarks as for curve 1.
						0.006188	23.039	
						0.004033	23.045	
						0.002174	23.047	
						0.0002863	23.034	
5	65-G(D)	323	Kestin, J. and Nagashima, A.	0.051	293.2	0.006377	20.864	Same remarks as for curve 1.
						0.004948	20.874	
						0.003269	20.883	
						0.001563	20.894	
						0.0002145	20.879	
6	65-G(D)	323	Kestin, J. and Nagashima, A.	0.000	293.2	0.01025	19.606	Same remarks as for curve 1.
						0.006924	19.601	
						0.005250	19.602	
						0.003550	19.609	
						0.002124	19.603	
						0.0009969	19.620	
						0.0001753	19.597	
7	65-G(D)	323	Kestin, J. and Nagashima, A.	0.741	303.2	0.02237	30.751	Same remarks as for curve 1.
						0.01950	30.745	
						0.01584	30.715	
						0.01306	20.718	
						0.009696	30.699	
						0.006486	30.673	
						0.003520	30.649	
						0.000677	30.639	
						0.000691	30.665	
8	65-G(D)	323	Kestin, J. and Nagashima, A.	0.567	303.2	0.01757	29.291	Same remarks as for curve 1.
						0.01576	29.277	
						0.01309	29.292	
						0.01052	29.286	
						0.008510	29.267	
						0.005450	29.275	
						0.002652	29.251	
						0.000565	29.244	
						0.000562	29.267	
9	65-G(D)	323	Kestin, J. and Nagashima, A.	0.350	303.2	0.01361	26.851	Same remarks as for curve 1.
						0.01166	26.855	
						0.009811	26.874	
						0.007903	26.864	
						0.005887	26.852	
						0.003708	26.844	
						0.001963	26.857	
						0.0004068	26.846	
10	65-G(D)	323	Kestin, J. and Nagashima, A.	0.154	303.2	0.009158	23.589	Same remarks as for curve 1.
						0.007769	23.595	
						0.006513	23.586	
						0.005217	23.590	
						0.003901	23.597	
						0.002579	23.606	
						0.001323	23.605	
						0.0002735	23.586	

TABLE 65-G(D)E.　EXPERIMENTAL VISCOSITY DATA AS A FUNCTION OF DENSITY FOR GASEOUS
HELIUM-NEON MIXTURES　(continued)

Cur. No.	Fig. No.	Ref. No.	Author(s)	Mole Fraction of Ne	Temp. (K)	Density (g cm^{-3})	Viscosity (N s m^{-2} x 10^{-6})	Remarks
11	65-G(D)	323	Kestin, J. and Nagashima, A.	0.051	303.2	0.006724	21.361	Same remarks as for curve 1.
						0.005448	21.381	
						0.004114	21.382	
						0.002836	21.389	
						0.001519	21.391	
						0.0002074	21.377	
12	65-G(D)	323	Kestin, J. and Nagashima, A.	0.000	303.2	0.009806	20.074	Same remarks as for curve 1.
						0.007700	20.082	
						0.006650	20.074	
						0.005581	20.068	
						0.004496	20.069	
						0.003424	20.077	
						0.002324	20.088	
						0.001566	20.082	
						0.0008202	20.085	
						0.0001717	20.080	

TABLE 65-G(D)S. SMOOTHED VISCOSITY VALUES AS A FUNCTION OF DENSITY FOR GASEOUS HELIUM-NEON MIXTURES

Density (g cm^{-3})	Mole Fraction of Neon					
	0.000 (293.2 K) [Ref. 323]	0.051 (293.2 K) [Ref. 323]	0.154 (293.2 K) [Ref. 323]	0.350 (293.2 K) [Ref. 323]	0.567 (293.2 K) [Ref. 323]	0.741 (293.2 K) [Ref. 323]
0.00000	19.600	20.888	23.038	26.173	28.540	29.940
0.00125	19.604	20.890	23.048			
0.00250	19.605	20.890	23.050	26.210	28.560	29.960
0.00375	19.605	20.883	23.051			
0.00500	19.604	20.878	23.050	26.225	28.578	29.970
0.00625	19.607	20.870	23.045			
0.00750	19.610	20.860	23.046	26.221	28.583	29.980
0.00875	19.610	20.850	23.048			
0.01000	19.607	20.840	23.042	26.215	28.590	29.990
0.01125	19.600					
0.01250				26.203	28.595	29.993
0.01500				26.190	28.602	30.000
0.01750					28.612	30.010
0.02000					28.620	30.025
0.02250						30.030
0.02500						30.032

Density (g cm^{-3})	Mole Fraction of Neon					
	0.000 (303.2 K) [Ref. 323]	0.051 (303.2 K) [Ref. 323]	0.154 (303.2 K) [Ref. 323]	0.350 (303.2 K) [Ref. 323]	0.567 (303.2 K) [Ref. 323]	0.741 (303.2 K) [Ref. 323]
0.00000	20.070	21.373	23.573	26.846	29.248	30.630
0.00125	20.078	21.390	23.600			
0.00250	20.079	21.391	23.602	26.860	29.262	30.651
0.00375	20.080	21.390	23.600			
0.00500	20.080	21.386	23.601	26.862	29.270	30.672
0.00625	20.085	21.372	23.598			
0.00750	20.083	21.345	23.598	26.868	29.275	30.685
0.00875	20.076		23.595			
0.01000	20.070		23.590	26.865	29.280	30.700
0.01250				26.860	29.295	30.718
0.01500				26.850	29.300	30.732
0.01750					29.299	30.750
0.02000						30.755
0.02250						30.750

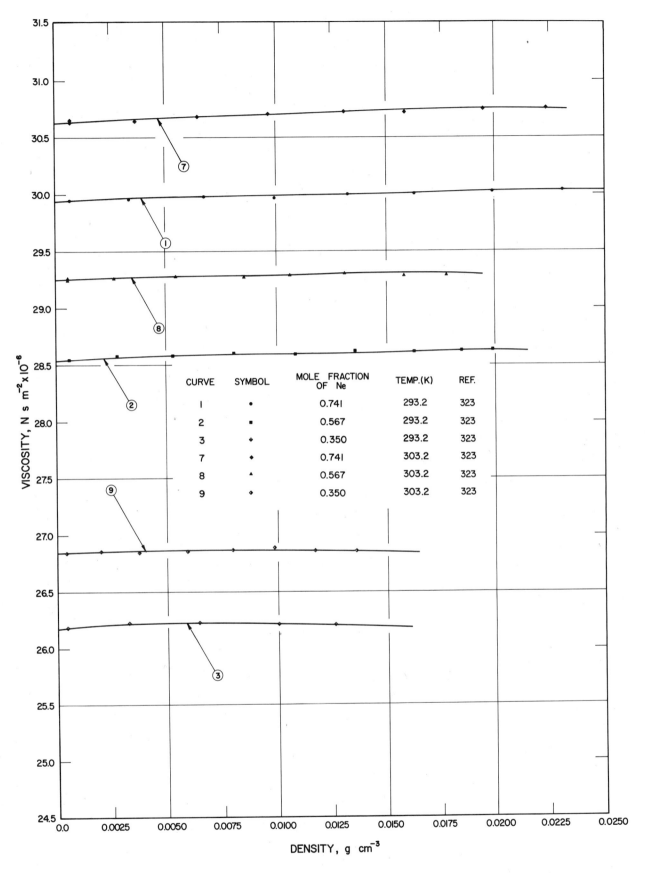

CURVE	SYMBOL	MOLE FRACTION OF Ne	TEMP.(K)	REF.
1	●	0.741	293.2	323
2	■	0.567	293.2	323
3	◆	0.350	293.2	323
7	◆	0.741	303.2	323
8	▲	0.567	303.2	323
9	◆	0.350	303.2	323

FIGURE 65-G(D). VISCOSITY DATA AS A FUNCTION OF DENSITY
FOR GASEOUS HELIUM-NEON MIXTURES

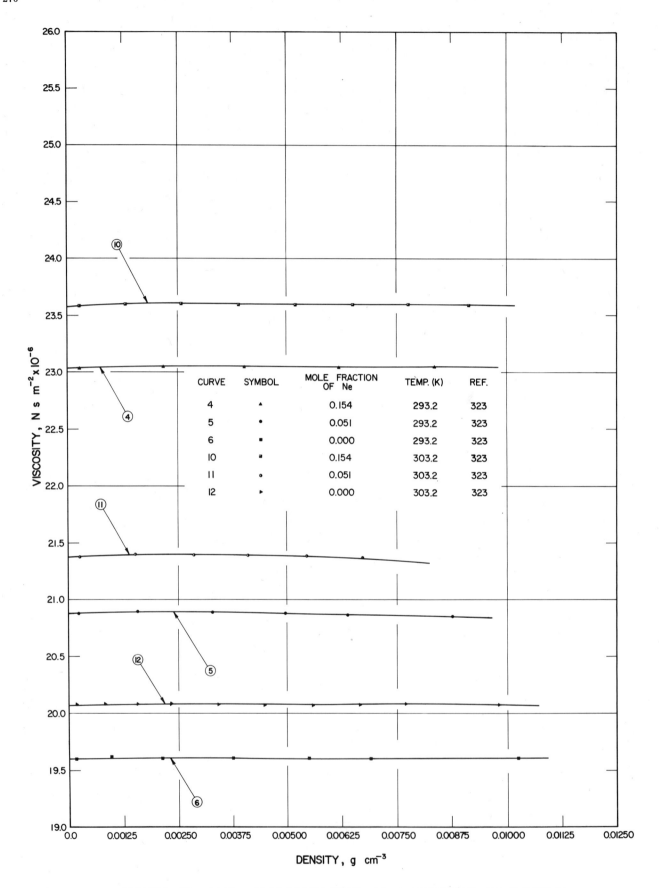

FIGURE 65-G(D). VISCOSITY DATA AS A FUNCTION OF DENSITY
FOR GASEOUS HELIUM-NEON MIXTURES (continued)

TABLE 66-G(C)E.　EXPERIMENTAL VISCOSITY DATA AS A FUNCTION OF COMPOSITION FOR GASEOUS
HELIUM-XENON MIXTURES

Cur. No.	Fig. No.	Ref. No.	Author(s)	Temp. (K)	Pressure (mm Hg)	Mole Fraction of Xe	Viscosity (N s m^{-2} x 10^{-6})	Remarks
1	66-G(C)	324	Thornton, E.	291.2	700	1.000	22.4	Xe: 99-100 pure, balance Kr, He:
						0.898	22.9	spectroscopically pure; modified
						0.792	23.2	Rankine viscometer, relative meas-
						0.687	23.7	urements; uncertainties: mixture
						0.594	24.2	composition ±0.3%, viscosity ±1.0%;
						0.494	24.5	L$_1$ = 0.199%, L$_2$ = 0.295%, L$_3$ =
						0.401	24.9	0.760%.
						0.304	25.2	
						0.201	25.2	
						0.169	24.8	
						0.063	23.2	
						0.000	19.4	

TABLE 66-G(C)S.　SMOOTHED VISCOSITY VALUES AS A FUNCTION OF COMPOSITION FOR GASEOUS
HELIUM-XENON MIXTURES

Mole Fraction of Xe	291.2 K [Ref. 9]
0.00	19.40
0.05	22.43
0.10	24.39
0.15	24.86
0.20	25.11
0.25	25.21
0.30	25.20
0.35	25.07
0.40	24.90
0.45	24.72
0.50	24.53
0.55	24.38
0.60	24.13
0.65	23.92
0.70	23.70
0.75	23.48
0.80	23.26
0.85	23.04
0.90	22.83
0.95	22.61
1.00	22.40

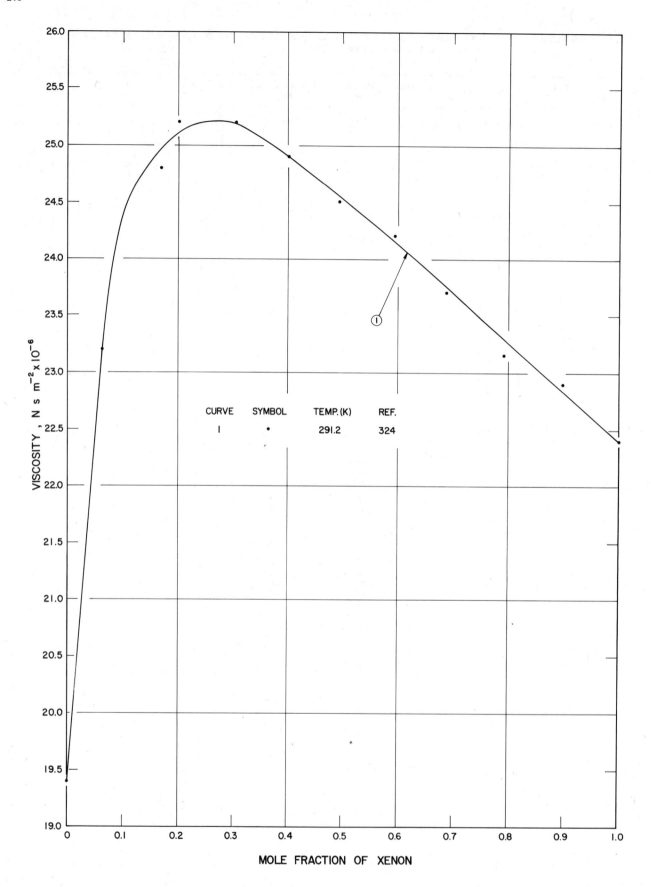

FIGURE 66-G (C). VISCOSITY DATA AS A FUNCTION OF COMPOSITION
FOR GASEOUS HELIUM- XENON MIXTURES

TABLE 67-G(C)E. EXPERIMENTAL VISCOSITY DATA AS A FUNCTION OF COMPOSITION FOR GASEOUS KRYPTON-NEON MIXTURES

Cur. No.	Fig. No.	Ref. No.	Author(s)	Temp. (K)	Pressure (mm Hg)	Mole Fraction of Kr	Viscosity (N s m^{-2} x 10^{-6})	Remarks
1	67-G(C)	278	Thornton, E.	291.2	700	1.000	24.9	Kr: 99-100 pure, balance Xe, Ne: spectroscopically pure; modified Rankine viscometer, relative measurements; uncertainties: mixture composition $\pm 0.3\%$, viscosity $\pm 1.0\%$; $L_1 = 0.248\%$, $L_2 = 0.336\%$, $L_3 = 0.662\%$.
						0.889	25.5	
						0.797	26.4	
						0.647	27.5	
						0.533	28.0	
						0.438	28.7	
						0.339	29.4	
						0.229	30.3	
						0.111	31.0	
						0.065	31.2	
						0.000	31.3	

TABLE 67-G(C)S. SMOOTHED VISCOSITY VALUES AS A FUNCTION OF COMPOSITION FOR GASEOUS KRYPTON-NEON MIXTURES

Mole Fraction of Kr	291.2 K [Ref. 278]
0.00	31.29
0.05	31.23
0.10	31.05
0.15	30.76
0.20	30.44
0.25	30.10
0.30	29.76
0.35	29.40
0.40	29.03
0.45	28.68
0.50	28.33
0.55	27.99
0.60	27.65
0.65	27.31
0.70	26.97
0.75	26.62
0.80	26.29
0.85	25.45
0.90	25.60
0.95	25.25
1.00	24.90

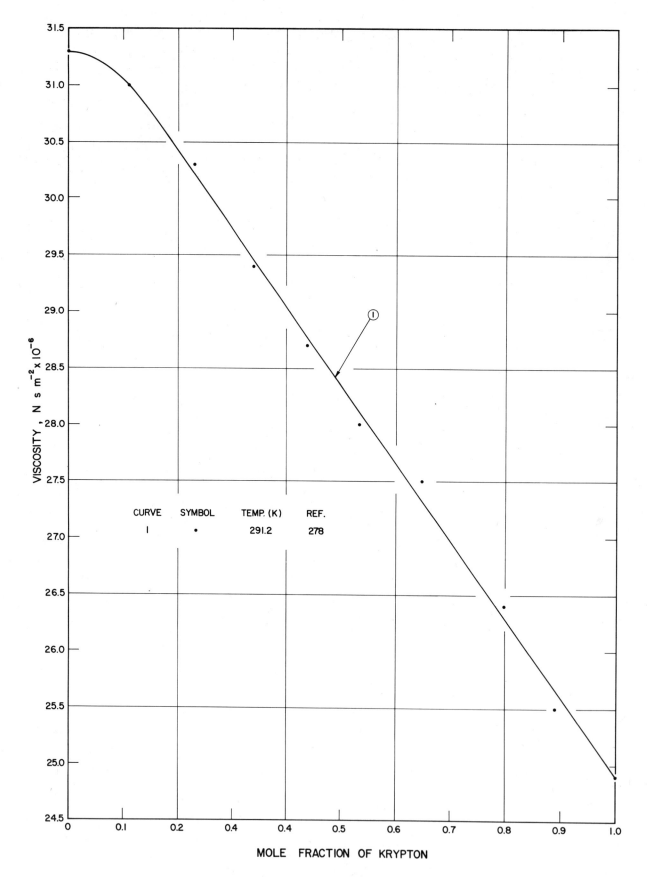

FIGURE 67- G (C). VISCOSITY DATA AS A FUNCTION OF COMPOSITION
FOR GASEOUS KRYPTON - NEON MIXTURES

TABLE 68-G(C)E.   EXPERIMENTAL VISCOSITY DATA AS A FUNCTION OF COMPOSITION FOR GASEOUS KRYPTON-XENON MIXTURES

Cur. No.	Fig. No.	Ref. No.	Author(s)	Temp. (K)	Pressure (mm Hg)	Mole Fraction of Xe	Viscosity (N s m^{-2} x 10^{-6})	Remarks
1	68-G(C)	324	Thornton, E.	291.2	700	1.000	22.5	Xe: 99-100 pure, balance Kr, Kr: 99-100 pure, balance Xe; modified Rankine viscometer, relative measurements; uncertainties: mixture composition ±0.3%, viscosity ±1.0%; L_1 = 0.729%, L_2 = 1.263%, L_3 = 2.418%.
						0.876	22.8	
						0.786	22.9	
						0.693	23.3	
						0.595	23.3	
						0.491	23.7	
						0.393	23.8	
						0.296	24.0	
						0.201	24.3	
						0.115	24.5	
						0.000	24.7	

TABLE 68-G(C)S.   SMOOTHED VISCOSITY VALUES AS A FUNCTION OF COMPOSITION FOR GASEOUS KRYPTON-XENON MIXTURES

Mole Fraction of Xe	291.2 K [Ref. 324]
0.00	24.70
0.05	24.60
0.10	24.48
0.15	24.39
0.20	24.29
0.25	24.17
0.30	24.06
0.35	23.95
0.40	23.84
0.45	23.73
0.50	23.62
0.55	23.51
0.60	23.40
0.65	23.29
0.70	23.17
0.75	23.06
0.80	23.95
0.85	22.84
0.90	22.73
0.95	22.61
1.00	22.50

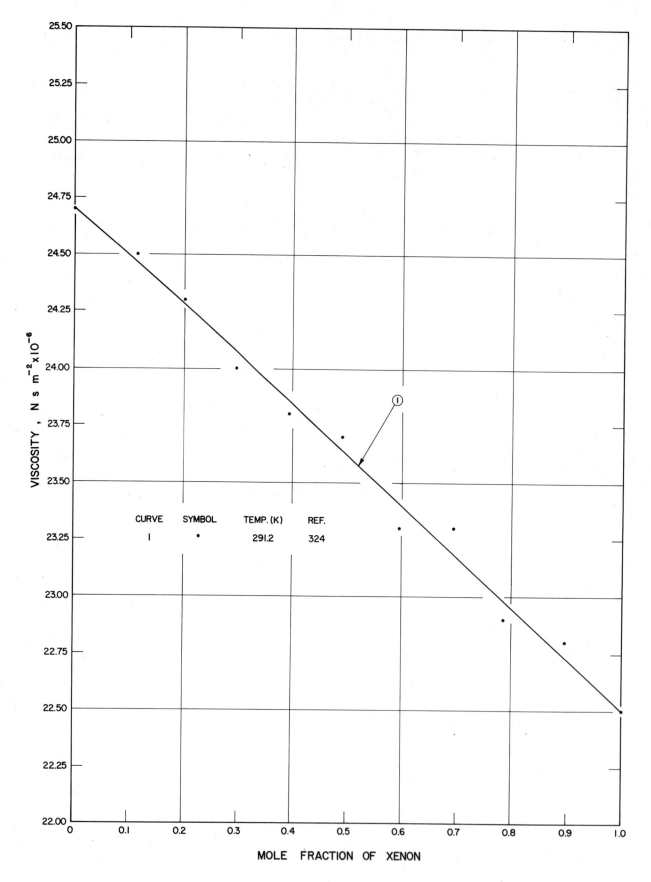

FIGURE 68-G(C). VISCOSITY DATA AS A FUNCTION OF COMPOSITION
FOR GASEOUS KRYPTON-XENON MIXTURES

TABLE 69-G(C)E.   EXPERIMENTAL VISCOSITY DATA AS A FUNCTION OF COMPOSITION FOR GASEOUS NEON-XENON MIXTURES

Cur. No.	Fig. No.	Ref. No.	Author(s)	Temp. (K)	Pressure (mm Hg)	Mole Fraction of Xe	Viscosity (N s m^{-2} x 10^{-6})	Remarks
1	69-G(C)	324	Thornton, E.	291.2	700	1.000	22.4	Xe: 99-100 pure, balance Kr, Ne: spectroscopically pure; modified Rankine viscometer, relative measurements; uncertainties: mixture composition ±0.3%, viscosity ±1.0%; $L_1 = 0.208\%$, $L_2 = 0.282\%$, $L_3 = 0.487\%$.
						0.903	23.2	
						0.794	24.0	
						0.594	25.8	
						0.393	27.8	
						0.285	29.1	
						0.199	29.9	
						0.103	30.6	
						0.000	31.0	

TABLE 69-G(C)S.   SMOOTHED VISCOSITY VALUES AS A FUNCTION OF COMPOSITION FOR GASEOUS NEON-XENON MIXTURES

Mole Fraction of Xe	291.2 K [Ref. 324]
0.00	31.00
0.05	30.83
0.10	30.60
0.15	30.32
0.20	29.97
0.25	29.53
0.30	29.01
0.35	28.43
0.40	27.84
0.45	27.29
0.50	26.79
0.55	26.31
0.60	25.84
0.65	25.38
0.70	24.94
0.75	24.50
0.80	24.06
0.85	23.64
0.90	23.22
0.95	22.81
1.00	22.40

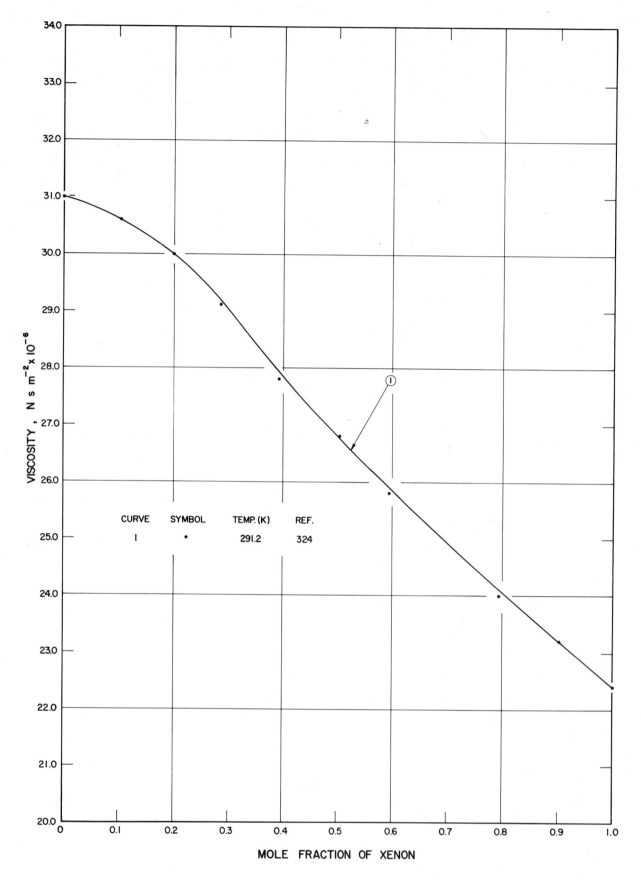

FIGURE 69-G(C). VISCOSITY DATA AS A FUNCTION OF COMPOSITION
FOR GASEOUS NEON-XENON MIXTURES

TABLE 70-G(D)E. EXPERIMENTAL VISCOSITY DATA AS A FUNCTION OF DENSITY FOR GASEOUS ARGON–CARBON DIOXIDE MIXTURES

Cur. No.	Fig. No.	Ref. No.	Author(s)	Mole Fraction of CO_2	Temp. (K)	Density (g cm^{-3})	Viscosity (N s m^{-2} x 10^{-6})	Remarks
1	70-G(D)	326	Kestin, J., Kobayashi, Y., and Wood, R.T.	0.9172	293.2	0.05135 0.02857 0.009322 0.001873	15.713 15.453 15.309 15.273	CO_2: 99.8 pure, Ar: 99.999 pure; oscillating disk viscometer; uncertainties: mixture composition ±0.002%, viscosity ±0.1%, viscosity ratios ±0.04%.
2	70-G(D)	326	Kestin, J., et al.	0.8425	293.2	0.05063 0.02867 0.009132 0.001859	16.307 16.038 15.874 15.826	Same remarks as for curve 1.
3	70-G(D)	326	Kestin, J., et al.	0.6339	293.2	0.04770 0.02744 0.008969 0.001807	17.940 17.645 17.459 17.386	Same remarks as for curve 1.
4	70-G(D)	326	Kestin, J., et al.	0.5398	293.2	0.04665 0.02658 0.008859 0.001798	18.682 18.396 18.191 18.123	Same remarks as for curve 1.
5	70-G(D)	326	Kestin, J., et al.	0.3324	293.2	0.04300 0.02633 0.008625 0.001765	20.271 20.023 19.806 19.728	Same remarks as for curve 1.
6	70-G(D)	326	Kestin, J., et al.	0.2675	293.2	0.04472 0.02551 0.008857 0.001753	20.826 20.532 20.321 20.229	Same remarks as for curve 1.
7	70-G(D)	326	Kestin, J., et al.	0.0000	293.2	0.04271 0.03806 0.03318 0.02941 0.02496 0.02106 0.01671 0.01252 0.008196 0.004983 0.001707	22.861 22.783 22.708 22.648 22.584 22.536 22.471 22.419 22.363 22.322 22.274	Same remarks as for curve 1.
8	70-G(D)	326	Kestin, J., et al.	1.0000	303.2	0.05057 0.04483 0.03924 0.03471 0.02871 0.02207 0.01865 0.01379 0.009020 0.005350 0.001803	15.585 15.504 15.447 15.392 15.327 15.277 15.254 15.216 15.194 15.172 15.157	Same remarks as for curve 1.
9	70-G(D)	326	Kestin, J., et al.	0.9172	303.2	0.04757 0.02784 0.009001 0.001766	16.211 15.978 15.815 15.781	Same remarks as for curve 1.
10	70-G(D)	326	Kestin, J., et al.	0.8425	303.2	0.04742 0.02750 0.008861 0.001751	16.810 16.555 16.390 16.339	Same remarks as for curve 1.
11	70-G(D)	326	Kestin, J., et al.	0.6339	303.2	0.04575 0.02624 0.008794 0.001759	18.484 18.203 18.014 17.946	Same remarks as for curve 1.
12	70-G(D)	326	Kestin, J., et al.	0.5398	303.2	0.04424 0.02624 0.008739 0.001776	19.252 18.968 18.757 18.687	Same remarks as for curve 1.
13	70-G(D)	326	Kestin, J., et al.	0.3324	303.2	0.04249 0.02545 0.008540 0.001706	20.871 20.622 20.403 20.330	Same remarks as for curve 1.

TABLE 70-G(D)E. EXPERIMENTAL VISCOSITY DATA AS A FUNCTION OF DENSITY FOR GASEOUS
ARGON-CARBON DIOXIDE MIXTURES (continued)

Cur. No.	Fig. No.	Ref. No.	Author(s)	Mole Fraction of CO_2	Temp. (K)	Density (g cm^{-3})	Viscosity (N s m^{-2} x 10^{-6})	Remarks
14	70-G(D)	326	Kestin, J., et al.	0.2675	303.2	0.04191	21.409	Same remarks as for curve 1.
						0.02593	21.163	
						0.008334	20.925	
						0.001695	20.845	
15	70-G(D)	326	Kestin, J., et al.	0.0000	303.2	0.04098	23.497	Same remarks as for curve 1.
						0.03660	23.425	
						0.03257	23.361	
						0.02845	23.297	
						0.02410	23.236	
						0.02022	23.177	
						0.01532	23.116	
						0.01206	23.070	
						0.008055	23.018	
						0.004849	22.971	
						0.001650	22.920	

TABLE 70-G(D)S.   SMOOTHED VISCOSITY VALUES AS A FUNCTION OF DENSITY FOR GASEOUS
ARGON-CARBON DIOXIDE MIXTURES

Density (g cm⁻³)	Mole Fraction of Carbon Dioxide						
	0.0000 (293.2 K) [Ref. 326]	0.2675 (293.2 K) [Ref. 326]	0.3324 (293.2 K) [Ref. 326]	0.5398 (293.2 K) [Ref. 326]	0.6339 (293.2 K) [Ref. 326]	0.8425 (293.2 K) [Ref. 326]	0.9172 (293.2 K) [Ref. 326]
0.0025	22.282						
0.0050	22.320	20.308	19.764	18.160	17.420	15.842	15.294
0.0100	22.387	20.330	19.820	18.220	17.460	15.880	15.320
0.0150	22.450	20.380	19.880	18.262	17.508	15.910	15.342
0.0200	22.520	20.440	19.942	18.320	17.550	15.950	15.380
0.0250	22.586	20.511	20.000	18.378	17.600	15.998	15.422
0.0300	22.658	20.580	20.075	18.440	17.660	16.050	15.478
0.0350	22.730	20.660	20.148	18.500	17.720	16.113	15.525
0.0400	22.820	20.738	20.226	18.579	17.800	16.174	15.582
0.0450	22.908	20.820	20.318	18.660	17.898	16.238	15.640
0.0500				18.758	18.010	16.300	15.700

Density (g cm⁻³)	Mole Fraction of Carbon Dioxide							
	0.0000 (303.2 K) [Ref. 326]	0.2675 (303.2 K) [Ref. 326]	0.3324 (303.2 K) [Ref. 326]	0.5398 (303.2 K) [Ref. 326]	0.6339 (303.2 K) [Ref. 326]	0.8425 (303.2 K) [Ref. 326]	0.9172 (303.2 K) [Ref. 326]	1.0000 (303.2 K) [Ref. 326]
0.0025	22.944							
0.0050	22.980	20.880	20.378	18.715	17.970	16.360	15.796	15.159
0.0100	23.040	20.940	20.422	18.758	18.100	16.400	15.820	15.179
0.0150	23.102	21.008	20.490	18.812	18.074	16.438	15.860	15.220
0.0200	23.172	21.080	20.558	18.880	18.140	16.480	15.900	15.260
0.0250	23.250	21.160	20.620	18.941	18.198	16.520	15.942	15.294
0.0300	23.320	21.240	20.680	19.020	18.260	16.578	15.995	15.338
0.0350	23.399	21.310	20.758	19.100	18.326	16.634	16.050	15.382
0.0400	23.470	21.380	20.830	19.180	18.400	16.700	16.115	15.440
0.0450				19.260	18.478	16.778	16.180	15.510
0.0500								15.580

288

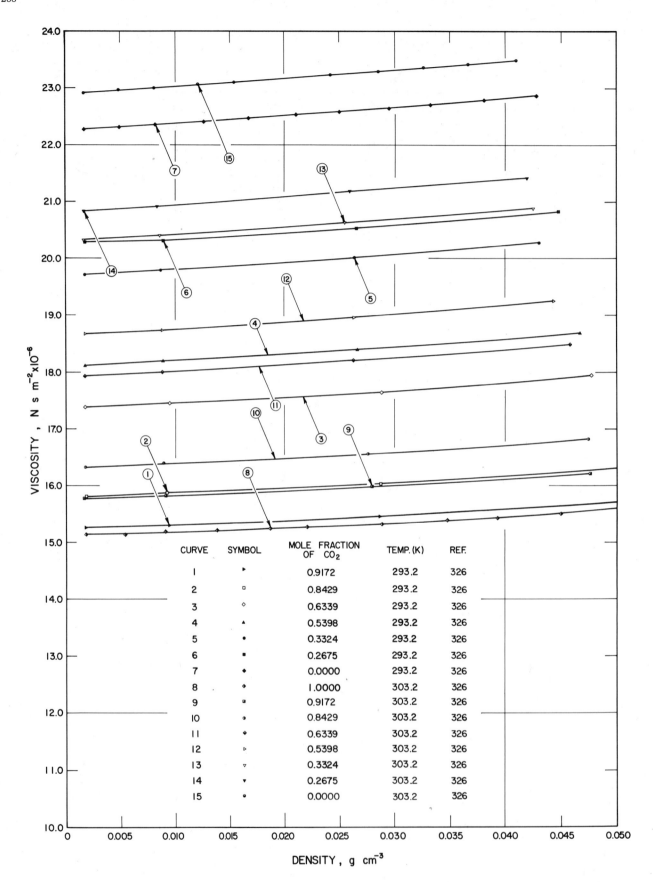

CURVE	SYMBOL	MOLE FRACTION OF CO_2	TEMP. (K)	REF.
1	▸	0.9172	293.2	326
2	◻	0.8429	293.2	326
3	◇	0.6339	293.2	326
4	▲	0.5398	293.2	326
5	•	0.3324	293.2	326
6	■	0.2675	293.2	326
7	◆	0.0000	293.2	326
8	◆	1.0000	303.2	326
9	▫	0.9172	303.2	326
10	◦	0.8429	303.2	326
11	◆	0.6339	303.2	326
12	▷	0.5398	303.2	326
13	▽	0.3324	303.2	326
14	▼	0.2675	303.2	326
15	•	0.0000	303.2	326

FIGURE 70-G(D). VISCOSITY DATA AS A FUNCTION OF DENSITY
FOR GASEOUS ARGON - CARBON DIOXIDE MIXTURES

TABLE 71-G(C)E.　EXPERIMENTAL VISCOSITY DATA AS A FUNCTION OF COMPOSITION FOR GASEOUS ARGON-HYDROGEN MIXTURES

Cur. No.	Fig. No.	Ref. No.	Author(s)	Temp. (K)	Pressure (atm)	Mole Fraction of Ar	Viscosity $(N \, s \, m^{-2} \times 10^{-6})$	Remarks
1	71-G(C)	226	Trautz, M. and Ludewigs, W.	293		1.0000	22.11	Ar: Linde Co., impurities 0.2 N_2;
						0.7058	21.40	H_2: made by electrolysis; capillary
						0.5543	20.56	method; $L_1 = 0.052\%$, $L_2 = 0.098\%$,
						0.3485	18.57	$L_3 = 0.215\%$.
						0.0000	8.75	
2	71-G(C)	226	Trautz, M. and Ludewigs, W.	373		1.0000	26.84	Same remarks as for curve 1 except
						0.7058	25.86	$L_1 = 0.080\%$, $L_2 = 0.179\%$, $L_3 =$
						0.5543	24.88	0.400%.
						0.3485	22.38	
						0.0000	10.29	
3	71-G(C)	226	Trautz, M. and Ludewigs, W.	473		1.0000	32.08	Same remarks as curve 1 except
						0.7058	30.70	$L_1 = 0.000\%$, $L_2 = 0.000\%$, $L_3 =$
						0.5543	29.48	0.000%.
						0.3485	26.36	
						0.0000	12.11	
4	71-G(C)	226	Trautz, M. and Ludewigs, W.	523		1.0000	34.48	Same remarks as for curve 1 except
						0.7058	33.10	$L_1 = 0.000\%$, $L_2 = 0.000\%$, $L_3 =$
						0.5543	31.64	0.000%.
						0.3485	28.26	
						0.0000	12.96	

TABLE 71-G(C)S.　SMOOTHED VISCOSITY VALUES AS A FUNCTION OF COMPOSITION FOR GASEOUS ARGON-HYDROGEN MIXTURES

Mole Fraction of Ar	293 K [Ref. 226]	373 K [Ref. 226]	473 K [Ref. 226]	523 K [Ref. 226]
0.00	8.75	10.29	12.11	12.96
0.05	9.08	12.83	15.30	16.40
0.10	13.00	15.00	18.10	19.50
0.15	14.60	17.20	20.50	22.08
0.20	15.96	18.90	22.50	24.16
0.25	17.02	20.30	24.04	25.80
0.30	17.90	21.48	25.38	27.20
0.35	18.64	22.41	26.44	28.34
0.40	19.22	23.22	27.38	29.34
0.45	19.71	23.90	28.20	30.20
0.50	20.16	24.48	28.88	30.87
0.55	20.38	24.90	29.44	31.60
0.60	20.90	25.30	29.90	32.18
0.65	21.80	25.61	30.32	32.68
0.70	21.40	25.90	30.68	33.07
0.75	21.60	26.15	30.90	33.40
0.80	21.75	26.28	31.22	33.70
0.85	21.84	26.48	31.48	33.92
0.90	21.99	26.61	31.70	34.16
0.95	22.10	26.77	31.90	34.32
1.00	22.11	26.84	32.08	34.48

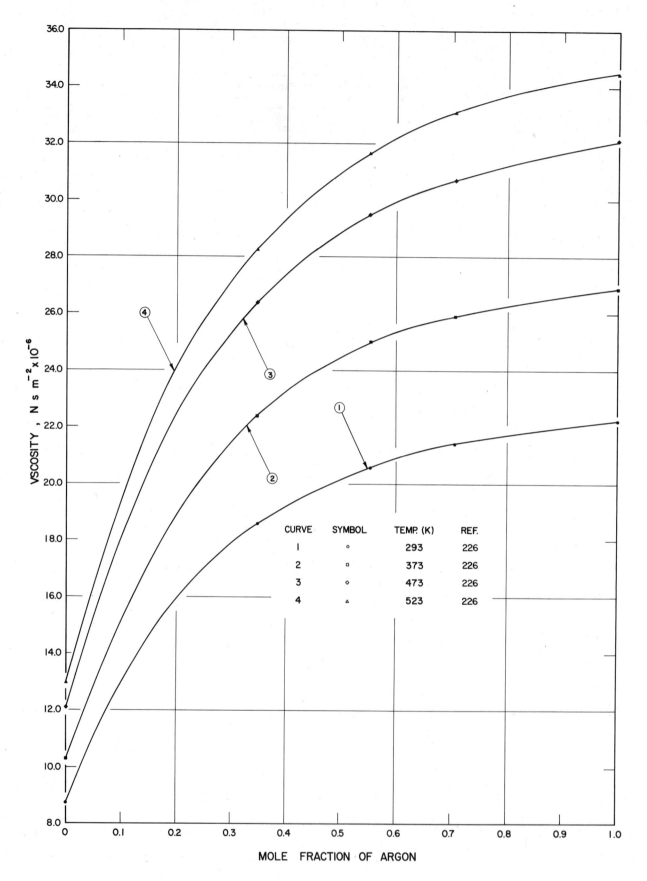

CURVE	SYMBOL	TEMP. (K)	REF.
1	○	293	226
2	□	373	226
3	◇	473	226
4	△	523	226

FIGURE 71-G(C). VISCOSITY DATA AS A FUNCTION OF COMPOSITION
FOR GASEOUS ARGON-HYDROGEN MIXTURES

TABLE 71-G(D)E.   EXPERIMENTAL VISCOSITY DATA AS A FUNCTION OF DENSITY FOR GASEOUS ARGON-HYDROGEN MIXTURES

Cur. No.	Fig. No.	Ref. No.	Author(s)	Mole Fraction of Ar	Temp. (K)	Density (g cm^3 x 10^{-4})	Viscosity (N s m^{-2} x 10^{-6})	Remarks
1	71-G(D)	327	Van Lierde, J.	0.361	286.0	0.109	17.31	Oscillating disk viscometer; original data reported as a function of pressure, density calculated from pressure using ideal gas equation.
						0.0175	14.28	
						0.00638	10.50	
						0.00231	6.23	
						0.000817	2.98	
						0.000290	1.11	
						0.000123	0.52	
						0.0000517	0.23	
2	71-G(D)	327	Van Lierde, J.	1.000	287.0	0.323	21.31	Same remarks as for curve 1.
						0.0532	18.72	
						0.0137	13.38	
						0.00616	9.16	
						0.00306	5.98	
						0.00154	3.56	
						0.000719	1.83	
						0.000365	0.99	
						0.000188	0.47	
						0.0000950	0.27	
3	71-G(D)	327	Van Lierde, J.	0.000	287.4	0.0201	8.47	Same remarks as for curve 1.
						0.000948	4.82	
						0.000341	2.79	
						0.000146	1.49	
						0.0000606	0.69	
						0.0000270	0.32	
4	71-G(D)	327	Van Lierde, J.	0.856	288.2	0.499	20.18	Same remarks as for curve 1.
						0.0624	18.57	
						0.0188	14.15	
						0.00791	9.71	
						0.00351	6.88	
						0.00176	4.09	
						0.000880	2.31	
						0.000458	1.20	
						0.000233	0.69	
						0.000119	0.35	
						0.0000609	0.20	
5	71-G(D)	327	Van Lierde, J.	0.545	288.2	0.202	19.50	Same remarks as for curve 1.
						0.0301	16.49	
						0.0105	12.55	
						0.00443	8.58	
						0.00261	6.15	
						0.00169	4.62	
						0.000839	2.71	
						0.000426	1.46	
						0.000225	0.80	
						0.000114	0.42	
						0.0000588	0.22	
6	71-G(D)	327	Van Lierde, J.	0.361	288.2	0.157	18.15	Same remarks as for curve 1.
						0.0231	17.22	
						0.0123	16.06	
						0.00487	13.44	
						0.00149	8.57	
						0.00128	7.70	
						0.000601	4.82	
						0.000242	2.25	
						0.000121	1.66	
7	71-G(D)	327	Van Lierde, J.	0.000	288.2	0.0145	8.55	Same remarks as for curve 1.
						0.00559	8.28	
						0.00280	7.93	
						0.00110	6.80	
						0.000907	6.71	
						0.000315	4.65	
						0.000183	3.38	
						0.000119	2.71	
						0.0000923	2.29	
						0.0000567	1.46	
						0.0000337	0.88	
						0.0000136	0.35	
8	71-G(D)	327	Van Lierde, J.	0.546	290.2	0.184	19.38	Same remarks as for curve 1.
						0.0385	18.63	
						0.0205	17.65	
						0.0110	16.23	
						0.00582	13.07	
						0.00318	11.72	
						0.00170	8.51	
						0.000914	5.87	
						0.000481	3.85	
						0.000265	2.10	

TABLE 71-G(D)E. EXPERIMENTAL VISCOSITY DATA AS A FUNCTION OF DENSITY FOR GASEOUS ARGON-HYDROGEN MIXTURES (continued)

Cur. No.	Fig. No.	Ref. No.	Author(s)	Mole Fraction of Ar	Temp. (K)	Density (g cm^{3}x10^{-4})	Viscosity (N s m^{-2}x10^{-6})	Remarks
9	71-G(D)	327	Van Lierde, J.	1.000	290.9	0.0429	20.50	Same remarks as for curve 1.
						0.0152	18.60	
						0.00543	13.96	
						0.00396	12.56	
						0.00203	8.93	
						0.00125	6.41	
						0.000670	4.07	
						0.000415	2.71	
						0.000160	1.12	
10	71-G(D)	327	Van Lierde, J.	0.856	291.5	0.106	21.15	Same remarks as for curve 1.
						0.0432	20.47	
						0.0151	18.16	
						0.00488	13.43	
						0.00191	8.32	
						0.00108	5.82	
						0.000672	4.08	
						0.000320	2.15	
						0.000143	0.87	

TABLE 71-G(D)S. SMOOTHED VISCOSITY VALUES AS A FUNCTION OF DENSITY FOR GASEOUS ARGON-HYDROGEN MIXTURES

Density (g cm^{-3}x10^{-4})	Mole Fraction of Argon							
	0.361 (288.2 K) [Ref. 327]	0.000 (288.2 K) [Ref. 327]	0.546 (290.2 K) [Ref. 327]	0.856 (291.5 K) [Ref. 327]	1.000 (287.0 K) [Ref. 327]	0.000 (287.4 K) [Ref. 327]	0.856 (288.2 K) [Ref. 327]	0.545 (288.2 K) [Ref. 327]
0.010	15.56	8.26	15.84	16.92	11.84	7.85	11.29	12.36
0.015	16.48	8.59	17.08	18.08	13.87	8.24	13.21	13.73
0.020	17.02	8.76	17.64	18.80	15.11	8.48	14.39	14.58
0.025	17.30		18.00	19.34	16.02	8.58	15.27	15.20
0.030	17.46		18.28	19.74	16.74		15.95	15.69
0.035	17.56		18.52	20.04	17.32		16.52	16.09
0.040	17.63		18.70	20.34	17.80		17.01	16.44
0.045	17.68		18.88	20.50	18.19		17.43	16.75
0.050	17.74		19.00	20.68	18.52		17.82	17.01
0.060	17.83		19.17	20.91	19.03		18.45	17.47
0.070	17.92		19.28	21.04	19.43		18.95	17.84
0.080	17.97		19.32	21.12	19.77		19.33	18.15
0.090	18.02		19.35	21.12	20.06			18.40
0.100	18.06		19.36	21.12	20.30			18.61
0.125	18.12		19.37		20.72			19.00
0.150	18.14		19.38		20.95			19.23
0.175			19.39		21.07			19.36
0.200			19.39		21.14			19.51
0.225					21.23			
0.250					21.28			
0.275					21.31			
0.300					21.32			

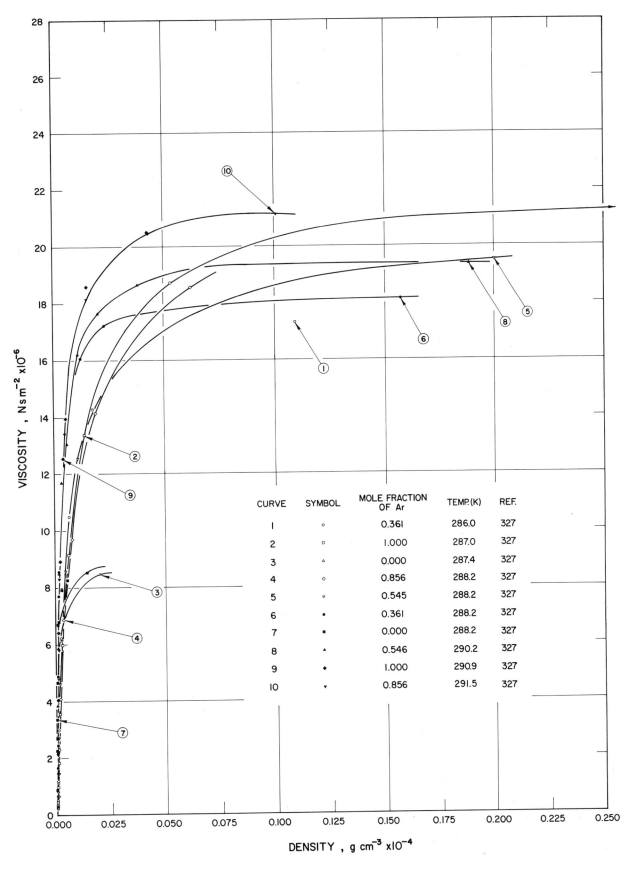

CURVE	SYMBOL	MOLE FRACTION OF Ar	TEMP.(K)	REF.
1	○	0.361	286.0	327
2	□	1.000	287.0	327
3	△	0.000	287.4	327
4	◇	0.856	288.2	327
5	▽	0.545	288.2	327
6	●	0.361	288.2	327
7	■	0.000	288.2	327
8	▲	0.546	290.2	327
9	◆	1.000	290.9	327
10	▼	0.856	291.5	327

FIGURE 71-G(D). VISCOSITY DATA AS A FUNCTION OF DENSITY FOR GASEOUS ARGON-HYDROGEN MIXTURES

TABLE 72-G(D)E.   EXPERIMENTAL VISCOSITY DATA AS A FUNCTION OF DENSITY FOR GASEOUS ARGON–NITROGEN MIXTURES

Cur. No.	Fig. No.	Ref. No.	Author(s)	Mole Fraction of Ar	Temp. (K)	Density $(g\ cm^{-3})$	Viscosity $(N\ s\ m^{-2} \times 10^{-6})$	Remarks
1	72-G(D)	328	DiPippo, R., Kestin, J., and Oguchi, K.	1.000	293.2	0.03864 0.03847 0.02520 0.02508 0.008352 0.001746	22.796 22.789 22.595 22.601 22.386 22.284	Ar: 99.99 pure, N_2: 99.999 pure; oscillating disk viscometer; uncertainties: error ± 0.1% and precision ± 0.05%.
2	72-G(D)	328	DiPippo, R., et al.	0.8010	293.2	0.03644 0.02368 0.007877 0.001640	21.907 21.731 21.534 21.430	Same remarks as for curve 1.
3	72-G(D)	328	DiPippo, R., et al.	0.6138	293.2	0.03405 0.02218 0.007336 0.001504	21.026 20.862 20.669 20.599	Same remarks as for curve 1.
4	72-G(D)	328	DiPippo, R., et al.	0.4054	293.2	0.03159 0.03154 0.02054 0.006835 0.001417	20.013 20.011 19.857 19.683 19.616	Same remarks as for curve 1.
5	72-G(D)	328	DiPippo, R., et al.	0.2263	293.2	0.02952 0.01920 0.006394 0.001330	19.103 18.959 18.791 18.722	Same remarks as for curve 1.
6	72-G(D)	328	DiPippo, R., et al.	1.000	303.2	0.03749 0.02432 0.008008 0.001656	23.429 23.245 23.033 22.938	Same remarks as for curve 1.
7	72-G(D)	328	DiPippo, R., et al.	0.8010	303.2	0.03512 0.02281 0.007566 0.001583	22.512 22.336 22.136 22.050	Same remarks as for curve 1.
8	72-G(D)	328	DiPippo, R., et al.	0.6138	303.2	0.03294 0.02149 0.007081 0.001487	21.604 21.437 21.256 21.182	Same remarks as for curve 1.
9	72-G(D)	328	DiPippo, R., et al.	0.4054	303.2	0.03036 0.01986 0.006618 0.001369	20.544 20.405 20.232 20.159	Same remarks as for curve 1.
10	72-G(D)	328	DiPippo, R., et al.	0.2263	303.2	0.02854 0.01842 0.006164 0.001276	19.606 19.468 19.308 19.241	Same remarks as for curve 1.

TABLE 72-G(D)S.   SMOOTHED VISCOSITY VALUES AS A FUNCTION OF DENSITY FOR GASEOUS ARGON-NITROGEN MIXTURES

Density (g cm^{-3})	Mole Fraction of Argon				
	0.2263 (293.2 K) [Ref. 328]	0.4054 (293.2 K) [Ref. 328]	0.6138 (293.2 K) [Ref. 328]	0.8010 (293.2 K) [Ref. 328]	1.0000 (293.2 K) [Ref. 328]
0.0025	18.730	19.638	20.610	21.445	22.292
0.0050	18.750	19.686	20.640	21.480	22.340
0.0075	18.770	19.732	20.670	21.510	22.371
0.0100	18.792	19.773	20.700	21.548	22.409
0.0125	18.821	19.812			
0.0150	18.855	19.850	20.767	21.620	22.472
0.0175		19.880			
0.0200	18.930	19.910	20.830	21.690	22.532
0.0250	19.020	19.960	20.900	21.760	22.590
0.0300	19.118	19.999	20.970	21.829	22.665
0.0350			21.040	21.885	22.739
0.0400					22.810

Density (g cm^{-3})	Mole Fraction of Argon				
	0.2263 (303.2 K) [Ref. 328]	0.4054 (303.2 K) [Ref. 328]	0.6138 (303.2 K) [Ref. 328]	0.8010 (303.2 K) [Ref. 328]	1.0000 (303.2 K) [Ref. 328]
0.0025	19.270	20.178		22.069	22.951
0.0050	19.310	20.226	21.190	22.100	22.990
0.0075	19.348	20.265	21.230	22.130	
0.0100	19.382	20.300	21.270	22.158	23.060
0.0125	19.420	20.340	21.308	22.192	
0.0150	19.450	20.370	21.345	22.225	23.130
0.0200	19.512	20.425	21.422	22.290	23.195
0.0250	19.570	20.489	21.496	22.360	23.260
0.0300	19.620	20.542	21.565	22.430	23.325
0.0350			21.630	22.501	23.390
0.0375					23.423

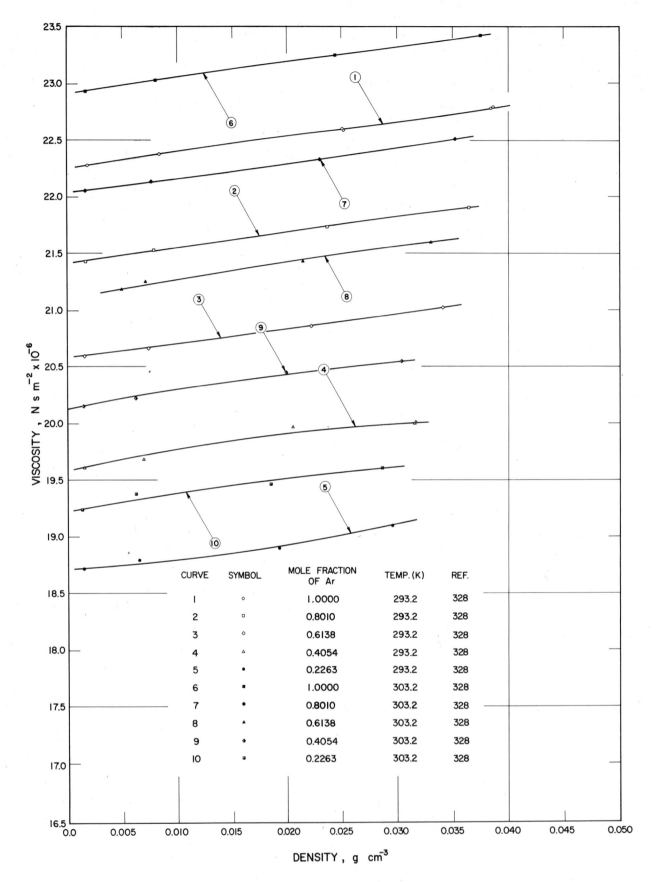

CURVE	SYMBOL	MOLE FRACTION OF Ar	TEMP.(K)	REF.
1	○	1.0000	293.2	328
2	□	0.8010	293.2	328
3	◇	0.6138	293.2	328
4	△	0.4054	293.2	328
5	●	0.2263	293.2	328
6	■	1.0000	303.2	328
7	◆	0.8010	303.2	328
8	▲	0.6138	303.2	328
9	◆	0.4054	303.2	328
10	▪	0.2263	303.2	328

DENSITY , g cm^{-3}

VISCOSITY , N s m^{-2} x10^{-6}

FIGURE 72 - G (D).   VISCOSITY DATA AS A FUNCTION OF DENSITY
FOR GASEOUS ARGON - NITROGEN MIXTURES

TABLE 73-G(D)E.   EXPERIMENTAL VISCOSITY DATA AS A FUNCTION OF DENSITY FOR GASEOUS
HELIUM-CARBON DIOXIDE MIXTURES

Cur. No.	Fig. No.	Ref. No.	Author(s)	Mole Fraction of CO_2	Temp. (K)	Density (g cm^{-3})	Viscosity (N s m^{-2} x 10^{-6})	Remarks
1	73-G(D)	328	DiPippo, R., Kestin, J., and Oguchi, K.	1.000	303.2	0.04633 0.02880 0.009107 0.009101 0.001854	15.495 15.323 15.199 15.201 15.167	CO_2: 99.8 pure, He: 99.995 pure; oscillating disk viscometer; error ±0.1%, precision ±0.05%.
2	73-G(D)	328	DiPippo, R., et al.	0.8626	303.2	0.03913 0.02470 0.02464 0.007935 0.001634	16.059 15.922 15.922 15.822 15.787	Same remarks as for curve 1.
3	73-G(D)	328	DiPippo, R., et al.	0.6655	303.2	0.02988 0.02581 0.02297 0.01909 0.01518 0.01002 0.006215 0.003702 0.001277	17.023 16.991 16.964 16.935 16.907 16.885 16.856 16.843 16.823	Same remarks as for curve 1.
4	73-G(D)	328	DiPippo, R., et al.	0.5095	303.2	0.02322 0.01494 0.004931 0.001021	17.962 17.899 17.845 17.812	Same remarks as for curve 1.
5	73-G(D)	328	DiPippo, R., et al.	0.3554	303.2	0.01705 0.01107 0.007366 0.003670 0.003671 0.002204 0.000764	18.992 18.957 18.938 18.922 18.916 18.908 18.894	Same remarks as for curve 1.
6	73-G(D)	328	DiPippo, R., et al.	0.2580	303.2	0.0131 0.008596 0.002886 0.000599	19.673 19.651 19.623 19.597	Same remarks as for curve 1.
7	73-G(D)	328	DiPippo, R., et al.	0.1961	303.2	0.01093 0.007135 0.004758 0.002374 0.001421 0.000492	20.058 20.043 20.031 20.023 20.015 20.002	Same remarks as for curve 1.
8	73-G(D)	328	DiPippo, R., et al.	0.0819	303.2	0.006669 0.006621 0.006612 0.004371 0.004342 0.001466 0.000307	20.477 20.467 20.469 20.465 20.458 20.459 20.435	Same remarks as for curve 1.
9	73-G(D)	328	DiPippo, R., et al.	0.0530	303.2	0.005570 0.003673 0.001231 0.000255	20.444 20.437 20.434 20.416	Same remarks as for curve 1.
10	73-G(D)	328	DiPippo, R., et al.	0.0414	303.2	0.005193 0.003373 0.001140 0.000238	20.401 20.403 20.392 20.387	Same remarks as for curve 1.
11	73-G(D)	328	DiPippo, R., et al.	0.000	303.2	0.003670 0.002377 0.002344 0.000802 0.000167	20.084 20.091 20.089 20.093 20.083	Same remarks as for curve 1.
12	73-G(D)	328	DiPippo, R., et al.	1.000	293.2	0.04871 0.03010 0.009414 0.001922	14.979 14.810 14.694 14.670	Same remarks as for curve 1.
13	73-G(D)	328	DiPippo, R., et al.	0.8626	293.2	0.04093 0.02565 0.008169 0.001678	15.538 15.413 15.315 15.289	Same remarks as for curve 1.

TABLE 73-G(D)E.   EXPERIMENTAL VISCOSITY DATA AS A FUNCTION OF DENSITY FOR GASEOUS HELIUM-CARBON DIOXIDE MIXTURES (continued)

Cur. No.	Fig. No.	Ref. No.	Author(s)	Mole Fraction of CO_2	Temp. (K)	Density $(g\ cm^{-3})$	Viscosity $(N\ s\ m^{-2} x\ 10^{-6})$	Remarks
14	73-G(D)	328	DiPippo, R., et al.	0.6655	293.2	0.03237	16.522	Same remarks as for curve 1.
						0.03109	16.509	
						0.01977	16.422	
						0.01295	16.387	
						0.006453	16.350	
						0.006440	16.353	
						0.006434	16.353	
						0.006433	16.349	
						0.001337	16.324	
15	73-G(D)	328	DiPippo, R., et al.	0.5095	293.2	0.02396	17.444	Same remarks as for curve 1.
						0.01554	17.379	
						0.005109	17.332	
						0.001061	17.301	
16	73-G(D)	328	DiPippo, R., et al.	0.3554	293.2	0.01765	18.477	Same remarks as for curve 1.
						0.01139	18.448	
						0.007629	18.424	
						0.003800	18.404	
						0.000796	18.387	
17	73-G(D)	328	DiPippo, R., et al.	0.2580	293.2	0.01375	19.165	Same remarks as for curve 1.
						0.008923	19.136	
						0.002976	19.112	
						0.000623	19.087	
18	73-G(D)	328	DiPippo, R., et al.	0.1961	293.2	0.01129	19.549	Same remarks as for curve 1.
						0.007352	19.533	
						0.002456	19.518	
						0.000510	19.487	
19	73-G(D)	328	DiPippo, R., et al.	0.0819	293.2	0.006884	19.990	Same remarks as for curve 1.
						0.004507	19.986	
						0.001513	19.978	
						0.001512	19.976	
						0.000315	19.960	
						0.000314	19.960	
20	73-G(D)	328	DiPippo, R., et al.	0.0530	293.2	0.005809	19.958	Same remarks as for curve 1.
						0.005809	19.955	
						0.003797	19.953	
						0.001270	19.943	
						0.000266	19.939	
21	73-G(D)	328	DiPippo, R., et al.	0.0414	293.2	0.005365	19.921	Same remarks as for curve 1.
						0.003509	19.919	
						0.001176	19.911	
						0.000245	19.895	

TABLE 73-G(D)S.   SMOOTHED VISCOSITY VALUES AS A FUNCTION OF DENSITY FOR GASEOUS
HELIUM-CARBON DIOXIDE MIXTURES

Density (g cm^{-3})	Mole Fraction of Carbon Dioxide				
	0.0414 (293.2 K) [Ref. 328]	0.0530 (293.2 K) [Ref. 328]	0.0819 (293.2 K) [Ref. 328]	0.1961 (293.2 K) [Ref. 328]	0.2580 (293.2 K) [Ref. 328]
0.0010	19.900	19.940	19.966	19.501	
0.0020	19.905	19.942	19.970	19.507	19.092
0.0030	19.910	19.948	19.972	19.510	
0.0040	19.915	19.950	19.978	19.515	19.110
0.0050	19.920	19.952	19.981	19.520	19.120
0.0060	19.925	19.960	19.986		19.125
0.0070			19.990	19.528	
0.0080			19.998	19.530	19.136
0.0090				19.538	
0.0100				19.541	19.148
0.0120				19.553	19.155
0.0140					19.163
0.0150					19.168

Density (g cm^{-3})	Mole Fraction of Carbon Dioxide				
	0.3554 (293.2 K) [Ref. 328]	0.5090 (293.2 K) [Ref. 328]	0.6655 (293.2 K) [Ref. 328]	0.8626 (293.2 K) [Ref. 328]	1.0000 (293.2 K) [Ref. 328]
0.0025	18.399	17.310	16.332	15.292	
0.0050	18.410	17.320	16.346	15.301	14.680
0.0075	18.422	17.332	16.358		
0.0100	18.435	17.345	16.368	15.325	14.698
0.0125	18.448	17.354			
0.0150	18.460	17.370	16.385	15.350	14.720
0.0175	18.470	17.389			
0.0200	18.481	17.410	16.410	15.380	14.747
0.0225		17.430			
0.0250		17.457	16.440	15.412	14.775
0.0300			16.500	15.449	14.810
0.0325			16.530		
0.0350				15.486	14.846
0.0400				15.530	14.888
0.0425				15.555	
0.0450					14.935

TABLE 73-G(D)S. SMOOTHED VISCOSITY VALUES AS A FUNCTION OF DENSITY FOR GASEOUS HELIUM–CARBON DIOXIDE MIXTURES (continued)

Density (g cm^{-3})	Mole Fraction of Carbon Dioxide					
	0.0000 (303.2 K) [Ref. 328]	0.0414 (303.2 K) [Ref. 328]	0.0530 (303.2 K) [Ref. 328]	0.0819 (303.2 K) [Ref. 328]	0.1961 (303.2 K) [Ref. 328]	0.2580 (303.2 K) [Ref. 328]
0.0010	20.090	20.389	20.420	20.440	20.008	19.599
0.0020	20.088	20.390	20.425	20.441	20.010	19.601
0.0030	20.087	20.392	20.430	20.448	20.018	19.612
0.0040	20.086	20.397	20.438	20.452	20.020	19.620
0.0050	20.091	20.399	20.440	20.459	20.028	19.630
0.0060		20.401	20.449	20.463	20.030	19.640
0.0070				20.47	20.035	19.649
0.0080					20.040	19.660
0.0090					20.042	19.662
0.0100					20.050	19.673
0.0110					20.053	

Density (g cm^{-3})	Mole Fraction of Carbon Dioxide				
	0.3554 (303.2 K) [Ref. 328]	0.5095 (303.2 K) [Ref. 328]	0.6655 (303.2 K) [Ref. 328]	0.8626 (303.2 K) [Ref. 328]	1.0000 (303.2 K) [Ref. 328]
0.0020	18.905				
0.0025		17.818	16.831	15.790	
0.0040	18.918				
0.0050	18.920	17.832	16.850	15.802	15.172
0.0060	18.928				
0.0075		17.850	16.865		
0.0080	18.939				
0.0100	18.948	17.868	16.880	15.835	15.200
0.0120	18.958				
0.0125		17.885	16.895		
0.0140	18.970				
0.0150	18.975	17.902	16.910	15.862	15.225
0.0170	18.990				
0.0175		17.920			
0.0200		17.940	16.940	15.892	15.255
0.0225		17.955			
0.0250			16.978	15.925	15.290
0.0300			17.018	15.970	15.330
0.0350				16.015	15.371
0.0400				16.063	15.422
0.0450					15.479

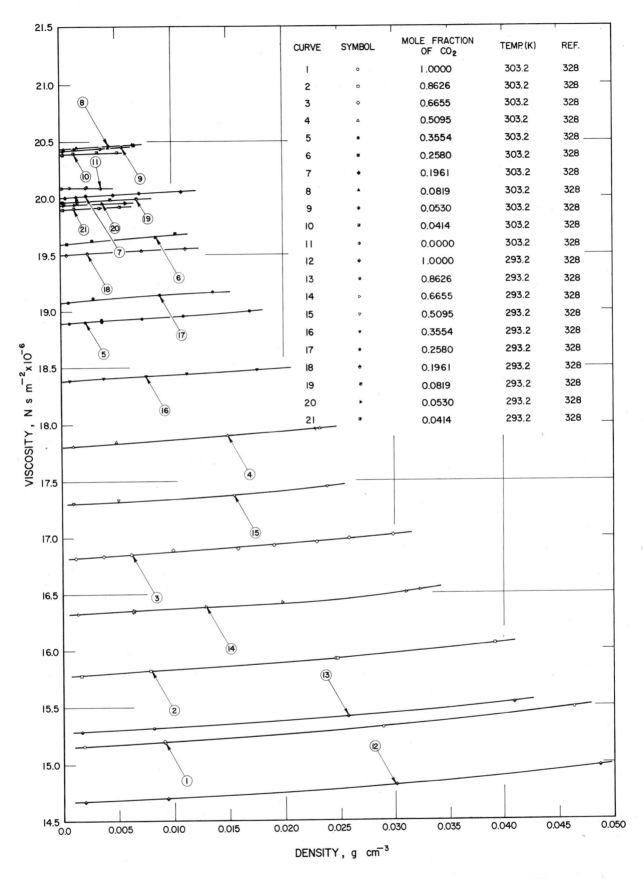

CURVE	SYMBOL	MOLE FRACTION OF CO_2	TEMP.(K)	REF.
1	○	1.0000	303.2	328
2	□	0.8626	303.2	328
3	◇	0.6655	303.2	328
4	△	0.5095	303.2	328
5	●	0.3554	303.2	328
6	■	0.2580	303.2	328
7	◆	0.1961	303.2	328
8	▲	0.0819	303.2	328
9	◆	0.0530	303.2	328
10	▫	0.0414	303.2	328
11	◦	0.0000	303.2	328
12	◆	1.0000	293.2	328
13	●	0.8626	293.2	328
14	▷	0.6655	293.2	328
15	▽	0.5095	293.2	328
16	▼	0.3554	293.2	328
17	●	0.2580	293.2	328
18	◆	0.1961	293.2	328
19	▪	0.0819	293.2	328
20	▶	0.0530	293.2	328
21	▪	0.0414	293.2	328

FIGURE 73-G(D). VISCOSITY DATA AS A FUNCTION OF DENSITY
FOR GASEOUS HELIUM-CARBON DIOXIDE MIXTURES

TABLE 74-G(C)E. EXPERIMENTAL VISCOSITY DATA AS A FUNCTION OF COMPOSITION FOR GASEOUS HELIUM-HYDROGEN MIXTURES

Cur. No.	Fig. No.	Ref. No.	Author(s)	Temp. (KP	Pressure (atm)	Mole Fraction of He	Viscosity (N s m^{-2} x 10^{-6})	Remarks
1	74-G(C)	74	Gille, A.	273.2		1.00000	18.925	He: spectroscopically pure, H$_2$: spectroscopically pure, electrolosis of sulfuric acid; capillary method; r_{18} = 0.0060482 ±3 cm; accuracy of η±0.02%; L$_1$ = 0.260%, L$_2$ = 0.349%, L$_3$ = 0.621%.
						0.96094	18.500	
						0.89569	17.596	
						0.86400	17.327	
						0.75087	16.032	
						0.59716	14.306	
						0.39857	12.267	
						0.18807	10.165	
						0.00000	8.410	
2	74-G(C)	74	Gille, A.	288.2		1.00000	19.611	Same remarks as for curve 1 except L$_1$ = 0.107%, L$_2$ = 0.146%, L$_3$ = 0.241%.
						0.96094	19.133	
						0.89569	18.319	
						0.86400	17.846	
						0.75087	16.528	
						0.59716	14.769	
						0.39857	12.652	
						0.18807	10.548	
						0.00000	8.776	
3	74-G(C)	327	van Lierde, J.	291.7		0.000	8.81	Oscillating disk viscometer; L$_1$ = 0.173%, L$_2$ = 0.255%, L$_3$ = 0.569%.
						0.189	10.57	
						0.353	12.02	
						0.503	13.43	
						0.565	13.97	
						0.683	15.36	
						0.811	16.86	
						1.000	19.69	
4	74-G(C)	221	Trautz, M. and Binkele, H.E.	293.0		1.0000	19.74	He: Linde Co., commercial grade, 99-99.5 purity; capillary method; r = 0.2019 mm; accuracy <±0.4%; L$_1$ = 0.115%, L$_2$ = 0.187%, L$_3$ = 0.398%.
						0.4480	13.17	
						0.3931	12.52	
						0.3082	11.66	
						0.0000	8.75	
5	74-G(C)	221	Trautz, M. and Binkele, H.E.	373.0		1.0000	23.20	Same remarks as for curve 4 except L$_1$ = 0.053%, L$_2$ = 0.084%, L$_3$ = 0.135%.
						0.4480	15.51	
						0.3931	14.78	
						0.3082	13.83	
						0.0000	10.29	
6	74-G(C)	74	Gille, A.	373.2		1.00000	23.408	Same remarks as for curve 1 except L$_1$ = 0.309%, L$_2$ = 0.414%, L$_3$ = 0.686%.
						0.96094	22.807	
						0.89569	22.032	
						0.86400	21.555	
						0.75087	19.860	
						0.59716	17.847	
						0.39857	15.174	
						0.18807	12.646	
						0.00000	10.450	
7	74-G(C)	221	Trautz, M. and Binkele, H.E.	473.0		1.0000	27.15	Same remarks as for curve 4 except L$_1$ = 0.103%, L$_2$ = 0.187%, L$_3$ = 0.404%.
						0.4480	18.17	
						0.3931	17.28	
						0.3082	16.19	
						0.0000	12.11	
8	74-G(C)	221	Trautz, M. and Binkele, H.E.	523.0		1.0000	29.03	Same remarks as for curve 4 except L$_1$ = 0.000%, L$_2$ = 0.000%, L$_3$ = 0.000%.
						0.4480	19.39	
						0.3931	18.52	
						0.3082	17.32	
						0.0000	19.96	

TABLE 74-G(C)S.   SMOOTHED VISCOSITY VALUES AS A FUNCTION OF COMPOSITION FOR GASEOUS
HELIUM–HYDROGEN MIXTURES

Mole Fraction of H₂	273.2 K [Ref. 74]	288.2 K [Ref. 74]	291.7 K [Ref. 327]	293.0 K [Ref. 221]	373.0 K [Ref. 221]	373.2 K [Ref. 74]	473.0 K [Ref. 221]	523.0 K [Ref. 221]
0.00	8.41	8.78	8.81	8.75	10.29	10.45	12.11	12.96
0.05	8.86	9.26	9.28	9.20	10.89	11.00	12.79	13.70
0.10	9.32	9.79	9.74	9.67	11.46	11.56	13.45	14.39
0.15	9.78	10.23	10.19	10.10	12.01	12.14	14.10	15.09
0.20	10.26	10.64	10.64	10.57	12.59	12.72	14.79	15.77
0.25	10.75	11.20	11.10	11.07	13.18	13.32	15.40	16.48
0.30	11.24	11.68	11.54	11.58	13.75	13.93	16.10	17.18
0.35	11.74	12.18	11.98	12.10	14.30	14.66	16.77	17.90
0.40	12.24	12.68	12.42	12.64	14.90	15.18	17.48	18.62
0.45	12.76	13.18	12.89	13.18	15.50	15.91	18.15	19.40
0.50	13.26	13.70	13.38	13.70	16.12	16.46	18.88	20.19
0.55	13.78	14.28	13.89	14.24	16.77	17.11	19.60	20.99
0.60	14.31	14.78	14.42	14.80	17.40	17.78	20.39	21.80
0.65	14.86	15.34	14.99	15.39	18.08	18.44	21.19	22.66
0.70	15.40	15.90	15.56	15.99	18.75	19.12	22.02	23.52
0.75	15.96	16.50	16.14	16.60	19.45	19.81	22.83	24.40
0.80	16.50	17.09	16.72	17.22	20.17	20.50	23.68	25.32
0.85	17.06	17.70	17.39	17.85	20.90	21.22	24.55	26.14
0.90	17.64	18.32	18.08	18.50	21.65	21.94	25.40	27.20
0.95	18.25	18.95	18.85	19.15	22.42	22.69	26.28	28.12
1.00	18.91	19.61	19.69	19.76	23.20	23.40	27.15	29.03

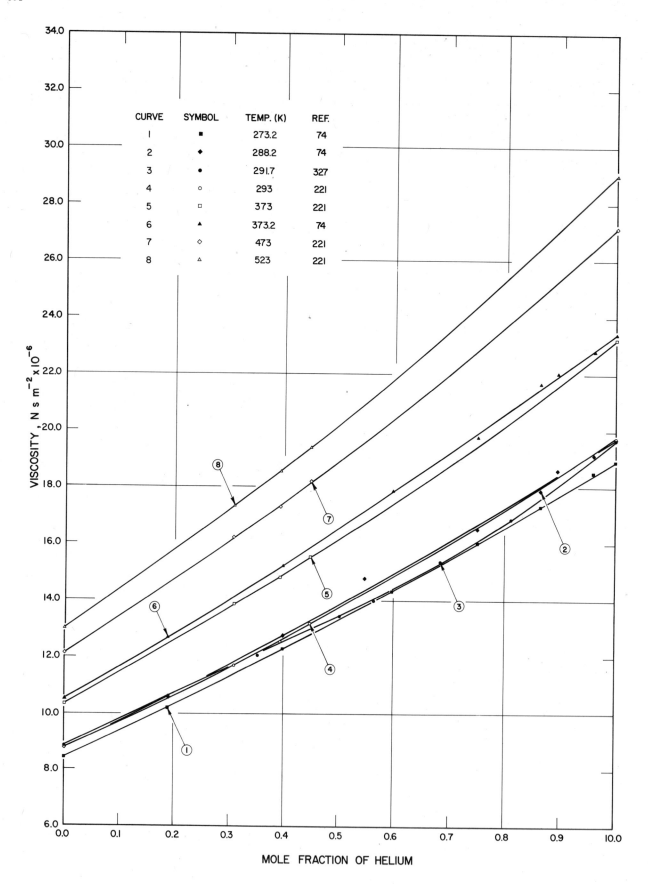

FIGURE 74-G(C).   VISCOSITY DATA AS A FUNCTION OF COMPOSITION
FOR GASEOUS HELIUM-HYDROGEN MIXTURES

TABLE 74-G(D)E. EXPERIMENTAL VISCOSITY DATA AS A FUNCTION OF DENSITY FOR GASEOUS HELIUM-HYDROGEN MIXTURES

Cur. No.	Fig. No.	Ref. No.	Author(s)	Mole Fraction of H_2	Temp. (K)	Density (g cm^{-3})	Viscosity (N s m^{-2} x 10^{-6})	Remarks
1	74-G(D)	329	Kestin, J. and Yata, J.	0.8596	293.2	0.000782 0.000464 0.000169	17.819 17.817 17.809	He: 99.995 pure, H_2: 99.999 pure; oscillating disk viscometer; error ±0.1%, precision ±0.05%.
2	74-G(D)	329	Kestin, J. and Yata, J.	0.8533	293.2	0.000782 0.000468 0.000171	17.739 17.737 17.739	Same remarks as for curve 1.
3	74-G(D)	329	Kestin, J. and Yata, J.	0.8488	293.2	0.000778 0.000464 0.000171	17.681 17.681 17.678	Same remarks as for curve 1.
4	74-G(D)	329	Kestin, J. and Yata, J.	0.8429	293.2	0.000775 0.000462 0.000167	17.637 17.631 17.618	Same remarks as for curve 1.
5	74-G(D)	329	Kestin, J. and Yata, J.	0.8325	293.2	0.000766 0.000460 0.000166	17.469 17.471 17.460	Same remarks as for curve 1.
6	74-G(D)	329	Kestin, J. and Yata, J.	0.7737	293.2	0.003369 0.002217 0.000737 0.000154	16.728 16.740 16.740 16.732	Same remarks as for curve 1.
7	74-G(D)	329	Kestin, J. and Yata, J.	0.6286	293.2	0.003273 0.002029 0.000677 0.000141	15.070 15.077 15.077 15.064	Same remarks as for curve 1.
8	74-G(D)	329	Kestin, J. and Yata, J.	0.5196	293.2	0.002919 0.001912 0.000634 0.000133	13.855 13.856 13.862 13.856	Same remarks as for curve 1.
9	74-G(D)	329	Kestin, J. and Yata, J.	0.2629	293.2	0.002412 0.001591 0.000529 0.000110	11.252 11.246 11.243 11.241	Same remarks as for curve 1.
10	74-G(D)	329	Kestin, J. and Yata, J.	0.8596	303.2	0.000757 0.000449 0.000166	18.247 18.240 18.239	Same remarks as for curve 1.
11	74-G(D)	329	Kestin, J. and Yata, J.	0.8533	303.2	0.000756 0.000449 0.000165	18.172 18.173 18.163	Same remarks as for curve 1.
12	74-G(D)	329	Kestin, J. and Yata, J.	0.8488	303.2	0.000752 0.000445 0.000163	18.112 18.113 18.104	Same remarks as for curve 1.
13	74-G(D)	329	Kestin, J. and Yata, J.	0.8429	303.2	0.000737 0.000444 0.000163	18.062 18.064 18.054	Same remarks as for curve 1.
14	74-G(D)	329	Kestin, J. and Yata, J.	0.8325	303.2	0.000741 0.000445 0.000162	17.898 17.891 17.893	Same remarks as for curve 1.
15	74-G(D)	329	Kestin, J. and Yata, J.	0.7737	303.2	0.003371 0.002128 0.000711 0.000149	17.126 17.132 17.131 17.129	Same remarks as for curve 1.
16	74-G(D)	329	Kestin, J. and Yata, J.	0.6286	303.2	0.003076 0.001959 0.000657 0.000186 0.000138	15.433 15.433 15.439 15.435 15.431	Same remarks as for curve 1.
17	74-G(D)	329	Kestin, J. and Yata, J.	0.5196	303.2	0.002856 0.001841 0.000614 0.000129	14.186 14.188 14.192 14.204	Same remarks as for curve 1.
18	74-G(D)	329	Kestin, J. and Yata, J.	0.5196	303.2	0.002386 0.002386 0.001525 0.000512 0.000108	11.518 11.516 11.518 11.517 11.493	Same remarks as for curve 1.

TABLE 74-G(D)S.   SMOOTHED VISCOSITY DATA AS A FUNCTION OF DENSITY FOR GASEOUS HELIUM–HYDROGEN MIXTURES

Density (g cm^{-3})	Mole Fraction of Hydrogen								
	0.2629 (293.2 K) [Ref. 329]	0.5196 (293.2 K) [Ref. 329]	0.6286 (293.2 K) [Ref. 329]	0.7737 (293.2 K) [Ref. 329]	0.8325 (293.2 K) [Ref. 329]	0.8429 (293.2 K) [Ref. 329]	0.8488 (293.2 K) [Ref. 329]	0.8533 (293.2 K) [Ref. 329]	0.8596 (293.2 K) [Ref. 329]
0.00010					17.455	17.614	17.677	17.738	17.894
0.00020					17.462	17.620	17.679	17.739	17.893
0.00025	11.241	13.863	15.067						
0.00030					17.466	17.625	17.680	17.739	17.893
0.00040					17.470	17.630	17.682	17.738	17.892
0.00050	11.244	13.860	15.075	16.738	17.472	17.633	17.682	17.738	17.892
0.00060					17.472	17.636	17.682	17.738	17.895
0.00070					17.471	17.637	17.682	17.739	17.898
0.00075	11.245		15.080						
0.00080					17.470	17.636	17.682	17.739	17.900
0.00090						17.636	17.682	17.739	17.902
0.00100	11.247	13.857	15.084	16.741			17.681	17.738	
0.00125	11.248	13.856							
0.00150	11.249	13.856	15.085	16.743					
0.00175	11.250								
0.00200	11.250	13.856	15.083	16.740					
0.00225	11.254								
0.00250	11.255	13.856	15.080	16.737					
0.00300		13.855	15.074	16.732					
0.00350		13.855	15.066	16.728					
0.00400				16.718					

Density (g cm^{-3})	Mole Fraction of Hydrogen								
	0.2629 (303.2 K) [Ref. 329]	0.5196 (303.2 K) [Ref. 329]	0.6286 (303.2 K) [Ref. 329]	0.7737 (303.2 K) [Ref. 329]	0.8325 (303.2 K) [Ref. 329]	0.8429 (303.2 K) [Ref. 329]	0.8488 (303.2 K) [Ref. 329]	0.8533 (303.2 K) [Ref. 329]	0.8596 (303.2 K) [Ref. 329]
0.00005					17.894	18.046		18.161	18.240
0.00010					17.894	18.050	18.102	18.162	18.240
0.00015					17.894	18.053			
0.00020					17.893	18.056	18.106	18.164	18.240
0.00025	11.503	14.200	15.435	17.128				18.166	18.240
0.00030					17.892	18.062	18.111	18.168	18.240
0.00040					17.892	18.064	18.113	18.170	18.241
0.00050	11.513	14.197	15.439	17.133	17.892	18.065	18.114	18.172	18.242
0.00060					17.895	18.064	18.114	18.171	18.245
0.00070					17.898	18.062	18.114	18.166	18.247
0.00075	11.520	14.195	15.441						
0.00080					17.900	18.061	18.112	18.160	18.248
0.00100	11.519	14.192	15.442	17.137					
0.00125	11.518	14.190							
0.00150	11.520	14.187	15.442	17.138					
0.00175	11.520								
0.00200	11.520	14.187	15.440	17.134					
0.00250	11.518	14.187	15.439	17.133					
0.00300		14.190	15.432	17.130					
0.00350			15.429	17.126					
0.00400				17.123					

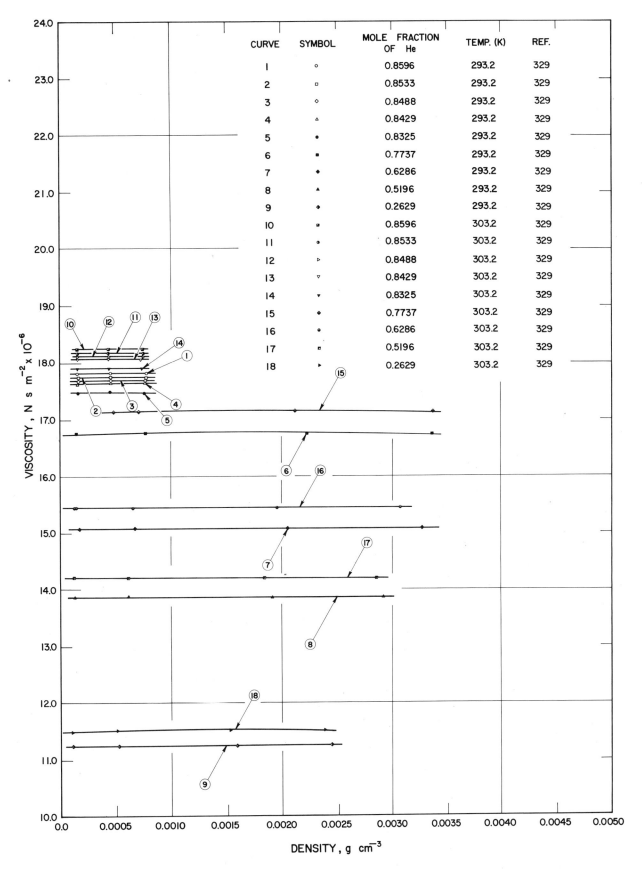

FIGURE 74-G(D). VISCOSITY DATA AS A FUNCTION OF DENSITY
FOR GASEOUS HELIUM-HYDROGEN MIXTURES

TABLE 75-G(D)E.   EXPERIMENTAL VISCOSITY DATA AS A FUNCTION OF DENSITY FOR GASEOUS HELIUM-NITROGEN MIXTURES

Cur. No.	Fig. No.	Ref. No.	Author(s)	Mole Fraction of N_2	Temp. (K)	Density (g cm^{-3})	Viscosity (N s m^{-2} x 10^{-6})	Remarks
1	75-G(D)	330	Kao, J.T.F. and Kobayashi, R.	0.0000	183.15	0.00264	14.244	N_2: 99.997 pure, He: 99.999 pure; capillary tube viscometer; error ±0.137%.
						0.00524	14.211	
						0.01032	14.220	
						0.01523	14.225	
						0.02000	14.272	
						0.02912	14.342	
						0.04590	14.564	
2	75-G(D)	330	Kao, J.T.F. and Kobayashi, R.	0.1283	183.15	0.00465	14.329	Same remarks as for curve 1.
						0.00923	14.337	
						0.01819	14.433	
						0.02689	14.510	
						0.03536	14.606	
						0.05168	14.816	
						0.08191	15.363	
3	75-G(D)	330	Kao, J.T.F. and Kobayashi, R.	0.4029	183.15	0.00907	13.655	Same remarks as for curve 1.
						0.01805	13.750	
						0.03572	13.978	
						0.05295	14.280	
						0.06968	14.641	
						0.10152	15.335	
						0.15837	17.032	
4	75-G(D)	330	Kao, J.T.F. and Kobayashi, R.	0.8412	183.15	0.01640	12.443	Same remarks as for curve 1.
						0.03336	12.700	
						0.06884	13.371	
						0.10594	14.287	
						0.14383	15.328	
						0.21751	17.909	
						0.33551	23.702	
5	75-G(D)	330	Kao, J.T.F. and Kobayashi, R.	1.0000	183.15	0.01921	11.904	Same remarks as for curve 1.
						0.03962	12.284	
						0.08435	13.230	
						0.13436	14.558	
						0.18829	16.167	
						0.29279	20.656	
						0.42992	29.987	
6	75-G(D)	330	Kao, J.T.F. and Kobayashi, R.	0.0000	223.15	0.00217	16.241	Same remarks as for curve 1.
						0.00431	16.239	
						0.00852	16.239	
						0.01261	16.248	
						0.01661	16.248	
						0.02431	16.276	
						0.03868	16.411	
						0.05499	16.644	
						0.06967	16.958	
7	75-G(D)	330	Kao, J.T.F. and Kobayashi, R.	0.1283	223.15	0.00385	16.377	Same remarks as for curve 1.
						0.00760	16.415	
						0.01499	16.450	
						0.02219	16.487	
						0.02923	16.534	
						0.04282	16.649	
						0.06829	16.955	
						0.09692	17.506	
						0.12190	18.180	
8	75-G(D)	330	Kao, J.T.F. and Kobayashi, R.	0.2540	223.15	0.00552	16.183	Same remarks as for curve 1.
						0.01089	16.202	
						0.02146	16.281	
						0.03174	16.361	
						0.04172	16.490	
						0.06082	16.774	
						0.09582	17.417	
						0.13424	18.341	
						0.16764	19.410	
9	75-G(D)	330	Kao, J.T.F. and Kobayashi, R.	0.4029	223.15	0.00747	15.802	Same remarks as for curve 1.
						0.01478	15.855	
						0.02921	15.996	
						0.04326	16.165	
						0.05693	16.368	
						0.08309	16.831	
						0.13070	17.891	
						0.18195	19.404	

TABLE 75-G(D)E.   EXPERIMENTAL VISCOSITY DATA AS A FUNCTION OF DENSITY FOR GASEOUS
HELIUM-NITROGEN MIXTURES (continued)

Cur. No.	Fig. No.	Ref. No.	Author(s)	Mole Fraction of N_2	Temp. (K)	Density (g cm^{-3})	Viscosity (N s m^{-2} x 10^{-6})	Remarks
10	75-G(D)	330	Kao, J.T.F. and Kobayashi, R.	0.6909	223.15	0.01139	15.016	Same remarks as for curve 1.
						0.02278	15.136	
						0.04557	15.470	
						0.06835	15.785	
						0.09061	16.229	
						0.13358	17.359	
						0.16856	19.878	
						0.28851	23.155	
						0.34873	26.673	
11	75-G(D)	330	Kao, J.T.F. and Kobayashi, R.	0.8412	223.15	0.01322	14.578	Same remarks as for curve 1.
						0.02678	14.744	
						0.05406	15.222	
						0.08156	15.759	
						0.10875	16.380	
						0.16227	17.826	
						0.25658	21.193	
						0.34615	26.105	
						0.41108	31.060	
12	75-G(D)	330	Kao, J.T.F. and Kobayashi, R.	1.0000	223.15	0.01530	14.229	Same remarks as for curve 1.
						0.03143	14.471	
						0.06437	15.055	
						0.09848	15.858	
						0.13324	16.785	
						0.20190	19.005	
						0.31894	24.152	
						0.41796	31.080	
						0.48364	37.511	
13	75-G(D)	330	Kao, J.T.F. and Kobayashi, R.	0.0000	273.15	0.00357	18.719	Same remarks as for curve 1.
						0.00699	18.718	
						0.01038	18.736	
						0.01370	18.737	
						0.02015	18.755	
						0.03233	18.859	
						0.04637	19.013	
						0.05926	19.213	
						0.07115	19.487	
14	75-G(D)	330	Kao, J.T.F. and Kobayashi, R.	0.0525	273.15	0.00464	18.814	Same remarks as for curve 1.
						0.00918	18.807	
						0.01362	18.782	
						0.01797	18.810	
						0.02640	18.892	
						0.04233	18.988	
						0.06035	19.178	
						0.07692	19.509	
						0.09213	19.875	
15	75-G(D)	330	Kao, J.T.F. and Kobayashi, R.	0.1283	273.15	0.00621	18.890	Same remarks as for curve 1.
						0.01229	18.941	
						0.01823	18.989	
						0.02405	19.043	
						0.03534	19.120	
						0.05667	19.395	
						0.08120	19.740	
						0.10336	20.141	
						0.12314	20.631	
16	75-G(D)	330	Kao, J.T.F. and Kobayashi, R.	0.2540	273.15	0.00891	18.748	Same remarks as for curve 1.
						0.01759	18.757	
						0.02606	18.826	
						0.03431	18.910	
						0.05021	19.125	
						0.07973	20.300	
						0.11292	20.300	
						0.14230	21.010	
						0.16868	21.927	
17	75-G(D)	330	Kao, J.T.F. and Kobayashi, R.	0.4029	273.15	0.01206	18.376	Same remarks as for curve 1.
						0.02384	18.513	
						0.03533	18.641	
						0.04653	18.801	
						0.06807	19.129	
						0.10783	19.911	
						0.15179	21.143	
						0.19014	22.371	
						0.22368	23.604	

310

TABLE 75-G(D)E. EXPERIMENTAL VISCOSITY DATA AS A FUNCTION OF DENSITY FOR GASEOUS
HELIUM-NITROGEN MIXTURES (continued)

Cur. No.	Fig. No.	Ref. No.	Author(s)	Mole Fraction of N_2	Temp. (K)	Density (g cm^{-3})	Viscosity (N s m^{-2} x 10^{-6})	Remarks
18	75-G(D)	330	Kao, J.T.F. and Kobayashi, R.	0.5450	273.15	0.01526	17.935	Same remarks as for curve 1.
						0.02996	18.109	
						0.04447	18.396	
						0.05864	18.591	
						0.08588	19.141	
						0.13588	20.339	
						0.19027	21.973	
						0.23664	23.869	
						0.27631	25.760	
19	75-G(D)	330	Kao, J.T.F. and Kobayashi, R.	0.6909	273.15	0.01861	17.564	Same remarks as for curve 1.
						0.03682	17.846	
						0.05482	18.148	
						0.07245	18.470	
						0.10643	19.174	
						0.16855	20.879	
						0.23486	23.207	
						0.28973	25.825	
						0.33520	28.444	
20	75-G(D)	330	Kao, J.T.F. and Kobayashi, R.	0.8412	273.15	0.01080	17.064	Same remarks as for curve 1.
						0.02160	17.237	
						0.04321	17.575	
						0.06456	17.899	
						0.08562	18.336	
						0.12638	19.307	
						0.20064	21.940	
						0.27785	25.035	
						0.33926	28.097	
						0.38842	31.680	
21	75-G(D)	330	Kao, J.T.F. and Kobayashi, R.	1.0000	273.15	0.02499	17.020	Same remarks as for curve 1.
						0.05072	17.246	
						0.07608	17.756	
						0.10178	18.358	
						0.15146	19.756	
						0.24133	23.066	
						0.33054	27.563	
						0.39744	32.340	
						0.44897	36.500	
22	75-G(D)	331	Makavetskas, R.A., Popov, V.N., and Tsederberg, N.V.	0.565	284.7	0.1370	20.96	Gas purities are not specified; capillary flow type viscometer; uncertainties are better than 4.5%; data corrected for thermal diffusion; original data reported as a function of pressure, density calculated from pressure through interpolation and extrapolation of P-V-T data of Witonsky and Miller [370].
						0.1093	20.45	
						0.0834	19.93	
						0.0534	19.36	
						0.0252	18.78	
						0.00850	18.49	
23	75-G(D)	331	Makavetskas, R.A., et al.	0.222	285.6	0.0690	20.88	Same remarks as for curve 22.
						0.0572	20.63	
						0.0407	20.32	
						0.0282	20.12	
						0.0133	20.05	
						0.00450	19.95	
24	75-G(D)	331	Makavetskas, R.A., et al.	0.412	285.6	0.1061	20.60	Same remarks as for curve 22.
						0.0849	20.27	
						0.0627	20.01	
						0.0421	19.71	
						0.0209	19.39	
						0.00670	19.22	
25	75-G(D)	331	Makavetskas, R.A., et al.	0.778	287.0	0.1673	21.75	Same remarks as for curve 22.
						0.1410	21.14	
						0.1041	20.16	
						0.0693	19.42	
						0.0325	18.62	
						0.0109	18.32	
26	75-G(D)	326	Kestin, J., Kobayashi, Y., and Wood, R.T.	0.7949	293.2	0.02479	18.400	N_2: 99.999 pure, He: 99.995 pure; oscillating disk viscometer; uncertainties: mixture composition $\pm 0.002\%$, viscosity $\pm 0.1\%$, viscosity ratios $\pm 0.04\%$.
						0.02184	18.360	
						0.01935	18.315	
						0.01444	18.260	
						0.009588	18.195	
						0.004686	18.145	
						0.002150	18.120	
						0.0009693	18.103	

TABLE 75-G(D)E.   EXPERIMENTAL VISCOSITY DATA AS A FUNCTION OF DENSITY FOR GASEOUS
HELIUM-NITROGEN MIXTURES (continued)

Cur. No.	Fig. No.	Ref. No.	Author(s)	Mole Fraction of N_2	Temp. (K)	Density (g cm^{-3})	Viscosity (N s m^{-2} x 10^{-6})	Remarks
27	75-G(D)	326	Kestin, J., et al.	0.7251	293.2	0.02273	18.556	Same remarks as for curve 26.
						0.01339	18.434	
						0.004466	18.331	
						0.0009080	18.294	
28	75-G(D)	326	Kestin, J., et al.	0.5005	293.2	0.01523	19.090	Same remarks as for curve 26.
						0.01013	19.030	
						0.003339	18.976	
						0.0006932	18.950	
29	75-G(D)	326	Kestin, J., et al.	0.2900	293.2	0.01153	19.612	Same remarks as for curve 26.
						0.009068	19.596	
						0.006792	19.583	
						0.004543	19.570	
						0.002276	19.562	
						0.0004648	19.542	
30	75-G(D)	326	Kestin, J., et al.	0.1682	293.2	0.007468	19.820	Same remarks as for curve 26.
						0.005176	19.809	
						0.001678	19.800	
						0.0003455	19.787	
31	75-G(D)	326	Kestin, J., et al.	0.1308	293.2	0.007443	19.860	Same remarks as for curve 26.
						0.004337	19.844	
						0.001462	19.831	
32	75-G(D)	326	Kestin, J., et al.	0.0361	293.2	0.004997	19.743	Same remarks as for curve 26.
						0.002979	19.746	
						0.001011	19.744	
						0.0002106	19.739	
33	75-G(D)	326	Kestin, J., et al.	1.0000	303.2	0.02854	18.394	Same remarks as for curve 26.
						0.02586	18.353	
						0.02409	18.322	
						0.02271	18.304	
						0.01972	18.262	
						0.01687	18.211	
						0.01564	18.200	
						0.01341	18.172	
						0.01125	18.143	
						0.009584	18.117	
						0.007259	18.091	
						0.005623	18.068	
						0.003425	18.046	
						0.001187	18.017	
						0.001160	18.011	
34	75-G(D)	326	Kestin, J., et al.	0.7949	303.2	0.02364	18.842	Same remarks as for curve 26.
						0.01383	18.707	
						0.004652	18.612	
						0.0009468	18.567	
35	75-G(D)	326	Kestin, J., et al.	0.7251	303.2	0.02142	19.019	Same remarks as for curve 26.
						0.01305	18.902	
						0.004434	18.803	
						0.0008901	18.757	
36	75-G(D)	326	Kestin, J., et al.	0.5005	303.2	0.01538	19.567	Same remarks as for curve 26.
						0.009445	19.508	
						0.003219	19.443	
						0.0006615	19.419	
37	75-G(D)	326	Kestin, J., et al.	0.3129	303.2	0.01175	20.026	Same remarks as for curve 26.
						0.006869	19.997	
						0.001924	19.973	
						0.0004675	19.957	
38	75-G(D)	326	Kestin, J., et al.	0.1686	303.2	0.008024	20.275	Same remarks as for curve 26.
						0.004794	20.261	
						0.001461	20.250	
						0.0003333	20.242	
39	75-G(D)	326	Kestin, J., et al.	0.1682	303.2	0.008071	20.301	Same remarks as for curve 26.
						0.004864	20.277	
						0.001660	20.262	
						0.0003363	20.246	
40	75-G(D)	326	Kestin, J., et al.	0.1308	303.2	0.007066	20.352	Same remarks as for curve 26.
						0.004250	20.315	
						0.001437	20.304	
						0.0002947	20.285	
41	75-G(D)	326	Kestin, J., et al.	0.0361	303.2	0.004806	20.204	Same remarks as for curve 26.
						0.002907	20.205	
						0.0009751	20.206	
						0.0002022	20.202	

TABLE 75-G(D)E.   EXPERIMENTAL VISCOSITY DATA AS A FUNCTION OF DENSITY FOR GASEOUS
HELIUM-NITROGEN MIXTURES (continued)

Cur. No.	Fig. No.	Ref. No.	Author(s)	Mole Fraction of N_2	Temp. (K)	Density (g cm^{-3})	Viscosity (N s m^{-2} x 10^{-6})	Remarks
42	75-G(D)	326	Kestin, J., et al.	0.0000	303.2	0.003927	20.074	Same remarks as for curve 26.
						0.003691	20.075	
						0.003198	20.072	
						0.002767	20.077	
						0.002378	20.069	
						0.002023	20.073	
						0.001203	20.077	
						0.0007979	20.071	
						0.0004885	20.070	
						0.0001656	20.067	
43	75-G(D)	330	Kao, J.T.F. and Kobayashi, R.	0.0000	323.15	0.00150	20.867	Same remarks as for curve 1.
						0.00299	20.817	
						0.00593	20.807	
						0.00883	20.812	
						0.01167	20.810	
						0.01721	20.804	
						0.02777	20.832	
						0.04009	20.975	
						0.05152	21.175	
						0.06219	21.353	
44	75-G(D)	330	Kao, J.T.F. and Kobayashi, R.	0.1283	323.15	0.00265	21.078	Same remarks as for curve 1.
						0.00527	21.133	
						0.01043	21.176	
						0.01549	21.207	
						0.02046	21.211	
						0.03013	21.228	
						0.04851	21.364	
						0.06986	21.689	
						0.08953	22.041	
						0.10756	22.523	
45	75-G(D)	330	Kao, J.T.F. and Kobayashi, R.	0.4029	323.15	0.00513	20.630	Same remarks as for curve 1.
						0.01020	20.670	
						0.02018	20.738	
						0.02992	20.807	
						0.03944	20.940	
						0.05781	21.246	
						0.09205	21.776	
						0.13055	22.696	
						0.16483	23.682	
						0.19540	24.744	
46	75-G(D)	330	Kao, J.T.F. and Kobayashi, R.	0.8412	323.15	0.00911	19.404	Same remarks as for curve 1.
						0.01818	19.482	
						0.03615	19.721	
						0.05384	20.020	
						0.07121	20.363	
						0.10476	21.023	
						0.16650	22.765	
						0.23315	25.059	
						0.28896	27.686	
						0.33563	30.290	
47	75-G(D)	330	Kao, J.T.F. and Kobayashi, R.	1.0000	323.15	0.01057	18.958	Same remarks as for curve 1.
						0.02113	19.116	
						0.04220	19.399	
						0.06307	19.772	
						0.08365	20.236	
						0.12357	21.168	
						0.19674	23.512	
						0.27386	26.779	
						0.33612	30.274	
						0.38657	33.860	
48	75-G(D)	331	Makavetskas, R.A., Popov, V.N., and Tsederberg, N.V.	0.778	588.8	0.0938	31.95	Same remarks as for curve 22.
						0.0726	31.42	
						0.0543	31.03	
						0.0344	30.62	
						0.0174	30.35	
						0.00680	30.12	
49	75-G(D)	331	Makavetskas, R.A., et al.	0.412	590.2	0.0575	31.57	Same remarks as for curve 22.
						0.0446	31.38	
						0.0328	31.20	
						0.0216	31.01	
						0.00960	30.80	
						0.00430	30.70	

TABLE 75-G(D)E.   EXPERIMENTAL VISCOSITY DATA AS A FUNCTION OF DENSITY FOR GASEOUS
HELIUM-NITROGEN MIXTURES (continued)

Cur. No	Fig. No.	Ref. No.	Author(s)	Mole Fraction of N_2	Temp. (K)	Density (g cm^{-3})	Viscosity (N s m^{-2} x 10^{-6})	Remarks
50	75-G(D)	331	Makavetskas, R.A., et al.	0.222	604.1	0.0372	32.46	Same remarks as for curve 22.
						0.0282	32.32	
						0.0203	32.21	
						0.0137	32.12	
						0.00650	32.04	
						0.00210	31.99	
51	75-G(D)	331	Makavetskas, R.A., et al.	0.565	604.8	0.0655	31.27	Same remarks as for curve 22.
						0.0523	31.05	
						0.0381	30.81	
						0.0250	30.43*	
						0.0118	30.31	
						0.00400	30.16	
52	75-G(D)	331	Makavetskas, R.A., et al.	0.222	822.8	0.0288	39.74	Same remarks as for curve 22.
						0.0220	39.62	
						0.0162	39.55**	
						0.0104	39.46	
						0.00530	39.39	
						0.00210	39.31	
53	75-G(D)	331	Makavetskas, R.A., et al.	0.412	873.2	0.0417	40.22	Same remarks as for curve 22.
						0.0299	40.12	
						0.0224	40.04	
						0.0146	39.95	
						0.00710	39.86	
						0.00290	39.81	
54	75-G(D)	331	Makavetskas, R.A., et al.	0.778	901.6	0.0648	40.58	Same remarks as for curve 22.
						0.0507	40.41	
						0.0367	39.97	
						0.0242	39.78	
						0.0115	39.53	
						0.00460	39.41	
55	75-G(D)	331	Makavetskas, R.A., et al.	0.565	952.6	0.0420	41.08	Same remarks as for curve 22.
						0.0333	40.80	
						0.0241	40.67	
						0.0160	40.44	
						0.00760	40.27	
						0.00280	40.21	

*Original table in the translation gives 40.43, which is believed to be in error.
**Original table gives 39.35, which is believed to be in error.

TABLE 75-G(D)S.   SMOOTHED VISCOSITY VALUES AS A FUNCTION OF DENSITY FOR GASEOUS HELIUM-NITROGEN MIXTURES

Density (g cm^{-3})	Mole Fraction of Nitrogen				
	0.0000 (183.2 K) [Ref. 330]	0.1283 (183.2 K) [Ref. 330]	0.4029 (183.2 K) [Ref. 330]	0.8412 (183.2 K) [Ref. 330]	1.0000 (183.2 K) [Ref. 330]
0.0050	14.223				
0.0100	14.202	14.342	13.662		
0.0150	14.202				
0.0200		14.437	13.775		
0.0250	14.286			12.600	12.000
0.0300	14.354	14.544			
0.0350	14.420				
0.0400	14.488	14.662	14.032		
0.0450	14.552				
0.0500	14.617	14.790	14.175	13.020	12.480
0.0600		14.943	14.320		
0.0700		15.123			
0.0750		15.222			
0.0800		15.324	14.680		
0.0850		15.428			
0.1000			15.278	14.130	13.605
0.1200			15.915		
0.1400			16.495		
0.1500				15.520	14.998
0.1600			17.079		
0.2000				17.200	16.620
0.2500				19.300	18.650
0.3000				21.790	21.140
0.3250				23.120	
0.3500					24.080
0.4000					27.600
0.4250					29.500

Density (g cm^{-3})	Mole Fraction of Nitrogen						
	0.0000 (223.2 K) [Ref. 330]	0.1283 (223.2 K) [Ref. 330]	0.2540 (223.2 K) [Ref. 330]	0.4029 (223.2 K) [Ref. 330]	0.6909 (223.2 K) [Ref. 330]	0.8412 (223.2 K) [Ref. 330]	1.0000 (223.2 K) [Ref. 330]
0.0100	16.242						
0.0125		16.440	16.218	15.835			
0.0200	16.264						
0.0250	16.287	16.505	16.308	15.950	15.180		
0.0300	16.318						
0.0375		16.600					
0.0400	16.418						
0.0500	16.563	16.720	16.595	16.260	15.510	15.170	14.810
0.0600	16.731						
0.0625		16.870					
0.0700	16.960						
0.0750	17.099	17.070	17.015	16.675	15.960		
0.1000		17.532	17.506	17.184	16.500	16.190	15.885
0.1250		18.270	18.090	17.749	17.110		
0.1500			18.815	18.432	17.840	17.492	17.230
0.1750				19.180			
0.2000					19.490	19.080	17.910
0.2500					21.410	20.920	20.910
0.3000					23.780	23.310	23.208
0.3500					26.800	26.320	26.000
0.4000						30.195	29.600
0.4500							34.060
0.4750							36.600

TABLE 75-G(D)S.   SMOOTHED VISCOSITY VALUES AS A FUNCTION OF DENSITY FOR GASEOUS HELIUM-NITROGEN MIXTURES (continued)

Density (g cm⁻³)	Mole Fraction of Nitrogen								
	0.0000 (273.15 K) [Ref. 330]	0.0525 (273.15 K) [Ref. 330]	0.1283 (273.15 K) [Ref. 330]	0.2540 (273.15 K) [Ref. 330]	0.4029 (273.15 K) [Ref. 330]	0.5450 (273.15 K) [Ref. 330]	0.6909 (273.15 K) [Ref. 330]	0.8412 (273.15 K) [Ref. 330]	1.0000 (273.15 K) [Ref. 330]
0.0100	18.714	18.800							
0.0125			18.940	18.742	18.380				
0.0200	18.754	18.830							
0.0250	18.794	18.859	19.030	18.810	18.520	18.090	17.680	17.280	
0.0300	18.838	18.892							
0.0375			19.140						
0.0400	18.930	18.969							
0.0500	19.054	19.056	19.270	19.110	18.840	18.470	18.080	17.690	17.240
0.0600	19.246	19.174							
0.0625			19.412						
0.0700	19.452	19.354							
0.0750			19.574	19.460	19.240	18.920			
0.0800	19.676	19.581							
0.0875			19.773						
0.0900		19.820							
0.1000			20.030	19.940	19.740	19.480	19.050	18.720	18.000
0.1125			20.342						
0.1250			20.720	20.540	20.358				
0.1500				21.247	21.078	20.800	20.380	20.175	19.200
0.1750				22.162	21.860				
0.2000					22.572	22.400	22.000	21.900	20.908
0.2250					23.531				
0.2500						24.450	24.020	23.800	23.200
0.2750						25.680			
0.3000							26.450	25.998	25.820
0.3250							27.830		
0.3500								28.720	28.750
0.3750								30.270	
0.4000									32.100
0.4500									35.800

Density (g cm⁻³)	Mole Fraction of Nitrogen						
	0.0361 (293.2 K) [Ref. 326]	0.1308 (293.2 K) [Ref. 326]	0.1682 (293.2 K) [Ref. 326]	0.2900 (293.2 K) [Ref. 326]	0.5005 (293.2 K) [Ref. 326]	0.7251 (293.2 K) [Ref. 326]	0.7449 (293.2 K) [Ref. 326]
0.00100	19.744						
0.00125		19.830	19.795	19.545	18.955		
0.00200	19.748						
0.00250	19.750	19.834	19.801	19.552	18.964	18.310	18.122
0.00300	19.750						
0.00375		19.843	19.807	19.560			
0.00400	19.749						
0.00500	19.743	19.850	19.811	19.568	18.985	18.336	18.146
0.00625		19.855	19.815	19.575			
0.00750		19.860	19.820	19.582	19.006	18.364	18.168
0.00875				19.591			
0.01000				19.600	19.033	18.343	18.202
0.01125				19.608			
0.01250					19.060	18.423	18.232
0.01500					19.087	18.455	18.264
0.01625					19.104		
0.01750						18.486	18.296
0.02000						18.520	18.330
0.02250						18.553	18.366
0.02500							18.401

TABLE 75-G(D)S.   SMOOTHED VISCOSITY VALUES AS A FUNCTION OF DENSITY FOR GASEOUS HELIUM-NITROGEN MIXTURES (continued)

Density (g cm^{-3})	Mole Fraction of Nitrogen									
	0.0000 (303.2 K) [Ref. 326]	0.0361 (303.2 K) [Ref. 326]	0.1308 (303.2 K) [Ref. 326]	0.1682 (303.2 K) [Ref. 326]	0.1686 (303.2 K) [Ref. 326]	0.3129 (303.2 K) [Ref. 326]	0.5005 (303.2 K) [Ref. 326]	0.7251 (303.2 K) [Ref. 326]	0.7949 (303.2 K) [Ref. 326]	1.0000 (303.2 K) [Ref. 326]
0.00050	20.071									
0.00100	20.073	20.245			20.249					
0.00125						19.970	19.400			
0.00150	20.075		20.292	20.255						
0.00200	20.077	20.246			20.253					
0.00250	20.078	20.246	20.303	20.265	20.255	19.980	19.410	18.780	18.584	18.032
0.00300	20.077	20.246			20.256					
0.00350	20.076		20.312	20.267						
0.00375						19.988	19.450			
0.00400	20.075	20.245			20.260					
0.00450			20.323	20.275						
0.00500		20.245		20.280	20.265	19.992	19.463	18.803	18.610	18.065
0.00550			20.332	20.283						
0.00600					20.267					
0.00625						19.998				
0.00650			20.342	20.290						
0.00700					20.270					
0.00750			20.351	20.295		20.005	19.455	18.840	18.635	18.096
0.00800					20.275					
0.00850			20.363	20.300						
0.00875						20.013				
0.01000						20.020	19.515	18.868	18.662	18.128
0.01125						20.028				
0.01250						20.034	19.540	18.897	18.690	18.160
0.01500							19.565	18.928	18.721	18.195
0.01625							19.577			
0.01750										
0.02000								18.970	18.756	18.228
0.02250								18.992	18.792	18.262
0.02500								19.027	18.828	18.297
										18.335

Density (g cm^{-3})	Mole Fraction of Nitrogen				
	0.0000 (323.15 K) [Ref. 330]	0.1283 (323.15 K) [Ref. 330]	0.4029 (323.15 K) [Ref. 330]	0.8412 (323.15 K) [Ref. 330]	1.0000 (323.15 K) [Ref. 330]
0.005	20.818	21.120			
0.010	20.805	21.174			
0.020	20.808	21.210			
0.025	20.824		20.780	19.620	19.110
0.030	20.856	21.226			
0.040	20.976	21.288			
0.050	21.147	21.378	21.040	19.985	19.305
0.060	21.316	21.512			
0.065	21.400				
0.070		21.690			
0.075			21.422		
0.080		21.875			
0.090		22.049			
0.100			21.940	20.970	20.550
0.125			22.565		
0.150			23.250	22.270	21.920
0.175			24.030		
0.200			24.915	23.870	23.620
0.250				25.800	25.650
0.300				28.280	28.080
0.325				29.680	
0.350					31.190
0.375					33.000

TABLE 75-G(D)S. SMOOTHED VISCOSITY VALUES AS A FUNCTION OF DENSITY FOR GASEOUS HELIUM–NITROGEN MIXTURES (continued)

Density (g cm^{-3})	Mole Fraction of Nitrogen					
	0.565 (284.7 K) [Ref. 331]	0.412 (285.6 K) [Ref. 331]	0.222 (285.6 K) [Ref. 331]	0.778 (287.0 K) [Ref. 331]	0.778 (588.8 K) [Ref. 331]	0.412 (590.2 K) [Ref. 331]
0.005	18.36	19.12	19.92	18.19	30.05	30.71
0.010	18.47	19.21	20.00	18.24	30.18	30.80
0.015	18.58	19.29	20.08	18.37	30.31	30.89
0.020	18.69	19.38	20.15	18.47	30.42	30.98
0.025	18.80	19.46	20.21	18.56	30.53	31.07
0.030	18.90	19.53	20.28	18.66	30.63	31.16
0.035	19.00	19.61	20.34	18.72	30.73	31.24
0.040	19.12	19.69	20.40	18.84	30.81	31.32
0.045	19.22	19.76	20.46	18.93	30.90	31.39
0.050	19.32	19.84	20.52	19.04	30.98	31.46
0.055	19.43	19.92	20.57	19.13	31.05	31.52
0.060	19.52	19.99	20.62	19.22	31.12	31.60
0.065	19.62	20.06	20.67	19.32	31.20	31.65
0.070	19.72	20.13	20.71	19.43	31.27	31.72
0.075	19.81	20.20		19.53	31.34	
0.080	19.89	20.28		19.65	31.42	
0.085	20.00	20.34		19.76	31.50	
0.090	20.08	20.40		19.86	31.56	
0.095	20.17	20.47		19.96	31.64	
0.100	20.26	20.52		20.08	31.71	
0.105	20.36	20.58		20.19		
0.110	20.45	20.64		20.28		
0.115	20.54			20.39		
0.120	20.63			20.49		
0.125	20.71			20.59		
0.130	20.80			20.69		
0.135	20.89			20.78		
0.140	20.98			20.87		

Density (g cm^{-3})	Mole Fraction of Nitrogen					
	0.222 (604.1 K) [Ref. 331]	0.565 (604.8 K) [Ref. 331]	0.222 (822.8 K) [Ref. 331]	0.412 (873.2 K) [Ref. 331]	0.778 (901.6 K) [Ref. 331]	0.565 (952.6 K) [Ref. 331]
0.005	32.02	30.16	39.36	39.82	39.40	40.20
0.010	32.09	30.28	39.45	39.87	39.50	40.32
0.015	32.16	30.39	39.54	39.96	39.60	40.44
0.020	32.22	30.49	39.61	40.01	39.69	40.56
0.025	32.28	30.59	39.68	40.07	39.78	40.67
0.030	32.36	30.68	39.74	40.12	39.87	40.78
0.035	32.42	30.78	39.80	40.17	39.96	40.88
0.040	32.48	30.87		40.23	40.04	40.99
0.045	32.55	30.95		40.28	40.12	41.09
0.050	32.61	31.03		40.32	40.20	41.20
0.055		31.10			40.27	
0.060		31.17			40.34	
0.065		31.24				
0.070		31.29				

318

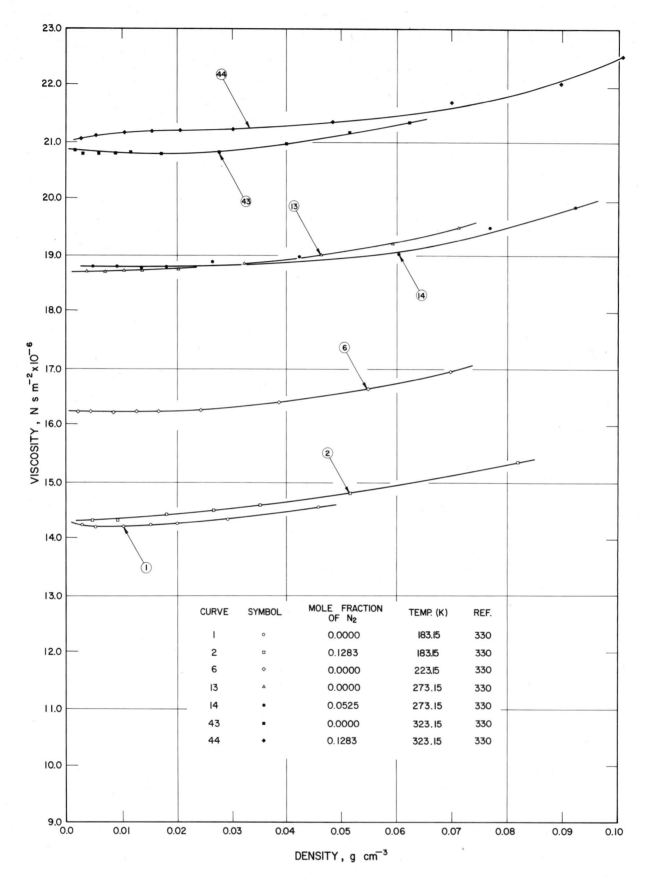

FIGURE 75-G(D). VISCOSITY DATA AS A FUNCTION OF DENSITY
FOR GASEOUS HELIUM-NITROGEN MIXTURES

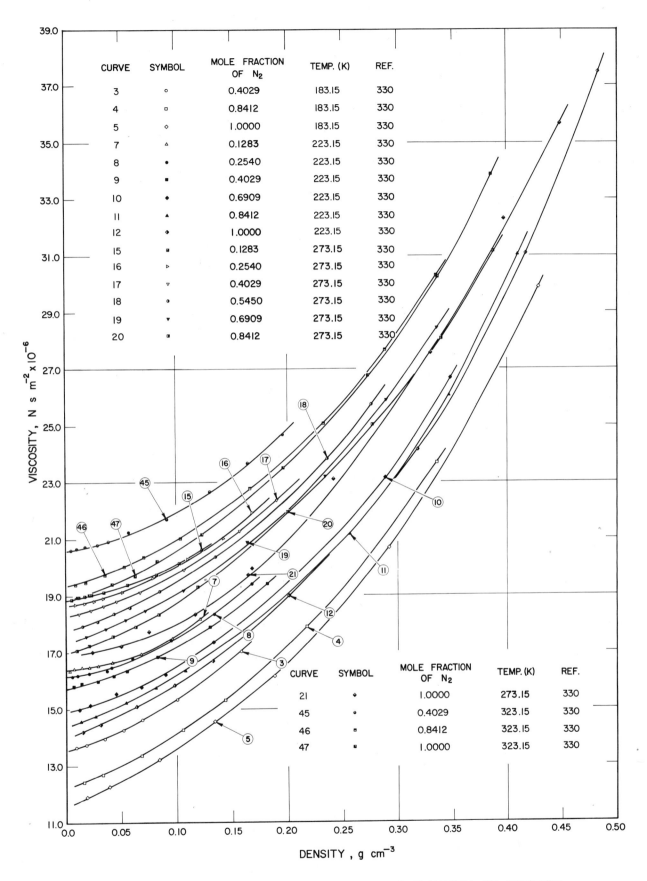

CURVE	SYMBOL	MOLE FRACTION OF N_2	TEMP. (K)	REF.
3	○	0.4029	183.15	330
4	□	0.8412	183.15	330
5	◇	1.0000	183.15	330
7	△	0.1283	223.15	330
8	●	0.2540	223.15	330
9	■	0.4029	223.15	330
10	◆	0.6909	223.15	330
11	▲	0.8412	223.15	330
12	◆	1.0000	223.15	330
15	◓	0.1283	273.15	330
16	▷	0.2540	273.15	330
17	▽	0.4029	273.15	330
18	◔	0.5450	273.15	330
19	▼	0.6909	273.15	330
20	◨	0.8412	273.15	330

CURVE	SYMBOL	MOLE FRACTION OF N_2	TEMP. (K)	REF.
21	◆	1.0000	273.15	330
45	◒	0.4029	323.15	330
46	◧	0.8412	323.15	330
47	◼	1.0000	323.15	330

FIGURE 75-G(D). VISCOSITY DATA AS A FUNCTION OF DENSITY
FOR GASEOUS HELIUM-NITROGEN MIXTURES (continued)

320

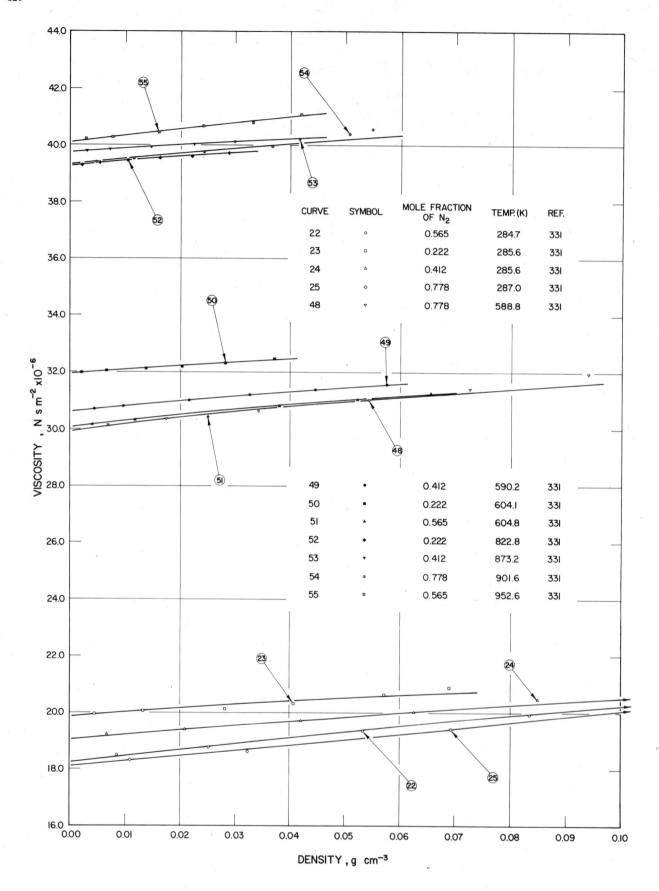

FIGURE 75 - G (D).   VISCOSITY DATA AS A FUNCTION OF DENSITY
FOR GASEOUS HELIUM-NITROGEN MIXTURES (continued)

321

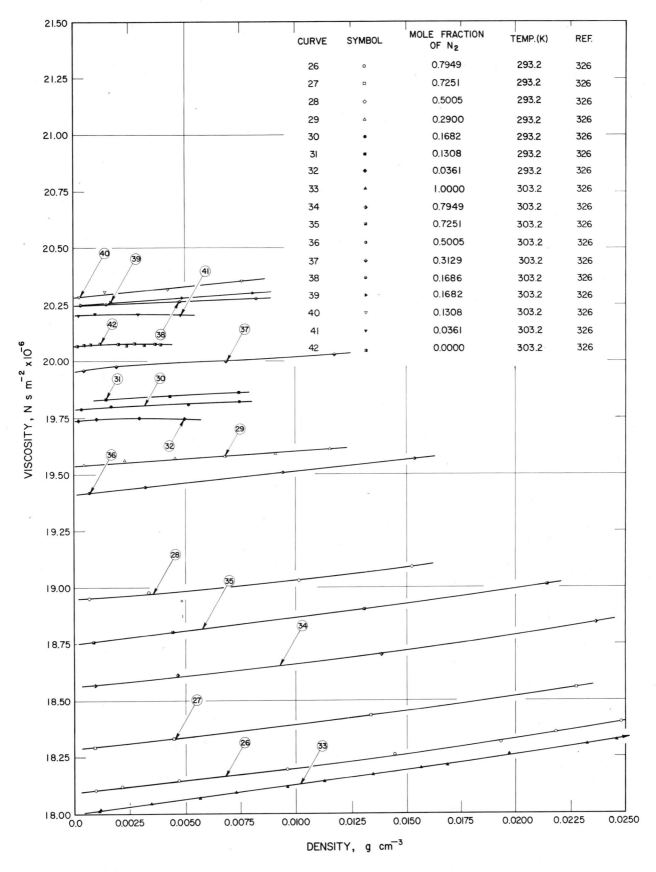

FIGURE 75-G(D). VISCOSITY DATA AS A FUNCTION OF DENSITY
FOR GASEOUS HELIUM-NITROGEN MIXTURES (continued)

TABLE 76-G(D)E.   EXPERIMENTAL VISCOSITY DATA AS A FUNCTION OF DENSITY FOR GASEOUS HELIUM-OXYGEN MIXTURES

Cur. No.	Fig. No.	Ref. No.	Author(s)	Mole Fraction of O_2	Temp. (K)	Density (g cm^{-3})	Viscosity (N s m^{-2} x10^{-6})	Remarks
1	76-G(D)	329	Kestin, J. and Yata, J.	1.0000	293.2	0.03319	20.764	O_2: 99.995 pure, He: 99.995 pure; oscillating disk viscometer; error ± 0.1% and precision ± 0.05%.
						0.02565	20.643	
						0.02023	20.577	
						0.01343	20.487	
						0.00667	20.406	
						0.00139	20.346	
2	76-G(D)	329	Kestin, J. and Yata, J.	0.7291	293.2	0.02609	21.230	Same remarks as for curve 1.
						0.02049	21.155	
						0.01528	21.099	
						0.01021	21.043	
						0.00509	20.997	
						0.00106	20.941	
3	76-G(D)	329	Kestin, J. and Yata, J.	0.5234	293.2	0.01844	21.503	Same remarks as for curve 1.
						0.01558	21.472	
						0.01166	21.448	
						0.007773	21.411	
						0.003882	21.372	
						0.000820	21.334	
4	76-G(D)	329	Kestin, J. and Yata, J.	0.4597	293.2	0.01722	21.573	Same remarks as for curve 1.
						0.01330	21.532	
						0.01056	21.511	
						0.007033	21.492	
						0.003515	21.450	
						0.000738	21.423	
5	76-G(D)	329	Kestin, J. and Yata, J.	0.3312	293.2	0.01378	21.580	Same remarks as for curve 1.
						0.01104	21.570	
						0.008308	21.551	
						0.005524	21.527	
						0.002766	21.515	
						0.000575	21.490	
6	76-G(D)	329	Kestin, J. and Yata, J.	0.1801	293.2	0.009210	21.248	Same remarks as for curve 1.
						0.007089	21.234	
						0.005606	21.234	
						0.003711	21.222	
						0.001883	21.216	
						0.000394	21.198	
7	76-G(D)	329	Kestin, J. and Yata, J.	0.1042	293.2	0.006841	20.798	Same remarks as for curve 1.
						0.005377	20.799	
						0.004295	20.796	
						0.002869	20.792	
						0.001438	20.783	
						0.000305	20.771	
8	76-G(D)	329	Kestin, J. and Yata, J.	0.0578	293.2	0.005660	20.378	Same remarks as for curve 1.
						0.004402	20.370	
						0.003470	20.375	
						0.003424	20.380	
						0.002329	20.375	
						0.001168	20.371	
						0.000243	20.366	
9	76-G(D)	329	Kestin, J. and Yata, J.	1.0000	303.2	0.03226	21.331	Same remarks as for curve 1.
						0.02459	21.227	
						0.01951	21.156	
						0.01292	21.072	
						0.00642	20.988	
						0.00133	20.918	
10	76-G(D)	329	Kestin, J. and Yata, J.	0.7291	303.2	0.02501	21.788	Same remarks as for curve 1.
						0.01965	21.724	
						0.01470	21.672	
						0.00985	21.611	
						0.00490	21.561	
						0.00102	21.513	
11	76-G(D)	329	Kestin, J. and Yata, J.	0.5234	303.2	0.01904	22.062	Same remarks as for curve 1.
						0.01504	22.035	
						0.01126	21.998	
						0.007497	21.972	
						0.003741	21.922	
						0.000777	21.895	
12	76-G(D)	329	Kestin, J. and Yata, J.	0.3312	303.2	0.01247	22.116	Same remarks as for curve 1.
						0.01013	22.106	
						0.008016	22.091	
						0.005343	22.079	
						0.002673	22.060	
						0.000562	22.043	

TABLE 76-G(D)E.   EXPERIMENTAL VISCOSITY DATA AS A FUNCTION OF DENSITY FOR GASEOUS HELIUM–OXYGEN MIXTURES (continued)

Cur. No.	Fig. No.	Ref. No.	Author(s)	Mole Fraction of O_2	Temp. (K)	Density (g cm^{-3})	Viscosity (N s m^{-2} 10^{-6})	Remarks
13	76–G(D)	329	Kestin, J. and Yata, J.	0.1801	303.2	0.008880	21.768	Same remarks as for curve 1.
						0.006852	21.757	
						0.005430	21.751	
						0.003607	21.743	
						0.001814	21.738	
						0.000379	21.724	
14	76–G(D)	329	Kestin, J. and Yata, J.	0.1042	303.2	0.006600	21.302	Same remarks as for curve 1.
						0.005222	21.304	
						0.004088	21.298	
						0.002760	21.297	
						0.001387	21.288	
						0.000287	21.265	
15	76–G(D)	329	Kestin, J. and Yata, J.	0.0578	303.2	0.005588	20.865	Same remarks as for curve 1.
						0.004494	20.873	
						0.003371	20.862	
						0.002251	20.869	
						0.001129	20.865	
						0.000235	20.845	
16	76–G(D)	329	Kestin, J. and Yata, J.	0.0000	303.2	0.003567	20.095	Same remarks as for curve 1.
						0.003014	20.102	
						0.002432	20.096	
						0.001605	20.096	
						0.000797	20.095	
						0.000169	20.078	

TABLE 76-G(D)S. SMOOTHED VISCOSITY VALUES AS A FUNCTION OF DENSITY FOR GASEOUS HELIUM-OXYGEN MIXTURES

Density (g cm^{-3})	Mole Fraction of Oxygen							
	0.0578 (293.2 K) [Ref. 329]	0.1042 (293.2 K) [Ref. 329]	0.1801 (293.2 K) [Ref. 329]	0.3312 (293.2 K) [Ref. 329]	0.4597 (293.2 K) [Ref. 329]	0.5234 (293.2 K) [Ref. 329]	0.7291 (293.2 K) [Ref. 329]	1.0000 (293.2 K) [Ref. 329]
0.00100	20.368	20.778	21.202					
0.00139								
0.00200	20.371	20.787	21.210	21.508				20.3460
0.00300	20.373	20.792	21.217		21.444			
0.00400	20.375	20.795	21.224	21.523				
0.00500	20.377	20.797	21.229	21.530	21.463	21.385	20.987	
0.00600	20.378	20.797	21.234	21.537	21.472			
0.00667								
0.00700		20.799			21.481			20.406
0.00800			21.242	21.549	21.490			
0.00900			21.246	21.555				
0.01000				21.561	21.509	21.432	21.043	
0.01100				21.567	21.518			
0.01200					21.527			
0.01300				21.578				
0.01343								
0.01400					21.545			20.486
0.01500				21.590	21.554			
0.02000						21.473	21.098	
0.02023						21.509	21.153	20.572
0.02500								
0.02565							21.208	
0.03319								20.645
								20.764

Density (g cm^{-3})	Mole Fraction of Oxygen							
	0.0000 (303.2 K) [Ref. 329]	0.0578 (303.2 K) [Ref. 329]	0.1042 (303.2 K) [Ref. 329]	0.1801 (303.2 K) [Ref. 329]	0.3312 (303.2 K) [Ref. 329]	0.5234 (303.2 K) [Ref. 329]	0.7291 (303.2 K) [Ref. 329]	1.0000 (303.2 K) [Ref. 329]
0.0000								
0.0005	20.091					21.878		
0.0010	20.095	20.856	21.280	21.728				
0.0020	20.097	20.861	21.289	21.734	22.058			
0.0025	20.098			21.736		21.900	21.531	20.935
0.0030	20.098	20.865	21.295					
0.0040	20.097	20.868	21.300		22.068			
0.0050		20.869	21.302	21.750	22.075	21.940	21.559	20.970
0.0060		20.870	21.304					
0.0070		20.870	21.305					
0.0075				21.762	22.091	21.967	21.587	21.003
0.0080				21.764				
0.0100				21.772	22.104	21.991	21.616	21.046
0.0125					22.104	22.013	21.645	21.069
0.0150						22.033	21.677	21.102
0.0175						22.052	21.702	
0.0200						22.069	21.730	21.167
0.0225							21.758	
0.0250							21.786	21.230
0.0300								21.294

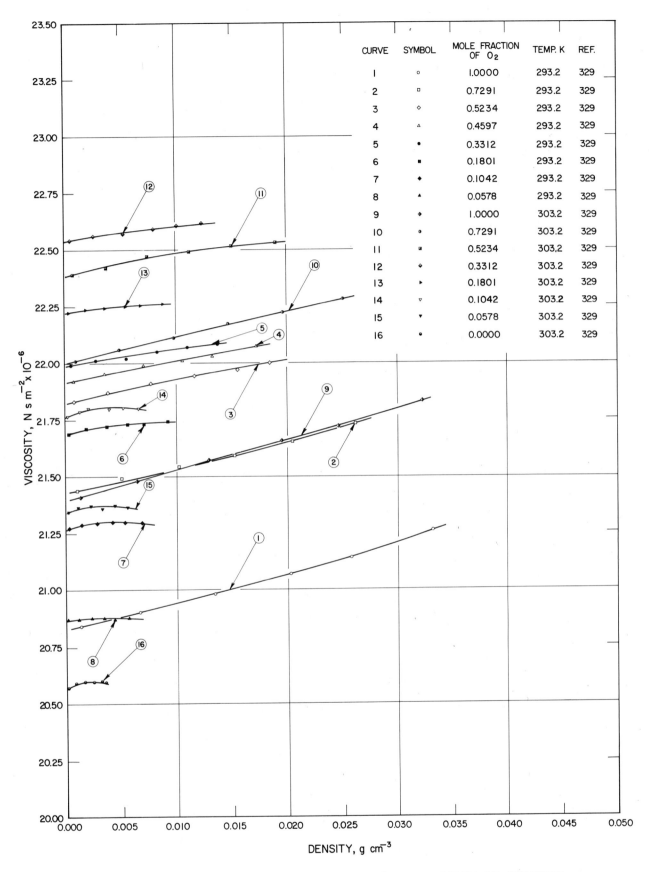

FIGURE 76-G(D). VISCOSITY DATA AS A FUNCTION OF DENSITY
FOR GASEOUS HELIUM- OXYGEN MIXTURES

TABLE 76-G(T)E.   EXPERIMENTAL VISCOSITY DATA AS A FUNCTION OF TEMPERATURE FOR GASEOUS
HELIUM-OXYGEN MIXTURES

Cur. No.	Fig. No.	Ref. No.	Author(s)	Mole Fraction of O$_2$	Pressure (atm)	Temp. (K)	Viscosity (N s m^{-2} x 10^{-6})	Remarks
1	76-G(T)	332	Johnson, C.A.	0.000	69.88	580.2	32.0	He: better than 99.95 pure, O$_2$: better than 99.6 pure; steady flow capillary viscometer; uncertainty ±1.0%.
					70.22	577.2	31.8	
					87.71	578.2	32.0	
2	76-G(T)	332	Johnson, C.A.	0.000	123.16	577.2	32.1	Same remarks as for curve 1.
					129.42	577.2	32.2	
					127.04	517.2	30.0	
					126.09	470.2	27.5	
					125.27	633.2	33.5	
					123.78	682.2	35.5	
					122.75	682.2	35.5	
3	76-G(T)	332	Johnson, C.A.	0.000	20.41	678.2	34.9	Same remarks as for curve 1.
					20.41	672.2	35.0	
					20.62	626.2	31.8	
					20.41	625.2	31.8	
					20.89	654.2	33.7	
					21.16	599.2	32.1	
					21.23	564.2	30.8	
					21.03	540.2	30.1	
					21.03	494.2	28.6	
					21.03	451.2	26.4	
4	76-G(T)	332	Johnson, C.A.	0.000	8.10	450.2	26.3	Same remarks as for curve 1.
					8.17	491.2	28.3	
					8.17	526.2	29.6	
5*	76-G(T)	332	Johnson, C.A.	0.000	21.23	828.2	21.3	Same remarks as for curve 1.
					21.23	352.2	22.2	
6	76-G(T)	332	Johnson, C.A.	0.000	70.09	460.2	30.5	Same remarks as for curve 1.
					74.03	559.2	30.6	
7	76-G(T)	332	Johnson, C.A.	0.000	38.72	539.2	30.1	Same remarks as for curve 1.
					38.31	554.2	30.7	
					36.88	582.2	32.2	
8	76-G(T)	332	Johnson, C.A.	0.180	132.49	683.2	39.0	Same remarks as for curve 1.
					131.94	642.2	38.6	
					131.46	614.2	36.4	
					131.40	583.2	34.8	
					131.19	550.2	33.6	
					130.24	518.2	32.0	
					129.63	485.2	30.8	
					129.22	463.2	29.7	
9	76-G(T)	332	Johnson, C.A.	0.180	87.30	463.2	29.6	Same remarks as for curve 1.
					86.96	493.2	31.8	
					86.62	529.2	32.4	
					84.85	476.2	34.1	
					84.17	611.2	36.0	
					83.15	707.2	39.6	
					80.97	643.2	37.1	
					80.43	679.2	38.8	
					80.16	607.2	35.4	
10	76-G(T)	332	Johnson, C.A.	0.180	45.93	704.2	39.3	Same remarks as for curve 1.
					45.59	667.2	37.8	
					45.18	637.2	36.4	
					44.91	613.2	35.8	
					44.57	578.2	34.1	
					44.30	557.2	33.2	
					43.89	525.2	32.0	
					43.55	493.2	30.5	
					43.07	465.2	29.6	
11	76-G(T)	332	Johnson, C.A.	0.531	68.52	512.2	32.8	Same remarks as for curve 1.
					68.73	485.2	31.4	
					68.86	555.2	34.5	
					68.97	586.2	35.4	
					69.34	615.2	36.6	
					69.17	636.2	37.4	
					70.02	701.2	39.5	
12	76-G(T)	332	Johnson, C.A.	0.531	45.11	582.2	35.4	Same remarks as for curve 1.
					44.84	696.2	39.4	
					44.71	664.2	38.6	
					44.50	628.2	36.5	
					44.23	602.2	36.2	
					43.55	555.2	33.7	
					43.21	527.2	32.4	
					43.01	497.2	31.4	

*Not shown in figure.

TABLE 76-G(T)E.   EXPERIMENTAL VISCOSITY DATA AS A FUNCTION OF TEMPERATURE FOR GASEOUS
HELIUM-OXYGEN MIXTURES (continued)

Cur. No.	Fig. No.	Ref. No.	Author(s)	Mole Fraction of O_2	Pressure (atm)	Temp. (K)	Viscosity (N s m^{-2} x 10^{-6})	Remarks
13	76-G(T)	332	Johnson, C.A.	0.717	127.45	468.2	30.4	Same remarks as for curve 1.
					127.04	497.2	31.8	
					123.91	542.2	33.6	
					123.23	582.2	35.3	
					123.16	625.2	36.9	
					122.41	652.2	38.0	
					122.01	717.2	39.6	
					121.57	688.2	39.3	
					120.92	658.2	37.9	
14	76-G(T)	332	Johnson, C.A.	0.717	87.78	656.2	37.8	Same remarks as for curve 1.
					87.51	624.2	36.6	
					87.17	598.2	35.4	
					86.96	570.2	34.4	
					86.42	545.2	33.6	
					86.15	719.2	40.0	
					85.47	687.2	38.9	
					84.58	533.2	32.8	
					84.17	494.2	31.4	
					83.63	475.2	30.6	
15	76-G(T)	332	Johnson, C.A.	0.717	44.91	471.2	29.6	Same remarks as for curve 1.
					44.91	498.2	31.3	
					44.64	542.2	32.8	
					44.23	566.2	33.9	
					44.09	594.2	34.8	
					43.69	618.2	36.0	
					43.28	718.2	39.2	
					42.66	681.2	38.1	
16	76-G(T)	332	Johnson, C.A.	1.000	99.42	570.2	35.0	Same remarks as for curve 1.
					99.42	569.2	35.1	
					98.33	519.2	32.8	
					97.92	470.2	30.5	
					97.92	494.2	31.6	
					98.73	547.2	33.4	
					98.39	598.2	35.6	
					99.07	625.2	36.7	
					99.48	648.2	37.6	
					99.48	672.2	38.7	
					100.57	699.2	39.7	
					102.18	724.2	40.5	
					95.88	567.2	34.4	
17	76-G(T)	332	Johnson, C.A.	1.000	50.15	566.2	34.1	Same remarks as for curve 1.
					49.81	597.2	34.8	
					51.17	570.2	33.6	
					51.44	596.2	34.6	
18	76-G(T)	332	Johnson, C.A.	1.000	100.16	329.2	26.4	Same remarks as for curve 1.
					99.76	353.2	27.2	
19	76-G(T)	332	Johnson, C.A.	1.000	52.53	475.2	29.6	Same remarks as for curve 1.
					51.92	516.2	31.4	
					51.71	549.2	33.2	
					51.37	594.2	34.8	
20	76-G(T)	332	Johnson, C.A.	1.000	51.10	627.2	35.9	Same remarks as for curve 1.
					50.90	722.2	39.8	
					50.56	688.2	38.5	
21	76-G(T)	332	Johnson, C.A.	1.000	128.33	465.2	30.3	Same remarks as for curve 1.
					128.06	503.2	33.5	
					127.65	531.2	33.1	
					127.31	571.2	34.3	
					124.39	703.2	36.7	
					123.57	648.2	37.1	
					123.37	603.2	35.5	

TABLE 76-G(T)S.   SMOOTHED VISCOSITY VALUES AS A FUNCTION OF TEMPERATURE FOR GASEOUS HELIUM-OXYGEN MIXTURES

Temp. K	Mole Fraction of Oxygen					
	0.000 (69.88-87.71 atm) [Ref. 329]	0.000 (122.75-129.42 atm) [Ref. 329]	0.000 (20.41-21.23 atm) [Ref. 329]	0.000 (8.10-8.17 atm) [Ref. 329]	0.000 (21.23 atm) [Ref. 329]	0.000 (36.88-38.72 atm) [Ref. 329]
355					22.19	
375					22.15	
400					22.10	
425					22.06	
450			26.75	26.58	22.01	
460		27.59	27.11	26.95	21.99	
475		28.13	27.65	27.50	21.96	
500		29.06	28.57	28.41	21.91	
525		29.98	29.49	29.33	21.86	29.80
550		30.90	30.41		21.82	30.72
575	31.74	31.82	31.32		21.77	31.64
600	32.64	32.73	32.23		21.72	
625		33.65	33.14		21.68	
650		34.57	34.07		21.63	
675		35.50	34.98		21.58	
700					21.54	
725					21.49	

Temp. K	Mole Fraction of Oxygen					
	0.180 (129.22-132.49 atm) [Ref. 329]	0.180 (80.16-87.30 atm) [Ref. 329]	0.180 (43.07-45.93 atm) [Ref. 329]	0.531 (68.52-70.02 atm) [Ref. 329]	0.531 (43.01-45.11 atm) [Ref. 329]	0.717 (120.92-127.45 atm) [Ref. 329]
460	29.84	29.66				
475	30.44	30.26	29.88			30.81
500	31.44	31.28	30.90	32.14	31.92	31.80
525	32.46	32.30	31.92	33.14	32.93	32.80
550	33.48	33.32	32.94	34.14	33.93	33.80
575	34.48	34.34	33.98	35.14	34.92	34.79
600	35.48	35.34	34.99	36.12	35.90	35.78
625	36.50	36.34	36.00	37.12	36.92	36.78
650	37.52	37.38	37.04	38.12	37.92	37.78
675	38.54	38.40	38.07	39.14	38.92	38.78
700		39.44	39.10			39.78

Temp. K	Mole Fraction of Oxygen				
	0.717 (83.63-87.78 atm) [Ref. 329]	0.717 (42.66-44.91 atm) [Ref. 329]	1.000 (95.88-102.18 atm) [Ref. 329]	1.000 49.81-51.49 atm) [Ref. 329]	1.000 51.37-52.53 atm) [Ref. 329]
475	30.48	30.01	30.76		29.79
500	31.48	31.02	31.74		30.82
525	32.50	32.05	32.74		31.84
550	33.52	33.08	33.74		32.88
575	34.52	34.10	34.73	33.87	33.91
600	35.53	35.12	35.71	34.89	34.93
625	36.54	36.12	36.71		
650	37.56	37.15	37.71		
675	38.59	38.19	38.72		
700	39.61	39.22	39.72		

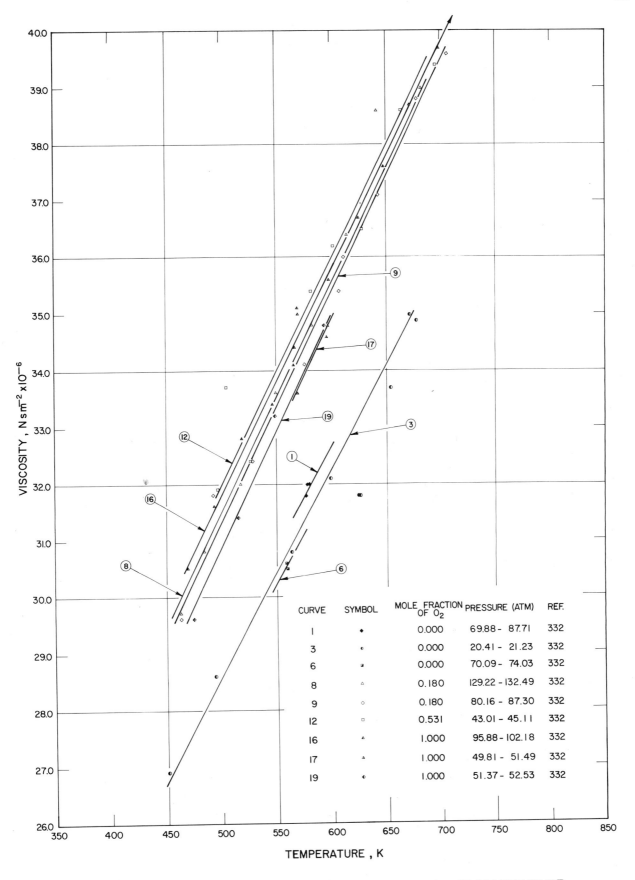

CURVE	SYMBOL	MOLE FRACTION OF O_2	PRESSURE (ATM)	REF.
1	◆	0.000	69.88 - 87.71	332
3	◐	0.000	20.41 - 21.23	332
6	▣	0.000	70.09 - 74.03	332
8	△	0.180	129.22 - 132.49	332
9	◇	0.180	80.16 - 87.30	332
12	□	0.531	43.01 - 45.11	332
16	▲	1.000	95.88 - 102.18	332
17	▵	1.000	49.81 - 51.49	332
19	◕	1.000	51.37 - 52.53	332

FIGURE 76-G(T). VISCOSITY DATA AS A FUNCTION OF TEMPERATURE
FOR GASEOUS HELIUM – OXYGEN MIXTURES

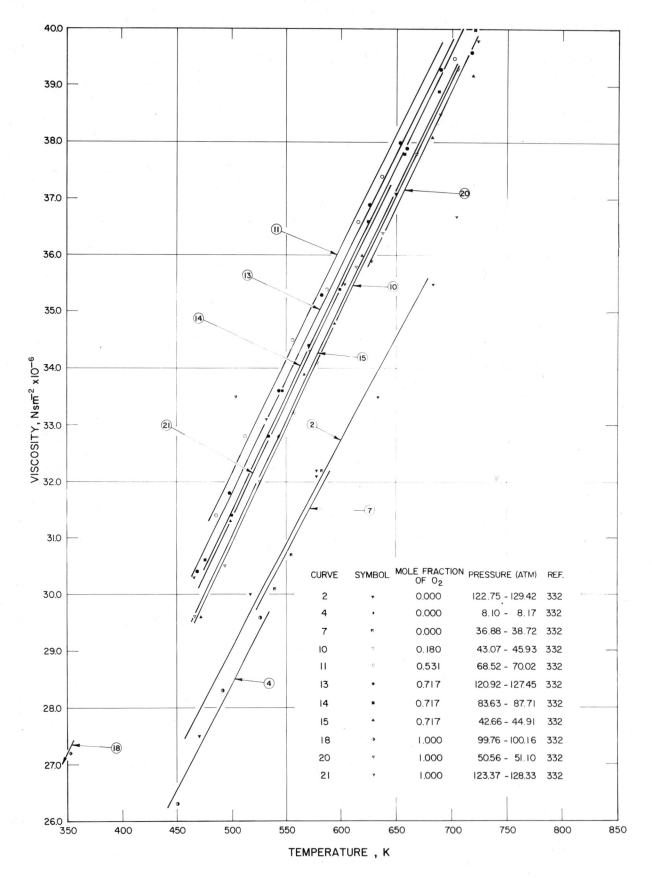

CURVE	SYMBOL	MOLE FRACTION OF O_2	PRESSURE (ATM)	REF.
2	▼	0.000	122.75 - 129.42	332
4	`	0.000	8.10 - 8.17	332
7	▫	0.000	36.88 - 38.72	332
10	▽	0.180	43.07 - 45.93	332
11	○	0.531	68.52 - 70.02	332
13	●	0.717	120.92 - 127.45	332
14	■	0.717	83.63 - 87.71	332
15	▲	0.717	42.66 - 44.91	332
18	`	1.000	99.76 - 100.16	332
20	▾	1.000	50.56 - 51.10	332
21	⸝	1.000	123.37 - 128.33	332

FIGURE 76 – G(T). VISCOSITY DATA AS A FUNCTION OF TEMPERATURE
FOR GASEOUS HELIUM-OXYGEN MIXTURES (continued)

TABLE 77-G(D)E.   EXPERIMENTAL VISCOSITY DATA AS A FUNCTION OF DENSITY FOR GASEOUS KRYPTON-CARBON DIOXIDE MIXTURES

Cur. No.	Fig. No.	Ref. No.	Author(s)	Mole Fraction of Kr	Temp. (K)	Density (g cm^{-3})	Viscosity (N s m^{-2} x 10^{-6})	Remarks
1	77-G(D)	329	Kestin, J. and Yata, J.	1.0000	293.2	0.07693 0.05416 0.01753 0.00367	25.762 25.488 25.122 25.000	Kr: 99.99 pure, CO$_2$: 99.8 pure; oscillating disk viscometer; error ±0.1%, precision ±0.05%.
2	77-G(D)	329	Kestin, J. and Yata, J.	0.7033	293.2	0.06535 0.04624 0.01512 0.00313	23.162 22.934 22.627 22.503	Same remarks as for curve 1.
2	77-G(D)	329	Kestin, J. and Yata, J.	0.4870	293.2	0.05778 0.04080 0.01336 0.00279	20.940 20.723 20.463 20.379	Same remarks as for curve 1.
4	77-G(D)	329	Kestin, J. and Yata, J.	0.2617	293.2	0.05139 0.03606 0.01152 0.00239	18.333 18.164 17.956 17.899	Same remarks as for curve 1.
5	77-G(D)	329	Kestin, J. and Yata, J.	0.0000	293.2	0.04393 0.03017 0.00941 0.00192	14.934 14.815 14.693 14.674	Same remarks as for curve 1.
6	77-G(D)	329	Kestin, J. and Yata, J.	1.0000	303.2	0.07382 0.05237 0.01702 0.00350	26.532 26.284 25.924 25.785	Same remarks as for curve 1.
7	77-G(D)	329	Kestin, J. and Yata, J.	0.7033	303.2	0.06314 0.04446 0.01454 0.00300	23.876 23.656 23.347 23.238	Same remarks as for curve 1.
8	77-G(D)	329	Kestin, J. and Yata, J.	0.4870	303.2	0.05587 0.03930 0.01287 0.00267	21.590 21.391 21.134 21.040	Same remarks as for curve 1.
9	77-G(D)	329	Kestin, J. and Yata, J.	0.2617	303.2	0.04897 0.03451 0.01113 0.00230	18.914 18.740 18.548 18.472	Same remarks as for curve 1.
10	77-G(D)	329	Kestin, J. and Yata, J.	0.0000	303.2	0.04178 0.02882 0.00906 0.00185	15.449 15.326 15.194 15.169	Same remarks as for curve 1.

TABLE 77-G(D)S.   SMOOTHED VISCOSITY VALUES AS A FUNCTION OF DENSITY FOR GASEOUS KRYPTON-CARBON DIOXIDE MIXTURES

Density (g cm^{-3})	Mole Fraction of Krypton				
	0.0000 (293.2 K) [Ref. 329]	0.2617 (293.2 K) [Ref. 329]	0.4870 (293.2 K) [Ref. 329]	0.7033 (293.2 K) [Ref. 329]	1.0000 (293.2 K) [Ref. 329]
0.005	14.680	17.918	20.394	22.523	25.020
0.010	14.700	17.941	20.434	22.580	25.062
0.015	14.722	17.976	20.475		
0.020	14.750	18.018	20.520	22.684	25.148
0.025	14.782	18.062	20.562	22.738	
0.030	14.820	18.110	20.610	22.780	25.242
0.035	14.860	18.162			
0.040	14.900	18.218	20.716	22.878	25.340
0.045	14.941	18.270			
0.050		18.322	20.838	22.982	25.442
0.055			20.900		
0.060				23.100	25.550
0.065				23.160	
0.070					25.678
0.075					25.740

Density (g cm^{-3})	Mole Fraction of Krypton				
	0.0000 (303.2 K) [Ref. 329]	0.2617 (303.2 K) [Ref. 329]	0.4870 (303.2 K) [Ref. 329]	0.7033 (303.2 K) [Ref. 329]	1.0000 (303.2 K) [Ref. 329]
0.005	15.180	18.494	21.068	23.262	25.790
0.010	15.198	18.528	21.108	23.300	25.840
0.015	15.220	18.564	21.150		25.890
0.020	15.250	18.708	21.199	23.390	25.940
0.025	15.286	18.750	21.244		25.968
0.030	15.329	18.798	21.300	23.490	26.044
0.035	15.380	18.746			
0.040	15.430	18.801	21.410	23.600	26.150
0.045		18.864			
0.050		18.930	21.534	23.720	26.260
0.055			21.596		
0.060				23.840	26.378
0.065				23.894	
0.070					26.482

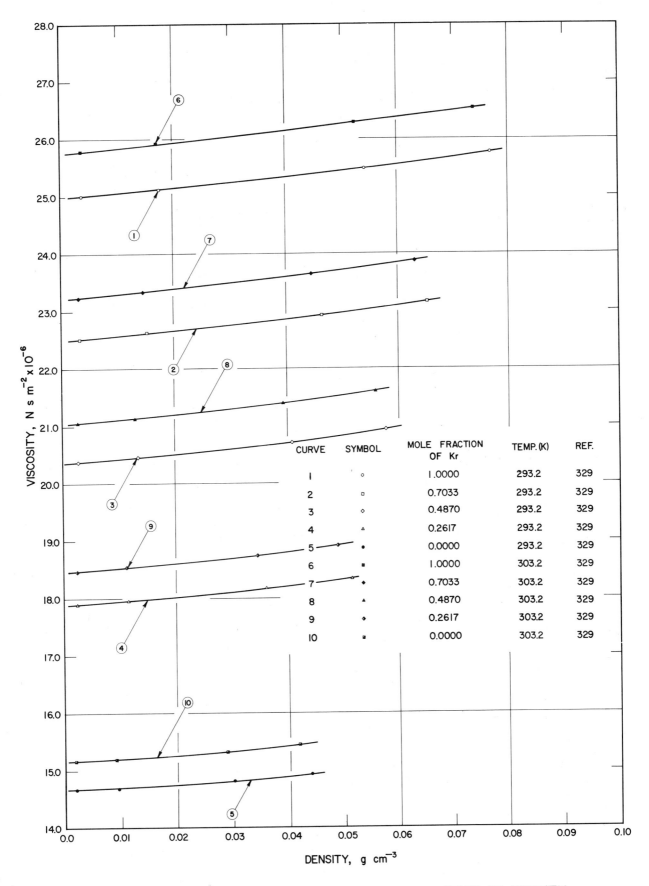

FIGURE 77-G(D). VISCOSITY DATA AS A FUNCTION OF DENSITY
FOR GASEOUS KRYPTON-CARBON DIOXIDE MIXTURES

TABLE 78-G(D)E. EXPERIMENTAL VISCOSITY DATA AS A FUNCTION OF DENSITY FOR GASEOUS
NEON–CARBON DIOXIDE MIXTURES

Cur. No.	Fig. No.	Ref. No.	Author(s)	Mole Fraction of CO_2	Temp. (K)	Density ($g\,cm^{-3}$)	Viscosity ($N\,s\,m^{-2} \times 10^{-6}$)	Remarks
1	78-G(D)	333	Breetveld, J.D., DiPippo, R., and Kestin, J.	1.0000	293.2	0.04877 0.02988 0.009435 0.001865	15.004 14.820 14.697 14.687	CO_2: 99.80 pure, Ne: 99.9925 pure; oscillating disk viscometer; precision ±0.1%.
2	78-G(D)	333	Breetveld, J.D., et al.	0.7938	293.2	0.04170 0.02577 0.008356 0.001653	17.240 17.069 16.956 16.919	Same remarks as for curve 1.
3	78-G(D)	333	Breetveld, J.D., et al.	0.5650	293.2	0.03233 0.02134 0.007090 0.001420	20.252 20.136 20.029 19.987	Same remarks as for curve 1.
4	78-G(D)	333	Breetveld, J.D., et al.	0.3797	293.2	0.02891 0.01833 0.006147 0.001233	23.289 23.191 23.089 23.049	Same remarks as for curve 1.
4	78-G(D)	333	Breetveld, J.D., et al.	0.2897	293.2	0.02646 0.01696 0.005651 0.001142	24.930 24.841 24.755 24.707	Same remarks as for curve 1.
6	78-G(D)	333	Breetveld, J.D., et al.	0.1238	293.2	0.02258 0.01450 0.004866 0.000964	28.483 28.420 28.367 28.315	Same remarks as for curve 1.
7	78-G(D)	333	Breetveld, J.D., et al.	1.0000	303.2	0.04590 0.02857 0.009066 0.001080	15.508 15.314 15.191 15.161	Same remarks as for curve 1.
8	78-G(D)	333	Breetveld, J.D., et al.	0.7938	303.2	0.03855 0.02480 0.008046 0.001598	17.756 17.609 17.472 17.443	Same remarks as for curve 1.
9	78-G(D)	333	Breetveld, J.D., et al.	0.5650	303.2	0.03262 0.02067 0.006775 0.001373	20.825 20.698 20.577 20.543	Same remarks as for curve 1.
10	78-G(D)	333	Breetveld, J.D., et al.	0.3797	303.2	0.02773 0.01794 0.006002 0.001192	23.891 23.814 23.693 23.646	Same remarks as for curve 1.
11	78-G(D)	333	Breetveld, J.D., et al.	0.2897	303.2	0.02518 0.01650 0.005359 0.001104	25.549 25.463 25.380 25.335	Same remarks as for curve 1.
12	78-G(D)	333	Breetveld, J.D., et al.	0.1238	303.2	0.02122 0.01409 0.004750 0.000942	29.149 29.101 29.019 28.971	Same remarks as for curve 1.
13	78-G(D)	333	Breetveld, J.D., et al.	0.0000	303.2	0.01827 0.01209 0.004074 0.000833	32.255 32.224 32.188 32.127	Same remarks as for curve 1.

TABLE 78-G(D)S. SMOOTHED VISCOSITY VALUES AS A FUNCTION OF DENSITY FOR GASEOUS NEON-CARBON DIOXIDE MIXTURES

Density (g cm^{-3})	Mole Fraction of Carbon Dioxide					
	0.1238 (293.2 K) [Ref. 333]	0.2897 (293.2 K) [Ref. 333]	0.3797 (293.2 K) [Ref. 333]	0.5650 (293.2 K) [Ref. 333]	0.7938 (293.2 K) [Ref. 333]	1.0000 (293.2 K) [Ref. 333]
0.0025	28.334	24.725				
0.0050	28.351	24.745	23.080	20.022	16.935	14.684
0.0075	28.368					
0.0100	28.386	24.766	23.120	20.054	16.965	14.696
0.0125	28.406					
0.0150	28.424	24.825	23.163	20.088	16.994	14.717
0.0175	28.443					
0.0200	28.462	24.870	23.205	20.125	17.025	14.743
0.0225	28.480					
0.0250	28.502	24.920	23.248	20.170	17.061	14.778
0.0300		24.972	23.300	20.224	17.118	14.820
0.0350		25.025	23.365	20.288	17.160	14.867
0.0375			23.401	20.323		
0.0400		25.082			17.218	14.914
0.0450					17.280	14.964
0.0500						15.020

Density (g cm^{-3})	Mole Fraction of Carbon Dioxide						
	0.0000 (303.2 K) [Ref. 333]	0.1238 (303.2 K) [Ref. 333]	0.2897 (303.2 K) [Ref. 333]	0.3797 (303.2 K) [Ref. 333]	0.5650 (303.2 K) [Ref. 333]	0.7938 (303.2 K) [Ref. 333]	1.0000 (303.2 K) [Ref. 333]
0.0025	32.158	28.989	25.348	23.658			
0.0050	32.184	29.011	25.370	23.677	20.565	17.453	15.175
0.0075	32.204	29.032	25.390	23.698			
0.0100	32.219	29.054	25.411	23.718	20.598	17.478	15.197
0.0125	32.231	29.075	25.430	23.741			
0.0150	32.242	29.096	25.450	23.766	20.642	17.515	15.223
0.0175	32.252	29.117	25.470				
0.0200	32.262	29.138	25.495	23.817	20.690	17.560	15.251
0.0225		29.160	25.520				
0.0250			25.548	23.866	20.743	17.610	15.285
0.0275				23.892			
0.0300					20.796	17.665	15.325
0.0350					20.853	17.718	15.375
0.0400					20.917	17.770	15.432
0.0425					20.955		
0.0450							15.495
0.0500							15.565

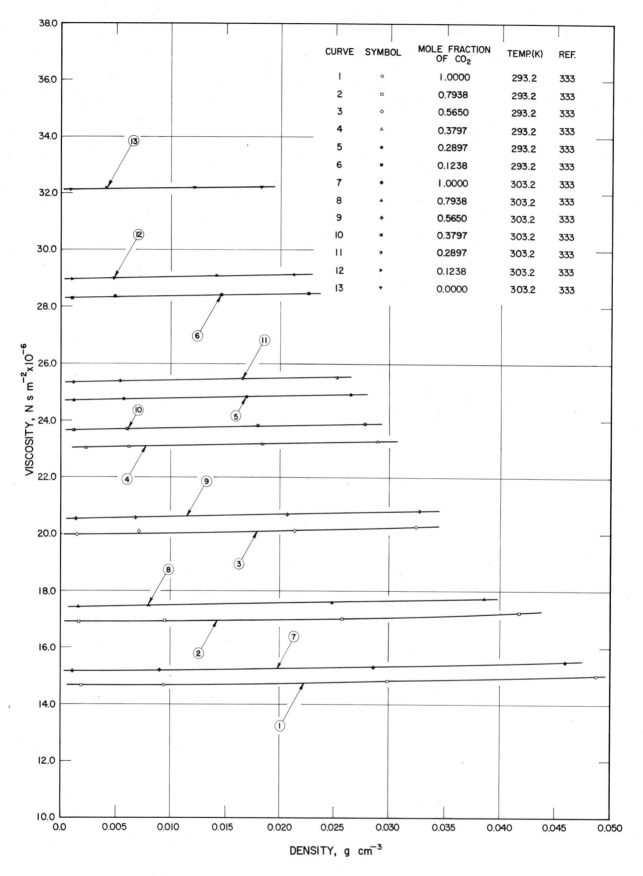

CURVE	SYMBOL	MOLE FRACTION OF CO₂	TEMP.(K)	REF.
1	○	1.0000	293.2	333
2	□	0.7938	293.2	333
3	◇	0.5650	293.2	333
4	△	0.3797	293.2	333
5	●	0.2897	293.2	333
6	■	0.1238	293.2	333
7	◆	1.0000	303.2	333
8	▲	0.7938	303.2	333
9	◆	0.5650	303.2	333
10	▣	0.3797	303.2	333
11	◉	0.2897	303.2	333
12	▶	0.1238	303.2	333
13	▼	0.0000	303.2	333

FIGURE 78-G (D). VISCOSITY DATA AS A FUNCTION OF DENSITY
FOR GASEOUS NEON-CARBON DIOXIDE MIXTURES

TABLE 79-G(C)E. EXPERIMENTAL VISCOSITY DATA AS A FUNCTION OF COMPOSITION FOR GASEOUS NEON-HYDROGEN MIXTURES

Cur. No.	Fig. No.	Ref. No.	Author(s)	Temp. (K)	Pressure (atm)	Mole Fraction of H_2	Viscosity (N s m^{-2} x 10^{-6})	Remarks
1	79-G(C)	327	van Lierde, J.	290.4		0.000	8.78	Oscillating disk viscometer; $L_1 = 0.729\%$, $L_2 = 1.263\%$, $L_3 = 2.418\%$.
						0.161	14.67	
						0.347	20.27	
						0.505	23.00	
						0.657	26.79	
						0.795	29.01	
						1.000	31.16	
2	79-G(C)	221	Trautz, M. and Binkele, H.E.	293.0		1.0000	30.92	Ne: Linde Co., commercial grade, 99-99.5 purity; capillary method, $v = 0.2019$ mm; accuracy $\pm 0.4\%$; $L_1 = 0.152\%$, $L_2 = 0.247\%$, $L_3 = 0.473\%$.
						0.7480	27.82	
						0.5391	24.27	
						0.2285	16.84	
						0.0000	8.75	
3	79-G(C)	221	Trautz, M. and Binkele, H.E.	373.0		1.0000	36.23	Same as for curve 2 except $L_1 = 0.152\%$, $L_2 = 0.246\%$, $L_3 = 0.467\%$.
						0.7480	32.69	
						0.5391	28.45	
						0.2285	19.81	
						0.0000	10.29	
4	79-G(C)	221	Trautz, M. and Binkele, H.E.	473.0		1.0000	42.20	Same as for curve 2 except $L_1 = 0.000\%$, $L_2 = 0.000\%$, $L_2 = 0.000\%$.
						0.7480	38.07	
						0.5391	33.27	
						0.2285	23.19	
						0.0000	12.11	
5	79-G(C)	221	Trautz, M. and Binkele, H.E.	523.0		1.0000	45.01	Same as for curve 2 except $L_1 = 0.079\%$, $L_2 = 0.176\%$, $L_3 = 0.393\%$.
						0.7480	40.54	
						0.5391	35.40	
						0.2285	24.76	
						0.0000	12.96	

TABLE 79-G(C)S. SMOOTHED VISCOSITY VALUES AS A FUNCTION OF COMPOSITION FOR GASEOUS NEON-HYDROGEN MIXTURES

Mole Fraction of H_2	290.4 K [Ref. 327]	293.0 K [Ref. 221]	373.0 K [Ref. 221]	473.0 K [Ref. 221]	523.0 K [Ref. 221]
0.00	8.78	8.75	10.32	12.11	12.96
0.05	10.78	10.07	12.70	14.64	15.78
0.10	12.60	12.61	14.90	17.18	18.44
0.15	14.30	14.34	16.90	19.61	21.00
0.20	15.88	16.00	18.80	21.90	23.42
0.25	17.32	17.60	20.68	24.01	25.64
0.30	18.64	19.00	22.19	26.01	27.70
0.35	19.90	20.30	23.68	27.81	29.60
0.40	21.10	21.45	25.05	29.42	31.34
0.45	22.18	22.50	26.38	30.90	32.91
0.50	23.42	23.51	27.60	32.26	34.40
0.55	24.51	24.50	28.72	33.60	35.75
0.60	25.59	25.44	29.82	34.85	37.10
0.65	26.54	26.34	30.88	36.01	38.35
0.70	27.47	27.16	31.82	37.11	39.52
0.75	28.31	27.94	32.71	38.10	40.70
0.80	29.08	28.65	33.50	39.00	41.72
0.85	29.72	29.30	34.25	39.88	42.70
0.90	30.30	29.90	34.90	40.70	43.60
0.95	30.78	30.44	35.56	41.48	44.40
1.00	31.16	30.92	36.40	42.20	45.01

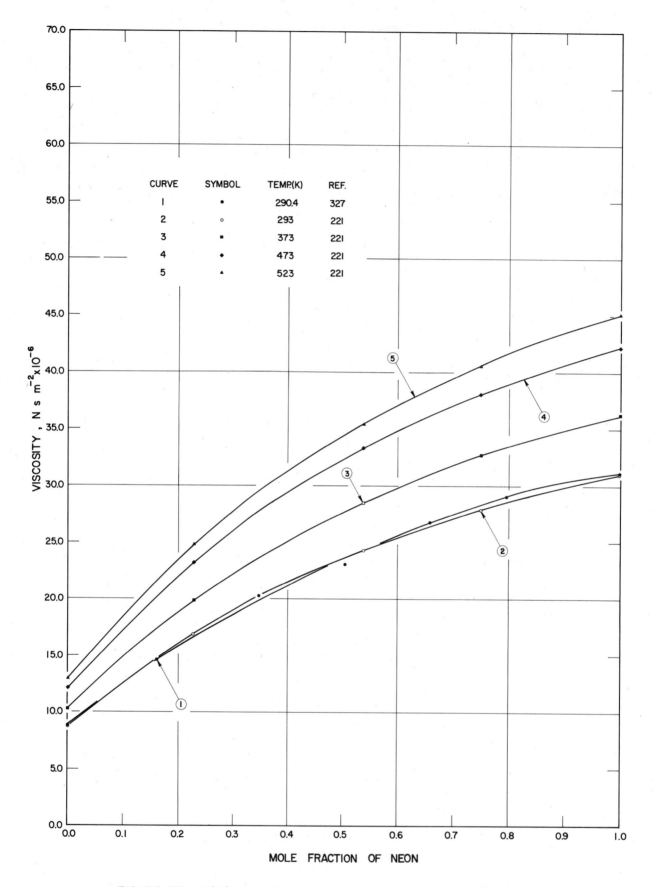

FIGURE 79-G(C). VISCOSITY DATA AS A FUNCTION OF COMPOSITION
FOR GASEOUS NEON-HYDROGEN MIXTURES

TABLE 80-G(D)E.   EXPERIMENTAL VISCOSITY DATA AS A FUNCTION OF DENSITY FOR GASEOUS NEON-NITROGEN MIXTURES

Cur. No.	Fig. No.	Ref. No.	Author(s)	Mole Fraction of N_2	Temp. (K)	Density (g cm^{-3})	Viscosity (N s m^{-2} x 10^{-6})	Remarks
1	80-G(D)	328	DiPippo, R., Kestin, J., and Oguchi, K.	0.7339	293.2	0.02490 0.01621 0.005375 0.001134	20.463 20.358 20.234 20.186	Oscillating disk viscometer; uncertainties: error ± 0.1% and precision ± 0.05%.
2	80-G(D)	328	DiPippo, R., et al.	0.4888	293.2	0.02293 0.01494 0.005006 0.001042	23.365 23.284 23.196 23.146	Same remarks as for curve 1.
3	80-G(D)	328	DiPippo, R., et al.	0.2479	293.2	0.02094 0.02094 0.01375 0.004606 0.000964	26.907 26.913 26.853 26.779 26.737	Same remarks as for curve 1.
4	80-G(D)	328	DiPippo, R., et al.	0.0000	293.2	0.01912 0.01666 0.01492 0.01251 0.004197 0.000879	31.539 31.531 31.523 31.506 31.441 31.400	Same remarks as for curve 1.
5	80-G(D)	328	DiPippo, R., et al.	1.0000	303.2	0.02605 0.02586 0.01697 0.005632 0.001178	18.366 18.362 18.234 18.077 18.025	Same remarks as for curve 1.
6	80-G(D)	328	DiPippo, R., et al.	0.7339	303.2	0.02403 0.02389 0.01566 0.005211 0.001090	20.972 20.967 20.879 20.762 20.704	Same remarks as for curve 1.
7	80-G(D)	328	DiPippo, R., et al.	0.4888	303.2	0.02215 0.01448 0.004825 0.001000	23.939 23.869 23.773 23.733	Same remarks as for curve 1.
8	80-G(D)	328	DiPippo, R., et al.	0.2479	303.2	0.02034 0.01327 0.004382 0.000930	27.557 27.491 27.418 27.386	Same remarks as for curve 1.
9	80-G(D)	328	DiPippo, R., et al.	0.0000	303.2	0.01848 0.01612 0.01448 0.01209 0.004061 0.000852	32.295 32.267 32.260 32.239 32.173 32.133	Same remarks as for curve 1.

TABLE 80-G(D)S.   SMOOTHED VISCOSITY VALUES AS A FUNCTION OF DENSITY FOR GASEOUS NEON–NITROGEN MIXTURES

Density (g cm⁻³)	Mole Fraction of Nitrogen			
	0.0000 (293.2 K) [Ref. 328]	0.2479 (293.2 K) [Ref. 328]	0.4888 (293.2 K) [Ref. 328]	0.7339 (293.2 K) [Ref. 328]
0.0025	31.424	26.755	23.160	20.205
0.0050	31.450	26.782	23.181	20.236
0.0075	31.473	26.807	23.210	20.266
0.0100	31.494	26.830	23.235	20.293
0.0125	31.512	26.852	23.263	20.322
0.0150	31.525	26.873	23.283	20.348
0.0175	31.534	26.889	23.306	20.376
0.0200	31.540	26.902	23.330	20.402
0.0225		26.912	23.353	20.428
0.0250		26.922	23.375	20.451

Density (g cm⁻³)	Mole Fraction of Nitrogen				
	0.0000 (303.2 K) [Ref. 328]	0.2479 (303.2 K) [Ref. 328]	0.4888 (303.2 K) [Ref. 328]	0.7339 (303.2 K) [Ref. 328]	1.0000 (303.2 K) [Ref. 328]
0.0025	32.154	27.400	23.742	20.726	18.035
0.0050	32.180	27.425	23.775	20.763	18.080
0.0075	32.207	27.448	23.800	20.797	18.120
0.0100	32.228	27.470	23.826	20.827	18.161
0.0125	32.248	27.492	23.850	20.855	18.200
0.0150	32.267	27.515	23.875	20.880	18.238
0.0175	32.285	27.536	23.900	20.905	18.277
0.0200	32.300	27.555	23.921	20.924	18.313
0.0225		27.575	23.943	20.951	18.344
0.0250		27.593	23.963	20.975	18.366

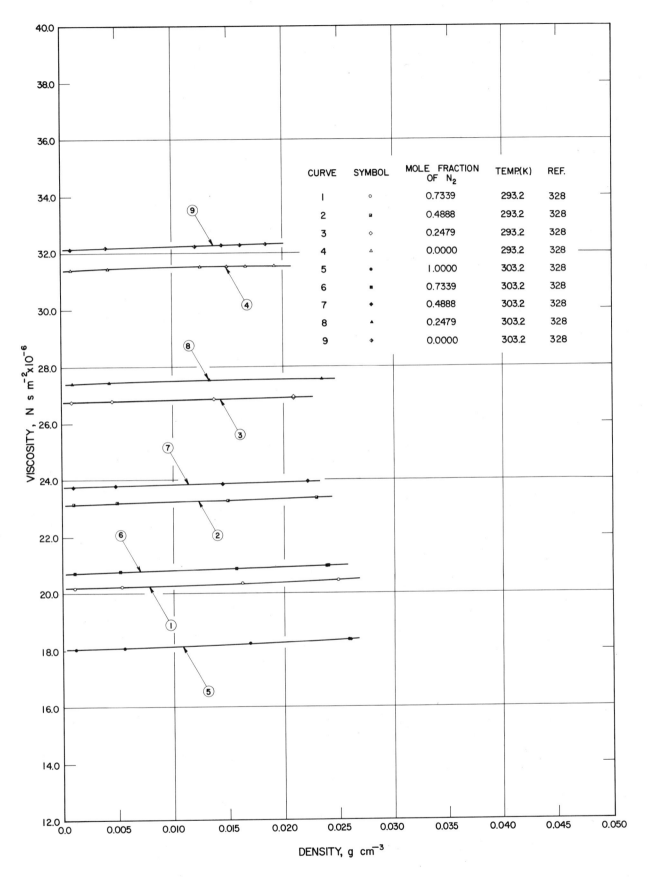

CURVE	SYMBOL	MOLE FRACTION OF N_2	TEMP(K)	REF.
1	○	0.7339	293.2	328
2	▫	0.4888	293.2	328
3	◇	0.2479	293.2	328
4	△	0.0000	293.2	328
5	●	1.0000	303.2	328
6	■	0.7339	303.2	328
7	◆	0.4888	303.2	328
8	▲	0.2479	303.2	328
9	◈	0.0000	303.2	328

VISCOSITY, N s m^{-2} x10^{-6}

DENSITY, g cm^{-3}

FIGURE 80-G (D). VISCOSITY DATA AS A FUNCTION OF DENSITY
FOR GASEOUS NEON-NITROGEN MIXTURES

TABLE 81-G(C)E.   EXPERIMENTAL VISCOSITY DATA AS A FUNCTION OF COMPOSITION FOR GASEOUS ARGON-AMMONIA MIXTURES

Cur. No.	Fig. No.	Ref. No.	Author(s)	Temp. (K)	Pressure (mm Hg)	Mole Fraction of Ar	Viscosity (N s m^{-2} x 10^{-6})	Remarks
1	81-G(C)	35	Chakraborti, P.K. and Gray, P.	298.2	243-142	1.000	22.54	Tank gases purified by distillation; capillary viscometer; relative measurements; accuracy ±1.0%; L_1 = 0.434%, L_2 = 0.592%, L_3 = 1.386%.
						0.852	20.72	
						0.785	19.85	
						0.691	18.74	
						0.595	17.56	
						0.501	16.44	
						0.386	15.04	
						0.274	13.52	
						0.172	12.23	
						0.054	10.67	
						0.000	10.16	
2	81-G(C)	35	Chakraborti, P.K. and Gray, P.	308.2	243-142	1.000	23.10	Same remarks as for curve 1 except L_1 = 0.648%, L_2 = 0.882%, L_3 = 2.182%.
						0.860	21.53	
						0.795	20.76	
						0.702	19.59	
						0.619	18.57	
						0.519	17.22	
						0.399	15.58	
						0.295	14.20	
						0.168	12.51	
						0.038	10.76	
						0.000	10.49	
3	81-G(C)	35	Chakraborti, P.K. and Gray, P.	353.2	243-142	1.000	25.71	Same remarks as for curve 1 except L_1 = 0.388%, L_2 = 0.474%, L_3 = 0.864%.
						0.860	23.94	
						0.684	21.62	
						0.594	20.37	
						0.491	18.90	
						0.381	17.28	
						0.278	15.81	
						0.184	14.52	
						0.053	12.62	
						0.000	11.98	

TABLE 81-G(C)S.   SMOOTHED VISCOSITY VALUES AS A FUNCTION OF COMPOSITION FOR GASEOUS ARGON-AMMONIA MIXTURES

Mole Fraction of Ar	298.2 K [Ref. 35]	308.2 K [Ref. 35]	353.2 K [Ref. 35]
0.00	10.08	10.49	11.98
0.05	10.78	11.12	12.68
0.10	11.41	11.79	13.39
0.15	12.04	12.41	14.08
0.20	12.68	13.04	14.78
0.25	13.30	13.70	15.48
0.30	13.92	14.34	16.18
0.35	14.59	14.98	16.83
0.40	15.22	15.60	17.58
0.45	15.84	16.24	18.22
0.50	16.48	16.88	18.91
0.55	17.10	17.51	19.60
0.60	17.70	18.16	20.30
0.65	18.31	18.78	21.00
0.70	18.92	19.40	21.68
0.75	19.52	20.04	22.39
0.80	20.12	20.68	23.06
0.85	20.72	21.28	23.72
0.90	21.32	21.88	24.41
0.95	21.90	22.48	25.08
1.00	22.54	23.10	25.71

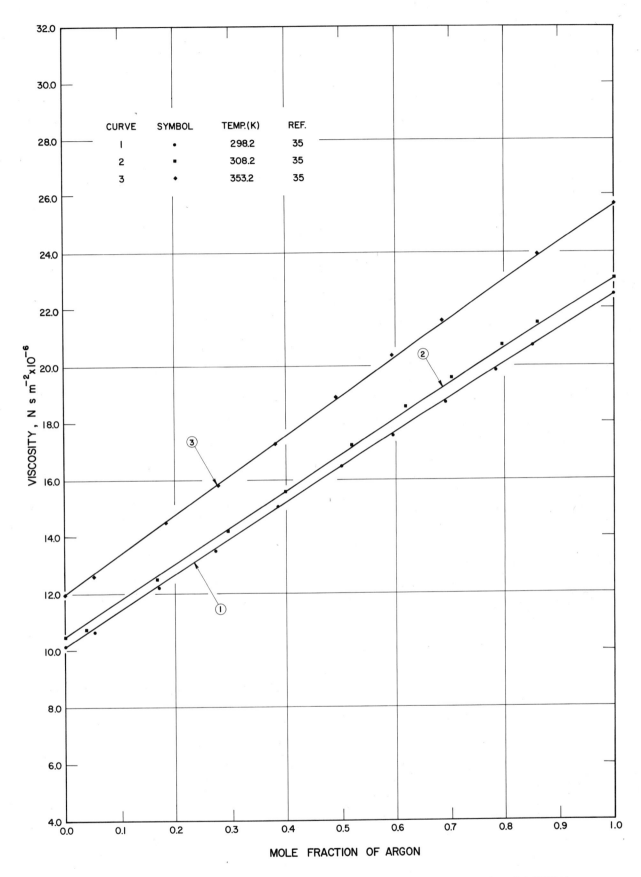

FIGURE 81-G(C).   VISCOSITY DATA AS A FUNCTION OF COMPOSITION
FOR GASEOUS ARGON-AMMONIA MIXTURES

TABLE 81–G(D)E.   EXPERIMENTAL VISCOSITY DATA AS A FUNCTION OF DENSITY FOR GASEOUS ARGON-AMMONIA MIXTURES

Cur No.	Fig. No.	Ref. No.	Author(s)	Mole Fraction of Ar	Temp. (K)	Density (g cm⁻³)	Viscosity (N s m⁻² x 10⁻⁶)	Remarks
1	81–G(D)	92	Iwasaki, H., Kestin, J., and Nagashima, A.	1.000	293.2	0.001684 0.009403 0.017944 0.034916 0.052123 0.069120 0.088147	22.275 22.362 22.462 22.681 22.954 23.221 23.572	Ar: 99.997 pure, NH_3: stored in liquid state at room temperature; oscillating disk viscometer; error ±1.5% to ±0.2% depending upon the composition being close to pure ammonia or argon respectively.
2	81–G(D)	92	Iwasaki, H., et al.	0.762	293.2	0.001459 0.002177 0.002872 0.004327 0.005806 0.007254 0.01016 0.01436 0.02206 0.02946 0.03592	20.093 20.103 20.081 20.106 20.136 20.128 20.171 20.240 20.355 20.442 20.531	Same remarks as for curve 1.
3	81–G(D)	92	Iwasaki, H., et al.	0.558	293.2	0.001266 0.001882 0.002515 0.003758 0.005044 0.006265 0.008847 0.01277 0.01862	17.630 17.672 17.737 17.737 17.757 17.800 17.850 17.892 17.940	Same remarks as for curve 1.
4	81–G(D)	92	Iwasaki, H., et al.	0.379	293.2	0.001081 0.001632 0.002211 0.003292 0.004419 0.005655 0.007073 0.008952 0.01029	15.473 15.479 15.492 15.499 15.504 15.509 15.524 15.526 15.518	Same remarks as for curve 1.
5	81–G(D)	92	Iwasaki, H., et al.	0.220	293.2	0.000939 0.001405 0.001903 0.003874 0.004860 0.005870 0.006795	13.588 13.598 13.609 13.616 13.592 13.613 13.601	Same remarks as for curve 1.
6	81–G(D)	92	Iwasaki, H., et al.	0.147	293.2	0.000883 0.001314 0.001788 0.002646 0.003627 0.004663 0.005642	12.155 12.162 12.170 12.155 12.169 12.160 12.114	Same remarks as for curve 1.
7	81–G(D)	92	Iwasaki, H., et al.	0.052	293.2	0.000786 0.001174 0.001562 0.002354 0.003171 0.004049 0.005141	10.910 10.906 10.906 10.886 10.884 10.855 10.771	Same remarks as for curve 1.
8	81–G(D)	92	Iwasaki, H., et al.	0.046	293.2	0.0002413 0.0003621 0.0004825 0.0007318 0.0009860 0.001248 0.001484	10.674 10.658 10.653 10.620 10.589 10.555 10.528	Same remarks as for curve 1.
9	81–G(D)	92	Iwasaki, H., et al.	0.000	293.2	0.0007844 0.001102 0.001466 0.002223 0.003008 0.003787 0.004602	9.882 9.865 9.847 9.808 9.774 9.734 9.695	Same remarks as for curve 1.

TABLE 81-G(D)E.   EXPERIMENTAL VISCOSITY DATA AS A FUNCTION OF DENSITY FOR GASEOUS ARGON-AMMONIA MIXTURES (continued)

Cur. No.	Fig. No.	Ref. No.	Author(s)	Mole Fraction of Ar	Temp. (K)	Density (g c m^{-3})	Viscosity (N s m^{-2} x 10^{-6})	Remarks
10	81-G(D)	92	Iwasaki, H., et al.	1.000	303.2	0.001611	22.944	Same remarks as for curve 1.
						0.009849	23.048	
						0.01808	23.136	
						0.03495	23.356	
						0.05235	23.628	
						0.06893	23.902	
						0.08567	24.206	
11	81-G(D)	92	Iwasaki, H., et al.	0.755	303.2	0.001439	20.981	Same remarks as for curve 1.
						0.002809	21.022	
						0.004253	21.038	
						0.008373	21.078	
						0.01554	21.168	
						0.02291	21.256	
						0.03046	21.361	
						0.03404	21.411	
12	81-G(D)	92	Iwasaki, H., et al.	0.532	303.2	0.005880	18.564	Same remarks as for curve 1.
						0.001258	18.494	
						0.002389	18.530	
						0.004004	18.558	
						0.007293	18.618	
						0.01030	18.623	
						0.01235	18.670	
13	81-G(D)	92	Iwasaki, H., et al.	0.330	303.2	0.001074	15.732	Same remarks as for curve 1.
						0.002052	15.740	
						0.003071	15.755	
						0.004139	15.763	
						0.006268	15.776	
						0.007374	15.778	
14	81-G(D)	92	Iwasaki, H., et al.	0.100	303.2	0.0008158	12.100	Same remarks as for curve 1.
						0.001593	12.102	
						0.002432	12.088	
						0.003282	12.072	
						0.005001	12.056	
						0.006248	12.021	
15	81-G(D)	92	Iwasaki, H., et al.	0.076	303.2	0.0007977	11.454	Same remarks as for curve 1.
						0.001149	11.441	
						0.001551	11.443	
						0.002358	11.423	
						0.003185	11.403	
						0.003820	11.388	
16	81-G(D)	92	Iwasaki, H., et al.	0.046	303.2	0.0007390	11.084	Same remarks as for curve 1.
						0.0007440	11.084	
						0.001112	11.070	
						0.001496	11.061	
						0.002267	11.039	
						0.003065	11.017	
						0.004039	10.983	
17	81-G(D)	92	Iwasaki, H., et al.	0.000	303.2	0.0007669	10.271	Same remarks as for curve 1.
						0.0007195	10.280	
						0.001063	10.256	
						0.001390	10.244	
						0.002120	10.213	
						0.002664	10.190	
						0.003598	10.148	
						0.004131	10.127	
						0.0007244	10.269	
						0.001390	10.242	
						0.002125	10.213	
						0.003604	10.147	
						0.004514	10.115	

TABLE 81-G(D)S.   SMOOTHED VISCOSITY VALUES AS A FUNCTION OF DENSITY FOR GASEOUS ARGON-AMMONIA MIXTURES

Density (g cm⁻³)	Mole Fraction of Argon								
	0.000 (293.2 K) [Ref. 92]	0.046 (293.2 K) [Ref. 92]	0.052 (293.2 K) [Ref. 92]	0.147 (293.2 K) [Ref. 92]	0.220 (293.2 K) [Ref. 92]	0.379 (293.2 K) [Ref. 92]	0.558 (293.2 K) [Ref. 92]	0.762 (293.2 K) [Ref. 92]	1.000 (293.2 K) [Ref. 92]
0.0005	9.893	10.646	10.912						
0.0010	9.870	10.586	10.909	12.559	13.590	15.470			
0.0012		10.561							
0.0015	9.847	10.525	10.906						
0.0018		10.490							
0.0020	9.823	10.466	10.900	12.158	13.600	15.482	17.658		
0.0025	9.799		10.892	12.159	13.601	15.490		20.092	
0.0030	9.775		10.878	12.159	13.605	15.495			
0.0035	9.749		10.860						
0.0040	9.723		10.838	12.155	13.610	15.505	17.740		
0.0045	9.692		10.812						
0.0050	9.670		10.782	12.138	13.613	15.510	17.772	20.122	
0.0060				12.106	13.615	15.519	17.800		
0.0070					13.610	15.520			
0.0080						15.523	17.842		
0.0090									
0.0100						15.522			
0.0120						15.520	17.872	20.184	22.368
0.0140							17.888		
0.0150							17.890	20.248	
0.0160							17.895		
0.0180							17.895		
0.0200									
0.0250								20.318	22.490
0.0300								20.390	22.552
0.0350								20.466	22.618
0.0400								20.546	22.760
0.0500									22.918
0.0600									23.075
0.0700									23.240
0.0750									23.330
0.0800									23.423
0.0900									23.602

Density (g cm⁻³)	Mole Fraction of Argon							
	0.000 (303.2 K) [Ref. 92]	0.046 (303.2 K) [Ref. 92]	0.076 (303.2 K) [Ref. 92]	0.100 (303.2 K) [Ref. 92]	0.330 (303.2 K) [Ref. 92]	0.532 (303.2 K) [Ref. 92]	0.755 (303.2 K) [Ref. 92]	1.000 (303.2 K) [Ref. 92]
0.0000								
0.0005	10.286							22.940
0.0007		11.089						
0.0010	10.262	11.080	11.452	12.100	15.732	18.490		
0.0015	10.240	11.064						
0.0020	10.218	11.050	11.434	12.094	15.745	18.518		
0.0025	10.196	11.034	11.423	1.2092	15.750	18.525	21.006	
0.0030	10.174	11.018	11.410	12.087	15.753			
0.0035	10.154	11.001	11.397					
0.0040	10.134	10.984	11.383	12.075	15.763	18.570		
0.0045	10.116							
0.0050				12.056	15.770	18.590	21.034	
0.0060				12.020	15.772	18.610		
0.0070					15.778	18.630		
0.0080						18.642		
0.0100						18.660	21.092	23.045
0.0120						18.668		
0.0150							21.153	
0.0200							21.215	23.151
0.0250							21.282	
0.0300							21.354	23.272
0.0350							21.425	
0.0400								23.415
0.0500								23.580
0.0600								23.753
0.0700								23.928
0.0800								24.105
0.0900								24.290

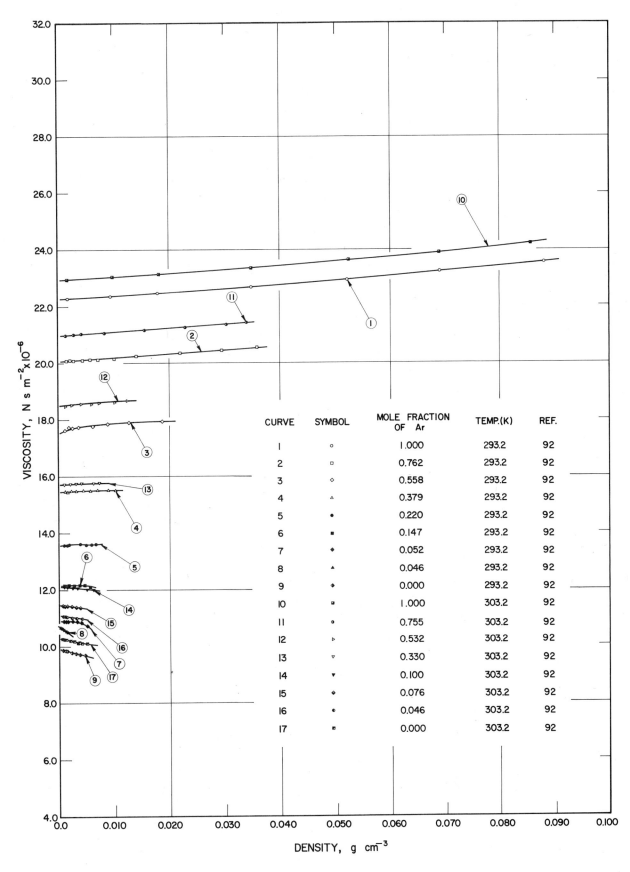

CURVE	SYMBOL	MOLE FRACTION OF Ar	TEMP.(K)	REF.
1	○	1.000	293.2	92
2	□	0.762	293.2	92
3	◇	0.558	293.2	92
4	△	0.379	293.2	92
5	●	0.220	293.2	92
6	■	0.147	293.2	92
7	◆	0.052	293.2	92
8	▲	0.046	293.2	92
9	◆	0.000	293.2	92
10	▫	1.000	303.2	92
11	◔	0.755	303.2	92
12	▷	0.532	303.2	92
13	▽	0.330	303.2	92
14	▼	0.100	303.2	92
15	◆	0.076	303.2	92
16	◦	0.046	303.2	92
17	▪	0.000	303.2	92

FIGURE 81-G(D).   VISCOSITY DATA AS A FUNCTION OF DENSITY
FOR GASEOUS ARGON – AMMONIA MIXTURES

TABLE 82-G(C)E. EXPERIMENTAL VISCOSITY DATA AS A FUNCTION OF COMPOSITION FOR GASEOUS ARGON-SULFUR DIOXIDE MIXTURES

Cur. No.	Fig. No.	Ref. No.	Author(s)	Temp. (K)	Pressure (atm)	Mole Fraction of SO$_2$	Viscosity (N s m^{-2} x 10^{-6})	Remarks
1	82-G C	35	Chakraborti, P.K. and Gray, P.	298.2	243-142	0.000	22.54	Gases purified by distillation; capillary viscometer, relative measurements; accuracy ± 1.0%; L$_1$ = 0.142%, L$_2$ = 0.256%, L$_3$ = 0.672%.
						0.191	20.07	
						0.250	19.44	
						0.314	18.68	
						0.404	17.74	
						0.500	16.85	
						0.612	15.81	
						0.720	14.97	
						0.830	14.13	
						0.954	13.31	
						1.000	13.17	
2	82-G(C)	35	Chakraborti, P.K. and Gray, P.	308.2	243-142	0.000	23.10	Same remarks as for curve 1 except L$_1$ = 0.194%, L$_2$ = 0.285%, L$_3$ = 0.676%.
						0.024	22.86	
						0.150	21.77	
						0.254	20.84	
						0.362	19.66	
						0.464	18.73	
						0.581	17.70	
						0.666	16.90	
						0.762	15.97	
						0.872	14.96	
						0.893	14.77	
						1.000	13.28	
3	82-G(C)	35	Chakraborti, P.K. and Gray, P.	353.2	243-142	0.000	25.71	Same remarks as for curve 1 except L$_1$ = 0.153%, L$_2$ = 0.244%, L$_3$ = 0.512%.
						0.043	25.50	
						0.163	24.34	
						0.264	23.37	
						0.387	22.13	
						0.483	21.13	
						0.586	20.11	
						0.687	19.03	
						0.781	17.86	
						0.885	16.65	
						0.920	16.29	
						1.000	15.23	

TABLE 82-G(C)S. SMOOTHED VISCOSITY VALUES AS A FUNCTION OF COMPOSITION FOR GASEOUS ARGON-SULFUR DIOXIDE MIXTURES

Mole Fraction of SO$_2$	298.2 K [Ref. 35]	308.2 K [Ref. 35]	353.2 K [Ref. 35]
0.00	22.45	23.10	25.71
0.05	21.88	22.64	25.32
0.10	21.22	22.16	24.91
0.15	20.58	21.69	24.46
0.20	19.96	21.22	24.00
0.25	19.39	20.74	23.52
0.30	18.82	20.28	23.04
0.35	18.28	19.82	22.53
0.40	17.77	19.36	22.00
0.45	17.27	18.90	21.49
0.50	16.80	18.44	20.97
0.55	16.34	17.98	20.45
0.60	15.95	17.52	19.92
0.65	15.48	17.05	19.36
0.70	15.08	16.60	18.80
0.75	14.70	16.16	18.24
0.80	14.34	15.70	17.67
0.85	14.01	15.22	17.07
0.90	13.70	14.80	16.46
0.95	13.42	14.20	15.85
1.00	13.17	13.28	15.23

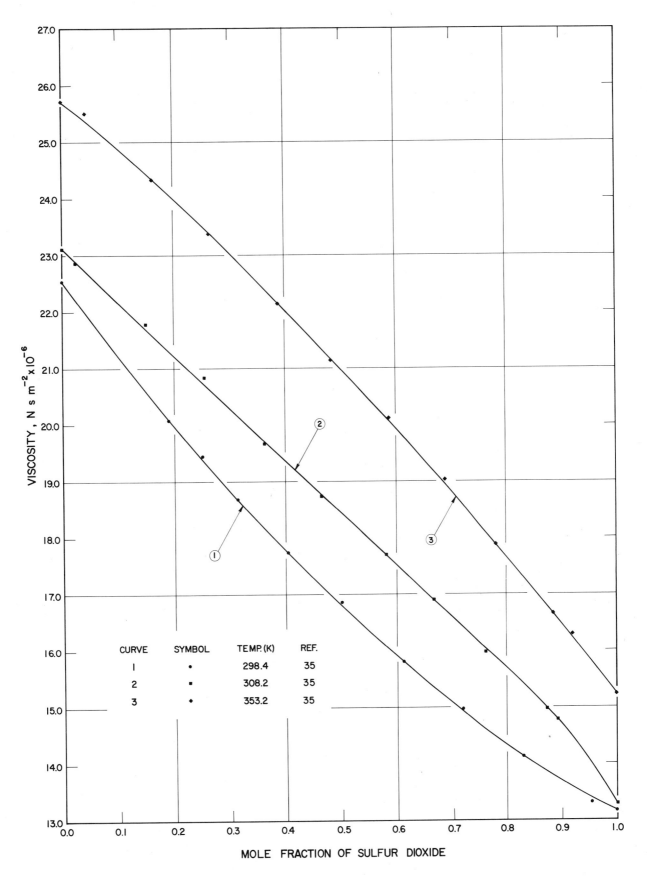

FIGURE 82-G(C). VISCOSITY DATA AS A FUNCTION OF COMPOSITION
FOR GASEOUS ARGON-SULFUR DIOXIDE MIXTURES

TABLE 83-L(C)E. EXPERIMENTAL VISCOSITY DATA AS A FUNCTION OF COMPOSITION FOR LIQUID BENZENE-CYCLOHEXANE MIXTURES

Cur. No.	Fig. No.	Ref. No.	Author(s)	Temp. (K)	Pressure (atm)	Mole Fraction of C_6H_{12}	Viscosity (N s m^{-2} x 10^{-6})	Remarks
1	83-L(C)	355	Ridgway, K. and Butler, P. A.	298.2		1.0000	869.0	Liquids supplied by British Drug Houses Ltd.; Ostwald viscometer; precision 0.1%; $L_1 = 0.017\%$, $L_2 = 0.039\%$, $L_3 = 0.105\%$.
						0.8718	762.2	
						0.7826	712.2	
						0.6636	659.9	
						0.5126	612.4	
						0.3530	587.9	
						0.2186	583.0	
						0.0967	592.6	
						0.0000	605.9	

TABLE 83-L(C)S. SMOOTHED VISCOSITY VALUES AS A FUNCTION OF COMPOSITION FOR LIQUID BENZENE CYCLOHEXANE MIXTURES

Mole Fraction of C_6H_{12}	298.2 K [Ref. 355]
0.00	605.9
0.05	598.4
0.10	592.4
0.15	587.5
0.20	584.0
0.25	581.4
0.30	580.6
0.35	582.3
0.40	588.7
0.45	598.1
0.50	609.5
0.55	622.5
0.60	637.8
0.65	655.0
0.70	776.0
0.75	697.5
0.80	721.0
0.85	746.5
0.90	776.5
0.95	822.6
1.00	869.0

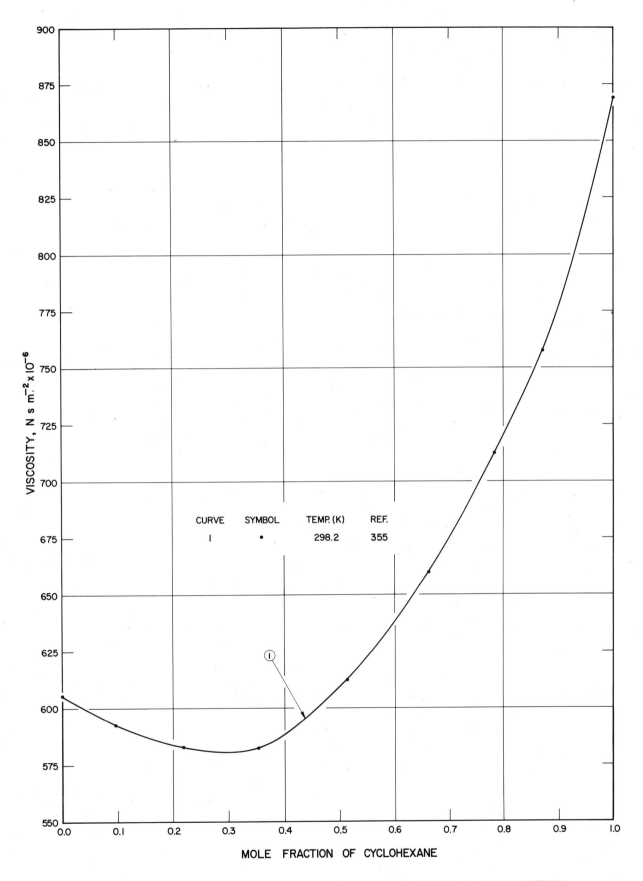

FIGURE 83 - L(C). VISCOSITY DATA AS A FUNCTION OF COMPOSITION
FOR LIQUID BENZENE-CYCLOHEXANE MIXTURES

TABLE 84-L(C)E.   EXPERIMENTAL VISCOSITY DATA AS A FUNCTION OF COMPOSITION FOR LIQUID
BENZENE-n-HEXANE MIXTURES

Cur. No.	Fig. No.	Ref. No.	Author(s)	Temp. (K)	Pressure (atm)	Mole Fraction of n-C_6H_{14}	Viscosity (N s m^{-2} x 10^{-6})	Remarks
1	84-L(C)	355	Ridgway, K. and Butler, P.A.	298.2		1.0000	300.8	Benzene: supplied by B.D.H. Ltd, n-Hexane: supplied by Phillips Petroleum Co.; Ostwald visco- meter; precision 0.1%; L_1 = 0.094%, L_2 = 0.177%, L_3 = 0.384%.
						0.8719	313.4	
						0.7335	327.0	
						0.5950	347.1	
						0.4296	382.2	
						0.2784	425.4	
						0.1189	513.9	
						0.0000	605.9	

TABLE 84-L(C)S.   SMOOTHED VISCOSITY VALUES AS A FUNCTION OF COMPOSITION FOR LIQUID
BENZENE-n-HEXANE MIXTURES

Mole Fraction of n-C_6H_{14}	298.2 K [Ref. 355]
0.00	605.9
0.05	565.8
0.10	527.0
0.15	492.5
0.20	462.5
0.25	438.0
0.30	417.5
0.35	401.6
0.40	388.6
0.45	377.2
0.50	366.4
0.55	356.4
0.60	347.2
0.65	339.0
0.70	331.5
0.75	325.0
0.80	319.2
0.85	314.2
0.90	309.5
0.95	305.2
1.00	300.8

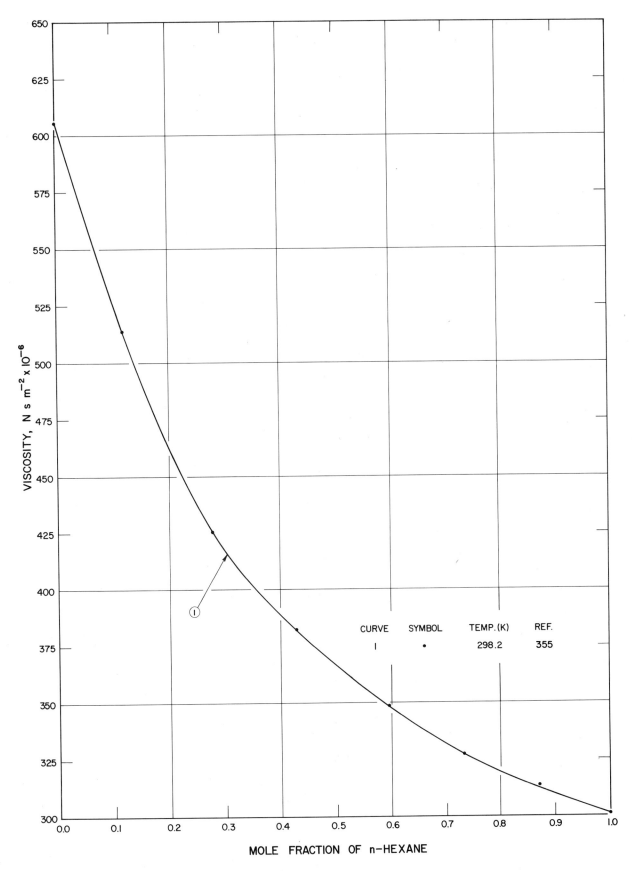

FIGURE 84-L(C). VISCOSITY DATA AS A FUNCTION OF COMPOSITION
FOR LIQUID BENZENE — n-HEXANE MIXTURES

TABLE 85-L(C)E.   EXPERIMENTAL VISCOSITY DATA AS A FUNCTION OF COMPOSITION FOR LIQUID
BENZENE–OCTAMETHYLCYCLOTETRASILOXANE MIXTURES

Cur. No.	Fig. No.	Ref. No.	Author(s)	Temp. (K)	Pressure (atm)	Mole Fraction of $[OSi(CH_3)_2]_4$	Viscosity ($N\ s\ m^{-2} \times 10^{-6}$)	Remarks
1	85-L(C)	360	Marsh, K.N.	291.2		0.0000	670.3	Benzene: A.R. grade shaken with
						0.0881	734.6	H_2SO_4 and washed with water, dried
						0.3511	1059.0	over $CaCl_2$ and Na, and then dis-
						0.5997	1493.0	tilled; Ostwald viscometer, relative
						0.7738	1885.0	measurements; $L_1 = 0.032\%$, $L_2 =$
						0.8529	2091.0	0.091%, $L_3 = 0.257\%$.
						0.9369	2328.0	
						1.0000	2520.0	
2	85-L(C)	360	Marsh, K.N.	298.2		0.0000	602.4	Same remarks as for curve 1 except
						0.0341	622.4	$L_1 = 0.167\%$, $L_2 = 0.279\%$, $L_3 =$
						0.0699	648.6	0.625%.
						0.1407	709.9	
						0.2235	794.4	
						0.2938	875.0	
						0.3751	981.4	
						0.4689	1113.0	
						0.6211	1363.0	
						0.6777	1466.0	
						0.7510	1608.0	
						0.8434	1804.0	
						0.8753	1880.0	
						0.9028	1939.0	
						0.9291	2010.0	
						1.0000	2190.0	
3	85-L(C)	360	Marsh, K.N.	308.2		0.0000	523.5	Same remarks as for curve 1 except
						0.0886	576.8	$L_1 = 0.093\%$, $L_2 = 0.173\%$, $L_3 =$
						0.3517	818.0	0.432%.
						0.6020	1127.0	
						0.7741	1390.0	
						0.8544	1527.0	
						0.9373	1682.0	
						1.0000	1806.0	
4	85-L(C)	360	Marsh, K.N.	318.2		0.0000	460.3	Same remarks as for curve 1 except
						0.0888	507.0	$L_1 = 0.205\%$, $L_2 = 0.311\%$, $L_3 =$
						0.3526	714.0	0.578%.
						0.6036	971.0	
						0.7763	1186.0	
						0.8562	1298.0	
						0.9134	1393.0	
						1.0000	1514.0	

TABLE 85-L(C)S.   SMOOTHED VISCOSITY VALUES AS A FUNCTION OF COMPOSITION FOR LIQUID
BENZENE-OCTAMETHYLCYCLOTETRASILOXANE MIXTURES

Mole Fraction of $[OSi(CH_3)_2]_4$	(291.2 K) [Ref. 360]	(298.2 K) [Ref. 360]	(308.2 K) [Ref. 360]	(318.2 K) [Ref. 360]
0.00	670.3	602.4	523.5	460.3
0.05	698.2	632.5	552.5	488.0
0.10	747.0	671.5	584.8	516.0
0.15	796.0	719.8	621.5	548.0
0.20	850.0	770.0	664.0	584.0
0.25	917.0	828.0	711.0	624.0
0.30	984.0	888.0	761.0	666.0
0.35	1059.0	950.0	816.0	712.5
0.40	1140.0	1018.8	871.0	760.0
0.45	1224.0	1090.0	930.0	810.0
0.50	1310.0	1164.0	992.0	830.0
0.55	1400.0	1244.0	1058.0	916.0
0.60	1500.0	1328.0	1124.0	971.0
0.65	1604.0	1417.5	1198.0	1030.0
0.70	1712.5	1511.0	1271.0	1090.0
0.75	1830.0	1609.0	1352.0	1156.0
0.80	1952.5	1718.0	1436.0	1221.0
0.85	2088.0	1824.5	1521.0	1291.5
0.90	2234.0	1942.0	1612.0	1364.0
0.95	2376.0	2062.0	1704.0	1438.0
1.00	2520.0	2190.0	1806.0	1514.0

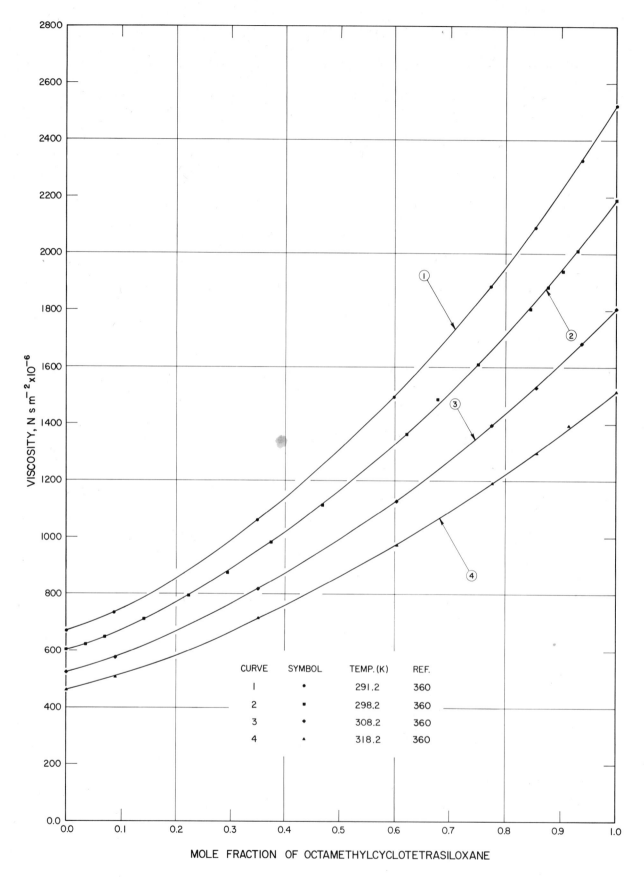

CURVE	SYMBOL	TEMP.(K)	REF.
1	•	291.2	360
2	▪	298.2	360
3	◆	308.2	360
4	▲	318.2	360

MOLE FRACTION OF OCTAMETHYLCYCLOTETRASILOXANE

FIGURE 85-L(C). VISCOSITY DATA AS A FUNCTION OF COMPOSITION
FOR LIQUID BENZENE – OCTAMETHYLCYCLOTETRASILOXANE MIXTURES

TABLE 86-G(D)E.   EXPERIMENTAL VISCOSITY DATA AS A FUNCTION OF DENSITY FOR GASEOUS
n-BUTANE-METHANE MIXTURES

Cur. No.	Fig. No.	Ref. No.	Author(s)	Mole Fraction of C_4H_{10}	Temp. (K)	Density (g cm^{-3})	Viscosity (N s m^{-2} x 10^{-6})	Remarks
1	86-G(D)	329	Kestin, J. and Yata, J.	1.0000	293.2	0.004445 0.003808 0.003239 0.002657	7.252 7.260 7.267 7.274	C_4H_{10}-CH_4: 99.99 pure; oscillating disk viscometer; calibrated with He and N_2 at 20 C; error ±0.1% and precision ±0.05%.
2	86-G(D)	329	Kestin, J. and Yata, J.	0.6447	293.2	0.002716 0.002452 0.002183 0.001903	8.128 8.133 8.131 8.131	Same remarks as for curve 1.
3	86-G(D)	329	Kestin, J. and Yata, J.	0.4579	293.2	0.003141 0.002578 0.002093 0.001541	8.726 8.726 8.723 8.722	Same remarks as for curve 1.
4	86-G(D)	329	Kestin, J. and Yata, J.	0.3026	293.2	0.004050 0.003045 0.002156 0.001251	9.352 9.348 9.339 9.335	Same remarks as for curve 1.
5	86-G(D)	329	Kestin, J. and Yata, J.	0.1568	293.2	0.006295 0.004455 0.002727 0.000983	10.092 10.064 10.042 10.026	Same remarks as for curve 1.
6	86-G(D)	329	Kestin, J. and Yata, J.	0.0000	293.2	0.01761 0.01384 0.01030 0.006809 0.003381 0.000701	11.321* 11.217* 11.137* 11.054 10.986 10.986	Same remarks as for curve 1.
7	86-G(D)	342	Dolan, J.P., Ellington, R.T., and Lee, A.L.	0.100	294.3	0.147 0.186 0.219	18.97 22.60 25.85	Capillary viscometer; maximum uncertainty of measurements ±0.5%; original data reported as a function of pressure, density calculated from pressure using volumetric data of Reamer et al. [369].
8	86-G(D)	329	Kestin, J. and Yata, J.	1.0000	303.2	0.005656 0.004027 0.002578	7.481 7.506 7.524	Same remarks as for curve 1.
9	86-G(D)	329	Kestin, J. and Yata, J.	0.6447	303.2	0.003032 0.002553 0.002197 0.001841	8.405 8.411 8.412 8.415	Same remarks as for curve 1.
10	86-G(D)	329	Kestin, J. and Yata, J.	0.4579	303.2	0.003490 0.002979 0.002206 0.001484	9.012 9.015 9.012 9.013	Same remarks as for curve 1.
11	86-G(D)	329	Kestin, J. and Yata, J.	0.3026	303.2	0.004335 0.004335 0.003986 0.003001 0.002122 0.001222	9.659 9.663 9.658 9.651 9.644 9.636	Same remarks as for curve 1.
12	86-G(D)	329	Kestin, J. and Yata, J.	0.1568	303.2	0.005153 0.003785 0.002238 0.000959	10.394 10.380 10.358 10.334	Same remarks as for curve 1.
13	86-G(D)	329	Kestin, J. and Yata, J.	0.0000	303.2	0.01506 0.003785 0.002238 0.000959	11.590* 10.380 10.358 10.334	Same remarks as for curve 1.
14	86-G(D)	342	Dolan, J.P., et al.	0.100	310.9	0.0351 0.0479 0.0612 0.0785 0.0962 0.114 0.132 0.197	11.89 12.37 13.00 13.91 15.07 16.26 17.75 23.87	Same remarks as for curve 7.
15	86-G(D)	342	Dolan, J.P., et al.	0.300	310.9	0.210 0.265 0.307	28.17 43.80 48.49	Same remarks as for curve 7.
16	86-G(D)	342	Dolan, J.P., et al.	0.500	310.9	0.326 0.397 0.441	48.81 70.95 76.74	Same remarks as for curve 7.

*Not shown in figure.

TABLE 86-G(D)E.   EXPERIMENTAL VISCOSITY DATA AS A FUNCTION OF DENSITY FOR GASEOUS
n–BUTANE-METHANE MIXTURES (continued)

Cur. No.	Fig. No.	Ref. No.	Author(s)	Mole Fraction of C_4H_{10}	Temp. (K)	Density (g cm⁻³)	Viscosity (N s m⁻² x 10⁻⁶)	Remarks
17	86-G(D)	343	Carmichael, L.T., Virginia, B., and Sage, B.H.	0.6060	310.9	0.433	75.757	Rotating cylinder viscometer; original data reported as a function of pressure, density calculated from pressure using volumetric data of Reamer et al. [369].
						0.433	77.328	
						0.433	77.708	
						0.433	77.968	
						0.438	76.866	
						0.438	76.869	
						0.455	77.231	
						0.455	77.279	
						0.457	77.633	
						0.457	78.467	
						0.461	80.642	
						0.461	80.914	
						0.461	81.112	
						0.462	79.795	
						0.462	79.500	
						0.462	79.885*	
						0.476	87.932	
						0.476	87.714	
						0.476	87.802*	
						0.476	86.986	
						0.476	87.154	
						0.476	87.044*	
						0.494	98.824	
						0.494	98.867*	
						0.494	98.823*	
						0.506	107.393*	
						0.506	107.520*	
						0.506	107.911*	
						0.514	113.280*	
						0.515	112.919*	
						0.515	112.712*	
						0.519	116.890*	
						0.519	116.855*	
						0.519	116.601*	
						0.519	117.219*	
18	86-G(D)	342	Dolan, J.P., et al.	0.100	344.3	0.0309	12.88	Same remarks as for curve 7.
						0.0418	13.20	
						0.0530	13.73	
						0.0672	14.42	
						0.0817	15.15	
						0.0962	16.02	
						0.111	16.86	
						0.165	21.29	
19	86-G(D)	342	Dolan, J.P., et al.	0.300	344.3	0.147	23.82	Same remarks as for curve 7.
						0.170	26.56	
						0.213	32.74	
						0.252	46.25	
20	86-G(D)	342	Dolan, J.P., et al.	0.500	344.3	0.253	46.50	Same remarks as for curve 7.
						0.317	53.49	
						0.365	58.44	
21	86-G(D)	342	Dolan, J.P., et al.	0.100	377.6	0.0373	14.21	Same remarks as for curve 7.
						0.0470	14.52	
						0.0592	15.02	
						0.0715	16.60	
						0.0838	16.25	
						0.0961	17.07	
						0.140	20.43	
						0.184	24.32	
						0.228	27.74	
22	86-G(D)	342	Dolan, J.P., et al.	0.300	377.6	0.0130	12.05	Same remarks as for curve 7.
						0.0265	12.30	
						0.0405	12.94	
						0.0548	13.48	
						0.0694	14.38	
						0.0879	15.81	
						0.107	17.50	
						0.125	19.61	
						0.144	21.53	
						0.181	26.14	
23	86-G(D)	342	Dolan, J.P., et al.	0.500	377.6	0.208	33.39	Same remarks as for curve 7.
						0.235	36.98	
						0.261	39.66	
						0.306	45.59	

*Not shown in figure.

TABLE 86-G(D)E.   EXPERIMENTAL VISCOSITY DATA AS A FUNCTION OF DENSITY FOR GASEOUS
n–BUTANE-METHANE MIXTURES (continued)

Cur. No.	Fig. No.	Ref. No.	Author(s)	Mole Fraction of C_4H_{10}	Temp. (K)	Density (g cm^{-3})	Viscosity (N s m^{-2} x 10^{-6})	Remarks
24	86-G(D)	343	Carmichael, L.T., et al.	0.6060	377.6	0.00220	10.926	Same remarks as for curve 17.
						0.00220	10.924*	
						0.00220	10.946	
						0.0458	11.940*	
						0.0458	11.942*	
						0.0458	11.950*	
						0.330	39.311	
						0.330	39.227*	
						0.330	39.065	
						0.363	45.463	
						0.363	45.830	
						0.363	46.078	
						0.363	46.110*	
						0.408	59.108	
						0.408	59.195*	
						0.408	59.285*	
						0.434	68.428	
						0.434	68.247	
						0.434	68.476*	
						0.446	73.541	
						0.446	73.689*	
						0.446	73.381	
						0.455	77.760	
						0.455	77.672*	
						0.455	77.774*	
25	86-G(D)	343	Carmichael, L.T., et al.	0.6060	444.3	0.00222	13.106	Same remarks as for curve 17.
						0.00222	13.113*	
						0.00222	13.124*	
						0.00222	13.184*	
						0.106	16.943	
						0.106	17.024*	
						0.106	17.146	
						0.240	28.782	
						0.240	28.606*	
						0.240	28.781*	
						0.283	34.455	
						0.283	34.590*	
						0.317	38.881	
						0.317	38.979*	
						0.360	47.684	
						0.360	47.722*	
						0.360	47.818	
						0.377	51.989	
						0.377	51.979*	
						0.377	52.213	
						0.389	53.731	
						0.389	53.963	
						0.389	53.962*	
26	86-G(D)	342	Dolan, J.P., et al.	0.750	444.3	0.204	17.70	Same remarks as for curve 7.
						0.249	22.72	
						0.283	28.10	
						0.309	33.44	
						0.346	37.62	
27		343	Carmichael, L.T., et al.	0.6060	477.6	0.106	16.698	Same remarks as for curve 17.
						0.106	16.738*	
						0.106	16.732*	
						0.198	24.284	
						0.198	25.072	
						0.198	25.986	
						0.278	33.810	
						0.278	33.919*	
						0.278	33.959*	
						0.323	41.919	
						0.323	41.771	
						0.323	42.010*	
						0.346	47.048	
						0.347	47.373	
						0.347	47.450*	
						0.355	48.772	
						0.358	49.867	
						0.360	50.350	

*Not shown in figure.

TABLE 86-G(D)S. SMOOTHED VISCOSITY VALUES AS A FUNCTION OF DENSITY FOR GASEOUS n-BUTANE-METHANE MIXTURES

Density (g cm^{-3})	Mole Fraction of n-Butane					
	0.0000 (293.2 K) [Ref. 329]	0.1568 (293.2 K) [Ref. 329]	0.3026 (293.2 K) [Ref. 329]	0.4579 (293.2 K) [Ref. 329]	0.6447 (293.2 K) [Ref. 329]	1.0000 (293.2 K) [Ref. 329]
0.00075		10.012				
0.00125	10.951	10.020	9.330	8.720		
0.00150			9.332	8.721		
0.00175		10.029	9.339	8.724	8.135	
0.00200			9.340	8.726	8.131	
0.00225		10.037	9.341	8.728	8.130	7.272
0.00250	10.968		9.343	8.727	8.128	7.270
0.00275		10.046	9.346	8.726	8.125	7.268
0.00300			9.349	8.725		7.267
0.00325		10.052	9.350	8.723		7.262
0.00350			9.350	8.721		7.261
0.00375	10.985	10.060	9.350			7.260
0.00400						7.258
0.00425		10.068				7.253
0.00450						7.250
0.00475		10.072				
0.00500	11.001					
0.00550		10.080				
0.00625	11.024					
0.00750	11.040					
0.00875	11.062					
0.01000	11.078					
0.01125	11.100					
0.01250	11.125					

Density (g cm^{-3})	Mole Fraction of n-Butane					
	0.0000 (303.2 K) [Ref. 329]	0.1568 (303.2 K) [Ref. 329]	0.3026 (303.2 K) [Ref. 329]	0.4579 (303.2 K) [Ref. 329]	0.6447 (303.2 K) [Ref. 329]	1.0000 (303.2 K) [Ref. 329]
0.00075			9.618			
0.00125	11.285	10.342	9.631	9.015		
0.00150				9.015	8.415	
0.00175		10.352	9.641	9.020	8.416	
0.00200				9.020	8.418	
0.00225		10.362	9.650	9.020	8.415	
0.00250	11.310			9.019	8.413	7.520
0.00275		10.370	9.654	9.018	8.410	
0.00300				9.016	8.409	7.519
0.00325		10.378	9.660		8.406	
0.00350				9.010	8.400	7.512
0.00375	11.329	10.386	9.660	9.009		7.510
0.00400						7.506
0.00425		10.390	9.660			
0.00450			9.658			7.499
0.00475		10.394				
0.00500	11.360					
0.00525		10.400				7.492
0.00600						7.482
0.00625	11.382					7.474
0.00750	11.410					
0.00875	11.435					
0.01000	11.460					
0.01125	11.490					
0.01250	11.535					

TABLE 86-G(D)S.   SMOOTHED VISCOSITY VALUES AS A FUNCTION OF DENSITY FOR GASEOUS
n-BUTANE-METHANE MIXTURES (continued)

Density (g cm^{-3})	Mole Fraction of n-Butane							
	0.100 (294.3 K) [Ref. 342]	0.100 (310.9 K) [Ref. 342]	0.300 (310.9 K) [Ref. 342]	0.500 (310.9 K) [Ref. 342]	0.606 (310.9 K) [Ref. 342]	0.100 (344.3 K) [Ref. 342]	0.300 (344.3 K) [Ref. 342]	0.500 (344.3 K) [Ref. 342]
0.04		12.09				13.21		
0.06		12.95				14.05		
0.08		13.97				15.07		
0.10		15.26				16.23		
0.12		16.75				17.59		
0.14	18.35	18.45				19.15		
0.16	20.08	20.24				20.83	25.39	
0.18	21.98	22.15					27.83	
0.20	23.94	24.19					30.47	
0.22	25.92		39.25				33.33	
0.24			41.23				36.25	
0.26			43.26					47.16
0.28			45.42					49.16
0.30			47.65					51.28
0.32								53.44
0.34				66.10				55.64
0.36				67.98				58.03
0.38				69.87				
0.40				71.90				
0.42				74.10				
0.44				76.53				
0.46					79.20			
0.48					89.40			
0.50					102.8			
0.52					117.4			

Density (g cm^{-3})	Mole Fraction of n-Butane						
	0.100 (377.6 K) [Ref. 342]	0.300 (377.6 K) [Ref. 342]	0.500 (377.6 K) [Ref. 342]	0.606 (377.6 K) [Ref. 342]	0.606 (444.3 K) [Ref. 342]	0.750 (444.3 K) [Ref. 342]	0.606 (477.6 K) [Ref. 342]
0.02		12.16		11.21	13.42		
0.04	14.44	12.88		11.77	14.02		
0.06	15.04	13.84		12.50	14.80		
0.08	16.02	15.08		13.43	15.73		
0.10	17.24	16.78		14.55	16.82		18.20
0.12	18.76	18.79		15.82	17.99		19.32
0.14	20.20	21.04		17.18	19.24		20.50
0.16	22.12	23.48		18.58	20.58		21.83
0.18	23.88	25.96		20.08	22.01		23.35
0.20	25.70			21.77	23.64		25.00
0.22	27.56		34.72	23.65	25.52	19.60	26.80
0.24			37.20	25.79	27.68	22.20	28.80
0.26			39.74	28.23	30.18	24.96	31.33
0.28			42.30	30.97	32.98	27.68	34.20
0.30			44.87	33.98	36.02	30.50	37.52
0.32				37.25	39.43	33.48	41.21
0.34				40.85	43.26	36.56	45.48
0.36				45.10	47.70		50.30
0.38				50.22	52.92		
0.40				56.39			
0.42				63.15			
0.44				70.72			

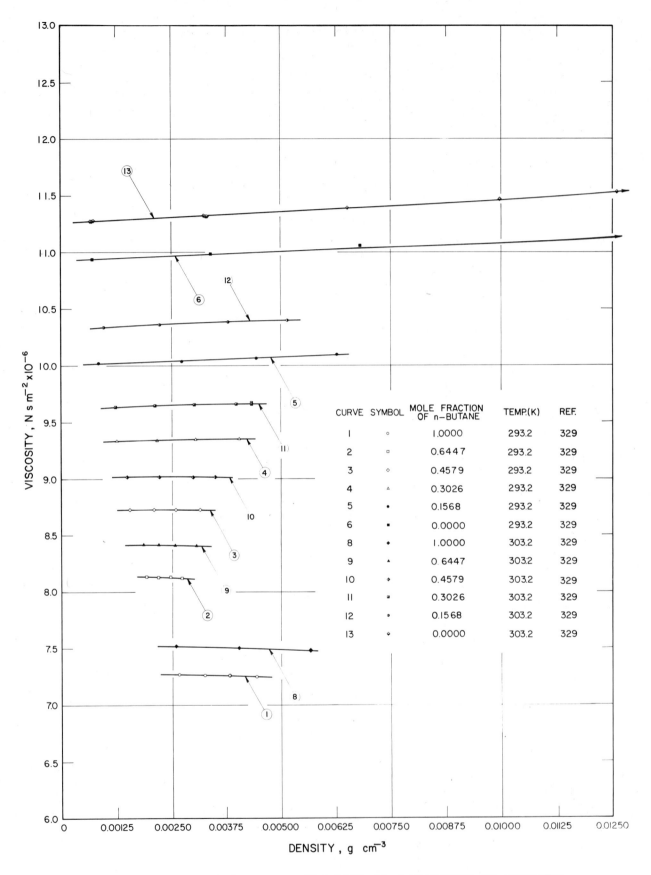

CURVE	SYMBOL	MOLE FRACTION OF n-BUTANE	TEMP.(K)	REF.
1	○	1.0000	293.2	329
2	□	0.6447	293.2	329
3	◇	0.4579	293.2	329
4	△	0.3026	293.2	329
5	•	0.1568	293.2	329
6	■	0.0000	293.2	329
8	◆	1.0000	303.2	329
9	▲	0.6447	303.2	329
10	◈	0.4579	303.2	329
11	◘	0.3026	303.2	329
12	◦	0.1568	303.2	329
13	◇	0.0000	303.2	329

FIGURE 86-G(D). VISCOSITY DATA AS A FUNCTION OF DENSITY
FOR GASEOUS n-BUTANE - METHANE MIXTURES

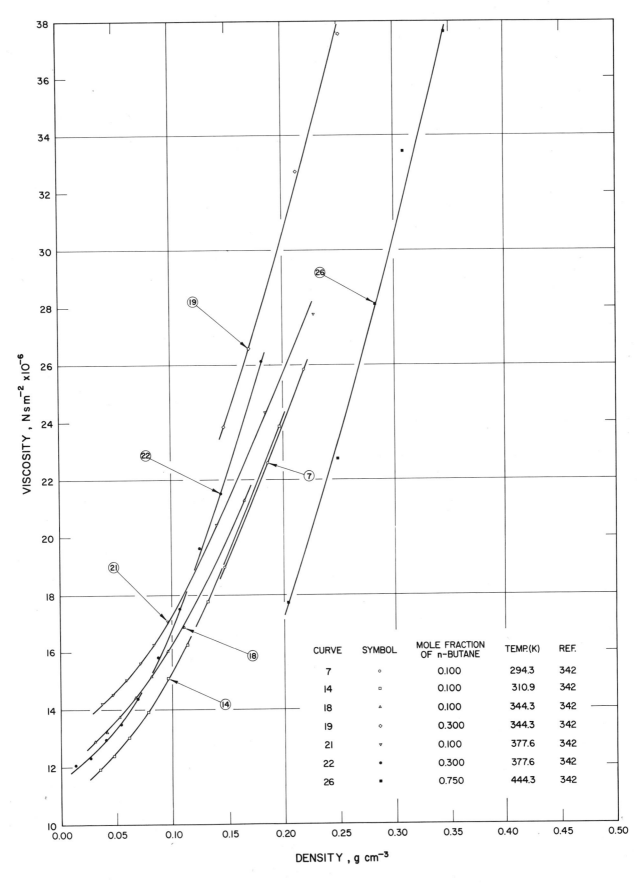

FIGURE 86-G(D). VISCOSITY DATA AS A FUNCTION OF DENSITY
FOR GASEOUS n-BUTANE-METHANE MIXTURES (continued)

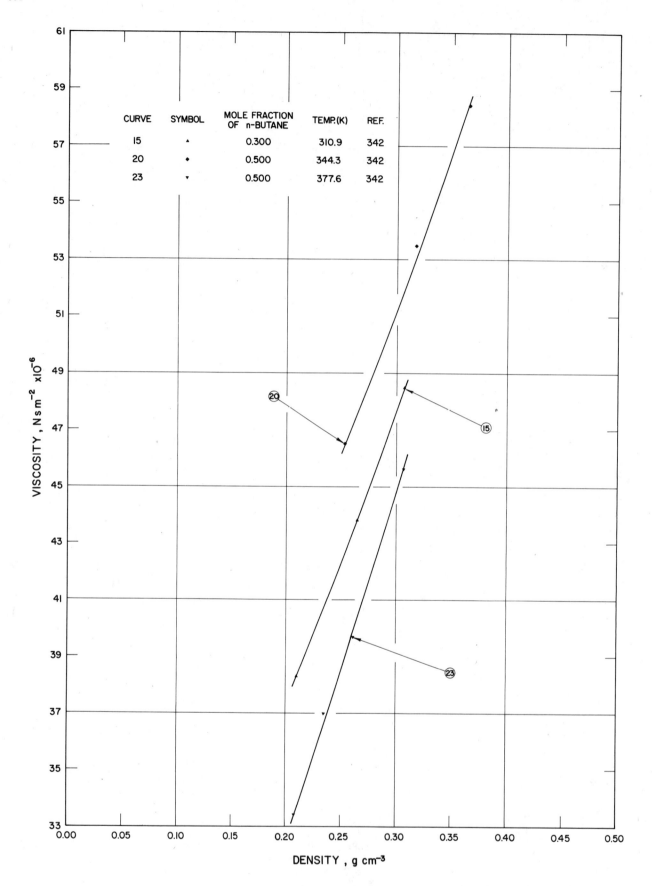

CURVE	SYMBOL	MOLE FRACTION OF n-BUTANE	TEMP.(K)	REF.
15	▲	0.300	310.9	342
20	♦	0.500	344.3	342
23	▼	0.500	377.6	342

VISCOSITY , N s m^{-2} x10^{-6}

DENSITY , g cm^{-3}

FIGURE 86-G(D). VISCOSITY DATA AS A FUNCTION OF DENSITY
FOR GASEOUS n-BUTANE-METHANE MIXTURES (continued)

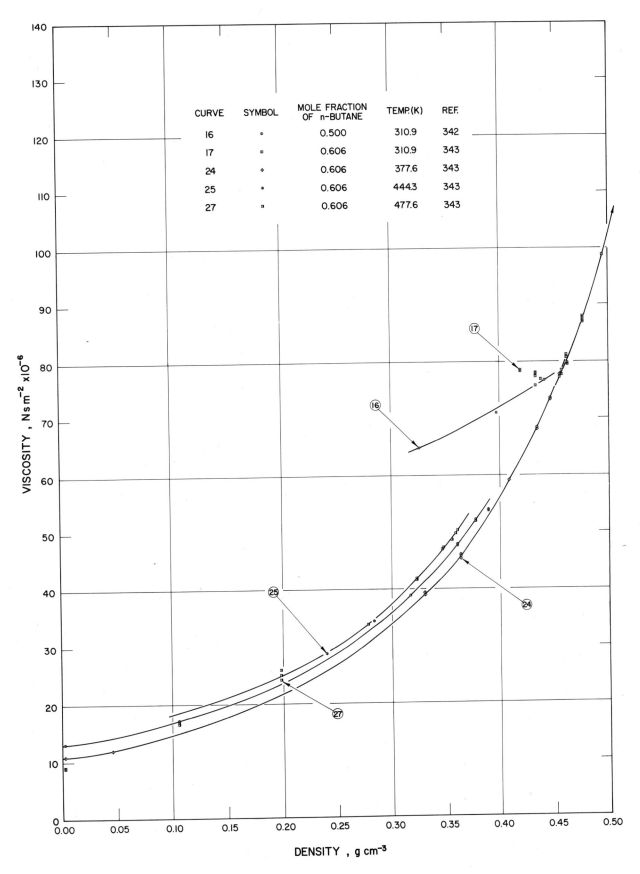

CURVE	SYMBOL	MOLE FRACTION OF n-BUTANE	TEMP.(K)	REF.
16	⊙	0.500	310.9	342
17	▥	0.606	310.9	343
24	◈	0.606	377.6	343
25	⊙	0.606	444.3	343
27	▫	0.606	477.6	343

FIGURE 86-G(D). VISCOSITY DATA AS A FUNCTION OF DENSITY FOR GASEOUS n-BUTANE-METHANE MIXTURES (continued)

TABLE 87-G(C)E.   EXPERIMENTAL VISCOSITY DATA AS A FUNCTION OF COMPOSITION FOR GASEOUS CARBON DIOXIDE-HYDROGEN MIXTURES

Cur. No.	Fig. No.	Ref. No.	Author(s)	Temp. (K)	Pressure (atm)	Mole Fraction of CO_2	Viscosity (N s m^{-2} x 10^{-6})	Remarks
1	87-G(C)	234	Trautz, M. and Kurz, F.	300.0		1.0000	14.93	CO_2: 99.966 pure, H_2: made by electrolysis; capillary method, d = 0.018 cm; L_1 = 0.033%, L_2 = 0.075%, L_3 = 0.200%.
						0.8821	15.02	
						0.8006	15.01	
						0.5871	15.06	
						0.4054	14.78	
						0.2150	13.76	
						0.1112	12.32	
						0.0000	8.91	
2	87-G(C)	234	Trautz, M. and Kurz, F.	400.0		1.0000	19.44	Same remarks as for curve 1 except L_1 = 0.045%, L_2 = 0.075%, L_3 = 0.154%.
						0.8821	19.51	
						0.8006	19.45	
						0.5871	19.33	
						0.4054	18.78	
						0.2150	17.13	
						0.1112	15.26	
						0.0000	10.81	
3	87-G(C)	234	Trautz, M. and Kurz, F.	500.0		1.0000	23.53	Same remarks as for curve 1 except L_1 = 0.017%, L_2 = 0.037%, L_3 = 0.099%.
						0.8821	23.60	
						0.8006	23.58	
						0.5871	23.21	
						0.4054	22.39	
						0.2150	20.26	
						0.1112	17.83	
						0.0000	12.56	
4	87-G(C)	234	Trautz, M. and Kurz, F.	550.0		1.0000	25.56	Same remarks as for curve 1 except L_1 = 0.275%, L_2 = 0.454%, L_3 = 1.097%.
						0.8821	25.54	
						0.8006	25.42	
						0.5871	25.06	
						0.4054	24.71	
						0.2150	21.73	
						0.1112	19.04	
						0.0000	13.41	
5	87-G(C)	337	Gururaja, G.J., Tirumarayanan, M.A., and Ramchandran, A.	300.7		1.000	14.990	Oscillating disk viscometer, calibrated to N_2; the viscosity of air, CO_2, and O_2 were measured at ambient temperature and pressure, the resulting precision was ± 1.0% of previous data.
				297.0		0.900	14.852	
				297.2		0.780	15.042	
				297.0		0.560	15.070	
				297.5		0.384	15.000	
				297.4		0.370	14.900	

TABLE 87-G(C)S.   SMOOTHED VISCOSITY VALUES AS A FUNCTION OF COMPOSITION FOR GASEOUS
CARBON DIOXIDE-HYDROGEN MIXTURES

Mole Fraction of CO_2	300 K [Ref. 234]	400 K [Ref. 234]	500 K [Ref. 234]	550 K [Ref. 234]
0.00	8.91	10.81	12.56	13.41
0.05	10.86	13.40	15.38	16.86
0.10	12.09	14.96	17.44	18.70
0.15	12.94	16.08	18.95	20.13
0.20	13.55	16.92	20.00	21.40
0.25	14.00	17.57	20.82	21.59
0.30	14.33	18.07	21.46	23.40
0.35	14.58	18.46	21.96	23.97
0.40	14.76	18.76	22.35	24.39
0.45	14.90	18.99	22.66	24.71
0.50	14.99	19.16	22.90	24.94
0.55	15.04	19.28	23.10	25.12
0.60	15.08	19.38	23.26	25.24
0.65	15.08	19.44	23.39	25.32
0.70	15.07	19.49	23.49	25.37
0.75	15.05	19.52	23.55	25.40
0.80	15.02	19.52	23.58	25.43
0.85	14.00	19.52	23.60	25.46
0.90	14.98	19.49	23.59	25.49
0.95	14.96	19.46	23.56	25.52
1.00	14.93	19.44	23.53	25.56

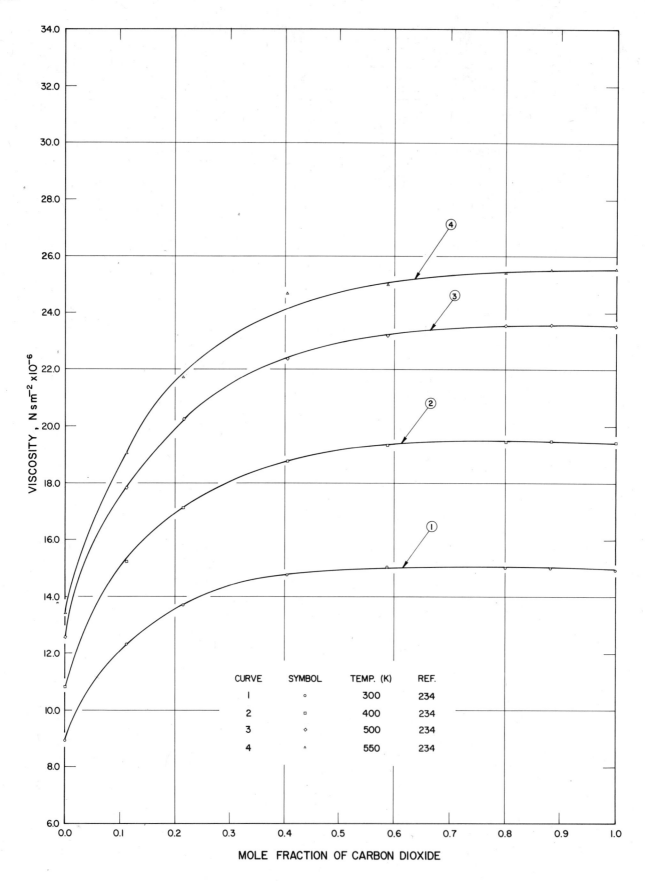

CURVE	SYMBOL	TEMP. (K)	REF.
1	○	300	234
2	□	400	234
3	◇	500	234
4	▵	550	234

MOLE FRACTION OF CARBON DIOXIDE

FIGURE 87-G(C). VISCOSITY DATA AS A FUNCTION OF COMPOSITION
FOR GASEOUS CARBON DIOXIDE – HYDROGEN MIXTURES

TABLE 88-G(D)E. EXPERIMENTAL VISCOSITY DATA AS A FUNCTION OF DENSITY FOR GASEOUS CARBON DIOXIDE-METHANE MIXTURES

Cur. No.	Fig. No.	Ref. No.	Author(s)	Mole Fraction of CO_2	Temp. (K)	Density (g cm^{-3})	Viscosity (N s m^{-2} x10^{-6})	Remarks
1	88-G(D)	335	DeWitt, K.J. and Thodos, G.	0.7570	50.1	0.0543	16.26	Gas purities not given as also an estimate of the accuracy; unsteady state transpiration type capillary viscometer.
						0.1254	18.01	
						0.3370	27.91	
						0.5126	41.98	
						0.6055	53.23	
						0.6609	61.36	
						0.7000	67.58	
						0.7298	72.61	
						0.7562	78.11	
						0.7763	82.16	
						0.7908	85.56	
2	88-G(D)	335	DeWitt, K.J. and Thodos, G.	0.5360	50.3	0.0444	15.53	Same remarks as for curve 1.
						0.0987	16.96	
						0.2309	22.66	
						0.3519	31.16	
						0.4348	39.24	
						0.4891	46.08	
						0.5292	51.78	
						0.5586	55.93	
						0.5816	59.65	
						0.6030	63.13	
						0.6200	65.94	
3	88-G(D)	335	DeWitt, K.J. and Thodos, G.	0.2450	50.3	0.0315	14.11	Same remarks as for curve 1.
						0.0655	15.07	
						0.1472	18.72	
						0.2193	23.41	
						0.2773	28.50	
						0.3198	33.19	
						0.3510	37.03	
						0.3748	40.31	
						0.3952	43.21	
						0.4138	46.14	
						0.4293	48.59	
4	88-G(D)	335	DeWitt, K.J. and Thodos, G.	0.7570	100.4	0.0447	18.24	Same remarks as for curve 1.
						0.0930	19.22	
						0.2145	23.67	
						0.3416	30.30	
						0.4436	37.55	
						0.5173	44.47	
						0.5731	50.95	
						0.6152	56.59	
						0.6483	61.49	
						0.6788	66.25	
						0.7022	70.13	
5	88-G(D)	335	DeWitt, K.J. and Thodos, G.	0.5360	100.3	0.0363	17.38	Same remarks as for curve 1.
						0.0770	18.25	
						0.1653	21.48	
						0.2520	25.80	
						0.3280	31.02	
						0.3898	36.27	
						0.4368	41.23	
						0.4720	45.68	
						0.5020	49.66	
						0.5279	53.35	
						0.5518	56.74	
6	88-G(D)	335	DeWitt, K.J. and Thodos, G.	0.2450	100.5	0.0263	15.76	Same remarks as for curve 1.
						0.0547	16.49	
						0.1139	18.77	
						0.1697	21.71	
						0.2198	25.05	
						0.2599	28.41	
						0.2944	32.00	
						0.3217	35.02	
						0.3462	38.07	
						0.3665	40.87	
						0.3833	43.23	
7	88-G(D)	335	DeWitt, K.J. and Thodos, G.	0.7570	150.7	0.0409	20.27	Same remarks as for curve 1.
						0.0789	20.97	
						0.1686	23.77	
						0.2586	27.77	
						0.3424	32.25	
						0.4127	37.04	
						0.4715	41.90	
						0.5188	46.59	
						0.5582	51.12	
						0.5926	55.60	
						0.6219	59.64	

TABLE 88-G(D)E.   EXPERIMENTAL VISCOSITY DATA AS A FUNCTION OF DENSITY FOR GASEOUS CARBON DIOXIDE-METHANE MIXTURES (continued)

Cur. No.	Fig. No.	Ref. No.	Author(s)	Mole Fraction of CO_2	Temp. (K)	Density (g cm^{-3})	Viscosity (N s m^{-2} x 10^{-6})	Remarks
8	88-G(D)	335	DeWitt, K.J. and Thodos, G.	0.5360	149.6	0.0317	19.25	Same remarks as for curve 1.
						0.0646	19.88	
						0.1331	22.05	
						0.2014	25.11	
						0.2646	28.52	
						0.3200	32.20	
						0.3653	35.82	
						0.4027	39.38	
						0.4370	43.14	
						0.4643	46.55	
						0.4900	49.93	
9	88-G(D)	335	DeWitt, K.J. and Thodos, G.	0.2450	150.2	0.0227	17.39	Same remarks as for curve 1.
						0.0467	17.95	
						0.0946	19.64	
						0.1402	21.73	
						0.1826	24.08	
						0.2200	26.65	
						0.2512	29.26	
						0.2789	31.96	
						0.3020	34.54	
						0.3234	36.97	
						0.3417	39.22	
10	88-G(D)	335	DeWitt, K.J. and Thodos, G.	0.7570	200.4	0.0328	22.10	Same remarks as for curve 1.
						0.0700	22.70	
						0.1425	24.74	
						0.2140	27.61	
						0.2828	30.81	
						0.3465	34.38	
						0.3993	37.92	
						0.4432	41.37	
						0.4830	44.94	
						0.5199	48.59	
						0.5508	52.10	
11	88-G(D)	335	DeWitt, K.J. and Thodos, G.	0.5360	200.6	0.0273	21.01	Same remarks as for curve 1.
						0.0555	21.51	
						0.1114	23.15	
						0.1691	25.39	
						0.2228	28.03	
						0.2728	30.81	
						0.3144	33.57	
						0.3501	36.37	
						0.3845	39.35	
						0.4149	42.43	
						0.4413	45.42	
12	88-G(D)	335	DeWitt, K.J. and Thodos, G.	0.2450	199.6	0.0197	18.92	Same remarks as for curve 1.
						0.0410	19.46	
						0.0813	20.69	
						0.1200	22.37	
						0.1573	24.21	
						0.1905	26.08	
						0.2202	28.23	
						0.2461	30.36	
						0.2692	32.52	
						0.2898	34.70	
						0.3079	36.70	
13	88-G(D)	329	Kestin, J. and Yata, J.	0.8565	303.2	0.04465	15.308	CO_2: 99.8 pure, CH_4: 99.99 pure; oscillating disk viscometer; error ± 0.1% and precision ± 0.05%.
						0.02570	15.088	
						0.00822	14.957	
						0.00170	14.920	
14	88-G(D)	329	Kestin, J. and Yata, J.	0.6624	303.2	0.03712	14.866	Same remarks as for curve 13.
						0.02164	14.649	
						0.00705	14.507	
						0.00146	14.466	
15	88-G(D)	329	Kestin, J. and Yata, J.	0.4806	303.2	0.03115	14.282	Same remarks as for curve 13.
						0.01823	14.072	
						0.00598	13.925	
						0.00124	13.881	
16	88-G(D)	329	Kestin, J. and Yata, J.	0.3257	303.2	0.02612	13.628	Same remarks as for curve 13.
						0.01544	13.434	
						0.00509	13.281	
						0.00106	13.237	

TABLE 88-G(D)E. EXPERIMENTAL VISCOSITY DATA AS A FUNCTION OF DENSITY FOR GASEOUS CARBON DIOXIDE-METHANE MIXTURES

Cur. No.	Fig. No.	Ref. No.	Author(s)	Mole Fraction of CO_2	Temp. (K)	Density (g cm^{-3})	Viscosity (N s m^{-2} x10^{-6})	Remarks
17	88-G(D)	329	Kestin, J. and Yata, J.	0.0000	303.2	0.01506	11.590	Same remarks as for curve 13.
						0.01262	11.526	
						0.009971	11.462	
						0.006518	11.387	
						0.003302	11.318	
						0.003294	11.318	
						0.003248	11.322	
						0.000733	11.276	
						0.000718	11.268	
						0.000669	11.267	
18	88-G(D)	329	Kestin, J. and Yata, J.	1.0000	303.2	0.04178	15.449	Same remarks as for curve 13.
						0.02882	15.326	
						0.00906	15.194	
						0.00185	15.169	
19	88-G(D)	329	Kestin, J. and Yata, J.	0.8565	293.2	0.04705	14.819	Same remarks as for curve 13.
						0.02665	14.585	
						0.00852	14.469	
						0.00176	14.433	
20	88-G(D)	329	Kestin, J. and Yata, J.	0.6624	293.2	0.03918	14.406	Same remarks as for curve 13.
						0.02246	14.177	
						0.00727	14.036	
						0.00149	14.003	
21	88-G(D)	329	Kestin, J. and Yata, J.	0.4806	293.2	0.03256	13.851	Same remarks as for curve 13.
						0.01890	13.636	
						0.00619	13.484	
						0.00128	13.448	
22	88-G(D)	329	Kestin, J. and Yata, J.	0.3257	293.2	0.02477	13.183	Same remarks as for curve 13.
						0.01605	13.026	
						0.00529	12.873	
						0.00110	12.826	
23	88-G(D)	329	Kestin, J. and Yata, J.	0.0000	293.2	0.01761	11.321	Same remarks as for curve 13.
						0.01384	11.217	
						0.01030	11.137	
						0.006809	11.054	
						0.003381	10.986	
						0.000701	10.936	
24	88-G(D)	329	Kestin, J. and Yata, J.	1.0000	293.2	0.04393	14.939	Same remarks as for curve 13.
						0.03017	14.815	
						0.00941	14.693	
						0.00192	14.674	

TABLE 88-G(D)S. SMOOTHED VISCOSITY VALUES AS A FUNCTION OF DENSITY FOR GASEOUS CARBON DIOXIDE-METHANE MIXTURES

Density (g cm⁻³)	Mole Fraction of Carbon Dioxide					
	0.0000 (293.2 K) [Ref. 329]	0.3257 (293.2 K) [Ref. 329]	0.4806 (293.2 K) [Ref. 329]	0.6624 (293.2 K) [Ref. 329]	0.8565 (293.2 K) [Ref. 329]	1.0000 (293.2 K) [Ref. 329]
0.0025	10.970	12.840	13.460	14.010		14.678
0.0050	11.012	12.870	13.470	14.020	14.450	14.680
0.0075	11.060	12.900	13.505	14.034		14.690
0.0100	11.112	12.930	13.530	14.050	14.478	14.700
0.0125	11.172	12.968				
0.0150	11.240	13.002	13.586	14.094	14.505	14.725
0.0175	11.320	13.046				
0.0200		13.090	13.650	14.140	14.535	14.752
0.0225		13.132				
0.0250		13.180	13.720	14.205	14.572	14.280
0.0300			13.810	14.280	14.620	14.812
0.0325			13.850			
0.0350				14.350	14.670	14.853
0.0375				14.387		
0.0400					14.730	14.900
0.0425						14.923
0.0450					14.800	

Density (g cm⁻³)	Mole Fraction of Carbon Dioxide					
	0.0000 (303.2 K) [Ref. 329]	0.3257 (303.2 K) [Ref. 329]	0.4806 (303.2 K) [Ref. 329]	0.6624 (303.2 K) [Ref. 329]	0.8565 (303.2 K) [Ref. 329]	1.0000 (303.2 K) [Ref. 329]
0.0020	11.294					
0.0025		13.252	13.890	14.476	14.925	15.170
0.0040	11.330					
0.0050	11.350	13.282	13.910	14.490	14.938	15.180
0.0075		13.318	13.935	14.508		
0.0080	11.420					
0.0100	11.460	13.350	13.962	14.530	14.967	15.190
0.0120	11.508					
0.0125		13.385	13.992			
0.0140	11.560					
0.0150	11.589	13.421	14.025	14.570	14.998	15.228
0.0175		13.461	14.060			
0.0200		13.502	14.099	14.622	15.034	15.258
0.0225		13.549				
0.0250		13.599	14.172	14.690	15.075	15.290
0.0300			14.260	14.760	15.128	15.330
0.0350				14.830	15.180	15.375
0.0375				14.865		
0.0400					15.243	15.420
0.0425						15.450
0.0450					15.310	

TABLE 88-G(D)S. SMOOTHED VISCOSITY VALUES AS A FUNCTION OF DENSITY FOR GASEOUS CARBON DIOXIDE-METHANE MIXTURES (continued)

Density (g cm^{-3})	Mole Fraction of Carbon Dioxide					
	0.7570 (323.3 K) [Ref. 335]	0.2450 (323.5 K) [Ref. 335]	0.5360 (323.5 K) [Ref. 335]	0.5360 (373.5 K) [Ref. 335]	0.7570 (373.6 K) [Ref. 335]	0.2450 (373.7 K) [Ref. 335]
0.020						15.67
0.050		14.60	15.70	17.60	18.32	16.40
0.100	17.39	16.48	17.08	18.95	19.45	18.06
0.120						18.95
0.150		19.00	18.85	20.80	21.00	20.50
0.175		20.50				
0.200	20.60	22.18	21.00	23.00	22.90	23.70
0.250		26.15	23.70	25.70	25.25	27.65
0.300	25.10	31.04	27.00	28.95	27.78	32.55
0.330						36.00
0.350		38.10	30.90	32.80		38.60
0.400	31.55	43.85	35.46	37.40	34.20	
0.420		47.00				
0.450			40.90	42.85		
0.500	40.50		47.40	49.30	42.70	
0.550			54.70	56.40		
0.600	52.50		62.50		54.50	
0.700	67.50				69.50	
0.750	76.50					
0.800	87.50					

Density (g cm^{-3})	Mole Fraction of Carbon Dioxide					
	0.5360 (422.8 K) [Ref. 335]	0.2450 (423.4 K) [Ref. 335]	0.7570 (423.9 K) [Ref. 335]	0.2450 (472.8 K) [Ref. 335]	0.7570 (473.6 K) [Ref. 335]	0.5360 (473.8 K) [Ref. 335]
0.020		17.20				
0.040		17.74				
0.050	19.60	18.04	20.40		22.34	21.50
0.100	20.09	19.90	21.53		24.41	22.78
0.150	22.82	22.26			25.00	24.60
0.200	25.03	26.70	24.89	19.00	27.00	26.80
0.250	28.08	30.60	27.22		29.28	29.50
0.280		33.70				
0.300	30.09	35.90	29.70		31.70	32.50
0.320		38.10				
0.350	34.58				34.60	36.30
0.400	39.10		36.02	19.45	37.95	40.08
0.420						42.90
0.440						45.02
0.450	44.70				42.00	
0.500			44.48	19.70	46.60	
0.550			50.00		51.82	
0.600			56.50			
0.800				20.06		
1.000				20.14		
1.200				20.24		
1.400				22.33		
1.500				23.80		
1.600				24.30		
1.800				25.84		
2.000				26.70		
2.300				28.95		
2.500				30.60		
2.800				33.70		
3.000				35.90		

374

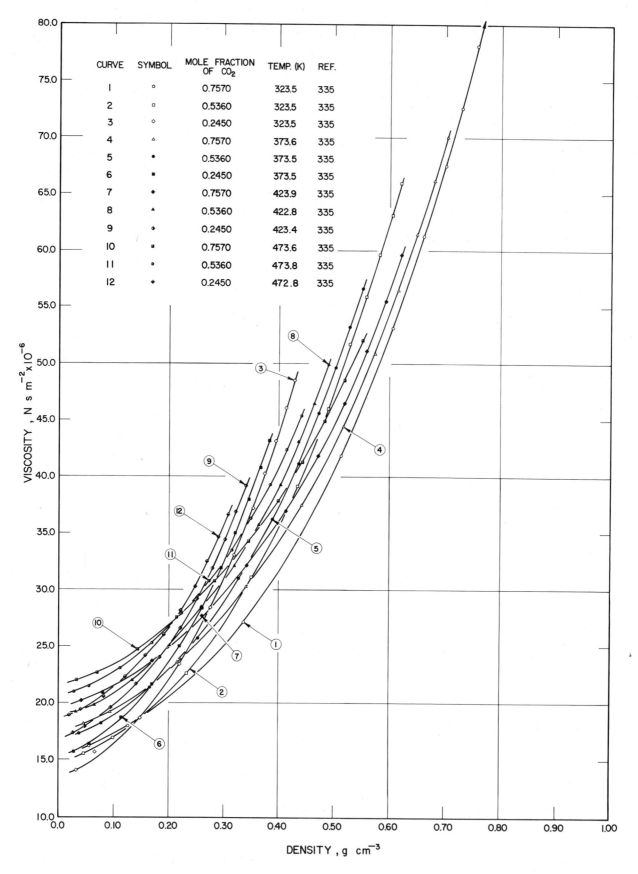

FIGURE 88-G(D). VISCOSITY DATA AS A FUNCTION OF DENSITY
FOR GASEOUS CARBON DIOXIDE-METHANE MIXTURES

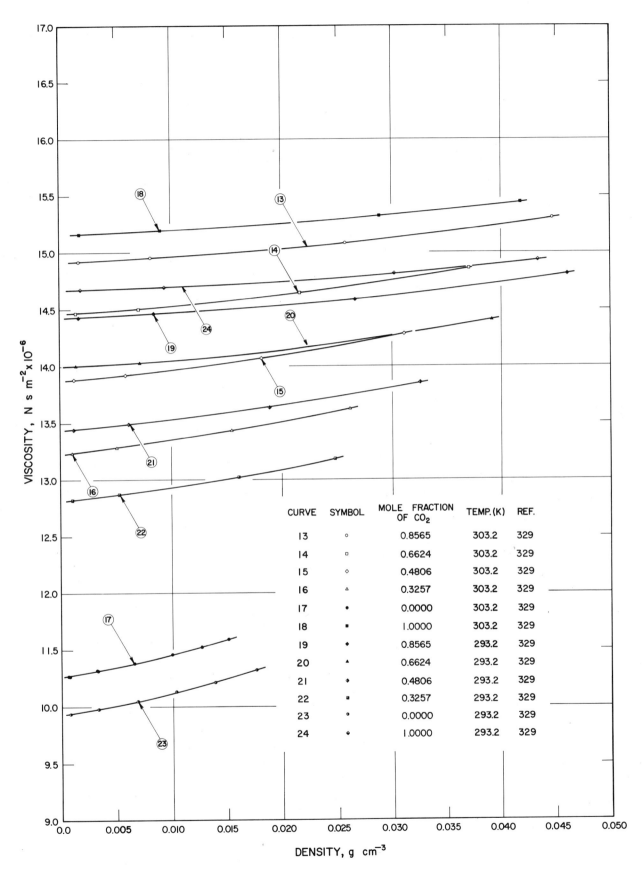

CURVE	SYMBOL	MOLE FRACTION OF CO_2	TEMP.(K)	REF.
13	○	0.8565	303.2	329
14	□	0.6624	303.2	329
15	◇	0.4806	303.2	329
16	△	0.3257	303.2	329
17	●	0.0000	303.2	329
18	■	1.0000	303.2	329
19	◆	0.8565	293.2	329
20	▲	0.6624	293.2	329
21	◆	0.4806	293.2	329
22	▣	0.3257	293.2	329
23	◦	0.0000	293.2	329
24	◈	1.0000	293.2	329

FIGURE 88-G(D).   VISCOSITY DATA AS A FUNCTION OF DENSITY
FOR GASEOUS CARBON DIOXIDE-METHANE MIXTURES (continued)

TABLE 89-G(C)E.   EXPERIMENTAL VISCOSITY DATA AS A FUNCTION OF COMPOSITION FOR GASEOUS CARBON DIOXIDE-NITROGEN MIXTURES

Cur. No.	Fig. No.	Ref. No.	Author(s)	Temp. (K)	Pressure (atm)	Mole Fraction of CO_2	Viscosity (N s m^{-2} x 10^{-6})	Remarks
1	89-G(C)	337	Gururaja, G.J., Tirunarayanan, M.A., and Ramchandran, A.	300.7		1.000	14.990	Oscillating disk viscometer, calibrated with nitrogen; estimated accuracy 1.0%; L_1 = 0.389%, L_2 = 0.564%, L_3 = 1.216%.
				297.0		0.800	15.270	
				297.9		0.750	15.690	
				297.2		0.580	16.100	
				296.6		0.326	16.720	
				297.2		0.277	16.920	
				297.0		0.226	17.010	
				298.2		0.000	17.796	

TABLE 89-G(C)S.   SMOOTHED VISCOSITY VALUES AS A FUNCTION OF COMPOSITION FOR GASEOUS CARBON DIOXIDE-NITROGEN MIXTURES

Mole Fraction of CO_2	(297.2 K) [Ref. 337]
0.00	17.80
0.05	17.63
0.10	17.46
0.15	17.30
0.20	17.13
0.25	16.97
0.30	16.81
0.35	16.65
0.40	16.50
0.45	16.35
0.50	16.21
0.55	16.03
0.60	15.95
0.65	15.82
0.70	15.70
0.75	15.58
0.80	15.46
0.85	15.34
0.90	15.22
0.95	15.10
1.00	15.00

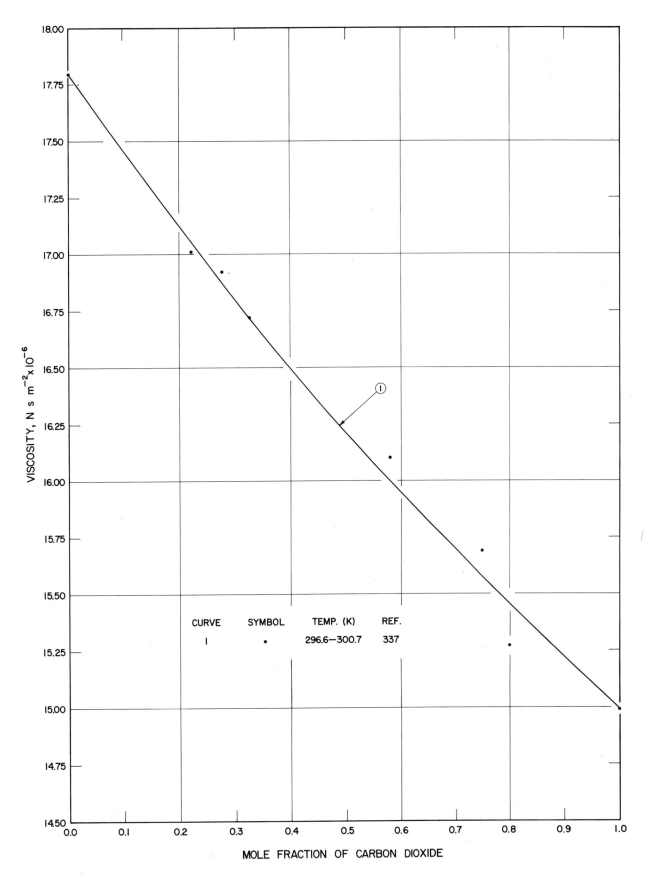

FIGURE 89-G(C). VISCOSITY DATA AS A FUNCTION OF COMPOSITION
FOR GASEOUS CARBON DIOXIDE - NITROGEN MIXTURES

TABLE 89-G(D)E.   EXPERIMENTAL VISCOSITY DATA AS A FUNCTION OF DENSITY FOR GASEOUS
CARBON DIOXIDE-NITROGEN MIXTURES

Cur. No.	Fig. No.	Ref. No.	Author(s)	Mole Fraction of CO_2	Temp. (K)	Density (g cm^{-3})	Viscosity (N s m^{-2} x10^{-6})	Remarks
1	89-G(D)	336	Kestin, J. and Leidenfrost, W.	0.9044	293.2	0.04244	15.290	CO_2: 99.695 pure, N_2: 99.999 pure; oscillating disk viscometer; uncertainties: mixture composition $\pm 1\%$, viscosity $\pm 0.05\%$.
						0.03499	15.181	
						0.02786	15.095	
						0.02094	15.041	
						0.01435	14.993	
						0.007940	14.960	
						0.001773	14.937	
2	89-G(D)	336	Kestin, J. and Leidenfrost, W.	0.7870	293.2	0.03966	15.697	Same remarks as for curve 1.
						0.03284	15.596	
						1.02622	15.534	
						0.01979	15.467	
						0.01358	15.424	
						0.007561	15.378	
						0.001659	15.350	
3	89-G(D)	336	Kestin, J. and Leidenfrost, W.	0.6568	293.2	0.03681	16.130	Same remarks as for curve 1.
						0.03062	16.046	
						0.02455	15.969	
						0.01860	15.897	
						0.01278	15.841	
						0.007111	15.764	
						0.001611	15.712	
4	89-G(D)	336	Kestin, J. and Leidenfrost, W.	0.5054	293.2	0.03383	16.623	Same remarks as for curve 1.
						0.02827	16.543	
						0.02270	16.465	
						0.01728	16.392	
						0.01190	16.302	
						0.006680	16.243	
						0.001510	16.184	
5	89-G(D)	336	Kestin, J. and Leidenfrost, W.	0.3752	293.2	0.03141	17.016	Same remarks as for curve 1.
						0.02624	16.948	
						0.02115	16.883	
						0.01614	16.825	
						0.01115	16.777	
						0.006251	16.729	
						0.001426	16.696	
6	89-G(D)	336	Kestin, J. and Leidenfrost, W.	0.2333	293.2	0.02898	17.425	Same remarks as for curve 1.
						0.01958	17.314	
						0.01038	17.226	
						0.001336	17.211	
7	89-G(D)	336	Kestin, J. and Leidenfrost, W.	0.1060	293.2	0.02674	17.708	Same remarks as for curve 1.
						0.02248	17.662	
						0.01819	17.638	
						0.01394	17.582	
						0.009681	17.535	
						0.005451	17.496	
						0.001249	17.440	
8	89-G(D)	326	Kestin, J., Kobayashi, Y., and Wood, R.T.	1.0000	293.2	0.05252	15.071	CO_2: 99.8 pure, N_2: 99.999 pure; oscillating disk viscometer; uncertainties: mixture composition $\pm 0.002\%$, viscosity $\pm 0.1\%$, viscosity ratio $\pm 0.04\%$.
						0.04810	15.019	
						0.04183	14.937	
						0.04167	14.946	
						0.03568	14.874	
						0.02973	14.821	
						0.02403	14.772	
						0.01942	14.752	
						0.01425	14.716	
						0.009412	14.686	
						0.005584	14.680	
						0.001908	14.673	
9	89-G(D)	326	Kestin, J., et al.	0.8131	293.2	0.04933	15.768	Same remarks as for curve 8.
						0.02714	15.471	
						0.008714	15.313	
						0.001757	15.278	
10	89-G(D)	326	Kestin, J., et al.	0.6882	293.2	0.03997	16.077	Same remarks as for curve 8.
						0.02490	15.888	
						0.008152	15.727	
						0.001638	15.681	
11	89-G(D)	326	Kestin, J., et al.	0.5057	293.2	0.03766	16.728	Same remarks as for curve 8.
						0.02316	16.543	
						0.007567	16.313	
						0.001526	16.264	

TABLE 89-G(D)E.   EXPERIMENTAL VISCOSITY DATA AS A FUNCTION OF DENSITY FOR GASEOUS CARBON DIOXIDE-NITROGEN MIXTURES (continued)

Cur. No.	Fig. No.	Ref. No.	Author(s)	Mole Fraction of CO_2	Temp. (K)	Density (g cm^{-3})	Viscosity (N s m^{-2} x 10^{-6})	Remarks
12	89-G(D)	326	Kestin, J., Kobayashi, Y., and Wood, R.T.	0.3101	293.2	0.03480	17.300	Same remarks as for curve 8.
						0.02088	17.085	
						0.006891	16.891	
						0.001392	16.825	
13	89-G(D)	326	Kestin, J., et al.	0.1607	293.2	0.01473	17.380	Same remarks as for curve 8.
						0.006411	17.266	
						0.001272	17.213	
14	89-G(D)	326	Kestin, J., et al.	0.0738	293.2	0.03050	17.827	Same remarks as for curve 8.
						0.01844	17.645	
						0.006614	17.485	
						0.001279	17.423	
15	89-G(D)	326	Kestin, J., et al.	1.0000	304.2	0.05079	15.637	Same remarks as for curve 8.
						0.04342	15.541	
						0.03933	15.494	
						0.03331	15.428	
						0.02486	15.353	
						0.01854	15.300	
						0.01380	15.269	
						0.009012	15.238	
						0.005331	15.223	
						0.001832	15.208	
16	89-G(D)	326	Kestin, J., et al.	0.8131	304.2	0.04466	16.275	Same remarks as for curve 8.
						0.02591	16.033	
						0.008352	15.874	
						0.001691	15.826	
17	89-G(D)	326	Kestin, J., et al.	0.3101	304.2	0.03311	17.820	Same remarks as for curve 8.
						0.02009	17.633	
						0.006568	17.423	
						0.001344	17.360	
18	89-G(D)	326	Kestin, J., et al.	0.0738	304.2	0.02997	18.348	Same remarks as for curve 8.
						0.01760	18.158	
						0.005870	18.002	
						0.001201	17.944	
19	89-G(D)	326	Kestin, J., et al.	0.0000	304.2	0.02852	18.446	Same remarks as for curve 8.
						0.01681	18.268	
						0.005617	18.123	
						0.001153	18.066	

TABLE 89-G(D)S. SMOOTHED VISCOSITY VALUES AS A FUNCTION OF DENSITY FOR GASEOUS CARBON DIOXIDE-NITROGEN MIXTURES

Density (g cm^{-3})	Mole Fraction of Carbon Dioxide						
	0.0738 (293.2 K) [Ref. 326]	0.1060 (293.2 K) [Ref. 336]	0.1607 (293.2 K) [Ref. 326]	0.2333 (293.2 K) [Ref. 336]	0.3101 (293.2 K) [Ref. 326]	0.3752 (293.2 K) [Ref. 336]	0.5054 (293.2 K) [Ref. 336]
0.0020			17.222				
0.0025	17.440	17.460		17.215	16.840	16.708	16.192
0.0040			17.240				
0.0050	17.470	17.491	17.250	17.210	16.870	16.722	16.221
0.0060			17.260				
0.0075	17.500	17.520		17.213		16.740	16.252
0.0080			17.285				
0.0100	17.532	17.546	17.311	17.225	16.930	16.760	16.286
0.0120			17.340				
0.0125		17.570		17.242			
0.0140			17.372				
0.0150	17.608	17.600	17.390	17.260	16.990	16.812	16.351
0.0200	17.670	17.644		17.318	17.070	16.870	16.420
0.0250	17.748	17.682		17.375	17.150	16.930	16.490
0.0275		17.710		17.408			
0.0300	17.820			17.437	17.230	16.990	16.560
0.0325	17.860					17.022	
0.0350					17.310		16.630

Density (g cm^{-3})	Mole Fraction of Carbon Dioxide						
	0.5057 (293.2 K) [Ref. 326]	0.6568 (293.2 K) [Ref. 336]	0.6882 (293.2 K) [Ref. 326]	0.7870 (293.2 K) [Ref. 336]	0.8131 (293.2 K) [Ref. 326]	0.9044 (293.2 K) [Ref. 336]	1.0000 (293.2 K) [Ref. 326]
0.0025	16.280	15.726	15.692	15.355		14.938	
0.0050	16.309	15.749	15.710	15.365	15.296	14.950	
0.0075	16.338	15.773	15.728				14.682
0.0100	16.362	15.800	15.746	15.380	15.324	14.972	14.701
0.0150	16.423	15.858	15.790	15.390	15.360	15.000	14.725
0.0200	16.485	15.918	15.835	15.430	15.400	15.033	14.755
0.0250	16.550	15.975	15.884	15.700	15.450	15.072	14.790
0.0300	16.618	16.040	15.949	15.520	15.502	15.120	14.821
0.0350	16.898	16.108	16.011	15.570	15.564	15.182	14.868
0.0375	16.725	16.140					
0.0400			16.078	15.630	15.628	15.258	14.920
0.0425						15.296	
0.0450					15.699		14.968
0.0500					15.770		15.040

TABLE 89-G(D)S.   SMOOTHED VISCOSITY VALUES AS A FUNCTION OF DENSITY FOR GASEOUS
CARBON DIOXIDE-NITROGEN MIXTURES (continued)

Density (g cm^{-3})	Mole Fraction of Carbon Dioxide				
	0.0000 (304.2 K) [Ref. 326]	0.0738 (304.2 K) [Ref. 326]	0.3101 (304.2 K) [Ref. 326]	0.8131 (304.2 K) [Ref. 326]	1.0000 (304.2 K) [Ref. 326]
0.0025	18.078	17.965	17.370		
0.0050	18.110	17.992	17.391	15.852	15.220
0.0075	18.139				
0.0100	18.170	18.060	17.466	15.892	15.242
0.0125		18.092			
0.0150	18.232	18.130	17.548	15.938	15.275
0.0200	18.310	18.190	17.620	15.980	15.315
0.0225		18.225			
0.0250	18.260	18.260	17.699	16.029	15.355
0.0275			17.732		
0.0300	18.348	18.340	17.774	16.087	15.399
0.0325			17.812		
0.0350				16.149	15.446
0.0400				16.212	15.502
0.0450				16.280	15.562
0.0500					15.627

382

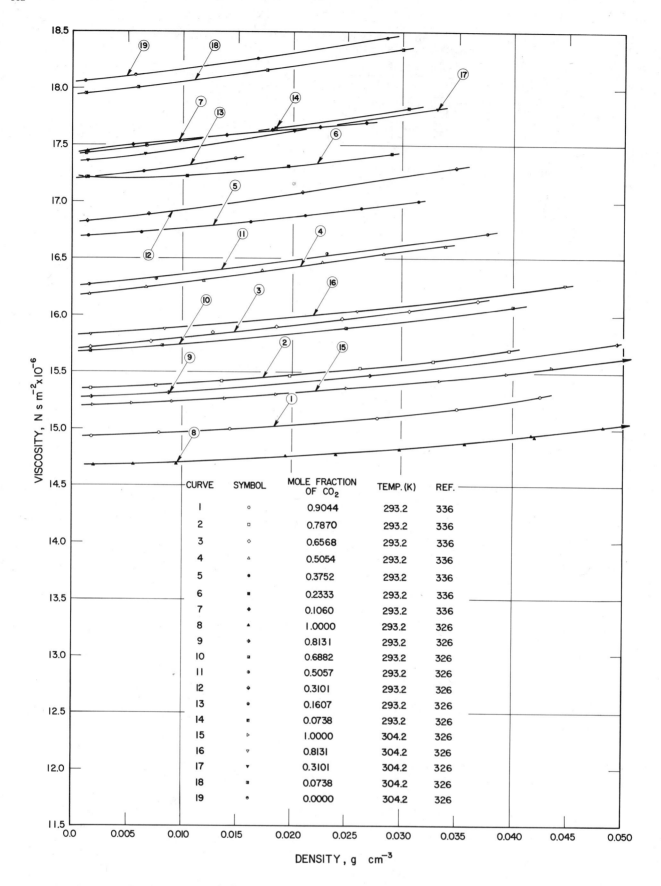

CURVE	SYMBOL	MOLE FRACTION OF CO_2	TEMP. (K)	REF.
1	○	0.9044	293.2	336
2	□	0.7870	293.2	336
3	◇	0.6568	293.2	336
4	△	0.5054	293.2	336
5	●	0.3752	293.2	336
6	■	0.2333	293.2	336
7	◆	0.1060	293.2	336
8	▲	1.0000	293.2	326
9	◈	0.8131	293.2	326
10	▣	0.6882	293.2	326
11	◉	0.5057	293.2	326
12	◆	0.3101	293.2	326
13	◉	0.1607	293.2	326
14	▥	0.0738	293.2	326
15	▷	1.0000	304.2	326
16	▽	0.8131	304.2	326
17	▼	0.3101	304.2	326
18	▣	0.0738	304.2	326
19	◉	0.0000	304.2	326

FIGURE 89-G(D). VISCOSITY DATA AS A FUNCTION OF DENSITY
FOR GASEOUS CARBON DIOXIDE-NITROGEN MIXTURES

TABLE 90-G(C)E.   EXPERIMENTAL VISCOSITY DATA AS A FUNCTION OF COMPOSITION FOR GASEOUS CARBON DIOXIDE-NITROUS OXIDE MIXTURES

Cur. No.	Fig. No.	Ref. No.	Author(s)	Temp. (K)	Pressure (atm)	Mole Fraction of CO_2	Viscosity ($N\ s\ m^{-2} \times 10^{-6}$)	Remarks
1	90-G(C)	234	Trautz, M. and Kurz, F.	300		1.0000	14.88	CO_2: 99.966 pure, N_2O: 1.3 parts per 1000; capillary method, d = 0.018 cm; $L_1 = 0.023\%$, $L_2 = 0.041\%$, $L_3 = 0.074\%$.
						0.5976	14.94	
						0.3967	14.95	
						0.1903	14.90	
						0.1087	14.95	
						0.0000	14.93	
2	90-G(C)	234	Trautz, M. and Kurz, F.	400		1.0000	19.93	Same remarks as for curve 1 except $L_1 = 0.103\%$, $L_2 = 0.148\%$, $L_3 = 0.308\%$.
						0.5976	19.50	
						0.3967	19.50	
						0.8003	19.48	
						0.1903	19.41	
						0.1087	19.45	
						0.0000	19.44	
3	90-G(C)	234	Trautz, M. and Kurz, F.	500		1.0000	23.55	Same remarks as for curve 1 except $L_1 = 0.051\%$, $L_2 = 0.114\%$, $L_3 = 0.296\%$.
						0.8003	23.57	
						0.5976	23.65	
						0.3967	23.65	
						0.1903	23.58	
						0.1087	23.58	
						0.0000	23.53	
4	90-G(C)	234	Trautz, M. and Kurz, F.	550		1.0000	25.55	Same remarks as for curve 1 except $L_1 = 0.206\%$, $L_2 = 0.259\%$, $L_3 = 0.353\%$.
						0.8003	25.55	
						0.5976	25.62	
						0.3967	25.64	
						0.1903	25.51	
						0.1087	25.55	
						0.0000	25.65	

TABLE 90-G(C)S.   SMOOTHED VISCOSITY VALUES AS A FUNCTION OF COMPOSITION FOR GASEOUS CARBON DIOXIDE-NITROUS OXIDE MIXTURES

Mole Fraction of CO_2	300.0 K [Ref. 234]	400.0 K [Ref. 234]	500.0 K [Ref. 234]	550.0 K [Ref. 234]
0.00	14.93	19.44	23.53	25.65
0.05	14.94	19.45	23.55	25.64
0.10	14.94	19.46	23.56	25.63
0.15	14.94	19.46	23.58	25.62
0.20	14.94	19.47	23.59	25.60
0.25	14.94	19.48	23.61	25.58
0.30	14.94	19.47	23.62	25.57
0.35	14.94	19.46	23.64	25.56
0.40	14.94	19.46	23.66	25.55
0.45	14.94	19.46	23.66	25.54
0.50	14.95	19.46	23.70	25.54
0.55	14.95	19.46	23.70	25.53
0.60	14.95	19.46	23.66	25.53
0.65	15.95	19.46	23.66	25.53
0.70	14.94	19.46	23.66	25.53
0.75	14.94	19.46	23.65	25.53
0.80	14.92	19.45	23.64	25.53
0.85	14.92	19.45	23.62	25.54
0.90	14.91	19.44	23.60	25.54
0.95	14.90	19.43	23.58	25.54
1.00	14.88	19.43	23.55	25.55

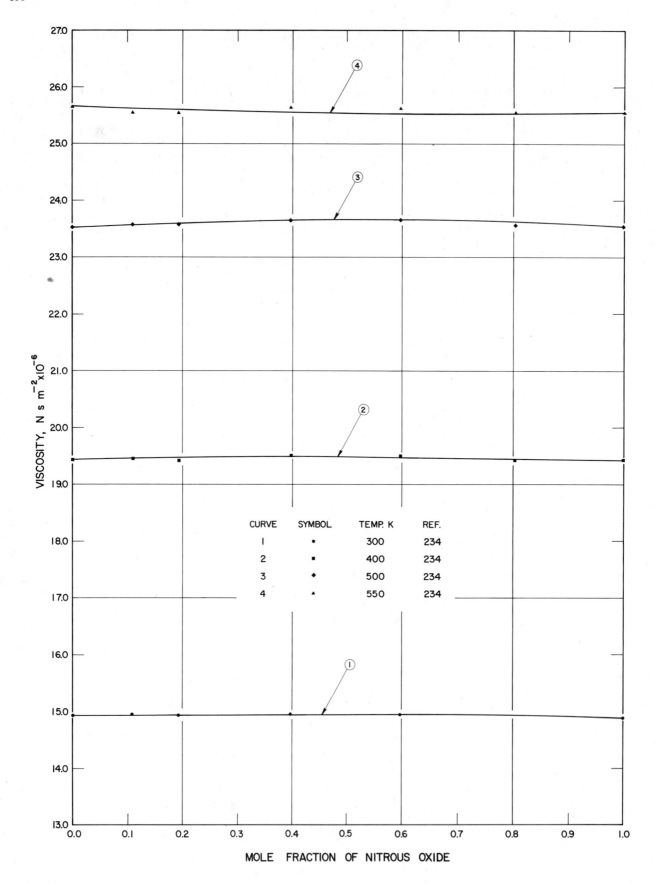

FIGURE 90-G(C). VISCOSITY DATA AS A FUNCTION OF COMPOSITION
FOR GASEOUS CARBON DIOXIDE-NITROUS OXIDE MIXTURES

TABLE 91-G(C)E.   EXPERIMENTAL VISCOSITY DATA AS A FUNCTION OF COMPOSITION FOR GASEOUS
CARBON DIOXIDE-OXYGEN MIXTURES

Cur. No.	Fig. No.	Ref. No.	Author(s)	Temp. (K)	Pressure (atm)	Mole Fraction of CO_2	Viscosity (N s m^{-2} x 10^{-6})	Remarks
1	91-G(C)	337	Gururaja, G.J., Tirumarayanan, M.A., and Ramchandran, A.	300.7	0.966	1.000	14.990	Oscillating disk viscometer, relative measurements; accuracy about 1.0%; $L_1 = 0.911\%$, $L_2 = 1.320\%$, $L_3 = 2.714\%$.
				297.6		0.917	15.420	
				297.6		0.800	15.950	
				298.2		0.710	16.600	
				298.2		0.560	17.710	
				298.2		0.339	18.450	
				298.2		0.306	18.600	
				297.4		0.195	18.950	
				302.6		0.000	20.800	

TABLE 91-G(C)S.   SMOOTHED VISCOSITY VALUES AS A FUNCTION OF COMPOSITION FOR GASEOUS
CARBON DIOXIDE-OXYGEN MIXTURES

Mole Fraction of CO_2	300 K [Ref. 337]
0.00	20.80
0.05	20.44
0.10	20.09
0.15	19.74
0.20	19.41
0.25	19.08
0.30	18.76
0.35	18.45
0.40	18.15
0.45	17.86
0.50	17.68
0.55	17.30
0.60	17.02
0.65	16.75
0.70	16.50
0.75	16.23
0.80	15.98
0.85	15.73
0.90	15.48
0.95	15.23
1.00	14.90

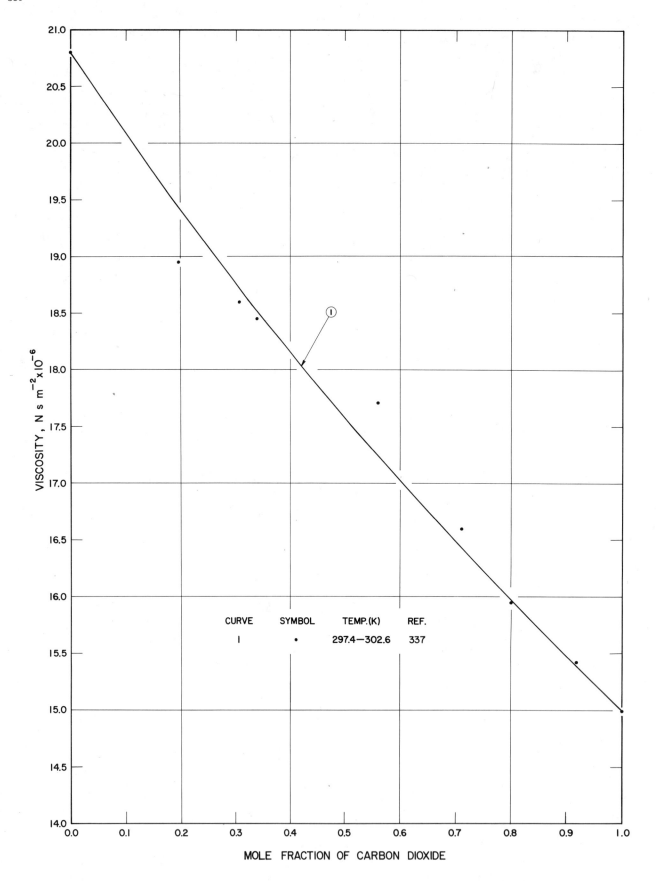

FIGURE 91-G(C). VISCOSITY DATA AS A FUNCTION OF COMPOSITION
FOR GASEOUS CARBON DIOXIDE-OXYGEN MIXTURES

TABLE 92-G(C)E.   EXPERIMENTAL VISCOSITY DATA AS A FUNCTION OF COMPOSITION FOR GASEOUS
CARBON DIOXIDE-PROPANE MIXTURES

Cur. No.	Fig. No.	Ref. No.	Author(s)	Temp. (K)	Pressure (atm)	Mole Fraction of C_3H_8	Viscosity (N s m^{-2} x 10^{-6})	Remarks
1	92-G(C)	234	Trautz, M. and Kurz, F.	300		1.0000	8.17	C_3H_8: 100 pure, CO_2: 99.966 pure;
						0.8106	9.26	capillary method, d = 0.018 cm;
						0.5975	10.58	L_1 = 0.027%, L_2 = 0.051%, L_3 =
						0.4224	11.74	0.122%.
						0.2117	13.26	
						0.0000	14.93	
2	92-G(C)	234	Trautz, M. and Kurz, F.	400		1.0000	10.70	Same remarks as for curve 1 except
						0.8106	12.13	L_1 = 0.063%, L_2 = 0.096%, L_3 =
						0.5975	13.83	0.174%.
						0.4224	15.33	
						0.2117	17.30	
						0.0000	19.44	
3	92-G(C)	234	Trautz, M. and Kurz, F.	500		1.0000	13.08	Same remarks as for curve 1 except
						0.8106	14.61	L_1 = 0.014%, L_2 = 0.027%, L_3 =
						0.5975	16.70	0.060%.
						0.4224	18.56	
						0.2117	20.93	
						0.0000	23.53	
4	92-G(C)	234	Trautz, M. and Kurz, F.	550		1.0000	14.22	Same remarks as for curve 1 except
						0.8106	16.01	L_1 = 0.007%, L_2 = 0.018%, L_3 =
						0.5975	18.15	0.044%.
						0.4224	20.10	
						0.2117	22.67	
						0.0000	25.56	

TABLE 92-G(C)S.   SMOOTHED VISCOSITY VALUES AS A FUNCTION OF COMPOSITION FOR GASEOUS
CARBON DIOXIDE-PROPANE MIXTURES

Mole Fraction of C_3H_8	300.0 K [Ref. 234]	400.0 K [Ref. 234]	500.0 K [Ref. 234]	550.0 K [Ref. 234]
0.00	14.93	19.44	23.54	25.56
0.05	14.52	18.89	22.90	24.81
0.10	14.11	18.38	22.28	24.12
0.15	13.72	17.88	21.67	23.46
0.20	13.34	17.38	21.07	22.81
0.25	12.98	16.91	20.48	22.19
0.30	12.62	16.42	19.91	21.58
0.35	12.26	15.98	19.35	20.98
0.40	11.91	15.53	18.80	20.36
0.45	11.56	15.09	18.28	19.80
0.50	11.22	14.66	17.74	19.24
0.55	10.89	14.23	17.21	18.68
0.60	10.57	13.80	16.64	18.14
0.65	10.25	13.39	16.08	17.62
0.70	9.94	13.00	15.68	17.12
0.75	9.63	12.61	15.20	16.62
0.80	9.33	12.22	14.83	16.13
0.85	9.04	11.84	14.29	15.64
0.90	8.75	11.46	13.87	15.16
0.95	8.47	11.09	13.47	14.70
1.00	8.18	10.72	13.08	14.22

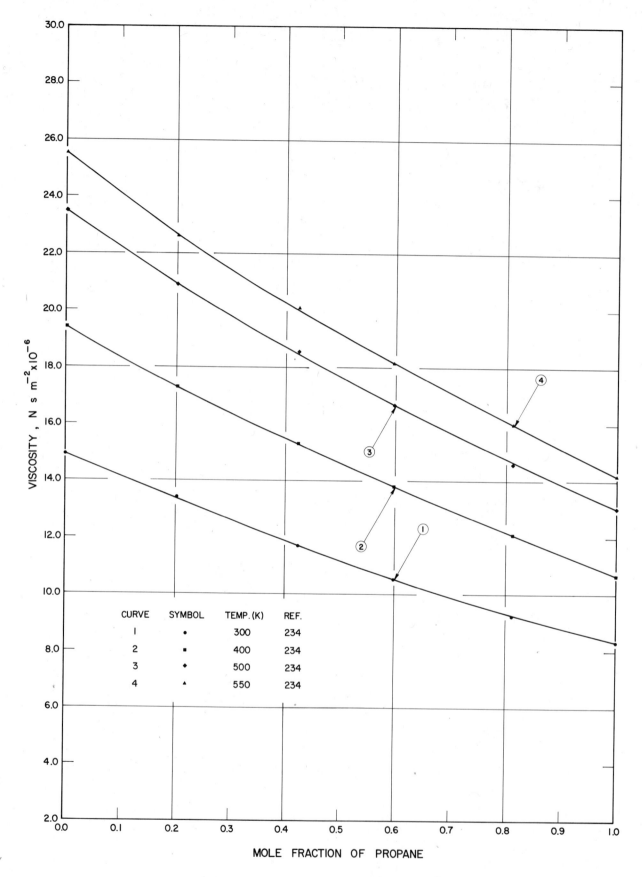

FIGURE 92-G(C). VISCOSITY DATA AS A FUNCTION OF COMPOSITION
FOR GASEOUS CARBON DIOXIDE-PROPANE MIXTURES

TABLE 93-G(C)E. EXPERIMENTAL VISCOSITY DATA AS A FUNCTION OF COMPOSITION FOR GASEOUS CARBON MONOXIDE-ETHYLENE MIXTURES

Cur. No.	Fig. No.	Ref. No.	Author(s)	Temp. (K)	Pressure (atm)	Mole Fraction of C_2H_4	Viscosity ($N \, s \, m^{-2} \times 10^{-6}$)	Remarks
1	93-G(C)	227	Trautz, M. and Melster, A.	300		0.0000 0.2632 0.4354 0.8062 1.0000	17.76 15.53 14.02 11.35 10.33	Capillary method, r = 0.2019 mm; L_1 = 0.348%, L_2 = 0.561%, L_3 = 1.041%.
2	93-G(C)	227	Trautz, M. and Melster, A.	400		0.0000 0.2632 0.4354 0.8062 1.0000	21.83 19.43 17.63 14.60 13.42	Same remarks as for curve 1 except L_1 = 0.191%, L_2 = 0.327%, L_3 = 0.674%.
3	93-G(C)	227	Trautz, M. and Melster, A.	500		0.0000 0.2632 0.4354 0.8062 1.0000	25.48 22.79 20.98 17.60 16.22	Same remarks as for curve 1 except L_1 = 0.081%, L_2 = 0.144%, L_3 = 0.308%.
4	93-G(C)	227	Trautz, M. and Melster, A.	550		0.0000 0.2632 0.4354 0.8062 1.0000	27.14 24.33 22.40 19.00 17.53	Same remarks as for curve 1 except L_1 = 0.127%, L_2 = 0.210%, L_3 = 0.413%.

TABLE 93-G(C)S. SMOOTHED VISCOSITY VALUES AS A FUNCTION OF COMPOSITION FOR GASEOUS CARBON MONOXIDE-ETHYLENE MIXTURES

Mole Fraction of C_2H_4	300.0 K [Ref. 227]	400.0 K [Ref. 227]	500.0 K [Ref. 227]	550.0 K [Ref. 227]
0.00	17.76	21.83	25.48	27.14
0.05	17.28	21.34	24.97	26.58
0.10	16.80	20.85	24.44	26.00
0.15	16.35	20.38	23.91	25.47
0.20	15.91	19.90	23.40	24.90
0.25	15.48	19.41	22.96	24.38
0.30	15.08	18.94	22.35	23.82
0.35	14.69	18.45	21.83	22.80
0.40	14.29	18.00	21.35	22.30
0.45	13.92	17.53	20.85	21.80
0.50	13.51	17.09	20.38	21.80
0.55	13.18	16.65	19.91	21.30
0.60	12.80	16.20	19.45	20.86
0.65	12.46	15.80	18.99	20.46
0.70	12.12	15.40	18.51	19.92
0.75	11.80	15.02	18.07	19.52
0.80	11.48	14.66	17.65	19.08
0.85	11.17	14.32	17.28	18.66
0.90	10.87	14.00	16.90	18.28
0.95	10.60	13.70	16.55	17.90
1.00	10.33	13.42	16.22	17.53

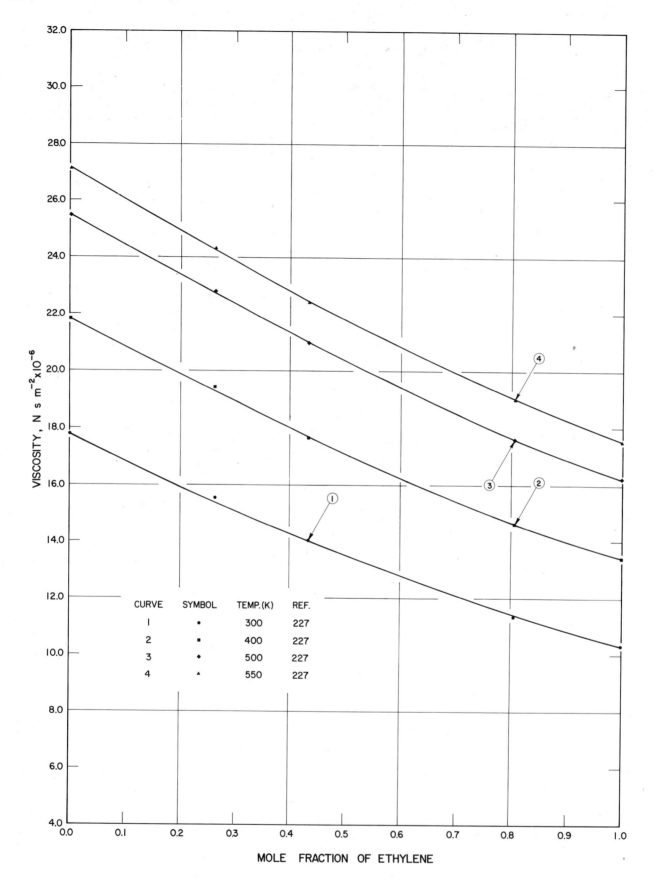

FIGURE 93-G(C). VISCOSITY DATA AS A FUNCTION OF COMPOSITION
FOR GASEOUS CARBON MONOXIDE-ETHYLENE MIXTURES

TABLE 94-G(C)E.   EXPERIMENTAL VISCOSITY DATA AS A FUNCTION OF COMPOSITION FOR GASEOUS
CARBON MONOXIDE-HYDROGEN MIXTURES

Cur. No.	Fig. No.	Ref. No.	Author(s)	Temp. (K)	Pressure (atm)	Mole Fraction of CO	Viscosity $(N\ s\ m^{-2} \times 10^{-6})$	Remarks
1	94-G(C)	327	Van Lierde, J.	293.3		0.000	8.84	Oscillating disk viscometer; $L_1 =$
						0.119	12.03	0.070%, $L_2 = 0.106\%$, $L_3 = 0.225\%$.
						0.191	13.28	
						0.274	14.46	
						0.386	15.62	
						0.494	16.30	
						0.613	16.86	
						1.000	17.68	

TABLE 94-G(C)S.   SMOOTHED VISCOSITY VALUES AS A FUNCTION OF COMPOSITION FOR GASEOUS
CARBON MONOXIDE-HYDROGEN MIXTURES

Mole Fraction of CO	293.3 K [Ref. 327]
0.00	8.84
0.05	10.42
0.10	11.66
0.15	12.64
0.20	13.46
0.25	14.16
0.30	14.78
0.35	15.28
0.40	15.70
0.45	16.06
0.50	16.36
0.55	16.61
0.60	16.82
0.65	17.00
0.70	17.15
0.75	17.28
0.80	17.39
0.85	17.48
0.90	17.56
0.95	17.63
1.00	17.68

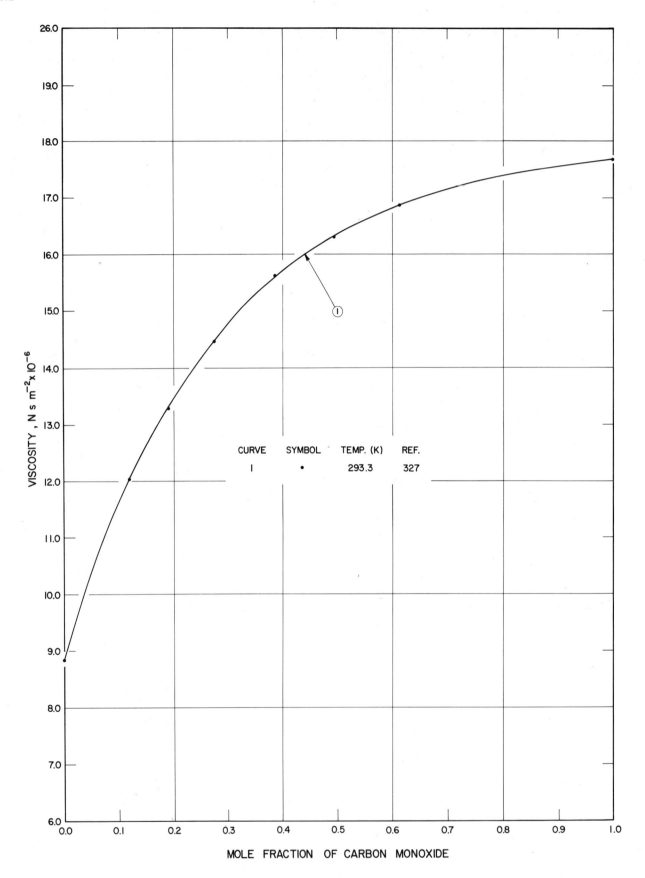

FIGURE 94-G(C). VISCOSITY DATA AS A FUNCTION OF COMPOSITION
FOR GASEOUS CARBON MONOXIDE-HYDROGEN MIXTURES

TABLE 95-L(T)E. EXPERIMENTAL VISCOSITY DATA AS A FUNCTION OF TEMPERATURE FOR LIQUID CARBON MONOXIDE-NITROGEN MIXTURES

Cur. No.	Fig. No.	Ref. No.	Author(s)	Temp. (K)	Pressure	Mole Fraction of N_2	Viscosity (N s m^{-2} x 10^{-6})	Remarks
1	95-L(C)	344	Gerf, S. F. and Galkov, G. I.	73.2 75.2 77.8 82.8 90.1 99.6 111.6 129.6		0.000	224.0 203.0 186.0 165.0 146.0 116.0 100.0 66.0	Mixture analysis ±0.2%; oscillating cylinder viscometer; n accuracy ±3%.
2	95-L(C)	344	Gerf, S. F. and Galkov, G. I.	76.4 82.0 90.1 100.8 111.6		0.252	183.0 151.0 132.0 109.0 89.0	Same remarks as for curve 1.
3	95-L(C)	344	Gerf, S. F. and Galkov, G. I.	77.2 83.0 90.1 100.0 111.6		0.453	171.0 147.0 127.0 108.0 86.0	Same remarks as for curve 1.
4	95-L(C)	344	Gerf, S. F. and Galkov, G. I.	81.0 90.1 111.6		0.687	153.0 123.0 84.0	Same remarks as for curve 1.

TABLE 95-L(T)S. SMOOTHED VISCOSITY VALUES AS A FUNCTION OF TEMPERATURE FOR LIQUID CARBON MONOXIDE-NITROGEN MIXTURES

Temp. (K)	Mole Fraction of Nitrogen			
	0.000 [Ref. 344]	0.252 [Ref. 344]	0.453 [Ref. 344]	0.687 [Ref. 344]
75	203			
80	178	162	156	
85	160	145	140	137
90	145.5	132	128	123
95	132.5	121	117	112.5
100	121	111	107.5	103
105	111	101	98	95
110	101	92	89	87
115	92			
120	83			
125	75			
130	67			

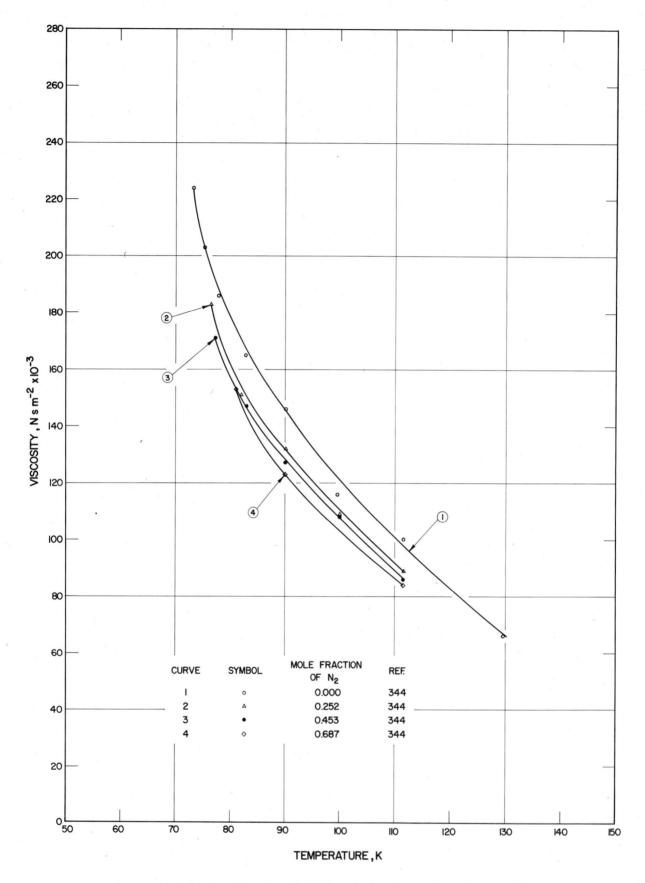

FIGURE 95-L(T). VISCOSITY DATA AS A FUNCTION OF TEMPERATURE
FOR LIQUID CARBON MONOXIDE-NITROGEN MIXTURES

TABLE 95-G(C)E.   EXPERIMENTAL VISCOSITY DATA AS A FUNCTION OF COMPOSITION FOR GASEOUS
CARBON MONOXIDE-NITROGEN MIXTURES

Cur. No.	Fig. No.	Ref. No.	Author(s)	Temp. (K)	Pressure	Mole Fraction of N_2	Viscosity ($N\ s\ m^{-2} \times 10^{-6}$)	Remarks
1	95-G(C)	227	Trautz, M. and Melster, A.	300		1.0000	17.81	Capillary method, r = 0.2019 mm; L_1 = 0.080%, L_2 = 0.103%, L_3 = 0.169%.
						0.8154	17.82	
						0.6030	17.81	
						0.3432	17.75	
						0.1629	17.74	
						0.0000	17.76	
2	95-G(C)	227	Trautz, M. and Melster, A.	400		1.0000	21.90	Same remarks as for curve 1 except L_1 = 0.069%, L_2 = 0.125%, L_3 = 0.275%.
						0.8154	21.86	
						0.6030	21.83	
						0.3432	21.91	
						0.1629	21.84	
						0.0000	21.83	
3	95-G(C)	227	Trautz, M. and Melster, A.	500		1.0000	25.60	Same remarks as for curve 1 except L_1 = 0.054%, L_2 = 0.080%, L_3 = 0.157%.
						0.8154	25.60	
						0.6030	25.58	
						0.3432	25.49	
						0.1629	25.51	
						0.0000	25.48	
4	95-G(C)	227	Trautz, M. and Melster, A.	550		1.0000	27.27	Same remarks as for curve 1 except L_1 = 0.079%, L_2 = 0.100%, L_3 = 0.147%.
						0.8154	27.21	
						0.6030	27.19	
						0.3432	27.22	
						0.1629	27.19	
						0.0000	27.14	

TABLE 95-G(C)S.   SMOOTHED VISCOSITY VALUES AS A FUNCTION OF COMPOSITION FOR GASEOUS
CARBON MONOXIDE-NITROGEN MIXTURES

Mole Fraction of N_2	300 K [Ref. 227]	400 K [Ref. 227]	500 K [Ref. 227]	550 K [Ref. 227]
0.00	17.76	21.83	25.48	27.14
0.05	17.76	21.84	25.50	27.15
0.10	17.76	21.84	25.50	27.15
0.15	17.76	21.84	25.51	27.15
0.20	17.76	21.84	25.52	27.16
0.25	17.76	21.85	25.52	27.16
0.30	17.76	21.85	25.53	27.16
0.35	17.76	21.85	25.53	27.16
0.40	17.76	21.85	25.54	27.17
0.45	17.77	21.86	25.54	27.18
0.50	17.77	21.86	25.54	27.18
0.55	17.78	21.86	25.55	27.19
0.60	17.78	21.86	25.55	27.20
0.65	17.78	21.86	25.56	27.21
0.70	17.79	21.87	25.56	27.22
0.75	17.79	21.88	25.56	27.23
0.80	17.79	21.88	25.57	27.24
0.85	17.80	21.88	25.58	27.24
0.90	17.80	21.88	25.58	27.25
0.95	17.80	21.89	25.59	27.26
1.00	17.81	21.90	25.60	27.27

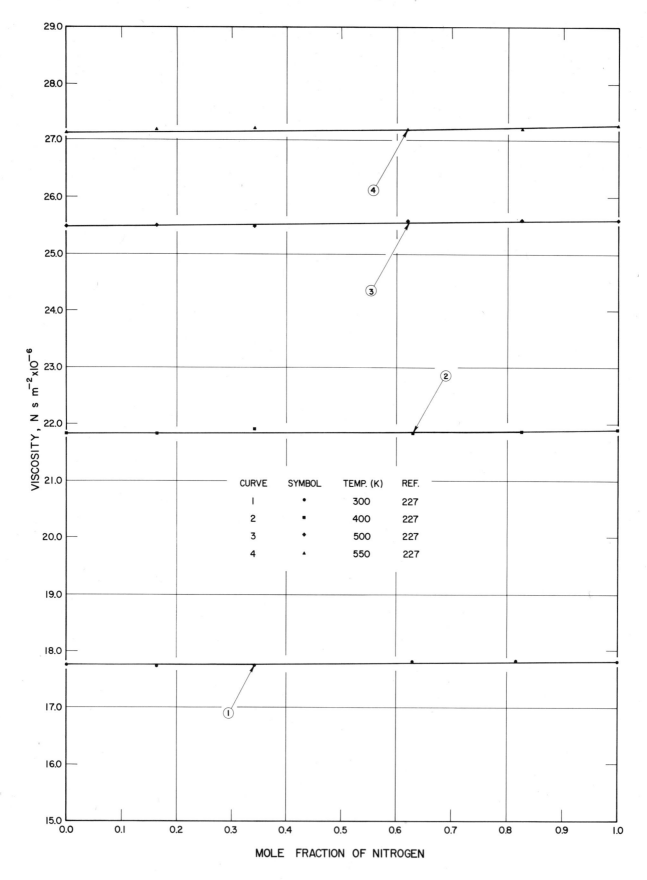

FIGURE 95-G(C). VISCOSITY DATA AS A FUNCTION OF COMPOSITION
FOR GASEOUS CARBON MONOXIDE-NITROGEN MIXTURES

TABLE 96-G(C)E. EXPERIMENTAL VISCOSITY DATA AS A FUNCTION OF COMPOSITION FOR GASEOUS CARBON MONOXIDE-OXYGEN MIXTURES

Cur. No.	Fig. No.	Ref. No.	Author(s)	Temp. (K)	Pressure (atm)	Mole Fraction of O_2	Viscosity (N s m^{-2} x 10^{-6})	Remarks
1	96-G(C)	227	Trautz, M. and Melster, A.	300		0.0000	17.76	Capillary method, r = 0.2019 mm; L_1 = 0.045%, L_2 = 0.077%, L_3 = 0.158%.
						0.2337	18.41	
						0.4201	19.00	
						0.7733	19.98	
						1.0000	20.57	
3	96-G(C)	227	Trautz, M. and Melster, A.	400		0.0000	21.83	Same remarks as for curve 1 except L_1 = 0.065%, L_2 = 0.103%, L_3 = 0.167%.
						0.2337	22.68	
						0.4201	23.43	
						0.7733	24.82	
						1.0000	25.68	
3	96-G(C)	227	Trautz, M. and Melster, A.	500		0.0000	25.48	Same remarks as for curve 1 except L_1 = 0.036%, L_2 = 0.058%, L_3 = 0.110%.
						0.2337	26.50	
						0.4201	27.41	
						0.7733	29.08	
						1.0000	30.17	

TABLE 96-G(C)S. SMOOTHED VISCOSITY VALUES AS A FUNCTION OF COMPOSITION FOR GASEOUS CARBON MONOXIDE-OXYGEN MIXTURES

Mole Fraction of O_2	300 K [Ref. 227]	400 K [Ref. 227]	500 K [Ref. 227]
0.00	17.76	21.83	25.48
0.05	17.90	22.01	25.69
0.10	18.05	22.20	25.90
0.15	18.19	22.40	26.12
0.20	18.33	22.58	26.34
0.25	18.48	22.78	26.58
0.30	18.62	22.96	26.81
0.35	18.76	23.16	27.05
0.40	18.91	23.34	27.28
0.45	19.06	23.54	27.52
0.50	19.21	23.72	27.76
0.55	19.35	23.92	28.00
0.60	19.50	24.12	28.24
0.65	19.64	24.30	28.47
0.70	19.78	24.50	28.71
0.75	19.92	24.69	28.95
0.80	20.07	24.88	29.20
0.85	20.19	25.08	29.44
0.90	20.32	25.28	29.68
0.95	20.45	25.47	29.92
1.00	20.57	25.68	30.17

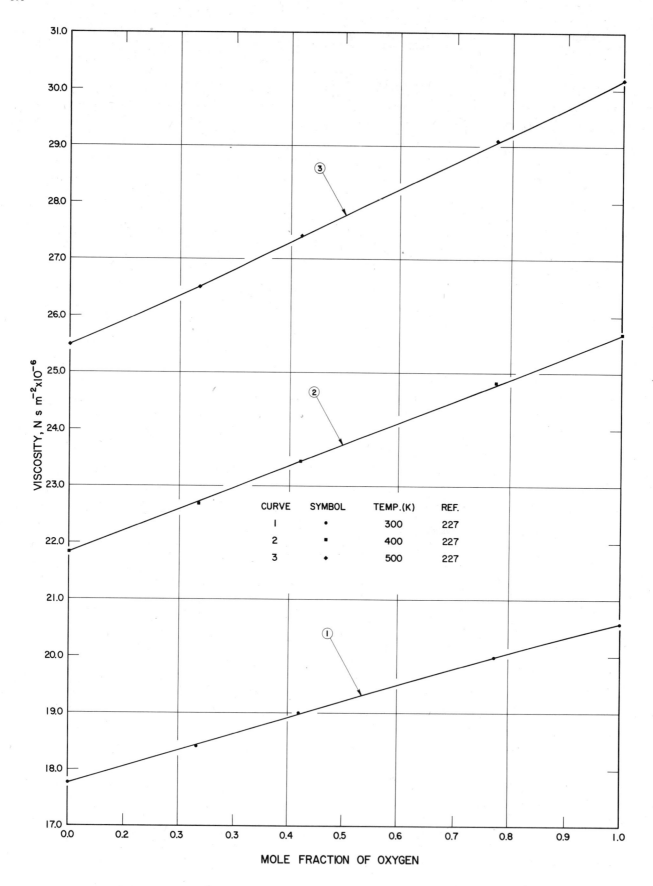

FIGURE 96-G(C). VISCOSITY DATA AS A FUNCTION OF COMPOSITION
FOR GASEOUS CARBON MONOXIDE-OXYGEN MIXTURES

TABLE 97-L(C)E. EXPERIMENTAL VISCOSITY DATA AS A FUNCTION OF COMPOSITION FOR LIQUID
CARBON TETRACHLORIDE-OCTAMETHYLCYCLOTETRASILOXANE MIXTURES

Cur. No.	Fig. No.	Ref. No.	Author(s)	Temp. (K)	Pressure (atm)	Mole Fraction of $[OSi(CH_3)_2]_4$	Viscosity (N s m^{-2} x 10^{-6})	Remarks
1	97-L(C)	360	Marsh, K.N.	291.2		0.0000	1001.0	Ostwald viscometer, relative measurements; L_1 = 0.497%, L_2 = 0.809%, L_3 = 1.783%.
						0.1780	1256.0	
						0.3227	1448.0	
						0.5718	1798.0	
						0.7258	2036.0	
						0.8618	2268.0	
						0.9815	2488.0	
						1.0000	2520.0	
2	97-L(C)	360	Marsh, K.N.	298.2		0.0000	901.0	Same remarks as for curve 1 except L_1 = 0.331%, L_2 = 0.501%, L_3 = 1.359%.
						0.1089	1044.0	
						0.1965	1140.0	
						0.2890	1245.0	
						0.4288	1407.0	
						0.5841	1595.0	
						0.6590	1694.0	
						0.8443	1950.0	
						0.9264	2073.0	
						0.9773	2147.0	
						1.0000	2190.0	
3	97-L(C)	360	Marsh, K.N.	308.2		0.0000	781.0	Same remarks as for curve 1 except L_1 = 0.316%, L_2 = 0.469%, L_3 = 0.942%.
						0.1756	964.0	
						0.3239	1101.0	
						0.5732	1339.0	
						0.7290	1493.0	
						0.8636	1646.0	
						0.9817	1786.0	
						1.0000	1806.0	
4	97-L(C)	360	Marsh, K.N.	318.2		0.0000	686.6	Same remarks as for curve 1 except L_1 = 0.433%, L_2 = 0.699%, L_3 = 1.687%.
						0.1779	844.0	
						0.3249	956.0	
						0.5816	1148.0	
						0.7307	1270.0	
						0.8652	1388.0	
						0.9821	1498.0	
						1.0000	1514.0	

TABLE 97-L(C)S. SMOOTHED VISCOSITY VALUES AS A FUNCTION OF COMPOSITION FOR LIQUID
CARBON TETRACHLORIDE-OCTAMETHYLCYCLOTETRASILOXANE MIXTURES

Mole Fraction of $[OSi(CH_3)_2]_4$	291.2 K [Ref. 360]	298.2 K [Ref. 360]	308.2 K [Ref. 360]	318.2 K [Ref. 360]
0.00	1001.0	901.0	781.0	686.6
0.05	1068.0	963.2	834.0	726.5
0.10	1131.5	1020.0	880.0	768.0
0.15	1196.0	1078.8	930.0	808.0
0.20	1261.5	1135.0	979.0	828.0
0.25	1330.0	1194.0	1036.0	888.0
0.30	1400.0	1252.0	1076.0	928.0
0.35	1479.0	1312.5	1124.0	968.5
0.40	1540.0	1371.5	1172.5	1008.0
0.45	1612.0	1432.5	1222.0	1048.0
0.50	1690.0	1496.0	1274.0	1088.0
0.55	1764.0	1559.5	1322.5	1129.0
0.60	1844.0	1622.0	1372.0	1169.0
0.65	1924.5	1688.0	1422.0	1211.0
0.70	2004.8	1757.0	1475.0	1252.0
0.75	2088.0	1821.5	1526.0	1293.0
0.80	2172.0	1891.0	1580.0	1334.0
0.85	2260.0	1964.0	1638.0	1379.0
0.90	2347.5	2039.0	1692.0	1424.0
0.95	2436.0	2112.5	1751.0	1470.0
1.00	2520.0	2190.0	1806.0	1514.0

400

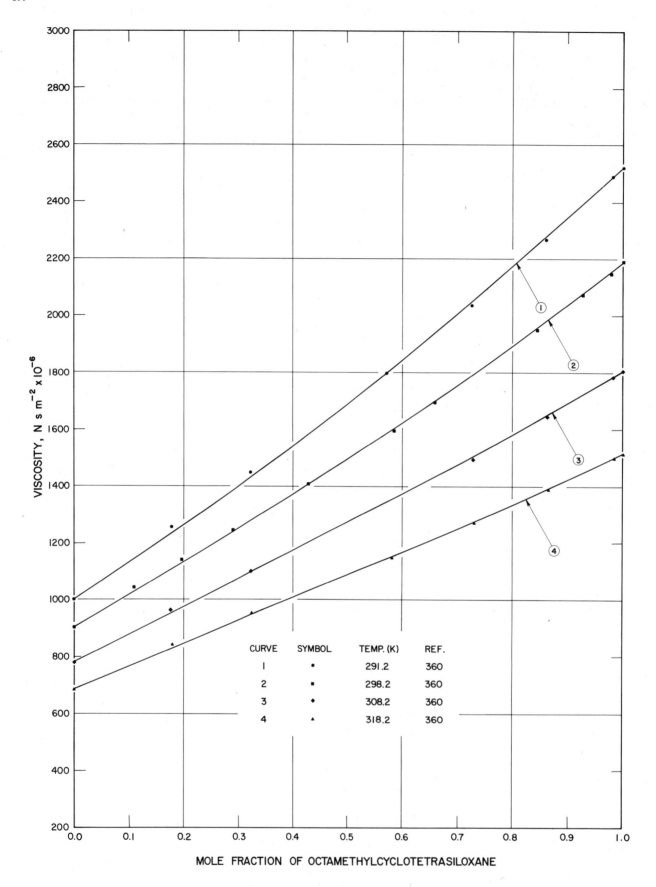

CURVE	SYMBOL	TEMP. (K)	REF.
1	•	291.2	360
2	■	298.2	360
3	♦	308.2	360
4	▲	318.2	360

MOLE FRACTION OF OCTAMETHYLCYCLOTETRASILOXANE

FIGURE 97-L(C).   VISCOSITY DATA AS A FUNCTION OF COMPOSITION
FOR LIQUID CARBON TETRACHLORIDE – OCTAMETHYLCYCLOTETRASILOXANE MIXTURES

TABLE 98-G(D)E.   EXPERIMENTAL VISCOSITY DATA AS A FUNCTION OF DENSITY FOR GASEOUS CARBON TETRAFLUORIDE-METHANE MIXTURES

Cur. No.	Fig. No.	Ref. No.	Author(s)	Mole Fraction of CF_4	Temp. (K)	Density (g cm^{-3})	Viscosity (N s m^{-2} x 10^{-6})	Remarks
1	98-G(D)	338	DeWitt, K.J. and Thodos, G.	1.0000	323.3	0.1259	20.57	Unsteady state transpiration type capillary viscometer; purity of the gases and accuracy of the data not specified.
						0.2803	23.93	
						0.4312	28.49	
						0.5743	34.14	
						0.6891	40.08	
						0.7870	46.12	
						0.8572	51.30	
						0.9162	56.24	
						0.9628	60.68	
						1.0042	64.95	
						1.0394	68.95	
						1.0702	72.68	
2	98-G(D)	338	DeWitt, K.J. and Thodos, G.	0.7330	323.5	0.0977	19.57	Same remarks as for curve 1.
						0.2040	22.01	
						0.3140	25.13	
						0.4252	29.29	
						0.5194	33.63	
						0.5975	38.02	
						0.6609	42.21	
						0.7126	46.13	
						0.7552	49.61	
						0.7929	53.01	
						0.8281	56.61	
						0.8568	60.60	
3	98-G(D)	338	DeWitt, K.J. and Thodos, G.	0.5390	323.4	0.0765	18.66	Same remarks as for curve 1.
						0.1553	20.45	
						0.2457	23.15	
						0.3290	26.32	
						0.4003	29.55	
						0.4693	33.33	
						0.5255	36.96	
						0.5714	40.27	
						0.6083	43.12	
						0.6432	46.10	
						0.6731	49.11	
						0.6991	51.96	
4	98-G(D)	338	DeWitt, K.J. and Thodos, G.	0.2500	323.3	0.0463	16.35	Same remarks as for curve 1.
						0.0956	17.79	
						0.1481	19.49	
						0.1993	21.63	
						0.2459	23.98	
						0.2903	26.61	
						0.3291	29.29	
						0.3626	31.95	
						0.3892	34.32	
						0.4139	36.51	
						0.4363	38.73	
						0.4566	41.09	
5	98-G(D)	338	DeWitt, K.J. and Thodos, G.	1.0000	373.4	0.1003	22.55	Same remarks as for curve 1.
						0.2125	24.71	
						0.3239	27.52	
						0.4339	30.96	
						0.5289	34.77	
						0.6161	38.69	
						0.6878	42.47	
						0.7516	46.26	
						0.8040	49.81	
						0.8510	53.29	
						0.8898	56.46	
						0.9298	59.74	
6	98-G(D)	338	DeWitt, K.J. and Thodos, G.	0.7330	373.8	0.0808	21.74	Same remarks as for curve 1.
						0.1611	23.18	
						0.2504	25.48	
						0.3261	27.85	
						0.4038	30.65	
						0.4737	33.70	
						0.5336	36.68	
						0.5854	39.59	
						0.6315	42.39	
						0.6699	45.04	
						0.7065	47.82	
						0.7380	50.43	

TABLE 98-G(D)E.   EXPERIMENTAL VISCOSITY DATA AS A FUNCTION OF DENSITY FOR GASEOUS
CARBON TETRAFLUORIDE-METHANE MIXTURES (continued)

Cur. No.	Fig. No.	Ref. No.	Author(s)	Mole Fraction of CF_4	Temp. (K)	Density (g cm^{-3})	Viscosity (N s m^{-2} x 10^{-6})	Remarks
7	98-G(D)	338	DeWitt, K.J. and Thodos, G.	0.5390	373.9	0.0556	20.50	Same remarks as for curve 1.
						0.1275	21.98	
						0.1941	23.75	
						0.2598	25.88	
						0.3197	28.09	
						0.3749	30.55	
						0.4236	32.97	
						0.4682	35.39	
						0.5079	37.89	
						0.5435	40.37	
						0.5735	42.65	
						0.6018	45.00	
8	98-G(D)	338	DeWitt, K.J. and Thodos, G.	0.2500	373.5	0.0383	18.27	Same remarks as for curve 1.
						0.0793	19.14	
						0.1207	20.49	
						0.1586	21.87	
						0.1998	23.64	
						0.2326	25.25	
						0.2660	27.09	
						0.2964	28.99	
						0.3218	30.73	
						0.3466	32.62	
						0.3688	34.46	
						0.3900	36.39	
9	98-G(D)	338	DeWitt, K. J. and Thodos, G.	1.0000	422.9	0.0878	24.43	Same remarks as for curve 1.
						0.1782	26.12	
						0.2667	28.20	
						0.3520	30.55	
						0.4339	33.36	
						0.5096	36.20	
						0.5747	39.00	
						0.6323	41.74	
						0.6856	44.61	
						0.7331	47.37	
						0.7727	49.92	
						0.8141	52.77	
10	98-G(D)	338	DeWitt, K.J. and Thodos, G.	0.7330	422.3	0.0676	23.61	Same remarks as for curve 1.
						0.1359	24.77	
						0.2102	26.51	
						0.2705	28.20	
						0.3346	30.21	
						0.3949	32.43	
						0.4480	34.67	
						0.4974	36.95	
						0.5409	39.09	
						0.5768	41.30	
						0.6163	43.51	
						0.6505	45.78	
11	98-G(D)	338	DeWitt, K.J. and Thodos, G.	0.5390	422.3	0.0555	22.66	Same remarks as for curve 1.
						0.1105	23.58	
						0.1645	24.97	
						0.2169	26.50	
						0.2674	28.13	
						0.3142	29.93	
						0.3578	31.78	
						0.3984	33.62	
						0.4353	35.52	
						0.4680	37.41	
						0.4973	39.24	
						0.5254	41.12	
12	98-G(D)	338	DeWitt, K.J. and Thodos, G.	0.2500	423.8	0.0301	19.94	Same remarks as for curve 1.
						0.0681	20.71	
						0.1028	21.74	
						0.1353	22.83	
						0.1676	24.07	
						0.1972	25.35	
						0.2255	26.71	
						0.2517	28.11	
						0.2760	29.52	
						0.2990	30.98	
						0.3211	32.51	
						0.3391	33.85	

TABLE 98-G(D)E.   EXPERIMENTAL VISCOSITY DATA AS A FUNCTION OF DENSITY FOR GASEOUS
CARBON TETRAFLUORIDE-METHANE MIXTURES (continued)

Cur. No.	Fig. No.	Ref. No.	Author(s)	Mole Fraction of CF$_4$	Temp. (K)	Density (g cm^{-3})	Viscosity (N s m^{-2} x10^{-6})	Remarks
13	98-G(D)	338	DeWitt, K.J. and Thodos, G.	1.0000	473.9	0.0775	26.39	Same remarks as for curve 1.
						0.1531	27.68	
						0.2310	29.26	
						0.3041	31.17	
						0.3736	33.37	
						0.4383	35.52	
						0.4964	37.72	
						0.5511	39.97	
						0.6007	42.25	
						0.6420	44.30	
						0.6830	46.49	
						0.7252	48.93	
14	98-G(D)	338	DeWitt, K.J. and Thodos, G.	0.7330	473.4	0.0597	25.50	Same remarks as for curve 1.
						0.1182	26.55	
						0.1814	27.73	
						0.2348	29.11	
						0.2896	30.67	
						0.3418	32.39	
						0.3901	34.19	
						0.4327	35.92	
						0.4733	37.63	
						0.5074	39.27	
						0.5451	41.25	
						0.5762	42.99	
15	98-G(D)	338	DeWitt, K.J. and Thodos, G.	0.5390	474.1	0.0488	24.44	Same remarks as for curve 1.
						0.0965	25.35	
						0.1446	26.28	
						0.1878	27.51	
						0.2318	28.69	
						0.2725	30.21	
						0.3115	31.68	
						0.3482	33.15	
						0.3807	34.63	
						0.4129	36.23	
						0.4414	37.76	
						0.4707	39.45	
16	98-G(D)	338	DeWitt, K.J. and Thodos, G.	0.2500	472.6	0.0296	21.54	Same remarks as for curve 1.
						0.0597	22.31	
						0.0890	23.04	
						0.1181	23.94	
						0.1457	24.92	
						0.1714	25.93	
						0.1964	27.02	
						0.2204	28.16	
						0.2425	29.31	
						0.2627	30.44	
						0.2830	31.66	
						0.3007	32.81	

TABLE 98-G(D)S.   SMOOTHED VISCOSITY VALUES AS A FUNCTION OF DENSITY FOR GASEOUS CARBON TETRAFLUORIDE-METHANE MIXTURES

Density (g cm^{-3})	Mole Fraction of Carbon Tetrafluoride							
	0.2500 (323.3 K) [Ref. 338]	1.0000 (323.3 K) [Ref. 338]	0.5390 (323.4 K) [Ref. 338]	0.7330 (323.5 K) [Ref. 338]	1.0000 (373.4 K) [Ref. 338]	0.2500 (373.5 K) [Ref. 338]	0.7330 (373.8 K) [Ref. 338]	0.5390 (373.9 K) [Ref. 338]
0.050	16.43		18.10			18.55		20.40
0.100	17.90		19.13		22.60	19.90	22.09	21.35
0.150	19.62	22.08		19.64		21.55		
0.200	21.72	24.02	21.72	22.00	24.45	23.55	24.07	23.90
0.250	24.20					26.05	25.40	25.52
0.300	27.23	24.45	25.08	24.75	26.90	29.10	26.90	27.40
0.350	30.90					32.75		
0.400	35.23	27.42	29.42	28.26	29.88	37.50	30.44	31.70
0.450	40.44							
0.500		32.09	35.03	32.65	33.48		34.90	37.30
0.600		35.05	42.32	37.20	37.85		40.32	40.30
0.700		40.08	51.18	45.00	43.00		47.10	
0.750							51.47	
0.800		47.10		53.80	49.28			
0.850				59.72				
0.900		54.75			57.13			
0.950		59.35						

Density (g cm^{-3})	Mole Fraction of Carbon Tetrafluoride							
	0.5390 (422.3 K) [Ref. 338]	0.7330 (422.3 K) [Ref. 338]	1.0000 (422.9 K) [Ref. 338]	0.2500 (423.8 K) [Ref. 338]	0.2500 (472.6 K) [Ref. 338]	0.7330 (473.4 K) [Ref. 338]	1.0000 (473.9 K) [Ref. 338]	0.5390 (474.1 K) [Ref. 338]
0.025				19.80	21.48			
0.050	22.58	23.48	23.92	20.32	22.05	25.38		24.43
0.075					22.66			
0.100	23.40	24.15	24.72	21.60	23.40	26.20	26.75	25.37
0.125					24.22			
0.150	24.30			23.30	25.10	27.14		26.50
0.175					26.08			
0.200	25.92	26.20	26.62	25.44	27.17	28.25	28.60	27.80
0.225				26.70	28.33			
0.250	28.00	27.58		28.00	29.65	29.55	29.70	29.39
0.275					31.10			
0.300	29.32	29.10	29.10	31.00	32.80	31.08	31.05	31.20
0.350				33.88		32.70		33.25
0.400	33.70	32.65	32.08			34.53	34.20	35.58
0.450	35.77					36.58		38.24
0.500	39.38	37.00	35.80			38.87	37.80	
0.550						41.50		
0.600		42.48	40.20					42.17
0.650		45.72						
0.700			45.25				47.50	
0.725							48.92	
0.800			51.48					

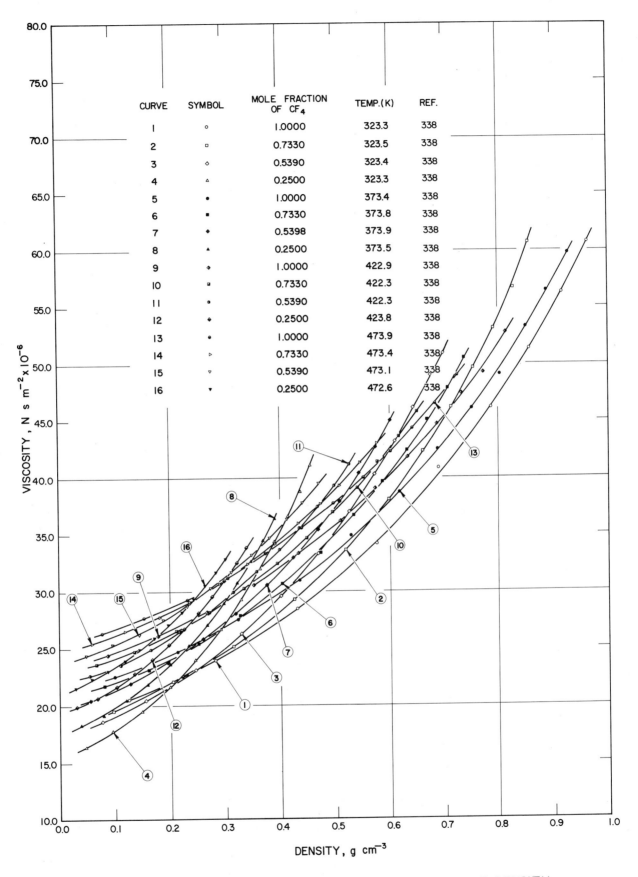

The following data table appears within the figure:

CURVE	SYMBOL	MOLE FRACTION OF CF_4	TEMP.(K)	REF.
1	○	1.0000	323.3	338
2	□	0.7330	323.5	338
3	◇	0.5390	323.4	338
4	△	0.2500	323.3	338
5	●	1.0000	373.4	338
6	■	0.7330	373.8	338
7	◆	0.5398	373.9	338
8	▲	0.2500	373.5	338
9	◈	1.0000	422.9	338
10	◨	0.7330	422.3	338
11	◉	0.5390	422.3	338
12	◆	0.2500	423.8	338
13	●	1.0000	473.9	338
14	▷	0.7330	473.4	338
15	▽	0.5390	473.1	338
16	▼	0.2500	472.6	338

FIGURE 98 - G(D). VISCOSITY DATA AS A FUNCTION OF DENSITY FOR GASEOUS CARBON TETRAFLUORIDE-METHANE MIXTURES

TABLE 99-G(C)E. EXPERIMENTAL VISCOSITY DATA AS A FUNCTION OF COMPOSITION FOR GASEOUS
CARBON TETRAFLUORIDE-SULFUR HEXAFLUORIDE MIXTURES

Cur. No.	Fig. No.	Ref. No.	Author(s)	Temp. (K)	Pressure (mm Hg)	Mole Fraction of SF$_6$	Viscosity (N s m^{-2} x 10^{-6})	Remarks
1	99-G(C)	339	Raw, C.J.G. and Tang, H.	303.1		1.000	15.90	SF$_6$: 95 pure, CF$_4$: 99 pure, gases further purified by vacuum distillation; transpiration type capillary flow constant volume gas viscometer, relative measurements; accuracy ± 1.0%; L$_1$ = 0.152%, L$_2$ = 0.260%, L$_3$ = 0.560%.
						0.743	15.99	
						0.509	16.15	
						0.246	16.43	
						0.000	17.67	
2	99-G(C)	339	Raw, C.J.G. and Tang, H.	313.1		1.000	16.36	Same remarks as for curve 1 except L$_1$ = 0.187%, L$_2$ = 0.304%, L$_3$ = 0.576%.
						0.743	16.46	
						0.509	16.59	
						0.246	16.89	
						0.000	18.17	
3	99-G(C)	339	Raw, C.J.G. and Tang, H.	329.1		1.000	17.06	Same remarks as for curve 1 except L$_1$ = 0.093%, L$_2$ - 0.165%, L$_3$ = 0.351%.
						0.743	17.17	
						0.509	17.30	
						0.246	17.59	
						0.000	18.94	
4	99-G(C)	339	Raw, C.J.G. and Tang, H.	342.0		1.000	17.59	Same remarks as for curve 1 except L$_1$ = 0.249%, L$_2$ = 0.420%, L$_3$ = 0.857%.
						0.743	17.71	
						0.509	17.89	
						0.246	18.16	
						0.000	19.57	

TABLE 99-G(C)S. SMOOTHED VISCOSITY VALUES AS A FUNCTION OF COMPOSITION FOR GASEOUS
CARBON TETRAFLUORIDE-SULFUR HEXAFLUORIDE MIXTURES

Mole Fraction of SF$_6$	303.1 K [Ref. 339]	313.1 K [Ref. 339]	329.1 K [Ref. 339]	342.0 K [Ref. 339]
0.00	17.67	18.17	18.94	19.57
0.05	17.33	17.83	18.57	19.16
0.10	17.05	17.54	18.25	18.81
0.15	16.80	17.27	17.98	18.54
0.20	16.60	17.05	17.76	18.32
0.25	16.43	16.85	17.58	18.15
0.30	16.31	16.75	17.44	18.02
0.35	16.23	16.66	17.35	17.92
0.40	16.16	16.59	17.27	17.84
0.45	16.11	16.54	17.22	17.79
0.50	16.07	16.50	17.19	17.74
0.55	16.03	16.47	17.16	17.73
0.60	16.01	16.45	17.15	17.69
0.65	16.00	16.43	17.13	17.67
0.70	15.98	16.41	17.12	17.66
0.75	15.96	16.46	17.11	17.64
0.80	15.95	16.39	17.10	17.63
0.85	15.93	16.38	17.09	17.62
0.90	15.92	16.37	17.07	17.61
0.95	15.91	16.36	17.07	17.60
1.00	15.90	16.36	17.06	17.59

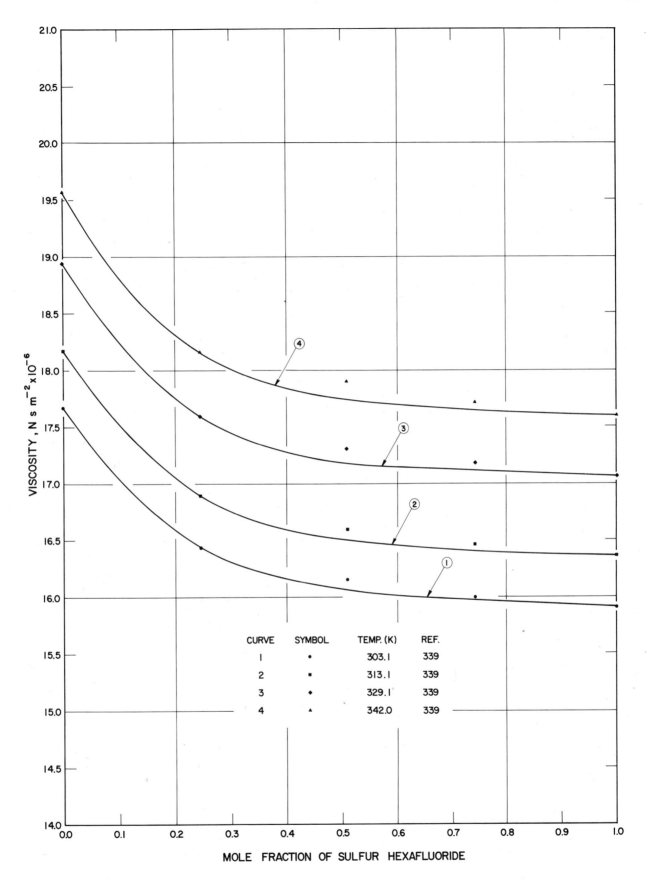

FIGURE 99-G(C). VISCOSITY DATA AS A FUNCTION OF COMPOSITION
FOR GASEOUS CARBON TETRAFLUORIDE-SULFUR HEXAFLUORIDE MIXTURES

TABLE 100-L(C)E. EXPERIMENTAL VISCOSITY DATA AS A FUNCTION OF COMPOSITION FOR LIQUID
CYCLOHEXANE - n-HEXANE MIXTURES

Cur. No.	Fig. No.	Ref. No.	Author(s)	Temp. (K)	Pressure (atm)	Mole Fraction of C_6H_{14}	Viscosity (N s m^{-2} x 10^{-6})	Remarks
1	100-L(C)	355	Ridgway, K. and Butler, P. A.	298.2		1.0000	300.8	Cyclohexane: supplied by B. D. H. and n-Hexane by Phillips Petroleum Co.; Ostwald viscometer; precision 0.1%; L_1 = 0.000%, L_2 = 0.000%, L_3 = 0.000%.
						0.8286	340.5	
						0.7258	367.0	
						0.5502	423.4	
						0.4127	484.6	
						0.2480	588.7	
						0.0966	734.7	
						0.0000	869.0	

TABLE 100-L(C)S. SMOOTHED VISCOSITY VALUES AS A FUNCTION OF COMPOSITION FOR LIQUID
CYCLOHEXANE - n-HEXANE MIXTURES

Mole Fraction of C_6H_{14}	298.2 K [Ref. 355]
0.00	869.0
0.05	797.0
0.10	731.0
0.15	673.8
0.20	625.4
0.25	585.0
0.30	550.0
0.35	519.6
0.40	491.8
0.45	467.2
0.50	445.0
0.55	425.0
0.60	406.5
0.65	389.9
0.70	374.6
0.75	360.8
0.80	347.8
0.85	335.8
0.90	323.8
0.95	312.5
1.00	300.8

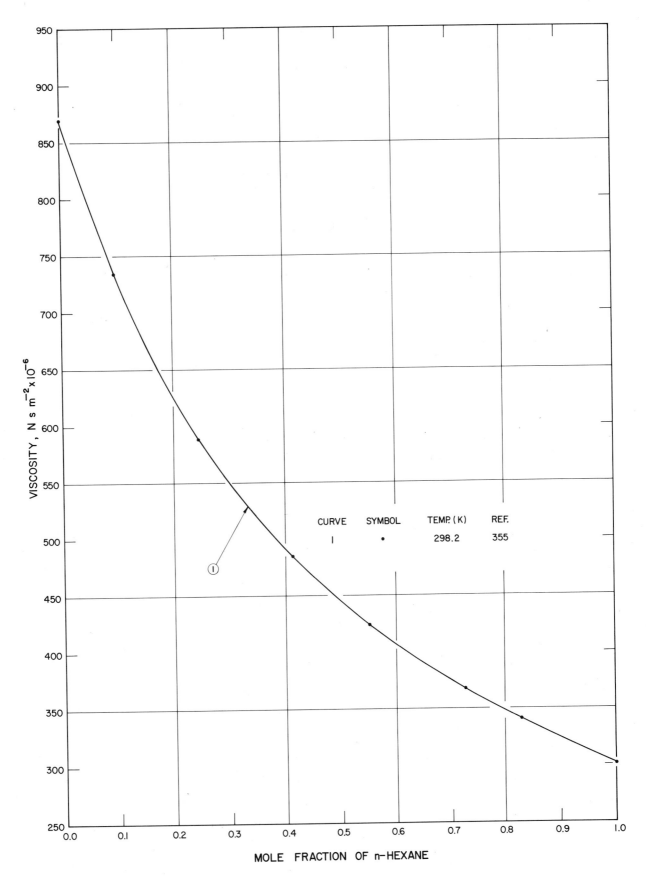

FIGURE 100-L(C). VISCOSITY DATA AS A FUNCTION OF COMPOSITION
FOR LIQUID CYCLOHEXANE - n-HEXANE MIXTURES

TABLE 101-L(D)E.　EXPERIMENTAL VISCOSITY DATA AS A FUNCTION OF DENSITY FOR LIQUID
n-DECANE-METHANE MIXTURES

Cur. No.	Fig. No.	Ref. No.	Author(s)	Mole Fraction of $n-C_{10}H_{22}$	Temp. (K)	Density $(g\,cm^{-3})$	Viscosity $(N\,s\,m^{-2} \times 10^{-6})$	Remarks
1	101-L(D)	353	Lee, A. L., Gonzalez, M. H., and Eakin, B. E.	0.700	311.0	0.6838	544.49	n-Decane: 99 pure, methane: 99.6 pure, 0.1 nitrogen and remainder as ethane, propane, n-butane, and carbon dioxide.
						0.6874	560.71	
						0.6911	578.91	
						0.6943	589.28	
						0.6975	611.01	
						0.7001	631.55	
						0.7031	659.72	
						0.7052	673.01	
2	101-L(D)	353	Lee, A. L., et al.	0.500	311.0	0.6453	435.17	Same remarks as for curve 1.
						0.6483	453.02	
						0.6520	470.17	
						0.6560	483.04	
						0.6616	513.47	
3	101-L(D)	353	Lee, A. L., et al.	0.300	311.0	0.5712	268.34	Same remarks as for curve 1.
						0.5808	279.33	
						0.5887	323.59	
4	101-L(D)	353	Lee, A. L., et al.	0.700	344.0	0.6556	369.44	Same remarks as for curve 1.
						0.6584	374.32	
						0.6616	380.07	
						0.6671	401.88	
						0.6719	418.23	
						0.6788	446.48	
						0.6855	479.80	
5	101-L(D)	353	Lee, A. L., et al.	0.500	344.0	0.6240	330.75	Same remarks as for curve 1.
						0.6296	342.60	
						0.6353	356.86	
						0.6420	378.83	
6	101-L(D)	353	Lee, A. L., et al.	0.300	344.0	0.5384	197.75	Same remarks as for curve 1.
						0.5459	213.25	
						0.5575	227.54	
						0.5674	231.88	
7	101-L(D)	353	Lee, A. L., et al.	0.700	378.0	0.6258	281.97	Same remarks as for curve 1.
						0.6289	286.14	
						0.6313	290.89	
						0.6339	295.95	
						0.6368	304.25	
						0.6388	309.01	
						0.6420	316.26	
						0.6459	326.56	
						0.6492	338.89	
						0.6535	350.18	
						0.6579	365.14	
8	101-L(D)	353	Lee, A. L., et al.	0.500	378.0	0.5952	260.38	Same remarks as for curve 1.
						0.6000	270.00	
						0.6055	274.33	
						0.6137	295.91	
9	101-L(D)	353	Lee, A. L., et al.	0.300	378.0	0.5092	168.48	Same remarks as for curve 1.
						0.5231	182.94	
						0.5351	197.12	
10	101-L(D)	353	Lee, A. L., et al.	0.700	411.0	0.5963	214.84	Same remarks as for curve 1.
						0.6008	221.82	
						0.6048	226.50	
						0.6074	231.41	
						0.6098	238.69	
						0.6123	241.75	
						0.6150	247.33	
						0.6200	256.77	
						0.6251	267.69	
11	101-L(D)	353	Lee, A. L., et al.	0.500	411.0	0.5614	195.13	Same remarks as for curve 1.
						0.5682	204.29	
						0.5744	210.69	
						0.5802	218.11	
						0.5894	231.79	
12	101-L(D)	353	Lee, A. L., et al.	0.300	411.0	0.4691	129.56	Same remarks as for curve 1.
						0.4790	137.54	
						0.4960	149.56	
						0.5098	155.43	
13	101-L(D)	353	Lee, A. L., et al.	0.500	444.0	0.5119	141.16	Same remarks as for curve 1.
						0.5193	144.95	
						0.5346	152.64	
						0.5426	159.70	

TABLE 101-L(D)S. SMOOTHED VISCOSITY VALUES AS A FUNCTION OF DENSITY FOR LIQUID n-DECANE-METHANE MIXTURES

Density (g cm^{-3})	Mole Fraction of n-Decane						
	0.700 (311.0 K) [Ref. 353]	0.500 (311.0 K) [Ref. 353]	0.300 (311.0 K) [Ref. 353]	0.700 (344.0 K) [Ref. 353]	0.500 (344.0 K) [Ref. 353]	0.300 (344.0 K) [Ref. 353]	0.700 (378.0 K) [Ref. 353]
0.540						205.5	
0.545						212.0	
0.550						219.0	
0.555						226.5	
0.560							
0.565							
0.570			264.5				
0.575			276.0				
0.580			288.5				
0.585			301.5				
0.590			316.0				
0.625					331.5		
0.630					344.0		288.0
0.635					357.0		300.5
0.640					371.5		314.0
0.645		434.5					328.0
0.650		456.5					342.5
0.655		481.0		365.0			358.0
0.660		506.5		381.0			
0.665				398.0			
0.670				416.0			
0.675				435.5			
0.680				456.0			
0.685	545.0			477.0			
0.690	572.5						
0.695	599.5						
0.700	630.4						
0.705	664.0						

Density (g cm^{-3})	Mole Fraction of n-Decane					
	0.500 (378.0 K) [Ref. 353]	0.300 (378.0 K) [Ref. 353]	0.700 (411.0 K) [Ref. 353]	0.500 (411.0 K) [Ref. 353]	0.300 (411.0 K) [Ref. 353]	0.500 (444.0 K) [Ref. 353]
0.470					129.5	
0.475					132.5	
0.480					135.5	
0.485					139.0	
0.490					142.5	
0.495					146.5	
0.500					150.5	
0.505					155.0	
0.510		169.0				140.0
0.515		173.5				142.5
0.520		178.5				145.0
0.525		184.5				148.0
0.530		190.5				151.0
0.535		197.0				154.5
0.540						158.0
0.545						161.5
0.560				193.5		
0.565				199.0		
0.570				205.0		
0.575				211.0		
0.580				217.5		
0.585				224.5		
0.590				232.0		
0.595	259.5		212.0			
0.600	268.0		219.5			
0.605	277.5		228.5			
0.610	287.5		232.0			
0.615	298.0		246.0			
0.620			256.0			
0.625			266.5			

412

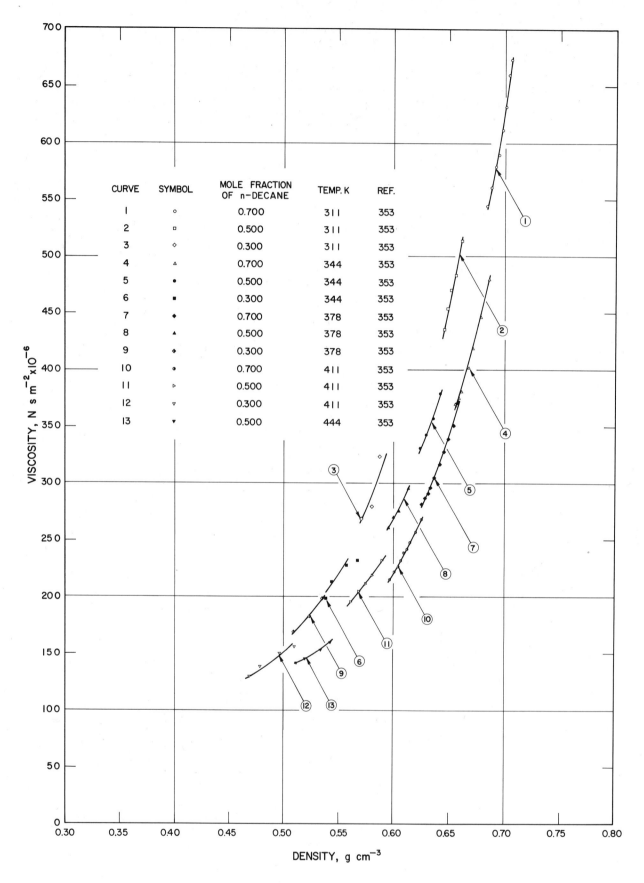

FIGURE 101-L(D). VISCOSITY DATA AS A FUNCTION OF DENSITY
FOR LIQUID n-DECANE - METHANE MIXTURES

TABLE 102-G(C)E.   EXPERIMENTAL VISCOSITY DATA AS A FUNCTION OF COMPOSITION FOR GASEOUS DEUTERIUM-HYDROGEN MIXTURES

Cur. No.	Fig. No.	Ref. No.	Author(s)	Temp. (K)	Pressure (mm Hg)	Mole Fraction of D_2	Viscosity ($N s m^{-2} \times 10^{-6}$)	Remarks
1	102-G(C)	179	Rietveld, A.O., Van Itterbeek, A., and Velds, C.A.	14.4	4-11	0.000 0.269 0.504 0.760 1.000	0.79 0.85 0.90 0.94 1.00	Hydrogen obtained from vapors over liquid hydrogen and then purified by condensation; oscillating disk viscometer; relative measurements; error: ±3% at low temperatures and ±2% at high temperatures; L_1 = 0.000%, L_2 = 0.000%, L_3 = 0.000%.
2	102-G(C)	179	Rietveld, A.O., et al.	20.4	4-11	0.000 0.334 0.677 1.000	1.08 1.19 1.29 1.37	Same remarks as for curve 1 except L_1 = 0.000%, L_2 = 0.000%, L_3 = 0.000%.
3	102-G(C)	179	Rietveld, A.O., et al.	71.5	14-40	0.000 0.248 0.502 0.749 1.000	3.24 3.58 3.90 4.16 4.44	Same remarks as for curve 1 except L_1 = 0.208%, L_2 = 0.330%, L_3 = 0.562%.
4	102-G(C)	179	Rietveld, A.O., et al.	90.1	14-40	0.000 0.262 0.502 0.745 1.000	3.86 4.31 4.68 5.00 5.33	Same remarks as for curve 1 except L_1 = 0.114%, L_2 = 0.184%, L_3 = 0.339%.
5	102-G(C)	179	Rietveld, A.O., et al.	196.0	14-40	0.000 0.251 0.497 0.753 1.000	6.75 7.51 8.17 8.80 9.36	Same remarks as for curve 1 except L_1 = 0.165%, L_2 = 0.263%, L_3 = 0.453%.
6	102-G(C)	179	Rietveld, A.O., et al.	229.0	14-40	0.000 0.248 0.505 0.755 1.000	7.57 8.38 9.15 9.78 10.43	Same remarks as for curve 1 except L_1 = 0.063%, L_2 = 0.102%, L_3 = 0.194%.
7	102-G(C)	179	Rietveld, A.O., et al.	293.1	14-40	0.000 0.246 0.507 0.753 1.000	8.86 9.84 10.78 11.56 12.30	Same remarks as for curve 1 except L_1 = 0.019%, L_2 = 0.035%, L_3 = 0.074%.

TABLE 102-G(C)S.   SMOOTHED VISCOSITY VALUES AS A FUNCTION OF COMPOSITION FOR GASEOUS DEUTERIUM-HYDROGEN MIXTURES

Mole Fraction of D_2	14.4 K [Ref. 179]	20.4 K [Ref. 179]	71.5 K [Ref. 179]	90.1 K [Ref. 179]	196.0 K [Ref. 179]	229.0 K [Ref. 179]	293.1 K [Ref. 179]
0.00	0.79	1.08	3.24	3.86	6.75	7.57	8.86
0.05	0.80	1.10	3.31	3.96	6.90	7.74	9.08
0.10	0.82	1.12	3.38	4.05	7.05	7.90	9.28
0.15	0.83	1.14	3.43	4.14	7.20	8.06	9.44
0.20	0.84	1.16	3.50	4.22	7.34	8.22	9.68
0.25	0.85	1.18	3.56	4.31	7.49	8.37	9.86
0.30	0.86	1.19	3.63	4.38	7.64	8.53	10.05
0.35	0.87	1.20	3.70	4.46	7.78	8.69	10.23
0.40	0.88	1.22	3.87	4.54	7.92	8.84	10.41
0.45	0.89	1.24	3.84	4.61	8.05	8.99	10.58
0.50	0.90	1.24	3.90	4.68	8.18	9.24	10.75
0.55	0.91	1.26	3.96	4.75	8.32	9.28	10.92
0.60	0.92	1.27	4.01	4.82	8.44	9.40	11.08
0.65	0.93	1.28	4.07	4.89	8.52	9.54	11.25
0.70	0.94	1.30	4.14	4.96	8.70	9.64	11.41
0.75	0.95	1.31	4.18	5.02	8.83	9.79	11.56
0.80	0.96	1.32	4.24	5.20	8.95	9.92	11.71
0.85	0.97	1.34	4.29	5.16	9.06	10.04	11.86
0.90	0.98	1.35	4.35	5.22	9.18	10.28	12.02
0.95	1.00	1.36	4.40	5.28	9.28	10.30	12.18
1.00	1.00	1.37	4.44	5.33	9.36	10.43	12.30

414

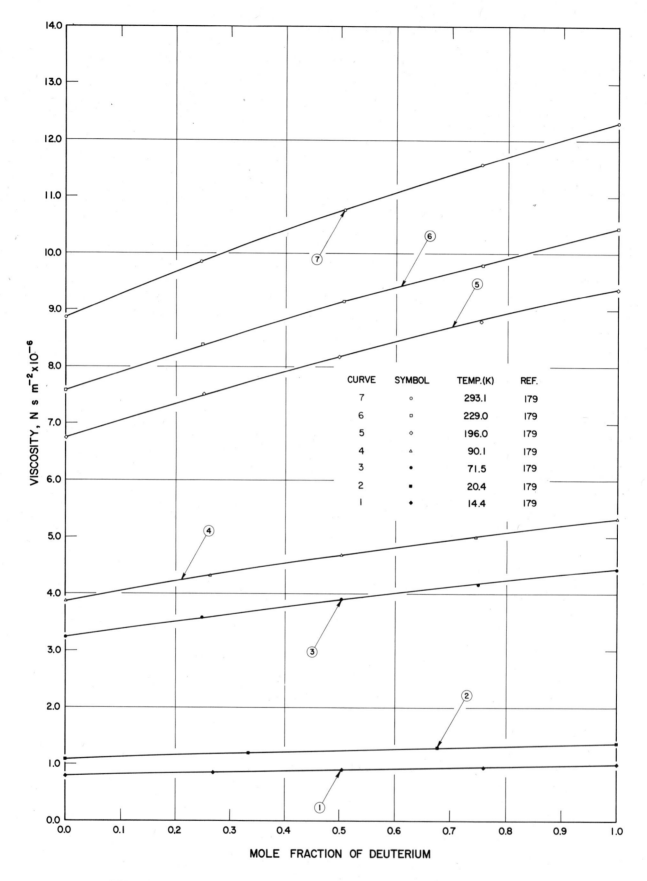

CURVE	SYMBOL	TEMP.(K)	REF.
7	○	293.1	179
6	□	229.0	179
5	◇	196.0	179
4	△	90.1	179
3	●	71.5	179
2	■	20.4	179
1	◆	14.4	179

VISCOSITY, N s m^{-2} x 10^{-6}

MOLE FRACTION OF DEUTERIUM

FIGURE 102-G(C). VISCOSITY DATA AS A FUNCTION OF COMPOSITION
FOR GASEOUS DEUTERIUM-HYDROGEN MIXTURES

TABLE 103-G(C)E.   EXPERIMENTAL VISCOSITY DATA AS A FUNCTION OF COMPOSITION FOR GASEOUS DEUTERIUM-HYDROGEN DEUTERIDE MIXTURES

Cur. No.	Fig. No.	Ref. No.	Author(s)	Temp. (K)	Pressure (mm Hg)	Mole Fraction of D_2	Viscosity (N s m^{-2} x 10^{-6})	Remarks
1	103-G(C)	179	Rietveld, A.O., Van Itterbeek, A., and Velds, C.A.	14.4	4-11	0.000 0.261 0.497 0.716 1.000	0.91 0.94 0.97 0.99 1.00	D_2: purity not specified, HD: 95 purity, rest being H_2 and D_2; oscillating disk viscometer; error in relative measurements ±3% at low temperatures and ±2% at high temperatures; $L_1 = 0.000\%$, $L_2 = 0.000\%$, $L_3 = 0.000\%$.
2	103-G(C)	179	Rietveld, A.O., et al.	20.4	4-11	0.000 0.242 0.503 0.751 1.000	1.27 1.31 1.34 1.38 1.41	Same remarks as for curve 1 except $L_1 = 0.000\%$, $L_2 = 0.000\%$, $L_3 = 0.000\%$.
3	103-G(C)	179	Rietveld, A.O., et al.	71.5	14-40	0.000 0.254 0.507 0.755 1.000	3.93 4.06 4.20 4.34 4.48	Same remarks as for curve 1 except $L_1 = 0.000\%$, $L_2 = 0.000\%$, $L_3 = 0.000\%$,
4	103-G(C)	179	Rietveld, A.O., et al.	90.1	14-40	0.000 0.238 0.492 0.749 1.000	4.74 4.90 5.07 5.25 5.40	Same remarks as for curve 1 except $L_1 = 0.280\%$, $L_2 = 0.626\%$, $L_3 = 1.400\%$.
5	103-G(C)	179	Rietveld, A.O., et al.	196.0	14-40	0.000 0.249 0.500 0.750 1.000	8.22 8.52 8.83 9.12 9.40	Same remarks as for curve 1 except $L_1 = 0.000\%$, $L_2 = 0.000\%$, $L_3 = 0.000\%$.
6	103-G(C)	179	Rietveld, A.O., et al.	229.0	14-40	0.000 0.249 0.495 0.755 1.000	9.10 9.46 9.80 10.16 10.48	Same remarks as for curve 1 except $L_1 = 0.000\%$, $L_2 = 0.000\%$, $L_3 = 0.000\%$.
7	103-G(C)	179	Rietveld, A.O., et al.	293.1	14-40	0.000 0.258 0.509 0.736 1.000	10.75 11.17 11.60 11.99 12.40	Same remarks as for curve 1 except $L_1 = 0.000\%$, $L_2 = 0.000\%$, $L_3 = 0.000\%$.

TABLE 103-G(C)S.   SMOOTHED VISCOSITY VALUES AS A FUNCTION OF COMPOSITION FOR GASEOUS DEUTERIUM-HYDROGEN DEUTERIDE MIXTURES

Mole Fraction of D_2	14.4 K [Ref. 179]	20.4 K [Ref. 179]	71.5 K [Ref. 179]	90.1 K [Ref. 179]	196.0 K [Ref. 179]	229.0 K [Ref. 179]	293.1 K [Ref. 179]
0.00	0.91	1.27	3.93	4.74	8.22	9.10	10.75
0.05	0.92	1.28	3.96	4.78	8.28	9.18	10.82
0.10	0.92	1.30	3.98	4.82	8.34	9.26	10.90
0.15	0.92	1.30	4.00	4.86	8.40	9.33	10.98
0.20	0.92	1.30	4.04	4.88	8.46	9.40	11.07
0.25	0.93	1.32	4.06	4.92	8.53	9.48	11.18
0.30	0.94	1.32	4.10	4.95	8.50	9.55	11.24
0.35	0.95	1.32	4.12	4.98	8.65	9.62	11.33
0.40	0.96	1.33	4.15	5.01	8.72	9.68	11.42
0.45	0.98	1.34	4.18	5.04	8.78	9.75	11.50
0.50	0.98	1.34	4.20	5.08	8.82	9.87	11.60
0.55	0.98	1.35	4.24	5.12	8.90	9.90	11.68
0.60	0.99	1.36	4.26	5.15	8.96	9.96	11.77
0.65	1.00	1.37	4.29	5.18	9.02	10.04	11.85
0.70	1.00	1.38	4.32	5.22	9.07	10.10	11.94
0.75	1.00	1.38	4.34	5.25	9.12	10.16	12.02
0.80	1.00	1.39	4.37	5.29	9.18	10.22	12.10
0.85	1.00	1.39	4.40	5.32	9.24	10.29	12.18
0.90	1.00	1.40	4.43	5.35	9.30	10.36	12.25
0.95	1.00	1.40	4.46	5.38	9.35	10.42	12.32
1.00	1.00	1.41	4.48	5.40	9.40	10.48	12.40

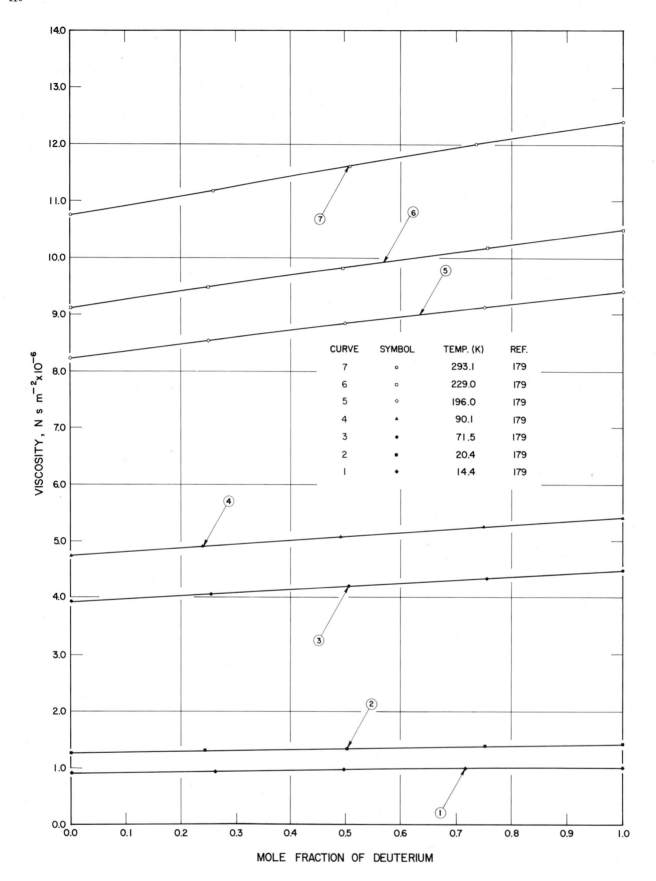

CURVE	SYMBOL	TEMP. (K)	REF.
7	○	293.1	179
6	□	229.0	179
5	◇	196.0	179
4	▲	90.1	179
3	●	71.5	179
2	■	20.4	179
1	◆	14.4	179

MOLE FRACTION OF DEUTERIUM

VISCOSITY, N s m^{-2}x10^{-6}

FIGURE 103-G(C).   VISCOSITY DATA AS A FUNCTION OF COMPOSITION
FOR GASEOUS DEUTERIUM-HYDROGEN DEUTERIDE MIXTURES

TABLE 104-L(T)E. EXPERIMENTAL VISCOSITY DATA AS A FUNCTION OF TEMPERATURE FOR LIQUID ETHANE-ETHYLENE MIXTURES

Cur. No.	Fig. No.	Ref. No.	Author(s)	Mole Fraction of C_2H_6	Pressure	Temp. (K)	Viscosity $(N\ s\ m^{-2} \times 10^{-6})$	Remarks
1	104-L(T)	70	Gerf, S. F. and Galkov, G. I.	0.000		105.0	66.0	Gas purity 99.8%; capillary method; accuracy ±1%.
						105.3	65.2	
						108.0	60.0	
						110.4	55.3	
						129.8	33.4	
						138.4	28.2	
						148.8	23.1	
						156.8	19.7	
						168.2	10.4	
2	104-L(T)	70	Gerf, S. F. and Galkov, G. I.	0.180		102.6	73.9	Same remarks as for curve 1.
						104.8	66.5	
						107.8	60.4	
						109.7	56.0	
						110.0	55.2	
						111.2	53.7	
						146.7	23.4	
						152.7	21.1	
						157.4	19.6	
						160.8	18.5	
3	104-L(T)	70	Gerf, S. F. and Galkov, G. I.	0.576		102.0	73.4	Same remarks as for curve 1.
						104.8	65.4	
						107.8	59.4	
						109.7	55.7	
						111.2	54.1	
						145.0	25.3	
						154.3	21.6	
						156.7	20.9	
4	104-L(T)	70	Gerf, S. F. and Galkov, G. I.	1.000		101.2	87.8	Same remarks as for curve 1.
						103.3	78.7	
						105.7	72.9	
						108.0	67.5	
						111.1	63.4	
						111.4	61.5	
						149.5	27.7	
						150.3	27.1	
						150.8	27.0	
						159.8	23.6	
						160.1	22.5	
						166.8	20.7	
						167.3	20.3	

TABLE 104-L(T)S. SMOOTHED VISCOSITY VALUES AS A FUNCTION OF TEMPERATURE FOR LIQUID ETHANE-ETHYLENE MIXTURES

Temp. (K)	Mole Fraction of C_2H_6			
	0.000 [Ref. 70]	0.180 [Ref. 70]	0.576 [Ref. 70]	1.000 [Ref. 70]
105	65.7	66.1	64.6	74.8
110	55.7	54.7	55.7	64.5
120	42.8	47.5	43.2	49.5
130	33.2	41.9	34.2	39.6
140	27.1	33.3	27.8	32.6
150	22.5	26.7	23.3	27.3
160	18.5	22.1	19.8	23.1
170	14.8	18.8	16.8	19.4

418

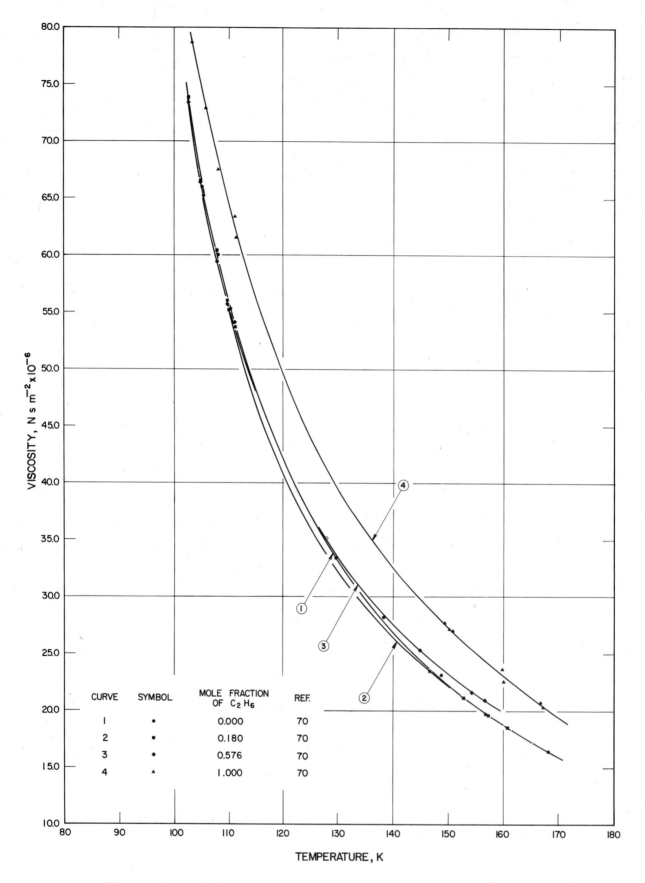

CURVE	SYMBOL	MOLE FRACTION OF C_2H_6	REF.
1	•	0.000	70
2	■	0.180	70
3	♦	0.576	70
4	▲	1.000	70

FIGURE 104-L(T). VISCOSITY DATA AS A FUNCTION OF TEMPERATURE
FOR LIQUID ETHANE-ETHYLENE MIXTURES

TABLE 105-G(C)E. EXPERIMENTAL VISCOSITY DATA AS A FUNCTION OF COMPOSITION FOR GASEOUS ETHANE-HYDROGEN MIXTURES

Cur. No.	Fig. No.	Ref. No.	Author(s)	Temp. (K)	Pressure	Mole Fraction of C_2H_6	Viscosity (N s m^{-2} x 10^{-6})	Remarks
1	105-G(C)	229	Trautz, M. and Sorg, K. G.	293. 0		1. 0000	9. 09	Capillary method; precision ± 0. 05%; L_1 = 0. 910%, L_2 = 1. 761%, L_3 = 3. 519%.
						0. 5500	9. 87	
						0. 1485	9. 93	
						0. 0000	8. 76	
2	105-G(C)	229	Trautz, M. and Sorg, K. G.	373. 0		1. 0000	11. 42	Same remarks as for curve 1 except L_1 = 0. 281%, L_2 = 0. 408%, L_3 = 0. 694%.
						0. 5500	12. 08	
						0. 1485	11. 89	
						0. 0000	10. 33	
3	105-G(C)	229	Trautz, M. and Sorg, K. G.	473. 0		1. 0000	14. 09	Same remarks as for curve 1 except L_1 = 0. 603%, L_2 = 1. 105%, L_3 = 2. 200%.
						0. 5500	14. 67	
						0. 1485	14. 12	
						0. 0000	12. 13	
4	105-G(C)	229	Trautz, M. and Sorg, K. G.	523. 0		1. 0000	15. 26	Same remarks as for curve 1 except L_1 = 0. 235%, L_2 = 0. 469%, L_3 = 0. 939%.
						0. 5500	15. 83	
						0. 1485	15. 11	
						0. 0000	12. 96	

TABLE 105-G(C)S. SMOOTHED VISCOSITY VALUES AS A FUNCTION OF COMPOSITION FOR GASEOUS ETHANE-HYDROGEN MIXTURES

Mole Fraction of C_2H_6	293. 0 K [Ref. 229]	373. 0 K [Ref. 229]	473. 0 K [Ref. 229]	523. 0 K [Ref. 229]
0. 00	8. 76	10. 33	12. 13	12. 96
0. 05	9. 25	10. 96	13. 06	13. 96
0. 10	9. 61	11. 46	13. 70	14. 64
0. 15	9. 92	11. 81	14. 12	15. 12
0. 20	10. 09	12. 04	14. 34	15. 44
0. 25	10. 20	12. 18	14. 66	15. 65
0. 30	10. 26	12. 26	14. 82	15. 79
0. 35	10. 30	12. 30	14. 94	15. 88
0. 40	10. 30	12. 29	15. 00	15. 94
0. 45	10. 30	12. 25	15. 04	15. 98
0. 50	10. 28	12. 20	15. 04	15. 98
0. 55	10. 23	12. 14	15. 00	15. 98
0. 60	10. 18	12. 08	14. 94	15. 96
0. 65	10. 08	12. 02	14. 88	15. 92
0. 70	9. 96	11. 96	14. 82	15. 86
0. 75	9. 85	11. 90	14. 72	15. 78
0. 80	9. 72	11. 83	14. 62	15. 70
0. 85	9. 58	11. 75	14. 51	15. 60
0. 90	9. 42	11. 66	14. 38	15. 50
0. 95	9. 25	11. 55	14. 23	15. 38
1. 00	9. 09	11. 42	14. 06	15. 26

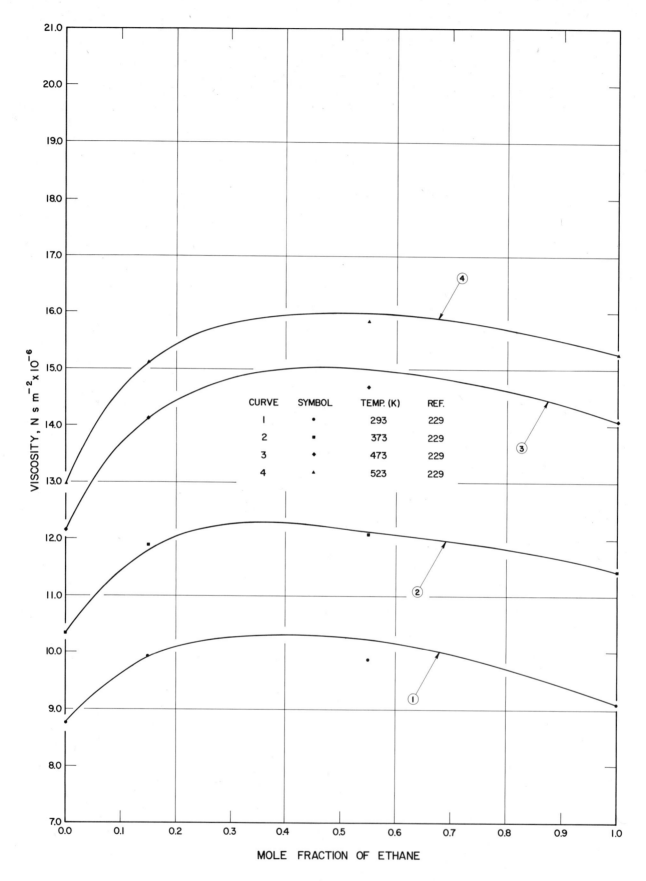

FIGURE 105 – G(C).   VISCOSITY DATA AS A FUNCTION OF COMPOSITION
FOR GASEOUS ETHANE – HYDROGEN MIXTURES

TABLE 106-G(C)E.   EXPERIMENTAL VISCOSITY DATA AS A FUNCTION OF COMPOSITION FOR GASEOUS
ETHANE-METHANE MIXTURES

Cur. No.	Fig. No.	Ref. No.	Author(s)	Temp. (K)	Pressure	Mole Fraction of C_2H_6	Viscosity $(N \ s \ m^{-2} \times 10^{-6})$	Remarks
1	106-G(C)	229	Trautz, M. and Sorg, K.G.	293.0		1.0000	9.09	CH_4: I. G. Farben, 99.9 pure; capillary method; precision $\pm 0.05\%$; $L_1 = 0.020\%$, $L_2 = 0.031\%$, $L_3 = 0.055\%$.
						0.8097	9.38	
						0.5126	9.86	
						0.1884	10.46	
						0.0000	10.87	
2	106-G(C)	229	Trautz, M. and Sorg, K.G.	373.0		1.0000	11.42	Same remarks as for curve 1 except $L_1 = 0.000\%$, $L_2 = 0.000\%$, $L_3 = 0.000\%$.
						0.8097	11.74	
						0.5126	12.26	
						0.1884	12.88	
						0.0000	13.31	
3	106-G(C)	229	Trautz, M. and Sorg, K.G.	473.0		1.0000	14.09	Same remarks as for curve 1 except $L_1 = 0.004\%$, $L_2 = 0.009\%$, $L_3 = 0.020\%$.
						0.8097	14.42	
						0.5126	14.96	
						0.1884	15.62	
						0.0000	16.03	
4	106-G(C)	229	Trautz, M. and Sorg, K.G.	523.0		1.0000	15.26	Same remarks as for curve 1 except $L_1 = 0.000\%$, $L_2 = 0.000\%$, $L_3 = 0.000\%$.
						0.8097	15.60	
						0.5126	16.14	
						0.1884	16.82	
						0.0000	17.25	

TABLE 106-G(C)S.   SMOOTHED VISCOSITY VALUES AS A FUNCTION OF COMPOSITION FOR GASEOUS
ETHANE-METHANE MIXTURES

Mole Fraction of C_2H_6	293.0 K [Ref. 229]	373.0 K [Ref. 229]	473.0 K [Ref. 229]	523.0 K [Ref. 229]
0.00	10.87	13.31	16.03	17.25
0.05	10.76	13.18	15.92	17.13
0.10	10.64	13.07	15.81	17.02
0.15	10.54	12.96	15.70	16.90
0.20	10.44	12.86	15.60	16.80
0.25	10.34	12.75	15.48	16.68
0.30	10.24	12.65	15.38	16.57
0.35	10.15	12.56	15.27	16.46
0.40	10.06	12.46	15.17	16.36
0.45	9.97	12.36	15.08	16.26
0.50	9.88	12.26	14.98	16.16
0.55	9.80	12.18	14.88	16.06
0.60	9.71	12.10	14.80	15.96
0.65	9.63	12.00	14.70	15.88
0.70	9.55	11.92	14.61	15.79
0.75	9.46	11.84	14.52	15.70
0.80	9.38	11.75	14.43	15.72
0.85	9.31	11.66	14.34	15.52
0.90	9.23	11.58	14.26	15.44
0.95	9.16	11.50	14.17	15.35
1.00	9.09	11.42	14.09	15.26

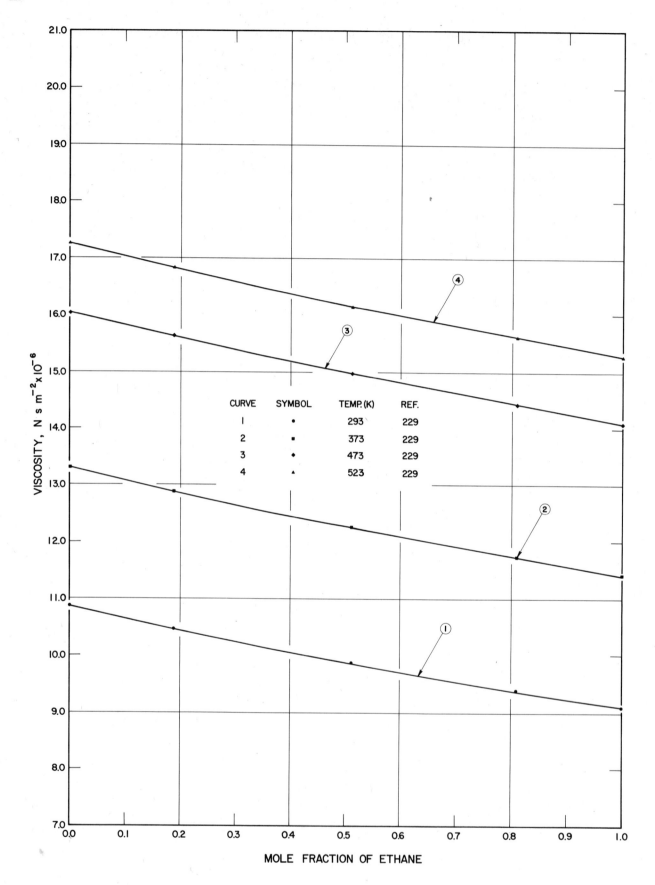

FIGURE 106-G(C). VISCOSITY DATA AS A FUNCTION OF COMPOSITION
FOR GASEOUS ETHANE-METHANE MIXTURES

TABLE 107-G(C)E.   EXPERIMENTAL VISCOSITY DATA AS A FUNCTION OF COMPOSITION FOR GASEOUS ETHANE-PROPANE MIXTURES

Cur. No.	Fig. No.	Ref. No.	Author(s)	Temp. (K)	Pressure	Mole Fraction of C_3H_8	Viscosity (N s m^{-2} x 10^{-6})	Remarks
1	107-G(E)	229	Trautz, M. and Sorg, K. G.	293.0		1.0000	8.01	C_3H_8: I. G. Farben, 99.9 pure; capillary method; precision ±0.05%; $L_1 = 0.167\%$, $L_2 = 0.286\%$, $L_3 = 0.591\%$.
						0.8474	8.15	
						0.7437	8.28	
						0.5673	8.41	
						0.0000	9.09	
2	107-G(E)	229	Trautz, M. and Sorg, K. G.	373.0		1.0000	10.08	Same remarks as for curve 1 except $L_1 = 0.039\%$, $L_2 = 0.086\%$, $L_3 = 0.193\%$.
						0.8474	10.25	
						0.7437	10.39	
						0.5673	10.58	
						0.0000	11.42	
3	106-G(E)	229	Trautz, M. and Sorg, K. G.	473.0		1.0000	12.53	Same remarks as for curve 1 except $L_1 = 0.217\%$, $L_2 = 0.314\%$, $L_3 = 0.620\%$.
						0.8474	12.72	
						0.7437	12.98	
						0.5673	13.13	
						0.0000	14.09	
4	107-G(E)	229	Trautz, M. and Sorg, K. G.	523.0		1.0000	13.63	Same remarks as for curve 1 except $L_1 = 0.184\%$, $L_2 = 0.297\%$, $L_3 = 0.558\%$.
						0.8474	13.82	
						0.7437	14.01	
						0.5673	14.25	
						0.0000	15.26	

TABLE 107-G(C)S.   SMOOTHED VISCOSITY VALUES AS A FUNCTION OF COMPOSITION FOR GASEOUS ETHANE-PROPANE MIXTURES

Mole Fraction of C_3H_8	293.0 K [Ref. 229]	373.0 K [Ref. 229]	473.0 K [Ref. 229]	523.0 K [Ref. 229]
0.00	9.09	11.42	14.09	15.26
0.05	9.03	11.32	14.00	15.18
0.10	8.97	11.24	13.90	15.10
0.15	8.92	11.17	13.82	15.02
0.20	8.86	11.10	13.73	14.94
0.25	8.80	11.03	13.65	14.86
0.30	8.75	10.96	13.56	14.78
0.35	8.70	10.89	13.49	14.70
0.40	8.64	10.82	13.41	14.62
0.45	8.58	10.78	13.33	14.54
0.50	8.53	10.68	13.26	14.45
0.55	8.48	10.62	13.18	14.36
0.60	8.42	10.25	13.11	14.26
0.65	8.37	10.48	13.04	14.17
0.70	8.32	10.42	12.96	14.08
0.75	8.26	10.36	12.89	14.02
0.80	8.22	10.30	12.82	13.94
0.85	8.16	10.24	12.74	13.86
0.90	8.11	10.18	12.67	13.80
0.95	8.06	10.14	12.60	13.72
1.00	8.01	10.08	12.53	13.63

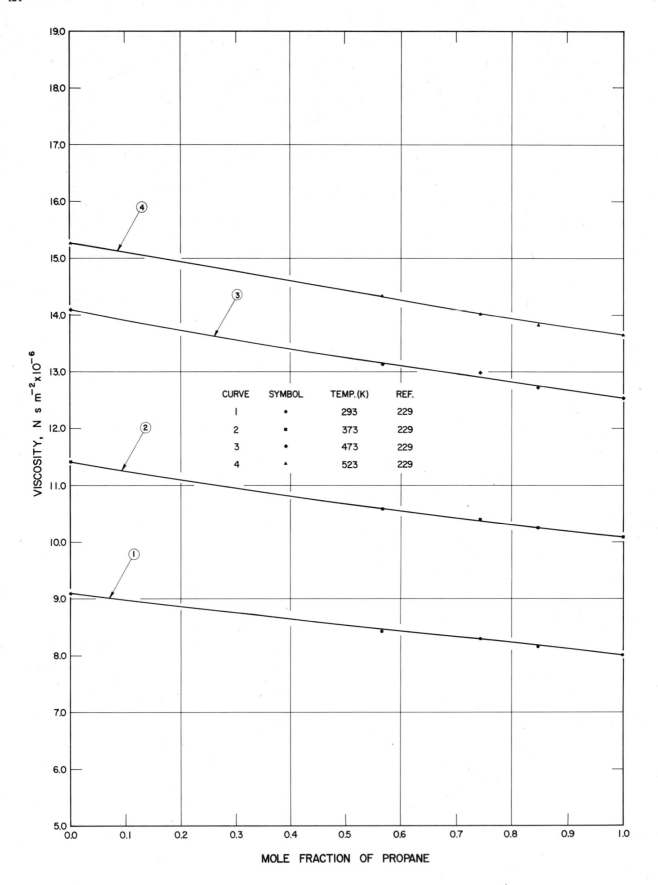

FIGURE 107 - G(C). VISCOSITY DATA AS A FUNCTION OF COMPOSITION
FOR GASEOUS ETHANE - PROPANE MIXTURES

TABLE 108-G(C)E. EXPERIMENTAL VISCOSITY DATA AS A FUNCTION OF COMPOSITION FOR GASEOUS ETHYLENE-HYDROGEN MIXTURES

Cur. No.	Fig. No.	Ref. No.	Author(s)	Temp. (K)	Pressure (atm)	Mole Fraction of C_2H_6	Viscosity $(N\ s\ m^{-2} \times 10^{-6})$	Remarks
1	108-G(C)	230	Trautz, M. and Stauf, F.W.	195.2		1.0000	7.18	Capillary method; accuracy estimated at $< \pm 4\%$ for pure gases; $L_1 = 0.167\%$, $L_2 = 0.312\%$, $L_3 = 0.706\%$.
						0.8082	7.31	
						0.6444	7.54	
						0.5087	7.64	
						0.2501	7.72	
						0.0000	6.70	
2	108-G(C)	230	Trautz, M. and Stauf, F.W.	233.2		1.0000	8.18	Same remarks as for curve 1 except $L_1 = 0.583\%$, $L_2 = 0.916\%$, $L_3 = 2.133\%$.
						0.8082	8.39	
						0.6444	8.52	
						0.5129	8.62	
						0.2501	8.66	
						0.1638	8.62	
						0.0000	7.40	
3	108-G(C)	230	Trautz, M. and Stauf, F.W.	272.2		1.0000	9.43	Same remarks as for curve 1 except $L_1 = 0.273\%$, $L_2 = 0.449\%$, $L_3 = 1.083\%$.
						0.8082	9.59	
						0.6444	9.85	
						0.5129	9.98	
						0.2501	9.96	
						0.1638	9.75	
						0.0000	8.30	
4	108-G(C)	230	Trautz, M. and Stauf, F.W.	293.2		1.0000	10.12	Same remarks as for curve 1 except $L_1 = 0.233\%$, $L_2 = 0.444\%$, $L_3 = 1.020\%$.
						0.8107	10.39	
						0.7033	10.53	
						0.5173	10.67	
						0.2160	10.60	
						0.0000	8.73	
5	108-G(C)	230	Trautz, M. and Stauf, F.W.	328.2		1.0000	11.22	Same remarks as for curve 1 except $L_1 = 0.375\%$, $L_2 = 0.573\%$, $L_3 = 1.157\%$.
						0.8707	11.54	
						0.7033	11.64	
						0.5173	11.73	
						0.2160	11.56	
						0.0000	9.43	
6	108-G(C)	230	Trautz, M. and Stauf, F.W.	373.2		1.0000	12.64	Same remarks as for curve 1 except $L_1 = 0.142\%$, $L_2 = 0.293\%$, $L_3 = 0.702\%$.
						0.8107	12.91	
						0.7033	12.98	
						0.5173	13.11	
						0.2114	12.78	
						0.0000	10.30	
7	108-G(C)	230	Trautz, M. and Stauf, F.W.	423.2		1.0000	14.08	Same remarks as for curve 1 except $L_1 = 0.098\%$, $L_2 = 0.126\%$, $L_3 = 0.222\%$.
						0.8043	14.32	
						0.7201	14.41	
						0.5197	14.63	
						0.2114	14.09	
						0.0000	11.23	
8	108-G(C)	230	Trautz, M. and Stauf, F.W.	473.2		1.0000	15.47	Same remarks as for curve 1 except $L_1 = 0.032\%$, $L_2 = 0.058\%$, $L_3 = 0.127\%$.
						0.8043	15.68	
						0.7201	15.74	
						0.5197	15.88	
						0.2114	15.29	
						0.0000	12.11	
9	108-G(C)	230	Trautz, M. and Stauf, F.W.	523.2		1.0000	16.81	Same remarks as for curve 1 except $L_1 = 0.000\%$, $L_2 = 0.000\%$, $L_3 = 0.000\%$.
						0.8043	16.94	
						0.7201	16.99	
						0.5116	17.09	
						0.2114	16.27	
						0.0000	12.94	

TABLE 108-G(C)S. SMOOTHED VISCOSITY VALUES AS A FUNCTION OF COMPOSITION FOR GASEOUS ETHYLENE-HYDROGEN MIXTURES

Mole Fraction of C_2H_4	195.2 K [Ref. 230]	233.2 K [Ref. 230]	272.2 K [Ref. 230]	293.2 K [Ref. 230]	328.2 K [Ref. 230]	373.2 K [Ref. 230]	423.2 K [Ref. 230]	473.2 K [Ref. 230]	523.2 K [Ref. 230]
0.00	6.70	7.40	8.30	8.73	9.43	10.30	11.23	12.11	12.94
0.05	7.06	7.82	9.02	9.52	10.28	11.38	12.76	13.42	14.36
0.10	7.32	8.15	9.44	10.02	10.84	12.04	13.41	14.22	15.22
0.15	7.51	8.38	9.72	10.34	11.23	12.47	13.80	14.80	15.79
0.20	7.64	8.55	9.89	10.56	11.50	12.76	14.06	15.22	16.20
0.25	7.72	8.66	9.98	10.69	11.69	12.94	14.26	15.51	16.50
0.30	7.76	8.73	10.03	10.77	11.81	13.06	14.40	15.70	16.72
0.35	7.76	8.76	10.04	10.80	11.86	13.12	14.51	15.82	16.89
0.40	7.74	8.76	10.04	10.82	11.88	13.15	14.57	15.88	17.00
0.45	7.70	8.74	10.02	10.82	11.86	13.16	14.60	15.90	17.09
0.50	7.65	8.69	10.00	10.80	11.82	13.14	14.62	15.90	17.09
0.55	7.60	8.65	9.96	10.75	11.78	13.12	14.60	15.89	17.10
0.60	7.54	8.60	9.92	10.70	11.72	13.09	14.58	15.87	17.08
0.65	7.51	8.56	9.88	10.63	11.68	13.05	14.53	15.83	17.10
0.70	7.46	8.51	9.82	10.56	11.63	13.01	14.47	15.78	17.00
0.75	7.42	8.46	9.76	10.48	11.57	12.96	14.41	15.74	16.97
0.80	7.37	8.40	9.70	10.51	11.50	12.90	14.34	15.68	16.94
0.85	7.33	8.36	9.64	10.33	11.44	12.84	14.28	15.63	16.90
0.90	7.29	8.31	9.58	10.26	11.44	12.78	14.22	15.58	16.88
0.95	7.24	8.25	9.51	10.18	11.29	12.71	14.15	15.52	16.84
1.00	7.18	8.18	9.43	10.12	11.22	12.64	14.08	15.47	16.81

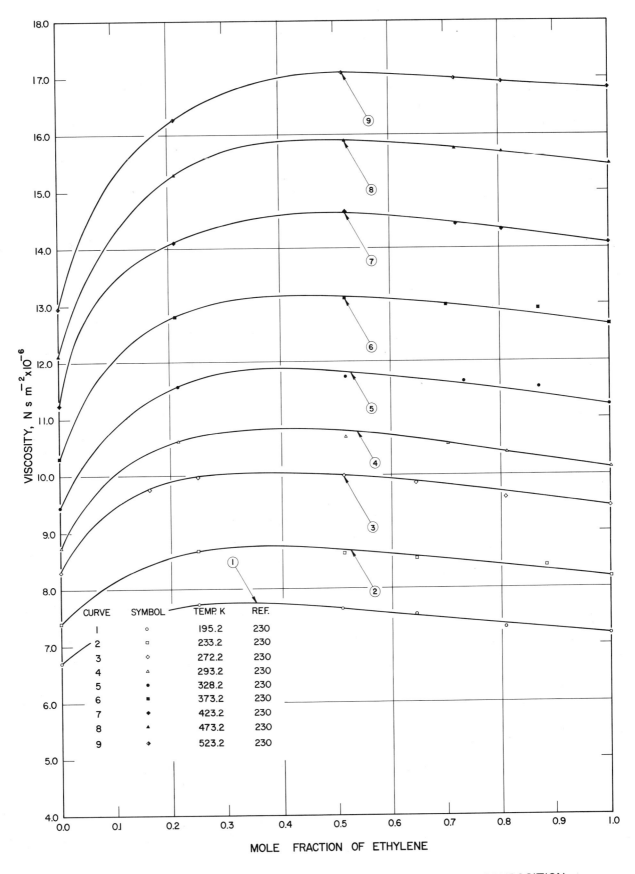

CURVE	SYMBOL	TEMP. K	REF.
1	○	195.2	230
2	□	233.2	230
3	◇	272.2	230
4	△	293.2	230
5	●	328.2	230
6	■	373.2	230
7	◆	423.2	230
8	▲	473.2	230
9	◈	523.2	230

MOLE FRACTION OF ETHYLENE

VISCOSITY, N s m^{-2} x10^{-6}

FIGURE 108-G(C). VISCOSITY DATA AS A FUNCTION OF COMPOSITION
FOR GASEOUS ETHYLENE-HYDROGEN MIXTURES

TABLE 109-L(T)E.   EXPERIMENTAL VISCOSITY DATA AS A FUNCTION OF TEMPERATURE FOR LIQUID ETHYLENE-METHANE MIXTURES

Cur. No.	Fig. No.	Ref. No.	Author(s)	Mole Fraction of C_2H_4	Pressure (atm)	Temp. (K)	Viscosity ($N \ s \ m^{-2} \times 10^{-6}$)	Remarks
1	109-L(T)	70	Gerf, S. F. and Galkov, G. I.	1.000		105.0	66.0	Gas purity 99.8; capillary method; accuracy $\pm 1\%$.
						105.3	65.2	
						108.0	60.0	
						110.4	55.3	
						129.8	33.4	
						138.4	28.2	
						148.8	23.1	
						156.8	19.7	
						168.2	16.4	
2	109-L(T)	70	Gerf, S. F. and Galkov, G. I.	0.230		92.6	31.1	Same remarks as for curve 1.
						94.9	28.1	
						99.2	24.0	
						101.1	22.4	
						104.1	20.5	
						109.4	18.2	
						111.0	17.0	
3	109-L(T)	70	Gerf, S. F. and Galkov, G. I.	0.398		93.7	40.6	Same remarks as for curve 1.
						95.1	38.2	
						97.5	34.4	
						99.5	32.7	
						102.6	29.9	
						105.2	26.3	
						107.2	25.3	
						111.2	22.6	
4	109-L(T)	70	Gerf, S. F. and Galkov, G. I.	0.590		96.6	48.8	Same remarks as for curve 1.
						98.9	43.8	
						102.6	38.2	
						104.9	35.2	
						107.8	32.5	
						111.2	29.5	
5	109-L(T)	70	Gerf, S. F. and Galkov, G. I.	0.763		98.9	59.5	Same remarks as for curve 1.
						101.5	52.6	
						104.1	48.2	
						106.4	45.1	
						108.4	42.1	
						111.1	39.0	
6	109-L(T)	70	Gerf, S. F. and Galkov, G. I.	0.000		94.4	18.7	Same remarks as for curve 1.
						98.3	16.2	
						102.4	14.4	
						108.8	12.5	
						111.2	11.9	
7	109-L(T)	70	Gerf, S. F. and Galkov, G. I.	0.196		133.4	109.0	Same remarks as for curve 1.
					0.190	152.4	89.0	
					0.196	173.6	73.0	
					0.200	184.0	62.0	
8	109-L(T)	70	Gerf, S. F. and Galkov, G. I.	0.555		143.6	163.0	Same remarks as for curve 1.
					0.590	161.4	131.0	
					0.588	183.0	108.0	
					0.584	199.0	82.0	
					0.590	205.4	80.0	
					0.605	217.8	68.0	
9	109-L(T)	70	Gerf, S. F. and Galkov, G. I.	0.730		149.8	193.0	Same remarks as for curve 1.
					0.750	165.0	165.0	
					0.750	179.4	151.0	
					0.750	196.8	129.0	
					0.796	214.6	114.0	
					0.796	238.2	100.0	
10	109-L(T)	70	Gerf, S. F. and Galkov, G. I.	1.000		183.8	135.0	Same remarks as for curve 1.
						204.0	115.0	
						226.4	92.0	
						252.2	72.0	
						273.1	64.0	

TABLE 109-L(T)S.   SMOOTHED VISCOSITY VALUES AS A FUNCTION OF TEMPERATURE FOR LIQUID
ETHYLENE-METHANE MIXTURES

T, K	Mole Fraction of Methane								
	0.000 [70]	0.195 [70]	0.230 [70]	0.398 [70]	0.585 [70]	0.590 [70]	0.745 [70]	0.763 [70]	1.000 [70]
92			32.10						
93				41.80					
94	18.98								
95	18.30			38.52					
96	17.65		27.00			50.30			69.70
98	16.50			34.41		45.30			
100	15.50		23.40	31.40		41.60		56.10	65.98
102	14.65			29.80		38.80		51.80	
104	13.90			27.80		36.30		48.40	
105	13.60		20.10	27.00		35.20		46.90	
106	13.30					34.15		45.42	
107								44.10	
108	12.70		18.50	24.70		32.20		42.80	
109								41.55	
110	12.20		17.60	23.40		30.50		40.50	56.40
112	11.80		16.75	22.10		28.90		38.12	
114						27.40			
120									44.00
130		112.40							33.20
140		101.81							27.00
150		91.48			149.75		193.10		22.60
160		81.79			133.20		173.10		18.98
170		73.10			118.90		158.15		15.98
175									14.70
180		65.10			106.55		147.65		139.05
190					95.30		135.15		128.60
200					85.10		126.40		118.20
210					75.40		118.60		108.10
220					66.20		111.45		98.35
230							104.95		89.10
240							98.75		78.80
250							92.90		73.40
260									67.15
270									61.60

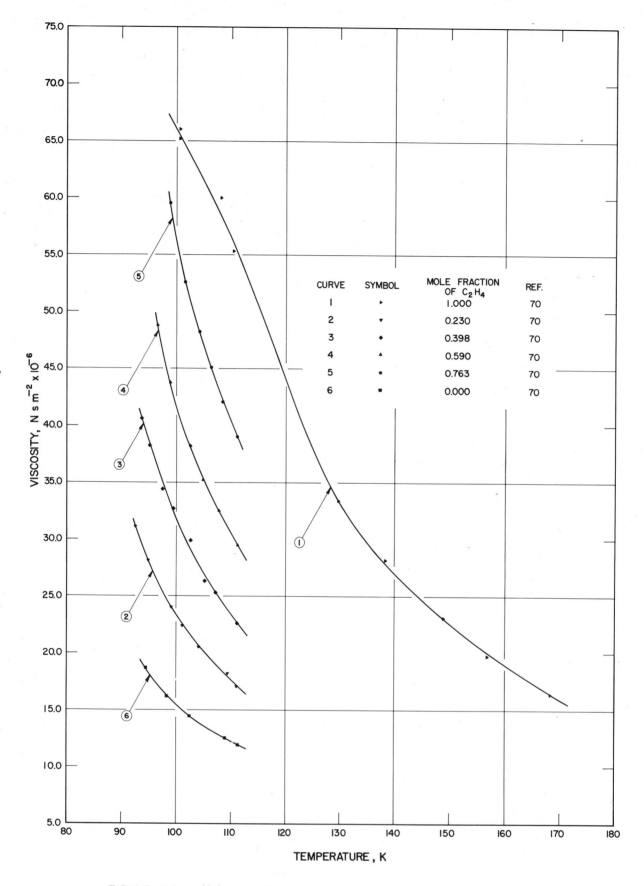

FIGURE 109 – L(T). VISCOSITY DATA AS A FUNCTION OF TEMPERATURE
FOR LIQUID ETHYLENE – METHANE MIXTURES

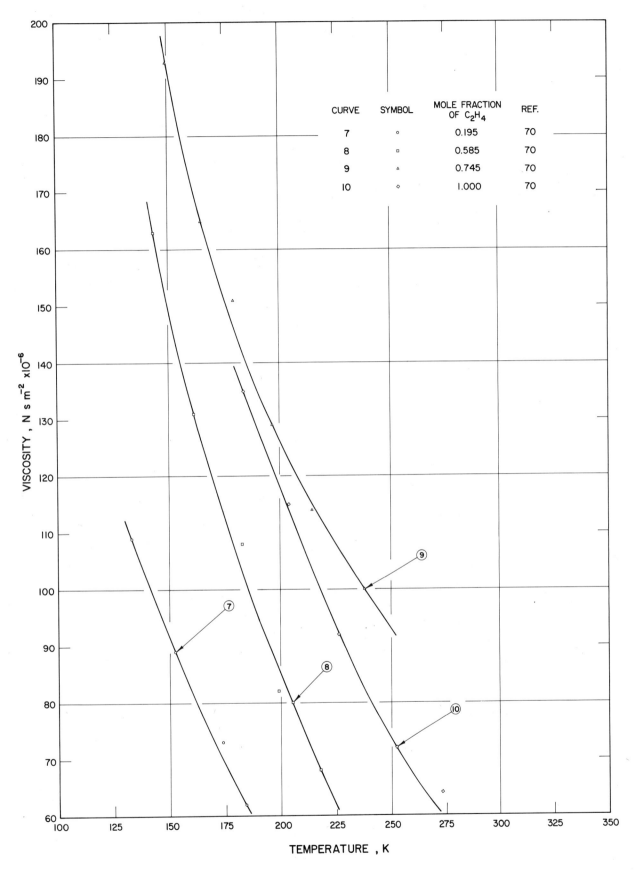

CURVE	SYMBOL	MOLE FRACTION OF C_2H_4	REF.
7	○	0.195	70
8	□	0.585	70
9	△	0.745	70
10	◇	1.000	70

VISCOSITY , $N s m^{-2} \times 10^{-6}$

TEMPERATURE , K

FIGURE 109 - L (T). VISCOSITY DATA AS A FUNCTION OF TEMPERATURE
FOR LIQUID ETHYLENE-METHANE MIXTURES (continued)

432

TABLE 110-G(C)E.   EXPERIMENTAL VISCOSITY DATA AS A FUNCTION OF COMPOSITION FOR GASEOUS
ETHYLENE-NITROGEN MIXTURES

Cur. No.	Fig. No.	Ref. No.	Author(s)	Temp. (K)	Pressure	Mole Fraction of C_2H_4	Viscosity (N s m^{-2} x 10^{-6})	Remarks
1	110-G(C)	227	Trautz, M. and Melster, A.	300.0		0.0000	17.81	Capillary method, r = 0.2019 mm; L_1 = 0.558%, L_2 = 0.913%, L_3 = 1.765%.
						0.2405	15.74	
						0.5695	13.08	
						0.7621	11.69	
						1.0000	10.33	
2	110-G(C)	227	Trautz, M. and Melster, A.	400.0		0.0000	21.90	Same remarks as for curve 1 except L_1 = 0.296%, L_2 = 0.481%, L_3 = 0.915%.
						0.2405	19.56	
						0.5695	16.55	
						0.7621	14.91	
						1.0000	13.48	
3	110-G(C)	227	Trautz, M. and Melster, A.	500.0		0.0000	25.60	Same remarks as for curve 1 except L_1 = 0.260%, L_2 = 0.533%, L_3 = 1.186%.
						0.2405	22.92	
						0.5695	19.63	
						0.7621	17.86	
						1.0000	16.22	
4	110-G(C)	227	Trautz, M. and Melster, A.	550.0		0.0000	27.27	Same remarks as for curve 1 except L_1 = 0.494%, L_2 = 0.796%, L_3 = 1.588%.
						0.2405	24.53	
						0.5695	21.08	
						0.7621	19.21	
						1.0000	17.53	

TABLE 110-G(C)S.   SMOOTHED VISCOSITY VALUES AS A FUNCTION OF COMPOSITION FOR GASEOUS
ETHYLENE-NITROGEN MIXTURES

Mole Fraction of C_2H_4	300.0 K [Ref. 227]	400.0 K [Ref. 227]	500.0 K [Ref. 227]	550.0 K [Ref. 227]
0.00	17.81	21.90	25.60	27.27
0.05	17.31	21.35	24.95	26.68
0.10	16.82	20.88	24.38	26.12
0.15	16.38	20.35	23.78	25.55
0.20	15.92	19.85	23.20	25.00
0.25	15.50	19.35	22.60	24.48
0.30	15.08	18.85	22.18	23.93
0.35	14.68	18.38	21.55	23.40
0.40	14.30	17.90	21.00	22.90
0.45	13.90	17.48	20.50	22.40
0.50	13.55	17.00	20.00	21.90
0.55	13.20	16.58	19.55	21.42
0.60	12.88	16.15	19.10	20.96
0.65	12.55	15.75	18.68	20.50
0.70	12.25	15.39	18.32	20.05
0.75	11.95	15.00	17.95	19.61
0.80	11.68	14.68	17.60	19.19
0.85	11.40	14.35	17.30	18.78
0.90	11.17	14.05	17.00	18.35
0.95	10.90	13.76	16.70	17.95
1.00	10.33	13.48	16.22	17.53

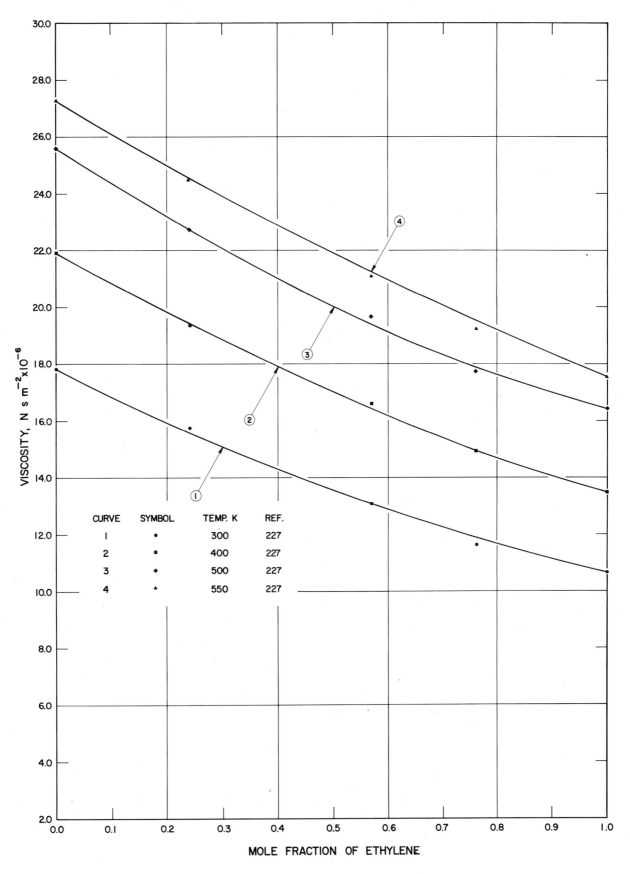

FIGURE 110 - G(C). VISCOSITY DATA AS A FUNCTION OF COMPOSITION
FOR GASEOUS ETHYLENE - NITROGEN MIXTURES

TABLE 111-G(C)E. EXPERIMENTAL VISCOSITY DATA AS A FUNCTION OF COMPOSITION FOR GASEOUS ETHYLENE-OXYGEN MIXTURES

Cur. No.	Fig. No.	Ref. No.	Author(s)	Temp. (K)	Pressure (atm)	Mole Fraction of C_2H_4	Viscosity (N s m^{-2} x 10^{-6})	Remarks
1	111-G(C)	227	Trautz, M. and Melster, A.	293.0		1.0000	20.19	Capillary method, r = 0.2019 mm; L_1 = 0.050%, L_2 = 0.092%, L_3 = 0.198%.
						0.8694	18.54	
						0.5855	15.29	
						0.2297	11.98	
						0.0000	10.10	
2	111-G(C)	227	Trautz, M. and Melster, A.	323.0		1.0000	21.81	Same remarks as for curve 1 except L_1 = 0.010%, L_2 = 0.022%, L_3 = 0.050%.
						0.8694	20.04	
						0.5855	16.58	
						0.2297	13.08	
						0.0000	11.07	
3	111-G(C)	227	Trautz, M. and Melster, A.	373.0		1.0000	24.33	Same remarks as for curve 1 except L_1 = 0.014%, L_2 = 0.030%, L_3 = 0.068%.
						0.8694	22.43	
						0.5855	18.65	
						0.2297	14.79	
						0.0000	12.62	

TABLE 111-G(C)S. SMOOTHED VISCOSITY VALUES AS A FUNCTION OF COMPOSITION FOR GASEOUS ETHYLENE-OXYGEN MIXTURES

Mole Fraction of C_2H_4	293.0 K [Ref. 227]	323.0 K [Ref. 227]	373.0 K [Ref. 227]
0.00	10.12	11.07	12.62
0.05	10.32	11.50	13.10
0.10	10.92	11.95	13.56
0.15	11.32	12.39	14.04
0.20	11.72	12.82	14.52
0.25	12.25	13.28	15.00
0.30	12.60	13.72	15.52
0.35	13.05	14.20	16.02
0.40	13.48	14.68	16.55
0.45	13.95	15.18	17.10
0.50	14.42	15.68	17.65
0.55	14.93	16.20	18.22
0.60	15.45	16.75	18.83
0.65	16.00	17.32	19.48
0.70	16.55	17.92	20.12
0.75	17.12	18.52	20.80
0.80	17.70	19.16	21.48
0.85	18.32	19.82	22.18
0.90	18.95	20.48	22.88
0.95	19.58	21.15	23.60
1.00	20.19	21.81	24.33

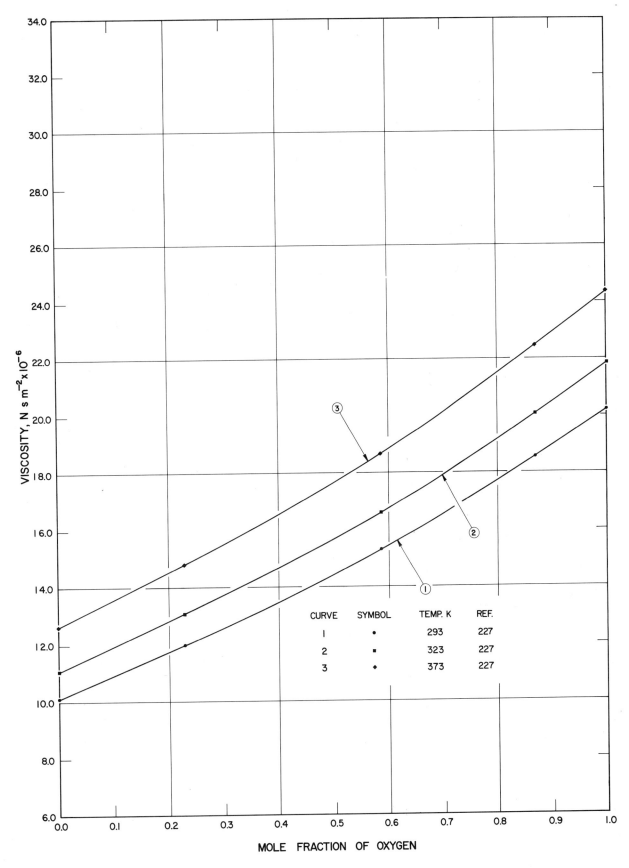

FIGURE III-G(C).   VISCOSITY DATA AS A FUNCTION OF COMPOSITION
FOR GASEOUS ETHYLENE-OXYGEN MIXTURES

TABLE 112-G(C)E. EXPERIMENTAL VISCOSITY DATA AS A FUNCTION OF COMPOSITION FOR GASEOUS n-HEPTANE-NITROGEN MIXTURES

Cur. No.	Fig. No.	Ref. No.	Author(s)	Temp. (K)	Pressure (atm)	Mole Fraction of n-C_7H	Viscosity (N s m^{-2} x 10^{-6})	Remarks
1	112-G(C)	307	Carmichael, L.T. and Sage, B.H.	310.9	0.398 0.781 1.000	1.0000 0.4848 0.0000	6.24 10.40 18.36	n-C_7H_{16}: 99.89 pure, N_2: 99.996 pure; oscillating cylinder visco-meter, calibrated with He; error ±1%, precision ±0.5%.
2	112-G(C)	307	Carmichael, L.T. and Sage, B.H.	344.3	0.398 2.574 1.000	1.0000 0.1471 0.0000	6.94 15.46 19.84	Same remarks as for curve 1.

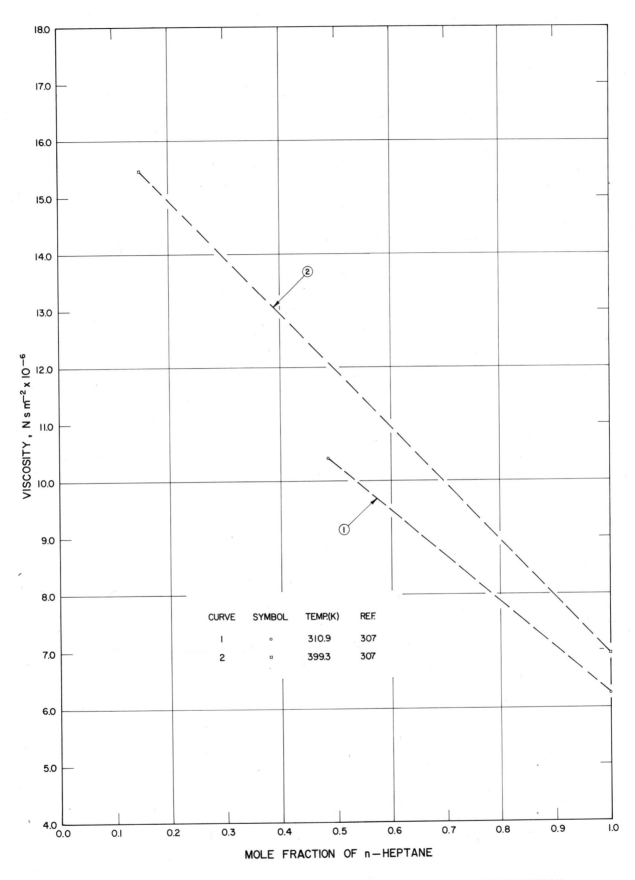

CURVE	SYMBOL	TEMP.(K)	REF.
1	○	310.9	307
2	□	399.3	307

MOLE FRACTION OF n−HEPTANE

FIGURE II2−G(C). VISCOSITY DATA AS A FUNCTION OF COMPOSITION
FOR GASEOUS n-HEPTANE − NITROGEN MIXTURES

TABLE 113-G(C)E.   EXPERIMENTAL VISCOSITY DATA AS A FUNCTION OF COMPOSITION FOR GASEOUS
HEXADECAFLUORO-n-HEPTANE - 2, 2, 4-TRIMETHYLPENTANE MIXTURES

Cur. No.	Fig. No.	Ref. No.	Author(s)	Temp. (K)	Pressure (mm Hg)	Mole Fraction of n-C$_7$H$_{16}$	Viscosity (N s m^{-2} x 10^{-6})	Remarks
1	113-G(C)	354	Lewis, J. E.	303.2	47	1.0000	8.2860	Hexadecafluoro-n-heptane: 99.95
						0.8941	8.1539	±0.02%, 2, 2, 4-trimethylpentane:
						0.6992	7.6732	99.98±0.02%; oscillating disk
						0.4830	6.9922	viscometer calibrated with air;
						0.3658	6.5046	L$_1$ = 0.128%, L$_2$ = 0.181%, L$_3$ =
						0.1550	5.6001	0.356%.
						0.0000	4.7861	
2	113-G(C)	354	Lewis, J. E.	323.2	40	1.0000	8.8168	Same remarks as for curve 1 except
						0.8941	8.6595	L$_1$ = 0.701%, L$_2$ = 0.925%, L$_3$ =
						0.6992	8.2324	1.441%.
						0.4830	7.5230	
						0.3658	6.8805	
						0.1550	5.8692	
						0.0000	5.1308	
3	113-G(C)	354	Lewis, J. E.	333.2	40	1.0000	9.0076	Same remarks as for curve 1 except
						0.4830	7.6423	L$_1$ = 0.000%, L$_2$ = 0.000%, L$_3$ =
						0.0000	5.3205	0.000%.

TABLE 113-G(C)S.   SMOOTHED VISCOSITY VALUES AS A FUNCTION OF COMPOSITION FOR GASEOUS
HEXADECAFLUORO-n-HEPTANE - 2, 2, 4-TRIMETHYLPENTANE MIXTURES

Mole Fraction of n-C$_7$H$_{16}$	303.2 K [Ref. 354]	323.2 K [Ref. 354]	333.2 K [Ref. 354]
0.00	4.786	5.131	5.321
0.05	5.007	5.407	5.590
0.10	5.332	5.670	5.850
0.15	5.508	5.930	6.110
0.20	5.818	6.175	6.360
0.25	6.040	6.415	6.612
0.30	6.255	6.650	6.850
0.35	6.460	6.870	7.082
0.40	6.660	7.108	7.300
0.45	6.858	7.320	7.520
0.50	7.045	7.510	7.710
0.55	7.220	7.577	7.900
0.60	7.390	7.880	8.070
0.65	7.545	8.042	8.230
0.70	7.690	8.190	8.380
0.75	7.820	8.325	8.548
0.80	7.940	8.450	8.650
0.85	8.042	8.560	8.770
0.90	8.132	8.665	8.870
0.95	8.218	8.748	8.970
1.00	8.286	8.816	9.008

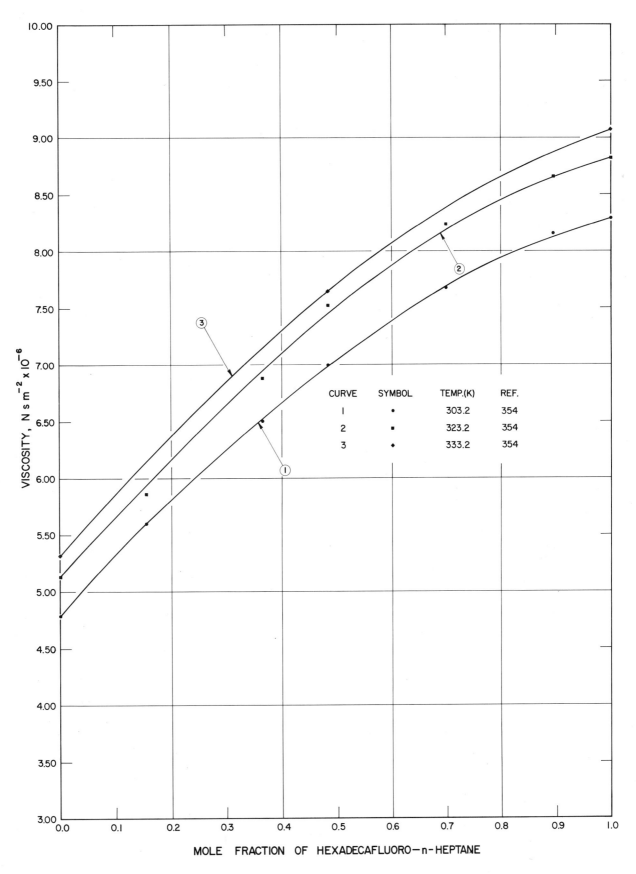

FIGURE 113-G(C). VISCOSITY DATA AS A FUNCTION OF COMPOSITION
FOR GASEOUS HEXADECAFLUORO-n-HEPTANE-
2,2,4-TRIMETHYLPENTANE MIXTURES

TABLE 114-G(C)E.   EXPERIMENTAL VISCOSITY DATA AS A FUNCTION OF COMPOSITION FOR GASEOUS HYDROGEN-HYDROGEN DEUTERIDE MIXTURES

Cur. No.	Fig. No.	Ref. No.	Author(s)	Temp. (K)	Pressure (mm Hg)	Mole Fraction of HD	Viscosity $(\text{N s m}^{-2} \times 10^{-6})$	Remarks
1	114-G(C)	179	Rietveld, A.O., Van Itterbeek, A., and Velds, C.A.	14.4	4-11	0.000	0.79	H_2: obtained from vapors over liquid hydrogen and then purified by condensation; oscillating disk viscometer, relative measurements; uncertainties ±3% at low temperatures and ±2% at high temperatures; $L_1 = 0.480\%$, $L_2 = 0.759\%$, $L_3 = 1.235\%$.
						0.254	0.82	
						0.501	0.84	
						0.757	0.87	
						1.000	0.88	
2	114-G(C)	179	Rietveld, A.O., et al.	20.4	4-11	0.000	1.11	Same remarks as for curve 1 except $L_1 = 0.333\%$, $L_2 = 0.745\%$, $L_3 = 1.667\%$.
						0.240	1.15	
						0.505	1.18	
						0.754	1.21	
						1.000	1.25	
3	114-G(C)	179	Rietveld, A.O., et al.	71.5	15-40	0.000	3.26	Same remarks as for curve 1 except $L_1 = 0.208\%$, $L_2 = 0.240\%$, $L_3 = 0.347\%$.
						0.250	3.45	
						0.499	3.62	
						0.749	3.79	
						1.000	3.95	
4	114-G(C)	179	Rietveld, A.O., et al.	90.1	15-40	0.000	3.92	Same remarks as for curve 1 except $L_1 = 0.268\%$, $L_2 = 0.508\%$, $L_3 = 1.113\%$.
						0.253	4.17	
						0.499	4.36	
						0.741	4.53	
						1.000	4.75	
5	114-G(C)	179	Rietveld, A.O., et al.	196.0	15-40	0.000	6.70	Same remarks as for curve 1 except $L_1 = 0.000\%$, $L_2 = 0.000\%$, $L_3 = 0.000\%$.
						0.236	7.07	
						0.496	7.48	
						0.746	7.81	
						1.000	8.16	
6	114-G(C)	179	Rietveld, A.O., et al.	229.0	15-40	0.000	7.45	Same remarks as for curve 1 except $L_1 = 0.000\%$, $L_2 = 0.000\%$, $L_3 = 0.000\%$.
						0.196	7.84	
						0.497	8.31	
						0.748	8.72	
						1.000	9.10	
7	114-G(C)	179	Rietveld, A.O., et al.	293.1	15-40	0.000	8.83	Same remarks as for curve 1 except $L_1 = 0.142\%$, $L_2 = 0.204\%$, $L_3 = 0.391\%$.
						0.241	9.28	
						0.498	9.80	
						0.748	10.20	
						1.000	10.69	

TABLE 114-G(C)S.   SMOOTHED VISCOSITY VALUES AS A FUNCTION OF COMPOSITION FOR GASEOUS HYDROGEN-HYDROGEN DEUTERIDE MIXTURES

Mole Fraction of HD	14.4 K [Ref. 179]	20.4 K [Ref. 179]	71.5 K [Ref. 179]	90.1 K [Ref. 179]	196.0 K [Ref. 179]	229.0 K [Ref. 179]	293.1 K [Ref. 179]
0.00	0.79	1.11	3.26	3.92	6.70	7.45	8.82
0.05	0.79	1.13	3.31	3.98	6.79	7.54	8.92
0.10	0.80	1.14	3.35	4.03	6.86	7.64	9.01
0.15	0.80	1.15	3.39	4.09	6.94	7.74	9.10
0.20	0.81	1.15	3.43	4.13	7.01	7.84	9.20
0.25	0.81	1.16	3.46	4.17	7.09	7.92	9.30
0.30	0.82	1.17	3.50	4.22	7.17	8.00	9.40
0.35	0.82	1.18	3.50	4.26	7.24	8.08	9.50
0.40	0.83	1.19	3.57	4.30	7.33	8.16	9.59
0.45	0.83	1.19	3.60	4.33	7.40	8.24	9.68
0.50	0.83	1.20	3.63	4.37	7.48	8.31	9.78
0.55	0.83	1.21	3.66	4.41	7.56	8.40	9.88
0.60	0.83	1.21	3.70	4.46	7.63	8.48	9.97
0.65	0.84	1.21	3.73	4.50	7.70	8.58	10.06
0.70	0.84	1.21	3.77	4.55	7.76	8.65	10.15
0.75	0.84	1.22	3.80	4.59	7.83	8.74	10.24
0.80	0.84	1.22	3.83	4.63	7.90	8.80	10.32
0.85	0.84	1.23	3.86	4.65	7.96	8.88	10.42
0.90	0.84	1.24	3.89	4.70	8.03	8.96	10.50
0.95	0.84	1.24	3.90	4.73	8.10	9.03	10.60
1.00	0.88	1.25	3.94	4.75	8.16	9.10	10.69

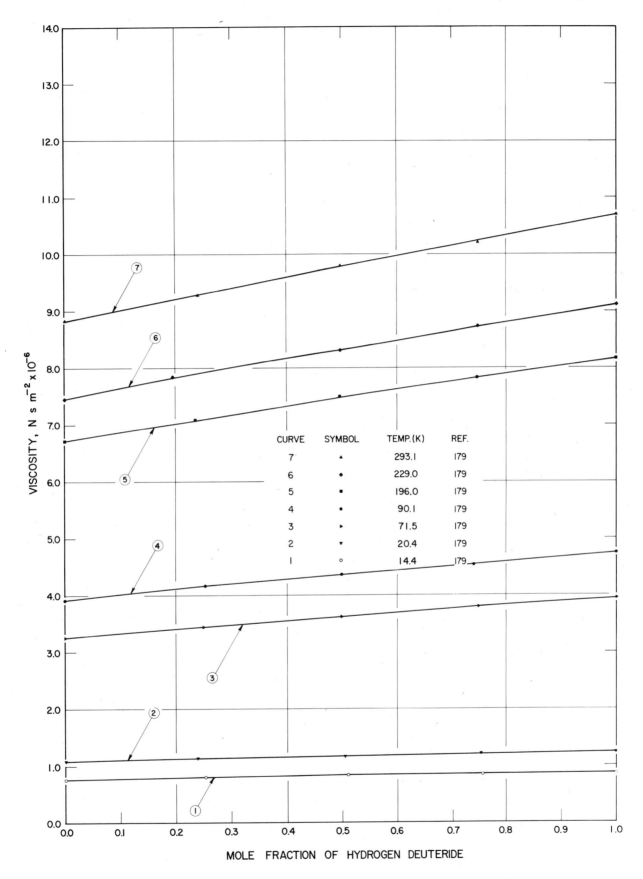

CURVE	SYMBOL	TEMP.(K)	REF.
7	▲	293.1	179
6	◆	229.0	179
5	■	196.0	179
4	●	90.1	179
3	▶	71.5	179
2	▼	20.4	179
1	○	14.4	179

MOLE FRACTION OF HYDROGEN DEUTERIDE

VISCOSITY, N s m^{-2} x 10^{-6}

FIGURE 114- G(C). VISCOSITY DATA AS A FUNCTION OF COMPOSITION
FOR GASEOUS HYDROGEN —HYDROGEN DEUTERIDE MIXTURES

TABLE 115-G(C)E.   EXPERIMENTAL VISCOSITY DATA AS A FUNCTION OF COMPOSITION FOR GASEOUS HYDROGEN-METHANE MIXTURES

Cur. No.	Fig. No.	Ref. No.	Author(s)	Temp. (K)	Pressure (atm)	Mole Fraction of CH_4	Viscosity (N s m^{-2} x 10^{-6})	Remarks
1	115-G(C)	229	Trautz, M. and Sorg, K. G.	293.0		1.0000	10.87	CH_4: I.G. Farben, 99.9 pure; capillary method; precision ±0.05%; $L_1 = 0.028\%$, $L_2 = 0.048\%$, $L_3 = 0.092\%$.
						0.7192	10.99	
						0.5145	10.98	
						0.3978	10.86	
						0.0777	9.55	
						0.0000	8.76	
2	115-G(C)	1	Adzumi, H.	293.2		0.0000	9.24	H_2: electrolysis of water, dried and traces of oxygen removed by passing over red hot copper; measurements relative to air; $L_1 = 0.313\%$, $L_2 = 0.560\%$, $L_3 = 1.287\%$.
						0.2083	10.62	
						0.3909	10.74	
						0.4904	11.10	
						0.6805	11.24	
						1.0000	11.25	
3	115-G(C)	1	Adzumi, H.	333.2		0.0000	10.08	Same remarks as for curve 2 except $L_1 = 0.261\%$, $L_2 = 0.391\%$, $L_3 = 0.784\%$.
						0.2083	11.60	
						0.3909	11.90	
						0.4904	12.34	
						0.6805	12.54	
						1.0000	12.55	
4	115-G(C)	229	Trautz, M. and Sorg, K. G.	373.0		1.0000	13.31	Same remarks as for curve 1 except $L_1 = 0.032\%$, $L_2 = 0.056\%$, $L_3 = 0.115\%$.
						0.7192	13.37	
						0.5145	13.28	
						0.3978	13.06	
						0.0777	11.32	
						0.0000	10.33	
5	115-G(C)	1	Adzumi, H.	373.2		0.0000	10.90	Same remarks as for curve 2 except $L_1 = 0.233\%$, $L_2 = 0.308\%$, $L_3 = 0.501\%$.
						0.2083	12.71	
						0.3909	13.12	
						0.4904	13.59	
						0.6805	13.80	
						1.0000	13.80	
6	115-G(C)	229	Trautz, M. and Sorg, K. G.	473.0		1.0000	16.03	Same remarks as for curve 1 except $L_1 = 0.029\%$, $L_2 = 0.050\%$, $L_3 = 0.110\%$.
						0.7192	16.02	
						0.5145	15.87	
						0.3978	15.51	
						0.0777	13.38	
						0.0000	12.13	
7	115-G(C)	229	Trautz, M. and Sorg, K. G.	523.0		1.0000	17.25	Same remarks as for curve 1 except $L_1 = 0.024\%$, $L_2 = 0.035\%$, $L_3 = 0.053\%$.
						0.7192	17.18	
						0.5145	16.99	
						0.3978	16.62	
						0.0777	14.23	
						0.0000	12.96	

TABLE 115-G(C)S.   SMOOTHED VISCOSITY VALUES AS A FUNCTION OF COMPOSITION FOR GASEOUS HYDROGEN-METHANE MIXTURES

Mole Fraction of CH_4	293.0 K [Ref. 229]	293.2 K [Ref. 1]	333.2 K [Ref. 1]	373.0 K [Ref. 229]	373.2 K [Ref. 1]	473.0 K [Ref. 229]	523.0 K [Ref. 229]
0.00	8.76	9.24	10.08	10.33	10.92	12.13	12.96
0.05	9.31	9.50	10.37	11.00	11.26	13.00	13.85
0.10	9.72	9.79	10.65	11.55	11.56	13.64	14.52
0.15	10.04	9.99	10.92	11.97	11.38	14.13	15.03
0.20	10.30	10.26	11.17	12.32	12.20	14.50	15.46
0.25	10.50	10.46	11.40	12.58	12.49	14.83	15.82
0.30	10.66	10.63	11.63	12.79	12.76	15.10	16.26
0.35	10.78	10.78	11.84	12.95	13.01	15.34	16.40
0.40	10.87	10.91	12.03	13.08	13.25	15.54	16.63
0.45	10.94	11.01	12.19	13.18	13.42	15.70	16.81
0.50	10.98	11.10	12.32	13.26	13.55	15.83	16.95
0.55	11.00	11.16	12.41	13.31	13.66	15.92	17.05
0.60	11.01	11.20	12.48	13.35	13.24	15.97	17.12
0.65	11.01	11.24	12.53	13.37	13.79	16.00	17.15
0.70	11.00	11.25	12.55	13.38	13.80	16.02	17.18
0.75	10.99	11.25	12.55	13.38	13.80	16.03	17.20
0.80	10.97	11.25	12.55	13.38	13.80	16.04	17.22
0.85	10.95	11.25	12.55	13.37	13.80	16.04	17.23
0.90	10.93	11.25	12.55	13.36	13.80	16.04	17.24
0.95	10.90	11.24	12.55	13.33	13.80	16.03	17.24
1.00	10.87	11.25	12.55	13.31	13.80	16.03	17.24

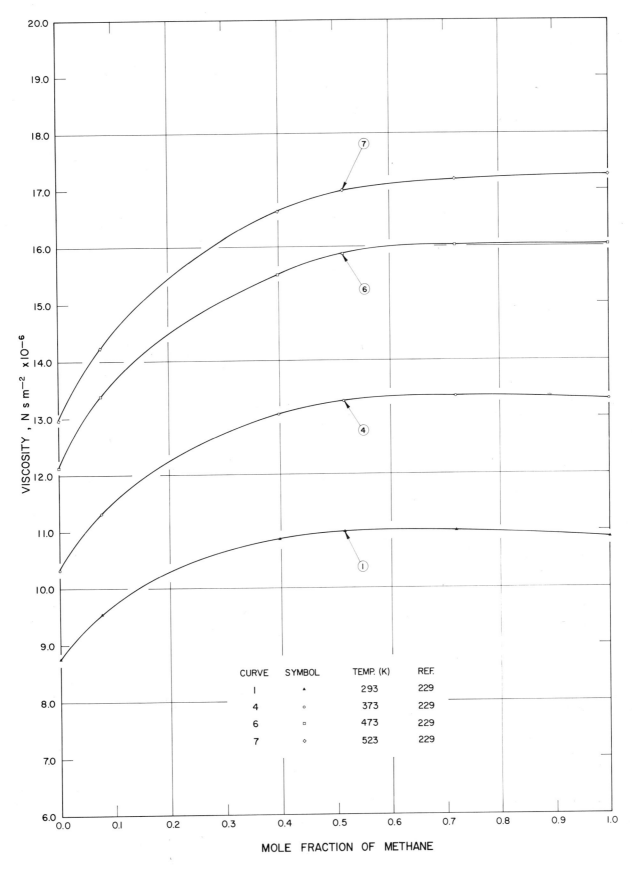

FIGURE II5 − G(C). VISCOSITY DATA AS A FUNCTION OF COMPOSITION
FOR GASEOUS HYDROGEN − METHANE MIXTURES

444

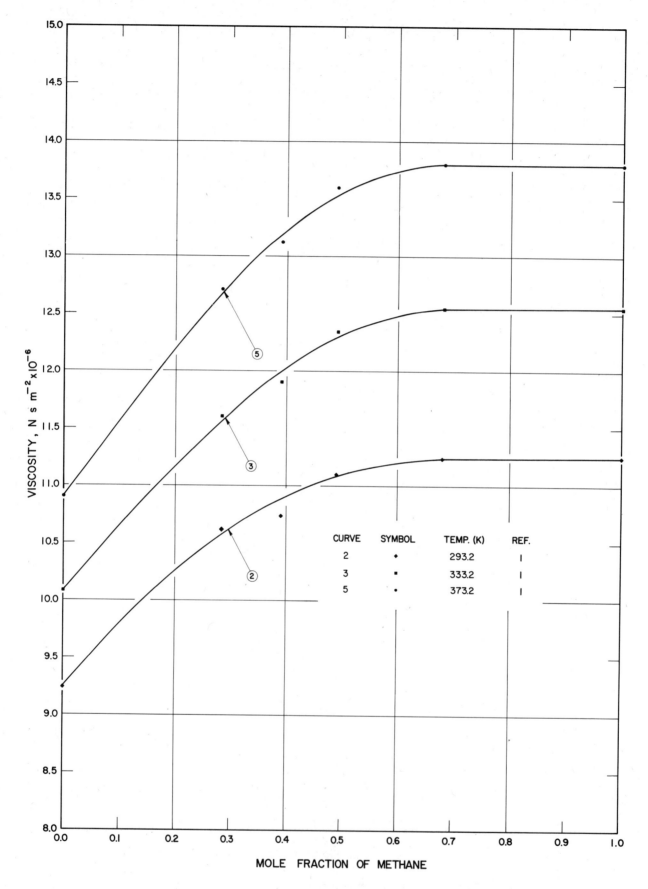

FIGURE II5 – G(C). VISCOSITY DATA AS A FUNCTION OF COMPOSITION
FOR GASEOUS HYDROGEN-METHANE MIXTURES (continued)

TABLE 116-G(C)E.   EXPERIMENTAL VISCOSITY DATA AS A FUNCTION OF COMPOSITION FOR GASEOUS HYDROGEN-NITRIC OXIDE MIXTURES

Cur. No.	Fig. No.	Ref. No.	Author(s)	Temp. (K)	Pressure (mm Hg)	Mole Fraction of NO	Viscosity $(N\ s\ m^{-2} \times 10^{-6})$	Remarks
1	116-G(C)	340	Alfons, K. and Walter, R.	273.2		0.0000	8.49	Modified Rankine type viscometer, calibrated with respect to air; $L_1 = 0.480\%$, $L_2 = 0.743\%$, $L_3 = 1.255\%$.
						0.1975	14.17	
						0.2299	14.52	
						0.2835	14.67	
						0.4508	15.95	
						0.7045	17.20	
						0.8503	17.50	
						1.0000	17.97	
2	116-G(C)	334	Strauss, W.A. and Edse, R.	293.2	751.64	1.0000	18.61	Capillary flow viscometer, relative measurements; $L_1 = 0.447\%$, $L_2 = 0.674\%$, $L_3 = 1.669\%$.
					751.96	0.8947	18.36	
					752.24	0.7932	18.02	
					752.49	0.6900	17.69	
					752.49	0.6204	17.36	
					753.05	0.4891	16.75	
					753.37	0.3926	16.04	
					753.48	0.2944	15.12	
					753.48	0.1931	14.01	
					752.98	0.1002	11.87	
					751.64	0.0000	8.88	
3	116-G(C)	334	Strauss, W.A. and Edse, R.	293.2	750.98	0.0000	9.01	Same remarks as for curve 2.
					751.28	0.0510	10.57	
					750.96	0.1499	12.97	
					751.12	0.2506	14.54	
					751.08	0.3425	15.85	
					751.32	0.4423	16.52	
					751.23	0.5393	17.19	
					751.23	0.6416	17.63	
					751.13	0.7453	18.04	
					751.23	0.8430	18.56	
					751.21	0.9524	18.62	
					751.15	1.0000	18.61	

TABLE 116-G(C)S.   SMOOTHED VISCOSITY VALUES AS A FUNCTION OF COMPOSITION FOR GASEOUS HYDROGEN-NITRIC OXIDE MIXTURES

Mole Fraction of NO	273.2 K [Ref. 340]	293.2 K [Ref. 334]
0.00	8.49	8.88
0.05	11.25	10.53
0.10	12.54	11.90
0.15	13.38	13.00
0.20	14.02	13.88
0.25	14.54	14.60
0.30	14.98	15.20
0.35	15.35	15.70
0.40	15.67	16.12
0.45	15.95	16.50
0.50	16.22	16.84
0.55	16.45	17.22
0.60	16.67	17.38
0.65	16.87	17.62
0.70	17.05	17.82
0.75	17.22	18.00
0.80	17.39	18.16
0.85	17.54	18.30
0.90	17.69	18.42
0.95	17.83	18.52
1.00	17.97	18.61

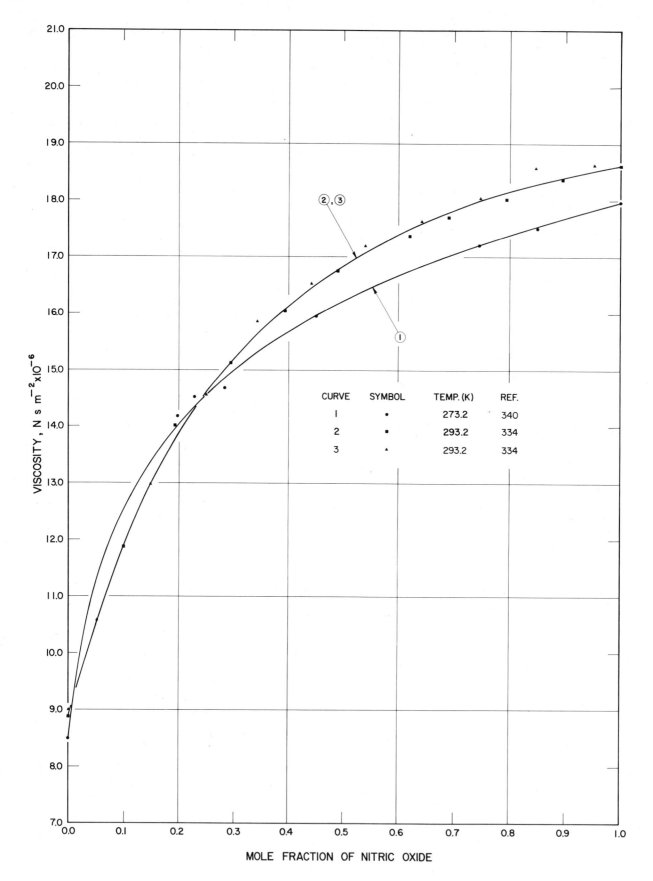

FIGURE 116 – G(C). VISCOSITY DATA AS A FUNCTION OF COMPOSITION
FOR GASEOUS HYDROGEN – NITRIC OXIDE MIXTURES

TABLE 117-G(C)E. EXPERIMENTAL VISCOSITY DATA AS A FUNCTION OF COMPOSITION FOR GASEOUS HYDROGEN–NITROGEN MIXTURES

Cur. No.	Fig. No.	Ref. No.	Author(s)	Temp. (K)	Pressure (mm Hg)	Mole Fraction of N_2	Viscosity ($N \ s \ m^{-2} \times 10^{-6}$)	Remarks
1	117-G(C)	252, 377	Van Itterbeek, A., Van Paemel, O., and Van Lierde, J.	82.2		0.000	3.62	Oscillating disk viscometer; accuracy of results not mentioned; $L_1 = 0.143\%$, $L_2 = 0.248\%$, $L_3 = 0.586\%$.
						0.160	4.73	
						0.351	5.09	
						0.441	5.19	
						0.620	5.33	
						0.759	5.37	
						1.000	5.44	
2	117-G(C)	252, 377	Van Itterbeek, A., et al.	90.2		0.000	3.92	Same remarks as for curve 1 except $L_1 = 0.141\%$, $L_2 = 0.290\%$, $L_3 = 0.789\%$.
						0.160	5.23	
						0.351	6.04	
						0.441	6.20	
						0.620	6.29	
						0.759	6.40	
						0.866	6.45	
						1.000	6.51	
3	117-G(C)	252, 377	Van Itterbeek, A., et al.	291.1		0.000	8.77	Same remarks as for curve 1 except $L_1 = 0.438\%$, $L_2 = 0.650\%$, $L_3 = 1.161\%$.
						0.160	12.51	
						0.441	15.60	
						0.620	16.60	
						0.759	16.77	
						0.866	17.42	
						1.000	17.52	
4	117-G(C)	252, 377	Van Itterbeek, A., et al.	291.2		0.000	8.82	Same remarks as for curve 1 except $L_1 = 0.065\%$, $L_2 = 0.106\%$, $L_3 = 0.246\%$.
						0.136	12.16	
						0.187	13.05	
						0.296	14.52	
						0.400	15.44	
						0.517	16.13	
						0.690	16.84	
						1.000	17.46	
5	117-G(C)	341	Pal, A.K. and Barua, A.K.	307.2	<100	0.0000	9.075	N_2 and H_2: better than 99.5 pure; oscillating disk viscometer, relative measurements; data agree with the literature values within 1.0%; $L_1 = 0.309\%$, $L_2 = 0.668\%$, $L_3 = 1.741\%$.
						0.2000	13.847	
						0.3991	15.958	
						0.5100	16.704	
						0.5794	16.995	
						0.7977	17.709	
						1.0000	18.163	
6	117-G(C)	341	Pal, A.K. and Barua, A.K.	325.4	<100	0.0000	9.445	Same remarks as for curve 5 except $L_1 = 0.303\%$, $L_2 = 0.588\%$, $L_3 = 1.452\%$.
						0.2000	14.254	
						0.3991	16.450	
						0.5100	17.300	
						0.5794	17.701	
						0.7977	18.500	
						1.0000	19.087	
7	117-G(C)	341	Pal, A.K. and Barua, A.K.	373.2	<100	0.0000	10.423	Same remarks as for curve 5 except $L_1 = 0.041\%$, $L_2 = 0.087\%$, $L_3 = 0.219\%$.
						0.2000	15.231	
						0.3991	18.120	
						0.5100	19.027	
						0.5794	19.501	
						0.7977	20.577	
						1.0000	21.012	
8	117-G(C)	341	Pal, A.K. and Barua, A.K.	422.7	<100	0.0000	11.490	Same remarks as for curve 5 except $L_1 = 0.211\%$, $L_2 = 0.378\%$, $L_3 = 0.915\%$.
						0.2005	16.470	
						0.3988	19.500	
						0.4996	20.636	
						0.5988	21.358	
						0.8002	22.258	
						1.0000	23.009	
9	117-G(C)	341	Pal, A.K. and Barua, A.K.	478.2	<100	0.0000	12.640	Same remarks as for curve 5 except $L_1 = 0.142\%$, $L_2 = 0.285\%$, $L_3 = 0.701\%$.
						0.2005	17.652	
						0.3988	21.399	
						0.4996	22.400	
						0.5988	23.130	
						0.8002	24.376	
						1.0000	25.259	

TABLE 117-G(C)S. SMOOTHED VISCOSITY VALUES AS A FUNCTION OF COMPOSITION FOR GASEOUS HYDROGEN-NITROGEN MIXTURES

Mole Fraction of N_2	82.2 K [Ref. 252]	90.2 K [Ref. 252]	291.1 K [Ref. 252]	291.2 K [Ref. 252]	307.2 K [Ref. 341]	325.4 K [Ref. 341]	373.2 K [Ref. 341]	422.7 K [Ref. 341]	478.2 K [Ref. 341]
0.00	3.62	3.92	8.77	8.82	9.07	9.94	10.42	11.49	12.64
0.05	4.08	4.38	10.14	10.26	10.23	10.64	11.66	12.72	13.81
0.10	4.33	4.80	11.33	11.45	11.38	11.80	12.90	13.95	15.02
0.15	4.68	5.16	12.36	12.44	12.56	12.99	14.10	15.14	16.27
0.20	4.86	5.48	13.20	13.24	13.61	14.05	15.23	16.30	16.50
0.25	4.98	5.72	13.88	13.96	14.42	14.90	16.21	17.38	18.70
0.30	5.06	5.90	14.44	14.55	15.07	15.58	16.97	18.28	19.71
0.35	5.12	6.02	14.92	15.04	15.58	16.11	17.60	19.04	20.56
0.40	5.16	6.14	15.32	15.42	15.99	16.55	18.12	19.68	21.45
0.45	5.21	6.20	15.66	15.77	16.33	16.94	18.56	20.20	21.84
0.50	5.26	6.25	15.96	16.05	16.62	17.27	18.96	20.62	22.33
0.55	5.29	6.29	16.20	16.30	16.86	17.54	19.32	21.00	22.76
0.60	5.32	6.32	16.43	16.52	17.07	17.80	19.62	21.29	23.16
0.65	5.35	6.36	16.62	16.70	17.26	18.00	19.90	21.56	23.50
0.70	5.37	6.38	16.79	16.86	17.43	18.19	20.15	21.81	23.82
0.75	5.38	6.42	16.94	17.00	17.58	18.36	20.36	22.04	24.11
0.80	5.40	6.44	17.06	17.11	17.72	18.52	20.55	22.24	24.38
0.85	5.41	6.46	17.18	17.21	17.84	18.70	20.70	22.44	24.62
0.90	5.42	6.48	17.30	17.29	17.96	18.82	20.82	22.63	24.84
0.95	5.43	6.48	17.50	17.38	18.06	18.96	20.93	22.82	25.06
1.00	5.44	6.50	17.52	17.46	18.16	19.09	21.01	23.01	25.27

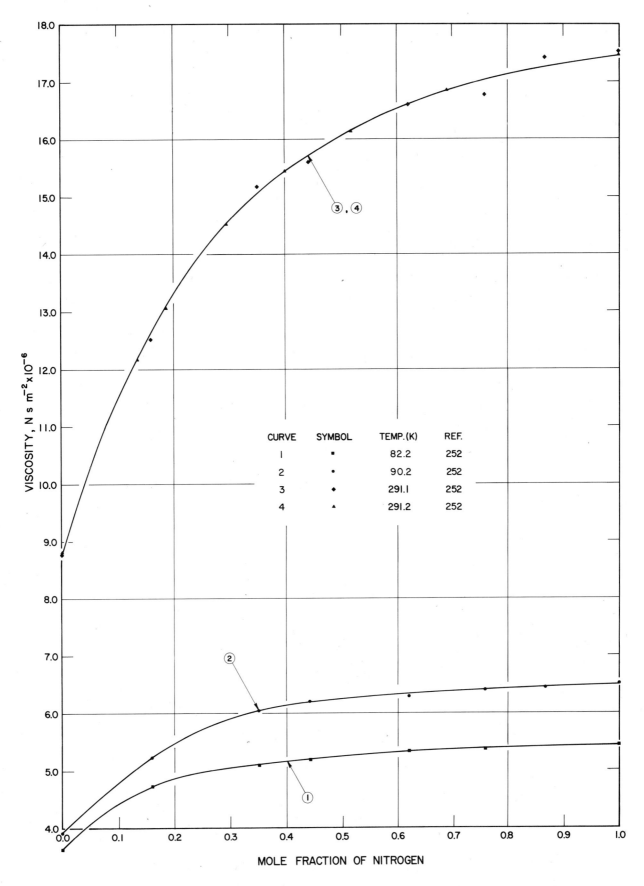

FIGURE 117-G(C). VISCOSITY DATA AS A FUNCTION OF COMPOSITION
FOR GASEOUS HYDROGEN – NITROGEN MIXTURES

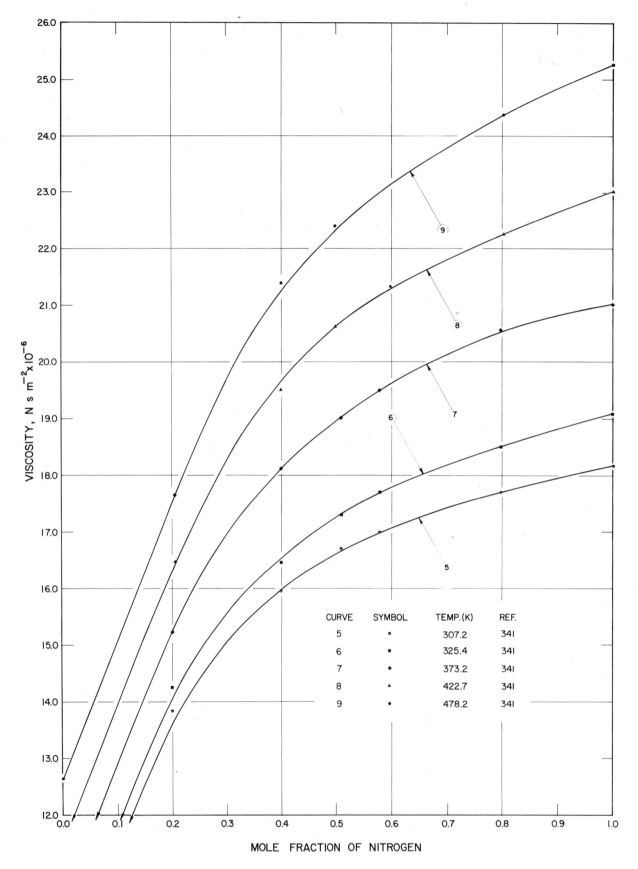

CURVE	SYMBOL	TEMP.(K)	REF.
5	•	307.2	341
6	■	325.4	341
7	♦	373.2	341
8	▲	422.7	341
9	•	478.2	341

MOLE FRACTION OF NITROGEN

FIGURE 117 – G(C). VISCOSITY DATA AS A FUNCTION OF COMPOSITION
FOR GASEOUS HYDROGEN-NITROGEN MIXTURES (continued)

TABLE 117-G(D)E. EXPERIMENTAL VISCOSITY DATA AS A FUNCTION OF DENSITY FOR GASEOUS
HYDROGEN-NITROGEN MIXTURES

Cur. No.	Fig. No.	Ref. No.	Author(s)	Mole Fraction of N$_2$	Temp. (K)	Density (g cm$^{-3}\cdot 10^{-4}$)	Viscosity (N s m^{-2} x 10^{-6})	Remarks
1	117-G(D)	327	Van Lierde, J.	1.000	90.2	0.598	6.46	Oscillating disk viscometer; original data reported as a function of pressure, density calculated from pressure using ideal gas equation.
						0.195	6.43	
						0.0383	6.28	
						0.0243	6.10	
						0.0111	5.96	
						0.00436	5.23	
						0.00276	4.48	
						0.00141	3.45	
						0.000799	2.32	
						0.000355	1.20	
2	117-G(D)	327	Van Lierde, J.	1.000	90.2	0.746	6.56	Same remarks as for curve 1.
						0.0736	6.32	
						0.0270	6.11	
						0.00987	5.76	
3	117-G(D)	327	Van Lierde, J.	0.866	90.2	0.415	6.64	Same remarks as for curve 1.
						0.0321	6.12	
						0.00934	5.64	
						0.00471	5.05	
						0.00185	3.80	
						0.000681	2.13	
						0.000319	1.24	
4	117-G(D)	327	Van Lierde, J.	0.866	90.2	0.480	6.41	Same remarks as for curve 1.
						0.0249	6.11	
						0.0129	5.71	
						0.00528	5.69	
						0.00151	3.64	
						0.000786	2.19	
5	117-G(D)	327	Van Lierde, J.	0.759	90.2	0.387	6.80	Same remarks as for curve 1.
						0.0255	6.30	
						0.0106	5.54	
						0.00549	4.80	
						0.00221	3.97	
						0.000580	1.96	
6	117-G(D)	327	Van Lierde, J.	0.759	90.2	0.426	6.30	Same remarks as for curve 1.
						0.0227	6.21	
						0.00725	5.44	
						0.00406	4.86	
						0.00198	4.21	
						0.00100	3.26	
7	117-G(D)	327	Van Lierde, J.	0.759	90.2	0.484	6.45	Same remarks as for curve 1.
						0.0334	6.06	
						0.00782	5.43	
						0.00329	4.44	
						0.00187	3.85	
						0.000724	2.69	
						0.000294	1.47	
8	117-G(D)	327	Van Lierde, J.	0.620	90.2	0.419	6.44	Same remarks as for curve 1.
						0.0398	6.62	
						0.0219	6.40	
						0.00455	5.26	
						0.00228	4.49	
						0.00135	3.79	
						0.000958	3.63	
9	117-G(D)	327	Van Lierde, J.	0.441	90.2	0.264	6.20	Same remarks as for curve 1.
						0.0235	6.04	
						0.00743	5.38	
						0.00306	4.71	
						0.00137	3.94	
						0.000878	3.31	
						0.000535	2.73	
						0.000247	1.56	
10	117-G(D)	327	Van Lierde, J.	0.351	90.2	0.277	5.95	Same remarks as for curve 1.
						0.0272	5.90	
						0.00914	5.44	
						0.00539	5.34	
						0.00318	4.86	
						0.00183	4.39	
						0.000791	3.35	
						0.000626	2.73	
						0.000327	2.16	

TABLE 117-G(D)E.   EXPERIMENTAL VISCOSITY DATA AS A FUNCTION OF DENSITY FOR GASEOUS
HYDROGEN-NITROGEN MIXTURES (continued)

Cur. No.	Fig. No.	Ref. No.	Author(s)	Mole Fraction of N_2	Temp. (K)	Density (g cm$^{-3} \cdot 10^{-4}$)	Viscosity (N s m^{-2} x 10^{-6})	Remarks
11	117-G(D)	327	Van Lierde, J.	0.1600	90.2	0.126	5.27	Same remarks as for curve 1.
						0.121	5.13	
						0.00950	5.03	
						0.00219	4.10	
						0.00160	4.09	
						0.000517	2.62	
						0.000312	2.19	
						0.000171	1.19	
12	117-G(D)	327	Van Lierde, J.	0.0000	90.2	0.0322	3.72	Same remarks as for curve 1.
						0.00874	3.93	
						0.00155	3.58	
						0.000355	2.50	
						0.000182	1.74	
						0.0000688	1.03	
						0.0000269	0.64	
13	117-G(D)	329	Kestin, J. and Yata, J.	0.8407	293.2	0.02318	17.600	N_2: 99.999 pure, H_2: 99.999 pure; oscillating disk visco-meter; accuracy ± 0.1% and precision ± 0.05%.
						0.01482	17.488	
						0.004968	17.365	
						0.001046	17.310	
14	117-G(D)	329	Kestin, J. and Yata, J.	0.6721	293.2	0.02025	17.121	Same remarks as for curve 13.
						0.01217	17.019	
						0.004055	16.926	
						0.000860	16.888	
15	117-G(D)	329	Kestin, J. and Yata, J.	0.4879	293.2	0.01527	16.234	Same remarks as for curve 13.
						0.009171	16.159	
						0.003059	16.100	
						0.000637	16.071	
16	117-G(D)	329	Kestin, J. and Yata, J.	0.2750	293.2	0.009253	14.420	Same remarks as for curve 13.
						0.005694	14.391	
						0.001898	14.318	
						0.000399	14.332	
17	117-G(D)	329	Kestin, J. and Yata, J.	0.1627	293.2	0.006159	12.802	Same remarks as for curve 13.
						0.003856	12.781	
						0.001297	12.759	
						0.000273	12.744	
18	117-G(D)	329	Kestin, J. and Yata, J.	0.0961	293.2	0.004411	11.473	Same remarks as for curve 13.
						0.002774	11.465	
						0.000938	11.445	
						0.000200	11.438	
19	117-G(D)	329	Kestin, J. and Yata, J.	0.0000	293.2	0.001936	8.829	Same remarks as for curve 13.
						0.001913	8.831	
						0.001582	8.826	
						0.001242	8.825	
						0.0008333	8.834	
						0.0004137	8.829	
						0.0000876	8.827	
20	117-G(D)	329	Kestin, J. and Yata, J.	1.0000	303.2	0.02648	18.367	Same remarks as for curve 13.
						0.02152	18.291	
						0.01701	18.163	
						0.01130	18.098	
						0.005650	18.036	
21	117-G(D)	329	Kestin, J. and Yata, J.	0.8407	303.2	0.02259	18.045	Same remarks as for curve 13.
						0.01445	17.939	
						0.00480	17.824	
						0.00102	17.782	
22	117-G(D)	329	Kestin, J. and Yata, J.	0.6721	303.2	0.01847	17.544	Same remarks as for curve 13.
						0.01176	17.464	
						0.003919	17.381	
						0.000815	17.351	
23	117-G(D)	329	Kestin, J. and Yata, J.	0.4879	303.2	0.01409	16.640	Same remarks as for curve 13.
						0.01409	16.636	
						0.008761	16.582	
						0.002947	16.520	
						0.000609	16.490	
24	117-G(D)	329	Kestin, J. and Yata, J.	0.2750	303.2	0.008755	14.786	Same remarks as for curve 13.
						0.005509	14.754	
						0.001841	14.720	
						0.000396	14.706	
25	117-G(D)	329	Kestin, J. and Yata, J.	0.1627	303.2	0.006000	13.120	Same remarks as for curve 13.
						0.003750	13.108	
						0.001255	13.088	
						0.000268	13.067	

TABLE 117-G(D)E.   EXPERIMENTAL VISCOSITY DATA AS A FUNCTION OF DENSITY FOR GASEOUS
HYDROGEN-NITROGEN MIXTURES (continued)

Cur. No.	Fig. No.	Ref. No.	Author(s)	Mole Fraction of N_2	Temp. (K)	Density (g cm^{-3} · 10^{-1})	Viscosity (N s m^{-2} x10^{-6})	Remarks
26	117-G(D)	329	Kestin, J. and Yata, J.	0.0961	303.2	0.004300	11.768	Same remarks as for curve 13.
						0.002700	11.748	
						0.000901	11.732	
						0.000192	11.726	
27	117-G(D)	329	Kestin, J. and Yata, J.	0.0000	303.2	0.001891	9.039	Same remarks as for curve 13.
						0.001209	9.031	
						0.0004042	9.027	
						0.0000847	9.025	

TABLE 117-G(D)S. SMOOTHED VISCOSITY VALUES AS A FUNCTION OF DENSITY FOR GASEOUS HYDROGEN-NITROGEN MIXTURES

Density (g cm^{-3} x 10^{-4})	Mole Fraction of Nitrogen					
	0.0000 (90.2 K) [Ref. 327]	0.1600 (90.2 K) [Ref. 327]	0.3510 (90.2 K) [Ref. 327]	0.4410 (90.2 K) [Ref. 327]	0.6200 (90.2 K) [Ref. 327]	1.0000 (90.2 K) [Ref. 327]
0.005	3.645	4.978	5.261			
0.010	3.679	5.038	5.760			5.906
0.015	3.695	5.063	5.825	5.917		6.060
0.020	3.706	5.081	5.864	6.001		6.144
0.025	3.712	5.095	5.890	6.052		6.200
0.030	3.717	5.106	5.904	6.086		6.238
0.035	3.720	5.115	5.911	6.110		6.265
0.040	3.721	5.122	5.917	6.127		6.286
0.045		5.130	5.920	6.139		6.297
0.050		5.133	5.922	6.146		6.304
0.075		5.156	5.930	6.155	6.261	6.328
0.100		5.178	5.932	6.162	6.279	6.350
0.125		5.199	5.938	6.170	6.291	6.372
0.150		5.220	5.940	6.177	6.309	6.394
0.175			5.945	6.182	6.322	6.415
0.200			5.948	6.190	6.339	6.435
0.250			5.950	6.198	6.362	6.474
0.300			5.952		6.390	6.511
0.350					6.411	6.549
0.400					6.436	6.581
0.450						6.619

Density (g cm^{-3} x 10^{-4})	Mole Fraction of Nitrogen						
	0.0000 (293.2 K) [Ref. 329]	0.0961 (293.2 K) [Ref. 329]	0.1627 (293.2 K) [Ref. 329]	0.2750 (293.2 K) [Ref. 329]	0.4879 (293.2 K) [Ref. 329]	0.6721 (293.2 K) [Ref. 329]	0.8407 (293.2 K) [Ref. 329]
0.00050	8.822	11.440					
0.00100	8.822	11.444	12.757				
0.00125				14.342	16.071	16.899	
0.00150	8.820	11.450	12.762				
0.00200	8.820	11.459	12.778				
0.00250	8.820	11.460	12.780	14.360	16.093	16.900	17.330
0.00300		11.468	12.781				
0.00350		11.472	12.790				
0.00375				14.378	16.110		
0.00400		11.478	12.794				
0.00450		11.479	12.798				
0.00500		11.479	12.800	14.391	16.124	16.932	17.362
0.00600			12.800				
0.00625				14.400			
0.00750				14.410	16.158	16.968	17.400
0.00875				14.420			
0.01000					16.180	17.000	17.438
0.01125					16.192		
0.01250					16.206	17.028	17.460
0.01500					16.230	17.055	17.490
0.01750						17.082	17.520
0.02000						17.118	17.560

TABLE 117-G(D)S.   SMOOTHED VISCOSITY VALUES AS A FUNCTION OF DENSITY FOR GASEOUS HYDROGEN-NITROGEN MIXTURES (continued)

Density (g cm^{-3} x 10^{-4})	Mole Fraction of Nitrogen							
	0.0000 (303.2 K) [Ref. 329]	0.0961 (303.2 K) [Ref. 329]	0.1627 (303.2 K) [Ref. 329]	0.2750 (303.2 K) [Ref. 329]	0.4879 (303.2 K) [Ref. 329]	0.6721 (303.2 K) [Ref. 329]	0.8407 (303.2 K) [Ref. 329]	1.0000 (303.2 K) [Ref. 329]
0.00050	9.021	11.724						
0.00100	9.028	11.736	13.080	14.720				
0.00125					16.500	17.358		
0.00150	9.031	11.740						
0.00200	9.037	11.744	13.098	14.736				
0.00250	9.040	11.748	13.100	14.744	16.520	17.370	17.800	18.021
0.00300		11.750	13.108	14.744				
0.00350		11.754						
0.00375					16.538			
0.00400		11.758	13.118	14.756				
0.00450		11.760	13.118					
0.00500		11.760	13.120	14.760	16.548	17.401	17.824	18.035
0.00600			13.122	14.772				
0.00625					16.562			
0.00700				14.780				
0.00750				14.781	16.578	17.438	17.850	18.050
0.00850				14.788				
0.01000					16.600	17.458	17.878	18.078
0.01250					16.628	17.478	17.909	18.102
0.01375					16.640			
0.01500						17.510	17.940	18.140
0.01750						17.540	17.978	18.180
0.01850						17.558		
0.02000							18.002	
0.02250							18.038	18.276
0.02500								18.328

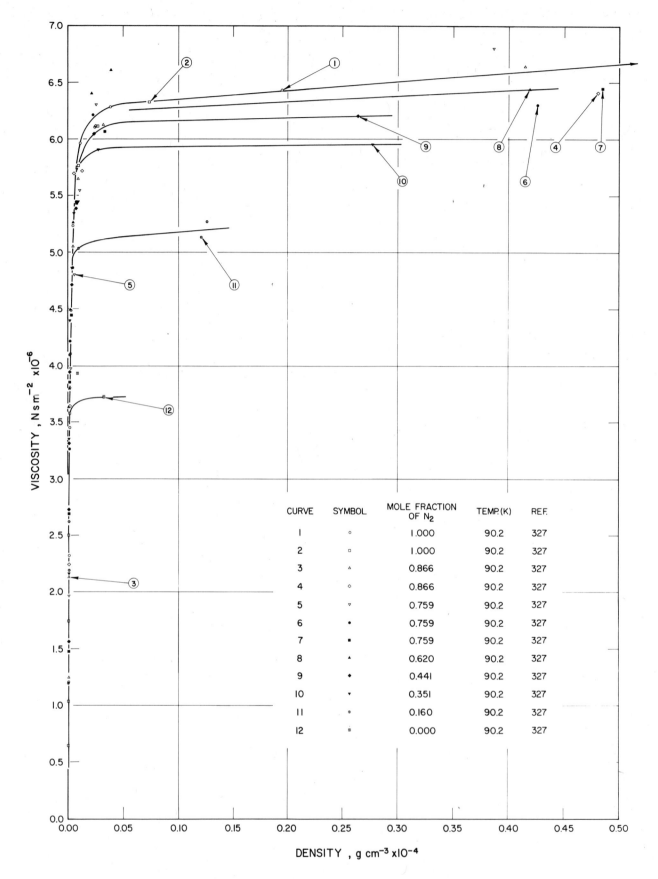

CURVE	SYMBOL	MOLE FRACTION OF N_2	TEMP.(K)	REF.
1	○	1.000	90.2	327
2	□	1.000	90.2	327
3	△	0.866	90.2	327
4	◇	0.866	90.2	327
5	▽	0.759	90.2	327
6	•	0.759	90.2	327
7	■	0.759	90.2	327
8	▲	0.620	90.2	327
9	◆	0.441	90.2	327
10	▼	0.351	90.2	327
11	○	0.160	90.2	327
12	⊡	0.000	90.2	327

DENSITY , g cm^{-3} x10^{-4}

FIGURE 117–G(D). VISCOSITY DATA AS A FUNCTION OF DENSITY
FOR GASEOUS HYDROGEN – NITROGEN MIXTURES

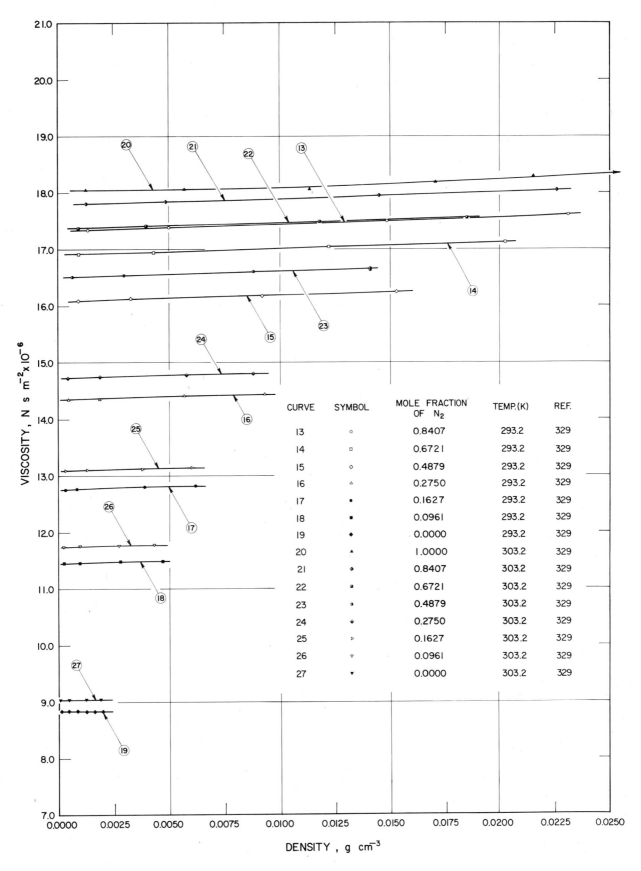

CURVE	SYMBOL	MOLE FRACTION OF N_2	TEMP.(K)	REF.
13	○	0.8407	293.2	329
14	□	0.6721	293.2	329
15	◇	0.4879	293.2	329
16	△	0.2750	293.2	329
17	●	0.1627	293.2	329
18	■	0.0961	293.2	329
19	◆	0.0000	293.2	329
20	▲	1.0000	303.2	329
21	◈	0.8407	303.2	329
22	▣	0.6721	303.2	329
23	◉	0.4879	303.2	329
24	◆	0.2750	303.2	329
25	▷	0.1627	303.2	329
26	▽	0.0961	303.2	329
27	▼	0.0000	303.2	329

FIGURE 117-G(D). VISCOSITY DATA AS A FUNCTION OF DENSITY
FOR GASEOUS HYDROGEN-NITROGEN MIXTURES (continued)

TABLE 118-G(C)E. EXPERIMENTAL VISCOSITY DATA AS A FUNCTION OF COMPOSITION FOR GASEOUS HYDROGEN-NITROUS OXIDE MIXTURES

Cur. No.	Fig. No.	Ref. No.	Author(s)	Temp. (K)	Pressure	Mole Fraction of N_2O	Viscosity ($N \ s \ m^{-2} \times 10^{-6}$)	Remarks
1	118-G(C)	234	Trautz, M. and Kurz, F.	300.0		1.0000	14.88	N_2O: 1.3 p per 1000, H_2: made by
						0.6011	14.81	electrolysis; capillary method,
						0.4039	14.51	d = 0.018 cm; L_1 = 0.027%, L_2 =
						0.2143	13.48	0.044%, L_3 = 0.083%.
						0.0000	8.91	
2	118-G(C)	234	Trautz, M. and Kurz, F.	400.0		1.0000	19.43	Same remarks as for curve 1 except
						0.6011	19.07	L_1 = 0.000%, L_2 = 0.000%, L_3 =
						0.4039	18.49	0.000%.
						0.2143	16.84	
						0.0000	10.81	
3	118-G(C)	234	Trautz, M. and Kurz, F.	500.0		1.0000	23.55	Same remarks as for curve 1 except
						0.6011	22.92	L_1 = 0.002%, L_2 = 0.004%, L_3 =
						0.4039	22.06	0.009%.
						0.2143	19.90	
						0.0000	12.56	
4	118-G(C)	234	Trautz, M. and Kurz, F.	550.0		1.0000	25.55	Same remarks as for curve 1 except
						0.6011	24.77	L_1 = 0.028%, L_2 = 0.063%, L_3 =
						0.4039	23.76	0.140%.
						0.2143	21.37	
						0.0000	13.41	

TABLE 118-G(C)S. SMOOTHED VISCOSITY VALUES AS A FUNCTION OF COMPOSITION FOR GASEOUS HYDROGEN-NITROUS OXIDE MIXTURES

Mole Fraction of N_2O	300.0 K [Ref. 234]	400.0 K [Ref. 234]	500.0 K [Ref. 234]	550.0 K [Ref. 234]
0.00	8.91	10.81	12.56	13.41
0.05	10.54	12.86	14.82	16.30
0.10	11.81	14.58	16.95	18.30
0.15	12.70	15.78	18.54	19.86
0.20	13.32	16.64	19.62	21.10
0.25	13.77	17.28	20.49	22.02
0.30	14.09	17.79	21.14	22.74
0.35	14.33	18.18	21.64	23.30
0.40	14.52	18.48	22.04	23.74
0.45	14.64	18.70	22.35	24.08
0.50	14.73	18.86	22.58	24.36
0.55	14.79	18.97	22.76	24.58
0.60	14.83	19.06	22.92	24.74
0.65	14.85	19.15	23.06	24.92
0.70	14.87	19.22	23.18	25.05
0.75	14.88	19.28	23.28	25.16
0.80	14.89	19.33	23.36	25.26
0.85	14.89	19.37	23.44	25.35
0.90	14.89	19.40	23.48	25.42
0.95	14.89	19.42	23.53	25.48
1.00	14.88	19.43	23.55	25.55

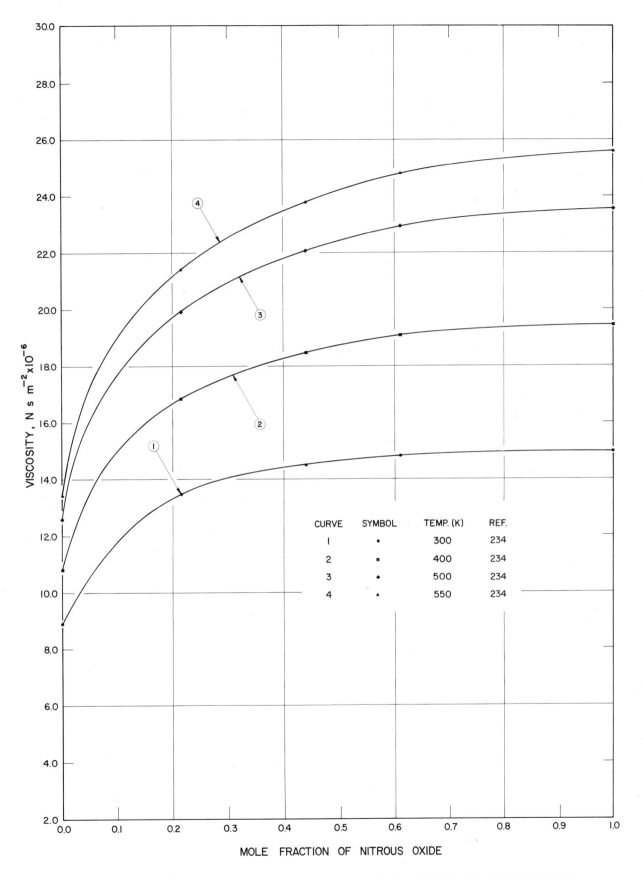

FIGURE 118-G(C). VISCOSITY DATA AS A FUNCTION OF COMPOSITION
FOR GASEOUS HYDROGEN-NITROUS OXIDE MIXTURES

TABLE 119-G(C)E. EXPERIMENTAL VISCOSITY DATA AS A FUNCTION OF COMPOSITION FOR GASEOUS HYDROGEN–OXYGEN MIXTURES

Cur. No.	Fig. No.	Ref. No.	Author(s)	Temp. (K)	Pressure (mm Hg)	Mole Fraction of O_2	Viscosity (N s m^{-2} x 10^{-6})	Remarks
1	119-G(C)	334	Strauss, W. A. and Edse, R.	293.2	741.1	0.000	8.78	Capillary flow viscometer, relative measurements; $L_1 = 0.245\%$, $L_2 = 0.495\%$, $L_3 = 1.767\%$.
					742.6	0.052	10.45	
					744.1	0.100	12.04	
					745.7	0.153	13.39	
					747.2	0.206	14.63	
					748.7	0.255	15.55	
					750.5	0.278	17.28	
					752.2	0.359	17.13	
					752.9	0.406	17.78	
					751.6	0.447	18.10	
					750.4	0.493	18.50	
					749.3	0.543	18.87	
					749.3	0.591	19.32	
					748.4	0.651	19.45	
					748.1	0.700	19.65	
					747.6	0.748	19.81	
					747.0	0.795	19.98	
					746.4	0.847	20.07	
					745.8	0.895	20.20	
					745.2	0.955	20.26	
					744.7	1.000	20.24	
2	119-G(C)	327	Van Lierde, J.	293.6		0.000	8.85	Oscillating disk viscometer; $L_1 = 0.093\%$, $L_2 = 0.197\%$, $L_3 = 0.498\%$.
						0.161	14.09	
						0.273	16.15	
						0.380	17.39	
						0.527	18.68	
						0.670	19.54	
						1.000	20.40	
3	119-G(C)	227	Trautz, M. and Melster, A.	300.0		1.0000	20.57	Capillary method, R = 0.2019 mm; $L_1 = 0.487\%$, $L_2 = 0.686\%$, $L_3 = 1.220\%$.
						0.8165	20.19	
						0.6055	19.25	
						0.3970	17.84	
						0.2192	14.94	
						0.0000	8.89	
4	119-G(C)	227	Trautz, M. and Melster, A.	400.0		1.0000	25.68	Same remarks as for curve 4 except $L_1 = 0.260\%$, $L_2 = 0.373\%$, $L_3 = 0.642\%$.
						0.8165	25.07	
						0.6055	23.81	
						0.3970	21.92	
						0.2192	18.58	
						0.0000	10.87	
5	119-G(C)	227	Trautz, M. and Melster, A.	500.0		1.0000	30.17	Same remarks as for curve 4 except $L_1 = 0.179\%$, $L_2 = 0.316\%$, $L_3 = 0.641\%$.
						0.8165	29.50	
						0.6055	27.90	
						0.3970	25.56	
						0.2192	21.58	
						0.0000	12.59	
6	119-G(C)	227	Trautz, M. and Melster, A.	550.0		1.0000	32.20	Same remarks as for curve 4 except $L_1 = 0.196\%$, $L_2 = 0.356\%$, $L_3 = 0.774\%$.
						0.8165	31.47	
						0.6055	29.78	
						0.3970	27.33	
						0.2192	22.88	
						0.0000	13.81	

TABLE 119-G(C)S.   SMOOTHED VISCOSITY VALUES AS A FUNCTION OF COMPOSITION FOR GASEOUS HYDROGEN-OXYGEN MIXTURES

Mole Fraction of O_2	(293.2 K) [Ref. 334]	(293.6 K) [Ref. 327]	(300.0 K) [Ref. 227]	(400.0 K) [Ref. 227]	(500.0 K) [Ref. 227]	(550.0 K) [Ref. 227]
0.00	8.78	8.85	8.89	10.87	12.59	13.81
0.05	10.40	10.73	10.70	13.30	15.00	16.30
0.10	11.92	12.46	12.25	15.30	17.21	18.60
0.15	13.28	13.83	13.60	16.92	19.22	20.62
0.20	14.48	14.88	14.70	18.25	21.00	22.40
0.25	15.49	15.72	15.60	19.40	22.42	23.85
0.30	16.33	16.43	16.20	20.33	23.60	24.17
0.35	17.05	17.05	17.09	21.16	24.62	26.21
0.40	17.64	17.58	17.65	21.88	25.55	27.19
0.45	18.14	18.14	18.15	22.48	26.28	28.00
0.50	18.55	18.47	18.60	23.04	26.94	28.70
0.55	18.91	18.85	18.95	25.55	27.51	29.32
0.60	19.21	19.16	19.28	23.90	28.01	29.86
0.65	19.48	19.45	19.55	24.25	28.44	30.35
0.70	19.69	19.68	19.78	24.35	28.82	30.75
0.75	19.87	19.88	19.98	24.80	29.15	31.09
0.80	20.00	20.03	20.12	25.05	29.42	31.38
0.85	20.09	20.16	20.28	25.22	29.65	31.58
0.90	20.16	20.25	20.40	25.40	29.85	31.82
0.95	20.22	20.33	20.48	25.55	30.03	32.02
1.00	20.24	20.40	20.51	25.68	30.17	32.20

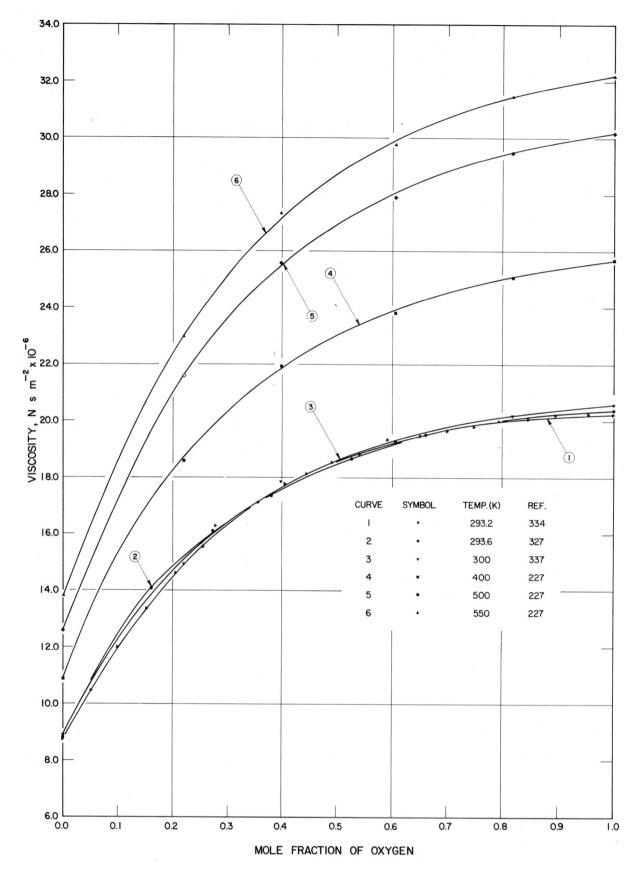

FIGURE 119-G(C). VISCOSITY DATA AS A FUNCTION OF COMPOSITION
FOR GASEOUS HYDROGEN - OXYGEN MIXTURES

TABLE 120-G(C)E.   EXPERIMENTAL VISCOSITY DATA AS A FUNCTION OF COMPOSITION FOR GASEOUS
HYDROGEN-PROPANE MIXTURES

Cur. No.	Fig. No.	Ref. No.	Author(s)	Temp. (K)	Pressure (atm)	Mole Fraction of C_3H_8	Viscosity (N s m^{-2} x 10^{-6})	Remarks
1	120-G(C)	340	Alfons, K. and Walter, R.	273.2		0.0000	8.601	Modified Rankine type viscometer,
						0.0313	8.900	calibrated with respect to air; $L_1 =$
						0.0785	9.390	0.510%, $L_2 = 0.777\%$, $L_3 = 2.036\%$.
						0.0891	9.500	
						0.1500	9.700	
						0.2218	9.600	
						0.3271	9.200	
						0.5182	8.700	
						0.6978	8.100	
						0.8037	7.700	
						1.0000	7.520	
2	120-G(C)	234	Trautz, M. and Kurz, F.	300.0		1.0000	8.17	C_3H_8: pure, H_2: made by electro-
						0.6296	8.74	lysis; capillary method, d = 0.018
						0.2118	9.85	cm; $L_1 = 0.0556\%$, $L_2 = 0.0939\%$,
						0.0775	9.70	$L_3 = 0.2250\%$.
						0.0000	8.91	
3	120-G(C)	234	Trautz, M. and Kurz, F.	400.0		1.0000	10.70	Same remarks as for curve 2 except
						0.6296	11.30	$L_1 = 0.052\%$, $L_2 = 0.123\%$, $L_3 =$
						0.2118	12.33	0.340%.
						0.0775	11.94	
						0.0000	10.81	
4	120-G(C)	234	Trautz, M. and Kurz, F.	500.0		1.0000	13.08	Same remarks as for curve 2 except
						0.6296	13.66	$L_1 = 0.070\%$, $L_2 = 0.121\%$, $L_3 =$
						0.2118	14.59	0.301%.
						0.0775	13.92	
						0.0000	12.56	
5	120-G(C)	234	Trautz, M. and Kurz, F.	550.0		1.0000	14.22	Same remarks as for curve 2 except
						0.6296	14.78	$L_1 = 0.033\%$, $L_2 = 0.066\%$, $L_3 =$
						0.2118	15.66	0.135%.
						0.0775	14.85	
						0.0000	13.47	

TABLE 120-G(S).   SMOOTHED VISCOSITY VALUES AS A FUNCTION OF COMPOSITION FOR GASEOUS
HYDROGEN-PROPANE MIXTURES

Mole Fraction of C_3H_8	(273.2 K) [Ref. 340]	(300.0 K) [Ref. 234]	(400.0 K) [Ref. 234]	(500.0 K) [Ref. 234]	(550.0 K) [Ref. 234]
0.00	8.60	8.89	10.81	12.56	13.47
0.05	9.05	9.50	11.69	13.60	14.45
0.10	9.53	9.80	12.10	14.14	15.10
0.15	9.65	9.90	12.28	14.46	15.50
0.20	9.62	9.88	12.32	14.68	15.66
0.25	9.48	9.78	12.29	14.62	15.62
0.30	9.33	9.64	12.18	14.56	15.55
0.35	9.17	9.48	12.02	14.44	15.45
0.40	9.00	9.31	11.82	14.25	15.34
0.45	8.84	9.15	11.66	14.06	15.22
0.50	8.67	9.02	11.54	13.92	15.10
0.55	8.51	8.90	11.44	13.80	14.97
0.60	8.37	8.80	11.34	13.70	14.84
0.65	8.23	8.69	11.25	13.60	14.72
0.70	8.10	8.58	11.15	13.50	14.62
0.75	7.98	8.48	11.04	13.40	14.51
0.80	7.86	8.39	10.95	13.31	14.41
0.85	7.76	8.32	10.88	13.23	14.34
0.90	7.67	8.27	10.81	13.16	14.30
0.95	7.59	8.22	10.75	13.12	14.26
1.00	7.52	8.17	10.70	13.07	14.22

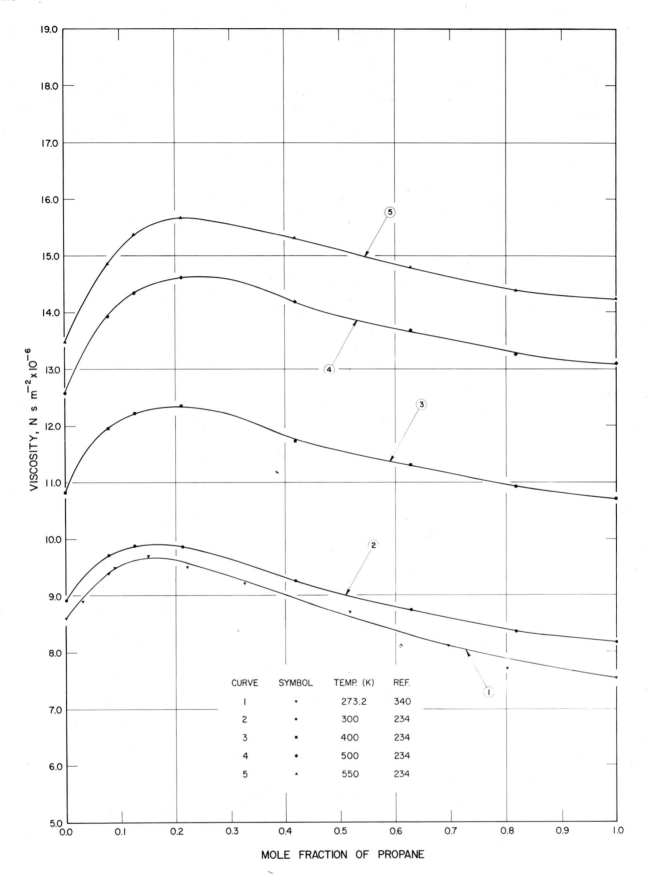

FIGURE 120-G(C).   VISCOSITY DATA AS A FUNCTION OF COMPOSITION
FOR GASEOUS HYDROGEN–PROPANE MIXTURES

TABLE 121-L(T)E.   EXPERIMENTAL VISCOSITY DATA AS A FUNCTION OF TEMPERATURE FOR LIQUID
METHANE-NITROGEN MIXTURES

Cur. No.	Fig. No.	Ref. No.	Author(s)	Mole Fraction of N_2	Pressure (atm)	Temp. (K)	Viscosity (N s m^{-2} x 10^{-6})	Remarks
1	121-L(T)	70	Gerf, S. F. and Galkov, G. I.	1.000		66.2 69.0 71.1 73.3 75.4 76.7 77.3	247.0 217.0 201.0 184.0 171.0 164.0 159.0	N_2 and CH_4: 99.8 pure; capillary flow viscometer, relative measurements.
2	121-L(T)	70	Gerf, S. F. and Galkov, G. I.	0.812		64.8 65.3 65.7 67.6 68.2 70.0 70.2 71.7 74.3 76.7 79.1 80.3	286.0 280.0 269.0 245.0 240.0 223.0 221.0 211.0 190.0 174.0 164.0 154.0	Same remarks as for curve 1.
3	121-L(T)	70	Gerf, S. F. and Galkov, G. I.	0.608		68.2 70.1 71.7 75.1 78.0 81.6 84.4	275.0 253.0 237.0 210.0 188.0 167.0 152.0	Same remarks as for curve 1.
4	121-L(**T**)	70	Gerf, S. F. and Galkov, G. I.	0.412		78.5 81.4 84.7 86.1	217.0 195.0 178.0 171.0	Same remarks as for curve 1.
5	121-L(T)	70	Gerf, S. F. and Galkov, G. I.	0.196		84.1 85.0 87.8 89.8	214.0 206.0 186.0 172.0	Same remarks as for curve 1.
6	121-L(T)	70	Gerf, S. F. and Galkov, G. I.	0.000		94.4 98.3 107.4 108.8 111.2	187.0 162.0 144.0 125.0 119.0	Same remarks as for curve 1.
7	121-L(T)	344	Gerf, S. F. and Galkov, G. I.	0.239		96.6 103.6 109.6 132.4 145.8	155.0 133.0 118.0 75.0 66.0	Oscillating cylinder; η accuracy ±3%, mixture analysis ±0.2%.
8	121-L(T)	70	Gerf, S. F. and Galkov, G. I.	0.494		93.4 96.4 103.8 110.8 138.8 139.4 146.8	133.0 125.0 113.0 103.0 64.0 64.0 58.0	Same remarks as for curve 7.
9	121-L(T)	70	Gerf, S. F. and Galkov, G. I.	0.727		96.2 102.5 110.3 128.8 137.2 146.5	116.0 103.0 93.0 68.0 57.0 44.0	Same remarks as for curve 7.

TABLE 121-L(T)S. SMOOTHED VISCOSITY VALUES AS A FUNCTION OF TEMPERATURE FOR LIQUID METHANE-NITROGEN MIXTURES

Temp. (K)	Mole Fraction of Nitrogen								
	0.000 [Ref. 70]	0.196 [Ref. 70]	0.412 [Ref. 70]	0.608 [Ref. 70]	0.812 [Ref. 70]	1.000 [Ref. 70]	0.239 [Ref. 344]	0.494 [Ref. 344]	0.727 [Ref. 344]
65.0					283.0				
66.0						248.0			
67.5				284.0	246.0	232.0			
70.0				254.8	223.0	208.4			
72.5				229.0	203.5	185.8			
75.0				210.4	185.5	173.0			
77.5				191.5	170.4	159.2			
78.0			220.0						
80.0			205.2	176.4	156.4	144.8			
82.0			192.2						
82.5			189.5	162.0	143.5	131.7			
83.0		220.0							
84.0			181.0						
85.0		216.4	176.0	149.2	131.0	119.0			
86.0		198.5	171.8						
87.5		187.8	164.4	134.8	119.5				
88.0		184.2							
90.0		177.2							
91.0	190.0	171.0							
94.0	190.0								
95.0	183.0						159.0	128.6	
96.0	176.0								
97.5	166.8								
98.0	163.9								
100.0	153.9						143.8	119.6	107.2
102.5	143.5								
104.0	138.0								
105.0	135.2						130.0	111.0	99.8
106.0	132.0								
107.5	128.0								
108.0	126.8								
110.0	121.8						117.2	102.5	93.2
112.0	117.0								
115.0							105.2	94.4	86.4
120.0							94.5	86.7	79.8
125.0							85.6	79.8	73.0
130.0							78.2	73.4	66.2
135.0							73.0	67.8	59.6
140.0							69.1	63.0	52.8
145.0							66.3	59.2	46.0
150.0							64.2		

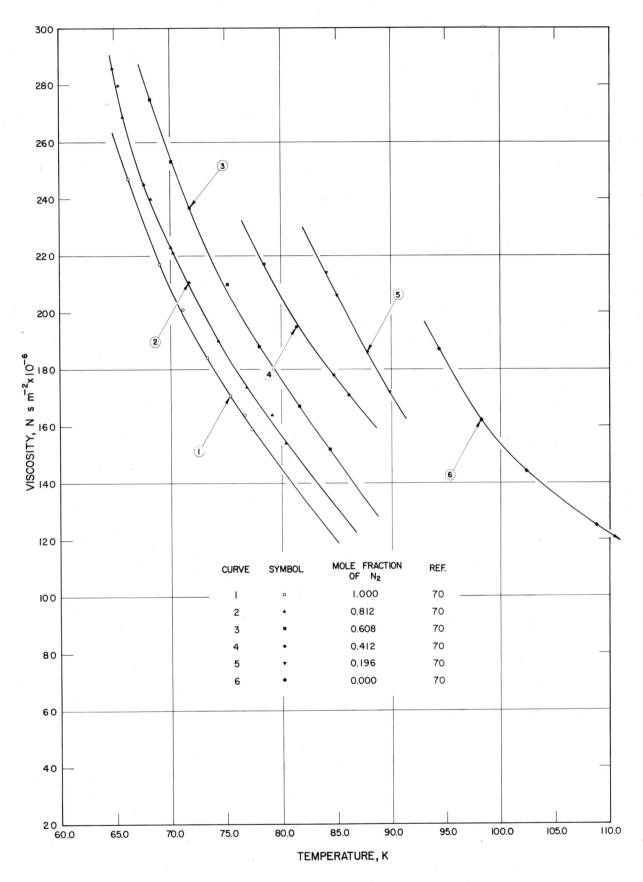

CURVE	SYMBOL	MOLE FRACTION OF N_2	REF.
1	○	1.000	70
2	▲	0.812	70
3	■	0.608	70
4	●	0.412	70
5	▼	0.196	70
6	◆	0.000	70

FIGURE 121-L(T). VISCOSITY DATA AS A FUNCTION OF TEMPERATURE
FOR LIQUID METHANE–NITROGEN MIXTURES

468

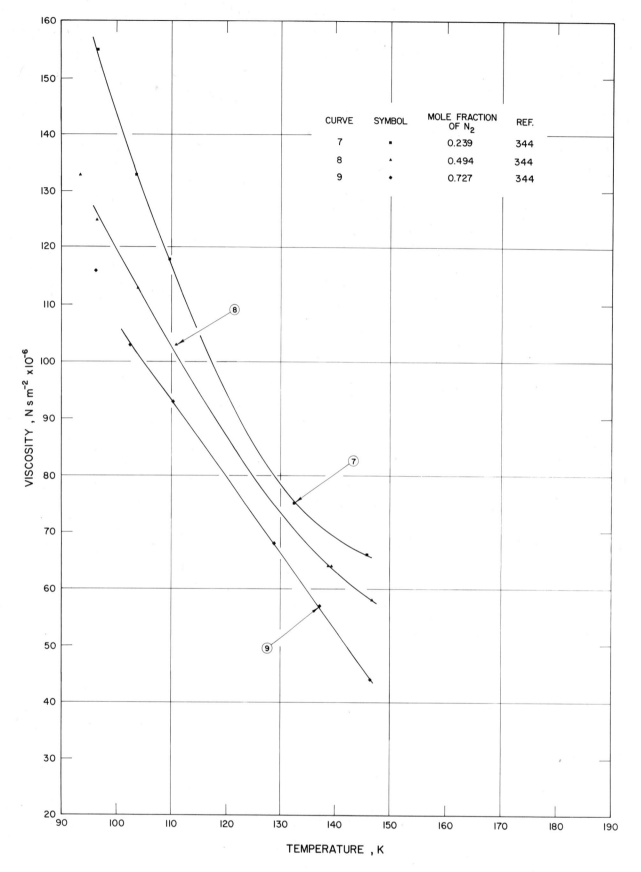

FIGURE 121-L(T). VISCOSITY DATA AS A FUNCTION OF TEMPERATURE
FOR LIQUID METHANE-NITROGEN MIXTURES (continued)

TABLE 121-G(D)E. EXPERIMENTAL VISCOSITY DATA AS A FUNCTION OF DENSITY FOR GASEOUS METHANE-NITROGEN MIXTURES

Cur. No.	Fig. No.	Ref. No.	Author(s)	Mole Fraction of N_2	Temp. (K)	Density (g cm^{-3})	Viscosity (N s m^{-2} x 10^{-6})	Remarks
1	121-G(D)	366	Gnezdilov, N. E. and Golubev, I. F.	0.722	273.2	0.0000	14.89	No purity specified for gases; composition analyzed by KhT-2M chrome-thermograph; capillary method; experimental error ±1%; original data reported as a function of pressure, density calculated from pressure using equations given by Miller et al. [375,376].
						0.0108	15.04	
						0.0217	15.10	
						0.0558	15.67	
						0.0558	15.63	
						0.115	17.08	
						0.173	18.93	
						0.173	19.02	
						0.222	21.07	
						0.222	21.29	
						0.264	23.48	
						0.264	23.60	
						0.298	25.52	
						0.330	27.83	
						0.330	27.96	
						0.355	30.00	
						0.355	30.08	
						0.384	31.85	
						0.408	33.56	
						0.408	33.68	
2	121-G(D)	366	Gnezdilov, N. E. and Golubev, I. F.	0.722	298.2	0.20982	15.90	Same remarks as for curve 1.
						0.0197	16.07	
						0.0501	16.54	
						0.101	17.79	
						0.151	19.34	
						0.193	21.04	
						0.236	22.80	
						0.270	24.69	
						0.270	24.98	
						0.301	26.69	
						0.329	28.51	
						0.351	30.07	
						0.351	30.38	
						0.364	31.80	
						0.364	32.07	
3	121-G(D)	366	Gnezdilov, N. E. and Golubev, I. F.	0.722	323.2	0.0181	16.99	Same remarks as for curve 1.
						0.0457	17.42	
						0.0910	18.48	
						0.136	20.00	
						0.136	20.18	
						0.175	21.68	
						0.213	23.12	
						0.246	24.50	
						0.275	26.03	
						0.299	27.41	
						0.299	27.69	
						0.324	29.20	
						0.344	30.61	
						0.344	30.81	
4	121-G(D)	366	Gnezdilov, N. E. and Golubev, I. F.	0.722	373.2	0.00780	18.66	Same remarks as for curve 1.
						0.0156	18.78	
						0.0389	19.14	
						0.0389	19.28	
						0.0773	20.00	
						0.114	20.92	
						0.146	22.11	
						0.178	23.24	
						0.208	24.42	
						0.235	25.62	
						0.258	26.93	
						0.280	28.16	
						0.299	29.55	
5	121-G(D)	366	Gnezdilov, N. E. and Golubev, I. F.	0.722	423.2	0.0137	20.54	Same remarks as for curve 1.
						0.0341	20.76	
						0.0675	21.49	
						0.0994	22.21	
						0.0994	22.30	
						0.128	23.10	
						0.156	24.05	
						0.182	25.08	
						0.206	26.06	
						0.227	27.16	
						0.248	28.18	
						0.267	29.30	
						0.267	29.36	

TABLE 121-G(D)E.   EXPERIMENTAL VISCOSITY DATA AS A FUNCTION OF DENSITY FOR GASEOUS
METHANE-NITROGEN MIXTURES (continued)

Cur. No.	Fig. No.	Ref. No.	Author(s)	Mole Fraction of N_2	Temp. (K)	Density (g cm^{-3})	Viscosity (N s m^{-2} x 10^{-6})	Remarks
6	121-G(D)	366	Gnezdilov, N.E. and Golubev, I.F.	0.722	473.2	0.0124	22.19	Same remarks as for curve 1.
						0.0303	22.39	
						0.0597	22.92	
						0.0880	23.62	
						0.113	24.26	
						0.113	24.36	
						0.138	25.12	
						0.161	26.04	
						0.183	26.82	
						0.203	27.74	
						0.223	28.33	
						0.244	29.41	
7	121-G(D)	366	Gnezdilov, N.E. and Golubev, I.F.	0.449	273.2	0.0189	13.44	Same remarks as for curve 1.
						0.0493	14.17	
						0.0493	13.94	
						0.102	15.61	
						0.158	17.84	
						0.204	20.42	
						0.204	20.51	
						0.242	22.68	
						0.242	22.82	
						0.271	25.32	
						0.302	27.69	
						0.302	27.50	
						0.319	29.83	
						0.337	31.60	
						0.337	31.73	
						0.350	33.27	
						0.350	33.32	
8	121-G(D)	366	Gnezdilov, N.E. and Golubev, I.F.	0.449	298.2	0.00848	14.19	Same remarks as for curve 1.
						0.0171	14.36	
						0.0438	14.93	
						0.0899	16.00	
						0.0899	16.21	
						0.136	17.78	
						0.175	19.94	
						0.212	21.90	
						0.245	23.87	
						0.272	25.95	
						0.291	27.86	
						0.310	29.56	
						0.310	29.71	
						0.326	31.46	
						0.326	31.50	
9	121-G(D)	366	Gnezdilov, N.E. and Golubev, I.F.	0.449	323.2	0.00780	15.09	Same remarks as for curve 1.
						0.0395	15.67	
						0.0809	16.65	
						0.0809	16.80	
						0.122	18.08	
						0.158	19.68	
						0.192	21.36	
						0.192	21.49	
						0.221	23.19	
						0.249	24.92	
						0.268	26.60	
						0.286	28.19	
						0.286	28.30	
						0.303	29.82	
						0.303	29.74	
10	121-G(D)	366	Gnezdilov, N.E. and Golubev, I.F.	0.449	373.2	0.00672	16.76	Same remarks as for curve 1.
						0.0134	16.87	
						0.0339	17.23	
						0.0682	18.01	
						0.101	19.01	
						0.130	20.21	
						0.160	21.55	
						0.187	22.91	
						0.212	24.14	
						0.212	24.26	
						0.231	25.62	
						0.248	26.85	
						0.248	27.00	
						0.263	28.14	
						0.263	28.21	

TABLE 121-G(D)E.   EXPERIMENTAL VISCOSITY DATA AS A FUNCTION OF DENSITY FOR GASEOUS
METHANE-NITROGEN MIXTURES (continued)

Cur. No.	Fig. No.	Ref. No.	Author(s)	Mole Fraction of N_2	Temp. (K)	Density (g cm^{-3})	Viscosity (N s m^{-2} x 10^{-6})	Remarks
11	121-G(D)	366	Gnezdilov, N. E. and Golubev, I. F.	0.449	423.2	0.0118	18.50	Same remarks as for curve 1.
						0.0296	18.82	
						0.0587	19.45	
						0.0872	20.28	
						0.113	21.11	
						0.137	22.16	
						0.160	23.30	
						0.182	24.44	
						0.203	25.46	
						0.219	26.49	
						0.234	27.51	
						0.234	27.64	
12	121-G(D)	366	Gnezdilov, N. E. and Golubev, I. F.	0.449	473.2	0.0106	20.07	Same remarks as for curve 1.
						0.0264	20.39	
						0.0520	20.86	
						0.0770	21.49	
						0.0990	22.24	
						0.121	23.12	
						0.141	23.96	
						0.161	24.86	
						0.180	25.79	
						0.199	26.63	
						0.215	27.65	

TABLE 121-G(D)S.   SMOOTHED VISCOSITY VALUES AS A FUNCTION OF DENSITY FOR GASEOUS METHANE-NITROGEN MIXTURES

Density (g cm^{-3})	Mole Fraction of Nitrogen					
	0.449 (273.2 K) [Ref. 366]	0.449 (298.2 K) [Ref. 366]	0.449 (323.2 K) [Ref. 366]	0.449 (373.2 K) [Ref. 366]	0.449 (423.2 K) [Ref. 366]	0.449 (423.2 K) [Ref. 366]
0.02	13.45	14.44	15.32	16.96	18.66	20.24
0.04	13.85	14.86	15.74	17.36	19.06	20.62
0.05	14.08	15.08	15.96	17.58	19.27	20.84
0.06	14.32	15.32	16.20	17.82	19.50	21.08
0.08	14.88	15.84	16.72	18.34	20.04	21.64
0.10	15.52	16.24	17.32	18.96	20.68	22.30
0.12	16.24	17.12	18.00	19.68	21.42	23.08
0.14	17.06	17.95	18.80	20.50	22.28	23.92
0.15	17.50	18.39	19.24	20.95	22.72	24.38
0.16	17.96	18.84	19.70	21.44	23.20	24.84
0.18	18.96	19.84	20.70	22.50	24.24	25.80
0.20	20.04	20.92	21.84	23.64	25.36	26.82
0.22	21.28	22.15	23.04	24.88	26.64	
0.24	22.60	23.48	24.24	26.28	27.94	
0.25	23.36	24.20	25.20	27.06		
0.26	24.16	24.96	25.98	27.90		
0.28	25.90	26.64	27.64			
0.30	27.68	28.60	29.48			
0.35	33.24					

Density (g cm^{-3})	Mole Fraction of Nitrogen					
	0.722 (273.2 K) [Ref. 366]	0.722 (298.2 K) [Ref. 366]	0.722 (323.2 K) [Ref. 366]	0.722 (373.2 K) [Ref. 366]	0.722 (423.2 K) [Ref. 366]	0.722 (423.2 K) [Ref. 366]
0.02	15.04	16.10	17.04	18.84	20.64	22.28
0.04	15.36	16.44	17.36	19.24	20.96	22.60
0.05	15.56	16.64	17.58	19.42	21.14	22.74
0.06	15.74	16.84	17.82	19.64	21.32	22.94
0.08	16.16	17.26	18.30	20.04	21.76	23.36
0.10	16.66	17.76	18.84	20.52	22.30	23.88
0.12	17.20	18.32	19.46	21.08	22.84	24.48
0.14	17.80	18.96	20.14	21.70	23.50	25.18
0.15	18.16	19.30	20.50	22.04	23.84	25.54
0.16	18.50	19.68	20.88	22.40	24.20	25.92
0.18	19.24	20.44	21.68	23.20	25.00	26.72
0.20	20.10	21.28	22.52	24.06	25.80	27.60
0.22	21.08	22.14	23.40	24.96	26.76	28.58
0.24	22.12	23.10	24.30	25.88	27.96	29.68
0.25	22.68	23.60	24.78	26.40		30.24
0.26	23.26	24.16	25.26	26.94		30.86
0.28	24.48	25.32	26.32	28.16		
0.30	25.80	26.60	27.54	29.56		
0.35	29.34	30.30	31.04			
0.40	33.04					

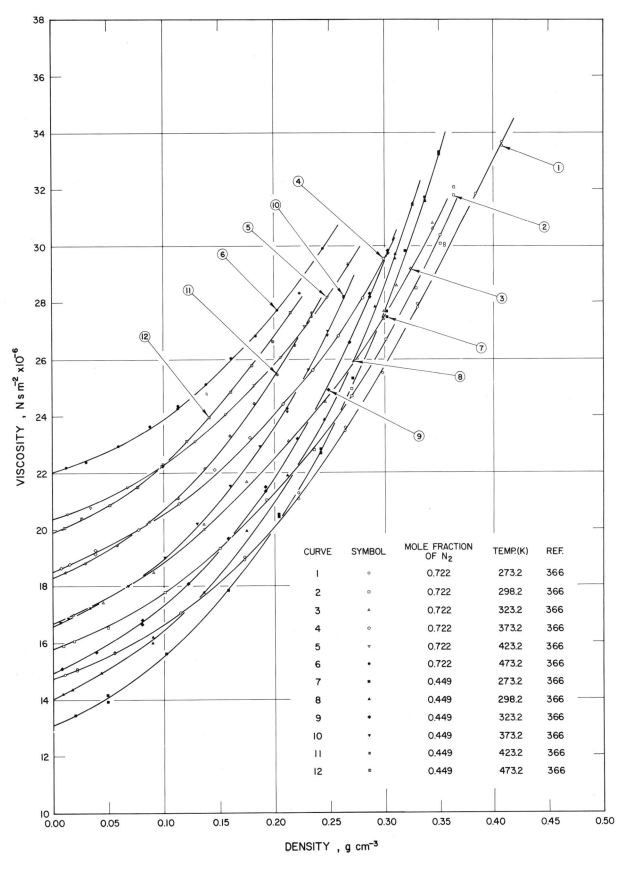

CURVE	SYMBOL	MOLE FRACTION OF N_2	TEMP.(K)	REF.
1	○	0.722	273.2	366
2	□	0.722	298.2	366
3	△	0.722	323.2	366
4	◇	0.722	373.2	366
5	▽	0.722	423.2	366
6	•	0.722	473.2	366
7	■	0.449	273.2	366
8	▲	0.449	298.2	366
9	◆	0.449	323.2	366
10	▼	0.449	373.2	366
11	◉	0.449	423.2	366
12	▥	0.449	473.2	366

FIGURE 121-G(D). VISCOSITY DATA AS A FUNCTION OF DENSITY FOR GASEOUS METHANE-NITROGEN MIXTURES

TABLE 122-G(C)E.   EXPERIMENTAL VISCOSITY DATA AS A FUNCTION OF COMPOSITION FOR GASEOUS METHANE-OXYGEN MIXTURES

Cur. No.	Fig. No.	Ref. No.	Author(s)	Temp. (K)	Pressure (mm Hg)	Mole Fraction of O_2	Viscosity (N s m^{-2} x 10^{-6})	Remarks
1	122-G(C)	334	Strauss, W. A. and Edse, R.	293.2	746.8	0.000	11.12	Capillary flow viscometer, relative measurements; $L_1 = 0.370\%$, $L_2 = 0.455\%$, $L_3 = 0.887\%$.
					747.4	0.051	11.59	
					748.2	0.099	12.15	
					749.0	0.142	12.56	
					750.1	0.198	13.16	
					751.0	0.251	13.74	
					750.7	0.296	14.13	
					750.6	0.349	14.52	
					750.2	0.501	15.92	
					750.2	0.549	16.30	
					750.3	0.597	16.75	
					749.9	0.647	17.18	
					749.4	0.702	17.62	
					749.3	0.765	18.31	
					748.9	0.799	18.39	
					748.9	0.849	18.86	
					748.6	0.898	19.29	
					748.3	0.951	19.71	
					748.0	1.000	20.04	
2	122-G(C)	334	Strauss, W. A. and Edse, R.	293.2	761.0	1.000	20.26	
					761.6	0.895	19.33	
					761.9	0.801	18.54	
					762.4	0.713	17.75	
					763.0	0.600	16.88	
					763.2	0.497	16.02	
					763.1	0.400	15.19	
					763.7	0.299	14.37	
					762.8	0.191	13.29	
					761.0	0.092	12.17	
					759.7	0.000	11.17	
3	122-G(C)	334	Strauss, W. A. and Edse, R.	293.2	759.7	0.000	11.05	
					760.3	0.048	11.63	
					761.9	0.147	12.70	
					763.9	0.244	13.87	
					763.5	0.353	14.72	
					763.0	0.504	15.88	
					763.3	0.554	16.52	
					762.8	0.656	17.33	
					762.3	0.747	18.13	
					761.8	0.860	18.98	
					761.3	0.951	19.70	
					761.0	1.000	20.02	

TABLE 122-G(C)S.   SMOOTHED VISCOSITY VALUES AS A FUNCTION OF COMPOSITION FOR GASEOUS METHANE-OXYGEN MIXTURES

Mole Fraction of O_2	(293.2 K) [Ref. 334]
0.00	11.12
0.05	11.68
0.10	12.22
0.15	12.72
0.20	13.25
0.25	13.76
0.30	14.12
0.35	14.66
0.40	15.12
0.45	15.54
0.50	15.97
0.55	16.38
0.60	16.81
0.65	17.22
0.70	17.65
0.75	18.05
0.80	18.45
0.85	18.89
0.90	19.30
0.95	19.70
1.00	20.11

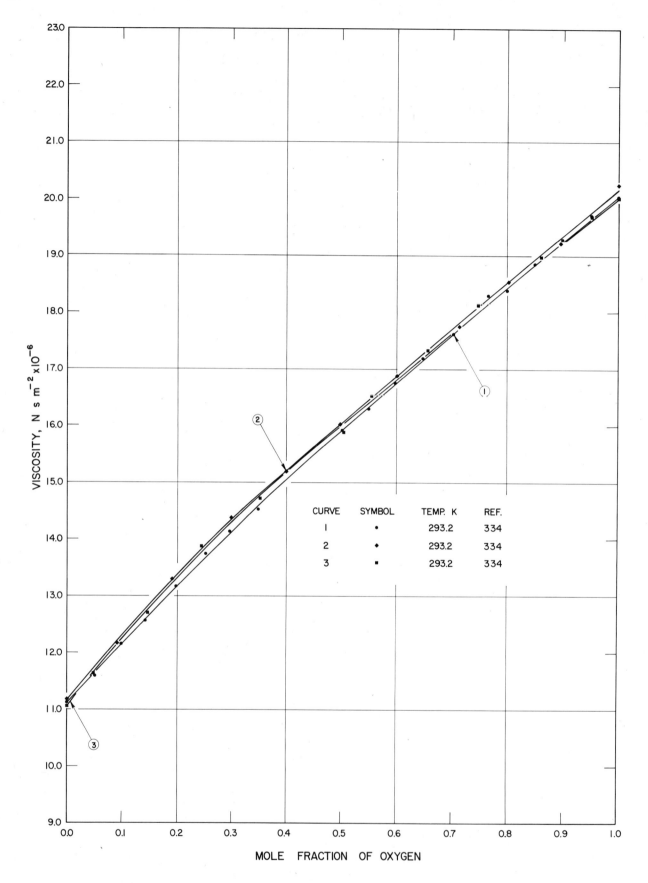

FIGURE 122-G(C). VISCOSITY DATA AS A FUNCTION OF COMPOSITION
FOR GASEOUS METHANE-OXYGEN MIXTURES

TABLE 123-L(D)E.   EXPERIMENTAL VISCOSITY DATA AS A FUNCTION OF DENSITY FOR LIQUID METHANE-PROPANE MIXTURES

Cur. No.	Fig. No.	Ref. No.	Author(s)	Mole Fraction of C_3H_8	Temp. (K)	Density $(g\,cm^{-3})$	Viscosity $(N\,s\,m^{-2}\,x\,10^{-6})$	Remarks
1	123-L(D)	72	Giddings, J.G., Kao, J.T.F., and Kobayashi, R.	1.0000	310.9	0.489	93.6	C_3H_8: research grade capillary tube viscometer; precision 0.25% excluding critical regions, error ±0.54%; original data reported as a function of pressure, density calculated from pressure using volumetric data of Reamer et al. [367], and Canjar and Manning [368].
						0.492	96.1	
						0.495	99.4	
						0.498	102.3	
						0.501	105.2	
						0.504	107.8	
						0.510	113.1	
						0.516	117.9	
						0.527	127.6	
						0.539	136.8	
						0.550	145.2	
						0.562	153.5	
						0.573	160.8	
2	123-L(D)	72	Giddings, J.G., et al.	0.7793	310.9	0.419	65.5	Same remarks as for curve 1.
						0.429	69.1	
						0.437	72.2	
						0.445	75.2	
						0.450	77.8	
						0.459	83.0	
						0.468	87.6	
						0.481	96.2	
						0.495	104.0	
						0.504	111.6	
						0.513	119.0	
						0.522	125.4	
3	123-L(D)	72	Giddings, J.G., et al.	0.6122	310.9	0.350	44.6	Same remarks as for curve 1.
						0.367	49.1	
						0.381	53.1	
						0.391	56.5	
						0.407	62.0	
						0.420	66.8	
						0.438	74.2	
4	123-L(D)	72	Giddings, J.G., et al.	0.3861	310.9	0.214	22.95	Same remarks as for curve 1.
						0.246	27.10	
						0.273	30.7	
						0.307	36.6	
						0.329	41.2	
						0.359	48.9	
						0.380	55.7	
5	123-L(D)	72	Giddings, J.G., et al.	1.0000	344.3	0.442	66.2	Same remarks as for curve 1.
						0.446	69.1	
						0.450	72.7	
						0.455	76.3	
						0.459	79.0	
						0.463	81.9	
						0.472	87.2	
						0.481	92.2	
						0.498	101.8	
						0.515	110.2	
						0.532	118.5	
						0.550	125.7	
						0.567	133.2	
6	123-L(D)	72	Giddings, J.G., et al.	0.7793	344.3	0.313	39.0	Same remarks as for curve 1.
						0.346	45.9	
						0.368	50.1	
						0.384	54.7	
						0.395	57.3	
						0.413	62.9	
						0.426	67.7	
						0.446	76.4	
						0.461	83.8	
						0.474	90.8	
						0.485	97.5	
						0.496	103.8	
7	123-L(D)	72	Giddings, J.G., et al.	0.6122	344.3	0.269	30.3	Same remarks as for curve 1.
						0.301	35.3	
						0.321	39.3	
						0.351	45.4	
						0.371	51.0	
						0.398	59.7	

TABLE 123-L(D)S.   SMOOTHED VISCOSITY VALUES AS A FUNCTION OF DENSITY FOR LIQUID
METHANE-PROPANE MIXTURES

Density (g cm⁻³)	Mole Fraction of Propane			
	1.000 (310.9 K) [Ref. 72]	0.7793 (310.9 K) [Ref. 72]	0.6122 (310.9 K) [Ref. 72]	0.3861 (310.9 K) [Ref. 72]
0.20				21.3
0.22				23.7
0.24				26.3
0.26				29.0
0.28				32.0
0.30				35.3
0.32				39.2
0.34				43.8
0.36			47.1	49.2
0.38			52.9	55.7
0.40			59.4	
0.41			63.0	
0.42		65.9	56.8	
0.43		69.2	70.7	
0.44		73.2	75.1	
0.45		77.8		
0.46		83.0		
0.47		88.9		
0.48		95.4		
0.49	92.4	101.9		
0.50	102.8	109.0		
0.51	113.0	116.3		
0.52	121.5	123.8		
0.53	129.5			
0.54	137.0			
0.55	143.8			
0.56	151.5			
0.57	159.7			

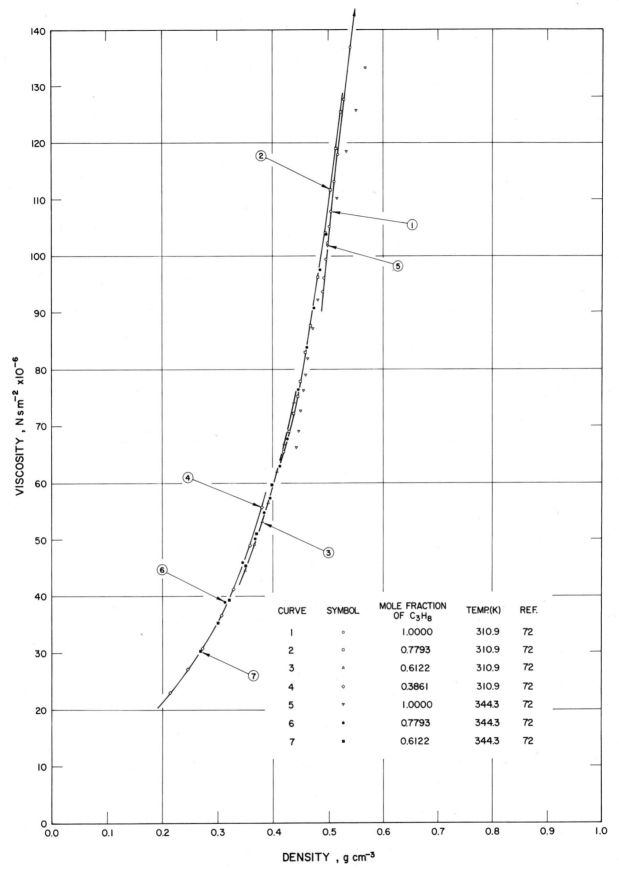

FIGURE 123-L(D). VISCOSITY DATA AS A FUNCTION OF DENSITY
FOR LIQUID METHANE – PROPANE MIXTURES

TABLE 123-G(C)E.   EXPERIMENTAL VISCOSITY DATA AS A FUNCTION OF COMPOSITION FOR GASEOUS
METHANE-PROPANE MIXTURES

Cur. No.	Fig. No.	Ref. No.	Author(s)	Temp. (K)	Pressure (atm)	Mole Fraction of C_3H_8	Viscosity ($N\ s\ m^{-2} \times 10^{-6}$)	Remarks
1	123-G(C)	229	Trautz, M. and Sorg, K.G.	293.0		1.0000	8.01	CH_4 and C_3H_8: I.G. Farben, 99.9 pure; capillary method; precision
						0.8341	8.31	$\pm\ 0.05\%$; $L_1 = 0.072\%$, $L_2 = 0.161\%$,
						0.6383	8.78	$L_3 = 0.360\%$.
						0.3684	9.48	
						0.0000	10.87	
2	123-G(C)	229	Trautz, M. and Sorg, K.G.	373.0		1.0000	10.08	Same remarks as for curve 1 except $L_1 = 0.125\%$, $L_2 = 0.218\%$, $L_3 =$
						0.8341	10.42	0.456%.
						0.6383	11.01	
						0.3684	11.82	
						0.0000	13.31	
3	123-G(C)	229	Trautz, M. and Sorg, K.G.	473.0		1.0000	12.53	Same remarks as for curve 1 except $L_1 = 0.119\%$, $L_2 = 0.266\%$, $L_3 =$
						0.8341	12.91	0.594%.
						0.6383	13.55	
						0.3684	14.41	
						0.0000	16.03	
4	123-G(C)	229	Trautz, M. and Sorg, K.G.	523.0		1.0000	13.63	Same remarks as for curve 1 except $L_1 = 0.072\%$, $L_2 = 0.160\%$, $L_3 =$
						0.8341	14.03	0.358%.
						0.6383	14.65	
						0.3684	15.53	
						0.0000	17.25	

TABLE 123-G(C)S.   SMOOTHED VISCOSITY VALUES AS A FUNCTION OF COMPOSITION FOR GASEOUS
METHANE-PROPANE MIXTURES

Mole Fraction of C_3H_8	(293.0 K) [Ref. 229]	(373.0 K) [Ref. 229]	(473.0 K) [Ref. 229]	(523.0 K) [Ref. 229]
0.00	10.86	13.31	16.03	17.25
0.05	10.64	13.10	15.81	17.02
0.10	10.42	12.90	15.58	16.79
0.15	10.22	12.64	15.36	16.55
0.20	10.04	12.48	15.13	16.31
0.25	9.86	12.29	14.92	16.08
0.30	9.69	12.10	14.71	15.84
0.35	9.54	11.92	14.51	15.61
0.40	9.40	11.73	14.31	15.40
0.45	9.26	11.56	14.12	15.18
0.50	9.14	11.39	13.94	14.98
0.55	9.01	11.23	13.76	14.80
0.60	8.88	11.07	13.60	14.63
0.65	8.76	10.92	13.44	14.48
0.70	8.64	10.77	13.28	14.34
0.75	8.52	10.63	13.14	14.20
0.80	8.42	10.51	13.01	14.08
0.85	8.32	10.38	12.88	13.95
0.90	8.20	10.27	12.76	13.84
0.95	8.10	10.17	12.64	13.73
1.00	8.00	10.08	12.53	13.62

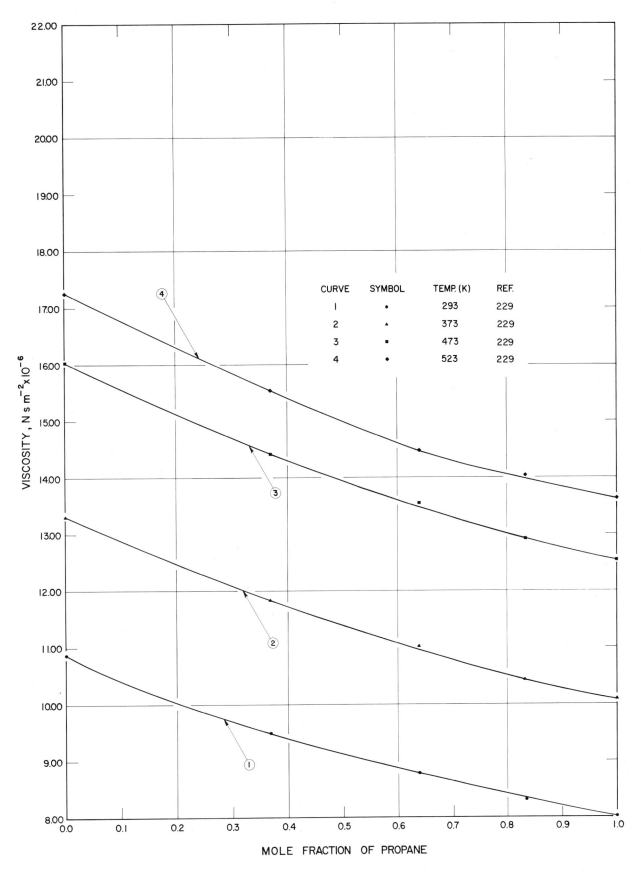

FIGURE 123-G(C). VISCOSITY DATA AS A FUNCTION OF COMPOSITION
FOR GASEOUS METHANE–PROPANE MIXTURES

TABLE 123-G(D)E.   EXPERIMENTAL VISCOSITY DATA AS A FUNCTION OF DENSITY FOR GASEOUS
METHANE-PROPANE MIXTURES

Cur. No.	Fig. No.	Ref. No.	Author(s)	Mole Fraction of C_3H_8	Temp. (K)	Density (g cm^{-3})	Viscosity (N s m^{-2} x 10^{-6})	Remarks
1	123-G(D)	72	Giddings, J.G., Kao, J.T.F., and Kobayashi, R.	1.0000	310.9	0.00175 0.0131	8.47 8.58	C_3H_8: research grade; capillary tube viscometer; precision 0.25% excluding critical regions, error ±0.54%; original data reported as a function of pressure, density calculated from pressure using volumetric data of Reamer et al. [367], and Canjar and Manning [368].
2	123-G(D)	72	Giddings, J.G., et al.	0.7793	310.9	0.00145 0.0122 0.0242	8.92 9.13 9.38	Same remarks as for curve 1.
3	123-G(D)	72	Giddings, J.G., et al.	0.6122	310.9	0.00130 0.00995 0.0201	9.3 9.5 9.7	Same remarks as for curve 1.
4	123-G(D)	72	Giddings, J.G., et al.	0.3861	310.9	0.00120 0.00773 0.0154 0.0244 0.0334 0.0445 0.0554	9.96 10.13 10.34 10.56 10.82 11.12 11.60	Same remarks as for curve 1.
5	123-G(D)	72	Giddings, J.G., et al.	0.2090	310.9	0.000750 0.00595 0.0122 0.0189 0.0256 0.0332 0.0408 0.0568 0.0745 0.0985 0.124 0.150 0.174 0.214 0.245 0.285 0.311 0.332	10.72 10.80 10.91 11.03 11.24 11.41 11.64 12.32 13.12 14.45 16.05 17.97 20.1 24.2 28.0 34.3 39.5 44.2	Same remarks as for curve 1.
6	123-G(D)	72	Giddings, J.G., et al.	0.0000	310.9	0.000630 0.00432 0.00873 0.0132 0.0178 0.0225 0.0272 0.0370 0.0470 0.0600 0.0732 0.0861 0.0998 0.125 0.149 0.188 0.217 0.240 0.258 0.274	11.62 11.68 11.79 11.90 12.02 12.16 12.31 12.65 13.03 13.68 14.22 14.94 15.71 17.52 19.3 22.8 26.1 29.2 31.8 34.2	Same remarks as for curve 1.
7	123-G(D)	72	Giddings, J.G., et al.	1.0000	344.3	0.00158 0.0115 0.0252 0.0432	9.35 9.53 9.79 10.25	Same remarks as for curve 1.
8	123-G(D)	72	Giddings, J.G., et al.	0.7793	344.3	0.00151 0.0103 0.0206 0.0347 0.0487	9.88 10.06 10.27 10.69 11.11	Same remarks as for curve 1.

TABLE 123-G(D)E.   EXPERIMENTAL VISCOSITY DATA AS A FUNCTION OF DENSITY FOR GASEOUS METHANE-PROPANE MIXTURES (continued)

Cur. No.	Fig. No.	Ref. No.	Author(s)	Mole Fraction of C_3H_8	Temp. (K)	Density $(g\,cm^{-3})$	Viscosity $(N\,s\,m^{-2} \times 10^{-6})$	Remarks
9	123-G(D)	72	Giddings, J.G., et al.	0.6122	344.3	0.00123	10.3	Same remarks as for curve 1.
						0.00870	10.4	
						0.0174	10.6	
						0.0279	10.8	
						0.0386	11.1	
						0.0523	11.6	
						0.0658	12.2	
						0.103	14.2	
						0.154	17.7	
						0.221	24.1	
10	123-G(D)	72	Giddings, J.G., et al.	0.3861	344.3	-0.000981	10.96	Same remarks as for curve 1.
						-0.00675	11.12	
						-0.0136	11.31	
						-0.0211	11.50	
						0.0287	11.70	
						-0.0370	11.93	
						0.0455	12.24	
						0.0643	12.91	
						0.0852	13.91	
						0.115	15.46	
						0.146	17.58	
						0.176	20.08	
						0.202	22.7	
						0.243	27.5	
						0.275	31.9	
						0.316	39.2	
						0.342	45.4	
						0.363	51.0	
						0.381	56.1	
11	123-G(D)	72	Giddings, J.G., et al.	0.2090	344.3	-0.000770	11.74	Same remarks as for curve 1.
						-0.000504	11.84	
						0.0109	11.96	
						-0.0167	12.07	
						0.0225	12.22	
						-0.0287	12.37	
						0.0348	12.56	
						0.0478	13.02	
						0.0615	14.09	
						0.0794	14.42	
						0.0983	15.43	
						0.117	16.60	
						0.136	17.90	
						0.170	20.7	
						0.200	23.6	
						0.244	28.9	
						0.276	33.5	
						0.299	37.7	
12	123-G(D)	72	Giddings, J.G., et al.	0.0000	344.3	0.000569	12.69	Same remarks as for curve 1.
						0.00389	12.75	
						0.00783	12.84	
						0.0118	12.94	
						0.0159	13.05	
						0.0199	13.15	
						0.0241	13.28	
						0.0324	13.55	
						0.0410	13.86	
						0.0513	14.29	
						0.0627	14.79	
						0.0731	15.33	
						0.0845	15.89	
						0.106	17.12	
						0.126	18.42	
						0.161	21.23	
						0.190	23.94	
						0.213	26.5	
						0.232	29.1	
						0.248	31.3	

484

TABLE 123-G(D)E. EXPERIMENTAL VISCOSITY DATA AS A FUNCTION OF DENSITY FOR GASEOUS
METHANE-PROPANE MIXTURES (continued)

Cur. No.	Fig. No.	Ref. No.	Author(s)	Mole Fraction of C_3H_8	Temp. (K)	Density (g cm^{-3})	Viscosity (N s m^{-2} x 10^{-6})	Remarks
13	123-G(D)	72	Giddings, J.G., et al.	1.0000	377.6	0.00144	10.28	Same remarks as for curve 1.
						0.0102	10.48	
						0.0218	10.65	
						0.0352	10.94	
						0.0517	11.54	
						0.0728	12.4	
						0.105	13.9	
						0.304	35.6	
						0.350	43.6	
						0.375	49.8	
						0.391	54.4	
						0.404	58.1	
						0.414	61.4	
						0.430	67.5	
						0.443	72.5	
						0.465	81.8	
14	123-G(D)	72	Giddings, J.G., et al.	0.7793	377.6	0.00133	10.83	Same remarks as for curve 1.
						0.00905	10.98	
						0.0181	11.16	
						0.0291	11.40	
						0.0401	11.75	
						0.0542	12.10	
						0.0683	12.50	
						0.106	14.6	
						0.159	18.5	
						0.231	25.8	
						0.279	32.1	
						0.309	37.1	
						0.330	41.4	
						0.360	47.7	
						0.381	53.1	
						0.409	62.0	
						0.428	69.3	
						0.444	76.2	
						0.457	82.5	
						0.468	88.7	
15	123-G(D)	72	Giddings, J.G., et al.	0.6122	377.6	0.00114	11.30	Same remarks as for curve 1.
						0.00774	11.38	
						0.0155	11.53	
						0.0244	11.73	
						0.0333	11.96	
						0.0435	12.29	
						0.0537	12.65	
						0.0773	13.64	
						0.105	15.1	
						0.144	17.7	
						0.184	21.2	
						0.220	24.9	
						0.249	28.6	
						0.291	34.9	
						0.320	40.4	
						0.358	49.0	
						0.383	55.7	
						0.402	61.5	
16	123-G(D)	72	Giddings, J.G., et al.	0.3861	377.6	0.000904	11.95	Same remarks as for curve 1.
						0.00615	12.09	
						0.0123	12.23	
						0.0189	12.39	
						0.0254	12.56	
						0.0325	12.75	
						0.0396	12.96	
						0.0548	13.50	
						0.0710	14.19	
						0.0920	15.32	
						0.114	16.53	
						0.136	17.99	
						0.158	19.65	
						0.197	25.08	
						0.229	26.5	
						0.276	32.8	
						0.307	38.3	
						0.332	43.4	
						0.352	48.2	

TABLE 123-G(D)E. EXPERIMENTAL VISCOSITY DATA AS A FUNCTION OF DENSITY FOR GASEOUS
METHANE-PROPANE MIXTURES (continued)

Cur. No.	Fig. No.	Ref. No.	Author(s)	Mole Fraction of C_3H_8	Temp. (K)	Density $(g\,cm^{-3})$	Viscosity $(N\,s\,m^{-2} \times 10^{-6})$	Remarks
17	123-G(D)	72	Giddings, J.G., et al.	0.2090	377.6	0.000724	12.68	Same remarks as for curve 1.
						0.00492	12.78	
						0.00984	12.90	
						0.0150	13.01	
						0.0201	13.14	
						0.0255	13.29	
						0.0308	13.44	
						0.0418	13.82	
						0.0532	14.28	
						0.0679	14.95	
						0.0828	15.68	
						0.0981	16.50	
						0.113	17.39	
						0.142	19.35	
						0.168	21.5	
						0.212	25.8	
						0.245	29.9	
						0.270	33.6	
						0.291	37.2	
						0.308	40.6	
18	123-G(D)	72	Giddings, J.G., et al.	0.0000	377.6	0.000518	13.70	Same remarks as for curve 1.
						0.00354	13.76	
						0.00711	13.84	
						0.0107	13.93	
						0.0143	14.03	
						0.0180	14.12	
						0.0216	14.23	
						0.0290	14.45	
						0.0365	14.71	
						0.0457	15.07	
						0.0553	15.47	
						0.0642	15.80	
						0.0740	16.36	
						0.0922	17.34	
						0.110	18.39	
						0.141	20.62	
						0.169	22.86	
						0.191	25.1	
						0.211	27.2	
						0.227	29.4	
19	123-G(D)	72	Giddings, J.G., et al.	0.7793	410.9	0.00120	11.80	Same remarks as for curve 1.
						0.00815	11.89	
						0.0163	11.99	
						0.0256	12.20	
						0.0349	12.48	
						0.0457	12.76	
						0.0565	13.20	
						0.0816	14.26	
						0.110	15.89	
						0.152	19.0	
						0.194	22.6	
						0.232	26.4	
						0.262	30.2	
						0.303	36.7	
						0.332	42.2	
						0.369	50.9	
						0.395	58.3	
						0.414	64.6	
						0.429	70.6	
						0.443	76.5	
20	123-G(D)	72	Giddings, J.G., et al.	0.6122	410.9	0.00104	12.25	Same remarks as for curve 1.
						0.00705	12.33	
						0.0141	12.44	
						0.0218	12.56	
						0.0295	12.76	
						0.0380	13.02	
						0.0465	13.29	
						0.0652	13.99	
						0.0853	14.97	
						0.112	16.64	
						0.140	18.71	
						0.169	21.0	
						0.195	23.3	
						0.239	28.0	
						0.273	32.6	
						0.318	40.7	
						0.348	44.9	
						0.371	53.3	

TABLE 123-G(D)E.   EXPERIMENTAL VISCOSITY DATA AS A FUNCTION OF DENSITY FOR GASEOUS
METHANE-PROPANE MIXTURES (continued)

Cur. No.	Fig. No.	Ref. No.	Author(s)	Mole Fraction of C_3H_8	Temp. (K)	Density (g cm^{-3})	Viscosity (N s m^{-2} x 10^{-6})	Remarks
21	123-G(D)	72	Giddings, J.G., et al.	0.3861	410.9	0.000744	12.92	Same remarks as for curve 1.
						0.00506	13.04	
						0.0112	13.20	
						0.0171	13.33	
						0.0229	13.47	
						0.0291	13.65	
						0.0353	13.83	
						0.0483	14.27	
						0.0617	14.74	
						0.0790	15.52	
						0.0966	16.41	
						0.114	17.51	
						0.131	18.69	
						0.165	21.3	
						0.195	23.9	
						0.242	29.2	
						0.276	34.0	
						0.303	38.6	
						0.324	43.0	
						0.341	47.0	
22	123-G(D)	72	Giddings, J.G., et al.	0.2090	410.9	0.000660	13.66	Same remarks as for curve 1.
						0.00449	13.73	
						0.00898	13.81	
						0.0145	13.90	
						0.0201	14.02	
						0.0255	14.17	
						0.0308	14.29	
						0.0374	14.61	
						0.0474	14.97	
						0.0599	15.50	
						0.0727	16.07	
						0.0854	16.69	
						0.0981	17.37	
						0.123	18.92	
						0.146	20.7	
						0.186	24.1	
						0.218	27.6	
						0.245	30.9	
						0.267	34.1	
						0.284	37.2	
23	123-G(D)	72	Giddings, J.G., et al.	0.0000	410.9	0.000476	14.65	Same remarks as for curve 1.
						0.00325	14.70	
						0.00651	14.78	
						0.00979	14.86	
						0.0131	14.94	
						0.0164	15.03	
						0.0197	15.12	
						0.0264	15.32	
						0.0329	15.54	
						0.0413	15.84	
						0.0496	16.17	
						0.0575	16.33	
						0.0661	16.91	
						0.0823	17.74	
						0.0979	18.61	
						0.126	20.49	
						0.152	22.4	
						0.174	24.3	
						0.193	26.1	
						0.210	28.0	

TABLE 123-G(D)S.   SMOOTHED VISCOSITY VALUES AS A FUNCTION OF DENSITY FOR GASEOUS METHANE-PROPANE MIXTURES

Density (g cm^{-3})	Mole Fraction of Methane							
	1.0000 (310.9 K) [Ref. 72]	0.7793 (310.9 K) [Ref. 72]	0.6122 (310.9 K) [Ref. 72]	0.3861 (310.9 K) [Ref. 72]	0.2090 (310.9 K) [Ref. 72]	0.0000 (310.9 K) [Ref. 72]	1.0000 (344.3 K) [Ref. 72]	0.7793 (344.3 K) [Ref. 72]
0.010	8.59	9.10	9.50	10.12	10.89	11.82	9.45	10.00
0.020	8.62	9.35	9.68	10.41	11.16	12.17	9.62	10.22
0.030		9.58	9.93	10.76	11.40	12.45	9.88	10.50
0.040				11.08	11.72	12.80	10.19	10.80
0.050				11.41	12.04	13.20	10.55	11.18
0.075					13.13	14.37		
0.100					14.44	15.70		
0.125					16.09	17.35		
0.150					17.99	19.25		
0.175					20.10	21.52		
0.200					22.61	24.22		
0.225					25.45	27.29		
0.250					28.70	30.72		
0.300					37.06			

Density (g cm^{-3})	Mole Fraction of Methane							
	0.6122 (344.3 K) [Ref. 72]	0.3861 (344.3 K) [Ref. 72]	0.2090 (344.3 K) [Ref. 72]	0.0000 (344.3 K) [Ref. 72]	1.0000 (377.6 K) [Ref. 72]	0.7793 (377.6 K) [Ref. 72]	0.6122 (377.6 K) [Ref. 72]	0.3861 (377.6 K) [Ref. 72]
0.010	10.47	11.19	11.92	12.96	10.40	11.00	11.42	12.20
0.020	10.66	11.41	12.20	13.19	10.60	11.20	11.70	12.50
0.030	10.95	11.72	12.47	13.49	10.90	11.42	11.98	12.80
0.040	11.21	12.02	12.80	13.82	11.18	11.70	12.20	13.05
0.050	11.57	12.39	13.18	14.20	11.48	12.00	12.56	13.40
0.075	12.61	13.40	14.22	15.36	12.42	13.00	13.50	14.40
0.100	14.00	14.61	15.58	16.72	13.70	14.30	14.80	15.70
0.125	15.59	16.11	17.13	18.39	15.10	15.90	16.38	17.30
0.150	17.40	17.89	18.98	20.25	16.70	17.80	18.20	19.05
0.175	19.45	20.00	21.11	22.44	18.60	19.90	20.34	21.10
0.200	21.82	22.45	23.60	25.05	20.84	22.40	22.80	23.40
0.225	24.60	25.24	26.42	28.08	23.40	25.20	25.62	26.10
0.250		28.38	29.67	31.55	26.35	28.20	28.94	29.20
0.300		36.20	37.90		33.83	35.84	36.60	37.00
0.350		47.20			43.60	45.40	47.00	47.70
0.400					57.00	59.10	60.90	
0.450					75.40	79.00		

Density (g cm^{-3})	Mole Fraction of						
	0.2090 (377.6 K) [Ref. 72]	0.0000 (377.6 K) [Ref. 72]	0.7793 (410.9 K) [Ref. 72]	0.6122 (410.9 K) [Ref. 72]	0.3861 (410.9 K) [Ref. 72]	0.2090 (410.9 K) [Ref. 72]	0.0000 (410.9 K) [Ref. 72]
0.010	12.90	13.98	11.98	12.40	13.20	13.92	14.90
0.020	13.20	14.22	12.18	12.60	13.40	14.20	15.20
0.030	13.50	14.60	12.40	12.90	13.70	14.40	15.50
0.040	13.80	14.90	12.60	13.18	14.00	14.70	15.80
0.050	14.20	15.30	12.90	13.42	14.40	15.10	16.20
0.075	15.30	16.40	13.90	14.42	15.40	16.18	17.30
0.100	16.60	17.70	15.20	15.80	16.70	17.50	18.70
0.125	18.20	19.30	16.90	17.60	18.30	19.20	20.40
0.150	20.00	21.20	18.80	19.42	20.10	21.00	22.30
0.175	22.20	23.40	20.80	21.50	22.20	23.20	24.40
0.200	24.60	26.20	23.18	23.80	24.60	25.60	26.70
0.225	27.40	29.20	25.70	26.40	27.20	28.40	29.30
0.250	30.60	32.60	28.60	29.40	30.20	31.70	
0.300	39.00		36.30	37.20	38.00		
0.350			46.20	48.00	49.50		
0.400			60.00				
0.450			79.62				

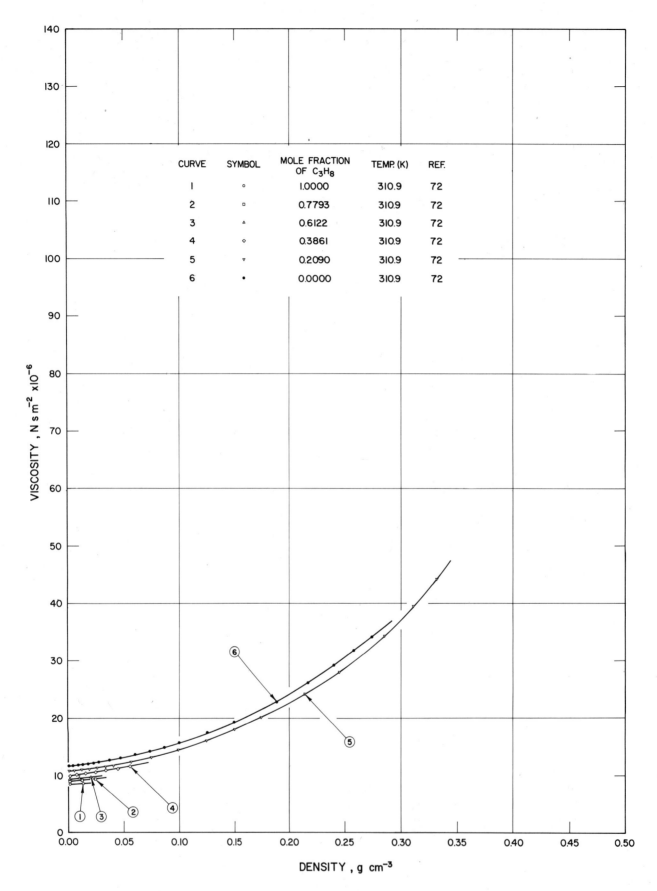

CURVE	SYMBOL	MOLE FRACTION OF C_3H_8	TEMP. (K)	REF.
1	○	1.0000	310.9	72
2	□	0.7793	310.9	72
3	△	0.6122	310.9	72
4	◇	0.3861	310.9	72
5	▽	0.2090	310.9	72
6	●	0.0000	310.9	72

FIGURE 123 – G(D). VISCOSITY DATA AS A FUNCTION OF DENSITY
FOR GASEOUS METHANE – PROPANE MIXTURES

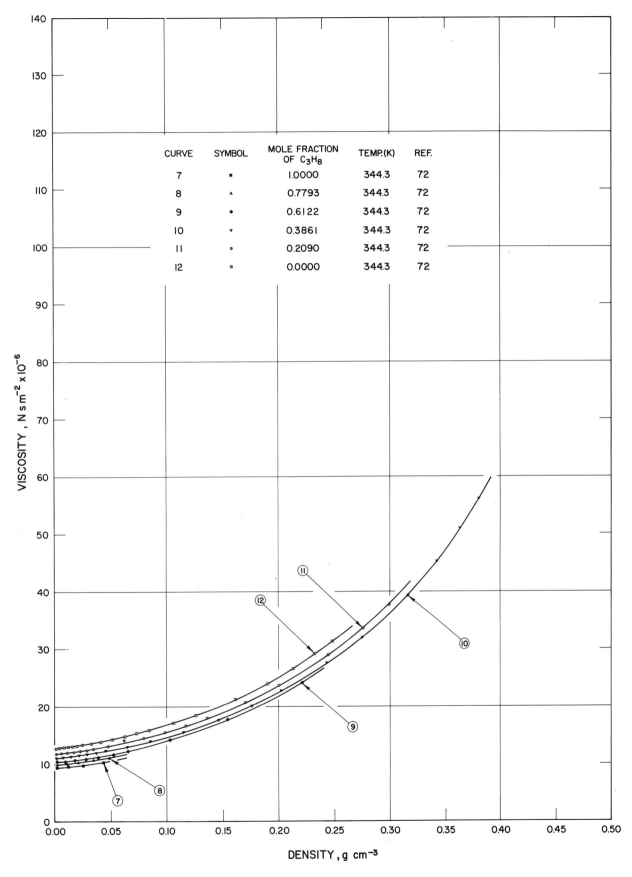

CURVE	SYMBOL	MOLE FRACTION OF C_3H_8	TEMP.(K)	REF.
7	▪	1.0000	344.3	72
8	▲	0.7793	344.3	72
9	◆	0.6122	344.3	72
10	▾	0.3861	344.3	72
11	◦	0.2090	344.3	72
12	▫	0.0000	344.3	72

VISCOSITY, N s m⁻² x 10⁻⁶

DENSITY, g cm⁻³

FIGURE 123–G(D).   VISCOSITY DATA AS A FUNCTION OF DENSITY
FOR GASEOUS METHANE-PROPANE MIXTURES (continued)

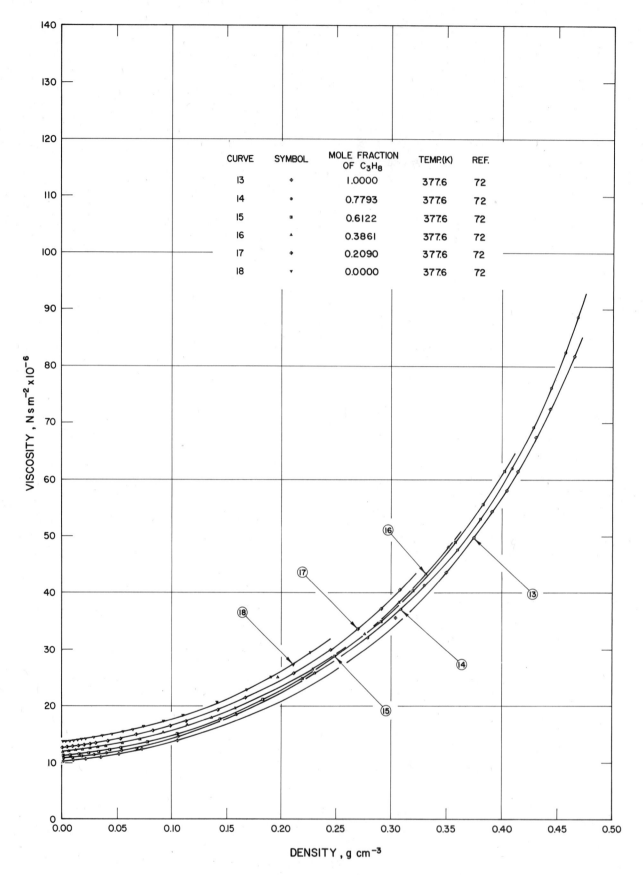

CURVE	SYMBOL	MOLE FRACTION OF C$_3$H$_8$	TEMP.(K)	REF.
13	◇	1.0000	377.6	72
14	○	0.7793	377.6	72
15	▫	0.6122	377.6	72
16	▲	0.3861	377.6	72
17	◆	0.2090	377.6	72
18	▼	0.0000	377.6	72

FIGURE 123 – G(D). VISCOSITY DATA AS A FUNCTION OF DENSITY
FOR GASEOUS METHANE-PROPANE MIXTURES (continued)

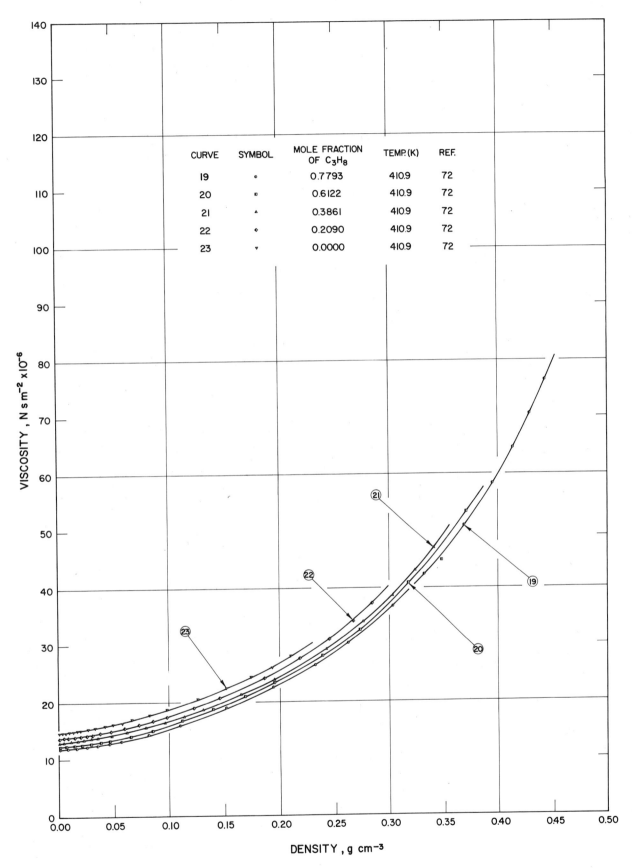

CURVE	SYMBOL	MOLE FRACTION OF C_3H_8	TEMP.(K)	REF.
19	◦	0.7793	410.9	72
20	▫	0.6122	410.9	72
21	▴	0.3861	410.9	72
22	◇	0.2090	410.9	72
23	▿	0.0000	410.9	72

FIGURE 123-G(D).   VISCOSITY DATA AS A FUNCTION OF DENSITY
FOR GASEOUS METHANE-PROPANE MIXTURES (continued)

TABLE 124-G(T)E. EXPERIMENTAL VISCOSITY DATA AS A FUNCTION OF TEMPERATURE FOR GASEOUS NITRIC OXIDE–NITROUS OXIDE MIXTURES

Cur. No.	Fig. No.	Ref. No.	Author(s)	Mole Fraction of N_2O	Pressure (atm)	Temp. (K)	Viscosity ($N \, s \, m^{-2} \times 10^{-6}$)	Remarks
1	124-G(T)	345	Hawksworth, W.A., Nourse, H.H.E., and Raw, C.J.G.	0.750		508.8	25.16	Gases were purified by vacuum distillation; capillary flow viscometer, calibrated with air; error ± 1.0%.
						580.8	28.20	
						633.0	29.85	
						660.4	30.82	
						692.8	32.32	
						734.3	33.40	
						778.5	35.18	
2	124-G(T)	345	Hawksworth, W.A., et al.	0.500		475.0	24.85	Same remarks as for curve 1.
						532.9	27.12	
						575.6	28.71	
						647.8	31.06	
						700.7	33.54	
						740.8	34.71	
						788.7	36.78	
3	124-G(T)	345	Hawksworth, W.A., et al.	0.250		510.4	27.32	Same remarks as for curve 1.
						575.1	30.21	
						597.9	30.99	
						645.7	32.56	
						699.8	34.25	
						742.2	36.11	
						785.0	37.40	
4	124-G(T)	345	Hawksworth, W.A., et al.	0.000		374.0	22.23	Same remarks as for curve 1.
						422.8	25.36	
						465.7	26.59	
						520.4	28.31	
						575.0	31.30	
						623.1	33.29	
						677.3	34.83	
						681.2	35.59	
						723.5	36.19	
						771.5	38.42	
						826.7	40.17	
						873.0	41.52	
						922.0	43.47	
						974.2	44.72	
						1023.2	45.48	
						1077.9	47.53	
						1174.8	49.86	
						1281.5	53.05	
5	124-G(T)	345	Hawksworth, W.A., et al.	1.000		429.2	20.66	Same remarks as for curve 1.
						530.1	24.84	
						582.2	26.39	
						636.0	28.75	
						684.8	30.14	
						739.9	32.63	
						793.2	33.95	
						886.6	39.19	
						916.6	43.58	
						943.2	48.30	
						956.4	52.83	
						977.8	55.59	
						1048.3	63.14	
						1174.8	73.60	
						1296.4	76.40	

TABLE 124-G(T)S. SMOOTHED VISCOSITY VALUES AS A FUNCTION OF TEMPERATURE FOR GASEOUS NITRIC OXIDE-NITROUS OXIDE MIXTURES

Temp. (K)	Mole Fraction of Nitrous Oxide				
	0.00 [Ref. 345]	0.25 [Ref. 345]	0.50 [Ref. 345]	0.75 [Ref. 345]	1.00 [Ref. 345]
375	22.32				
400	23.61				
450					21.50
500	28.30	26.88	26.78	24.81	23.50
525		27.90			
530			26.77		
550		28.94	27.60	26.78	
575		29.96			
600	32.24	30.96	29.49	28.68	27.28
650		32.35	31.40	30.52	
700	35.82	34.63	33.30	32.32	31.00
725		35.50	34.26		
750		36.32	35.24	34.18	
775		37.15	36.22		
800	39.22	37.92	37.20	35.94	34.57
900	42.42				41.12
1000	45.45				59.50
1100	48.30				68.58
1200	50.95				73.60
1300	53.51				76.40

494

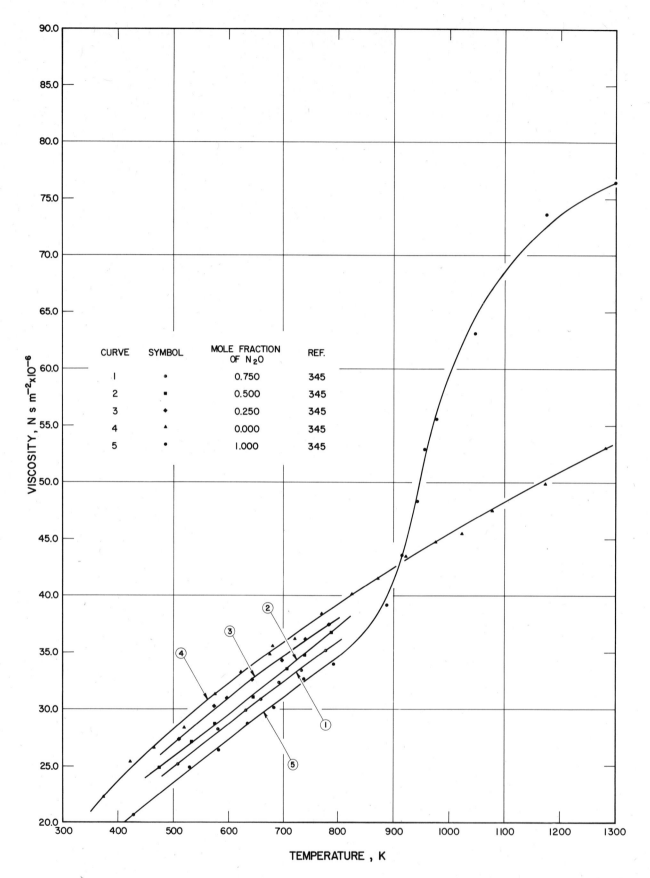

FIGURE 124-G(T). VISCOSITY DATA AS A FUNCTION OF TEMPERATURE
FOR GASEOUS NITRIC OXIDE – NITROUS OXIDE MIXTURES

TABLE 125-G(C)E. EXPERIMENTAL VISCOSITY DATA AS A FUNCTION OF COMPOSITION FOR GASEOUS NITRIC OXIDE-NITROGEN MIXTURES

Cur. No.	Fig. No.	Ref. No.	Author(s)	Temp. (K)	Pressure (atm)	Mole Fraction of NO	Viscosity (N s m^{-2} x 10^{-6})	Remarks
1	125-G(C)	315	Trautz, M. and Gabriel, E.	293.0		0.0000	17.47	NO: from solution of sodium nitrade;
						0.2674	17.78	capillary method; L_1 = 0.112%,
						0.5837	18.27	L_2 = 0.185%, L_3 = 0.368%.
						0.6948	18.33	
						1.0000	18.82	
2	125-G(C)	315	Trautz, M. and Gabriel, E.	373.0		0.0000	20.84	Same remarks as for curve 1 except
						0.2674	21.32	L_1 = 0.204%, L_2 = 0.282%, L_3 =
						0.5837	22.09	0.496%.
						0.6948	22.22	
						1.0000	22.72	

TABLE 125-G(C)S. SMOOTHED VISCOSITY VALUES AS A FUNCTION OF COMPOSITION FOR GASEOUS NITRIC OXIDE-NITROGEN MIXTURES

Mole Fraction of NO	293.0 K [Ref. 315]	373.0 K [Ref. 315]
0.00	17.47	20.84
0.05	17.53	20.95
0.10	17.58	21.11
0.15	17.64	21.16
0.20	17.70	21.26
0.25	17.76	21.36
0.30	17.82	21.46
0.35	17.89	21.55
0.40	17.95	21.65
0.45	18.02	21.74
0.50	18.09	21.83
0.55	18.16	21.91
0.60	18.23	22.01
0.65	18.30	22.10
0.70	18.37	22.19
0.75	18.49	22.28
0.80	18.52	22.37
0.85	18.60	22.45
0.90	18.67	22.54
0.95	18.75	22.63
1.00	18.82	22.72

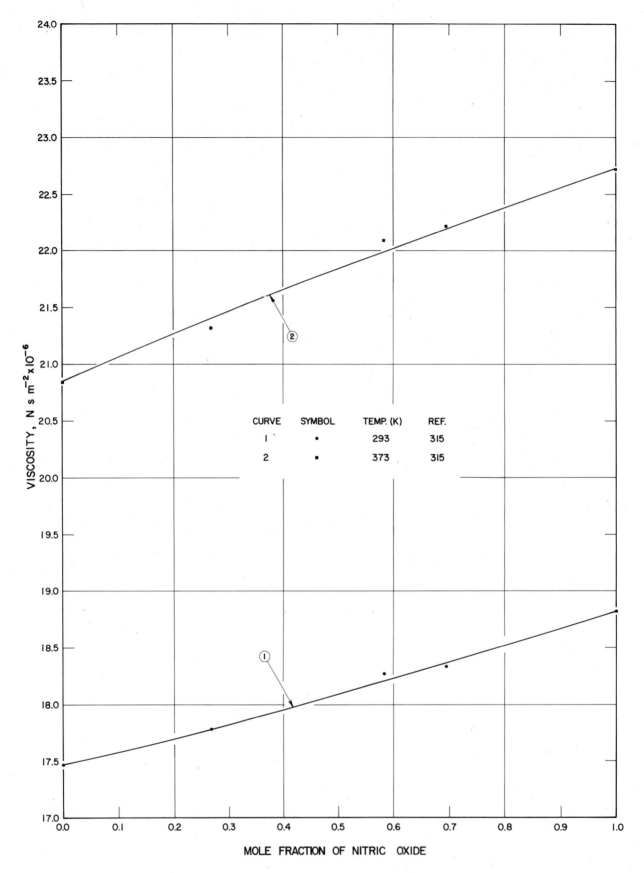

FIGURE 125-G(C). VISCOSITY DATA AS A FUNCTION OF COMPOSITION
FOR GASEOUS NITRIC OXIDE—NITROGEN MIXTURES

TABLE 126-G(C)E.   EXPERIMENTAL VISCOSITY DATA AS A FUNCTION OF COMPOSITION FOR GASEOUS
NITROGEN-OXYGEN MIXTURES

Cur. No.	Fig. No.	Ref. No.	Author(s)	Temp. (K)	Pressure (atm)	Mole Fraction of O_2	Viscosity (N s m^{-2} x 10^{-6})	Remarks
1	126-G(C)	337	Gururaja, G.J., Tirunarayanan, M.A., and Ramchandran, A.	298.2 297.6 298.4 298.5 298.2 298.2 298.2 302.6		0.000 0.132 0.256 0.410 0.510 0.660 0.760 1.000	17.796 17.850 18.450 18.855 19.100 19.650 19.750 20.800	No purity specified; oscillating disk viscometer, calibrated to N_2; viscosity measured at ambient temperature and pressure; precision was ±1.0% of previous data; L_1 = 0.486%, L_2 = 0.616%, L_3 = 1.102%.
2	126-G(C)	227	Trautz, M. and Melster, A.	300.0		1.0000 0.7592 0.4107 0.2178 0.0000	20.57 19.95 18.94 18.43 17.81	Capillary method, R = 0.2019 mm; L_1 = 0.051%, L_2 = 0.088%, L_3 = 0.190%.
3	126-G(C)	227	Trautz, M. and Melster, A.	400.0		1.0000 0.7592 0.4107 0.2178 0.0000	25.68 24.80 23.45 22.75 21.90	Same remarks as for curve 2 except L_1 = 0.061%, L_2 = 0.090%, L_3 = 0.154%.
4	126-G(C)	227	Trautz, M. and Melster, A.	500.0		1.0000 0.7592 0.4107 0.2178 0.0000	30.17 29.09 27.41 26.58 25.60	Same remarks as for curve 2 except L_1 = 0.066%, L_2 = 0.106%, L_3 = 0.226%.
5	126-G(C)	227	Trautz, M. and Melster, A.	550.0		1.0000 0.7592 0.4107 0.2178 0.0000	27.14 24.33 22.40 19.00 17.53	Same remarks as for curve 2 except L_1 = 1.842%, L_2 = 2.587%, L_3 = 4.859%.

TABLE 126-G(C)S.   SMOOTHED VISCOSITY VALUES AS A FUNCTION OF COMPOSITION FOR GASEOUS
NITROGEN-OXYGEN MIXTURES

Mole Fraction of O_2	297.6– 302.6 K [Ref. 337]	300.0 K [Ref. 227]	400.0 K [Ref. 227]	500.0 K [Ref. 227]	550.0 K [Ref. 227]
0.00	17.67	17.81	21.90	25.60	17.53
0.05	17.78	17.95	22.09	25.82	17.99
0.10	17.90	18.09	22.28	26.11	18.44
0.15	18.03	18.23	22.46	26.28	18.90
0.20	18.16	18.38	22.65	26.51	19.36
0.25	18.29	18.50	22.84	26.74	19.82
0.30	18.44	18.66	23.03	26.97	20.29
0.35	18.58	18.80	23.22	27.19	20.76
0.40	18.74	18.94	23.41	27.42	21.23
0.45	18.93	19.08	23.60	27.65	21.72
0.50	19.07	19.22	23.79	27.88	22.20
0.55	19.24	19.37	23.98	28.15	22.69
0.60	19.41	19.50	24.17	28.83	23.18
0.65	19.59	19.65	24.36	28.56	23.66
0.70	19.76	19.80	24.54	28.79	24.16
0.75	19.97	19.94	24.73	29.02	24.66
0.80	20.11	20.07	24.92	29.24	25.15
0.85	20.29	21.21	25.11		25.66
0.90	20.46	20.33	25.30	29.69	26.16
0.95	20.64	20.46	25.49		26.64
1.00	20.80	20.57	25.68	30.15	27.14

498

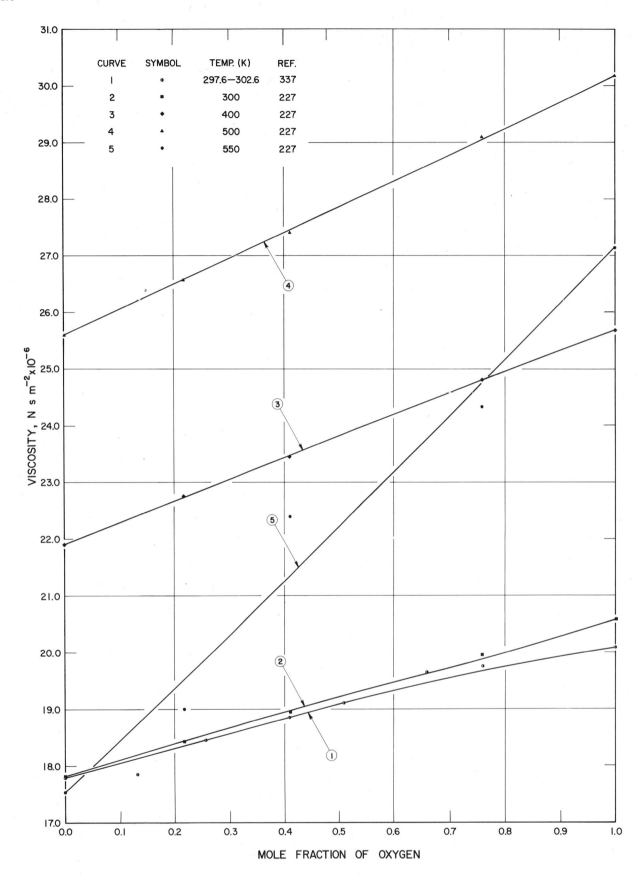

FIGURE 126-G(C). VISCOSITY DATA AS A FUNCTION OF COMPOSITION
FOR GASEOUS NITROGEN—OXYGEN MIXTURES

TABLE 127-G(C)E.   EXPERIMENTAL VISCOSITY DATA AS A FUNCTION OF COMPOSITION FOR GASEOUS
NITROUS OXIDE-PROPANE MIXTURES

Cur. No.	Fig. No.	Ref. No.	Author(s)	Temp. (K)	Pressure (atm)	Mole Fraction of C_3H_8	Viscosity (N s m^{-2} x 10^{-6})	Remarks
1	127-G(C)	234	Trautz, M. and Kurz, F.	300.0		1.0000	8.17	N_2O: 1.3 p per 1000, C_3H_8: 100 pure; capillary method; d = 0.018 cm; L_1 = 0.227%, L_2 = 0.446%, L_3 = 0.986%.
						0.7984	9.26	
						0.4171	11.67	
						0.2018	13.26	
						0.0000	14.88	
2	127-G(C)	234	Trautz, M. and Kurz, F.	400.0		1.0000	10.70	Same remarks as for curve 1 except L_1 = 0.037%, L_2 = 0.083%, L_3 = 0.187%.
						0.7984	12.13	
						0.4171	15.25	
						0.2018	17.25	
						0.0000	19.43	
3	127-G(C)	234	Trautz, M. and Kurz, F.	500.0		1.0000	13.08	Same remarks as for curve 1 except L_1 = 0.000%, L_2 = 0.000%, L_3 = 0.000%.
						0.7984	14.78	
						0.4171	18.54	
						0.2018	20.83	
						0.0000	23.55	
4	127-G(C)	234	Trautz, M. and Kurz, F.	550.0		1.0000	14.22	Same remarks as for curve 1 except L_1 = 0.000%, L_2 = 0.000%, L_3 = 0.000%.
						0.7984	16.10	
						0.4171	20.12	
						0.2018	22.71	
						0.0000	25.56	

TABLE 127-G(C)S.   SMOOTHED VISCOSITY VALUES AS A FUNCTION OF COMPOSITION OF GASEOUS
NITROUS OXIDE-PROPANE MIXTURES

Mole Fraction of C_3H_8	300.0 K [Ref. 234]	400.0 K [Ref. 234]	500.0 K [Ref. 234]	550.0 K [Ref. 234]
0.00	14.88	19.43	23.55	25.56
0.05	14.48	18.88	22.86	24.78
0.10	14.09	18.33	22.16	24.07
0.15	13.68	17.79	21.48	23.40
0.20	13.29	17.26	20.86	22.74
0.25	12.90	16.76	20.28	22.11
0.30	12.53	16.30	19.74	21.49
0.35	12.16	15.84	19.22	20.88
0.40	11.80	15.40	18.70	20.30
0.45	11.45	14.97	18.20	19.75
0.50	11.20	14.54	17.70	19.20
0.55	10.78	14.12	17.20	18.66
0.60	10.46	13.70	16.70	18.14
0.65	10.14	13.29	16.26	17.62
0.70	9.84	12.90	15.70	17.12
0.75	9.54	12.51	15.23	16.60
0.80	9.25	12.14	14.77	16.10
0.85	8.97	11.77	14.33	15.60
0.90	8.70	11.42	13.91	15.13
0.95	8.44	11.08	13.49	14.67
1.00	8.18	10.72	13.08	14.22

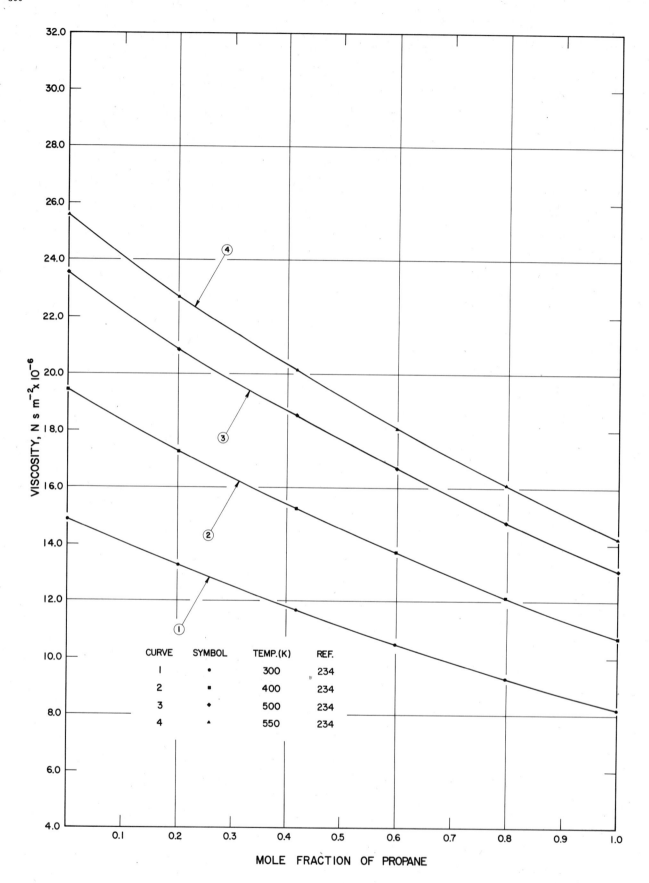

FIGURE 127-G(C). VISCOSITY DATA AS A FUNCTION OF COMPOSITION
FOR GASEOUS NITROUS OXIDE—PROPANE MIXTURES

TABLE 128-G(C)E. EXPERIMENTAL VISCOSITY DATA AS A FUNCTION OF COMPOSITION FOR GASEOUS CARBON DIOXIDE-HYDROGEN CHLORIDE MIXTURES

Cur. No	Fig. No.	Ref. No.	Author(s)	Temp. (K)	Pressure (atm)	Mole Fraction of CO_2	Viscosity (N s m^{-2} x 10^{-6})	Remarks
1	128-G(C)	346	Jung, J. and Schmick, H.	291.0		0.0000	14.26	Effusion method of Trautz and Weizel; L_1 = 0.018%, L_2 = 0.026%, L_3 = 0.041%.
						0.2000	14.53	
						0.4000	14.73	
						0.6000	14.83	
						0.8000	14.81	
						1.0000	14.64	
2	128-G(C)	346	Jung, J. and Schmick, H.	291.16		0.1000	14.59	Same remarks as for curve 1 except L_1 = 0.030%, L_2 = 0.060%, L_3 = 0.166%.
						0.2000	14.72	
						0.3000	14.83	
						0.4000	14.92	
						0.5000	14.99	
						0.6000	15.02	
						0.7000	15.03	
						0.8000	15.00	
						0.9000	14.95	

TABLE 128-G(C)S. SMOOTHED VISCOSITY VALUES AS A FUNCTION OF COMPOSITION FOR GASEOUS CARBON DIOXIDE-HYDROGEN CHLORIDE MIXTURES

Mole Fraction of CO_2	291.2 K [Ref. 346]
0.00	14.44
0.05	14.52
0.10	14.59
0.15	14.66
0.20	14.72
0.25	14.78
0.30	14.83
0.35	14.88
0.40	14.92
0.45	14.96
0.50	14.98
0.55	15.00
0.60	15.02
0.65	15.02
0.70	15.03
0.75	15.02
0.80	15.01
0.85	14.98
0.90	14.95
0.95	14.91
1.00	14.83

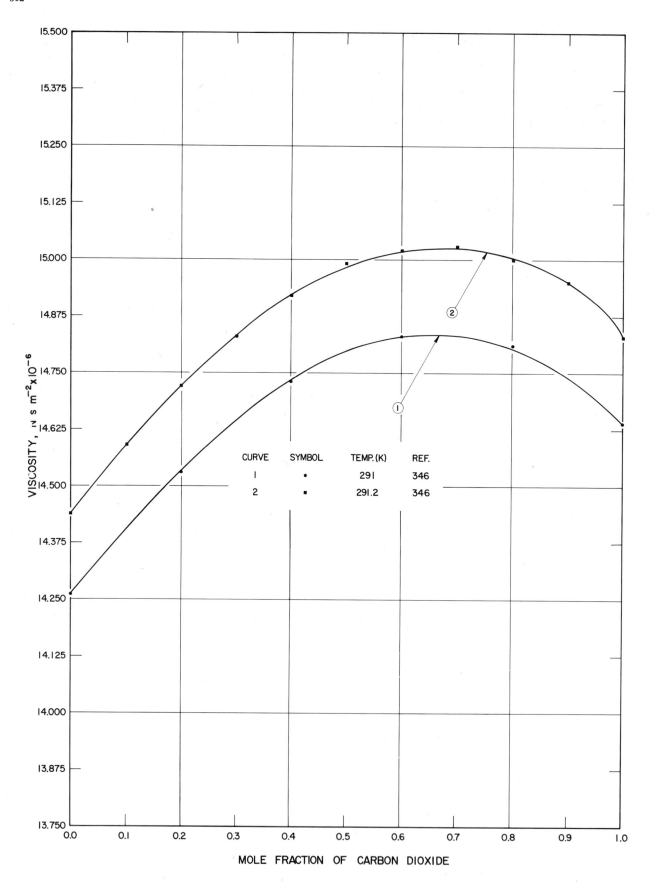

FIGURE 128-G(C). VISCOSITY DATA AS A FUNCTION OF COMPOSITION
FOR GASEOUS CARBON DIOXIDE—HYDROGEN CHLORIDE MIXTURES

TABLE 129-G(C)E. EXPERIMENTAL VISCOSITY DATA AS A FUNCTION OF COMPOSITION FOR GASEOUS CARBON DIOXIDE-SULFUR DIOXIDE MIXTURES

Cur. No.	Fig. No.	Ref. No.	Author(s)	Temp. (K)	Pressure (mm Hg)	Mole Fraction of SO_2	Viscosity ($N\ s\ m^{-2} \times 10^{-6}$)	Remarks
1	129-G(C)	346	Jung, G. and Schmick, H.	289.0		0.0000	14.58	Effusion method of Trautz and Weizel; $L_1 = 0.084\%$, $L_2 = 0.116\%$, $L_3 = 0.237\%$.
						0.2000	14.28	
						0.4000	13.88	
						0.6000	13.46	
						0.8000	12.99	
						1.0000	12.43	
2	129-G(C)	346	Jung, G. and Schmick, H.	289.0		0.900	12.88	Same remarks as for curve 1 except $L_1 = 0.073\%$, $L_2 = 0.096\%$, $L_3 = 0.195\%$.
						0.800	13.16	
						0.700	13.38	
						0.600	13.63	
						0.500	13.84	
						0.400	14.07	
						0.300	14.29	
						0.200	14.47	
						0.100	14.64	
3	129-G(C)	35	Chakraborti, P.K. and Gray, P.	298.2	243-142	0.000	14.80	Gases purified by distillation between liquid nitrogen traps; capillary flow method, relative measurements; precision $\pm 0.2\%$, accuracy 1.0%; $L_1 = 0.158\%$, $L_2 = 0.249\%$, $L_3 = 0.431\%$.
						0.008	14.79	
						0.152	14.73	
						0.179	14.71	
						0.277	14.62	
						0.389	14.54	
						0.424	14.40	
						0.503	14.33	
						0.596	14.15	
						0.655	14.10	
						0.712	13.99	
						0.783	13.78	
						0.822	13.67	
						0.972	13.18	
						1.000	13.17	
4	129-G(C)	35	Chakraborti, P.K. and Gray, P.	308.2	243-142	0.000	15.38	Same remarks as for curve 3 except $L_1 = 0.230\%$, $L_2 = 0.273\%$, $L_3 = 0.438\%$.
						0.041	15.37	
						0.177	15.23	
						0.269	15.10	
						0.396	14.77	
						0.509	14.58	
						0.608	14.36	
						0.697	14.20	
						0.782	13.96	
						0.866	13.77	
						1.000	13.28	
5	129-G(C)	35	Chakraborti, P.K. and Gray, P.	353.2	243-142	0.000	17.30	Same remarks as for curve 3 except $L_1 = 0.041\%$, $L_2 = 0.071\%$, $L_3 = 0.174\%$.
						0.048	17.20	
						0.182	17.02	
						0.288	16.85	
						0.388	16.68	
						0.500	16.45	
						0.598	16.23	
						0.694	16.03	
						0.792	15.79	
						0.878	15.56	
						1.000	15.23	

TABLE 129-G(C)S. SMOOTHED VISCOSITY VALUES AS A FUNCTION OF COMPOSITION FOR GASEOUS
CARBON DIOXIDE-SULFUR DIOXIDE MIXTURES

Mole Fraction of SO_2	289.0 K [Ref. 346]	289.0 K [Ref. 346]	298.2 K [Ref. 35]	308.2 K [Ref. 35]	353.2 K [Ref. 35]
0.00	14.580	14.77	14.80	15.38	17.30
0.05	14.508	14.70	14.78	15.33	17.23
0.10	14.440	14.62	14.76	15.27	17.21
0.15	14.365	14.55	14.73	15.21	17.08
0.20	14.280	14.46	14.69	15.14	17.00
0.25	14.200	14.37	14.65	15.07	16.92
0.30	14.110	14.28	14.60	14.99	16.83
0.35	14.012	14.18	14.54	14.91	16.74
0.40	14.913	14.09	14.48	14.82	16.64
0.45	13.810	13.98	14.41	14.73	16.55
0.50	13.700	13.88	14.33	14.63	16.45
0.55	13.590	13.76	14.25	14.53	16.35
0.60	13.475	13.64	14.16	14.41	16.24
0.65	13.355	13.52	14.06	14.29	16.13
0.70	13.235	13.40	13.95	14.17	16.01
0.75	13.111	13.28	13.84	14.03	15.89
0.80	12.980	13.16	13.72	13.90	15.77
0.85	12.855	13.02	13.59	13.59	15.64
0.90	12.720	12.88	13.46	13.46	15.51
0.95	12.580	12.74	13.32	13.32	15.37
1.00	12.440	12.60	13.17	13.28	15.23

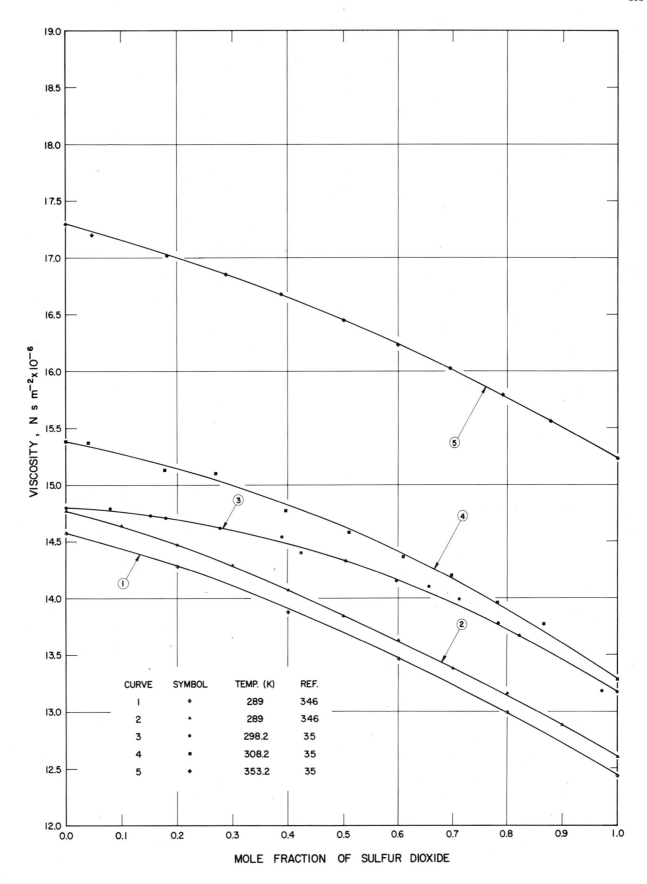

FIGURE 129-G(C). VISCOSITY DATA AS A FUNCTION OF COMPOSITION
FOR GASEOUS CARBON DIOXIDE—SULFUR DIOXIDE MIXTURES

TABLE 130-G(C)E. EXPERIMENTAL VISCOSITY DATA AS A FUNCTION OF COMPOSITION FOR GASEOUS CARBON TETRACHLORIDE-DICHLOROMETHANE MIXTURES

Cur. No.	Fig. No.	Ref. No.	Author(s)	Temp. (K)	Pressure (atm)	Mole Fraction of CCl_4	Viscosity (N s m^{-2} x 10^{-6})	Remarks
1	130-G(C)	292	Mueller, C.R. and Ignatowski, A.J.	293.15		0.0000	10.25	Oscillating disks; L_1 = 0.086%, L_2 = 0.140%, L_3 = 0.294%.
						0.1575	10.21	
						0.2015	10.16	
						0.4986	10.13	
						0.6876	10.00	
						0.8616	9.91	
						1.0000	9.82	
2	130-G(C)	292	Mueller, C.R. and Ignatowski, A.J.	353.26		0.0000	12.02	Same remarks as for curve 1 except L_1 = 0.283%, L_2 = 0.411%, L_3 = 0.700%.
						0.2261	12.12	
						0.6351	11.92	
						1.0000	11.60	
3	130-G(C)	292	Mueller, C.R. and Ignatowski, A.J.	413.43		0.0000	14.27	Same remarks as for curve 1 except L_1 = 0.332%, L_2 = 0.419%, L_3 = 0.728%.
						0.1615	14.25	
						0.2882	14.03	
						0.4738	14.11	
						0.7096	13.82	
						0.8739	13.68	
						1.0000	13.63	

TABLE 130-G(C)S. SMOOTHED VISCOSITY VALUES AS A FUNCTION OF COMPOSITION FOR GASEOUS CARBON TETRACHLORIDE-DICHLOROMETHANE MIXTURES

Mole Fraction of CCl_4	(293.15 K) [Ref. 292]	(353.26 K) [Ref. 292]	(413.43 K) [Ref. 292]
0.00	10.25	12.02	14.27
0.05	10.24	12.03	14.25
0.10	10.23	12.04	14.22
0.15	10.22	12.05	14.19
0.20	10.21	12.06	14.17
0.25	10.20	12.06	14.14
0.30	10.19	12.07	14.11
0.35	10.17	12.07	14.08
0.40	10.15	12.07	14.05
0.45	10.13	12.06	14.02
0.50	10.11	12.05	13.99
0.55	10.09	12.04	13.96
0.60	10.06	12.02	13.93
0.65	10.04	12.00	13.89
0.70	10.01	11.96	13.86
0.75	10.00	11.92	13.82
0.80	9.95	11.88	13.79
0.85	9.91	11.82	13.75
0.90	9.87	11.76	13.71
0.95	9.84	11.70	13.68
1.00	9.82	11.60	13.63

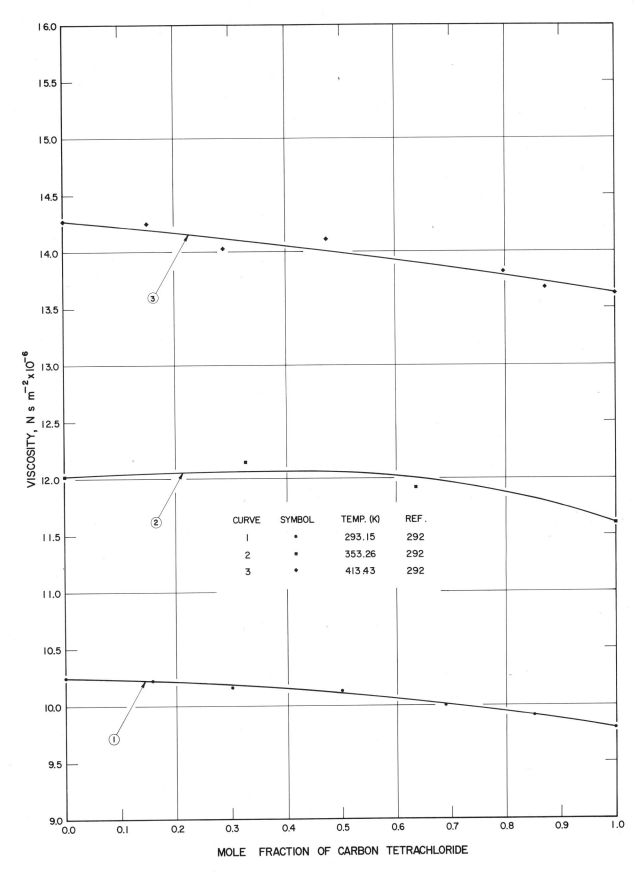

FIGURE 130-G(C). VISCOSITY DATA AS A FUNCTION OF COMPOSITION
FOR GASEOUS CARBON TETRACHLORIDE—DICHLOROMETHANE MIXTURES

TABLE 131-L(C)E.   EXPERIMENTAL VISCOSITY DATA AS A FUNCTION OF COMPOSITION FOR LIQUID
CARBON TETRACHLORIDE-ISOPROPYL ALCOHOL MIXTURES

Cur. No.	Fig. No.	Ref. No.	Author(s)	Temp. (K)	Pressure (atm)	Mole Fraction of CCl_4	Viscosity (N s m^{-2} x 10^{-6})	Remarks
1	131-L(C)	352	Katti, P.K. and Prakash, O.	313.2		1.000	739.0	Merck's isopropyl alcohol and B.D.H. carbon tetrachloride were further purified; Ostwald visco-meter; error ± 0.5%; $L_1 = 0.057\%$, $L_2 = 0.111\%$, $L_3 = 0.239\%$.
						0.885	729.1	
						0.780	733.6	
						0.675	754.8	
						0.579	781.2	
						0.500	817.5	
						0.398	874.6	
						0.315	935.5	
						0.255	986.8	
						0.121	1144.8	
						0.000	1330.0	

TABLE 131-L(C)S.   SMOOTHED VISCOSITY VALUES AS A FUNCTION OF COMPOSITION FOR LIQUID
CARBON TETRACHLORIDE-ISOPROPYL ALCOHOL MIXTURES

Mole Fraction of CCl_4	(313.2 K) [Ref. 352]
0.00	1330.0
0.05	1250.0
0.10	1175.0
0.15	1105.0
0.20	1044.0
0.25	992.0
0.30	946.4
0.35	908.0
0.40	874.0
0.45	844.0
0.50	817.5
0.55	794.8
0.60	775.5
0.65	760.0
0.70	746.8
0.75	737.9
0.80	731.8
0.85	729.0
0.90	729.2
0.95	733.2
1.00	739.0

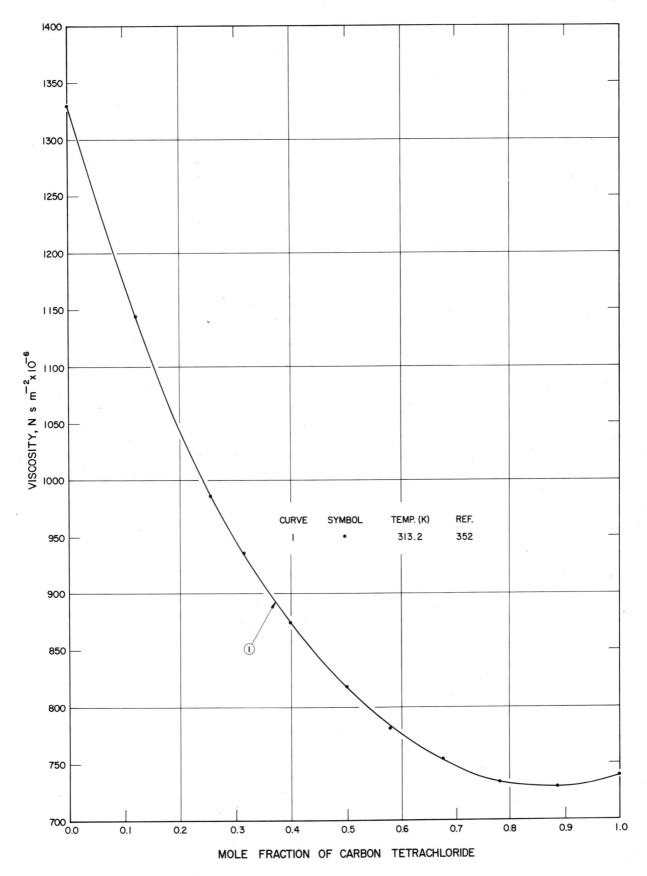

FIGURE 131-L(C). VISCOSITY DATA AS A FUNCTION OF COMPOSITION
FOR LIQUID CARBON TETRACHLORIDE—ISOPROPYL ALCOHOL MIXTURES

TABLE 132-L(C)E.   EXPERIMENTAL VISCOSITY DATA AS A FUNCTION OF COMPOSITION FOR LIQUID
CARBON TETRACHLORIDE-METHANOL MIXTURES

Cur. No.	Fig. No.	Ref. No.	Author(s)	Temp. (K)	Pressure (atm)	Mole Fraction of CCl$_4$	Viscosity (N s m^{-2} x 10^{-6})	Remarks
1	132-L(C)	352	Katti, P.K. and Prakash, O.	313.2		1.000	739.0	Merck's methanol and B.D.H. carbon tetrachloride were further purified before use; Ostwald vis- cometer; error ± 0.5%; L$_1$ = 0.022%, L$_2$ = 0.041%, L$_3$ = 0.099%.
						0.895	759.5	
						0.807	762.4	
						0.697	750.0	
						0.650	742.7	
						0.490	695.5	
						0.320	624.8	
						0.280	605.0	
						0.210	570.0	
						0.090	505.0	
						0.000	456.0	

TABLE 132-L(C)S.   SMOOTHED VISCOSITY VALUES AS A FUNCTION OF COMPOSITION FOR LIQUID
CARBON TETRACHLORIDE-METHANOL MIXTURES

Mole Fraction of CCl$_4$	(313.2 K) [Ref. 352]
0.00	456
0.05	484
0.10	511
0.15	539
0.20	565
0.25	591
0.30	615
0.35	638
0.40	660
0.45	680
0.50	698
0.55	715
0.60	730
0.65	742
0.70	751
0.75	758
0.80	762
0.85	762
0.90	759
0.95	751
1.00	739

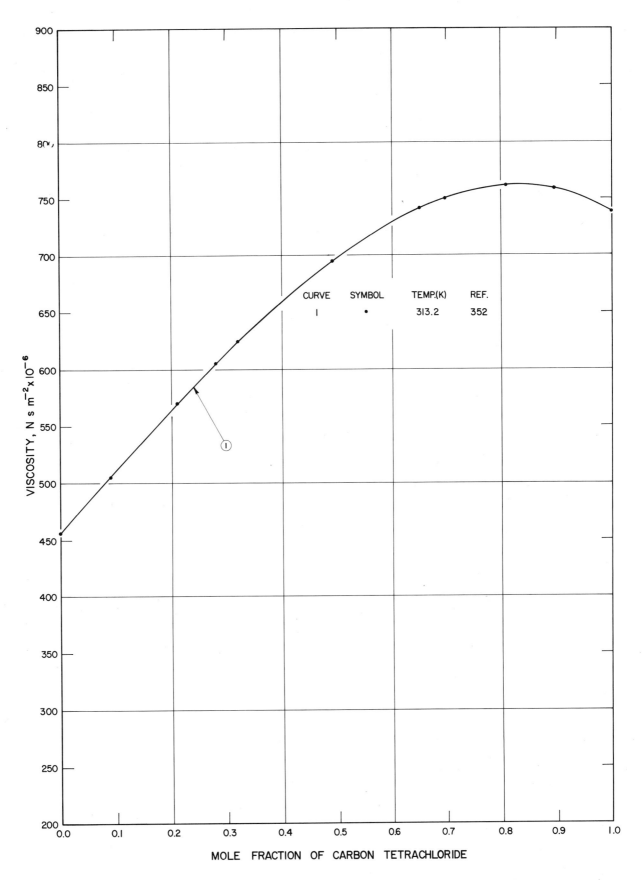

FIGURE 132-L(C). VISCOSITY DATA AS A FUNCTION OF COMPOSITION
FOR LIQUID CARBON TETRACHLORIDE-METHANOL MIXTURES

TABLE 133-L(C)E.   EXPERIMENTAL VISCOSITY DATA AS A FUNCTION OF COMPOSITION FOR LIQUID DIOXANE-BENZYL ACETATE MIXTURES

Cur. No.	Fig. No.	Ref. No.	Author(s)	Temp. (K)	Pressure (atm)	Mole Fraction of $C_9H_{10}O_2$	Viscosity (N s m^{-2} x 10^{-6})	Remarks
1	133-L(C)	351	Katti, P.K. and Chaudhri, M.M.	313.2		0.000	625.6	Liquids were purified (ref. J. Chem. Eng. Data, 9, 128, 1964); Ostwald viscometer; error ± 0.5%; L_1 = 0.851%, L_2 = 1.311%, L_3 = 2.683%.
						0.200	725.0	
						0.300	802.4	
						0.380	857.0	
						0.520	958.1	
						0.645	1060.2	
						0.748	1147.0	
						0.875	1233.3	
						1.000	1352.5	

TABLE 133-L(C)S.   SMOOTHED VISCOSITY VALUES AS A FUNCTION OF COMPOSITION FOR LIQUID DIOXANE-BENZYL ACETATE MIXTURES

Mole Fraction of $C_9H_{10}O_2$	(313.2 K) [Ref. 351]
0.00	626
0.05	655
0.10	684
0.15	714
0.20	742
0.25	772
0.30	802
0.35	833
0.40	864
0.45	896
0.50	930
0.55	965
0.60	1001
0.65	1038
0.70	1076
0.75	1116
0.80	1160
0.85	1204
0.90	1250
0.95	1300
1.00	1352

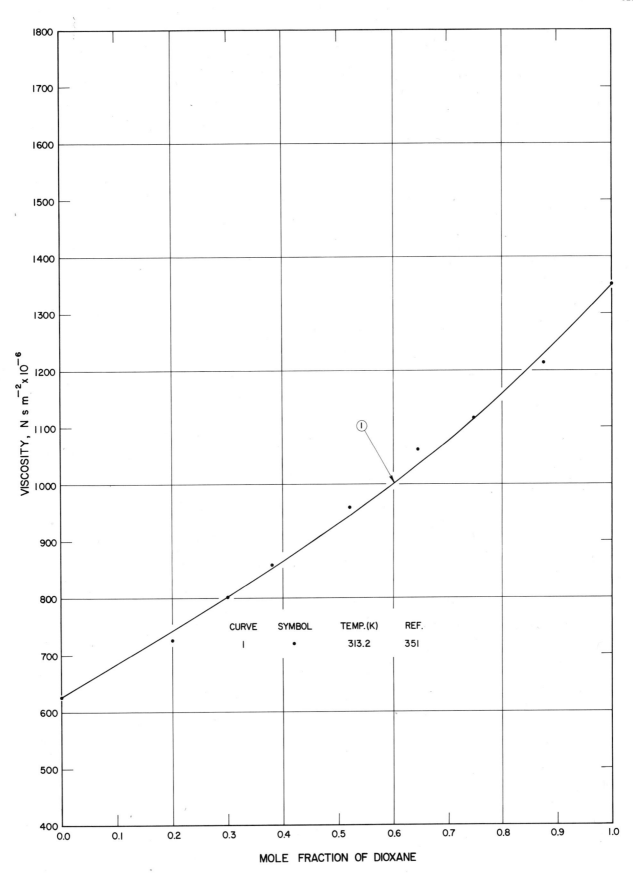

FIGURE 133 - L(C). VISCOSITY DATA AS A FUNCTION OF COMPOSITION
FOR LIQUID DIOXANE – BENZYL ACETATE MIXTURES

TABLE 134-G(C)E.   EXPERIMENTAL VISCOSITY DATA AS A FUNCTION OF COMPOSITION FOR GASEOUS
ETHYLENE-AMMONIA MIXTURES

Cur. No.	Fig. No.	Ref. No.	Author(s)	Temp. (K)	Pressure (atm)	Mole Fraction of C_2H_4	Viscosity (N s m^{-2} x 10^{-6})	Remarks
1	134-G(C)	222	Trautz, M. and Heberling, R.	293.2		0.0000	10.08	C_2H_4 obtained by chemical reaction;
						0.1133	10.01	NH_3: I. G. Farben, 99.997% pure,
						0.1929	10.13	chief impurities O_2, H_2, N_2; capil-
						0.3039	10.22	lary transpiration method, d =
						0.4828	10.30	0.04038 cm; experimental error
						0.7007	10.27	<3%; L_1 = 0.036%, L_2 = 0.067%,
						0.8904	10.15	L_3 = 0.148%.
						1.0000	9.82	
2	134-G(C)	222	Trautz, M. and Heberling, R.	373.2		0.0000	12.57	Same remarks as for curve 1 except
						0.1133	12.94	L_1 = 0.039%, L_2 = 0.061%, L_3 =
						0.1929	13.01	0.142%.
						0.3039	13.04	
						0.4828	13.03	
						0.7007	12.91	
						0.8904	12.69	
						1.0000	12.79	
3	134-G(C)	222	Trautz, M. and Heberling, R.	473.2		0.0000	15.41	Same remarks as for curve 1 except
						0.1133	16.47	L_1 = 0.029%, L_2 = 0.060%, L_3 =
						0.1929	16.48	0.152%.
						0.3039	16.39	
						0.4828	16.22	
						0.7007	15.95	
						0.8904	15.61	
						1.0000	16.46	
4	134-G(C)	222	Trautz, M. and Heberling, R.	523.2		0.0000	16.66	Same remarks as for curve 1 except
						0.1133	18.09	L_1 = 0.060%, L_2 = 0.077%, L_3 =
						0.1929	18.05	0.153%.
						0.3039	17.91	
						0.4828	17.64	
						0.7007	17.29	
						0.8904	16.89	
						1.0000	18.13	

TABLE 134-G(C)S.   SMOOTHED VISCOSITY VALUES AS A FUNCTION OF COMPOSITION FOR GASEOUS
ETHYLENE-AMMONIA MIXTURES

Mole Fraction of C_2H_4	(293.2 K) [Ref. 222]	(373.2 K) [Ref. 222]	(473.2 K) [Ref. 222]	(523.2 K) [Ref. 222]
0.00	9.82	12.79	16.46	18.13
0.05	9.91	12.87	16.47	18.12
0.10	9.99	12.93	16.47	18.10
0.15	10.06	12.97	16.47	18.08
0.20	10.13	13.08	16.45	18.04
0.25	10.18	13.03	16.43	17.99
0.30	10.22	13.04	16.40	17.94
0.35	10.25	13.04	16.36	17.87
0.40	10.28	13.05	16.31	17.80
0.45	10.29	13.04	16.26	17.72
0.50	10.30	13.02	16.20	17.64
0.55	10.30	13.01	16.15	17.56
0.60	10.30	12.98	16.09	17.48
0.65	10.29	12.95	16.02	17.39
0.70	10.27	12.91	15.95	17.30
0.75	10.23	12.87	15.88	17.20
0.80	10.22	12.82	15.79	17.10
0.85	10.19	12.76	15.70	16.99
0.90	10.16	12.70	15.61	16.88
0.95	10.12	12.63	15.50	16.77
1.00	10.08	12.56	15.41	16.65

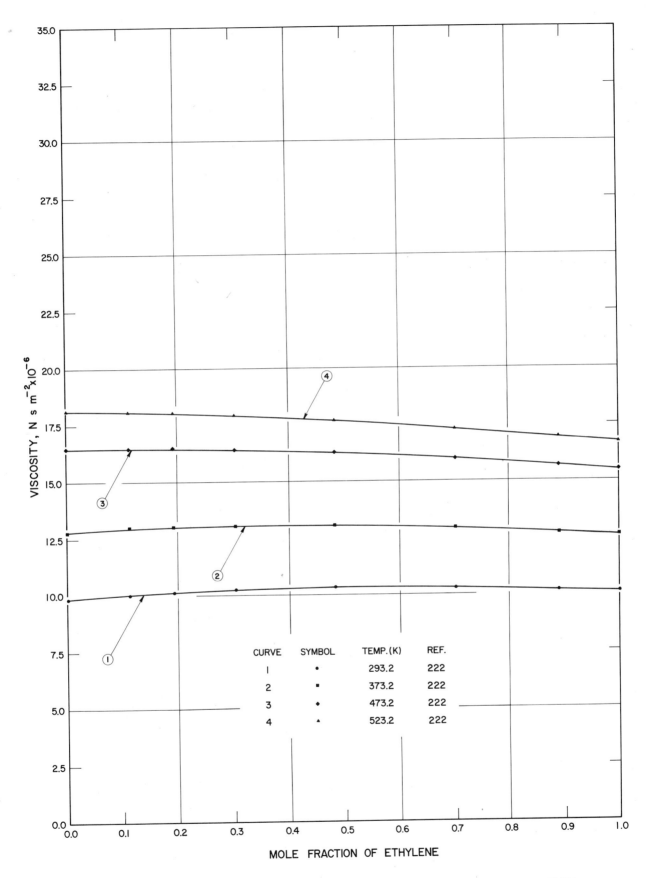

FIGURE 134-G(C). VISCOSITY DATA AS A FUNCTION OF COMPOSITION
FOR GASEOUS ETHYLENE—AMMONIA MIXTURES

TABLE 135-G(C)E.   EXPERIMENTAL VISCOSITY DATA AS A FUNCTION OF COMPOSITION FOR GASEOUS HYDROGEN-AMMONIA MIXTURES

Cur. No.	Fig. No.	Ref. No.	Author(s)	Temp. (K)	Pressure (mm Hg)	Mole Fraction of NH_3	Viscosity ($N\ s\ m^{-2} \times 10^{-6}$)	Remarks
1	135-G(C)	222	Trautz, M. and Heberling, R.	293.2		1.0000	9.82	H_2: by electrolysis of KOH on pure nickel electrodes; NH_3: I.G. Farben, 99.997% pure, chief impurities O_2, H_2, N_2; capillary transpiration method, $d = 0.04038$ cm; experimental error < 3%; $L_1 = 0.049\%$, $L_2 = 0.111\%$, $L_3 = 0.298\%$.
						0.9005	10.04	
						0.7087	10.47	
						0.5177	10.80	
						0.2975	10.87	
						0.2239	10.72	
						0.1082	10.11	
						0.0000	8.77	
2	135-G(C)	341	Pal, A.K. and Barua, A.K.	306.2	< 100	0.0000	9.055	H_2: 99.5 pure; oscillating disk viscometer, relative measurements; uncertainty in mixture composition $\pm 0.5\%$; data agree with available values in literature within 1%; $L_1 = 0.265\%$, $L_2 = 0.496\%$, $L_3 = 1.156\%$.
						0.1950	11.840	
						0.3990	12.381	
						0.5360	12.244	
						0.6770	12.000	
						0.8550	11.461	
						1.0000	10.590	
3	135-G(C)	341	Pal, A.K. and Barua, A.K.	327.2	< 100	0.0000	9.491	Same remarks as for curve 2 except $L_1 = 0.025\%$, $L_2 = 0.065\%$, $L_3 = 0.172\%$.
						0.1950	12.516	
						0.3990	13.071	
						0.5360	13.049	
						0.6770	12.758	
						0.8550	12.150	
						1.0000	11.375	
4	135-G(C)	341	Pal, A.K. and Barua, A.K.	371.2	< 100	0.0000	10.397	Same remarks as for curve 2 except $L_1 = 0.455\%$, $L_2 = 0.962\%$, $L_3 = 2.502\%$.
						0.1950	13.582	
						0.3990	14.579	
						0.5360	14.609	
						0.6770	14.504	
						0.8550	14.135	
						1.0000	13.001	
5	135-G(C)	222	Trautz, M. and Heberling, R.	373.2		1.0000	12.79	Same remarks as for curve 1 except $L_1 = 0.192\%$, $L_2 = 0.344\%$, $L_3 = 0.868\%$.
						0.9005	12.99	
						0.7087	13.33	
						0.5177	13.54	
						0.2975	13.29	
						0.2239	12.99	
						0.1082	12.04	
						0.0000	10.30	
6	135-G(C)	341	Pal, A.K. and Barua, A.K.	421.2	< 100	0.0000	11.458	Same remarks as for curve 2 except $L_1 = 1.230\%$, $L_2 = 2.097\%$, $L_3 = 5.123\%$.
						0.1400	14.917	
						0.4054	15.937	
						0.5170	16.030	
						0.6005	16.201	
						0.8042	15.991	
						1.0000	14.850	
7	135-G(C)	222	Trautz, M. and Heberling, R.	473.2		1.0000	16.46	Same remarks as for curve 1 except $L_1 = 0.038\%$, $L_2 = 0.064\%$, $L_3 = 0.122\%$.
						0.9005	16.60	
						0.7087	16.80	
						0.5177	16.76	
						0.2975	16.10	
						0.2239	15.60	
						0.1082	14.32	
						0.0000	12.11	
8	135-G(C)	341	Pal, A.K. and Barua, A.K.	479.2	< 100	0.0000	12.621	Same remarks as for curve 2 except $L_1 = 1.267\%$, $L_2 = 2.184\%$, $L_3 = 5.243\%$.
						0.1400	16.460	
						0.4054	17.719	
						0.5170	17.905	
						0.6005	18.020	
						0.8042	17.971	
						1.0000	17.002	
9	135-G(C)	222	Trautz, M. and Heberling, R.	523.2		1.0000	18.13	Same remarks as for curve 1 except $L_1 = 0.023\%$, $L_2 = 0.046\%$, $L_3 = 0.109\%$.
						0.9005	18.26	
						0.7087	18.39	
						0.5177	18.26	
						0.2975	17.40	
						0.2239	16.80	
						0.0000	12.96	

TABLE 135-(C)S. SMOOTHED VISCOSITY VALUES AS A FUNCTION OF COMPOSITION FOR GASEOUS HYDROGEN-AMMONIA MIXTURES

Mole Fraction of NH₃	(293.2 K) [Ref. 222]	(306.2 K) [Ref. 341]	(327.2 K) [Ref. 341]	(371.2 K) [Ref. 341]	(373.2 K) [Ref. 222]	(421.2 K) [Ref. 341]	(473.2 K) [Ref. 222]	(479.2 K) [Ref. 341]	(523.2 K) [Ref. 222]
0.00	8.77	9.06	9.49	10.40	10.30	11.46	12.11	12.62	12.96
0.05	9.53	10.32	10.89	11.47	11.22	12.68	13.24	13.97	14.29
0.10	10.04	11.06	11.67	12.41	11.92	13.61	14.19	15.01	15.25
0.15	10.39	11.54	12.18	13.12	12.46	14.32	14.89	15.78	15.99
0.20	10.63	11.87	12.54	13.63	12.84	14.85	15.40	16.37	16.57
0.25	10.79	12.10	12.78	14.00	13.12	15.26	15.81	16.83	17.04
0.30	10.87	12.24	12.94	14.27	13.33	15.56	16.12	17.19	17.42
0.35	10.91	12.33	13.04	14.45	13.46	15.78	16.36	17.48	17.71
0.40	10.90	12.39	13.08	14.56	13.52	15.93	16.50	17.70	17.94
0.45	10.88	12.39	13.08	14.64	13.56	16.04	16.65	17.87	18.10
0.50	10.82	12.36	13.07	14.66	13.55	16.10	16.76	18.00	18.23
0.55	10.76	12.29	13.03	14.66	13.52	16.12	16.80	18.08	18.31
0.60	10.68	12.19	12.96	14.60	13.47	16.10	16.82	18.10	18.36
0.65	10.59	12.07	12.85	14.52	13.42	16.05	16.82	18.09	18.38
0.70	10.50	11.92	12.72	14.39	13.35	15.97	16.79	18.03	18.40
0.75	10.40	11.75	12.56	14.23	13.27	15.86	16.76	17.93	18.39
0.80	10.29	11.56	12.38	14.04	13.19	15.70	16.72	17.80	18.37
0.85	10.18	11.35	12.16	13.81	13.10	15.52	16.66	17.63	18.34
0.90	10.07	11.12	11.92	13.56	13.01	15.32	16.60	17.44	18.28
0.95	9.96	10.86	11.66	13.28	12.90	15.10	16.54	17.22	18.22
1.00	9.80	10.59	11.38	13.00	12.68	14.85	16.46	17.00	18.13

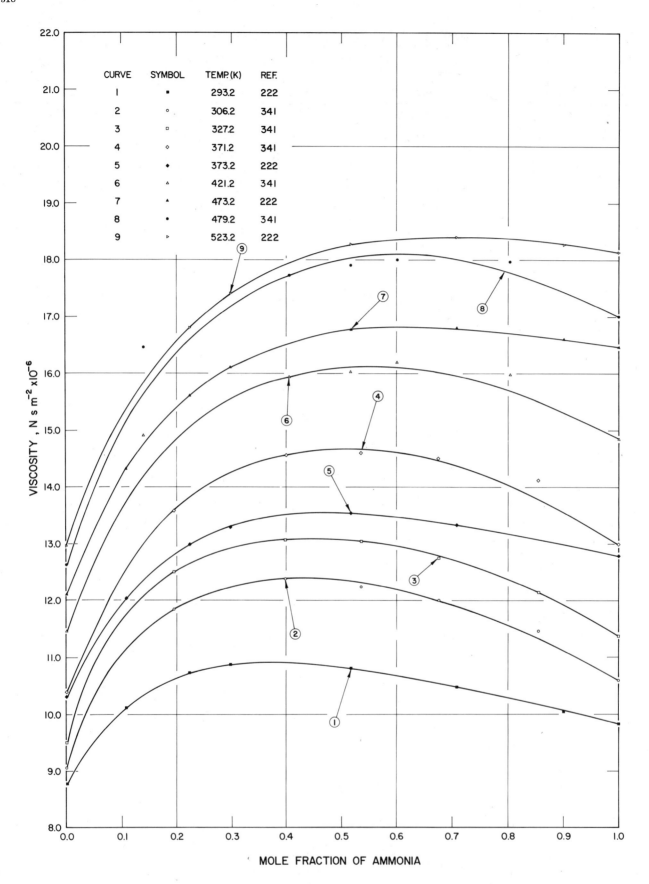

CURVE	SYMBOL	TEMP.(K)	REF.
1	■	293.2	222
2	○	306.2	341
3	□	327.2	341
4	◇	371.2	341
5	◆	373.2	222
6	△	421.2	341
7	▲	473.2	222
8	●	479.2	341
9	▷	523.2	222

VISCOSITY , N s m^{-2} x10^{-6}

MOLE FRACTION OF AMMONIA

FIGURE 135-G(C). VISCOSITY DATA AS A FUNCTION OF COMPOSITION
FOR GASEOUS HYDROGEN – AMMONIA MIXTURES

TABLE 136-G(C)E. EXPERIMENTAL VISCOSITY DATA AS A FUNCTION OF COMPOSITION FOR GASEOUS HYDROGEN-ETHYL ETHER MIXTURES

Cur. No.	Fig. No.	Ref. No.	Author(s)	Temp. (K)	Pressure (atm)	Mole Fraction of $(C_2H_5)_2O$	Viscosity (N s m^{-2} x 10^{-6})	Remarks
1	136-G(C)	226	Trautz, M. and Ludewigs, W.	288.16		1.0000 0.2650 0.1330 0.0000	7.29 9.00 9.37 8.68	$(C_2H_5)_2O$: no purity specified, H_2: made by electrolysis; capillary method; $L_1 = 0.000\%$, $L_2 = 0.000\%$, $L_3 = 0.000\%$.
2	136-G(C)	226	Trautz, M. and Ludewigs, W.	373.16		1.0000 0.2650 0.1330 0.0000	9.49 11.19 11.46 10.35	Same remarks as for curve 1 except $L_1 = 0.035\%$, $L_2 = 0.053\%$, $L_3 = 0.097\%$.
3	136-G(C)	226	Trautz, M. and Ludewigs, W.	423.15		1.0000 0.2650 0.1330 0.0000	10.70 12.52 12.62 11.34	Same remarks as for curve 1 except $L_1 = 0.000\%$, $L_2 = 0.000\%$, $L_3 = 0.000\%$.
4	136-G(C)	226	Trautz, M. and Ludewigs, W.	486.16		1.0000 0.2650 0.1330 0.0000	12.15 13.91 14.03 12.48	Same remarks as for curve 1 except $L_1 = 0.002\%$, $L_2 = 0.004\%$, $L_3 = 0.007\%$.

TABLE 136-G(C)S. SMOOTHED VISCOSITY VALUES AS A FUNCTION OF COMPOSITION FOR GASEOUS HYDROGEN-ETHYL ETHER MIXTURES

Mole Fraction of $(C_2H_5)_2O$	(288.16 K) [Ref. 226]	(373.16 K) [Ref. 226]	(423.15 K) [Ref. 226]	(486.16 K) [Ref. 226]
0.00	8.68	10.35	11.34	12.48
0.05	9.09	10.96	11.86	13.27
0.10	9.32	11.32	12.43	13.88
0.15	9.35	11.48	12.67	14.06
0.20	9.24	11.43	12.68	14.08
0.25	9.07	11.27	12.57	13.97
0.30	8.88	11.05	12.39	13.77
0.35	8.70	10.87	12.22	13.60
0.40	8.54	10.71	12.07	13.45
0.45	8.40	10.56	11.93	13.30
0.50	8.27	10.43	11.80	13.17
0.55	8.16	10.31	11.68	13.05
0.60	8.04	10.20	11.56	12.94
0.65	7.94	10.10	11.44	12.84
0.70	7.84	10.00	11.32	12.73
0.75	7.74	9.91	11.21	12.63
0.80	7.65	9.82	11.10	12.54
0.85	7.56	9.73	10.99	12.44
0.90	7.47	9.65	10.89	12.35
0.95	7.38	9.57	10.79	12.25
1.00	7.29	9.49	10.70	12.15

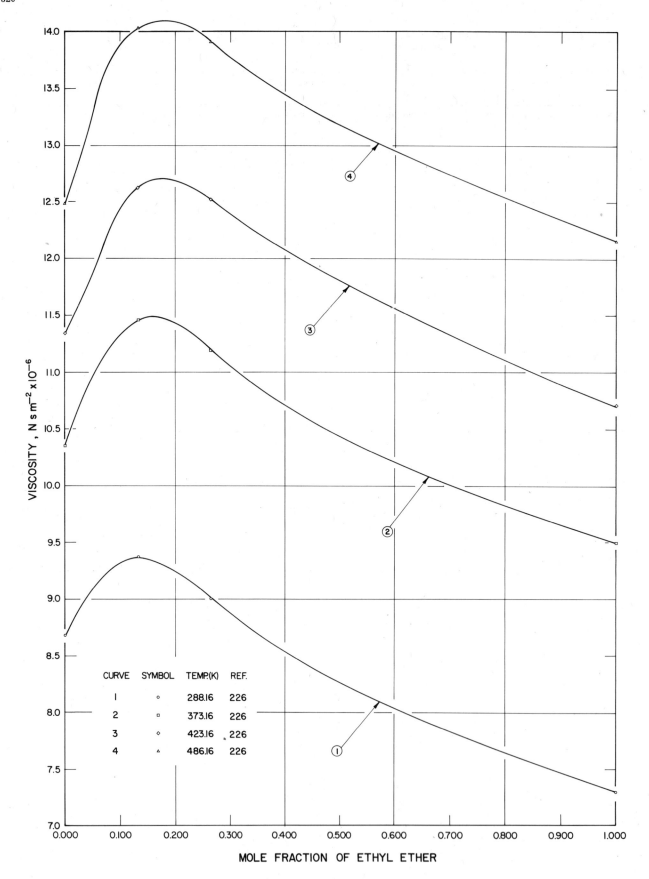

CURVE	SYMBOL	TEMP.(K)	REF.
1	○	288.16	226
2	□	373.16	226
3	◇	423.16	226
4	△	486.16	226

VISCOSITY , N s m^{-2} x10^{-6}

MOLE FRACTION OF ETHYL ETHER

FIGURE 136 - G (C). VISCOSITY DATA AS A FUNCTION OF COMPOSITION
FOR GASEOUS HYDROGEN − ETHYL ETHER MIXTURES

TABLE 137-G(C)E.   EXPERIMENTAL VISCOSITY DATA AS A FUNCTION OF COMPOSITION FOR GASEOUS HYDROGEN-HYDROGEN CHLORIDE MIXTURES

Cur. No.	Fig. No.	Ref. No.	Author(s)	Temp. (K)	Pressure (atm)	Mole Fraction of HCl	Viscosity (N s m^{-2} x 10^{-6})	Remarks
1	137-G(C)	228	Trautz, M. and Narath, A.	294.16		1.0000	14.37	Capillary method, d = 0.152 mm;
						0.8220	14.61	precision ± 2%; L$_1$ = 0.040%, L$_2$ =
						0.7179	14.69	0.058%, L$_3$ = 0.102%.
						0.5042	14.71	
						0.2031	13.42	
						0.0000	8.81	
2	137-G(C)	228	Trautz, M. and Narath, A.	327.16		1.0000	16.05	Same remarks as for curve 1 except
						0.8220	16.26	L$_1$ = 0.011%, L$_2$ = 0.028%, L$_3$ =
						0.7179	16.32	0.068%.
						0.5042	16.25	
						0.2031	14.72	
						0.0000	9.47	
3	137-G(C)	228	Trautz, M. and Narath, A.	372.16		1.0000	18.28	Same remarks as for curve 1 except
						0.8220	18.48	L$_1$ = 0.081%, L$_2$ = 0.161%, L$_3$ =
						0.7179	18.55	0.379%.
						0.5042	18.31	
						0.2031	16.29	
						0.0000	10.36	
4	137-G(C)	228	Trautz, M. and Narath, A.	427.16		1.0000	20.94	Same remarks as for curve 1 except
						0.8417	20.99	L$_1$ = 0.180%, L$_2$ = 0.338%, L$_3$ =
						0.6989	21.04	0.766%.
						0.5092	20.53	
						0.2409	18.66	
						0.0000	11.42	
5	137-G(C)	228	Trautz, M. and Narath, A.	473.16		1.0000	23.04	Same remarks as for curve 1 except
						0.8417	23.11	L$_1$ = 0.088%, L$_2$ = 0.154%, L$_3$ =
						0.6989	23.04	0.305%.
						0.5092	22.61	
						0.2409	20.24	
						0.0000	12.24	
6	137-G(C)	228	Trautz, M. and Narath, A.	523.16		1.0000	25.28	Same remarks as for curve 1 except
						0.7947	25.27	L$_1$ = 0.087%, L$_2$ = 0.123%, L$_3$ =
						0.6312	25.07	0.198%.
						0.5178	24.54	
						0.2991	22.81	
						0.0000	13.15	

TABLE 137-G(C)S.   SMOOTHED VISCOSITY VALUES AS A FUNCTION OF COMPOSITION FOR GASEOUS HYDROGEN-HYDROGEN CHLORIDE MIXTURES

Mole Fraction of HCl	(294.2 K) [Ref. 228]	(327.2 K) [Ref. 228]	(372.2 K) [Ref. 228]	(427.2 K) [Ref. 228]	(473.2 K) [Ref. 228]	(523.2 K) [Ref. 228]
0.00	8.81	9.41	10.36	11.42	12.24	13.15
0.05		12.57	13.08	14.10	15.20	16.48
0.10	12.58	13.45	14.50	16.28	17.41	18.89
0.15	13.04	14.16	15.52	17.49	18.77	20.34
0.20	13.41	14.70	16.25	18.24	19.70	21.38
0.25	13.73	15.12	16.80	18.74	20.33	22.18
0.30	14.00	15.44	17.22	19.10	20.81	22.84
0.35	14.23	15.70	17.56	19.42	21.24	23.36
0.40	14.40	15.91	17.82	19.72	21.62	23.82
0.45	14.54	16.07	18.04	20.00	21.96	24.18
0.50	14.62	16.18	18.20	20.24	22.25	24.48
0.55	14.70	16.26	18.33	20.44	22.50	24.72
0.60	14.74	16.32	18.41	20.62	22.70	24.92
0.65	14.74	16.34	18.46	20.79	22.85	25.09
0.70	14.72	16.33	18.48	20.88	22.97	25.19
0.75	14.68	16.31	18.48	20.96	23.05	25.26
0.80	14.64	16.28	18.47	20.99	23.10	25.32
0.85	14.60	16.24	18.44	20.98	23.11	25.34
0.90	14.53	16.18	18.40	20.98	23.10	25.34
0.95	14.46	16.12	18.33	20.96	23.08	25.33
1.00	14.38	16.05	18.28	20.44	23.04	25.28

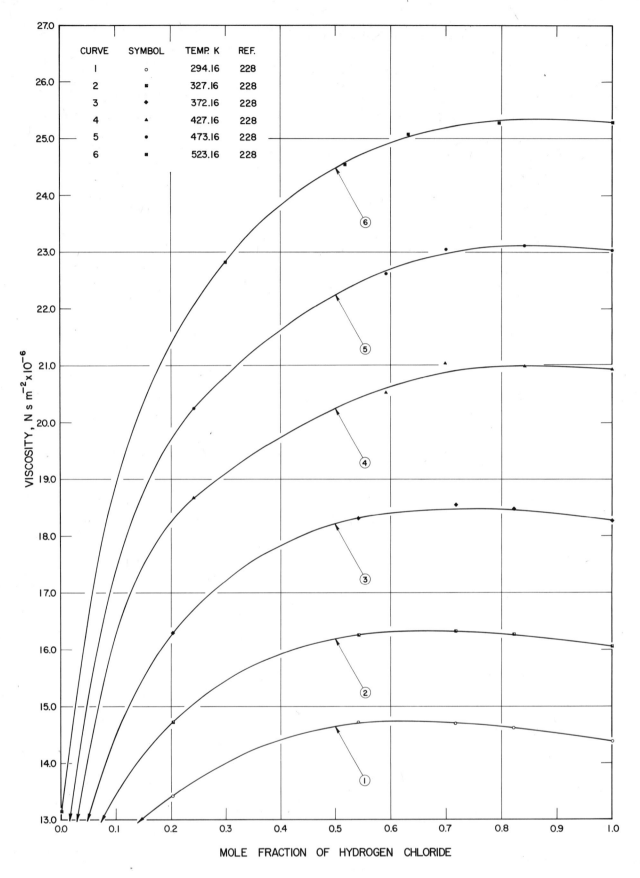

CURVE	SYMBOL	TEMP. K	REF.
1	○	294.16	228
2	■	327.16	228
3	◆	372.16	228
4	▲	427.16	228
5	●	473.16	228
6	■	523.16	228

VISCOSITY, N s m^{-2} x10^{-6}

MOLE FRACTION OF HYDROGEN CHLORIDE

FIGURE 137−G(C). VISCOSITY DATA AS A FUNCTION OF COMPOSITION
FOR GASEOUS HYDROGEN−HYDROGEN CHLORIDE MIXTURES

TABLE 138-G(C)E.   EXPERIMENTAL VISCOSITY DATA AS A FUNCTION OF COMPOSITION FOR GASEOUS
HYDROGEN-SULFUR DIOXIDE MIXTURES

Cur. No.	Fig. No.	Ref. No.	Author(s)	Temp. (K)	Pressure (mm Hg)	Mole Fraction of SO_2	Viscosity (N s m^{-2} x 10^{-6})	Remarks
1	138-G(C)	231	Trautz, M. and Weizel, W.	290.16		1.0000	12.59	Capillary method, D = 0.15 mm; accuracy: pure SO_2 ±0.2, pure H_2 ±0.8, for SO_2 <30% of mixture ±0.4; precision: ±0.3% for SO_2 <30% of mixture; L_1 = 0.116%, L_2 = 0.231%, L_3 = 0.587%.
						0.8215	12.93	
						0.5075	13.50	
						0.3903	13.70	
						0.2286	13.44	
						0.1676	13.04	
						0.0000	8.88	
2	138-G(C)	347	Pal, A.K. and Barua, A.K.	303.20	100	1.0000	13.301	H_2: 99.95 pure; oscillating disk viscometer, relative measurements; error ±1.0%; L_1 = 1.181%, L_2 = 1.624%, L_3 = 2.567%.
						0.8219	13.445	
						0.5957	13.501	
						0.4919	13.675	
						0.4059	13.701	
						0.2005	13.641	
						0.0000	9.000	
3	138-G(C)	231	Trautz, M. and Weizel, W.	318.16		1.0000	13.86	Same remarks as for curve 1 except L_1 = 0.234%, L_2 = 0.492%, L_3 = 1.288%.
						0.8028	14.25	
						0.5075	14.75	
						0.2963	14.94	
						0.2286	14.53	
						0.1676	14.10	
						0.0000	9.45	
4	138-G(C)	347	Pal, A.K. and Barua, A.K.	328.20	100	1.0000	14.402	Same remarks as for curve 2 except L_1 = 1.657%, L_2 = 2.196%, L_3 = 3.135%.
						0.7866	14.546	
						0.5975	14.721	
						0.4863	14.846	
						0.4000	14.801	
						0.2005	14.712	
						0.0000	9.560	
5	138-G(C)	231	Trautz, M. and Weizel, W.	343.16		1.0000	14.98	Same remarks as for curve 1 except L_1 = 0.408%, L_2 = 0.594%, L_3 = 1.333%.
						0.8028	15.35	
						0.6999	15.57	
						0.6175	15.74	
						0.4823	15.87	
						0.2963	15.96	
						0.2306	15.57	
						0.1676	15.00	
						0.1657	15.05	
						0.0000	9.94	
6	138-G(C)	231	Trautz, M. and Weizel, W.	365.16		1.0000	15.99	Same remarks as for curve 1 except L_1 = 0.225%, L_2 = 0.320%, L_3 = 0.675%.
						0.8028	16.33	
						0.6999	16.48	
						0.6175	16.75	
						0.4823	16.82	
						0.2306	16.40	
						0.1676	15.73	
						0.1657	15.77	
						0.0000	10.37	
7	138-G(C)	347	Pal, A.K. and Barua, A.K.	373.20	100	1.0000	16.890	Same remarks as for curve 2 except L_1 = 1.193%, L_2 = 1.637%, L_3 = 2.612%.
						0.7866	16.806	
						0.5975	16.795	
						0.4863	16.691	
						0.4000	16.595	
						0.2005	16.289	
						0.0000	10.470	
8	138-G(C)	231	Trautz, M. and Weizel, W.	397.16		1.0000	17.39	Same remarks as for curve 1 except L_1 = 0.242%, L_2 = 0.439%, L_3 = 0.953%.
						0.6760	17.97	
						0.4698	18.14	
						0.3265	18.01	
						0.1636	16.85	
						0.0000	11.02	
9	138-G(C)	347	Pal, A.K. and Barua, A.K.	423.20	100	1.0000	19.220	Same remarks as for curve 2 except L_1 = 0.342%, L_2 = 0.555%, L_3 = 1.278%.
						0.8110	19.203	
						0.6024	19.250	
						0.5023	19.252	
						0.4018	19.253	
						0.2000	17.788	
						0.0000	11.550	
10	138-G(C)	231	Trautz, M. and Weizel, W.	432.16		1.0000	18.97	Same remarks as for curve 1 except L_1 = 0.367%, L_2 = 0.498%, L_3 = 0.839%.
						0.6760	19.42	
						0.4698	19.60	
						0.3265	19.42	
						0.1676	18.03	
						0.1512	17.48	
						0.0000	11.67	

TABLE 138-G(C)E. EXPERIMENTAL VISCOSITY DATA AS A FUNCTION OF COMPOSITION FOR GASEOUS
HYDROGEN-SULFUR DIOXIDE MIXTURES (continued)

Cur. No.	Fig. No.	Ref. No.	Author(s)	Temp. (K)	Pressure (mm Hg)	Mole Fraction of SO_2	Viscosity (N s m^{-2} x 10^{-6})	Remarks
11	138-G(C)	231	Trautz, M. and Weizel, W.	472.16		1.0000	20.71	Same remarks as for curve 1 except $L_1 = 0.118\%$, $L_2 = 0.181\%$, $L_3 = 0.329\%$.
						0.6760	21.18	
						0.4905	21.21	
						0.3265	20.98	
						0.1512	19.53	
						0.0000	12.87	
12	138-G(C)	347	Pal, A.K. and Barua, A.K.	473.20	100	1.0000	21.150	Same remarks as for curve 2 except $L_1 = 0.109\%$, $L_2 = 0.159\%$, $L_3 = 0.348\%$.
						0.8110	21.411	
						0.6024	21.499	
						0.5023	21.540	
						0.4018	21.337	
						0.2000	19.472	
						0.0000	12.260	

TABLE 138-G(C)S. SMOOTHED VISCOSITY VALUES AS A FUNCTION OF COMPOSITION FOR GASEOUS
HYDROGEN-SULFUR DIOXIDE MIXTURES

Mole Fraction of SO_2	(373.2 K) [Ref. 347]	(397.2 K) [Ref. 231]	(423.2 K) [Ref. 347]	(432.2 K) [Ref. 231]	(472.2 K) [Ref. 231]	(473.2 K) [Ref. 347]
0.00	10.47	11.02	11.55	11.67	12.87	12.26
0.05	14.20	15.11	15.17	15.51	17.54	16.90
0.10	15.20	16.10	16.44	16.76	18.76	18.10
0.15	15.82	16.73	17.24	17.64	19.51	18.92
0.20	16.25	17.18	17.81	18.26	20.02	19.54
0.25	16.56	17.49	18.24	18.77	20.44	20.05
0.30	16.78	17.72	18.56	19.16	20.80	20.50
0.35	16.94	17.92	18.80	19.46	21.06	21.92
0.40	17.04	18.04	18.99	19.58	21.02	21.26
0.45	17.11	18.12	19.12	19.59	21.26	21.44
0.50	17.14	18.17	19.22	19.60	21.28	21.54
0.55	17.14	18.18	19.28	19.58	21.30	21.58
0.60	17.12	18.16	19.32	19.56	21.28	21.58
0.65	17.10	18.10	19.34	19.52	21.26	21.56
0.70	17.06	18.02	19.34	19.44	21.22	21.52
0.75	17.04	17.94	19.33	19.40	21.16	21.48
0.80	17.00	17.84	19.34	19.32	21.08	21.42
0.85	16.98	17.74	19.30	19.25	21.00	21.39
0.90	16.96	17.72	19.28	19.16	19.91	21.32
0.95	16.92	17.51	19.26	19.07	19.80	21.25
1.00	16.89	17.39	19.22	18.97	20.71	21.16

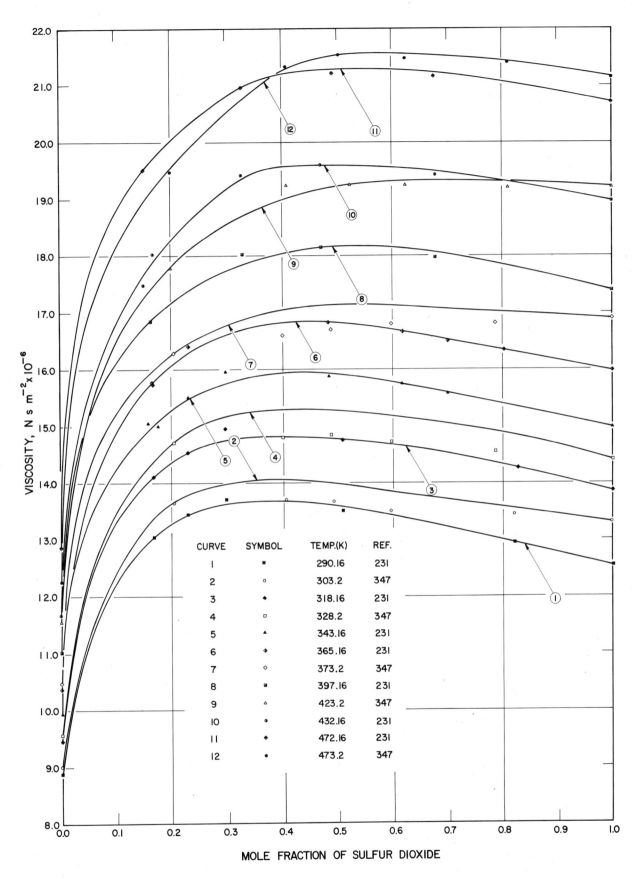

CURVE	SYMBOL	TEMP.(K)	REF.
1	■	290.16	231
2	○	303.2	347
3	◆	318.16	231
4	□	328.2	347
5	▲	343.16	231
6	◆	365.16	231
7	◇	373.2	347
8	▣	397.16	231
9	△	423.2	347
10	●	432.16	231
11	◆	472.16	231
12	●	473.2	347

MOLE FRACTION OF SULFUR DIOXIDE

FIGURE 138-G(C). VISCOSITY DATA AS A FUNCTION OF COMPOSITION
FOR GASEOUS HYDROGEN – SULFUR DIOXIDE MIXTURES

TABLE 139-G(C)E.   EXPERIMENTAL VISCOSITY DATA AS A FUNCTION OF COMPOSITION FOR GASEOUS
METHANE-AMMONIA MIXTURES

Cur. No.	Fig. No.	Ref. No.	Author(s)	Temp. (K)	Pressure (mm Hg)	Mole Fraction of NH_3	Viscosity $(N\ s\ m^{-2} \times 10^{-6})$	Remarks
1	139-G(C)	346	Jung, G. and Schmick, H.	287.66		0.0000	10.91	Effusion method of Trautz and Weizel; $L_1 = 0.118\%$, $L_2 = 0.210\%$, $L_3 = 0.616\%$.
						0.9000	10.08	
						0.8000	10.39	
						0.7000	10.61	
						0.6000	10.77	
						0.5000	10.91	
						0.4000	10.99	
						0.3000	11.05	
						0.2000	11.05	
						0.1000	10.99	
						1.0000	9.79	
2	139-G(C)	35	Chakraborti, P.K. and Gray, P.	298.2	243-142	0.0000	11.00	NH_3: purified by distillation between liquid nitrogen traps, CH_4: 99.8 pure; capillary viscometer, relative measurements; error 1.0% and precision $\pm$ 0.2%; $L_1 = 0.488\%$, $L_2 = 0.636\%$, $L_3 = 1.220\%$.
						0.7400	11.01	
						0.1970	11.09	
						0.3020	11.12	
						0.4040	11.27	
						0.4970	11.28	
						0.5910	11.18	
						0.7000	10.89	
						0.7950	10.71	
						0.8980	10.39	
						1.0000	10.16	
3	139-G(C)	35	Chakraborti, P.K. and Gray, P.	308.2	243-142	0.0000	11.38	Same remarks as for curve 2 except $L_1 = 0.279\%$, $L_2 = 0.343\%$, $L_3 = 0.666\%$.
						0.8000	11.34	
						0.1850	11.37	
						0.3060	11.40	
						0.4060	11.35	
						0.4990	11.30	
						0.5980	11.28	
						0.6970	11.29	
						0.7980	11.18	
						0.8710	10.96	
						1.0000	10.49	
4	139-G(C)	35	Chakraborti, P.K. and Gray, P.	353.2	243-142	0.0000	12.53	Same remarks as for curve 2 except $L_1 = 0.059\%$, $L_2 = 0.110\%$, $L_3 = 0.159\%$.
						0.4600	12.62	
						0.1780	12.77	
						0.2900	12.85	
						0.3940	12.88	
						0.4970	12.87	
						0.5960	12.80	
						0.6890	12.72	
						0.7780	12.58	
						0.8350	12.43	
						1.0000	11.98	

TABLE 139-G(C)S. SMOOTHED VISCOSITY VALUES AS A FUNCTION OF COMPOSITION FOR GASEOUS METHANE-AMMONIA MIXTURES

Mole Fraction of NH_3	(287.7 K) [Ref. 346]	(298.2 K) [Ref. 35]	(308.2 K) [Ref. 35]	(353.2 K) [Ref. 35]
0.00	10.91	11.00	11.38	12.53
0.05	10.96	11.05	11.38	12.61
0.10	10.99	11.09	11.38	12.68
0.15	11.02	11.13	11.39	12.74
0.20	11.03	11.16	11.39	12.79
0.25	11.04	11.13	11.38	12.82
0.30	11.04	11.19	11.38	12.85
0.35	11.02	11.19	11.38	12.87
0.40	10.99	11.19	11.37	12.88
0.45	10.95	11.17	11.37	12.88
0.50	10.90	11.14	11.36	12.87
0.55	10.84	11.11	11.35	12.85
0.60	10.77	11.05	11.33	12.82
0.65	10.69	10.99	11.30	12.77
0.70	10.55	10.90	10.25	12.71
0.75	10.49	10.81	11.19	12.63
0.80	10.37	10.70	11.10	12.54
0.85	10.24	10.58	11.00	12.43
0.90	10.10	10.44	10.86	12.30
0.95	9.95	10.30	10.69	12.15
1.00	9.79	10.16	10.50	11.98

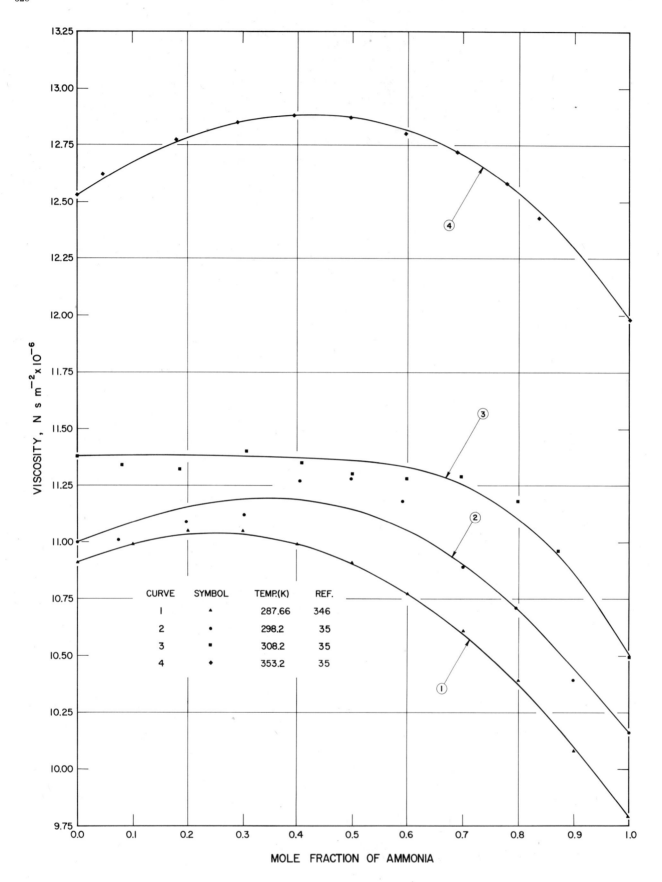

FIGURE 139-G(C). VISCOSITY DATA AS A FUNCTION OF COMPOSITION
FOR GASEOUS METHANE- AMMONIA MIXTURES

TABLE 140-G(C)E.   EXPERIMENTAL VISCOSITY DATA AS A FUNCTION OF COMPOSITION FOR GASEOUS METHANE–SULFUR DIOXIDE MIXTURES

Cur. No.	Fig. No.	Ref. No.	Author(s)	Temp. (K)	Pressure (mm Hg)	Mole Fraction of SO_2	Viscosity (N s m^{-2} x 10^{-6})	Remarks
1	140-G(C)	35	Chakraborti, P.K. and Gray, P.	308.2	142-243	0.000	11.38	SO_2: tank gas was purified by distillation between liquid nitrogen traps; capillary flow method, relative measurements; precision $\pm 0.2\%$ and accuracy 1.0%; L_1 = 0.096%, L_2 = 0.149%, L_3 = 0.398%.
						0.085	11.86	
						0.221	12.60	
						0.302	12.87	
						0.433	13.24	
						0.567	13.48	
						0.674	13.57	
						0.791	13.59	
						0.871	13.56	
						1.000	13.28	
2	140-G(C)	35	Chakraborti, P.K. and Gray, P.	353.2	142-243	0.000	12.53	Same remarks as for curve 1 except L_1 = 0.598%, L_2 = 0.896%, L_3 = 1.919%.
						0.146	13.60	
						0.260	13.86	
						0.392	14.69	
						0.478	14.91	
						0.590	15.12	
						0.681	15.23	
						0.871	15.23	
						1.000	15.21	

TABLE 140-G(C)S.   SMOOTHED VISCOSITY VALUES AS A FUNCTION OF COMPOSITION FOR GASEOUS METHANE–SULFUR DIOXIDE MIXTURES

Mole Fraction of SO_2	(308.2 K) [Ref. 35]	(353.2 K) [Ref. 35]
0.00	11.39	12.53
0.05	11.68	12.84
0.10	11.95	13.15
0.15	12.22	13.43
0.20	12.46	13.72
0.25	12.67	13.97
0.30	12.86	14.21
0.35	13.03	14.43
0.40	13.17	14.62
0.45	13.29	14.79
0.50	13.39	14.93
0.55	13.47	15.05
0.60	13.53	15.14
0.65	13.57	15.21
0.70	13.60	15.25
0.75	13.61	15.28
0.80	13.60	15.29
0.85	13.57	15.29
0.90	13.53	15.28
0.95	13.44	15.25
1.00	13.28	15.22

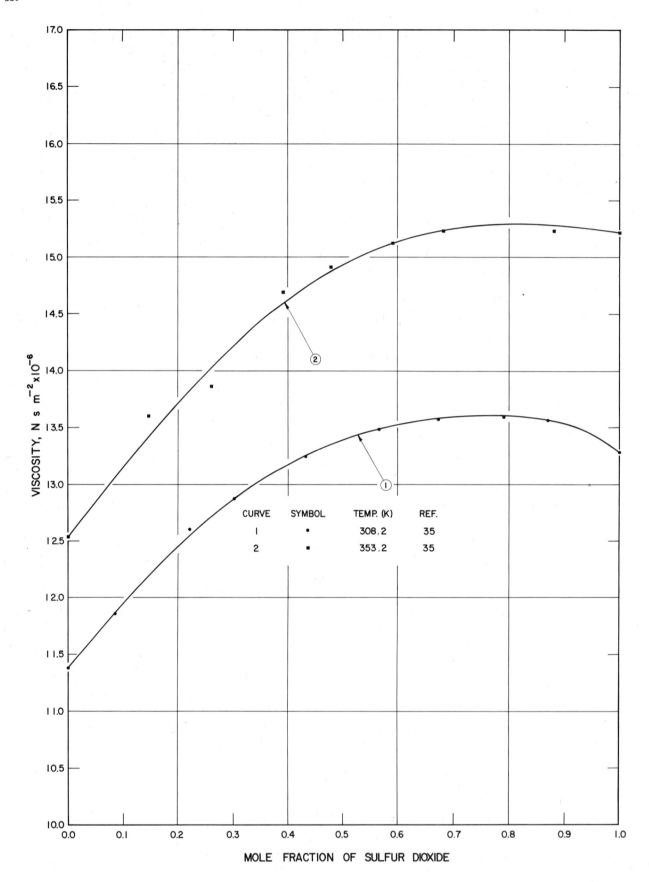

FIGURE 140-G(C). VISCOSITY DATA AS A FUNCTION OF COMPOSITION
FOR GASEOUS METHANE−SULFUR DIOXIDE MIXTRES

TABLE 141-G(C)E.   EXPERIMENTAL VISCOSITY DATA AS A FUNCTION OF COMPOSITION FOR GASEOUS
NITROGEN-AMMONIA MIXTURES

Cur. No.	Fig. No.	Ref. No.	Author(s)	Temp. (K)	Pressure (mm Hg)	Mole Fraction of N_2	Viscosity ($N\ s\ m^{-2} x 10^{-6}$)	Remarks
1	141-G(C)	222	Trautz, M. and Heberling, R.	293.2		0.0000 0.1117 0.2853 0.4362 0.7080 0.8889 1.0000	17.45 10.92 12.54 13.83 15.85 16.90 9.82	N_2: obtained by chemical reaction; NH_3: I.G. Farben, 99.997% pure, chief impurities O_2, H_2, N_2; capillary transpiration method, d = 0.04038 cm; experimental error <3%; $L_1 = 0.358\%$, $L_2 = 0.678\%$, $L_3 = 1.675\%$.
2	141-G(C)	347	Pal, A.K. and Barua, A.K.	297.2	100	0.0000 0.2036 0.4291 0.4973 0.5980 0.7993 1.0000	10.281 11.944 13.617 14.160 14.861 16.785 17.505	N_2: 99.95 pure; oscillating disk viscometer, relative measurements; error ± 1.0%; $L_1 = 1.016\%$, $L_2 = 1.564\%$, $L_3 = 3.726\%$.
3	141-G(C)	347	Pal, A.K. and Barua, A.K.	327.2	100	0.0000 0.2036 0.4291 0.4973 0.5980 0.7993 1.0000	11.372 13.640 15.171 15.805 16.703 17.937 19.130	Same remarks as for curve 2 except $L_1 = 0.321\%$, $L_2 = 0.499\%$, $L_3 = 1.102\%$.
4	141-G(C)	347	Pal, A.K. and Barua, A.K.	373.2	100	0.0000 0.2036 0.4291 0.4973 0.5980 0.7993 1.0000	13.075 15.495 17.010 17.734 18.508 19.892 21.010	Same remarks as for curve 2 except $L_1 = 0.379\%$, $L_2 = 0.464\%$, $L_3 = 0.694\%$.
5	141-G(C)	222	Trautz, M. and Heberling, R.	373.2		0.0000 0.1117 0.2853 0.4362 0.7080 0.8889 1.0000	20.85 13.98 15.69 17.10 19.20 20.31 12.79	Same remarks as for curve 1 except $L_1 = 0.057\%$, $L_2 = 0.107\%$, $L_3 = 0.209\%$.
6	141-G(C)	347	Pal, A.K. and Barua, A.K.	423.2	100	0.0000 0.2397 0.4080 0.5072 0.6015 0.7748 1.0000	14.928 17.611 19.003 19.901 20.375 21.672 23.050	Same remarks as for curve 2 except $L_1 = 0.824\%$, $L_2 = 1.288\%$, $L_3 = 2.509\%$.
7	141-G(C)	222	Trautz, M. and Heberling, R.	473.2		0.0000 0.1117 0.2853 0.4362 0.7080 0.8889 1.0000	24.62 17.68 19.46 20.85 22.96 24.08 16.46	Same remarks as for curve 1 except $L_1 = 0.677\%$, $L_2 = 0.913\%$, $L_3 = 1.733\%$.
8	141-G(C)	222	Trautz, M. and Heberling, R.	523.2		0.0000 0.1117 0.2853 0.4362 0.7080 0.8889 1.0000	26.27 19.39 21.12 22.50 24.60 25.72 18.13	Same remarks as for curve 1 except $L_1 = 0.299\%$, $L_2 = 0.475\%$, $L_3 = 0.974\%$.
9	141-G(C)	347	Pal, A.K. and Barua, A.K.	573.2	100	0.0000 0.2397 0.4080 0.5072 0.6015 0.7748 1.0000	16.798 19.572 20.785 21.520 22.211 23.625 25.225	Same remarks as for curve 2 except $L_1 = 1.124\%$, $L_2 = 1.569\%$, $L_3 = 3.359\%$.

TABLE 141-G(C)S.   SMOOTHED VISCOSITY VALUES AS A FUNCTION OF COMPOSITION FOR GASEOUS NITROGEN-AMMONIA MIXTURES

Mole Fraction of N_2	(293.2 K) [Ref. 222]	(297.2 K) [Ref. 347]	(327.2 K) [Ref. 347]	(373.2 K) [Ref. 222]	(373.2 K) [Ref. 347]	(423.2 K) [Ref. 347]	(473.2 K) [Ref. 222]	(523.2 K) [Ref. 222]	(573.2 K) [Ref. 222]
0.00	9.82	10.30	11.37	13.08	12.79	14.93	16.46	18.13	16.80
0.05	10.32	10.64	11.88	13.62	13.38	15.49	17.04	18.70	17.26
0.10	10.80	11.00	12.36	14.13	13.87	15.87	17.62	19.25	17.72
0.15	11.30	11.35	12.86	14.64	14.37	16.34	18.20	19.79	18.16
0.20	11.78	11.70	13.32	15.12	14.88	16.81	18.76	20.30	18.60
0.25	12.25	12.07	13.78	15.57	15.38	17.27	19.32	20.78	19.04
0.30	12.70	12.44	14.24	16.02	15.86	17.73	19.86	21.25	19.47
0.35	13.12	12.85	14.68	16.45	16.32	18.19	20.34	21.68	19.90
0.40	13.54	13.24	15.10	16.86	16.77	18.64	20.76	22.10	20.32
0.45	13.92	13.64	15.50	17.26	17.22	19.09	21.17	22.60	20.76
0.50	14.28	14.01	15.90	17.64	17.64	19.52	21.55	22.90	21.16
0.55	14.62	14.38	16.26	18.04	18.04	19.95	21.90	23.27	21.58
0.60	14.95	14.74	16.62	18.42	18.42	20.36	22.24	23.62	22.00
0.65	15.28	15.11	16.96	18.78	18.78	20.76	22.60	23.99	22.48
0.70	15.60	15.48	17.30	19.12	19.12	21.14	22.89	24.32	22.80
0.75	15.92	15.84	17.62	19.45	19.44	21.51	23.18	24.66	23.21
0.80	16.24	16.18	17.94	19.77	19.76	21.84	23.48	24.98	23.62
0.85	16.56	16.52	18.24	20.15	20.08	22.17	23.76	25.32	24.02
0.90	16.85	16.86	18.54	20.40	20.38	22.48	24.06	25.64	24.40
0.95	17.17	17.18	18.84	20.70	20.67	22.76	24.35	25.95	24.83
1.00	17.45	17.50	19.13	21.01	20.85	23.05	24.62	26.27	25.23

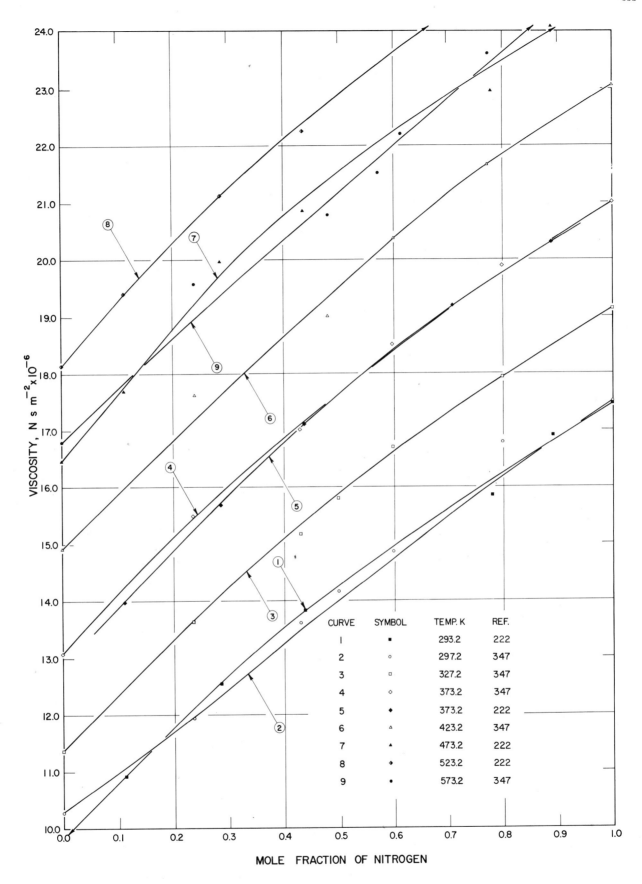

CURVE	SYMBOL	TEMP. K	REF.
1	■	293.2	222
2	○	297.2	347
3	□	327.2	347
4	◇	373.2	347
5	◆	373.2	222
6	△	423.2	347
7	▲	473.2	222
8	⬧	523.2	222
9	●	573.2	347

MOLE FRACTION OF NITROGEN

FIGURE 141-G(C). VISCOSITY DATA AS A FUNCTION OF COMPOSITION
FOR GASEOUS NITROGEN-AMMONIA MIXTURES

TABLE 142-G(C)E.　EXPERIMENTAL VISCOSITY DATA AS A FUNCTION OF COMPOSITION FOR GASEOUS NITROUS OXIDE-AMMONIA MIXTURES

Cur. No.	Fig. No.	Ref. No.	Author(s)	Temp. (K)	Pressure (mm Hg)	Mole Fraction of N_2O	Viscosity ($N \cdot s \cdot m^{-2} \times 10^{-6}$)	Remarks
1	142-G(C)	35	Chakraborti, P.K. and Gray, P.	298.2	142-243	1.000	14.86	Both gases were purified by distillation between liquid nitrogen traps; capillary flow viscometer, relative measurements; precision $\pm 0.2\%$, accuracy 1.0%; $L_1 = 0.421\%$, $L_2 = 0.535\%$, $L_3 = 1.075\%$.
						0.899	14.63	
						0.802	14.36	
						0.702	13.81	
						0.598	13.43	
						0.507	13.08	
						0.406	12.66	
						0.303	12.03	
						0.207	11.46	
						0.105	10.78	
						0.000	10.16	
2	142-G(C)	35	Chakraborti, P.K. and Gray, P.	308.2	142-243	1.000	15.38	Same remarks as for curve 1 except $L_1 = 0.142\%$, $L_2 = 0.192\%$, $L_3 = 0.430\%$.
						0.951	15.27	
						0.821	14.95	
						0.706	14.65	
						0.602	14.32	
						0.502	13.93	
						0.402	13.52	
						0.313	13.05	
						0.210	12.40	
						0.112	11.67	
						0.000	10.49	
3	142-G(C)	35	Chakraborti, P.K. and Gray, P.	353.2	142-243	1.000	17.30	Same remarks as for curve 1 except $L_1 = 0.180\%$, $L_2 = 0.228\%$, $L_3 = 0.408\%$.
						0.919	17.23	
						0.816	17.00	
						0.716	16.74	
						0.606	16.35	
						0.502	15.86	
						0.408	15.40	
						0.320	14.88	
						0.221	14.18	
						0.142	13.55	
						0.000	11.98	

TABLE 142-G(C)S.　SMOOTHED VISCOSITY VALUES AS A FUNCTION OF COMPOSITION FOR GASEOUS NITROUS OXIDE-AMMONIA MIXTURES

Mole Fraction of NO_2	(298.2 K) [Ref. 35]	(308.2 K) [Ref. 35]	(353.2 K) [Ref. 35]
0.00	10.16	10.49	12.00
0.05	10.48	11.03	12.58
0.10	10.82	11.51	13.10
0.15	11.14	11.94	13.58
0.20	11.46	12.32	14.02
0.25	11.78	12.68	14.41
0.30	12.08	13.00	14.76
0.35	12.36	13.28	15.10
0.40	12.63	13.53	15.39
0.45	12.88	13.76	15.66
0.50	13.12	13.97	15.90
0.55	13.34	14.16	16.13
0.60	13.55	14.33	16.33
0.65	13.76	14.49	16.51
0.70	13.95	14.64	16.67
0.75	14.13	14.78	16.81
0.80	14.30	14.91	16.93
0.85	14.45	15.03	17.04
0.90	14.60	15.17	17.13
0.95	14.73	15.29	17.22
1.00	14.86	15.40	17.30

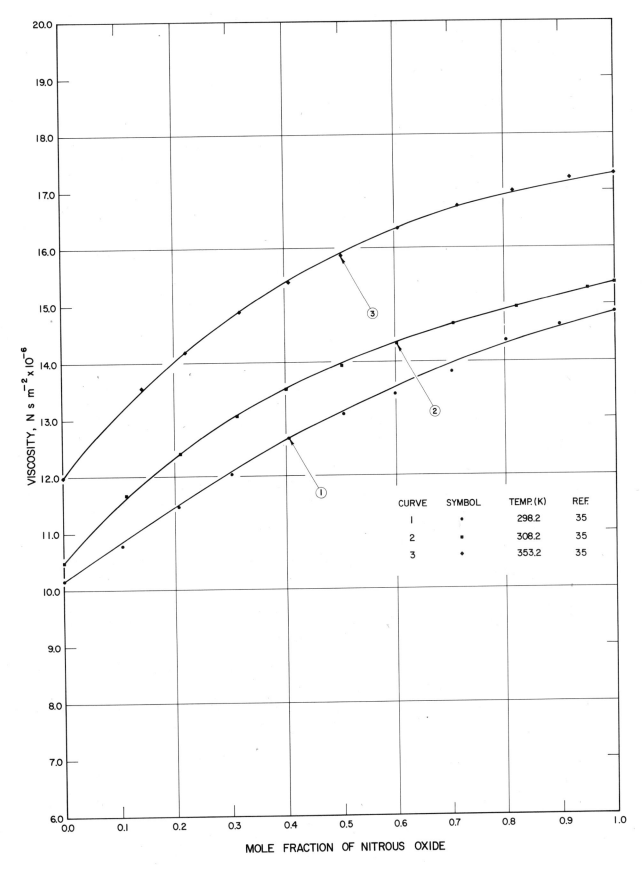

FIGURE 142-G(C). VISCOSITY DATA AS A FUNCTION OF COMPOSITION
FOR GASEOUS NITROUS OXIDE—AMMONIA MIXTURES

TABLE 143-G(C)E.   EXPERIMENTAL VISCOSITY DATA AS A FUNCTION OF COMPOSITION FOR GASEOUS NITROUS OXIDE-SULFUR DIOXIDE MIXTURES

Cur. No.	Fig. No.	Ref. No.	Author(s)	Temp. (K)	Pressure (mm Hg)	Mole Fraction of N_2O	Viscosity ($N\ s\ m^{-2} \times 10^{-6}$)	Remarks
1	143-G(C)	35	Chakraborti, P.K. and Gray, P.	298.2	142-243	0.000	14.86	SO_2 and N_2O: tank gases purified by distillation between liquid nitrogen traps; capillary viscometer, relative measurements; precision ± 0.2% and accuracy 1.0%; $L_1 = 0.319\%$, $L_2 = 0.470\%$, $L_3 = 1.345\%$.
						0.043	14.79	
						0.178	14.80	
						0.297	14.64	
						0.401	14.48	
						0.493	14.27	
						0.596	14.04	
						0.702	13.92	
						0.800	13.71	
						0.900	13.39	
						0.914	13.20	
						1.000	13.17	
2	143-G(C)	35	Chakraborti, P.K. and Gray, P.	308.2	142-243	0.000	15.38	Same remarks as for curve 1 except $L_1 = 0.753\%$, $L_2 = 1.011\%$, $L_3 = 1.841\%$.
						0.042	15.04	
						0.147	15.05	
						0.249	14.90	
						0.398	14.69	
						0.476	14.55	
						0.575	14.36	
						0.672	14.31	
						0.777	14.11	
						0.879	13.87	
						1.000	13.28	
3	143-G(C)	35	Chakraborti, P.K. and Gray, P.	353.2	142-243	0.000	17.30	Same remarks as for curve 1 except $L_1 = 0.107\%$, $L_2 = 0.147\%$, $L_3 = 0.289\%$.
						0.035	17.26	
						0.183	17.07	
						0.273	16.94	
						0.375	16.78	
						0.474	16.56	
						0.576	16.36	
						0.675	16.17	
						0.786	15.89	
						0.895	15.60	
						1.000	15.23	

TABLE 143-G(C)S.   SMOOTHED VISCOSITY VALUES AS A FUNCTION OF COMPOSITION FOR GASEOUS NITROUS OXIDE-SULFUR DIOXIDE MIXTURES

Mole Fraction of N_2O	(298.2 K) [Ref. 35]	(308.2 K) [Ref. 35]	(353.2 K) [Ref. 35]
0.00	14.88	15.38	17.30
0.05	14.86	15.31	17.24
0.10	14.83	15.24	17.18
0.15	14.80	15.16	17.12
0.20	14.76	15.08	17.05
0.25	14.71	15.00	16.98
0.30	14.65	14.92	16.90
0.35	14.57	14.83	16.82
0.40	14.49	14.73	16.73
0.45	14.40	14.63	16.64
0.50	14.30	14.53	16.54
0.55	14.21	14.43	16.44
0.60	14.10	14.31	16.33
0.65	14.00	14.20	16.22
0.70	13.89	14.08	16.09
0.75	13.77	13.96	15.96
0.80	13.66	13.84	15.83
0.85	13.54	13.71	15.69
0.90	13.42	13.58	15.54
0.95	13.29	13.43	15.39
1.00	13.17	13.28	15.23

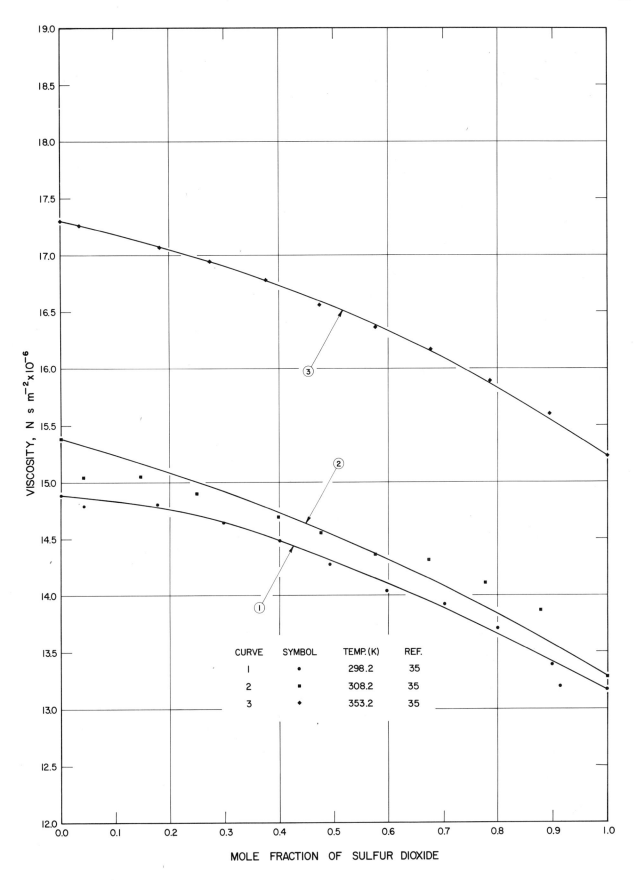

CURVE	SYMBOL	TEMP.(K)	REF.
1	•	298.2	35
2	■	308.2	35
3	♦	353.2	35

MOLE FRACTION OF SULFUR DIOXIDE

FIGURE 143-G(C). VISCOSITY DATA AS A FUNCTION OF COMPOSITION
FOR GASEOUS NITROUS OXIDE – SULFUR DIOXIDE MIXTURES

TABLE 144-G(C)E.   EXPERIMENTAL VISCOSITY DATA AS A FUNCTION OF COMPOSITION FOR GASEOUS OXYGEN-AMMONIA MIXTURES

Cur. No.	Fig. No.	Ref. No.	Author(s)	Temp. (K)	Pressure (atm)	Mole Fraction of O_2	Viscosity ($N\ s\ m^{-2} \times 10^{-6}$)	Remarks
1	144-G(C)	222	Trautz, M. and Heberling, R.	293.2		0.0000	20.23	O_2: by electrolysis of KOH on pure nickel electrodes; NH_3: I.G. Farben, 99.997% pure, chief impurities O_2, H_2, N_2; capillary transpiration method; d = 0.04038 cm; experimental error <3%; L_1 = 0.092%, L_2 = 0.203%, L_3 = 0.523%.
						0.1245	11.43	
						0.2921	13.50	
						0.5214	16.04	
						0.7014	17.83	
						0.8649	19.24	
						1.0000	9.82	
2	144-G(C)	222	Trautz, M. and Heberling, R.	373.2		0.0000	24.40	Same remarks as for curve 1 except L_1 = 0.012%, L_2 = 0.033%, L_3 = 0.086%.
						0.1245	14.59	
						0.2921	16.89	
						0.5214	19.72	
						0.7014	21.70	
						0.8649	23.26	
						1.0000	12.79	
3	144-G(C)	222	Trautz, M. and Heberling, R.	473.2		0.0000	29.02	Same remarks as for curve 1 except L_1 = 0.282%, L_2 = 0.440%, L_3 = 0.891%.
						0.1245	18.40	
						0.2921	20.85	
						0.5214	23.90	
						0.7014	26.04	
						0.8649	27.73	
						1.0000	16.46	

TABLE 144-G(C)S.   SMOOTHED VISCOSITY VALUES AS A FUNCTION OF COMPOSITION FOR GASEOUS OXYGEN-AMMONIA MIXTURES

Mole Fraction of O_2	(293.2 K) [Ref. 222]	(373.2 K) [Ref. 222]	(473.2 K) [Ref. 222]
0.00	9.82	12.79	16.46
0.05	10.48	13.52	17.26
0.10	11.12	14.24	18.05
0.15	11.75	14.95	18.80
0.20	12.38	15.64	19.55
0.25	13.00	16.32	20.28
0.30	13.61	17.00	20.96
0.35	14.20	17.64	21.65
0.40	14.78	18.28	22.29
0.45	15.32	18.89	22.92
0.50	15.84	19.48	23.53
0.55	16.34	20.06	24.13
0.60	16.81	20.62	24.69
0.65	17.26	21.26	25.28
0.70	17.70	21.70	25.80
0.75	18.15	22.20	26.33
0.80	18.59	22.68	26.86
0.85	19.01	23.15	27.20
0.90	19.42	23.58	27.92
0.95	19.82	24.00	28.48
1.00	20.23	24.40	29.02

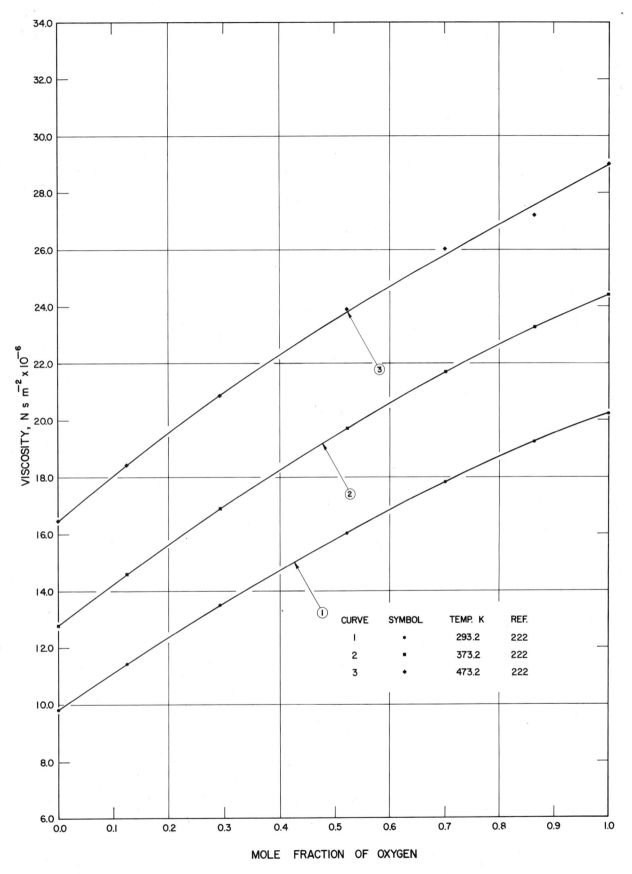

FIGURE 144-G(C). VISCOSITY DATA AS A FUNCTION OF COMPOSITION
FOR GASEOUS OXYGEN—AMMONIA MIXTURES

TABLE 145-G(C)E.   EXPERIMENTAL VISCOSITY DATA AS A FUNCTION OF COMPOSITION FOR GASEOUS
AMMONIA-METHYLAMINE MIXTURES

Cur. No.	Fig. No.	Ref. No.	Author(s)	Temp. (K)	Pressure (mm Hg)	Mole Fraction of CH_3NH_2	Viscosity (N s m^{-2} x 10^{-6})	Remarks
1	145-G(C)	348	Burch, L.G. and Raw, C.J.G.	273	110-220	1.000 0.750 0.500 0.250 0.000	8.71 8.89 9.00 9.09 9.20	CH_3NH_2: 98.0 pure, NH$_3$: 99.99 pure, gases were further purified by vacuum distillation; capillary flow viscometer, relative measurements; uncertainty ± 0.5%; L_1 = 0.000%, L_2 = 0.000%, L_3 = 0.000%.
2	145-G(C)	348	Burch, L.G. and Raw, C.J.G.	298	110-220	1.000 0.750 0.500 0.250 0.000	9.43 9.64 9.80 9.94 10.09	Same remarks as for curve 1
3	145-G(C)	348	Burch, L.G. and Raw, C.J.G.	323	110-220	1.000 0.750 0.500 0.250 0.000	10.15 10.40 10.60 10.79 10.99	Same remarks as for curve 1.
4	145-G(C)	348	Burch, L.G. and Raw, C.J.G.	348	110-220	1.000 0.750 0.500 0.250 0.000	10.88 11.15 11.40 11.64 11.89	Same remarks as for curve 1.
5	145-G(C)	348	Burch, L.G. and Raw, C.J.G.	373	110-220	1.000 0.750 0.500 0.250 0.000	11.61 11.91 12.21 12.50 12.79	Same remarks as for curve 1.
6	145-G(C)	348	Burch, L.G. and Raw, C.J.G.	423	110-220	1.000 0.750 0.500 0.250 0.000	13.07 13.45 13.85 14.22 14.60	Same remarks as for curve 1 except L_1 = 0.031%, L_2 = 0.068%, L_3 = 0.153%.
7	145-G(C)	348	Burch, L.G. and Raw, C.J.G.	473	110-220	1.000 0.750 0.500 0.250 0.000	14.66 15.10 15.55 16.02 16.47	Same remarks as for curve 1.
8	145-G(C)	348	Burch, L.G. and Raw, C.J.G.	523	110-220	1.000 0.750 0.500 0.250 0.000	16.11 16.63 17.15 17.72 18.25	Same remarks as for curve 1.
9	145-G(C)	348	Burch, L.G. and Raw, C.J.G.	573	110-220	1.000 0.750 0.500 0.250 0.000	17.56 18.19 18.50 19.43 20.03	Same remarks as for curve 1 except L_1 = 0.440%, L_2 = 0.574%, L_3 = 0.883%.
10	145-G(C)	348	Burch, L.G. and Raw, C.J.G.	623	110-220	1.000 0.750 0.500 0.250 0.000	19.01 19.70 20.41 21.07 21.81	Same remarks as for curve 1 except L_1 = 0.105%, L_2 = 0.159%, L_3 = 0.320%.
11	145-G(C)	348	Burch, L.G. and Raw, C.J.G.	673	110-220	1.000 0.750 0.500 0.250 0.000	20.48 21.28 22.05 22.83 23.60	Same remarks as for curve 1.

TABLE 145-G(C)S. SMOOTHED VISCOSITY VALUES AS A FUNCTION OF COMPOSITION FOR GASEOUS AMMONIA-METHYLAMINE MIXTURES

Mole Fraction of CH₃NH₂	(273.0 K) [Ref. 348]	(298.0 K) [Ref. 348]	(323.0 K) [Ref. 348]	(348.0 K) [Ref. 348]	(373.0 K) [Ref. 348]	(423.0 K) [Ref. 348]
0.00	9.20	10.09	10.99	11.89	12.79	14.60
0.05	9.18	10.04	10.94	11.84	12.74	14.53
0.10	9.16	10.01	10.91	11.79	12.68	14.46
0.15	9.14	9.98	10.87	11.74	12.63	14.38
0.20	9.12	9.96	10.83	11.69	12.57	14.30
0.25	9.10	9.93	10.79	11.64	12.50	14.22
0.30	9.08	9.90	10.75	11.59	12.46	14.15
0.35	9.06	9.88	10.72	11.54	12.39	14.08
0.40	9.04	9.86	10.68	11.49	12.33	14.00
0.45	9.01	9.84	10.64	11.44	12.27	13.93
0.50	9.00	9.81	10.60	11.40	12.21	13.85
0.55	8.97	9.78	10.56	11.35	12.16	13.77
0.60	8.95	9.75	10.52	11.30	12.10	13.70
0.65	8.92	9.72	10.48	11.25	12.04	13.62
0.70	8.90	6.69	10.44	11.20	11.98	13.54
0.75	8.90	9.65	10.40	11.15	11.91	13.45
0.80	8.87	9.61	10.35	11.10	11.86	13.38
0.85	8.83	9.58	10.31	11.05	11.80	13.30
0.90	8.80	9.54	10.26	11.00	11.74	13.23
0.95	8.76	9.49	10.22	10.94	11.68	13.16
1.00	8.71	9.43	10.15	10.88	11.61	13.09

Mole Fraction of CH₃NH₂	(473.0 K) [Ref. 348]	(523.0 K) [Ref. 348]	(573.0 K) [Ref. 348]	(623.0 K) [Ref. 348]	(673.0 K) [Ref. 348]
0.00	16.47	18.25	20.03	21.81	23.60
0.05	16.39	18.15	19.86	21.66	23.45
0.10	16.30	18.05	19.70	21.50	23.30
0.15	16.22	17.94	19.55	21.34	23.15
0.20	16.12	17.83	19.40	21.19	23.00
0.25	16.02	17.72	19.26	21.04	21.28
0.30	15.94	17.60	19.12	20.90	21.68
0.35	15.84	17.50	19.00	20.76	21.52
0.40	15.74	17.38	18.87	20.62	21.37
0.45	15.64	17.27	18.75	20.48	21.21
0.50	15.50	17.15	18.63	20.35	22.05
0.55	15.46	17.06	18.52	20.22	21.90
0.60	15.36	16.95	18.40	20.08	21.74
0.65	15.28	16.84	18.29	19.96	21.59
0.70	15.19	16.74	18.18	19.82	21.44
0.75	15.10	16.63	18.08	19.69	21.28
0.80	15.02	16.53	17.97	19.55	21.12
0.85	14.94	16.43	17.86	19.42	20.96
0.90	14.85	16.33	17.76	19.28	20.80
0.95	14.76	16.22	17.66	19.15	20.64
1.00	14.66	16.11	17.56	19.01	20.48

542

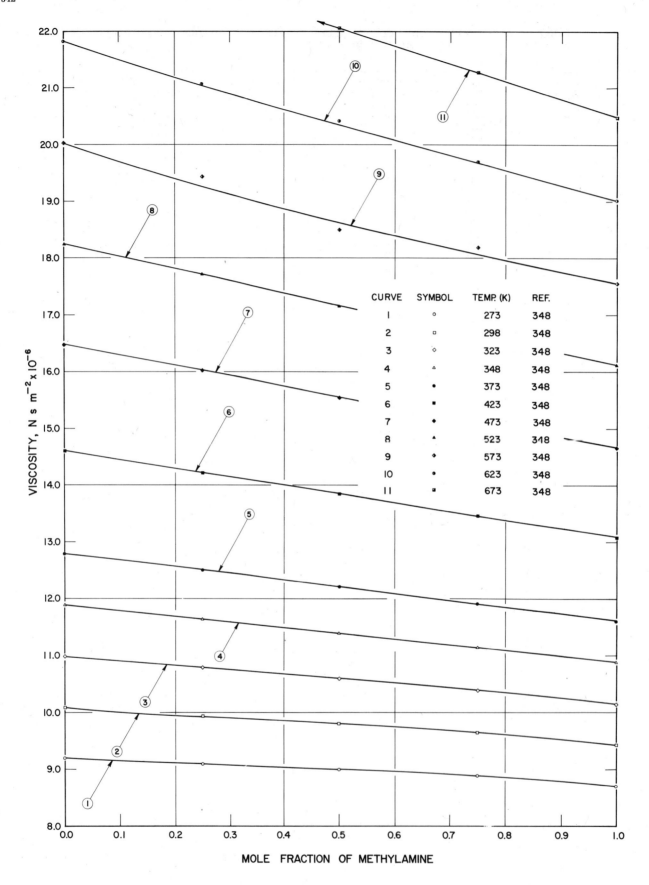

CURVE	SYMBOL	TEMP. (K)	REF.
1	o	273	348
2	□	298	348
3	◇	323	348
4	△	348	348
5	●	373	348
6	■	423	348
7	◆	473	348
8	▲	523	348
9	◆	573	348
10	●	623	348
11	■	673	348

MOLE FRACTION OF METHYLAMINE

FIGURE 145-G(C).   VISCOSITY DATA AS A FUNCTION OF COMPOSITION
FOR GASEOUS AMMONIA—METHYLAMINE MIXTURES

TABLE 146-L(C)E.   EXPERIMENTAL VISCOSITY DATA AS A FUNCTION OF COMPOSITION FOR GASEOUS
ANILINE-BENZYL ACETATE MIXTURES

Cur. No.	Fig. No.	Ref. No.	Author(s)	Temp. (K)	Pressure (atm)	Mole Fraction of Benzyl Acetate	Viscosity (N s m^{-2} x 10^{-6})	Remarks
1	146-L(C)	351	Katti, P.K. and Chaudhri, M.M.	303.2		0.000	3145.7	Liquids were purified (ref. J.
						0.125	2910.0	Chem. Eng. Data, 9, 128, 1964);
						0.300	2600.0	Ostwald viscometer; error ± 0.5%;
						0.435	2383.6	L$_1$ = 0.049%, L$_2$ = 0.085%, L$_3$ =
						0.495	2284.5	0.172%.
						0.605	2123.5	
						0.750	1928.5	
						0.850	1809.5	
						1.000	1652.4	

TABLE 146-L(C)S.   SMOOTHED VISCOSITY VALUES AS A FUNCTION OF COMPOSITION FOR GASEOUS
ANILINE-BENZYL ACETATE MIXTURES

Mole Fraction of Benzyl Acetate	(303.2 K) [Ref. 351]
0.00	3150.0
0.05	3055.0
0.10	2960.0
0.15	2867.5
0.20	2777.5
0.25	2690.0
0.30	2600.0
0.35	2517.5
0.40	2437.5
0.45	2357.5
0.50	2280.0
0.55	2205.0
0.60	2132.5
0.65	2060.0
0.70	1992.5
0.75	1927.5
0.80	1865.0
0.85	1807.5
0.90	1750.0
0.95	1700.0
1.00	1652.5

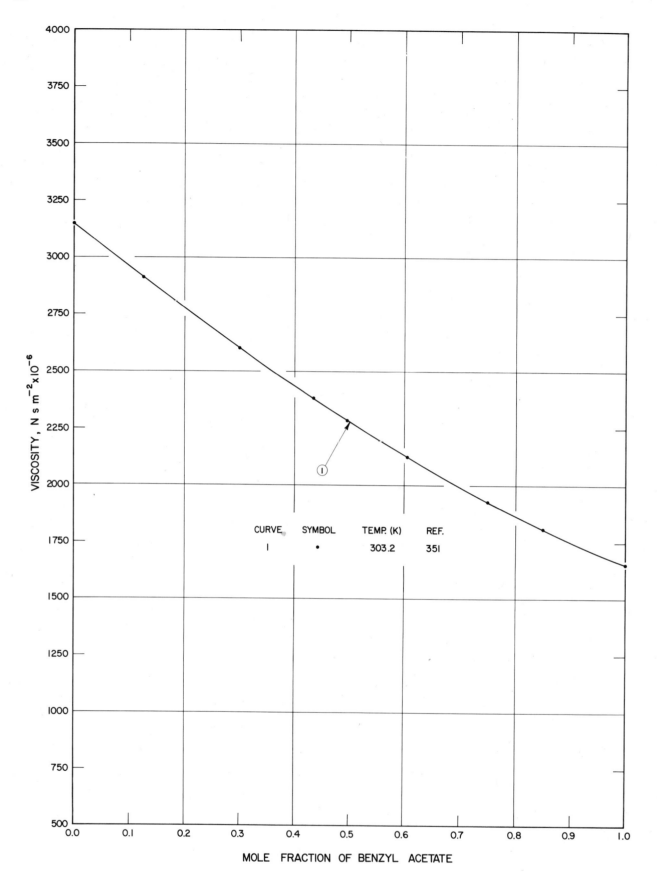

FIGURE 146-L(C). VISCOSITY DATA AS A FUNCTION OF COMPOSITION
FOR LIQUID ANILINE—BENZYL ACETATE MIXTURES

TABLE 147-L(C)E. EXPERIMENTAL VISCOSITY DATA AS A FUNCTION OF COMPOSITION FOR LIQUID
BENZYL ACETATE - META-CRESOL MIXTURES

Cur. No.	Fig. No.	Ref. No.	Author(s)	Temp. (K)	Pressure (atm)	Mole Fraction of Benzyl Acetate	Viscosity (N s m^{-2} x 10^{-6})	Remarks
1	147-L(C)	351	Katti, P.K. and Chaudhri, M.M.	313.2		0.000	6180.0	Liquids were purified (ref. J. Chem. Eng. Data, 9, 128, 1964); Ostwald viscometer; error ± 0.5%; L_1 = 0.0102%, L_2 = 0.0270%, L_3 = 0.071%.
						0.115	5113.7	
						0.272	3917.2	
						0.435	3060.0	
						0.620	2337.0	
						0.810	1764.6	
						1.000	1352.5	

TABLE 147-L(C)S. SMOOTHED VISCOSITY VALUES AS A FUNCTION OF COMPOSITION FOR LIQUID
BENZYL ACETATE - META-CRESOL MIXTURES

Mole Fraction of Benzyl Acetate	(313.2 K) [Ref. 351]
0.00	6180.0
0.05	5700.0
0.10	5250.0
0.15	4814.0
0.20	4420.0
0.25	4062.0
0.30	3748.0
0.35	3466.0
0.40	3216.0
0.45	2999.0
0.50	2785.0
0.55	2590.0
0.60	2410.0
0.65	2235.0
0.70	2075.0
0.75	1930.0
0.80	1792.0
0.85	1672.0
0.90	1560.0
0.95	1478.0
1.00	1352.5

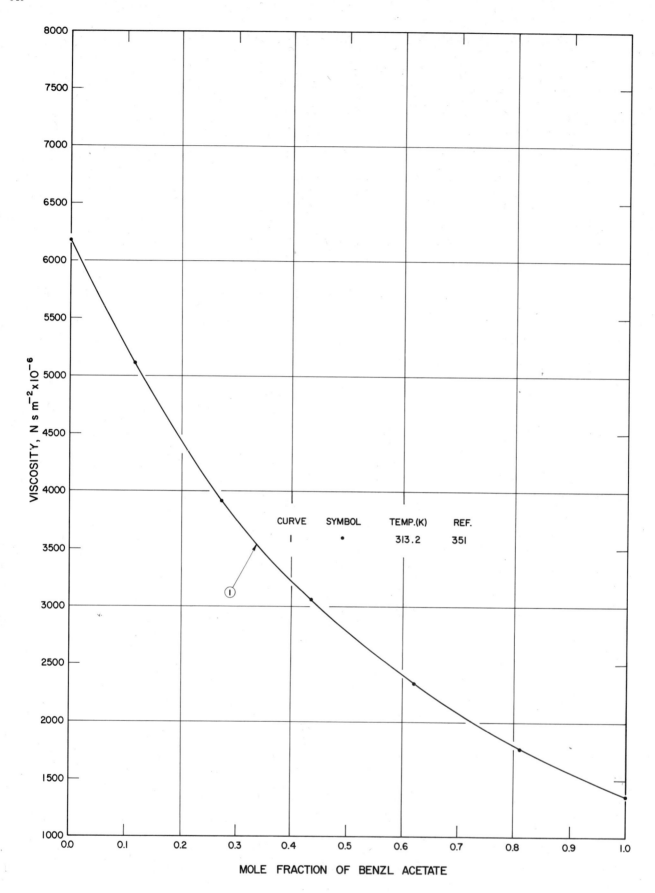

FIGURE 147-L(C). VISCOSITY DATA AS A FUNCTION OF COMPOSITION
FOR LIQUID BENZYL ACETATE—meta-CRESOL MIXTURES

TABLE 148-G(C)E.   EXPERIMENTAL VISCOSITY DATA AS A FUNCTION OF COMPOSITION FOR GASEOUS DIMETHYL ETHER-METHYL CHLORIDE MIXTURES

Cur. No.	Fig. No.	Ref. No.	Author(s)	Temp. (K)	Pressure (atm)	Mole Fraction of $(CH_3)_2O$	Viscosity $(N\ s\ m^{-2} \times 10^{-6})$	Remarks
1	148-G(C)	349	Chakraborti, P.K. and Gray, P.	308.2		0.000	9.66	Tank gases were purified by fractionation at liquid nitrogen temperature; capillary flow viscometer, relative measurements; precision $\pm 0.4\%$ and accuracy $\pm 1.0\%$; $L_1 = 0.083\%$, $L_2 = 0.128\%$, $L_3 = 0.291\%$.
						0.046	9.75	
						0.222	9.99	
						0.299	10.09	
						0.401	10.24	
						0.508	10.41	
						0.604	10.54	
						0.699	10.70	
						0.802	10.86	
						0.877	10.99	
						1.000	11.26	
2	148-G(C)	349	Chakraborti, P.K. and Gray, P.	353.2		0.000	10.98	Same remarks as for curve 1 except $L_1 = 0.082\%$, $L_2 = 0.109\%$, $L_3 = 0.258\%$.
						0.063	11.09	
						0.191	11.29	
						0.281	11.42	
						0.400	11.66	
						0.474	11.76	
						0.588	11.97	
						0.669	12.12	
						0.761	12.32	
						1.000	12.78	

TABLE 148-G(C)S.   SMOOTHED VISCOSITY VALUES AS A FUNCTION OF COMPOSITION FOR GASEOUS DIMETHYL ETHER-METHYL CHLORIDE MIXTURES

Mole Fraction of $(CH_3)_2O$	(308.2 K) [Ref. 349]	(353.2 K) [Ref. 349]
0.00	9.66	10.98
0.05	9.73	11.06
0.10	9.80	11.14
0.15	9.88	11.21
0.20	9.95	11.30
0.25	10.02	11.38
0.30	10.09	11.46
0.35	10.17	11.54
0.40	10.24	11.63
0.45	10.32	11.72
0.50	10.39	11.81
0.55	10.47	11.86
0.60	10.54	12.00
0.65	10.61	12.10
0.70	10.69	12.19
0.75	10.77	12.29
0.80	10.85	12.38
0.85	10.94	12.48
0.90	11.04	12.58
0.95	11.15	12.68
1.00	11.26	12.78

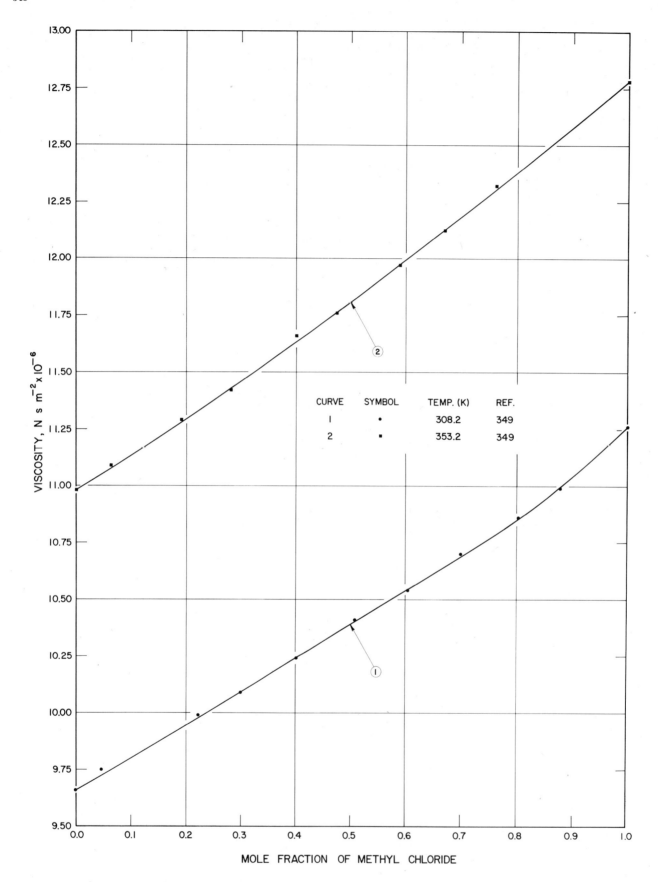

FIGURE 148-G(C).   VISCOSITY DATA AS A FUNCTION OF COMPOSITION
FOR GASEOUS DIMETHYL ETHER—METHYL CHLORIDE MIXTURES

TABLE 149-G(C)E.   EXPERIMENTAL VISCOSITY DATA AS A FUNCTION OF COMPOSITION FOR GASEOUS
DIMETHYL ETHER-SULFUR DIOXIDE MIXTURES

Cur. No.	Fig. No.	Ref. No.	Author(s)	Temp. (K)	Pressure (atm)	Mole Fraction of SO_2	Viscosity (N s m^{-2} x 10^{-6})	Remarks
1	149-G(C)	349	Chakraborti, P.K. and Gray, P.	308.2		0.000	9.66	Tank gases were purified by fractionation at liquid nitrogen temperature; capillary flow viscometer, relative measurements; precision ± 0.4% and accuracy ± 1.0%; L_1 = 0.279%, L_2 = 0.372%, L_3 = 0.807%.
						0.058	9.83	
						0.184	10.31	
						0.294	10.70	
						0.391	11.06	
						0.492	11.45	
						0.591	11.79	
						0.692	12.20	
						0.782	12.54	
						0.844	12.79	
						1.000	13.28	
2	149-G(C)	349	Chakraborti, P.K. and Gray, P.	353.2		0.000	10.98	Same remarks as for curve 1 except L_1 = 0.259%, L_2 = 0.369%, L_3 = 0.953%.
						0.049	11.14	
						0.190	11.69	
						0.279	12.04	
						0.389	12.53	
						0.504	13.05	
						0.570	13.33	
						0.648	13.77	
						0.748	14.10	
						0.866	14.64	
						1.000	15.23	

TABLE 149-G(C)S.   SMOOTHED VISCOSITY VALUES AS A FUNCTION OF COMPOSITION FOR GASEOUS
DIMETHYL ETHER-SULFUR DIOXIDE MIXTURES

Mole Fraction of SO_2	(308.2 K) [Ref. 349]	(353.2 K) [Ref. 349]
0.00	9.66	10.97
0.05	9.88	11.17
0.10	10.05	11.37
0.15	10.23	11.57
0.20	10.41	11.78
0.25	10.58	11.98
0.30	10.76	12.18
0.35	10.94	12.40
0.40	11.12	12.60
0.45	11.29	12.81
0.50	11.48	13.02
0.55	11.66	13.23
0.60	11.84	13.44
0.65	12.02	13.66
0.70	12.20	13.88
0.75	12.38	14.10
0.80	12.56	14.32
0.85	12.74	14.54
0.90	12.92	14.77
0.95	13.10	14.99
1.00	13.28	15.23

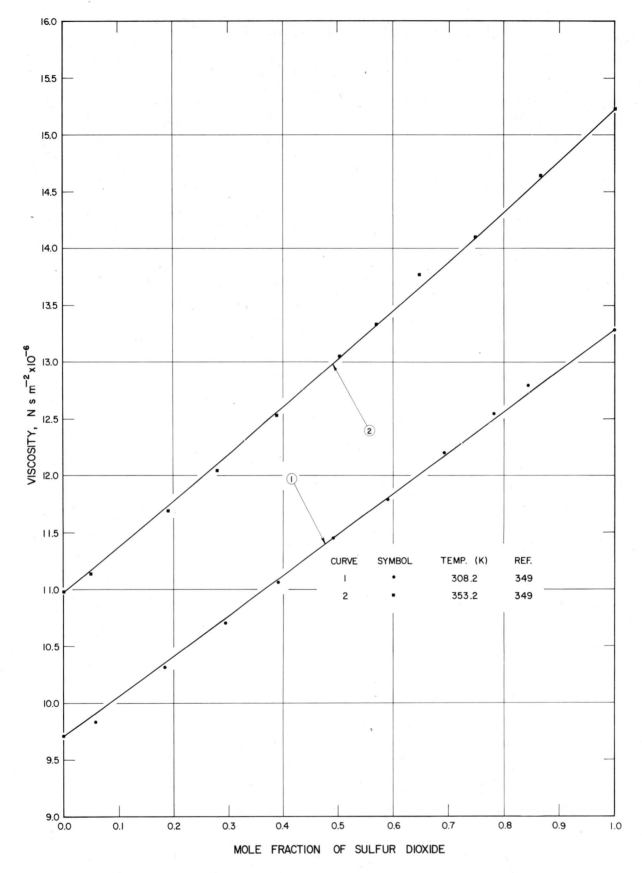

FIGURE 149-G(C). VISCOSITY DATA AS A FUNCTION OF COMPOSITION
FOR GASEOUS DIMETHYL ETHER−SULFUR DIOXIDE MIXTURES

TABLE 150-G(C)E.   EXPERIMENTAL VISCOSITY DATA AS A FUNCTION OF COMPOSITION FOR GASEOUS
METHYL CHLORIDE-SULFUR DIOXIDE MIXTURES

Cur. No.	Fig. No.	Ref. No.	Author(s)	Temp. (K)	Pressure (atm)	Mole Fraction of CH_3Cl	Viscosity (N s m^{-2} x 10^{-6})	Remarks
1	150-G(C)	349	Chakraborti, P.K. and Gray, P.	308.2		0.000	11.26	Tank gases were purified by fractionation at liquid nitrogen temperature; capillary flow viscometer, relative measurements; precision ± 0.4% and accuracy ± 1.0%; $L_1 = 0.332\%$, $L_2 = 0.466\%$, $L_3 = 0.537\%$.
						0.045	11.30	
						0.167	11.56	
						0.286	11.83	
						0.369	12.06	
						0.492	12.31	
						0.604	12.56	
						0.690	12.73	
						0.768	12.92	
						0.847	13.10	
						1.000	13.28	
2	150-G(C)	349	Chakraborti, P.K. and Gray, P.	353.2		0.000	12.78	Same remarks as for curve 1 except $L_1 = 0.092\%$, $L_2 = 0.127\%$, $L_3 = 0.233\%$.
						0.051	12.86	
						0.183	13.19	
						0.285	13.43	
						0.394	13.77	
						0.483	14.00	
						0.589	14.28	
						0.686	14.56	
						0.793	14.87	
						1.000	15.23	

TABLE 150-G(C)S.   SMOOTHED VISCOSITY VALUES AS A FUNCTION OF COMPOSITION FOR GASEOUS
METHYL CHLORIDE-SULFUR DIOXIDE MIXTURES

Mole Fraction of CH_3Cl	(308.2 K) [Ref. 349]	(353.2 K) [Ref. 349]
0.00	11.26	12.77
0.05	11.37	12.89
0.10	11.48	13.01
0.15	11.59	13.13
0.20	11.70	13.24
0.25	11.82	13.37
0.30	11.93	13.50
0.35	12.03	13.63
0.40	12.14	13.77
0.45	12.25	13.90
0.50	12.35	14.04
0.55	12.46	14.18
0.60	12.56	14.32
0.65	12.66	14.45
0.70	12.75	14.57
0.75	12.85	14.71
0.80	12.94	14.83
0.85	13.03	14.94
0.90	13.12	15.05
0.95	13.20	15.15
1.00	13.28	15.23

552

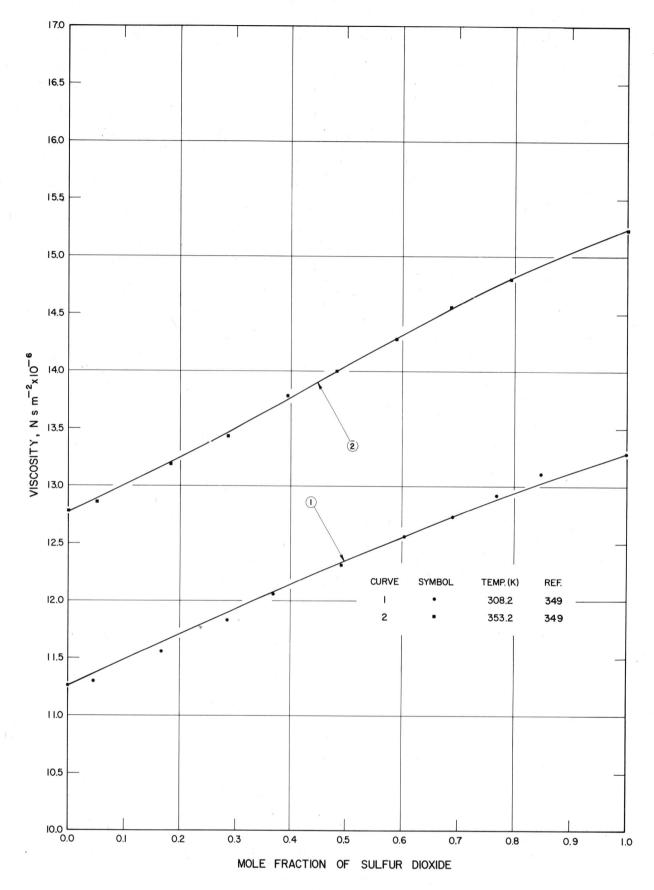

FIGURE 150-G(C). VISCOSITY DATA AS A FUNCTION OF COMPOSITION
FOR GASEOUS METHYL CHLORIDE—SULFUR DIOXIDE MIXTURES

TABLE 151-L(T). RECOMMENDED VISCOSITY VALUES FOR LIQUID REFRIGERANT 500

DISCUSSION

SATURATED LIQUID

Only one set of experimental data, by Phillips [371], was found in the literature.

They were fitted to the equation

$$\log \mu = A + B/T$$

from which recommended values were generated.

The deviations from the equation are within ± 1%, indicating good internal consistency, but we need independent determination to assess the accuracy.

RECOMMENDED VALUES

[Temperature, T, K; Viscosity, μ, N s m^{-2} · 10^{-3}]

SATURATED LIQUID	
T	μ
200	0.611
205	0.559
210	0.515
215	0.476
220	0.442
225	0.412
230	0.386
235	0.362
240	0.341
245	0.322
250	0.304
255	0.289
260	0.274
265	0.261
270	0.249
275	0.232
280	0.227
285	0.217
290	0.208
295	0.200
300	0.1915
305	0.1840
310	0.1768
315	0.1700
320	0.1632
325	0.1560
330	0.1482
335	0.1400
340	0.1313
345	0.1221
350	0.1124
355	0.1024
360	0.0916
365	0.0800
370	0.0670
375	0.0520
379*	0.0284

* Crit. Temp.

554

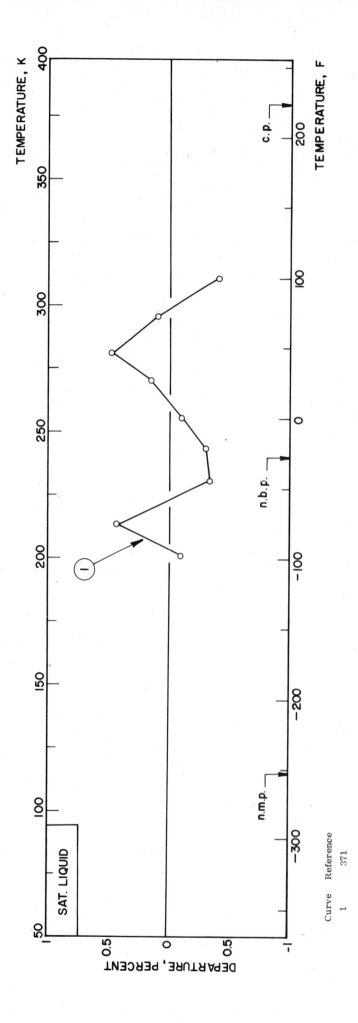

FIGURE 151-L(T). DEPARTURE PLOT FOR VISCOSITY OF LIQUID REFRIGERANT 500

TABLE 151-V(T). RECOMMENDED VISCOSITY VALUES FOR REFRIGERANT 500 VAPOR

DISCUSSION

SATURATED VAPOR

Recommended values for the viscosity of the saturated vapor were generated by means of the method of Stiel and Thodos[207]which make use of the excess viscosity concept. Excess viscosity, gotten from a reduced excess viscosity versus reduced temperature curve obtained with a number of other refrigerants, were combined with the recommended values for the 1 atm gas.

The accuracy is of about 5%, but the figure may be higher around the critical temperature.

RECOMMENDED VALUES

[Temperature, T, K, Viscosity, μ, 10^{-3} N s m^{-2}]

SATURATED VAPOR

T	μ
240	0.00959
250	0.01006
260	0.01052
270	0.01102
280	0.01152
290	0.01201
300	0.01258
310	0.01319
320	0.01381
330	0.01448
340	0.01532
350	0.0163
360	0.0173
370	0.0193
379*	0.0284

* Crit. Temp.

TABLE 151-G(T).   RECOMMENDED VISCOSITY VALUES FOR GASEOUS REFRIGERANT 500

DISCUSSION

GAS

Only one set of experimental data, by Latto [137], was found in the literature.

From the data, the function

$$\sigma^2 \Omega = \frac{K \sqrt{T}}{\mu}$$

was computed and then adjusted to a linear equation in 1/T from which the recommended values were generated.

The accuracy is thought to be ±2%.

RECOMMENDED VALUES

[Temperature, T, K, Viscosity, μ, 10^{-6} N s m^{-2}]

GAS

T	μ
240	9.57
250	10.02
260	10.46
270	10.91
280	11.35
290	11.79
300	12.2
310	12.7
320	13.1
330	13.5
340	13.9
350	14.4
360	14.8
370	15.2
380	15.6
390	16.0

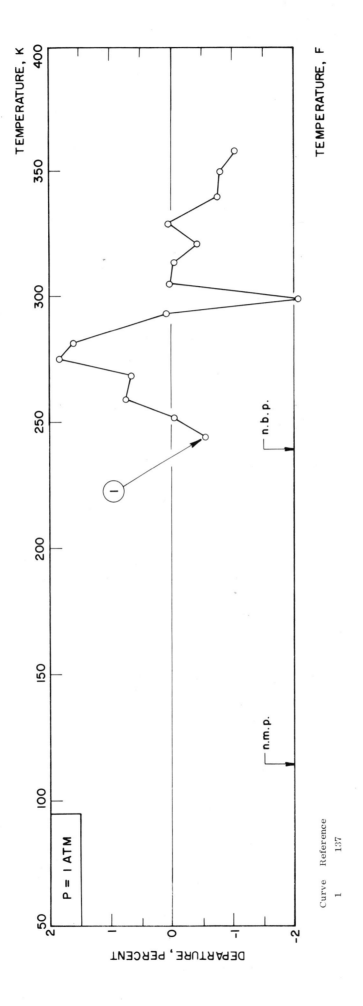

FIGURE 151-G(T). DEPARTURE PLOT FOR VISCOSITY OF GASEOUS REFRIGERANT 500

TABLE 152-L(T).   RECOMMENDED VISCOSITY VALUES FOR LIQUID REFRIGERANT 502

DISCUSSION

SATURATED LIQUID

Two sets of experimental data were found in the literature, those of
Gordon [79] and of Phillips [371]. They are in fair agreement.

The data below 270 K were least square fitted to an equation:

$$\log \mu = A + B/T$$

from which recommended values were generated. Above 270 K the curve was
drawn graphically.

The accuracy is thought to be better than ±5%.

RECOMMENDED VALUES

[Temperature, T, K, Viscosity, μ, 10^{-3} N s m^{-2}]

SATURATED LIQUID

T	μ
200	0.571
210	0.488
220	0.423
230	0.371
240	0.329
250	0.2949
260	0.2664
270	0.2395
280	0.2150
290	0.1940
300	0.1755
310	0.1585
320	0.1425
330	0.1265
340	0.1090
350	0.089
360*	0.030

* Crit. Temp.

559

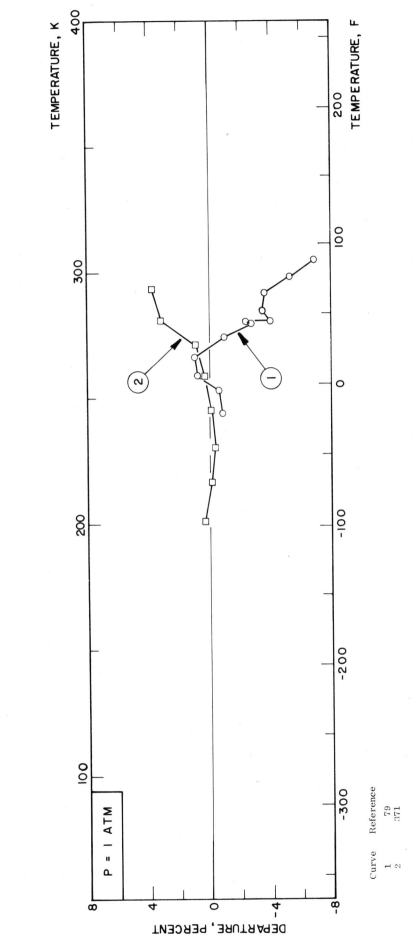

FIGURE 152-L(T). DEPARTURE PLOT FOR VISCOSITY OF LIQUID REFRIGERANT 502

TABLE 152-V(T). RECOMMENDED VISCOSITY VALUES FOR REFRIGERANT 502 VAPOR

DISCUSSION

SATURATED VAPOR

Recommended values for the viscosity of the saturated vapor were computed with the method of Stiel and Thodos [207] which makes use of the excess viscosity concept. From a graph of reduced excess viscosity versus reduced temperature, constructed with data for several refrigerants, excess viscosity was derived, and added to the recommended values.

The accuracy should be poor, about ±10% close to the boiling point, but may reach ±20% when approaching the critical point.

RECOMMENDED VALUES

[Temperature, T, K, Viscosity, μ, 10^{-3} N s m^{-2}]

SATURATED VAPOR

T	μ
220	0.00941
230	0.00990
240	0.01037
250	0.01084
260	0.01134
270	0.01185
280	0.01240
290	0.01301
300	0.0137
310	0.0143
320	0.0152
330	0.0161
340	0.0174
350	0.0194
360*	0.0304

* Crit. Temp.

TABLE 152-G(T). RECOMMENDED VISCOSITY VALUES FOR GASEOUS REFRIGERANT 502

RECOMMENDED VALUES

[Temperature, T, K, Viscosity, μ, 10^{-6} N s m^{-2}]

GAS

T	μ
220	9.41
230	9.86
240	10.30
250	10.75
260	11.19
270	11.62
280	12.05
290	12.48
300	12.90
310	13.32
320	13.74
330	14.15
340	14.55
350	14.96
360	15.36
370	15.75
380	16.15
390	16.54
400	16.93

DISCUSSION

GAS

Values for the viscosity of the gas were found in a manufacturer's technical note [281]. They were read from the curve, then checked by means of the equation:

$$\delta^2 \Omega = K \sqrt{\frac{T}{\mu}}$$

which was fitted to a linear equation in 1/T, and from which recommended values were generated.

These are estimated values, and there is no means to check their accuracy.

562

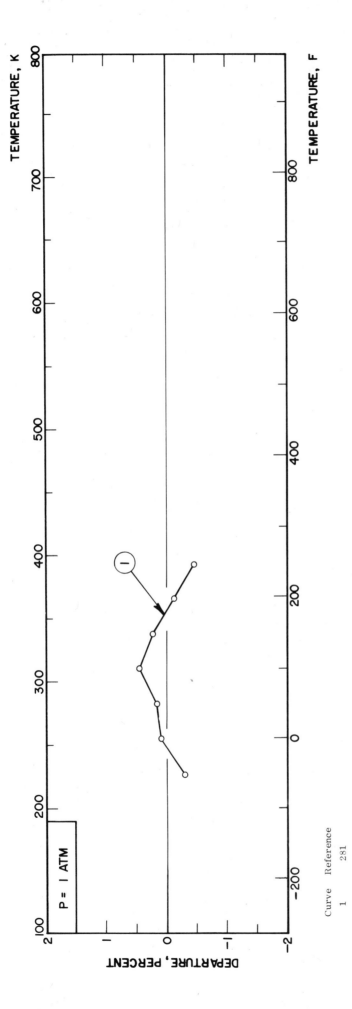

FIGURE 152-G(T). DEPARTURE PLOT FOR VISCOSITY OF GASEOUS REFRIGERANT 502

TABLE 153-L(T). RECOMMENDED VISCOSITY VALUES FOR LIQUID REFRIGERANT 503

DISCUSSION

SATURATED LIQUID

Only one set of experimental data, the results of Phillips [371], was found in the literature. The author gives an equation of the type:

$$\log \mu = A + B/T + CT + DT^2$$

which was adopted to generate recommended values. The deviations from the equation are small but there is no means to assess the accuracy. Values above 260 K were obtained by extrapolation and should be considered as tentative values.

RECOMMENDED VALUES

[Temperature, T, K: Viscosity, μ, N s m^{-2} · 10^{-3}]

SATURATED LIQUID

T	μ
180	0.341
185	0.317
190	0.295
195	0.275
200	0.257
205	0.240
210	0.225
215	0.210
220	0.197
225	0.1847
230	0.1732
235	0.1624
240	0.1523
245	0.1428
250	0.134
255	0.125
260	0.117
265	0.110
270	0.103
275	0.096
280	0.090
285	0.084
290	0.078

TABLE 153-L(T). DEPARTURE PLOT FOR VISCOSITY OF LIQUID REFRIGERANT 503

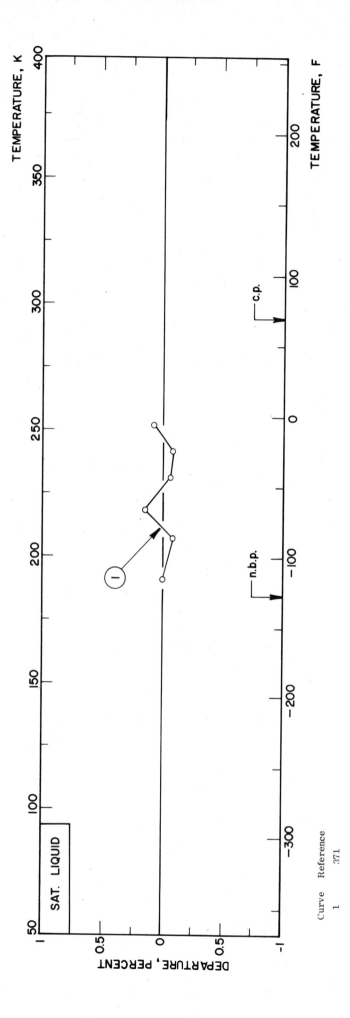

SAT. LIQUID

Curve   Reference
  1   371

TABLE 154-L(T). RECOMMENDED VISCOSITY VALUES FOR LIQUID REFRIGERANT 504

DISCUSSION

SATURATED LIQUID

Only one set of experimental data, the results of Phillips [371], was found in the literature. The author gives an equation of the type:

$$\log \mu = A + B/T + CT + DT^2$$

which was adopted, to generate recommended values. Values above 290 K, obtained from the equation, represent an extrapolation.

The deviations from the equation are small, but there is no means to assess accuracy.

RECOMMENDED VALUES

[Temperature, T, K; Viscosity, μ, N s m^{-2} · 10^{-3}]

SATURATED LIQUID

T	μ
200	0.477
205	0.433
210	0.396
215	0.364
220	0.334
225	0.319
230	0.294
235	0.276
240	0.260
245	0.245
250	0.232
255	0.220
260	0.208
265	0.197
270	0.187
275	0.1776
280	0.1683
285	0.1594
290	0.1507
295	0.1424
300	0.134
305	0.126
310	0.119
315	0.111
320	0.104

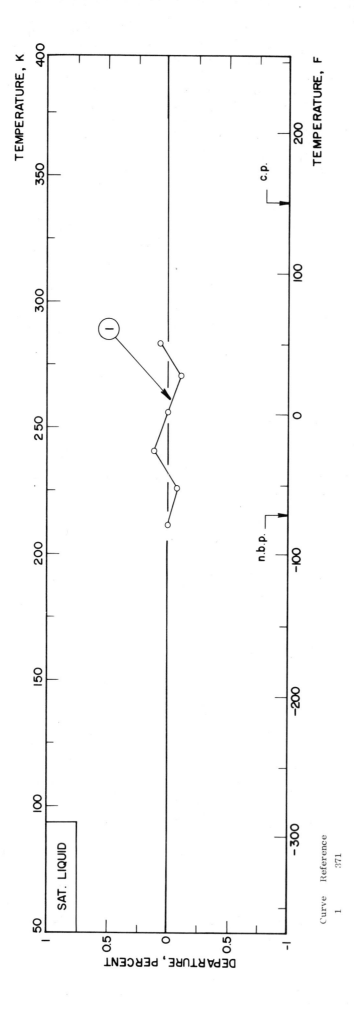

FIGURE 154-L(T). DEPARTURE PLOT FOR VISCOSITY OF LIQUID REFRIGERANT 504

TABLE 155-L(T)E. EXPERIMENTAL VISCOSITY DATA AS A FUNCTION OF TEMPERATURE FOR LIQUID SODIUM CHLORATE-SODIUM NITRATE MIXTURES

Cur. No.	Fig. No.	Ref. No.	Author(s)	Mole Fraction of NaClO₃	Pressure (atm)	Temp. (K)	Viscosity (N s m⁻² x 10⁻³)	Remarks
1	155-L(T)	358	Campbell, A.N. and Van Der Kouwe, E.T.	1.000		534.5	6.95	NaClO₃ and NaNO₃: Fisher reagent grade chemicals and dried at 130 C; maximum impurity in NaClO₃ was 0.005% of sodium bromate and in NaNO₃ was 0.0005% heavy metals; capillary flow viscometer; over-all accuracy of measurements ± 1.0%.
						534.8	6.93	
						537.4	6.74	
						537.7	6.70	
						539.5	6.59	
						541.7	6.46	
						545.5	6.19	
						548.5	5.98	
						550.5	5.86	
						553.9	5.71	
						556.0	5.58	
						559.2	5.43	
2	155-L(T)	358	Campbell, A.N. and Van Der Kouwe, E.T.	0.727		511.2	8.53	Same remarks as for curve 1.
						513.2	8.33	
						518.5	7.77	
						521.2	7.54	
						526.2	7.16	
						529.3	6.91	
						530.2	6.86	
						533.5	6.60	
						537.1	6.37	
						543.6	5.96	
						544.4	5.88	
						548.2	5.66	
						553.5	5.37	
						557.5	5.16	
						560.7	4.98	
3	155-L(T)	358	Campbell, A.N. and Van Der Kouwe, E.T.	0.515		511.2	7.56	Same remarks as for curve 1.
						514.9	7.24	
						519.4	6.92	
						522.7	6.66	
						526.5	6.38	
						528.6	6.24	
						530.5	6.07	
						531.7	6.14	
						535.6	5.77	
						541.0	5.47	
						546.0	5.22	
						550.7	5.00	
						555.7	4.79	
4	155-L(T)	358	Campbell, A.N. and Van Der Kouwe, E.T.	0.389		522.4	6.11	Same remarks as for curve 1.
						526.2	5.88	
						529.4	5.70	
						530.0	5.67	
						532.9	5.50	
						535.2	5.38	
						541.7	5.06	
						545.2	4.92	
						546.7	4.84	
						549.2	4.77	
						550.9	4.68	
						559.4	4.37	
						559.9	4.31	
						561.5	4.26	

TABLE 155-L(T)S. SMOOTHED VISCOSITY VALUES AS A FUNCTION OF TEMPERATURE FOR LIQUID
SODIUM CHLORATE–SODIUM NITRATE MIXTURES

Temp. (K)	Mole Fraction of Sodium Chlorate			
	0.389 [Ref. 358]	0.515 [Ref. 358]	0.727 [Ref. 358]	1.000 [Ref. 358]
510.0		7.650	8.665	
515.0		7.265	8.150	
520.0	6.250	6.890	7.670	
525.0	5.960	6.520	7.240	
530.0	5.670	6.175	6.850	
535.0	5.400	5.850	6.480	6.934
537.5				6.730
540.0	5.150	5.555	6.140	6.535
542.5				6.360
545.0	4.920	5.278	5.990	6.192
547.5				6.040
550.0	4.715	5.035	5.555	5.900
552.5				5.760
555.0	4.510	4.821	5.280	
557.5				5.510
560.0	4.320	4.620	5.020	5.390

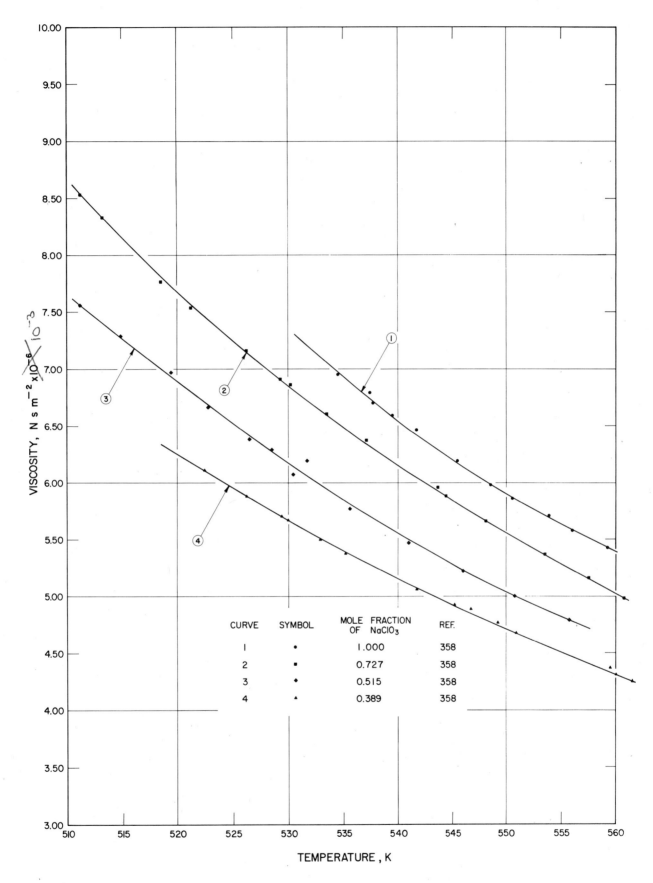

FIGURE 155-L(T). VISCOSITY DATA AS A FUNCTION OF TEMPERATURE
FOR LIQUID SODIUM CHLORATE—SODIUM NITRATE MIXTURES

TABLE 156-G(C)E.   EXPERIMENTAL VISCOSITY DATA AS A FUNCTION OF COMPOSITION FOR GASEOUS SULFUR DIOXIDE-SULFURYL FLUORIDE MIXTURES

Cur. No.	Fig. No.	Ref. No.	Author(s)	Temp. (K)	Pressure (atm)	Mole Fraction of SO_2F_2	Viscosity $(N\ s\ m^{-2} \times 10^{-6})$	Remarks
1	156-G(C)	350	Chang, K.C., Hesse, R.J., and Raw, C.J.G.	273		0.00	12.26	SO_2: 99.98 pure, SO_2F_2: 99.5% pure; constant volume transpiration type viscometer, relative measurements; $L_1 = 0.000\%$, $L_2 = 0.000\%$, $L_3 = 0.000\%$.
						0.25	12.89	
						0.50	13.43	
						0.75	13.88	
						1.00	14.13	
2	156-G(C)	350	Chang, K.C., et al.	323		0.00	14.42	Same remarks as for curve 1.
						0.25	15.12	
						0.50	15.69	
						0.75	16.06	
						1.00	16.22	
3	156-G(C)	350	Chang, K.C., et al.	373		0.00	16.52	Same remarks as for curve 1.
						0.25	17.27	
						0.50	17.86	
						0.75	18.16	
						1.00	18.28	
4	156-G(C)	350	Chang, K.C., et al.	423		0.00	18.62	Same remarks as for curve 1.
						0.25	19.40	
						0.50	19.97	
						0.75	20.23	
						1.00	20.29	
5	156-G(C)	350	Chang, K.C., et al.	473		0.00	20.69	Same remarks as for curve 1 except $L_1 = 0.006\%$, $L_2 = 0.013\%$, $L_3 = 0.028\%$.
						0.25	21.43	
						0.50	21.98	
						0.75	22.17	
						1.00	22.25	
6	156-G(C)	350	Chang, K.C., et al.	523		0.00	22.69	Same remarks as for curve 1 except $L_1 = 0.007\%$, $L_2 = 0.015\%$, $L_3 = 0.033\%$.
						0.25	23.43	
						0.50	23.93	
						0.75	24.13	
						1.00	24.22	
7	156-G(C)	350	Chang, K.C., et al.	573		0.00	24.68	Same remarks as for curve 1.
						0.25	25.35	
						0.50	25.82	
						0.75	26.03	
						1.00	26.14	
8	156-G(C)	350	Chang, K.C., et al.	623		0.00	26.61	Same remarks as for curve 1.
						0.25	27.21	
						0.50	27.66	
						0.75	27.87	
						1.00	28.01	
9	156-G(C)	350	Chang, K.C., et al.	673		0.00	28.45	Same remarks as for curve 1.
						0.25	29.07	
						0.50	29.46	
						0.75	29.68	
						1.00	29.83	

TABLE 156-G(C)S.   SMOOTHED VISCOSITY VALUES AS A FUNCTION OF COMPOSITION FOR GASEOUS
SULFUR DIOXIDE-SULFURYL FLUORIDE MIXTURES

Mole Fraction of SO_2F_2	(273.0 K) [Ref. 350]	(323.0 K) [Ref. 350]	(373.0 K) [Ref. 350]	(423.0 K) [Ref. 350]	(473.0 K) [Ref. 350]	(523.0 K) [Ref. 350]	(573.0 K) [Ref. 350]	(623.0 K) [Ref. 350]	(673.0 K) [Ref. 350]
0.00	12.26	14.42	16.52	18.62	20.69	22.69	24.68	26.61	28.45
0.05	12.39	14.57	16.59	18.80	20.84	22.87	24.83	26.78	28.60
0.10	12.52	14.72	16.85	18.96	21.00	23.04	24.97	26.89	28.73
0.15	12.65	14.86	17.00	19.12	21.15	23.19	25.11	26.99	28.85
0.20	12.77	14.99	17.14	19.26	21.30	23.33	25.23	27.10	28.96
0.25	12.89	15.12	17.27	19.40	21.44	23.43	25.35	27.21	29.07
0.30	13.01	15.24	17.41	19.54	21.57	23.58	25.46	27.32	29.16
0.35	13.12	15.37	17.53	19.66	21.70	23.69	25.57	27.42	29.25
0.40	13.23	15.48	17.65	19.78	21.81	23.78	25.66	27.51	29.33
0.45	13.33	15.59	17.76	19.88	21.90	23.87	25.75	27.59	29.40
0.50	13.43	15.69	17.86	19.97	21.98	23.94	25.82	27.66	29.46
0.55	13.54	15.78	17.93	20.05	22.04	24.00	25.88	27.72	29.52
0.60	13.63	15.86	18.00	20.11	22.09	24.04	25.93	27.76	29.56
0.65	13.73	15.94	18.07	20.16	22.12	24.08	25.97	27.80	29.61
0.70	13.81	16.00	18.12	20.20	22.15	24.11	26.00	27.83	29.65
0.75	13.88	16.06	18.16	20.23	22.17	24.13	26.03	27.87	29.68
0.80	13.94	16.11	18.20	20.25	22.19	24.15	26.06	27.90	29.71
0.85	14.00	16.14	18.23	20.27	22.21	24.17	26.08	27.93	29.75
0.90	14.05	16.17	18.25	20.28	22.30	24.19	26.10	27.96	29.78
0.95	14.09	16.20	18.27	20.29	22.34	24.20	26.12	27.99	29.80
1.00	14.13	16.22	18.28	20.29	22.25	24.22	26.14	28.01	29.83

572

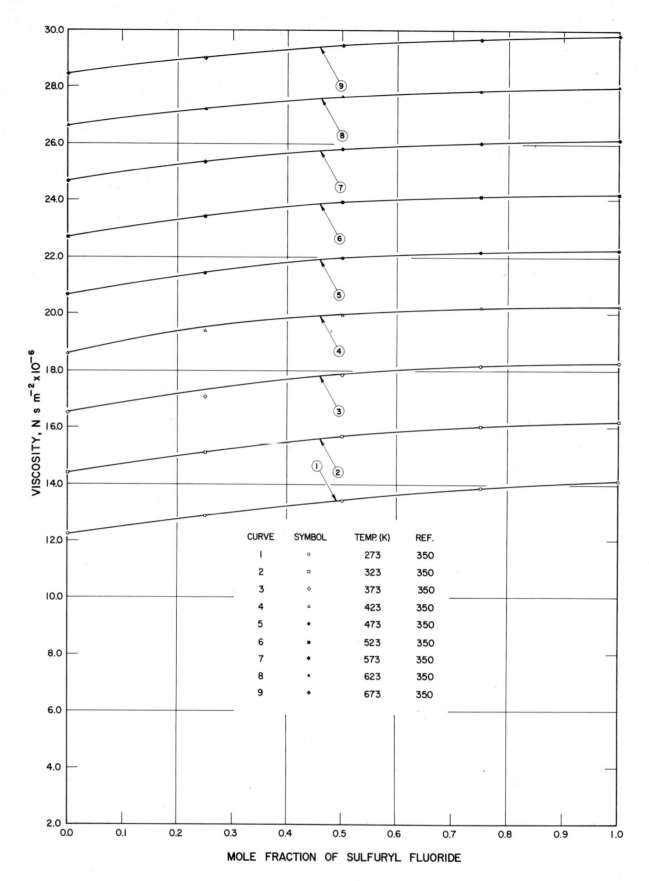

CURVE	SYMBOL	TEMP. (K)	REF.
1	○	273	350
2	□	323	350
3	◇	373	350
4	△	423	350
5	●	473	350
6	■	523	350
7	◆	573	350
8	▲	623	350
9	◆	673	350

MOLE FRACTION OF SULFURYL FLUORIDE

FIGURE 156-G(C). VISCOSITY DATA AS A FUNCTION OF COMPOSITION
FOR GASEOUS SULFUR DIOXIDE—SULFURYL FLUORIDE MIXTURES

TABLE 157-L(T)E.   EXPERIMENTAL VISCOSITY DATA AS A FUNCTION OF TEMPERATURE FOR LIQUID
IRON-CARBON MIXTURES

Cur. No.	Fig. No.	Ref. No.	Author(s)	Mole Fraction of Fe	Pressure (atm)	Temp. (K)	Viscosity (N s m^{-2} x 10^{-6}) x10^{-3}	Remarks
1	157-L(T)	356	Vatolin, N.V., Vostrayakov, A.A., and Esin, O.A.	0.9992		1823.2 1853.2 1873.2 1973.2	7.80 5.97 5.92 4.30	No purity specified for metals; oscillant crucible method; precision and accuracy not given.
2	157-L(T)	356	Vatolin, N.V., et al.	0.9980		1893.2 1943.2 1993.2	4.55 3.99 3.12	Same remarks as for curve 1.
3	157-L(T)	356	Vatolin, N.V., et al.	0.9975		1823.2 1853.2 1973.2	4.92 4.34 3.50	Same remarks as for curve 1.
4	157-L(T)	356	Vatolin, N.V., et al.	0.9960		1833.2 1853.2 1883.2 1953.2 1983.2	5.10 4.88 4.12 3.52 2.88	Same remarks as for curve 1.
5	157-L(T)	356	Vatolin, N.V., et al.	0.9936		1843.2 1903.2 1923.2 1953.2 1973.2	4.79 3.90 3.77 3.50 3.54	Same remarks as for curve 1.
6	157-L(T)	356	Vatolin, N.V., et al.	0.9870		1723.2 1793.2 1863.2 1973.2	7.76 5.93 4.60 3.54	Same remarks as for curve 1.
7	157-L(T)	356	Vatolin, N.V., et al.	0.9790		1713.2 1743.2 1763.2 1853.2 1873.2	6.94 6.41 6.25 4.79 4.70	Same remarks as for curve 1.
8	157-L(T)	356	Vatolin, N.V., et al.	0.9715		1623.2 1693.2 1723.2 1823.2 1873.2 1903.2	9.23 7.45 6.60 4.47 4.40 3.83	Same remarks as for curve 1.
9	157-L(T)	356	Vatolin, N.V., et al.	0.9580		1543.2 1703.2 1753.2 1823.2	8.60 5.75 4.89 3.39	Same remarks as for curve 1.
10	157-L(T)	356	Vatolin, N.V., et al.	0.9514		1633.2 1693.2 1763.2 1833.2 1873.2	4.06 2.42 2.45 2.03 1.36	Same remarks as for curve 1.

TABLE 157-L(T)S.   SMOOTHED VISCOSITY VALUES AS A FUNCTION OF TEMPERATURE FOR LIQUID IRON-CARBON MIXTURES

$N\,S\,m^{-2} \times 10^{-6}$

Temp. (K)	Mole Fraction of Iron				
	0.9992 [Ref. 356]	0.9980 [Ref. 356]	0.9975 [Ref. 356]	0.9960 [Ref. 356]	0.9936 [Ref. 356]
1820	7950		4970		
1825	7625		4875	5690	
1840	6820		4580		4852
1850	6440		4415	4920	4660
1870				4380	
1875	5740		4125	4275	4241
1880				4180	4170
1890	5430	4600			
1900	5250	4430	3920	3885	3940
1920	4950	4122			3765
1925	4875	4050	3760	3630	3730
1940	4675				3630
1950	4550	3715	3615	3425	3570
1960		3509		3355	
1970	4320				3470
1975		3420	3370	3260	3445
1980		3360		3240	3420
2000		3150			

Temp. (K)	Mole Fraction of Iron				
	0.9870 [Ref. 356]	0.9790 [Ref. 356]	0.9715 [Ref. 356]	0.9580 [Ref. 356]	0.9514 [Ref. 356]
1550				8480	
1575				7990	
1600				7515	
1625			9170	7030	
1630					4130
1650			8540	6570	3750
1660					3560
1675			7885	6110	3300
1690					3040
1700		7234	7215	5665	2875
1720					2590
1725	7710	6820	6570	5200	2525
1740					2340
1750	7040	6407	6020	4740	2232
1770					2035
1775	6370	6000	5530		
1800	5760	5606	5092	3902	
1825	5245	5228	4710	3540	
1840		5022			
1850	4800	4900	4380	3210	
1875	4450	4600	4110		
1900	4140		3870		
1925	3892				
1950	3690				
1975	3530				
1980	3415				

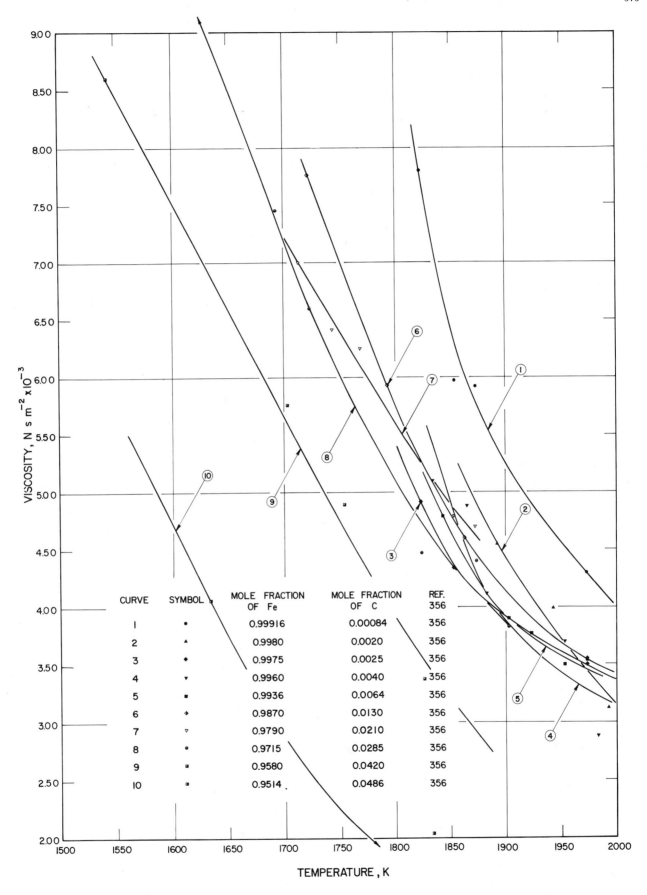

CURVE	SYMBOL	MOLE FRACTION OF Fe	MOLE FRACTION OF C	REF. 356
1	•	0.99916	0.00084	356
2	▲	0.9980	0.0020	356
3	◆	0.9975	0.0025	356
4	▼	0.9960	0.0040	356
5	■	0.9936	0.0064	356
6	◈	0.9870	0.0130	356
7	▽	0.9790	0.0210	356
8	⊙	0.9715	0.0285	356
9	▣	0.9580	0.0420	356
10	⊡	0.9514	0.0486	356

FIGURE 157-L(T). VISCOSITY DATA AS A FUNCTION OF TEMPERATURE
FOR LIQUID IRON—CARBON MIXTURES

TABLE 158-L(T)E.   EXPERIMENTAL VISCOSITY DATA AS A FUNCTION OF TEMPERATURE FOR LIQUID LEAD-TIN MIXTURES

Cur. No.	Fig. No.	Ref. No.	Author(s)	Mole Fraction of Pb	Pressure (atm)	Temp. (K)	Viscosity (N s m^{-2} x 10^{-3}) x 10^{-3}	Remarks
1	158-L(T)	357	Yao, T.P. and Kondic, V.	0.000		504.2	2.75	Sn: 99.9885 pure, Pb: 99.9962 pure; oscillating pendulum method, relative measurements.
						504.2	2.67	
						505.2	2.68	
						505.2	2.63	
						505.7	2.56	
						505.7	2.54	
						506.2	2.46	
						506.2	2.44	
						508.2	2.38	
						508.2	2.35	
						508.7	2.26	
						508.7	2.20	
						509.7	2.22	
						509.7	2.11	
						518.2	2.10	
						518.2	1.99	
						527.2	1.97	
						527.2	1.94	
						531.2	1.92	
						531.2	1.89	
						534.2	1.89	
						534.2	1.82	
						542.2	1.86	
						542.2	1.76	
						554.2	1.74	
						554.2	1.73	
						558.2	1.69	
						558.2	1.56	
						566.2	1.70	
						566.2	1.64	
						571.7	1.70	
						571.7	1.65	
						585.2	1.69	
						585.2	1.60	
						599.2	1.63	
						599.2	1.58	
						607.2	1.61	
						607.2	1.56	
						618.2	1.53	
						618.2	1.51	
						622.2	1.54	
						622.2	1.52	
						628.2	1.53	
						628.2	1.49	
						638.7	1.50	
						638.7	1.48	
						648.7	1.51	
						648.7	1.50	
						657.2	1.52	
						657.2	1.48	
						660.2	1.52	
						660.2	1.46	
2	158-L(T)	357	Yao, T.P. and Kondic, V.	0.025		503.2	8.54	Same remarks as for curve 1.
						513.2	2.05	
						515.2	1.98	
						525.7	1.78	
						540.2	1.74	
						549.2	1.65	
						566.2	1.63	
						572.2	1.57	
						624.2	1.51	
						666.2	1.44	
3	158-L(T)	357	Yao, T.P. and Kondic, V.	0.300		473.2	3.97	Same remarks as for curve 1.
						488.2	2.53	
						503.7	2.47	
						523.7	2.24	
						549.7	2.13	
						574.2	2.11	
						608.2	2.05	
						640.2	1.96	
						669.7	1.88	
						719.2	1.79	

TABLE 158-L(T)E. EXPERIMENTAL VISCOSITY DATA AS A FUNCTION OF TEMPERATURE FOR LIQUID LEAD-TIN MIXTURES (continued)

Cur. No.	Fig. No.	Ref. No.	Author(s)	Mole Fraction of Pb	Pressure (atm)	Temp. (K)	Viscosity (N s m^{-2} x 10^{-6}) × 10^{-3}	Remarks
4	158-L(T)	357	Yao, T.P. and Kondic, V.	0.382		460.2	3.75	Same remarks as for curve 1.
						461.2	3.10	
						467.2	2.77	
						473.7	2.66	
						484.7	2.21	
						494.7	2.06	
						507.2	1.99	
						524.7	2.06	
						541.7	2.02	
						550.7	2.41	
						572.2	2.28	
						591.2	2.32	
						630.7	2.26	
						719.2	2.14	
						725.7	2.12	
5	158-L(T)	357	Yao, T.P. and Kondic, V.	1.000		616.2	2.73	Same remarks as for curve 1.
						622.2	2.68	
						626.2	2.60	
						633.2	2.48	
						653.2	2.41	
						704.2	2.21	
						716.2	1.99	
						736.2	1.96	
						753.2	1.93	
						763.2	1.90	
						773.2	1.87	

TABLE 158-L(T)S. SMOOTHED VISCOSITY VALUES AS A FUNCTION OF TEMPERATURE FOR LIQUID LEAD-TIN MIXTURES

N s m^{-2} × 10^{-6}

Temp. (K)	Mole Fraction of Lead				
	0.000 [Ref. 357]	0.025 [Ref. 357]	0.300 [Ref. 357]	0.382 [Ref. 357]	1.000 [Ref. 357]
460.0				4150	
470.0			4950		
475.0				2550	
480.0			2945	2060	
500.0			2455	2060	
505.0	2620				
510.0	2015	4200			
520.0		1880			
525.0			2220	2040	
540.0	1835	1790			
550.0			2150	2230	
560.0	1707	1610			
575.0			2100	2310	
580.0	1640	1570			
600.0	1570	1540	2050	2300	2950
612.5					2780
620.0	1530	1510			
625.0			2000	2280	2630
637.5					2490
640.0	1510	1480			
650.0			1950	2250	2360
660.0	1490	1445			
675.0			1900	2220	2180
700.0			1850	2180	2180
725.0			1800		1990
750.0					1940
775.0					1890

578

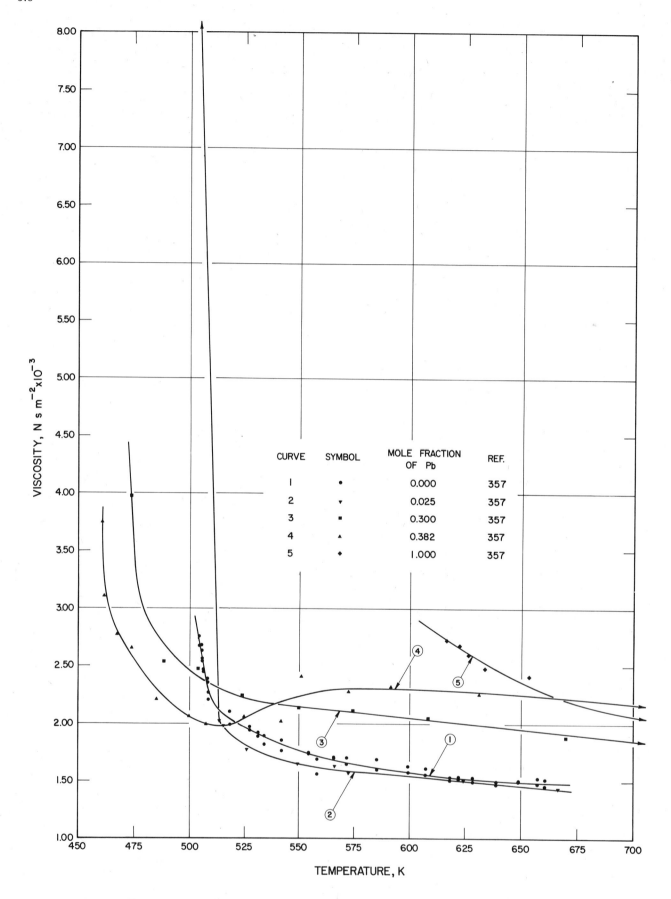

FIGURE 158-L(T). VISCOSITY DATA AS A FUNCTION OF TEMPERATURE
FOR LIQUID LEAD—TIN MIXTURES

5. TERNARY SYSTEMS

TABLE 159-G(C)E.   EXPERIMENTAL VISCOSITY DATA AS A FUNCTION OF COMPOSITION FOR GASEOUS ARGON-HELIUM-NEON MIXTURES

Cur. No.	Fig. No.	Ref. No.	Author(s)	Temp. (K)	Pressure (atm)	Mole Fraction of			Viscosity (N s m^{-2} x 10^{-6})	Remarks
						Ar	He	Ne		
1		223	Trautz, M. and Kipphan, K.F.	293.2		0.1615	0.5175	0.3210	26.02	Gas purity: Ar < 0.5% N2,
						0.1702	0.4746	0.3552	26.29	He < 1% Ne, Ne < 1% He;
						0.2382	0.5429	0.2189	25.04	method of Trautz and
						0.2670	0.1754	0.5576	27.40	Weizel, calibrated with
						0.3213	0.3594	0.3193	25.69	air.
						0.3333	0.2042	0.4625	26.55	
						0.4414	0.1883	0.3703	25.57	
						0.5851	0.1983	0.2166	24.11	
2		223	Trautz, M. and Kipphan, K.F.	373.2		0.1615	0.5175	0.3210	30.69	Same remarks as for
						0.1702	0.4746	0.3552	31.00	curve 1.
						0.2382	0.5429	0.2189	29.57	
						0.2670	0.1754	0.5576	32.37	
						0.3213	0.3594	0.3193	30.44	
						0.3333	0.2042	0.4625	31.47	
						0.4414	0.1883	0.3703	30.45	
						0.5851	0.1983	0.2166	28.86	
3		223	Trautz, M. and Kipphan, K.F.	473.2		0.1615	0.5175	0.3210	35.93	Same remarks as for
						0.2382	0.5429	0.2189	34.70	curve 1.
						0.2670	0.1754	0.5576	37.90	
						0.3213	0.3594	0.3193	35.74	
						0.3333	0.2042	0.4625	36.92	
						0.4414	0.1883	0.3703	35.82	
						0.5851	0.1983	0.2166	34.15	

581

TABLE 160-G(C)E. EXPERIMENTAL VISCOSITY DATA AS A FUNCTION OF COMPOSITION FOR GASEOUS ARGON-HELIUM-CARBON DIOXIDE MIXTURES

Cur. No.	Fig. No.	Ref. No.	Author(s)	Temp. (K)	Pressure (atm)	Mole Fraction of			Viscosity ($N \, s \, m^{-2} \times 10^{-6}$)	Remarks
						Ar	He	CO_2		
1		361	Strunk, M.R. and Fehsenfeld, G.D.	278.2		0.2295 0.1322	0.1630 0.6385	0.6075 0.2293	16.86 19.53	Ar: Matheson Co., specified purity 99.995, chief impurities O_2 and N_2, He: Matheson Co., specified purity 99.9, chief impurities N_2 and CO_2, CO_2: Matheson Co., specified purity 99.8, chief impurities N_2 and O_2; mixtures prepared according to Dalton's law of partial pressures; mixtures analyzed on mass spectrometer; rolling ball viscometer; experimental error $\pm 1.5\%$.
2		361	Strunk, M.R. and Fehsenfeld, G.D.	323.2		0.2295 0.1322	0.1630 0.6385	0.6075 0.2293	19.15 21.72	Same remarks as for curve 1.
3		361	Strunk, M.R. and Fehsenfeld, G.D.	363.2		0.2295 0.1322	0.1630 0.6385	0.6075 0.2293	21.03 23.48	Same remarks as for curve 1.

TABLE 161-G(C)E.   EXPERIMENTAL VISCOSITY DATA AS A FUNCTION OF COMPOSITION FOR GASEOUS
ARGON-HELIUM-METHANE MIXTURES

Cur. No.	Fig. No.	Ref. No.	Author(s)	Temp. (K)	Pressure (atm)	Mole Fraction of			Viscosity $(N\ s\ m^{-2} \times 10^{-6})$	Remarks
						Ar	He	CH_4		
1		361	Strunk, M.R. and Fehsenfeld, G.D.	278.2		0.3909	0.4597	0.1494	19.54	Ar: Matheson Co., specified purity 99.995, chief impurities O_2 and N_2, He: Matheson Co., specified purity 99.9, chief impurities N_2 and CO_2, CH_4: Matheson Co., specified purity 99.0, chief impurities CO_2, N_2, ethane, propane; mixtures prepared according to Dalton's law of partial pressures; mixtures analyzed on mass spectrometer; rolling ball viscometer; experimental error $\pm 1.5\%$.
						0.4510	0.1612	0.3878	16.82	
2		361	Strunk, M.R. and Fehsenfeld, G.D.	323.2		0.3909	0.4597	0.1494	22.05	Same remarks as for curve 1.
						0.4510	0.1612	0.3878	19.10	
3		361	Strunk, M.R. and Fehsenfeld, G.D.	363.2		0.3909	0.4597	0.1494	23.87	Same remarks as for curve 1.
						0.4510	0.1612	0.3878	20.91	

TABLE 162-G(C)E.   EXPERIMENTAL VISCOSITY DATA AS A FUNCTION OF COMPOSITION FOR GASEOUS
ARGON-CARBON DIOXIDE-METHANE MIXTURES

Cur. No.	Fig. No.	Ref. No.	Author(s)	Temp. (K)	Pressure (atm)	Mole Fraction of			Viscosity $(N\ s\ m^{-2} \times 10^{-6})$	Remarks
						Ar	CO_2	CH_4		
1		361	Strunk, M.R. and Fehsenfeld, G.D.	278.2		0.1498 0.4267	0.6396 0.1668	0.2106 0.4065	14.82 16.20	Ar: Matheson Co., specified purity 99.995, chief impurities O_2 and N_2, CO_2: Matheson Co., specified purity 99.8, chief impurities N_2 and O_2, CH_4: Matheson Co., specified purity 99.0, chief impurities CO_2, N_2, ethane, propane; mixtures prepared according to Dalton's law of partial pressures; mixtures analyzed on mass spectrometer; rolling ball viscometer; experimental error $\pm 1.5\%$.
2		361	Strunk, M.R. and Fehsenfeld, G.D.	323.2		0.1498 0.4267	0.6396 0.1668	0.2106 0.4065	17.02 18.42	Same remarks as for curve 1.
3		361	Strunk, M.R. and Fehsenfeld, G.D.	363.2		0.1498 0.4267	0.6396 0.1668	0.2106 0.4065	18.74 20.11	Same remarks as for curve 1.

TABLE 163-G(C)E.   EXPERIMENTAL VISCOSITY DATA AS A FUNCTION OF COMPOSITION FOR GASEOUS
CARBON DIOXIDE-HYDROGEN-OXYGEN MIXTURES

Cur. No.	Fig. No.	Ref. No.	Author(s)	Temp. (K)	Pressure (psia)	Mole Fraction of			Viscosity ($N \cdot s \cdot m^{-2} \times 10^{-6}$)	Remarks
						CO_2	H_2	O_2		
1		337	Gururaja, G.J., Tirunarayanan, M.A., and Kamchandran, A.	298.5	14.2	12.8	74.9	12.3	15.85	No purity specified; oscillating disc viscometer, calibrated to N_2; viscosity measured at ambient temperature and pressure; precision was ±1.0% of previous data.
				299.0		70.0	14.7	15.3	15.78	
				299.3		78.0	6.0	16.0	15.80	
				298.0		9.2	69.1	21.7	16.61	
				298.1		17.4	58.9	23.7	16.72	
				297.9		59.0	14.3	26.7	16.67	
				297.0		19.4	50.0	30.6	17.12	
				297.7		42.0	18.8	39.2	17.10	
				297.0		33.8	22.0	44.2	17.86	
				296.3		33.3	22.0	44.7	16.85	
				297.0		11.4	40.3	48.3	18.19	
				296.8		14.7	36.0	49.3	18.18	

TABLE 164-G(C)E. EXPERIMENTAL VISCOSITY DATA AS A FUNCTION OF COMPOSITION FOR GASEOUS CARBON DIOXIDE-NITROGEN-OXYGEN MIXTURES

Cur. No.	Fig. No.	Ref. No.	Author(s)	Temp. (K)	Pressure (atm)	Mole Fraction of			Viscosity $(N\ s\ m^{-2} x\ 10^{-6})$	Remarks
						CO_2	N_2	O_2		
1		337	Gururaja, G.J., Tirunarayana, M.A., and Ramchandran, A.	297.45		0.084	0.883	0.033	17.45	No purity specified; oscillating disc viscometer, calibrated to N_2; viscosity measured at ambient temperature and pressure; precision was ±1% of previous data.
				297.83		0.098	0.812	0.090	17.55	
				295.92		0.090	0.796	0.114	17.90	
				298.45		0.146	0.736	0.118	17.60	
				297.80		0.297	0.500	0.203	17.40	
				297.80		0.507	0.280	0.213	16.80	
				296.94		0.090	0.680	0.230	18.05	
				297.45		0.062	0.858	0.081	17.78	
				297.85		0.128	0.703	0.169	17.65	
				297.45		0.212	0.520	0.268	17.65	
				297.30		0.266	0.400	0.334	17.65	
2		363	Herning, F. and Zipperer, L.	293		0.086	0.891	0.023	17.56	Capillary method.
						0.133	0.828	0.039	17.49	
						0.062	0.831	0.107	17.93	

TABLE 164-G(T)E.   EXPERIMENTAL VISCOSITY DATA AS A FUNCTION OF TEMPERATURE FOR
GASEOUS CARBON DIOXIDE-NITROGEN-OXYGEN MIXTURES

Cur. No.	Fig. No.	Ref. No.	Author(s)	Mole Fraction of			Pressure (atm)	Temp. (K)	Viscosity ($N\ s\ m^{-2} \times 10^{-6}$)	Remarks
				CO_2	N_2	O_2				
1		364	Kenney, M.J., Sarjant, R.J., and Thring, M.W.	0.1062	0.8865	0.0073		317.2	18.44	Pure gases were obtained from commercial cylinders; relative capillary flow viscometer, calibrated for air; estimated maximum error ± 2.0%.
								561.2	27.23	
								687.2	31.21	
								813.7	35.36	
								1027.7	41.84	
								1160.7	45.16	
2		364	Kenney, M.J., et al.	0.1500	0.8450	0.005		305.2	17.63	Same remarks as for curve 1.
								548.7	27.33	
								737.2	32.20	
								904.7	38.05	
								1039.2	41.11	
								1134.2	44.13	
3		364	Kenney, M.J., et al.	0.1982	0.7954	0.0064		291.4	17.03	Same remarks as for curve 1.
								403.2	21.69	
								496.2	24.89	
								631.2	29.01	
								811.2	34.77	
								932.2	38.23	
								1047.2	41.87	
								1146.2	43.74	

TABLE 165-G(D)E.   EXPERIMENTAL VISCOSITY DATA AS A FUNCTION OF DENSITY FOR GASEOUS HYDROGEN-METHANE-NITROGEN MIXTURES

Cur. No.	Fig. No.	Ref. No.	Author(s)	Mole Fraction of			Temp. (K)	Density (g cm^{-3})	Viscosity (N s m^{-2} x10^{-6})	Remarks
				H$_2$	CH$_4$	N$_2$				
1		366	Gnezdilov, N.E. and Golubev, I.F.	0.577	0.218	0.205	273.2	0.01	12.60	No purity specified for gases; composition analyzed by means of Kh T-2M chrome-thermograph; capillary method; experimental error ± 1%; density calculated from data given.
								0.02	12.68	
								0.03	13.11	
								0.04	13.57	
								0.05	13.81	
								0.06	14.36	
								0.07	14.60	
								0.08	15.12	
								0.09	15.54	
								0.10	16.15	
								0.11	16.65	
								0.12	17.38	
								0.13	18.06	
								0.14	18.70	
								0.15	19.15	
								0.16	19.80	
								0.17	20.61	
								0.18	21.25	
								0.19	22.09	
								0.20	22.84	
								0.21	23.77	
								0.22	24.43	
2		366	Gnezdilov, N.E. and Golubev, I.F.	0.577	0.218	0.205	298.2	0.01	13.44	Same remarks as for curve 1.
								0.02	13.52	
								0.03	13.95	
								0.04	14.41	
								0.05	14.65	
								0.06	15.20	
								0.07	15.44	
								0.08	15.96	
								0.09	16.38	
								0.10	16.99	
								0.11	17.49	
								0.12	18.22	
								0.13	18.90	
								0.14	19.54	
								0.15	19.99	
								0.16	20.64	
								0.17	21.45	
								0.18	22.09	
								0.19	22.93	
								0.20	23.68	
								0.21	24.61	
								0.22	25.27	
3		366	Gnezdilov, N.E. and Golubev, I.F.	0.577	0.218	0.205	323.2	0.01	14.25	Same remarks as for curve 1.
								0.02	14.33	
								0.03	14.76	
								0.04	15.22	
								0.05	15.46	
								0.06	16.01	
								0.07	16.25	
								0.08	16.77	
								0.09	17.19	
								0.10	17.80	
								0.11	18.30	
								0.12	19.03	
								0.13	19.71	
								0.14	20.35	
								0.15	20.80	
								0.16	21.45	
								0.17	22.26	
								0.18	22.90	
								0.19	23.74	
								0.20	24.69	
								0.21	25.42	
								0.22	26.08	

TABLE 165-G(D)E.   EXPERIMENTAL VISCOSITY DATA AS A FUNCTION OF DENSITY FOR GASEOUS
HYDROGEN-METHANE-NITROGEN MIXTURES (continued)

Cur. No.	Fig. No.	Ref. No.	Author(s)	Mole Fraction of			Temp. (K)	Density $(g\ cm^{-3})$	Viscosity $(N\ s\ m^{-2} \times 10^{-6})$	Remarks
				H_2	CH_4	N_2				
4		366	Gnezdilov, N. E. and Golubev, I. F.	0.577	0.218	0.205	373.2	0.01	15.78	Same remarks as for curve 1.
								0.02	15.87	
								0.03	16.30	
								0.04	16.76	
								0.05	17.00	
								0.06	17.55	
								0.07	17.79	
								0.08	18.31	
								0.09	18.73	
								0.10	19.32	
								0.11	19.84	
								0.12	20.57	
								0.13	21.25	
								0.14	21.89	
								0.15	22.34	
								0.16	22.99	
								0.17	23.80	
								0.18	24.44	
								0.19	25.28	
								0.20	26.03	
								0.21	26.96	
								0.22	27.62	
5		366	Gnezdilov, N. E. and Golubev, I. F.	0.577	0.218	0.205	423.2	0.01	17.28	Same remarks as for curve 1.
								0.02	17.36	
								0.03	17.79	
								0.04	18.25	
								0.05	18.49	
								0.06	19.04	
								0.07	19.28	
								0.08	19.80	
								0.09	20.22	
								0.10	20.83	
								0.11	21.33	
								0.12	22.06	
								0.13	22.74	
								0.14	23.38	
								0.15	23.83	
								0.16	24.48	
								0.17	25.29	
								0.18	25.93	
								0.19	26.77	
								0.20	27.72	
								0.21	28.45	
								0.22	29.11	
6		366	Gnezdilov, N. E. and Golubev, I. F.	0.577	0.218	0.205	473.2	0.01	18.78	Same remarks as for curve 1.
								0.02	18.86	
								0.03	19.29	
								0.04	19.75	
								0.05	19.99	
								0.06	20.54	
								0.07	20.78	
								0.08	21.30	
								0.09	21.72	
								0.10	22.31	
								0.11	22.83	
								0.12	23.56	
								0.13	24.24	
								0.14	24.88	
								0.15	25.33	
								0.16	25.98	
								0.17	26.79	
								0.18	27.43	
								0.19	28.27	
								0.20	29.02	
								0.21	29.95	
								0.22	30.61	

TABLE 165-G(D)E.   EXPERIMENTAL VISCOSITY DATA AS A FUNCTION OF DENSITY FOR GASEOUS HYDROGEN-METHANE-NITROGEN MIXTURES (continued)

Cur. No.	Fig. No.	Ref. No.	Author(s)	Mole Fraction of			Temp. (K)	Density (g cm^{-3})	Viscosity (N s m^{-2} x 10^{-6})	Remarks
				H$_2$	CH$_4$	N$_2$				
7		366	Gnezdilov, N.E. and Golubev, I.F.	0.577	0.218	0.205	523.2	0.01	20.26	Same remarks as for curve 1.
								0.02	20.34	
								0.03	20.77	
								0.04	21.23	
								0.05	21.47	
								0.06	22.02	
								0.07	22.26	
								0.08	22.78	
								0.09	23.20	
								0.10	23.81	
								0.11	24.31	
								0.12	25.04	
								0.13	25.72	
								0.14	26.36	
								0.15	26.81	
								0.16	27.46	
								0.17	28.27	
								0.18	28.91	
								0.19	29.75	
								0.20	30.70	
								0.21	31.43	
								0.22	32.09	
8		366	Gnezdilov, N.E. and Golubev, I.F.	0.498	0.188	0.314	273.2	0.01	14.31	Same remarks as for curve 1.
								0.02	14.53	
								0.03	14.72	
								0.04	14.97	
								0.05	15.29	
								0.06	15.64	
								0.07	15.91	
								0.08	16.25	
								0.09	16.69	
								0.10	17.07	
								0.11	17.50	
								0.12	17.85	
								0.13	18.34	
								0.14	18.98	
								0.15	19.52	
								0.16	20.19	
								0.17	20.78	
								0.18	21.52	
								0.19	22.40	
								0.20	23.19	
								0.21	24.23	
								0.22	25.18	
9		366	Gnezdilov, N.E. and Golubev, I.F.	0.498	0.188	0.314	298.2	0.01	15.24	Same remarks as for curve 1.
								0.02	15.46	
								0.03	15.65	
								0.04	15.90	
								0.05	16.22	
								0.06	16.57	
								0.07	16.84	
								0.08	17.18	
								0.09	17.62	
								0.10	18.00	
								0.11	18.43	
								0.12	18.78	
								0.13	19.27	
								0.14	19.91	
								0.15	20.45	
								0.16	21.12	
								0.17	21.71	
								0.18	22.45	
								0.19	23.33	
								0.20	24.17	
								0.21	25.16	
								0.22	26.11	

TABLE 165-G(D)E.   EXPERIMENTAL VISCOSITY DATA AS A FUNCTION OF DENSITY FOR GASEOUS
HYDROGEN-METHANE-NITROGEN MIXTURES (continued)

Cur. No.	Fig. No.	Ref. No.	Author(s)	Mole Fraction of			Temp. (K)	Density (g cm^{-3})	Viscosity (N s m^{-2} x 10^{-6})	Remarks
				H$_2$	CH$_4$	N$_2$				
10		366	Gnezdilov, N.E. and Golubev, I.F.	0.498	0.188	0.314	323.2	0.01	16.11	Same remarks as for curve 1.
								0.02	16.33	
								0.03	16.52	
								0.04	16.77	
								0.05	17.09	
								0.06	17.44	
								0.07	17.71	
								0.08	18.05	
								0.09	18.49	
								0.10	18.87	
								0.11	19.30	
								0.12	19.65	
								0.13	20.14	
								0.14	20.78	
								0.15	21.32	
								0.16	21.99	
								0.17	22.58	
								0.18	23.32	
								0.19	24.20	
								0.20	24.99	
								0.21	26.03	
								0.22	26.98	
11		366	Gnezdilov, N.E. and Golubev, I.F.	0.498	0.188	0.314	373.2	0.01	17.79	Same remarks as for curve 1.
								0.02	18.01	
								0.03	18.20	
								0.04	18.45	
								0.05	18.77	
								0.06	19.12	
								0.07	19.39	
								0.08	19.73	
								0.09	20.17	
								0.10	20.55	
								0.11	20.98	
								0.12	21.33	
								0.13	21.82	
								0.14	22.46	
								0.15	23.00	
								0.16	23.67	
								0.17	24.25	
								0.18	25.00	
								0.19	25.88	
								0.20	26.67	
								0.21	27.71	
								0.22	28.66	
12		366	Gnezdilov, N.E. and Golubev, I.F.	0.498	0.188	0.314	423.2	0.01	19.36	Same remarks as for curve 1.
								0.02	19.58	
								0.03	19.77	
								0.04	20.02	
								0.05	20.34	
								0.06	20.69	
								0.07	20.96	
								0.08	21.30	
								0.09	21.74	
								0.10	22.12	
								0.11	22.55	
								0.12	22.90	
								0.13	23.39	
								0.14	24.03	
								0.15	24.57	
								0.16	25.24	
								0.17	25.83	
								0.18	26.57	
								0.19	27.45	
								0.20	28.24	
								0.21	29.28	
								0.22	30.23	

TABLE 165-G(D) E.   EXPERIMENTAL VISCOSITY DATA AS A FUNCTION OF DENSITY FOR GASEOUS
HYDROGEN-METHANE-NITROGEN MIXTURES (continued)

Cur. No.	Fig. No.	Ref. No.	Author(s)	Mole Fraction of H_2	CH_4	N_2	Temp. (K)	Density ($g\ cm^{-3}$)	Viscosity ($N\ s\ m^{-2} \times 10^{-6}$)	Remarks
13		366	Gnezdilov, N.E. and Golubev, I.F.	0.498	0.188	0.314	473.2	0.01	20.88	Same remarks as for curve 1.
								0.02	21.10	
								0.03	21.29	
								0.04	21.54	
								0.05	21.86	
								0.06	22.21	
								0.07	22.48	
								0.08	22.82	
								0.09	23.26	
								0.10	23.64	
								0.11	24.07	
								0.12	24.42	
								0.13	24.91	
								0.14	25.55	
								0.15	26.09	
								0.16	26.76	
								0.17	27.34	
								0.18	28.09	
								0.19	28.97	
								0.20	29.76	
								0.21	30.80	
								0.22	31.75	

TABLE 166-G(C)E.   EXPERIMENTAL VISCOSITY DATA AS A FUNCTION OF COMPOSITION FOR GASEOUS
DIMETHYL ETHER-METHYL CHLORIDE-SULFUR DIOXIDE MIXTURES

Cur. No.	Fig. No.	Ref. No.	Author(s)	Temp. (K)	Pressure (atm)	Mole Fraction of			Viscosity (N s m^{-2} x 10^{-6})	Remarks
						$(CH_3)_2O$	CH_3Cl	SO_2		
1		349	Chakraborti, P.K. and Gray, P.	308.2		26.3	25.6	48.1	12.08	$(CH_3)_2O$ and CH_3Cl in gas
						25.5	48.8	25.7	11.45	cylinders, SO_2 in syphons
						33.7	33.5	32.9	11.53	obtained from Matheson
						48.9	25.2	25.9	11.02	Co.; all purified by frac-
										tionation at liquid nitrogen
										temperature; capillary flow
										viscometer calibrated with
										air, Ar, N_2O, and CH_4;
										estimated maximum uncer-
										tainty is ±1.0% and pre-
										cision ±0.4%.
2		349	Chakraborti, P.K. and Gray, P.	353.2		25.3	25.5	49.2	13.86	Same remarks as for
						24.4	49.4	26.2	13.19	curve 1.
						33.3	33.1	33.6	13.26	
						50.1	25.0	24.9	12.69	

6. QUATERNARY SYSTEMS

TABLE 167-G(C)E.   EXPERIMENTAL VISCOSITY DATA AS A FUNCTION OF COMPOSITION FOR GASEOUS
ARGON-HELIUM-CARBON DIOXIDE-METHANE MIXTURES

Cur. No.	Fig. No.	Ref. No.	Author(s)	Temp. (K)	Pressure (atm)	Mole Fraction of				Viscosity (N s m⁻² x 10⁻⁶)	Remarks
						Ar	He	CO_2	CH_4		
1		361	Strunk, M.R. and Fehsenfeld, G.D.	278.2	1.0 1.0	0.1010 0.3823	0.1847 0.3166	0.3820 0.1095	0.3323 0.1916	14.88 18.42	Ar: Matheson Co., specified purity 99.995, chief impurities O_2 and N_2, He: Matheson Co., specified purity 99.9, chief impurities N_2 and CO_2, CO_2: Matheson Co., specified purity 99.8, chief impurities N_2 and O_2, CH_4: Matheson Co., specified purity 99.0, chief impurities CO_2, N_2, ethane, propane; mixtures prepared according to Dalton's law of partial pressures; mixtures analyzed on mass spectrometer; rolling ball viscometer; experimental error ±1.5%.
2		361	Strunk, M.R. and Fehsenfeld, G.D.	323.2	1.0 1.0	0.1010 0.3823	0.1847 0.3166	0.3820 0.1095	0.3323 0.1916	17.70 20.80	Same remarks as for curve 1.
3		361	Strunk, M.R. and Fehsenfeld, G.D.	363.2	1.0 1.0	0.1010 0.3823	0.1847 0.3166	0.3820 0.1095	0.3323 0.1916	18.70 22.72	Same remarks as for curve 1.

TABLE 168-G(T)E.   EXPERIMENTAL VISCOSITY DATA AS A FUNCTION OF TEMPERATURE FOR GASEOUS CARBON DIOXIDE-HYDROGEN-NITROGEN-OXYGEN MIXTURES

Cur. No.	Fig. No.	Ref. No.	Author(s)	CO$_2$	Mole Fraction of H$_2$	N$_2$	O$_2$	Pressure (atm)	Temp. (K)	Viscosity (N s m^{-2} x 10^{-6})	Remarks
1		362	Schmid, C.	0.1080	0.0220	0.8500	0.0200		300.5	18.27	Capillary method; error always less than 4%.
									415.5	23.19	
									524.5	27.15	
									654	31.76	
									814.5	36.65	
									973	41.17	
									1125.5	44.97	
									1279	48.56	

TABLE 169-G(D)E. EXPERIMENTAL VISCOSITY DATA AS A FUNCTION OF DENSITY FOR GASEOUS ETHANE-METHANE-NITROGEN-PROPANE MIXTURES

Cur. No.	Fig. No.	Ref. No.	Author(s)	C_2H_6	Mole Fraction of CH_4	N_2	C_3H_8	Temp. (K)	Density (g cm^{-3})	Viscosity (N s m^{-2} x 10^{-6})	Remarks
1		365	Carr, N.L.	0.257	0.735	0.006	0.002	298.5	0.0083	10.66	Mixtures simulated, all gases well dried, obtained commercially and subjected to spectroscopic analysis; capillary pyrex viscometer of Rankine type enclosed in a special high pressure bomb; maximum experimental error <2% in all cases, <1% in most cases.
									0.0500	11.95	
									0.0933	13.90	
									0.1315	15.91	
									0.1592	17.51	
									0.1811	18.65	
									0.2080	23.09	
2		365	Carr, N.L.	0.257	0.735	0.006	0.002	298.8	0.0084	10.42	Same remarks as for curve 1.
									0.0696	12.81	
									0.0696	12.76	
									0.2372	25.86	
									0.2592	28.77	
									0.2770	31.58	
3		365	Carr, N.L.	0.257	0.735	0.006	0.002	299.0	0.0086	10.69	Same remarks as for curve 1.
									0.2746	31.85	
									0.2976	35.42	
									0.3172	38.70	
									0.3384	42.64	
									0.3474	44.47	
									0.3554	46.20	
									0.3642	48.07	
									0.3741	49.76	
									0.3791	51.18	
									0.3906	54.30	
4		365	Carr, N.L.	0.257	0.735	0.006	0.002	299.5	0.0084	10.69	Same remarks as for curve 1.
									0.0084	10.58	
									0.0208	10.93	
									0.0411	11.44	
									0.0694	12.70	
									0.1202	15.27	
5		365	Carr, N.L.	0.036	0.956	0.003	0.005	302.7	0.0081	11.20	Same remarks as for curve 1.
									0.0991	16.05	
									0.2109	24.26	
									0.2288	26.09	
									0.2385	27.90	
									0.2602	31.23	
									0.2761	34.24	
									0.2870	37.03	
									0.2941	38.23	
6		365	Carr, N.L.	0.036	0.956	0.003	0.005	302.7	0.0112	11.21	Same remarks as for curve 1.
									0.0208	11.45	
									0.0384	12.15	
									0.0529	12.75	
									0.0832	14.22	
7		365	Carr, N.L.	0.036	0.956	0.003	0.005	302.8	0.0676	13.33	Same remarks as for curve 1.
									0.1468	17.91	
									0.1702	19.91	
									0.1934	21.85	
8		365	Carr, N.L.	0.257	0.735	0.006	0.002	338.8	0.0178	12.31	Same remarks as for curve 1.
									0.0412	13.25	
									0.0714	14.37	
									0.0887	15.43	
									0.1050	16.07	
									0.1423	18.21	
									0.1423	18.40	
									0.1719	20.29	
									0.2070	23.19	
									0.2270	24.97	
									0.2615	27.98	
									0.2894	30.97	
9		365	Carr, N.L.	0.036	0.956	0.003	0.005	377.6	0.0038	13.62	Same remarks as for curve 1.
									0.0197	13.92	
									0.0396	14.64	
									0.0582	15.48	
									0.0747	16.65	
									0.1178	18.74	
									0.1508	21.11	
									0.1893	24.80	
									0.2102	27.32	

TABLE 169-G(D)E.   EXPERIMENTAL VISCOSITY DATA AS A FUNCTION OF DENSITY FOR GASEOUS
ETHANE-METHANE-NITROGEN-PROPANE MIXTURES (continued)

Cur. No.	Fig. No.	Ref. No.	Author(s)	Mole Fraction of				Temp. (K)	Density (g cm^{-3})	Viscosity (N s m^{-2} x 10^{-6})	Remarks
				C$_2$H$_6$	CH$_4$	N$_2$	C$_3$H$_8$				
10		365	Carr, N. L.	0.036	0.956	0.003	0.005	397.9	0.0048	14.03	Same remarks as for curve 1.

7. MULTICOMPONENT SYSTEMS

TABLE 170-G(C)E. EXPERIMENTAL VISCOSITY DATA AS A FUNCTION OF COMPOSITION FOR GASEOUS ARGON-HELIUM-AIR-CARBON DIOXIDE MIXTURES

Cur. No.	Fig. No.	Ref. No.	Author(s)	Temp. (K)	Pressure (atm)	Ar	He	Air	CO$_2$	Viscosity (N s m^{-2} x 10^{-6})	Remarks
1		361	Strunk, M.R. and Fehsenfeld, G.D.	278.2	1.0 1.0	0.1875 0.2914	0.0964 0.4038	0.3254 0.2033	0.3907 0.1015	17.02 19.87	Ar: Matheson Co., specified purity 99.995, chief impurities O$_2$ and N$_2$, He: Matheson Co., specified purity 99.9, chief impurities N$_2$ and CO$_2$, Air: Matheson Co., 20.9 O$_2$, 79 N$_2$, 0.1 Ar, no CO$_2$, CO$_2$: Matheson Co., specified purity 99.8, chief impurities N$_2$ and O$_2$; mixtures prepared according to Dalton's law of partial pressures; mixtures analyzed on mass spectrometer; rolling ball viscometer; experimental error ±1.5%.
2		361	Strunk, M.R. and Fehsenfeld, G.D.	323.2	1.0 1.0	0.1875 0.2914	0.0964 0.4038	0.3254 0.2033	0.3907 0.1015	19.30 22.32	Same remarks as for curve 1.
3		361	Strunk, M.R. and Fehsenfeld, G.D.	363.2	1.0 1.0	0.1875 0.2914	0.0964 0.4038	0.3254 0.2033	0.3907 0.1015	21.64 24.48	Same remarks as for curve 1.

TABLE 171-G(C)E.   EXPERIMENTAL VISCOSITY DATA AS A FUNCTION OF COMPOSITION FOR GASEOUS
ARGON-HELIUM-AIR-METHANE MIXTURES

Cur. No	Fig. No.	Ref. No.	Author(s)	Temp. (K)	Pressure (atm)	Mole Fraction of				Viscosity ($N\ s\ m^{-2} \times 10^{-6}$)	Remarks
						Ar	He	Air	CH_4		
1		361	Strunk, M.R. and Fehsenfeld, G.D.	278.2	1.0	0.1869	0.3438	0.1088	0.3605	15.72	Ar: Matheson Co., specified purity 99.995, chief impurities O_2 and N_2, He: Matheson Co., specified purity 99. 9, chief impurities N_2 and CO_2, Air: Matheson Co., specified purity 20.9 O_2, 79 N_2, 0.1 Ar, no CO_2, CH_4: Matheson Co., specified purity 99.0, chief impurities CO_2, N_2, ethane, propane; mixtures analyzed on mass spectrometer; rolling ball viscometer; experimental error ±1.5%.
					1.0	0.2922	0.1948	0.4014	0.1116	18.69	
2		361	Strunk, M.R. and Fehsenfeld, G.D.	323.2	1.0	0.1869	0.3438	0.1088	0.3605	17.77	Same remarks as for curve 1.
					1.0	0.2922	0.1948	0.4014	0.1116	20.83	
3		361	Strunk, M.R. and Fehsenfeld, G.D.	363.2	1.0	0.1869	0.3438	0.1088	0.3605	19.63	Same remarks as for curve 1.
					1.0	0.2922	0.1948	0.4014	0.1116	22.92	

TABLE 172-G(C)E.   EXPERIMENTAL VISCOSITY DATA AS A FUNCTION OF COMPOSITION FOR GASEOUS ARGON-AIR-CARBON DIOXIDE MIXTURES

Cur. No.	Fig. No.	Ref. No.	Author(s)	Temp. (K)	Pressure (atm)	Mole Fraction of			Viscosity ($N \ s \ m^{-2} \times 10^{-6}$)	Remarks
						Ar	Air	CO_2		
1		361	Strunk, M.R. and Fehsenfeld, G.D.	278.2	1.0 1.0	0.4748 0.1915	0.3194 0.5225	0.2058 0.2860	18.78 17.44	Ar: Matheson Co., specified purity 99.995, chief impurities O_2 and N_2, Air: Matheson Co., 20.9 O_2, 79 N_2, 0.1 Ar, no CO_2, CO_2: Matheson Co., specified purity 98.8, chief impurities N_2 and O_2; mixtures prepared according to Dalton's law of partial pressures; mixtures analyzed on mass spectrometer; rolling ball viscometer; experimental error ±1.5%.
2		361	Strunk, M.R. and Fehsenfeld, G.D.	323.2	1.0 1.0	0.4748 0.1915	0.3194 0.5225	0.2058 0.2860	21.18 19.62	Same remarks as for curve 1.
3		361	Strunk, M.R. and Fehsenfeld, G.D.	363.2	1.0 1.0	0.4748 0.1915	0.3194 0.5225	0.2058 0.2860	23.53 21.80	Same remarks as for curve 1.

TABLE 173-G(C)E.   EXPERIMENTAL VISCOSITY DATA AS A FUNCTION OF COMPOSITION FOR GASEOUS ARGON-AIR-CARBON DIOXIDE-METHANE MIXTURES

Cur. No.	Fig. No.	Ref. No.	Author(s)	Temp. (K)	Pressure (atm)	Mole Fraction of				Viscosity $(N\ s\ m^{-2} \times 10^{-6})$	Remarks
						Ar	Air	CO_2	CH_4		
1		361	Strunk, M.R. and Fehsenfeld, G.C.	278.2	1.0 1.0	0.1026 0.3538	0.2242 0.3215	0.2966 0.2080	0.3766 0.1167	14.42 17.58	Ar: Matheson Co., specified purity 99.995, chief impurities O_2 and N_2, Air: Matheson Co., 20.9 O_2, 79 N_2, 0.1 Ar, no CO_2, CO_2: Matheson Co., specified purity 98.8, chief impurities N_2 and O_2, CH_4: Matheson Co., specified purity 99.0, chief impurities CO_2, N_2, ethane, propane; mixtures prepared according to Dalton's law of partial pressures; mixtures analyzed on mass spectrometer; rolling ball viscometer; experimental error ± 1.5%.
2		361	Strunk, M.R. and Fehsenfeld, G.C.	323.2	1.0 1.0	0.1026 0.3538	0.2242 0.3215	0.2966 0.2080	0.3766 0.1167	16.44 20.18	Same remarks as for curve 1.
3		361	Strunk, M.R. and Fehsenfeld, G.C.	363.2	1.0 1.0	0.1026 0.3538	0.2242 0.3215	0.2966 0.2080	0.3766 0.1167	18.31 22.10	Same remarks as for curve 1.

TABLE 174-G(C)E.   EXPERIMENTAL VISCOSITY DATA AS A FUNCTION OF COMPOSITION FOR GASEOUS HELIUM–AIR–CARBON DIOXIDE MIXTURES

Cur. No.	Fig. No.	Ref. No.	Author(s)	Temp. (K)	Pressure (atm)	Mole Fraction of			Viscosity (N s m^{-2} x 10^{-6})	Remarks
						He	Air	CO$_2$		
1		361	Strunk, M.R. and Fehsenfeld, G.D.	278.2	1.0 1.0	0.1714 0.4697	0.2353 0.3784	0.5933 0.1519	15.77 18.36	He: Matheson Co., specified purity 99.995, chief impurities O$_2$ and N$_2$, Air: Matheson Co., specified purity 20.9 O$_2$, 79 N$_2$, 0.1 Ar, no CO$_2$, CO$_2$: Matheson Co., specified purity 99.8, chief impurities N$_2$ and O$_2$; mixtures prepared according to Dalton's law of partial pressures; mixtures analyzed on mass spectrometer; rolling ball viscometer; experimental error ±1.5%.
2		361	Strunk, M.R. and Fehsenfeld, G.D.	323.2	1.0 1.0	0.1714 0.4697	0.2353 0.3784	0.5933 0.1519	17.86 20.40	Same remarks as for curve 1.
3		361	Strunk, M.R. and Fehsenfeld, G.D.	363.2	1.0 1.0	0.1714 0.4697	0.2353 0.3784	0.5933 0.1519	19.62 21.98	Same remarks as for curve 1.

TABLE 175-G(C)E. EXPERIMENTAL VISCOSITY DATA AS A FUNCTION OF COMPOSITION FOR GASEOUS HELIUM-AIR-CARBON DIOXIDE-METHANE MIXTURES

Cur. No.	Fig. No.	Ref. No.	Author(s)	Temp. (K)	Pressure (atm)	Mole Fraction of				Viscosity $(N\ s\ m^{-2} \times 10^{-6})$	Remarks
---	---	---	---	---	---	He	Air	CO_2	CH_4		
1		361	Strunk, M.R. and Fehsenfeld, G.C.	278.2	1.0 1.0	0.3992 0.0977	0.1183 0.4085	0.1869 0.2867	0.2956 0.2071	15.09 15.34	He: Matheson Co., specified purity 99.995, chief impurities O_2 and N_2, Air: Matheson Co., specified purity 20.9 O_2, 79 N_2, 0.1 Ar, no CO_2, CO_2: Matheson Co., specified purity 99.8, chief impurities N_2 and O_2, CH_4: Matheson Co., specified purity 99.0, chief impurities CO_2, N_2, ethane, propane; mixtures prepared according to Dalton's law of partial pressures; mixtures analyzed on mass spectrometer; rolling ball viscometer; experimental error ± 1.5%.
2		361	Strunk, M.R. and Fehsenfeld, G.C.	323.2	1.0 1.0	0.3992 0.0977	0.1183 0.4085	0.1869 0.2867	0.2956 0.2071	17.13 17.53	Same remarks as for curve 1.
3		361	Strunk, M.R. and Fehsenfeld, G.C.	363.2	1.0 1.0	0.3992 0.0977	0.1183 0.4085	0.1869 0.2867	0.2956 0.2071	18.66 19.06	Same remarks as for curve 1.

TABLE 176-G(C)E.   EXPERIMENTAL VISCOSITY DATA AS A FUNCTION OF COMPOSITION FOR GASEOUS HELIUM-AIR-METHANE MIXTURES

Cur. No.	Fig. No.	Ref. No.	Author(s)	Temp. (K)	Pressure (atm)	Mole Fraction of He	Air	CH_4	Viscosity (N s m^{-2} x 10^{-6})	Remarks
1		361	Strunk, M.R. and Fehsenfeld, G.D.	278.2	1.0 1.0	0.6055 0.2281	0.1675 0.6173	0.2270 0.1546	15.97 16.67	He: Matheson Co., specified purity 99.995, chief impurities O_2 and N_2, Air: Matheson Co., specified purity 20.9 O_2, 79 N_2, 0.1 Ar, no CO_2, CH_4: Matheson Co., specified purity 99.0, chief impurity CO_2, N_2, ethane, propane; mixtures prepared according to Dalton's law of partial pressures; mixtures analyzed on mass spectrometer; rolling ball viscometer; experimental error ± 1.5%.
2		361	Strunk, M.R. and Fehsenfeld, G.D.	323.2	1.0 1.0	0.6055 0.2281	0.1675 0.6173	0.2270 0.1546	17.80 18.51	Same remarks as for curve 1.
3		361	Strunk, M.R. and Fehsenfeld, G.D.	363.2	1.0 1.0	0.6055 0.2281	0.1675 0.6173	0.2270 0.1546	19.33 20.36	Same remarks as for curve 1.

TABLE 177-G(D)E.   EXPERIMENTAL VISCOSITY DATA AS A FUNCTION OF DENSITY FOR GASEOUS HELIUM-n-BUTANE-ETHANE-METHANE-NITROGEN-PROPANE-i-BUTANE MIXTURES

Cur. No.	Fig. No.	Ref. No.	Author(s)	Mole Fraction	Temp. (K)	Density (g cm^{-3})	Viscosity (N s m^{-2} x 10^{-6})	Remarks
1		365	Carr, N. L.	See footnote	299.7	0.0068	12.00	Mixtures simulated, all gases well dried, obtained commercially and subjected to spectroscopic analysis; capillary pyrex viscometer of Rankine type enclosed in a special high pressure bomb; maximum experimental error <2% in all cases, <1% in most cases.
						0.0190	12.36	
						0.0374	12.99	
						0.0637	13.89	
						0.0655	13.95	
						0.1010	15.56	
						0.1385	17.59	
						0.1743	19.87	
2		365	Carr, N. L.	See footnote	301.2	0.0068	11.99	Same remarks as for curve 1.
						0.1778	20.09	
						0.2042	22.26	
						0.2608	26.60	
						0.2914	30.60	
						0.3178	34.32	
						0.3453	37.64	
						0.3627	42.08	
						0.3764	45.17	
3		365	Carr, N. L.	See footnote	338.9	0.0066	13.30	Same remarks as for curve 1.
						0.0197	13.67	
						0.0240	13.97	
						0.1109	15.86	
						0.1375	17.19	
4		365	Carr, N. L.	See footnote	338.9	0.0070	13.40	Same remarks as for curve 1.
						0.0527	14.75	
						0.1368	17.37	
						0.1941	20.38	
						0.2200	23.54	
						0.2533	26.60	
						0.2711	29.28	
						0.2927	31.90	
						0.3318	36.17	

Mole Fractions: 0.008 He, 0.006 n-C_4H_{10}, 0.061 C_2H_6, 0.731 CH_4, 0.158 N_2, 0.034 C_3H_8, and 0.002 i-C_4H_{10}.

TABLE 178-L(T). RECOMMENDED VISCOSITY VALUES FOR LIQUID AIR

DISCUSSION

SATURATED LIQUID

Three sets of experimental data were found in the literature. They are those of Rudenko[188], Naiki[158] and Verschaffelt[256]. Only the data of Rudenko covers a substantial temperature range. They were fitted to an equation

$$\log \mu = A + B/T$$

The accuracy of the data is poor.

From 110 K to the critical temperature, recommended values were generated by drawing a smooth curve joining the value of viscosity at 110 K and the value at the critical point determined by the method of Jossi, Stiel and Thodos [100]. The accuracy of the correlation is of about ±15%.

RECOMMENDED VALUES

[Temperature, T, K; Viscosity, μ, 10^{-3} N s m^{-2}]

SATURATED LIQUID

T	μ
60	0.325
65	0.264
70	0.221
75	0.189
80	0.165
85	0.147
90	0.132
95	0.120
100	0.1101
105	0.1019
110	0.0949
115	0.0865
120	0.0750
125	0.0615
130	0.0420
133*	0.0207

* Crit. Temp.

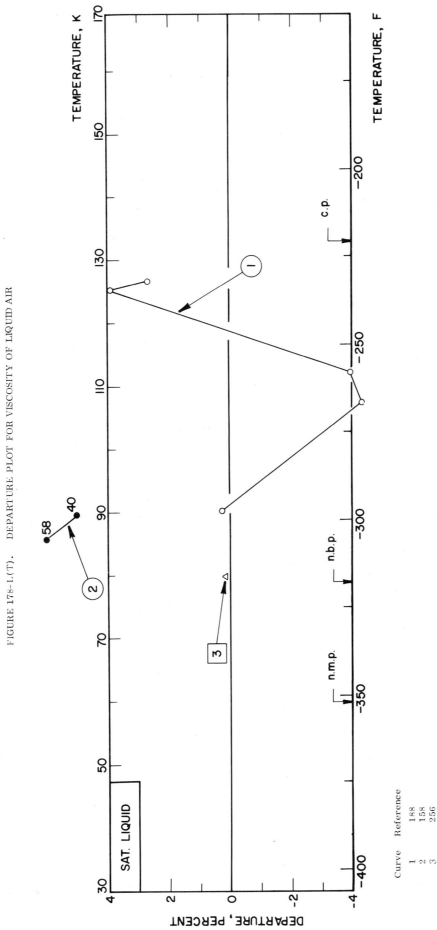

FIGURE 178-L(T).   DEPARTURE PLOT FOR VISCOSITY OF LIQUID AIR

TABLE 178-V(T). RECOMMENDED VISCOSITY VALUES FOR AIR VAPOR

DISCUSSION

RECOMMENDED VALUES

[Temperature, T, K; Viscosity, μ, 10^{-3} N s m^{-2}]

SATURATED VAPOR

SATURATED VAPOR

Recommended values of the viscosity of the saturated vapor were computed by the correlation technique of Jossi, Stiel and Thodos [100] using the recommended values of viscosity of the gas at 1 atmosphere and the density values given by Din [49]. The accuracy is of about ±5%.

T	μ
80	0.0055
85	0.0060
90	0.0065
95	0.0070
100	0.0075
105	0.0080
110	0.0086
115	0.0093
120	0.0102
125	0.0117
130	0.0143
133*	0.0207

* Crit. Temp.

TABLE 178-G(T).   RECOMMENDED VISCOSITY VALUES FOR GASEOUS AIR

RECOMMENDED VALUES

[Temperature, T, K; Viscosity, μ, 10^{-6} N s m^{-2}]

GAS

T	μ	T	μ	T	μ
		450	24.93	850	37.83
		460	25.32	860	38.10
		470	25.70	870	38.37
80	5.52	480	26.07	880	38.64
90	6.35	490	26.45	890	38.91
100	7.06	500	26.82	900	39.18
110	7.75	510	27.18	910	39.45
120	8.43	520	27.54	920	39.71
130	9.09	530	27.90	930	39.97
140	9.74	540	28.25	940	40.23
150	10.38	550	28.60	950	40.49
160	11.00	560	28.95	960	40.75
170	11.61	570	29.29	970	41.00
180	12.20	580	29.69	980	41.26
190	12.79	590	29.97	990	41.52
200	13.36	600	30.30	1000	41.77
210	13.92	610	30.63	1050	43.0
220	14.47	620	30.96	1100	44.2
230	15.01	630	31.28	1150	45.4
240	15.54	640	31.61	1200	46.5
250	16.06	650	31.93	1250	47.7
260	16.57	660	32.24	1300	48.8
270	17.07	670	32.56	1350	49.9
280	17.57	680	32.87	1400	50.9
290	18.05	690	33.18	1450	51.9
300	18.53	700	33.49	1500	53.0
310	19.00	710	33.79	1550	54.0
320	19.46	720	34.09	1600	54.9
330	19.92	730	34.39	1650	55.9
340	20.37	740	34.69	1700	56.9
350	20.81	750	34.98	1750	57.8
360	21.25	760	35.28	1800	58.7
370	21.68	770	35.57	1850	59.6
380	22.11	780	35.86	1900	60.5
390	22.52	790	36.15	1950	61.4
400	22.94	800	36.43	2000	62.3
410	23.35	810	36.72		
420	23.75	820	36.99		
430	24.15	830	37.27		
440	24.54	840	37.55		

DISCUSSION

GAS

The literature revealed a rather abundant experimental work on air: 51 sets of data were found. Most of them are in the normal range of temperature close to room temperature. At high temperature the results of Vasilesco [254-5] and Trautz [226, 228, 229, 231] appears the most consistent, while at low temperature, the data of Fortier [66], Johnson[97-8]and of Filipova [62] are in good agreement. At about 20 C, very precise measurements of Bond [17], Bearden [8], Rigden[183], Kellström[104-6] and Majumdar[146-7] were selected, and the curve was forced to fit these data.

The theoretical expression for viscosity:

$$\mu = \frac{K\sqrt{T}}{\sigma^2 \Omega} \tag{1}$$

was used to obtain $\sigma^2\Omega$ from the experimental data. A plot of $\sigma^2\Omega$ versus $1/T$ revealed that the equation:

$$\sigma^2\Omega = A + B/T\, e^{-C/T} \tag{2}$$

proposed by Keyes [121] was able to represent the data, which were least square fitted to such an equation. Values obtained from this equation (2) were used in equation (1) to generate the table of recommended values, which are thought to be accurate to ±2% in the whole range covered.

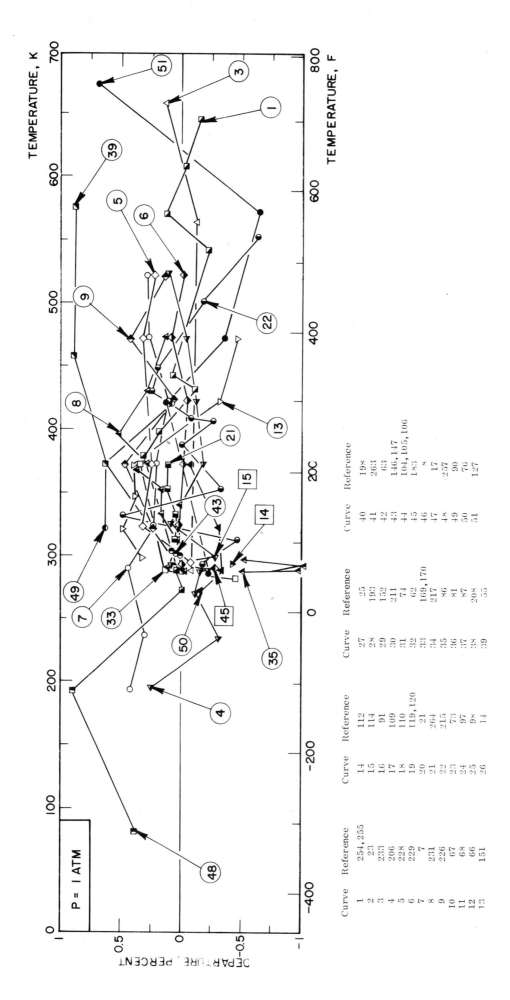

FIGURE 178-G(T).   DEPARTURE PLOT FOR VISCOSITY OF GASEOUS AIR

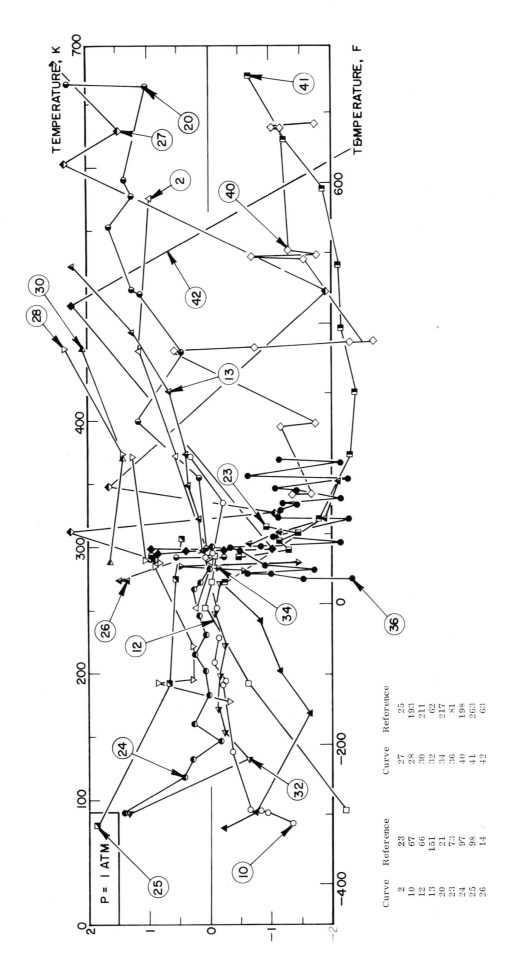

FIGURE 178-G(T).   DEPARTURE PLOT FOR VISCOSITY OF GASEOUS AIR (continued)

Curve	Reference		Curve	Reference
2	23		27	25
10	67		28	193
12	66		30	211
13	151		32	62
20	21		34	217
23	73		36	81
24	97		40	198
25	98		41	263
26	14		42	63

TABLE 179-G(C)E.   EXPERIMENTAL VISCOSITY DATA AS A FUNCTION OF COMPOSITION FOR GASEOUS
AIR-CARBON DIOXIDE MIXTURES

Cur. No.	Fig. No.	Ref. No.	Author(s)	Temp. (K)	Pressure (atm)	Mole Fraction of CO_2	Viscosity $(N\ s\ m^{-2} \times 10^{-6})$	Remarks
1	179-G(C)	346	Jung, G. and Schmick, H.	290		1.000	14.55	Effusion method of Trautz and Weizel; $L_1 = 0.042\%$, $L_2 = 0.076\%$, $L_3 = 0.162\%$.
						0.800	15.23	
						0.600	15.91	
						0.400	16.60	
						0.200	17.30	
						0.000	17.97	

TABLE 179-G(S).   SMOOTHED VISCOSITY VALUES AS A FUNCTION OF COMPOSITION FOR GASEOUS
AIR-CARBON DIOXIDE MIXTURES

Mole Fraction of CO_2	(290.0 K) [Ref. 346]
1.00	14.55
0.95	14.71
0.90	14.88
0.85	15.06
0.80	15.23
0.75	15.40
0.70	15.52
0.65	15.73
0.60	15.91
0.55	16.08
0.50	16.24
0.45	16.42
0.40	16.59
0.35	16.76
0.30	16.93
0.25	17.10
0.20	17.27
0.15	17.45
0.10	17.62
0.05	17.80
0.00	17.97

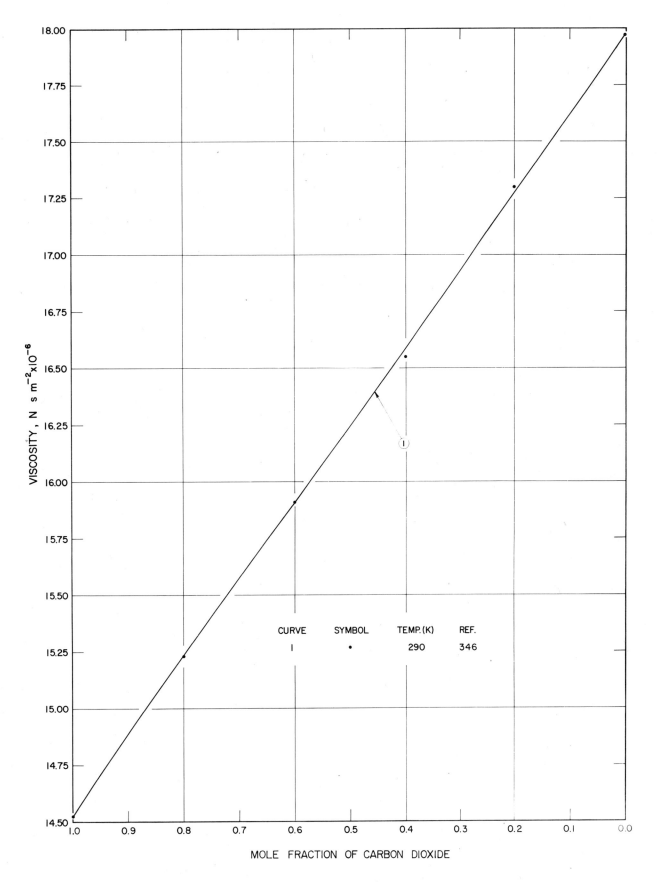

FIGURE 179-G(C). VISCOSITY DATA AS A FUNCTION OF COMPOSITION
FOR GASEOUS AIR-CARBON DIOXIDE MIXTURES

TABLE 180-G(C)E.   EXPERIMENTAL VISCOSITY DATA AS A FUNCTION OF COMPOSITION FOR GASEOUS AIR-CARBON DIOXIDE-METHANE MIXTURES

Cur. No.	Fig. No.	Ref. No.	Author(s)	Temp. (K)	Pressure (atm)	Mole Fraction of			Viscosity (N s m^{-2} x 10^{-6})	Remarks
						Air	CO$_2$	CH$_4$		
1		361	Strunk, M.R. and Fehsenfeld, G.D.	278.2	1.0 1.0	0.5022 0.2212	0.1195 0.5270	0.3783 0.2518	14.52 14.28	Air: Matheson Co., 20.9 O$_2$, 79 N$_2$, 0.1 Ar, no CO$_2$; CO$_2$: Matheson Co., specified purity 99.8%, chief impurities N$_2$ and O$_2$; CH$_4$: Matheson Co., specified purity 99.0%, chief impurities CO$_2$, N$_2$, ethane, propane; mixtures prepared according to Dalton's law of partial pressures; mixtures analyzed on mass spectrometer; rolling ball viscometer; experimental error ±1.5%.
2		361	Strunk, M.R. and Fehsenfeld, G.D.	323.2	1.0 1.0	0.5022 0.2212	0.1195 0.5270	0.3783 0.2518	16.52 16.34	Same remarks as for curve 1.
3		361	Strunk, M.R. and Fehsenfeld, G.D.	363.2	1.0 1.0	0.5022 0.2212	0.1195 0.5270	0.3783 0.2518	17.92 17.94	Same remarks as for curve 1.

TABLE 181-G(C)E.   EXPERIMENTAL VISCOSITY DATA AS A FUNCTION OF COMPOSITION FOR GASEOUS
AIR-METHANE MIXTURES

Cur. No.	Fig. No.	Ref. No.	Author(s)	Temp. (K)	Pressure (mm Hg)	Mole Fraction of Air	Viscosity (N s m^{-2} x 10^{-6})	Remarks
1	181-G(C)	334	Strauss, W.A. and Edse, R.	293.2	756.5	1.000	17.95	Capillary flow viscometer, relative measurements; L_1 = 0.397%, L_2 = 0.695%, L_3 = 2.007%.
					757.0	0.902	17.37	
					757.2	0.804	16.77	
					757.8	0.713	16.24	
					758.4	0.609	15.70	
					758.6	0.505	15.08	
					758.6	0.405	14.39	
					759.5	0.302	13.78	
					758.8	0.199	13.01	
					756.9	0.109	12.04	
					755.8	0.000	11.21	
2	181-G(C)	334	Strauss, W.A. and Edse, R.	293.2	755.7	0.000	11.09	Same remarks as for curve 1 except L_1 = 0.691%, L_2 = 1.612%, L_3 = 5.451%.
					756.3	0.045	11.47	
					757.8	0.150	12.39	
					757.9	0.253	13.34	
					759.2	0.354	13.99	
					758.8	0.441	14.55	
					759.1	0.559	16.25	
					758.3	0.654	15.97	
					757.6	0.749	16.54	
					757.5	0.854	17.05	
					756.5	0.949	17.63	
					756.2	1.000	17.96	
3	181-G(C)	334	Strauss, W.A. and Edse, R.	293.2	749.4	0.000	11.29	Same remarks as for curve 1 except L_1 = 0.526%, L_2 = 1.108%, L_3 = 3.516%.
					751.2	0.106	12.15	
					752.4	0.199	13.06	
					753.1	0.306	13.87	
					752.3	0.384	14.84	
					752.5	0.505	15.11	
					752.1	0.601	15.66	
					750.4	0.699	16.19	
					750.9	0.798	16.72	
					750.4	0.901	17.32	
					751.7	1.000	17.84	
4	181-G(C)	334	Strauss, W.A. and	293.2	750.0	1.000	17.97	Same remarks as for curve 1 except L_1 = 0.256%, L_2 = 0.353%, L_3 = 0.865%.
					750.1	0.946	17.58	
					750.7	0.852	17.06	
					751.3	0.747	16.51	
					752.0	0.636	15.91	
					752.3	0.553	15.33	
					752.5	0.442	14.67	
					752.9	0.348	14.09	
					753.6	0.252	13.53	
					751.7	0.152	12.59	
					750.1	0.048	11.69	
					749.5	0.000	11.28	

TABLE 181-G(C)S.   SMOOTHED VISCOSITY VALUES AS A FUNCTION OF COMPOSITION FOR GASEOUS
AIR-METHANE MIXTURES

Mole Fraction of Air	293.2 K [Ref. 334]
0.00	11.20
0.05	11.58
0.10	12.10
0.15	12.44
0.20	12.85
0.25	13.34
0.30	13.65
0.35	14.02
0.40	14.37
0.45	15.70
0.50	15.02
0.55	15.33
0.60	15.63
0.65	15.92
0.70	16.22
0.75	16.50
0.80	16.79
0.85	17.08
0.90	17.35
0.95	17.64
1.00	17.91

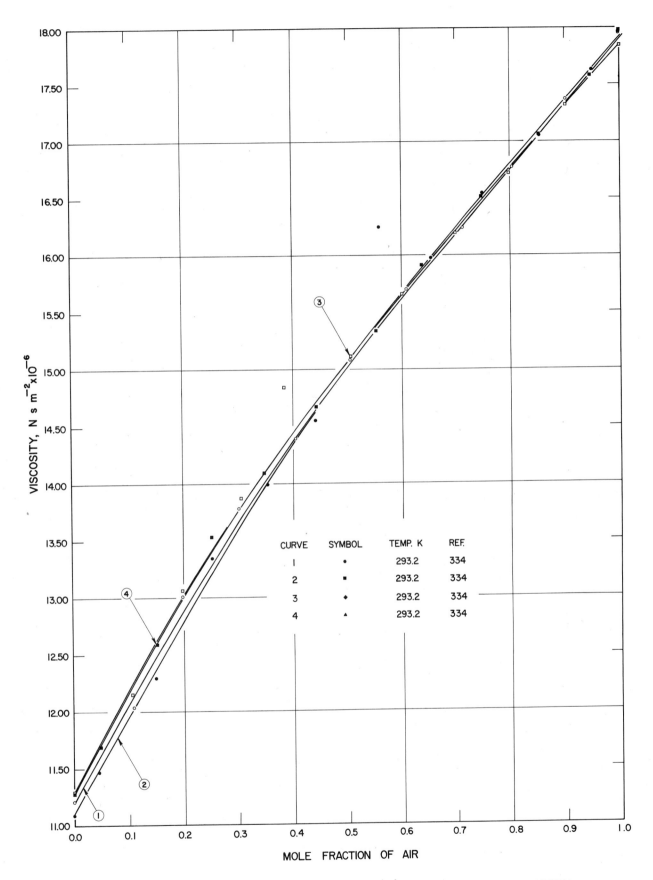

FIGURE 181-G(C). VISCOSITY DATA AS A FUNCTION OF COMPOSITION
FOR GASEOUS AIR-METHANE MIXTURES

TABLE 182-G(C)E.   EXPERIMENTAL VISCOSITY DATA AS A FUNCTION OF COMPOSITION FOR GASEOUS
CARBON DIOXIDE–CARBON MONOXIDE–HYDROGEN–METHANE–NITROGEN MIXTURES

Cur. No.	Fig. No.	Ref. No.	Author(s)	Temp. (K)	Mole Fraction of					Viscosity ($N \ s \ m^{-2} \times 10^{-6}$)	Remarks
					CO_2	CO	H_2	CH_4	N_2		
1		363	Herning, F. and Zipperer, L.	293.2	0.106	0.298	0.039	0.003	0.554	17.43	Capillary method.
					0.089	0.307	0.033	0.004	0.567	17.47	
					0.087	0.328	0.015	0.002	0.568	17.49	

TABLE 183-G(T)E. EXPERIMENTAL VISCOSITY DATA AS A FUNCTION OF TEMPERATURE FOR GASEOUS
CARBON DIOXIDE-CARBON MONOXIDE-HYDROGEN-METHANE-NITROGEN-OXYGEN MIXTURES

Cur. No.	Fig. No.	Ref. No.	Author(s)	Mole Fraction of						Temp. (K)	Viscosity $(N \ s \ m^{-2} \times 10^{-6})$	Remarks
				CO_2	CO	H_2	CH_4	N_2	O_2			
1		362	Schmid, C.	0.037	0.271	0.095	0.016	0.578	0.003	300.5	18.15	Capillary method; error always less than 4%.
										366.5	21.00	
										477	25.11	
										565.5	28.19	
										676.5	31.97	
										776	34.99	
										866	37.55	
										981	40.45	
										1070	42.86	
										1176	45.35	
										1282	47.92	

TABLE 184-G(C)E.　EXPERIMENTAL VISCOSITY DATA AS A FUNCTION OF COMPOSITION FOR GASEOUS CARBON DIOXIDE-CARBON MONOXIDE-HYDROGEN-METHANE-NITROGEN-OXYGEN – HEAVIER HYDROCARBONS MIXTURES

Cur. No.	Fig. No.	Ref. No.	Author(s)	Temp. (K)	*Mole Fraction of							Viscosity (N s m^{-2} x 10^{-6})
					CO_2	CO	H_2	CH_4	N_2	O_2	H.H.	
1		363	Herning, F. and Zipperer, L.	293.0	0.017	0.060	0.575	0.240	0.078	0.009	0.021	12.62
					0.021	0.057	0.530	0.243	0.117	0.009	0.023	13.04
					0.020	0.046	0.549	0.235	0.116	0.014	0.020	13.10
					0.033	0.038	0.513	0.296	0.100	0.006	0.014	13.22
					0.022	0.041	0.531	0.295	0.092	0.006	0.013	13.06
					0.022	0.040	0.523	0.299	0.094	0.010	0.012	13.07
					0.025	0.149	0.530	0.181	0.091	0.008	0.016	13.55
					0.048	0.264	0.172	0.026	0.482	0.003	0.005	17.14
					0.035	0.273	0.144	0.037	0.500	0.003	0.008	17.12
					0.031	0.286	0.177	0.042	0.450	0.005	0.009	17.15

*Capillary method used to obtain values.

TABLE 185-G(T)E.   EXPERIMENTAL VISCOSITY DATA AS A FUNCTION OF TEMPERATURE FOR GASEOUS
CARBON DIOXIDE–CARBON MONOXIDE–HYDROGEN–NITROGEN–OXYGEN MIXTURES

Cur. No.	Fig. No.	Ref. No.	Author(s)	Mole Fraction of					Temp. (K)	Viscosity ($N \ s \ m^{-2} \times 10^{-6}$)	Remarks
				CO_2	CO	H_2	N_2	O_2			
1		362	Schmid, C.	0.067	0.078	0.022	0.832	0.001	307.5	18.42	Capillary method: error always less than 4%.
									417	22.95	
									519	26.55	
									668	31.69	
									815	36.03	
									975	40.48	
									1116	44.01	
									1285	48.08	
2		362	Schmid, C.	0.064	0.003	0.007	0.890	0.030	314	19.04	Same remarks as for curve 1.
									368	21.44	
									518	27.06	
									695	33.30	
									820	37.02	
									974.5	41.13	
									1126	44.85	
									1287	48.95	
3		362	Schmid, C.	0.060	0.257	0.115	0.567	0.001	302	18.23	Same remarks as for curve 1.
									439	23.82	
									526	26.86	
									653	31.14	
									819	36.27	
									976	40.41	
									1126.3	44.00	
									1283	47.77	

TABLE 186-G(C)E.   EXPERIMENTAL VISCOSITY DATA AS A FUNCTION OF COMPOSITION FOR GASEOUS
AIR-AMMONIA MIXTURES

Cur. No.	Fig. No.	Ref. No.	Author(s)	Temp. (K)	Pressure (atm)	Mole Fraction of NH_3	Viscosity (N s m^{-2} x 10^{-6})	Remarks
1	186-G(C)	346	Jung, G. and Schmick, H.	288.7		0.000	9.88	Effusion method of Trautz and Weizel; $L_1 = 0.264\%$, $L_2 = 0.571\%$, $L_3 = 1.820\%$.
						0.100	11.00	
						0.200	12.03	
						0.300	13.06	
						0.400	14.03	
						0.500	14.92	
						0.600	15.75	
						0.700	16.18	
						0.800	17.13	
						0.900	17.64	
						1.000	18.10	

TABLE 186-G(C)S.   SMOOTHED VISCOSITY VALUES AS A FUNCTION OF COMPOSITION FOR GASEOUS
AIR-AMMONIA MIXTURES

Mole Fraction of NH_3	(288.7 K) [Ref. 346]
0.00	9.88
0.05	10.46
0.10	11.02
0.15	11.56
0.20	12.08
0.25	12.58
0.30	13.09
0.35	13.58
0.40	14.04
0.45	14.50
0.50	14.94
0.55	15.36
0.60	15.75
0.65	16.13
0.70	16.48
0.75	16.81
0.80	17.13
0.85	17.40
0.90	17.66
0.95	17.89
1.00	18.10

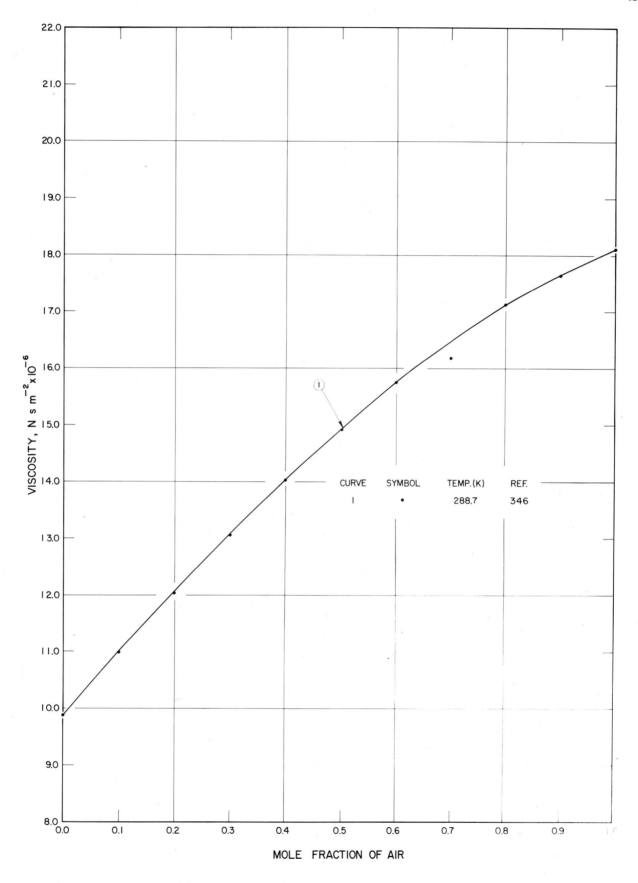

FIGURE 186-G(C). VISCOSITY DATA AS A FUNCTION OF COMPOSITION
FOR GASEOUS AIR-AMMONIA MIXTURES

TABLE 187-G(C)E.   EXPERIMENTAL VISCOSITY DATA AS A FUNCTION OF COMPOSITION FOR GASEOUS
AIR-HYDROGEN CHLORIDE MIXTURES

Cur. No.	Fig. No.	Ref. No.	Author(s)	Temp. (K)	Pressure (atm)	Mole Fraction of HCl	Viscosity (N s m^{-2} x 10^{-6})	Remarks
1	187-G(C)	346	Jung, G. and Schmick, H.	289.7		1.000	14.26	Effusion method of Trautz and
						0.900	14.89	Weizel; L$_1$ = 0.079%, L$_2$ = 0.109%,
						0.800	15.45	L$_3$ = 0.260%.
						0.700	15.92	
						0.600	16.38	
						0.500	16.78	
						0.400	17.15	
						0.300	17.49	
						0.200	17.78	
						0.100	18.00	
						0.000	18.18	
2	187-G(C)	346	Jung, G. and Schmick, H.	291.3		1.000	14.07	Same remarks as for curve 1 except
						0.800	15.35	L$_1$ = 0.169%, L$_2$ = 0.377%, L$_3$ =
						0.600	16.16	0.920%.
						0.400	16.93	
						0.200	17.55	
						0.000	17.94	

TABLE 187-G(C)S.   SMOOTHED VISCOSITY VALUES AS A FUNCTION OF COMPOSITION FOR GASEOUS
AIR-HYDROGEN CHLORIDE MIXTURES

Mole Fraction of HCl	(289.7 K) [Ref. 346]
0.00	18.19
0.05	18.11
0.10	18.01
0.15	17.90
0.20	17.77
0.25	17.64
0.30	17.49
0.35	17.34
0.40	17.17
0.45	16.99
0.50	16.80
0.55	16.40
0.60	16.38
0.65	16.26
0.70	15.92
0.75	15.67
0.80	15.41
0.85	15.15
0.90	14.87
0.95	14.58
1.00	14.27

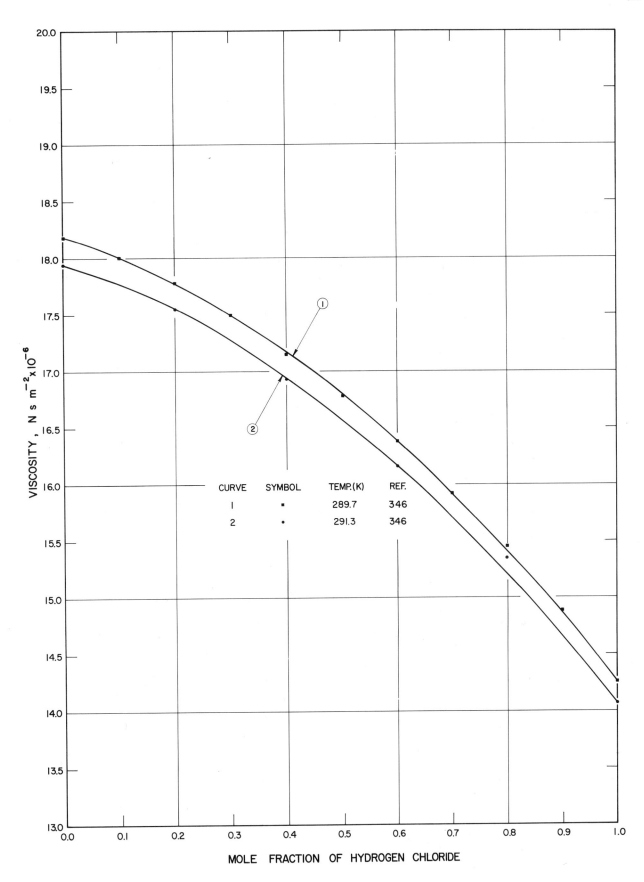

FIGURE 187-G(C). VISCOSITY DATA AS A FUNCTION OF COMPOSITION
FOR GASEOUS AIR-HYDROGEN CHLORIDE MIXTURES

TABLE 188-G(C)E.   EXPERIMENTAL VISCOSITY DATA AS A FUNCTION OF COMPOSITION FOR GASEOUS
AIR-HYDROGEN SULFIDE MIXTURES

Cur. No.	Fig. No.	Ref. No.	Author(s)	Temp. (K)	Pressure (atm)	Mole Fraction of H_2S	Viscosity (N s m^{-2} x 10^{-6})	Remarks
1	188-G(C)	346	Jung, G. and Schmick, H.	290.36		1.000	12.60	Effusion method of Trautz and Weizel; L_1 = 0.108%, L_2 = 0.167%, L_3 = 0.339%.
						0.900	13.31	
						0.800	14.03	
						0.700	14.69	
						0.600	15.35	
						0.500	16.03	
						0.400	16.55	
						0.300	17.09	
						0.200	17.55	
						0.100	17.95	
						0.000	18.27	

TABLE 188-G(C)S.   SMOOTHED VISCOSITY VALUES AS A FUNCTION OF COMPOSITION FOR GASEOUS
AIR-HYDROGEN SULFIDE MIXTURES

Mole Fraction of H_2S	(290.4 K) [Ref. 346]
0.00	18.27
0.05	18.13
0.10	19.95
0.15	17.74
0.20	17.53
0.25	17.30
0.30	17.06
0.35	16.81
0.40	16.55
0.45	16.28
0.50	16.00
0.55	15.70
0.60	15.40
0.65	15.07
0.70	14.74
0.75	14.39
0.80	14.03
0.85	13.68
0.90	13.31
0.95	12.95
1.00	12.60

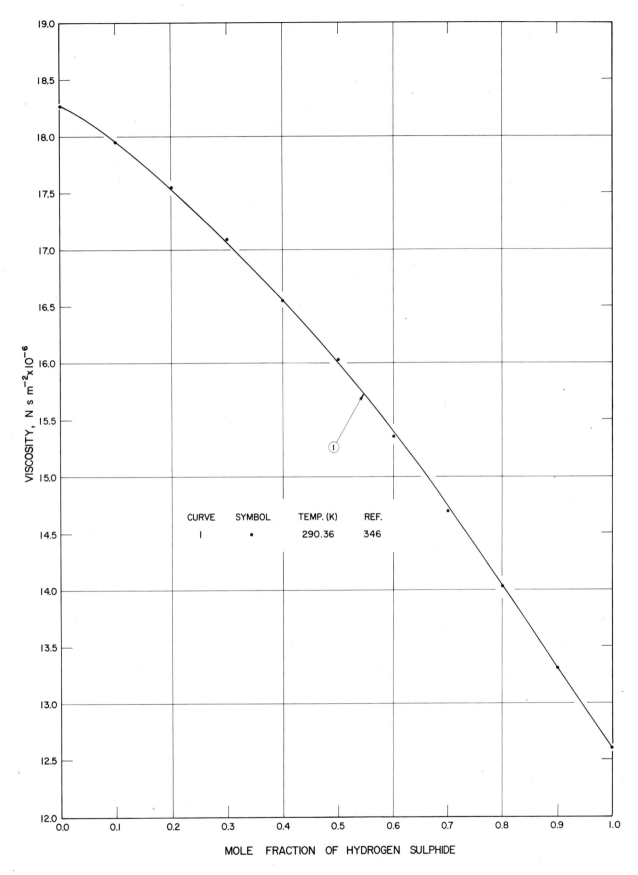

FIGURE 188-G(C). VISCOSITY DATA AS A FUNCTION OF COMPOSITION
FOR GASEOUS AIR–HYDROGEN SULFIDE MIXTURES

References to Data Sources

Ref. No.	TPRC No.	

1. 11499 — Adzumi, H., "The Flow of Gaseous Mixtures Through Capillaries. I. The Viscosity of Binary Gaseous Mixtures," Bull. Chem. Soc. Japan, 12, 199-226, 1937.

2. 9302 — Amdur, I. and Mason, E.A., "Properties of Gases at Very High Temperatures," Phys. Fluids, 1, 370-83, 1958.

3. 24839 — Andrussow, L., "Diffusion, Viscosity and Conductivity of Gases," 2nd ASME Symp. Thermophysical Properties, 279-87, 1962.

4. 60183 — Barker, J.A., Fock, W., and Smith, F., "Calculation of Gas Transport Properties and the Interaction of Argon Atoms," Phys. Fluids, 7, 897-903, 1964.

5. 18203 — Baron, J.D., Roof, J.G., and Wells, F.W., "Viscosity of Nitrogen, Methane, Ethane, and Propane at Elevated Temperature and Pressure," J. Chem. Eng. Data, 4, 283-8, 1959.

6. 33445 — Barua, A.K., Ross, J., and Afzal, M., "Viscosity of Hydrogen, Deuterium, Methane and Carbon Monoxide from -50 C to 150 C below 200 Atmospheres," Project Squid Tech. Rept. BRN-10-P, 21 pp., 1964. [AD 429 502]

7. 26015 — Baumann, P.B., "The Viscosity of Binary Mixtures of Hydrogen with Ether Vapor, Nitrogen and Carbon Monoxide," Heidelberg University Doctoral Dissertation, 52 pp., 1928.

8. 9871 — Bearden, J.A., "A Precision Determination of the Viscosity of Air," Phys. Rev., 56, 1023-40, 1939.

9. 5473 — Becker, E.W., Misenta, R., and Schmeissner, F., "Viscosity of Gaseous Helium-3 and Helium-4 between 1.3 K and 4.2 K. Quantum Statistics of the Gas-Kinetic Collision at Low Temperatures," Z. Physik, 137, 126-36, 1954.

10. 5474 — Becker, E.W. and Stehl, O., "Viscosity Difference Between Ortho- and Para-Hydrogen at Low Temperatures," Z. Physik, 133, 615-28, 1952.

11. 5475 — Becker, E.W. and Misenta, R., "Viscosity of HD and Helium-3 between 14 and 20 K," Z. Physik, 140, 535-9, 1955.

12. 23543 — Benning, A.F. and McHarness, R.C., "Thermodynamic Properties of Freon 114 Refrigerant CClF$_2$-CClF$_2$ with Addition of Other Physical Properties," E.I. DuPont de Nemours No. T-114 B, 11 pp., 1944.

13. 23546 — Benning, A.F. and McHarness, R.C., "Thermodynamic Properties of Freon-113 Trichlorotrifluoromethane CCl$_2$F-CClF$_2$, with Addition of Other Physical Properties," E.I. DuPont de Nemours No. T-113 A, 12 pp., 1938.

14. 10260 — Benning, A.F. and Markwood, W.H., "The Viscosities of Freon Refrigerants," Refrig. Eng., 37, 243-7, 1939.

15. 42454 — Bewilogua, L., Handstein, A., and Hoeger, H., "Measurement on Liquid Neon," Cryogenics, 6(1), 21-4, 1966.

16. — Bicher, L.B., Jr. and Katz, D.L., "Viscosities of Natural Gases," Ind. Eng. Chem., 35, p. 754, 1943.

17. 9377 — Bond, W.N., "Viscosity of Air," Nature, 137, p. 1031, 1936.

18. 24607 — Bonilla, C.F., Brooks, R.D., and Walker, P.L., "The Viscosity of Steam and Nitrogen at Atmospheric Pressure and High Temperatures," in Proc. of the General Discussion on Heat Transfer, The IME and the ASME, Section II, 167-73, 1951.

19. 30818 — Boon, J.P. and Thomaes, G., "The Viscosity of Liquefied Gases," Physica, 29, 208-14, 1963.

20. 41782 — Boon, J.P., Thomaes, G., and Legros, J.C., "The Principle of Corresponding States for the Viscosity of Simple Liquids," Physica, 33(3), 547-57, 1967.

21. 25394 — Braune, H., Basch, R., and Wentzel, W., "The Viscosity of Some Gases and Vapors. I. Air and Bromine," Z. Phys. Chem., Abt. A, 137, 176-92, 1928.

22. 7029 — Braune, H. and Linke, R., "The Viscosity of Gases and Vapors. III. Influence of the Dipole Moment on the Magnitude of the Sutherland Constant," Z. Physik. Chem., 148A, 195-215, 1930.

23. 10240 — Breitenbach, P., "On the Viscosity of Gases and Their Alteration with Temperature," Ann. Physik, 5(4), 140-65, 1901.

24. 4333 — Bresler, S.E. and Landerman, A., "Viscosity of the Liquid Methane and Deuteriomethane," J. Exptl. Theoret. Phys. (USSR), 10(2), 50-1, 1940.

25. 10284 — Bremond, P., "The Viscosities of Gases at High Temperatures," Comptes Rendus, 196, 1472-4, 1933.

26. 33759 — Bruges, E.A., Latto, B., and Ray, A.K., "New Correlations and Tables of the Coefficient of Viscosity of Water and Steam up to 1000 Bar and 1000 C," Int. J. Heat Mass Transfer, 9(5), 465-80, 1966.

27. — Bruges, E.A. and Gibson, M.R., "The Viscosity of Compressed Water to 10 Kilobar and Steam to 1500 C," 7th Int. Conf. on Steam, Tokyo Paper B-16, 1968.

28. 9360 — Buddenberg, J.W. and Wilke, C.R., "Viscosities of Some Mixed Gases," J. Phys. and Colloid Chem., 55, 1491-8, 1951.

Ref. No.	TPRC No.	
29	26122	Carmichael, L.T., Reamer, H.H., and Sage, B.H., "Viscosity of Ammonia at High Pressures," J. Chem. Eng. Data, 8, 400-4, 1963.
30	29494	Carmichael, L.T. and Sage, B.H., "Viscosity of Ethane at High Pressures," J. Chem. Eng. Data, 8, 94-8, 1963.
31	10334	Carmichael, L.T. and Sage, B.H., "Viscosity of Liquid Ammonia at High Pressures," Ind. Eng. Chem., 44, 2728-32, 1952.
32	26167	Carmichael, L.T. and Sage, B.H., "Viscosity of Hydrocarbons. N-Butane," J. Chem. Eng. Data, 8, 612-6, 1963.
33	37900	Carmichael, L.T., Berry, V., and Sage, B.H., "Viscosity of Hydrocarbons, Methane," J. Chem. Eng. Data, 10, 57-61, 1965.
34	10340	Carr, N.L., "Viscosities of Natural-Gas Components and Mixtures," Inst. Gas Technol. Res. Bull. 23, 59 pp., 1953.
35	34426	Chakraborti, P.K. and Gray, P., "Viscosities of Gaseous Mixtures Containing Polar Gases. Mixtures with One Polar Constituent," Trans. Faraday Soc., 61(11), 2422-34, 1965.
36	24608	Comings, E.W., "Recent Advances in the Use of High Pressures," Ind. Eng. Chem., 39(8), 948-52, 1947.
37	3371	Comings, E.W. and Egly, R.S., "Viscosity of Ethylene and of Carbon Dioxide under Pressure," Ind. Eng. Chem., 33, 1224-9, 1941.
38	24609	Comings, E.W., Mayland, B.J., and Egly, R.S., "Viscosity of Gases at High Pressures," Univ. Illinois Eng. Expt. Sta. Bull. 354, 68 pp., 1944.
39	10148	Coremans, J.M.J., Van Itterbeek, A., Beenakker, J.J.M., Knaap, H.F.P., and Zandbergen, P., "Viscosity of Gaseous Helium, Neon, Hydrogen, and Deuterium below 80 K," Kamerlingh onnes Lab. Leiden Neth. Physica, 24, 557-76, 1958.
40	56153	Coughlin, J., "The Vapor Viscosities of Refrigerants," Purdue Univ. M.S. Thesis, 49 pp., 1953.
41	5455	Craven, P.M. and Lambert, J.D., "The Viscosities of Organic Vapors," Proc. Roy. Soc. (London), A205, 439-49, 1951.
42	36450	Das Gupta, A. and Barua, A.K., "Calculation of the Viscosity of Ammonia at Elevated Pressures," J. Chem. Phys., 42(8), 2849-51, 1965.
43	42183	De Bock, A., Grevendonk, W., and Awouters, H., "Pressure Dependence of the Viscosity of Liquid Argon and Liquid Oxygen, Measured by Means of a Torsionally Vibrating Quartz Crystal," Physica, 34(1), 49-52, 1967.
44	47154	De Bock, A., Grevendonk, W., and Herreman, W., "Shear Viscosity of Liquid Argon," Physica, 37(2), 227-32, 1967.
45	3046	Guimaraes De Carvalho, H., "Variation of Viscosity of Gases with Temperature," Anais Assoc. Quim. Brasil, 4, 79-82, 1945.
46	29692	DiGeronimo, J.P., "Viscosity Correlations of n-Paraffin Hydrocarbons," Newark College of Engineering, Newark, N.J., M.S. Thesis, 41 pp., 1960.
47	9812	De Rocco, A.G. and Halford, J.O., "Intermolecular Potentials of Argon, Methane, and Ethane," J. Chem. Phys., 28, 1152-4, 1958.
48	36203	Diller, D.E., "Measurements of the Viscosity of Parahydrogen," J. Chem. Phys., 42, 2089-100, 1965.
49	28079	Din, F. (Editor), Thermodynamic Functions of Gases, Ed. Butterworths Scientific Publ., London, Current Edition. s.d.
50	40170	DiPippo, R., Kestin, J., and Whitelaw, J.H., "A High-Temperature Oscillating-Disk Viscometer," Physica, 32, 2064-80, 1966.
51	47413	DiPippo, R., "An Absolute Determination of the Viscosity of Seven Gases to High Temperatures," Brown Univ., Ph.D. Thesis, 106 pp., 1964. [Univ. Micr. 67-2231]
52	30399	Dolan, J.P., Starling, K.E., Lee, A.L., Eakin, B.E., and Ellington, R.T., "Liquid, Gas, and Dense-Fluid Viscosity of Butane," J. Chem. Eng. Data, 8, 396-9, 1963.
53	27243	Eakin, B.E., Starling, K.E., Dolan, J.P., and Ellington, R.T., "Liquid, Gas, and Dense Fluid Viscosity of Ethane," J. Chem. Eng. Data, 7, 33-6, 1962.
54	22432	Edwards, R.S., "The Effect of Temperature on the Viscosity of Neon," Proc. Roy. Soc. (London), A119, 578-90, 1928.
55	7637	Edwards, R.S. and Rankine, A.O., "The Effect of Temperature on the Viscosity of Air," Proc. Roy. Soc. (London), A117, 245-57, 1927.
56	22400	Edwards, R.S. and Worswick, B., "On the Viscosity of Ammonia Gas," Proc. Phys. Soc. London, 38, 16-23, 1925.

Ref. No.	TPRC No.	
57	22310	Eglin, J.M., "Coefficients of Viscosity and Slip of Carbon Dioxide by the Oil Drop Method, and the Law of Motion of an Oil Drop in Carbon Dioxide, Oxygen and Helium at Low Pressures," Phys. Rev., 22, 161-70, 1923.
58	47610	Eisele, E.H., "Determination of Dynamic Viscosity of Several Freon Compounds at Temperatures in the Range 200 F to -200 F," Purdue Univ. M.S. Thesis, 122 pp., 1965.
59	18277	Ellis, C.P. and Raw, C.J.G., "High Temperature Gas Viscosities. II. Nitrogen, Nitric Oxide, Boron Trifluoride, Silicon Tetrafluoride, and Sulfur Hexafluoride," J. Chem. Phys., 30, 574-6, 1959.
60	27768	Esipov, Yu. L. and Gagarin, V.I., "Specific Gravity and Viscosity of Furfural-Water Solutions," Gidrolizn i Lesokhim Prom., 15, 15-16, 1962.
61	3742	Felsing, W.A. and Blankenship, F., "Effect of Pressure on the Viscosity of C_2H_4," Proc. OKLA Acad. Sci., 24, 90-1, 1944.
62	22612, 17363	Filippova, G.P. and Ishkin, I.P., "The Viscosity of Air and Argon at Temperatures of from 0 to -183 C and Pressures of from 1 to 150 Atmospheres," Kislorod, 12(2), p. 38, 1959; English translation: RTS-1696, N61-15235, 3 pp., 1960.
63	10397	Fisher, W.J., "The Coefficients of Gas Viscosity. II," Phys. Rev., 28, 73-106, 1909.
64	24080	Flynn, G.P., Hanks, R.V., LeMaire, N.A., and Ross, J., "Viscosity of Nitrogen, Helium, Neon, and Argon from -78.5 to 100 C below 200 Atmospheres," J. Chem. Phys., 38, 154-62, 1963. [AD 294 401]
65	23179, 32116	Forster, S., "Viscosity Measurements in Liquid Neon, Argon, and Nitrogen," Monatsber. Deut. Akad. Wiss. Berlin, 5(10), 695-60, 1963; English translation: Cryogenics, 3, 176-7, 1963.
66	6735	Fortier, A., "Viscosity of Gases and Sutherland's Constant," Compt. Rend., 203, 711-2, 1936.
67	14832	Fortier, A., "The Viscosity of Air and Gases," Publ. Sci. et Tech. du Ministere de l'Air, No. 111, 74 pp., 1937.
68	6734	Fortier, A., "The Viscosity of Air and the Electronic Charge," Compt. Rend. Acad. Sci., 208, 506-7, 1939.
69	5437, 21126	Galkov, G.I. and Gerf, S.F., "Viscosity of Liquefied Pure Gases and their Mixtures II," J. Tech. Phys. (USSR), 11, 613-6, 1941; English translation: SLA 61-18003, 4 pp., 1961.
70	5434, 21037	Gerf, S.F. and Galkov, G.I., "Viscosity of Liquefied Pure Gases and Their Mixtures," J. Tech. Phys. (USSR), 10, 725-32, 1940; English translation: N61-18004, 8 pp., 1961.
71	33059	Giddings, J.G., "The Viscosity of Light Hydrocarbon Mixtures at High Pressures. The Methane-Propane System," Rice Univ., Houston, Texas, Ph.D. Thesis, 202 pp., 1964.
72	39467	Giddings, J.G., Kao, J.T.F., and Kobayashi, R., "Development of a High-Pressure Capillary-Tube Viscometer and its Application to Methane, Propane, and Their Mixtures in the Gaseous and Liquid Regions," J. Chem. Phys., 45, 578-86, 1966.
73	10396	Gilchrist, L., "An Absolute Determination of the Viscosity of Air," Phys. Rev., 1, 124-40, 1913.
74	8361	Gille, A., "The Coefficient of Viscosity for Mixtures of Helium and Hydrogen," Ann. Physik, 48, 799-837, 1915.
75	32097	Gnapp, J.I., "Extrapolation of Viscosity Data for Liquids," Newark College of Engineering, M.S. Thesis, 124 pp., 1961.
76	7215	Golubev, I.F., "The Viscosity of Gases and Gaseous Mixtures at High Pressures. I," J. Tech. Phys., USSR, 8, 1932-7, 1938.
77	42489	Golubev, I.F. and Gnezdilov, N.E., "Viscosity of Helium and Helium-Hydrogen Mixtures up to 250 and 500 kg/cm² Pressure," Gazov. Promy., 10(12), 38-42, 1965.
78	33226	Gonzalez, M.H. and Lee, A.L., "Viscosity of Isobutane," J. Chem. Eng. Data, 11, 357-9, 1966.
79	52196	Gordon, D.T., "The Measurements and Analysis of Liquid Viscosity Data for Eight Freon Refrigerants," Purdue Univ., M.S. Thesis, 108 pp., 1968.
80		Graham, T., "On the Motion of Gases," Phil. Trans. Roy. Soc. (London), 136, 573-632, 1846.
81	24638	Grindlay, J.H. and Gibson, A.H., "In the Frictional Resistance to the Flow of Air through a Pipe," Proc. Roy. Soc. (London), A 80, 114-39, 1908.
82	34248	Guevara, F.A. and Wagner, W.E., "Measurement of Helium and Hydrogen Viscosities to 2340 K," NASA LA-3319 and CFSTI N65-33510, 41 pp., 1965.
83	22850	Guenther, P., "Viscosity of Gases at Low Temperatures," Z. Physik. Chem., 110, 626-36, 1924.
84	22645	Guenther, P., "The Viscosity of Hydrogen at Low Temperatures," Sitz preuss Akad., 720-8, 1920.
85	48301	Hanley, H.J.M. and Childs, G.E., "Discrepancies between Viscosity Data for Simple Gases," Science, 159(3819), 1114-7, 1968.
86	10415	Hogg, J.L., "Viscosity of Air," Proc. Amer. Acad. Arts and Sci., 40, 611-26, 1905.
87	10405	Houston, W.V., "The Viscosity of Air," Phys. Rev., 52, 751-7, 1937.
88	39545	Huang, E.T.S., Swift, G.W., and Kurata, F., "Viscosities of Methane and Propane at Low Temperatures and High Pressures," A.I.Ch.E. J., 12(5), 932-6, 1966.

634

Ref. No.	TPRC No.	
89	9864	Ishida, Y., "Determination of Viscosities and of the Stokes-Millikan Law Constant by the Oil-Drop Method," Phys. Rev., 21, 550-63, 1923.
90	5416	Iwasaki, H., "Measurement of Viscosities of Gases at High Pressure. I. Viscosity of Air at 50, 100, and 150 C," Scientific Reports Research Inst., Tohoku Univ., Japan, 3, 247-57, 1951.
91	32457	Iwasaki, H. and Kestin, J., "The Viscosity of Argon-Helium Mixtures," Physica, 29, 1345-72, 1963.
92	33101	Iwasaki, H., Kestin, J., and Nagashima, A., "Viscosity of Argon-Ammonia Mixtures," J. Chem. Phys., 40, 2988-95, 1964.
93	27667	Iwasaki, H., Kestin, J., and Nagashima, A., "The Viscosity of Argon-Ammonia Mixtures," BRN-6-P, 29 pp., 1963. [AD 410 092]
94	24610	Iwasaki, H. and Takahashi, H., "The Viscosity of Methane at High Pressure," Kogyo Kagaku Zasshi Japan, 62(7), 918-21, 1959.
95	5415	Jackson, W.M., "Viscosities of the Binary Gas Mixtures, Methane-Carbon Dioxide and Ethylene-Argon," J. Phys. Chem., 60, 789-91, 1956.
96	3417	Johns, H.E., "The Viscosity of Liquid Hydrogen," Can. J. Research, 17A, 221-6, 1939.
97	1335	Johnston, H.L. and Grilly, E.R., "Viscosities of Carbon Monoxide, Helium, Neon, and Argon between 80 and 300 K. Coefficients of Viscosity," J. Phys. Chem., 46(8), 948-63, 1942.
98	5413	Johnston, H.L. and McCloskey, K.E., "Viscosities of Several Common Gases between 90 K and Room Temperature," J. Phys. Chem., 44, 1038-58, 1940.
99	8120	Johnston, H.L., Mattox, R.W., and Powers, R.W., "Viscosities of Air and Nitrogen at Low Pressures," NACA TN 2546, 22 pp., 1951.
100	27869	Jossi, J.A., Stiel, L.I., and Thodos, G., "The Viscosity of Pure Substances in the Dense Gaseous and Liquid Phases," A.I.Ch.E. J., 8, 59-63, 1962.
101	19608	Kanda, E., "Viscosity of Fluorine Gas at Low Temperatures," Bull. Chem. Soc. Japan, 12, 463-79, 1937.
102	12012	Keesom, W.H. and Keesom, P.H., "Viscosity of Hydrogen Vapor," Physica, 7, 29-32, 1940.
103	11917	Keesom, W.H. and MacWood, G.E., "The Viscosity of Hydrogen Vapor," Physica, 5, 749-52, 1938.
104	6721	Kesslstrom, G., "The Viscosity of Air under Pressure of 1-30 kg/cm², " Ark. Mat. Astr. Fys., 27A(23), 1-15, 1941.
105	5265	Kellstrom, G., "Note on a Paper 'A New Determination of the Viscosity of Air by the Rotating-Cylinder Method," Phil. Mag., 31, 466-70, 1941.
106	15811	Kellstrom, G., "Viscosity of Air and the Electronic Charge," Nature, 136, 682-3, 1935.
107	32662, 32663	Kesselman, P.M. and Chernyshev, S.K., "Thermal Properties of Some Hydrocarbons at High Temperatures," Teplofiz. Vys. Temp., 3, 700-7, 1965; English translation: High Temp., 3, 651-7, 1965.
108	15650	Kestin, J. and Leidenfrost, W., "Viscosity of Helium," Physica, 25, 537-55, 1959.
109	19284	Kestin, J. and Leidenfrost, W., "Absolute Determination of the Viscosity of 11 Gases Over a Range of Pressures," Physica, 25, 1033-62, 1959.
110	8346	Kestin, J. and Leidenfrost, W., "The Effect of Moderate Pressures on the Viscosity of Five Gases," in Thermodynamic and Transport Properties of Gases, Liquids and Solids Symp., ASME, N.Y., 321-38, 1959.
111	31968	Kestin, J. and Nagashima, A., "The Viscosity of the Isotopes of Hydrogen and their Intermolecular Force Potentials," Brown Univ., Providence, R.I., Rept. BRN-11-P, 23 pp., 1963. [AD 429 501]
112	1675	Kestin, J. and Pilarczyk, K., "Measurement of the Viscosity of Five Gases at Elevated Pressures by the Oscillating Disk Method," Trans. ASME, 76, 987-99, 1954.
113	32023	Kestin, J. and Richardson, P.D., "The Viscosity of Superheated Steam up to 275 C. A Refined Determination," J. Heat Transfer, 85C, 295-302, 1963.
114	6631	Kestin, J. and Wang, H.E., "The Viscosity of Five Gases. A Re-Evaluation," Trans. ASME, 80, 11-7, 1958.
115	6604	Kestin, J. and Wang, H.E., "The Viscosity of Five Gases: A Re-Evaluation," ASME 56-A-72, AFOSR TN 56-98, 23 pp., 1956. [AD 82 011]
116	20730	Kestin, J. and Wang, H.E., "Viscosity of Superheated Steam up to 270 Degrees," Physica, 26, 575-84, 1960.
117	30819	Kestin, J. and Whitelaw, J.H., "A Relative Determination of the Viscosity of Several Gases by the Oscillating Disk Method," Physica, 29, 335-56, 1963. [AD 287 471]
118	42615	Kestin, J. and Whitelaw, J.H., "Sixth International Conference on the Properties of Steam - Transport Properties of Water Substance," Trans. ASME (J. Engrg. Power), 88(1), 82-104, 1966.
119	33444	Kestin, J. and Whitelaw, J.H., "The Viscosity of Dry and Humid Air," Project Squid Tech. Rept. BRN-12-P, 31 pp., 1964. [AD 429 500]

Ref. No.	TPRC No.	
120	37448	Kestin, J. and Whitelaw, J.H., "The Viscosity of Dry and Humid Air," Int. J. Heat Mass Transfer, $\underline{7}$, 1245-55, 1964.
121	10637	Keyes, F.G., "The Heat Conductivity, Viscosity, Specific Heat and Prandtl Numbers for Thirteen Gases," Project Squid Tech. Rept. 37, 33 pp., 1952. [AD 167 173]
122	24649	Keyes, F.G., "Summary of Measured Thermal Conductivities and Values of Viscosities. Transport Properties in Gases," in Proc. Second Biennial Gas Dynamics Symposium, Northwestern Univ. Press, 51-4, 1958.
123	57313	Kinser, R.E., "Viscosity of Several Fluorinated Hydrocarbons in the Liquid Phase," Purdue Univ. M.S. Thesis, 54 pp., 1956.
124	7667	Kiyama, R. and Makita, T., "The Viscosity of Carbon Dioxide, Ammonia, Acetylene, Argon, and Oxygen under High Pressure," Rev. Phys. Chem. Japan, $\underline{22}$, 49-58, 1952.
125	2993	Kiyama, R. and Makita, T., "An Improved Viscometer for Compressed Gases and the Viscosity of Oxygen," Rev. Phys. Chem. Japan, $\underline{26}$(2), 70-4, 1956.
126	22179	Klemenc, A. and Remi, W., "Experimental Investigation of the Viscosity of Nitric Oxide, Propane and their Mixtures with Hydrogen," Monatsh. Chemie, $\underline{44}$, 307-16, 1924.
127	5397	Kompaneets, V.Ya., "Experimental Determination of the Viscosity of Gases and Gaseous Mixtures at High Temperatures," Sbornik Nauch. Rabot Leningrad Inst. Mekhanizatsii Sel'sk. Khoz., $\underline{9}$, 113-26, 1953.
128		Kopsch, W., "The Coefficient of Viscosity of Hydrogen and Argon at Low Temperatures," Halle, Germany, Dissertation, 1909.
129	28213	Krueger, S., "A Correlation of Viscosity of n-Paraffin Hydrocarbons," Newark College of Engrg., Newark, N.J., M.S. Thesis, 68 pp., 1963.
130	24631	Kuenen, J.P. and Visser, S.W., "The Viscosity of Normal Butane Vapor," Verslag Gewone Vergader. Afdeel. Natuurk., Ned. Akad. Wetenschap., $\underline{22}$, 336-43, 1913.
131	25376	Kundt. A. and Warburg, E., "The Viscosity and Thermal Conductivity of Rarefied Gases," Pogg. Ann., $\underline{155}$, 525-50, 1875.
132	2343	Kuss, E., "High-Pressure Research. II. Viscosity of Compressed Gases," Z. Angew Physik, $\underline{4}$, 203-7, 1952.
133	1117	Lambert, J.D., Cotton, K.J., Pailthorpe, M.W., Robinson, A.M., Scrivins, J., Vale, W.R.F., and Young, R.M., "Transport Properties of Gaseous Hydrocarbons," Proc. Roy. Soc. (London), $\underline{231}$, 280-90, 1955.
134	35257	Latto, B., "The Viscosity of Steam at Atmospheric Pressure," Mechanical Engr. Dept., Glasgow Univ. Ph.D. Thesis, 207 pp., 1965.
135	36265	Latto, B., "Viscosity of Steam at Atmospheric Pressure," Intern. J. Heat Mass Transfer, $\underline{8}$, 689-720, 1965.
136	59509	Latto, B. and Cal-Salvum, A.J., "Absolute Viscosity of CCl_2F_2 and $ChCl_2F$," J. Mech. Engng. Sci., $\underline{12}$(2), 135-42, 1970.
137	58068	Latto, B., Hesoun, P., and Asrani, S.C., "Absolute Viscosity and Molecular Parameter for R13, R500, R12, and R22," in 5th ASME Symp. on Thermophys. Properties, ASME, N.Y., 177-85, 1970.
138	5394	Lazarre, F. and Vodar, B., "Determination of the Viscosity of Nitrogen Compressed up to 3000 kg/cm^2," Compt. Rend. Acad. Sci., $\underline{243}$, 487-9, 1956.
139	5393	Leipunskii, O.I., "The Viscosity of Compressed Gases," Acta Phys. (USSR), $\underline{18}$, 172-82, 1943.
140	51487	Lilios, N., "The Viscosities of Several Liquid Refrigerants at Atmospheric Pressure," Purdue Univ. M.S. Thesis, 72 pp., 1957.
141	5388	Linke, R., "The Viscosities of the Freon Compounds and of MeCl in the Liquid and Gaseous States," Warme-Kalte-Tech., $\underline{44}$, 52-3, 1942.
142	5387	Lipkin, M.R., "Viscosity of Propane, Butane and Isobutane," Ind. Eng. Chem., $\underline{34}$, 976-8, 1942.
143	24630	McCullum, R.G., "High Temperature Viscosity Measurement of Fluorinated Hydrocarbon Compounds in the Vapor Phase," Purdue Univ. M.S. Thesis, 84 pp., 1958.
144	20993	McCoubrey, J.C. and Singh, N.M., "Intermolecular Forces in Quasi-Spherical Molecules. II," Trans. Faraday Soc., $\underline{55}$, 1826-30, 1959.
145	8867	McCoubrey, J.C. and Singh, N.M., "Intermolecular Forces in Quasi-Spherical Molecules," Trans. Faraday Soc., $\underline{53}$, 877-83, 1957.
146	5383	Majumdar, V.D. and Oka, V.S., "Atomic Function of Some Gases in the Light of Revised Viscosity Determinations," J. Univ. Bombay, $\underline{17A}$(5), 35-40, 1949.
147	7634	Majumdar, V.D. and Vajifdar, M.B., "Coefficient of Viscosity of Air," Proc. Ind. Acad. Sci., $\underline{8A}$, 171-8, 1938.
148	23177, 23178	Makavetskas, R.A., Popov, V.N., and Tsederberg, N.V., "Experimental Study of the Viscosity of Helium and Nitrogen," Teplofiz. Vys. Temp., $\underline{1}$(2), 191-7, 1963; English translation: High Temp., $\underline{1}$(2), 169-75, 1963.

636

Ref. No.	TPRC No.	
149	2492	Makita, T., "The Viscosity of Freons under Pressure," Rev. Phys. Chem. Japan, 24, 74-80, 1954.
150	5382	Makita, T., "Viscosity of Gases under High Pressure," Mem. Fac. Ind. Arts, Kyoto Tech. Univ. Sci. and Technol., No. 4, 19-35, 1955.
151	6611	Makita, T., "The Viscosity of Argon, Nitrogen and Air at Pressures up to 800 kg/cm to the Second Power," Rev. Phys. Chem. Japan, 27, 16-21, 1957.
152	24635	Markowski, H., "The Viscosity of Oxygen, Hydrogen, Chemical and Atmospheric Nitrogen and Its Change with Temperature," Ann. Physik, 14(4), 742-5, 1904.
153	4302	Mason, S.G. and Maass, O., "Measurement of Viscosity in the Critical Region. Ethylene," Can. J. Research, 18B, 128-37, 1940.
154	5377	Michels, A., Botzen, A., and Schuurman, W., "The Viscosity of Argon at Pressures up to 2000 Atmospheres," Physica, XX, 1141-8, 1954.
155	5375	Michels, A., Schipper, A.C.J., and Rintoul, W.H., "The Viscosity of Hydrogen and Deuterium at Pressures up to 2000 Atmospheres," Physica, 19, 1011-28, 1953.
156		Miyabe, K. and Nishikawa, K., "Correlation of Viscosity for Water and Water Vapor," 7th Int. Conf. on Prop. of Steam, Tokyo, Paper B-6, 1968.
157	19208	Monchick, L., "Collision Integrals for the Exponential Repulsive Potential," Phys. Fluids, 2, 695-700, 1959.
158	6728	Naiki, T., Hanai, T., and Shimizu, S., "Measurement of the Viscosity of Liquid Air," Bull. Inst. Chem. Research, Kyoto Univ., 31(1), 56-8, 1953.
159	21674	Nasini, A. and Rossi, C., "Viscosity of Rare Gases," Gazz. Chim. Ital., 58, 433-42, 1928.
160	57430	Neduzhii, I.A. and Khmara, Yu.I., "Experimental Investigation of the Liquid Viscosity of Propylene, Isobutylene, Butadiene-1,3, Toluol, and Cyclohexane," Teplofiz. Kharakteristiki Veshchestv. GSSSD Moskow, 158-60, 1968; English translation: TT69-55091, 158-60, 1970.
161	9016, 9137	Novikov, I.I., "Some Relationships for Viscosity and Thermal Conductivity of Liquids and Gases," Atomnaya energiya, 2, 468-9, 1957; English translation: J. Nucl. Energy, 6(4), p. 370, 1958.
162	9966	Onnes, H.K., Dorsman, C., and Weber, S., "The Viscosity of Gases at Low Temperatures. I. Hydrogen," Verslag Koninkl. ned Akad Wetenschap., 21, 1375-84, 1913.
163	41443	Gorrell, J.H., Jr. and Bubois, J.T., "Viscosity and Intermolecular Potentials of Hydrogen Sulphide, Sulphur Dioxide and Ammonia," Trans. Faraday Soc., 63, 347-54, 1967. [AD 656 156]
164	24613	Phillips, P., "The Viscosity of Carbon Dioxide," Proc. Roy. Soc. (London), 87, 48-61, 1912.
165	5305	Rietveld, A.O., Van Itterbeek, A., and Van den Berg, G.J., "Coefficient of Viscosity of Gases and Gas Mixtures at Low Temperatures," Physica, 19, 517-24, 1953.
166	57384	Phillips, T.W. and Murphy, K.P., "Liquid Viscosity of Halocarbons," J. Chem. Eng. Data, 15(2), 304-7, 1970.
167		Pinevich, G., "Viscosity of Water-Ammonia Solutions and of Liquid Ammonia," Refrig. Tech. (Moscow), 20, p. 30, 1948.
168	4307	Pleskov, V.A. and Igamberdyev, I., "Viscosity of Mixtures of Ammonia and Water at 20 C," J. Phys. Chem. USSR, 13, 701-2, 1939.
169	20633	Rankine, A.O., "On the Viscosities of the Gases of the Argon Group," Proc. Roy. Soc., 83, 516-25, 1910.
170	24614	Rankine, A.O., "On the Variation with Temperature of the Viscosities of the Gases of the Argon Group," Proc. Roy. Soc., 84, 181-92, 1910.
171	25389	Rankine, A.O., "The Viscosity of Gases of the Argon Group," Physik. Z., 11, 491-502, 1910.
172	25390	Rankine, A.O., "On the Variation with Temperature of the Viscosities of the Gases on the Argon Group," Physik. Z., 11, 745-52, 1910.
173	24615	Rankine, A.O., "One a Method of Measuring of Viscosity of Vapors of Volatile Liquids. An Application to Bromine," Proc. Roy. Soc., A88, 575-88, 1913.
174	22529	Rankine, A.O. and Smith, C.J., "The Viscosities and Molecular Dimensions of Methane, Sulfuretted Hydrogen and Cyanogen," Phil. Mag., 42, 615-20, 1921.
175	24643	Rappenecker, K., "The Viscosity Coefficients of Vapors and their Dependence on the Temperature," Z. Phys. Chem., 72, 695-722, 1910.
176	6918	Raw, C.J.G. and Ellis, C.P., "High-Temperature Gas Viscosities. I. Nitrous Oxide and and Oxygen," J. Chem. Phys., 28, 1198-200, 1958.
177	8577	Reed, J.F. and Rabinovitch, B.S., "Viscosities of Fluorinated Methyl Bromides and Chlorides," J. Chem. Eng. Data, 2, p. 75, 1957.
178	8808	Rietveld, A.O. and Van Itterbeek, A., "Viscosity of Mixtures of H_2 and HD between 300 and 14 K," Physica, 23, 838-42, 1957.

Ref. No.	TPRC No.	
179	15721	Rietveld, A.O., Van Itterbeek, A., and Velds, C.A., "Viscosity of Binary Mixtures of Hydrogen Isotopes and Mixtures of He and Ne," Physica, 25, 205-16, 1959.
180	7196	Rietveld, A.O. and Van Itterbeek, A., "Measurements of the Viscosity of Ne-A Mixtures between 300 and 70 K," Physica, 22, 785-90, 1956.
181	5305	Rietveld, A.O., Van Itterbeek, A., and Van den Berg, G.J., "Measurements on the Viscosity of Mixtures of He and Argon," Physica, 19, 517-24, 1953.
182	33954	Rigby, M. and Smith, E.B., "Viscosities of the Inert Gases," Trans. Faraday Soc., 62, 54-8, 1966.
183	7194	Rigden, P.J., "The Viscosity of Air, Oxygen, and Nitrogen," Phil. Mag., 25, 961-81, 1938.
184	50238	Riley, V.J., "The Viscosity of Liquid Freon 11 and Freon 22 at Temperatures to -110 C," Purdue Univ. M.S. Thesis, 97 pp., 1962.
185		Rivkin, S.L., "Equations of Thermal Conductivity and Dynamic Viscosity of Water Substance," 7th Int. Conference on Prop. of Steam, Tokyo, Paper B-10, 33 pp., 1968.
186	6673	Ross, J.F. and Brown, G.M., "Viscosities of Gases at High Pressures," Ind. Eng. Chem., 49, 2026-33, 1957.
187	32722 23676	Rudenko, N.S., "Viscosity of Liquid Hydrogen and Deuterium," Zh. Fiz. Khim., 37(12), 2761-2, 1963; English translation: Russ. J. Phys. Chem., 37(12), 1493-4, 1963.
188	5300	Rudenko, N.S., "Viscosity of Liquid O_2, N_2, CH_4, C_2H_4, and Air," J. Expt'l. Theor. Phys. USSR, 9, 1078-80, 1939.
189	9888	Rudenko, N.S. and Shubnikov, L.V., "The Viscosity of Liquid Nitrogen, Carbon Monoxide, Argon, and Oxygen in Dependency of Temperature," Phys. Z. Sowjetunion, 6, 470-7, 1934.
190	14159	Sage, B.H. and Lacey, W.N., "Effect of Pressure Upon Viscosity of Air, Methane and Two Natural Gases," Am. Inst. Mining Met. Engrs., Tech. Pub. 845, 16 pp., 1937.
191	11722	Sage, B.H., Yale, W.D., and Lacey, W.N., "Effect of Pressure on Viscosity of Butane and i-Butane," Ind. Eng. Chem., 31, 223-6, 1939.
192	11714	Sage, B.H. and Lacey, W.N., "Viscosity of Hydrocarbon Solutions. Viscosity of Liquid and Gaseous Propane," Ind. Eng. Chem., 30, 829-34, 1938.
193	31581	Schmitt, K., "The Viscosity of Some Gases and Gas Mixtures at Different Temperatures," Ann. Physik, 30, 393-410, 1909.
194	25383	Schultze, H., "The Viscosity of Argon and Its Change with Temperature," Ann. Physik, 5(4), 140-65, 1901.
195	25382	Schultze, H., "The Viscosity of Helium and Its Change with the Temperature," Ann. Physik, 6, 302-14, 1901.
196	20027	Senftleben, H., "Measurements of Physical Gas Constants," Arch. Eisenhuttenw., 31, 709-10, 1960.
197	11660, 10303	Shifrin, A.S., "Viscosity of Steam at Atmospheric Pressure," Teploenergetika, 6(9), 22-7, 1959; English translation: MDF-S-142, 10 pp., 1959.
198	7195	Shilling, W.G. and Laxton, A.E., "The Effect of Temperature on the Viscosity of Air," Phil. Mag., 10, 721-33, 1930.
199	21093	Shimotake, H., "Viscosity of Ammonia in the Dense-Phase Region Pressures Up to 5000 Lb./Sq. In. and Temperatures of 100, 150, and 200 Degrees," Univ. Micro. Publ. No. 60-4795, 84 pp., 1960.
200	5292	Smith, A.S. and Brown, G.G., "Correlating Fluid Viscosity," Ind. Eng. Chem., 35, 705-11, 1943.
201	16316	Smith, C.J., "XXIII. An Experimental Comparison of the Viscous Properties of (A) Carbon Dioxide and Nitrous Oxide, (B) Nitrogen and Carbon Monoxide," Proc. Phys. Soc. (London), 34, 155-64, 1922.
202	16910	Stakelbeck, H., "The Viscosities of Various Refrigerants in the Liquid and Vapor States and Their Dependence on Pressure and Temperature," Z. Ges. Kälte-Ind., 40, 33-40, 1933.
203	32009	Starling, K.E. and Ellington, R.T., "Viscosity Correlations for Nonpolar Dense Fluids," A.I.Ch.E. J., 10, 11-5, 1964.
204	25715	Starling, K.E., Eakin, B.E., and Ellington, R.T., "Liquid, Gas, and Dense-Fluid Viscosity of Propane," A.I.Ch.E. J., 6, 438-42, 1960.
205	9865	States, M.N., "The Coefficient of Viscosity of Helium and the Coefficients of Slip of Helium and Oxygen by the Constant Deflection Method," Phys. Rev., 21, 662-71, 1923.
206	26017	Stauf, F.W., "The Viscosity of Ethylene and Its Mixtures with Hydrogen," Heidelberg Univ. Doctoral Dissertation, 34 pp., 1927.
207	27872	Stiel, L.I. and Thodos, G., "The Viscosity of Polar Gases at Normal Pressures," A.I.Ch.E. J., 8, 229-32, 1962.
208	6729	Sutherland, B.P. and Maass, O., "Measurement of the Viscosity of Gases Over a Large Temperature Range," Can. J. Research, 6, 428-43, 1932.
209	25714	Swift, G.W., Lohrenz, J., and Kurata, F., "Liquid Viscosities Above the Normal Boiling Point for Methane, Ethane, Propane and n-Butane," A.I.Ch.E. J., 6, 415-9, 1960.

Ref. No.	TPRC No.	

210   Tanishita, I., Watanabe, K., and Oguchi, K., "Formulation of Viscosity for Water Substance as a Function of Temperature and Density," 7th Int. Conf. on Prop. of Steam, Tokyo, Paper B-7, 1968.

211   25392   Tanzler, P., "The Coefficient of Viscosity for Mixtures of Argon and Helium," Verhandl. deut. Physik. Ges., 8, 222-35, 1906.

212   25379   Thomsen, E., "The Viscosity of Gas Mixtures," Ann. Physik., 36, 815-33, 1911.

213   30962   Thornton, E. and Baker, W.A.D., "Viscosity and Thermal Conductivity of Binary Gas Mixtures: Argon-Neon, Argon-Helium, and Neon-Helium," Proc. Phys. Soc., 80, 1171-5, 1962.

214   24859   Thornton, E., "Viscosity of Binary Mixtures of Rare Gases," in Progr. in International Research on Thermodynamic and Transport Properties, Academic Press, 527-9, 1962.

215   12175   Titani, T., "Viscosity of Vapors of Organic Compounds. III," Bull. Chem. Soc. Japan, 8, 255-76, 1933.

216   21501   Titani, T., "Viscosity of Vapours of Organic Compounds, Part I," Bull. Inst. Phys. Chem. Res. Japan, 8, 433-60, 1929.

217   24636   Tomlinson, H., "The Coefficient of Viscosity of Air," Phil. Trans. Roy. Soc. (London), 177(2), 767-89, 1886.

218   33119   Trappeniers, N.J., Botzen, A., Van den Berg, H.R., and Van Oosten, J., "The Viscosity of Neon between 25 C and 75 C at Pressures up to 1800 Atmospheres," Physica, 30, 985-96, 1964.

219   5335   Trautz, M., "The Applicability of the Uniformity Principle of Kamerlingh-Onnes to the Estimation of Viscosities," J. Prakt. Chem., 162, 218-23, 1943.

220   15506   Trautz, M. and Baumann, P.B., "Viscosity, Heat Conductivity and Diffusion in Gaseous Mixtures. II. The Viscosities of Hydrogen, Nitrogen, and Hydrogen-Carbon Monoxide Mixtures," Ann. Physik., 2, 733-6, 1929.

221   8354   Trautz, M. and Binkele, H.E., "Viscosity, Heat Conductivity, and Diffusion in Gaseous Mixtures. VIII. The Viscosity of Hydrogen, Helium, Neon, Argon, and their Binary Mixtures," Ann. Physik, 5, 561-80, 1930.

222   6713   Trautz, M. and Heberling, R., "Viscosity, Heat Conductivity and Diffusion in Gas Mixtures. XVII. The Viscosity of Ammonia and its Mixtures with Hydrogen, Nitrogen, Oxygen, Ethylene," Ann. Physik., 10, 155-77, 1931.

223   13313   Trautz, M. and Kipphan, K.F., "Viscosity, Heat Conductivity and Diffusion in Gaseous Mixtures. IV. The Viscosity of Binary and Ternary Mixtures of Noble Gases," Ann. Physik, 2, 743-8, 1929.

224   13869   Trautz, M. and Husseini, I., "Viscosity, Heat Conductivity, and Diffusion in Gaseous Mixtures. XXVI. The Viscosity of Propylene and Beta-Butylene and of Their Mixtures with Helium of Hydrogen," Ann. Physik, 20, 121-6, 1934.

225   13870   Trautz, M. and Ruf, F., "Viscosity, Heat Conductivity and Diffusion in Gaseous Mixtures. XVII. The Viscosity of Chlorine and of Hydrogen Iodide. A Test of Methods of Viscosity Measurements on Corrosive Gases," Ann. Physik, 20, 127-34, 1934.

226   15507   Trautz, M. and Ludesigs, W., "Viscosity, Heat Conductivity, and Diffusion in Gaseous Mixtures. VI. Viscosity Determinations on Pure Gases by Direct Measurement and by Measurements on their Mixtures," Ann. Physik, 3, 409-28, 1929.

227   15509   Trautz, M. and Melster, A., "Viscosity, Heat Conductivity and Diffusion in Gaseous Mixtures. XI. The Viscosity of Hydrogen, Nitrogen, Carbon Monoxide, Ethylene, Oxygen Mixtures," Ann. Physik, 7(5), 409-26, 1930.

228   25381   Trautz, M. and Narath, A., "The Viscosity of Gas Mixtures," Ann. Physik, 79, 637-72, 1926.

229   21421   Trautz, M. and Sorg, K.G., "Viscosity, Heat Conductivity, and Diffusion in Gaseous Mixtures. XVI. The Viscosity of Hydrogen, Methane, Ethane, Propane, and their Binary Mixtures," Ann. Physik, 10, 81-96, 1931.

230   21409   Trautz, M. and Stauf, F.W., "Viscosity, Heat Conductivity, and Diffusion in Gaseous Mixtures. III. The Viscosity of Hydrogen-Ethylene Mixtures," Ann. Physik, 2(5), 737-42, 1929.

231   21402   Trautz, M. and Weizel, W., "Determination of the Viscosity of Sulfur Dioxide and its Mixtures with Hydrogen," Ann. Physik, 78, 305-69, 1925.

232   8358   Trautz, M. and Zimmerman, H., "The Viscosity, Heat Conductivity and Diffusion in Gaseous Mixtures. XXX. The Viscowity at Low Temperatures of Hydrogen, Helium and Neon and their Binary Mixtures Down to 90 Abs.," Ann. Physik, 22(5), 189-93, 1935.

233   8355   Trautz, M. and Zink, R., "Viscosity, Heat Conductivity and Diffusion in Gaseous Mixtures. XII. Viscosity of Gases at High Temperatures," Ann. Physik, 7(5), 427-52, 1930.

234   21419   Trautz, M. and Kwiz, F., "Viscosity, Heat Conductivity, and Diffusion in Gaseous Mixtures. XV. The Viscosity of Hydrogen, Nitrous Oxide, Carbon Dioxide, and Propane and their Binary Mixtures," Ann. Physik, 9, 981-1003, 1931.

235   24627   Tsui, C.Y., "Viscosity Measurements for Several Fluorinated Hydrocarbon Vapors at Elevated Pressures and Temperatures," Purdue Univ. M.S. Thesis, 95 pp., 1959.

Ref. No.	TPRC No.	
236	6824	Uchiyama, H., "Viscosity of Gases at Atmospheric Pressure," Chem. Eng. Japan, 19, 342-8, 1955.
237	13454	Van Cleave, A.B. and Maass, O., "The Variation of the Viscosity of Gases with Temperature Over a Large Temperature Range," Can. J. Research, 13B, 140-8, 1935.
238	25146	Van Dyke, K.S., "The Coefficients of Viscosity of Slip of Air and of Carbon Dioxide by the Rotating Cylinder Method," Phys. Rev., 21, 250-65, 1923.
239	5478	Van Itterbeek, A. and van Paemel, O., "Determination of the Viscosity of Liquid Hydrogen and Deuterium," Physica, 8, 133-43, 1941.
240	40172	Van Itterbeek, A., "Viscosity of Liquefied Gases at Pressures between 1 and 100 Atmospheres," Physica, 32(11), 2171-2, 1966.
241	33839	Van Itterbeek, A., "Viscosity of Liquefied Gases at Pressure Above 1 Atmosphere," Physica, 32(2), 489-93, 1966.
242	4766	Van Itterbeek, A., "Viscosity of Light and Heavy Methane Between 322 K and 90 K," Physica, 7, 831-7, 1940.
243	10275	Van Itterbeek, A. and Claes, A., "Viscosity of Gaseous Oxygen at Low Temperatures. Dependence on the Pressure," Physica, 3, 275-81, 1936.
244	11919	Van Itterbeek, A. and Claes, A., "The Viscosity of Hydrogen and Deuterium Gas Between 293 K and 14 K," Physica, 5, 938-44, 1938.
245	11684	Van Itterbeek, A. and Claes, A., "Viscosity of Light Hydrogen Gas and Deuterium Between 293 K and 14 K," Nature, 142, 793-4, 1938.
246	11923	Van Itterbeek, A. and Keesom, W.H., "Measurements on the Viscosity of Helium Gas Between 293 and 1.6 K," Physica, 5, 257-69, 1938.
247	3653	Van Itterbeek, A. and van Paemel, O., "Measurements of the Viscosity of Gases for Low Pressures at Room Temperature and at Low Temperatures," Physica, 7, 273-83, 1940.
248	9295	Van Itterbeek, A. and van Paemel, O., "Measurements on the Viscosity of Argon Gas at Room Temperature and Between 90 and 55 K," Physica, 5, 1009-12, 1938.
249	7177	Van Itterbeek, A. and Keesom, W.H., "Measurements of the Viscosity of Oxygen Gas at Liquid Oxygen Temperatures," Physica, 2, 97-103, 1935.
250	4315	Van Itterbeek, A. and van Paemel, O., "Measurements of the Viscosity of Neon, Hydrogen, Dueterium and Helium as a Function of the Temperature, between Room Temperature and Liquid-Hydrogen Temperatures," Physica, 7, 265-72, 1940.
251	5420	Van Itterbeek, A., Schapink, F.W., Van den Berg, G.J., and Van Beek, H.J.M., "Measurements of the Viscosity of Helium Gas at Liquid-Helium Temperatures as a Function of Temperature and Pressure," Physica, 19, 1158-62, 1953.
252	5419	Van Itterbeek, A., van Paemel, O., and Van Lierde, J., "The Viscosity of Gas Mixtures," Physica, 13, 88-95, 1947.
253	5312	Van Paemel, O., "Measurements and Theoretical Considerations Relating to the Viscosity of Gases and Condensed Gases," Verh Kon Vlaamsche Acad. Wetensch, Letteren Schoone Kunsten Belgie, Klasse Wetensch., 3(3), 3-59, 1941.
254	5279	Vasilesco, V., "Experimental Research on the Viscosity of Gases at High Temperatures," Ann. Phys., 20, 292-334, 1945.
255	8260	Vasilesco, V., "Experimental Research on the Viscosity of Gases at High Temperatures," Ann. Phys., 20, 137-76, 1945.
256	24620	Verschaffelt, J.E. and Nicaise, C., "The Viscosity of Liquefied Gases. IX. Preliminary Determination of the Viscosity of Liquid Hydrogen," Proc. Acad. Sci. (Amsterdam), 19, 1084-98, 1917.
257	8360	Vogel, H., "The Viscosity of Several Gases and its Temperature Dependence at Low Temperatures," Ann. Physik, 43, 1235-72, 1914.
258	24645	Volker, E., "The Viscosity of Carbon Dioxide and Hydrogen at Low Temperatures," Halle Univ. Doctoral Dissertation, 32 pp., 1910.
259	26019	Vukalovich, M.P., "Thermodynamic Properties of Water and Steam," V.E.B. Verlag Technik, Berlin, 245 pp., 1958.
260	24621	Warburg, E. and von Babo, L., "The Relation between Viscosity and Density of Liquid, Particularly Gaseous Liquid Bodies," Wied. Ann., 17, 390-427, 1882.
261	24628	Wellman, E.J., "Viscosity Determination for Several Fluorinated Hydrocarbon Vapors with a Rolling Ball Viscometer," Purdue Univ. Ph.D. Thesis, 103 pp., 1955. [Univ. Microfilms Publ. UM-13959]
262	28033	Wilbers, O.J., "Viscosity Measurements of Several Hydrocarbon Vapors at Low Temperatures," Purdue Univ. M.S. Thesis, 77 pp., 1961.
263	7633	Williams, F.A., "The Effect of Temperature on the Viscosity of Air," Proc. Roy. Soc. (London), A110, 141-67, 1926.

Ref. No.	TPRC No.	
264	3110	Wobser, R. and Muller, F., "The Viscosity of Gases and Vapors and the Measurement of Viscosity with the Hoppler Viscometer," Kolloid-Beihefte, 52, 165-276, 1941.
265	24629	Witzell, O.W. and Kamien, C.Z., "Viscosity of Refrigerants," ASHRAE J., 65, 663-74, 1959.
266	22277	Yen, K.L., "An Absolute Determination of the Coefficients of Viscosity of Hydrogen, Nitrogen and Oxygen," Phil. Mag., 38, 582-97, 1919.
267	27786	Zaloudik, P., "Viscosity Measurements with Hopplers Viscometer," Chem. Prumysl., 12, 81-3, 1962.
268	8847	Zhdanova, N.F., "Temperature Dependence of the Viscosity of Liquid Argon," Zhur. Eksptl. i Theoret. Fiz., 31(4), 724-5, 1956; English translation: Soviet Phys.-JETP, 4, 749-50, 1957.
269	26032	Ziegler, E., "Concerning the Thermal Conductivity of Ethane and Methane," Phil. Diss. Halle Univ., 39 pp., 1904.
270	24624	Zimmer, O., "The Viscosity of Ethylene and Carbon Monoxide and its Change at Low Temperatures," Halle Univ. Doctoral Dissertation, 30 pp., 1912.
271	1661	Hilsenrath, J. and Touloukian, Y.S., "The Viscosity, Thermal Conductivity, and Prandtl Number for Air, O_2, N_2, NO, H_2, CO, CO_2, H_2O, He, and A," The Transactions of the ASME, 76, 1967-85, 1954.
272		E. I. duPont de Nemours and Co., Inc., Methyl Chloride refrigerant Technical Note.
273	18993	Titani, T., "Viscosity of Vapors of Organic Compounds. II," Bull. Chem. Soc. Japan, 5, 98-108, 1930.
274	23548	Benning, A.F. and McHarness, R.C., "Thermodynamic Properties of Freon-11 Trichloromonofluoromethane (CCl_3F) with Addition of Other Physical Properties," E. I. duPont Technical Note T-11-B, 11 pp., 1938.
275	23552	E. I. duPont de Nemours and Co., Inc., "Properties and Applications of the Freon Fluorinated Hydrocarbons," Bulletin B-2, 11 pp., 1957.
276	26045	Pennsalt Chemicals, Isotron Controlled-Process Refrigerants," Leaflet, 3 pp., 1957.
277	60184	E. I. duPont de Nemours and Co., Inc., "Thermodynamic Properties of Freon-12 Refrigerant," Technical Note 12, 31 pp., 1956.
278	24144	Thornton, E., "Viscosity and Thermal Conductivity of Binary Gas Mixtures. Krypton-Argon, Krypton-Neon, and Krypton-Helium," Proc. Phys. Soc. (London), 77, 1166-9, 1961.
279	23543	E. I. duPont de Nemours and Co., Inc., "Thermodynamic Properties of Freon-114 Refrigerant $CClF_2$-$CClF_2$ with Addition of Other Physical Properties," Technical Note T-114, 11 pp., 1944.
280	60185	E. I. duPont de Nemours and Co., Inc., "Thermodynamic Properties of Freon C318 Refrigerant," Technical Note C-318, 35 pp., 1964.
281		E. I. duPont de Nemours and Co., Inc., "Transport Properties of Freon Fluorocarbons," Technical Note C-30, 23 pp., 1967.
282	30191	Huth, F., "The Viscosity of Liquid Neon," Cryogenics, 2(6), p. 368, 1962.
283	33752	Corruccini, R.J., "Properties of Liquid Hydrogen," Meeting Int. Inst. of Refrigeration, Comm. I, NBS, 53 pp., 1965.
284	5211	Swindells, J.F., Cole, J.R., and Godfrey, T.B., "Absolute Viscosity of Water at 20 Degrees," J. Research Natl. Bur. Stand. (Res. Paper No. 2279), 48, 1-31, 1952.
285	19391	Roscoe, R. and Bainbridge, W., "Viscosity Determination by the Oscillating Vessel Method. II. The Viscosity of Water at 20 Degrees," Proc. Phys. Soc., 72, 585-95, 1958.
286	22284	Ray, S., "Viscosity of Air in a Transverse Electric Field," Phil. Mag., 43, 1129-34, 1922.
287	22434	Nasini, A.G., "Molecular Dimensions of Organic Compounds. II. Viscosity of Vapors. Benzene, Toluene and Cyclohexane," Proc. Roy. Soc. (London), 123, 692-704, 1929.
288	10641	Spencer, A.N. and Trowbridge-Williams, J.L., "The Viscosity of Gaseous Boron Trifluoride," UKAEA and ASTIA IGR-R-CA-235, 8 pp., 1957. [AD 200 161]
289	6914	Raw, C.J.G., "Properties of the Boron Halides. V. The Intermolecular Force Constants of Boron Trifluoride," J. S. African Chem. Inst., 7, p. 20, 1954.
290	6897	Cooke, B.A. and MacKenzie, H.A.E., "Properties of the Boron Halides. I. Viscosity of Boron Trifluoride in the Range 20-200 Degrees," J. S. African Chem. Inst., 4, 123-9, 1951.
291	23835	Panchenkov, G.M., Makarov, A.V., Dyachenko, V.Ya., and Moiseev, V.D., "Viscosity of Boron Trifluoride," Vestnik Moskovskogo Univ. Seriya II, Khim., 17, 11-3, 1962.
292	18306	Mueller, C.R. and Ignatowski, A.J., "Equilibrium and Transport Properties of the Carbon Tetrachloride-Methylene Chloride System," J. Chem. Phys., 32, 1430-4, 1960.
293	20442	Sperry, E.H. and Mack, E., Jr., "The Collision Area of the Gaseous Carbon Tetrachloride Molecule," J. Am. Chem. Soc., 54(3), 904-7, 1932.
294	22340	Bleakney, W.M., "Measurements on the Vapor Viscosities of the Two Common Pentanes, Two Pentanes and Carbon Tetrachloride," Physics, 3, 123-36, 1932.

Ref. No.	TPRC No.	
295	25136	Rankine, A.O., "The Viscosities of Gaseous Chlorine and Bromine," Nature (London), 88, 469-70, 1912.
296	21416	Trautz, M. and Winterkorn, H., "Viscosity, Heat Conductivity and Diffusion of Gaseous Mixtures. XVIII. The Measurement of Viscosity in Corrosive Gases (Cl_2, HI)," Ann. Physik, 10(5), 511-28, 1931.
297	22230	Campetti, A., "Physical Constants of Chlorine Under the Action of Light," Nuovo Cimento, 17(1), 143-58, 1919.
298	13456	Van Cleave, A.B. and Maass, O., "The Viscosities of Deuterium-Hydrogen Mixtures," Can. J. Research, 13, p. 384, 1935.
299	29823	Amdur, I., "Viscosity of Deuterium," J. Am. Chem. Soc., 57, 588-9, 1935.
300	16908	Khalilov, K., "Viscosity of Liquids and Saturated Vapors at High Temperatures and Pressures," J. Exptl. Theoret. Phys. USSR, 9, 335-45, 1939.
301	18209	Reid, R.C. and Belenyessy, L.T., "Viscosity of Polar Vapor Mixtures," J. Chem. Eng. Data, 5, 150-1, 1960.
302	10407	Eucken, A., "On the Thermal Conductivity, Specific Heat and Viscosity of Gas," Physik Z., 14, p. 324, 1913.
303	24642	Pedersen, F.M., "The Influence of Molecular Structure Upon the Internal Friction of Certain Isometric Ether Gases," Phys. Rev., 25, 225-54, 1907.
304	5441	Franck, E.U. and Stober, W., "The Viscosity and Effective Molecular Diameter of Fluorine," Z. Naturforsch., 7, 822-3, 1952.
305		Melaven, R.M. and Mack, E., "The Collision Areas and Shapes of Carbon Chain Molecules in the Gaseous State: Normal Heptane, Normal Octane, Normal Nonane," J. Am. Chem. Soc., 54, 888-904, 1932.
306	30266	Agaev, N.A. and Golubev, I.F., "The Viscosities of Liquid and Gaseous n-Heptane and n-Octane at High Pressures and at Different Temperatures," Gazovaya Prom., 8, 50-3, 1963.
307	33779	Carmichael, L.T. and Sage, B.H., "Viscosity and Thermal Conductivity of Nitrogen-n-Heptane and Nitrogen-n-Octane Mixtures," A.I.Ch.E. J., 2(3), 559-62, 1966.
308	6059	McCoubrey, J.C., McCrea, J.N., and Ubbelohde, A.R., "The Configuration of Flexible Polymethylene Molecules in the Gas Phase," J. Chem. Soc., 1961-71, 1951.
309	22413	Harle, H., "Viscosities of the Hydrogen Halides," Proc. Roy. Soc. (London), A100, p. 429, 1922.
310	41442	Pal, A.K. and Barua, A.K., "Viscosity and Intermolecular Potentials of Hydrogen Sulphide, Sulphur Dioxide and Ammonia," Trans. Faraday Soc., 63(2), 341-6, 1967.
311	20635	Rankine, A.O., "On the Viscosities of the Vapor of Iodine," Proc. Roy. Soc. (London), 91A(8), p. 201, 1915.
312	29802	Clifton, D.G., "Measurements of the Viscosity of Krypton," J. Chem. Phys., 38, p. 1123, 1963.
313	36848	Trappeniers, N.J., Botzen, A., Van Oosten, J., and Van den Berg, H.R., "The Viscosity of Krypton between 25 and 75 C and at Pressures Up to 2000 Atm.," Physica, 31, p. 945, 1965.
314	13871	Trautz, M. and Freytag, A., "Viscosity, Heat Conductivity and Diffusion in Gaseous Mixtures. XXVIII. The Viscosity of Cl_2, NO, and NOCl. Viscosity During the Reaction 2 NO + Cl_2 = 2 NOCl," Ann. Physik., 20, 135-44, 1934.
315	21422	Trautz, M. and Gabriel, E., "Viscosity, Heat Conductivity and Diffusion in Gaseous Mixtures. XX. The Viscosity of Nitric Oxide and Its Mixtures with Nitrogen," Ann. Physik, 11, 606-10, 1931.
316	29728	Peter, S. and Wagner, E., "The Methodics of Accurate Viscosity Measurements with Capillary Viscometers. II. Influence of the Capillary Forces and of the Change of the Hydrostatic Pressure on the Measurement," Z. Physik. Chem., 17, 199-219, 1958.
317		Beer, H., "Heat Transfer in Dissociated Gases," Chem. Ing. Tech., 31(10), p. 1047, 1965.
318		Timrot, D.L., Serednickaja, M.A., and Traktueva, S.A., "Investigation of the Viscosity of Dissociating Nitrogen Tetraoxide by the Method of a Vibrating Disc," Teplofiz. Vys. Temp., 7(5), 885-92, 1969.
319	22329	Day, R.K., "Variation of the Vapor Viscosities of Pentane and Isopentane with Pressure by the Rotating Cylinder Method," Phys. Rev., 40, 281-90, 1932.
320	30480	McCoubrey, J.C. and Singh, N.M., "The Vapor-Phase Viscosities of the Pentanes," J. Phys. Chem., 67, 517-8, 1963.
321	21541	Stewart, W.W. and Maass, O., "The Coefficient of Viscosity of Sulphur Dioxide Over a Low Temperature Range," Can. J. Research, 6, p. 453, 1932.
322	25090	Smith, C.J., "The Viscosity and Molecular Dimensions of Sulfur Dioxide," Phil. Mag., 44, 508-11, 1922.
323	37222	Kestin, J. and Nagashima, A., "Viscosity of Neon-Helium and Neon-Argon Mixtures at 20 and 30 C," J. Chem. Phys., 40, 3648-54, 1964.
324	15622	Thornton, E., "Viscosity and Thermal Conductivity of Binary Gas Mixtures. Xenon-Krypton, Xenon-Argon, Xenon-Neon, and Xenon-Helium," Proc. Phys. Soc. (London), 76, 104-12, 1960.

Ref. No.	TPRC No.	
325	21675	Nasini, A.G. and Rossi, C., "Viscosity of Mixtures of Rare Gases. I," Gazz. Chim. Ital., _58_, 898-912, 1928.
326	39402	Kestin, J., Kobayashi, Y., and Wood, R.T., "The Viscosity of Four Binary, Gaseous Mixtures at 20 and 30 C," Physica, _32_(6), 1065-89, 1966.
327	5390	van Lierde, J., "Measurements of Thermal Diffusion and Viscosity of Certain Gas Mixtures at Low and Very Low Temperatures," Verh. Koninkl. Vlaam. Acad. Wetensch. Belg., Kl. Wetensch., _9_(24), 7-78, 1947.
328	34992	Di Pippo, R., Kestin, J., and Oguchi, K., "Viscosity of Three Binary Gaseous Mixtures," J. Chem. Phys., _46_(12), 4758-64, 1967.
329	48647	Kestin, J. and Yata, J., "Viscosity and Diffusion Coefficient of Six Binary Mixtures," J. Chem. Phys., _49_(11), 4780-91, 1968.
330	39148	Kao, J.T.F. and Kobayashi, R., "Viscosity of Helium and Nitrogen and Their Mixtures at Low Temperatures and Elevated Pressures," J. Chem. Phys., _47_(8), 2836-49, 1967.
331	31983	Makavetskas, R.A., Popov, V.N., and Tsederberg, N.V., "An Experimental Investigation of the Viscosity of Mixtures of Nitrogen and Helium," Teplofiz. Vys. Temp., _1_(3), 348-55, 1963.
332	16644	Johnson, C.A., "Viscosity of Gas Mixtures," SURI Ch.E. 273-566F3, AECU-3301, 119 pp., 1956.
333	39459	Breetveld, J.D., Di Pippo, R., and Kestin, J., "Viscosity and Binary Diffusion Coefficient of Neon-Carbon Dioxide Mixtures at 20 and 30 C," J. Chem. Phys., _45_(1), 124-6, 1966.
334	6650	Strauss, W.A. and Edse, R., "Measurements of the Viscosity of Gas Mixtures," WADC TR 57-484, 15 pp., 1957. [AD 142 082]
335	39179	DeWitt, K.J. and Thodos, G., "Viscosities of Binary Mixtures in the Dense Gaseous State: The Methane-Carbon Dioxide System," Can. J. Research, _44_(3), 148-51, 1966.
336	15651	Kestin, J. and Leidenfrost, W., "Effect of Pressure on the Viscosity of N_2-CO_2 Mixtures," Physica, _25_, 525-36, 1959.
337	39122	Gururaja, G.J., Tirunarayanan, M.A., and Ramchandran, A., "Dynamic Viscosity of Gas Mixtures," J. Chem. Eng. Data, _12_(4), 562-7, 1967.
338	39502	DeWitt, K.J. and Thodos, G., "Viscosities of Binary Mixtures in the Dense Gaseous State: The Methane-Tetrafluoromethane System," Physica, _32_(8), 1459-72, 1966.
339	30437	Kaw, C.J.G. and Tang, H., "Viscosity and Diffusion Coefficients of Gaseous Sulfur Hexafluoride-Carbon Tetrafluoride Mixtures," J. Chem. Phys., _39_(10), 2616-8, 1963.
340	22179	Alfons, K. and Walter, K., "Experimental Investigation of the Coefficients of Viscosity of Nitric Oxide, Propane and Their Mixtures with Hydrogen," Monatsh. Chemie, _44_, 307-16, 1924.
341	36791	Pal, A.K. and Barua, A.K., "Viscosity of Hydrogen-Nitrogen and Hydrogen-Ammonia Gas Mixtures," J. Chem. Phys., _47_(1), 216-8, 1967.
342	36031	Dolan, J.P., Ellington, R.T., and Lee, A.L., "Viscosity of Methane-Butane Mixtures," J. Chem. Eng. Data, _9_(4), 484-7, 1964.
343	40602	Carmichael, L.T., Virginia, B., and Sage, B.H., "Viscosity of a Mixture of Methane and n-Butane," J. Chem. Eng. Data, _12_(1), 44-7, 1967.
344	21156	Gerf, S.F. and Galkov, G.I., "Viscosity of Liquefied Pure Gases and Their Mixtures. III," J. Tech. Phys. (USSR), _11_, 801-8, 1941.
345	27488	Hawksworth, W.A., Nourse, H.H.E., and Raw, C.J.G., "High-Temperature Gas Viscosities. III. NO-N_2O Mixtures," J. Chem. Phys., _37_(4), 918-9, 1962.
346	22885	Jung, G. and Schmick, H., "The Influence of Molecular Attractive Forces on the Viscosity of Gas Mixtures," Z. Physik. Chem., _B7_, 130-47, 1930.
347	49917	Pal, A.K. and Barua, A.K., "Viscosity of Polar-Nonpolar Gas Mixtures," Indian J. Phys., _41_(10), 713-8, 1967.
348	34795	Burch, L.G. and Raw, C.J.G., "Transport Properties of Polar-Gas Mixtures. I. Viscosities of Ammonia-Methylamine Mixtures," J. Chem. Phys., _47_(8), 2798-801, 1967.
349	39381	Chakraborti, P.K. and Gray, P., "Viscosities of Gaseous Mixtures Containing Polar Gases: More Than One Polar Constituent," Trans. Faraday Soc., _62_(7), 1769-75, 1966.
350	57310	Chang, K.C., Hesse, R.J., and Raw, C.J.G., "Transport Properties of Polar Gas Mixtures SO_2 + SO_2F_2 Mixtures," Trans. Faraday Soc., _66_, 590-6, 1970.
351	37951	Katti, P.K. and Chaudhri, M.M., "Viscosities of Binary Mixtures of Benzyl Acetate with Dioxane, Aniline, and m-Cresol," J. Chem. Eng. Data, _9_(3), 442-3, 1964.
352		Katti, P.K. and Prakash, O., "Viscosities of Binary Mixtures of Carbon Tetrachloride with Methanol and Isopropyl Alcohol," J. Chem. Eng. Data, _11_(1), 46-7, 1966.
353	33143	Lee, A.L., Gonzalez, M.H., and Eakin, B.E., "Viscosity of Methane-n-Decane Mixtures," J. Chem. Eng. Data, _11_(3), 281-7, 1966.

Ref. No.	TPRC No.	
354	28616	Lewis, J.E., "Thermodynamic and Intermolecular Properties of Binary Liquid Systems," Purdue Univ. Ph.D. Dissertation, 151 pp., 1956.
355	34565	Ridgway, K. and Butler, P.A., "Some Physical Properties of the Ternary System Benzene-Cyclohexane-n-Hexane," J. Chem. Eng. Data, 12(4), 509-15, 1967.
356	32932	Vatolin, N.V., Vostryakov, A.A., and Esin, O.A., "Viscosity of Molten Ferrocarbon Alloys," Phys. Metals Metallography (USSR), 15(2), 53-8, 1963.
357	7466	Yao, T.P. and Kondic, V., "The Viscosity of Molten Tin, Lead, Zinc, Aluminum, and Some of Their Alloys," J. Inst. Metals, 81(1), 17-24, 1952.
358	48549	Campbell, A.N. and Van der Kouive, E.T., "Studies on the Thermodynamics and Conductances of Molten Salts and Their Mixtures. V. The Density, Change of Volume on Fusion, Viscosity, and Surface Tension of Sodium Chlorate and of Its Mixtures with Sodium Nitrate," Can. J. Chem., 46(8), 1279-86, 1968.
359	5415	Morrison-Jackson, W., "Viscosities of the Binary Gas Mixtures Methane-Carbon Dioxide and Ethylene-Argon," J. Phys. Chem., 60, 789-91, 1956.
360	48784	Marsh, K.N., "Mutual Diffusion in Octamethylcyclotetrasiloxane Mixtures," Trans. Faraday Soc., 64(4), 894-901, 1968.
361	40696	Strunk, M.R. and Fehsenfeld, G.D., "The Prediction of the Viscosity of Multicomponent Nonpolar Gaseous Mixtures at Atmospheric Pressure," Univ. of Missouri at Rolla, M.S. Thesis, 95 pp., 1964. [AD 18254]
362	5297	Schmid, C., "Viscosity of Gases and Gaseous Mixtures at High Temperatures," Gas-und Wasserfach, 85, 92-103, 1942.
363	14264	Herning, F. and Zipperer, L., "Calculation of the Viscosity of Technical Gas Mixtures from the Viscosity of the Individual Gases," Gas-und Wasserfach, 79, 49-54, 69-73, 1936.
364	5407	Kenney, M.J., Sarjant, R.J., and Thring, M.W., "The Viscosity of Mixtures of Gases at High Temperatures," Brit. J. Appl. Phys., 7(9), 324-9, 1956.
365	10340	Carr, N.L., "Viscosities of Natural Gas Components and Mixtures," Inst. Gas, Technol. Res. Bull. No. 23, June 1953.
366	40097	Gnezdilov, N.E. and Golubev, I.F., "Viscosity of Methane-Nitrogen and Methane-Nitrogen-Hydrogen Mixtures at Temperatures from 298 to 473 K and Pressures up to 490.3×10^5 N/m^2," Teploenergetika, 14(1), 89-90, 1967.
367		Reamer, H.H., Sage, B.H., and Lacey, W.N., "Phase Equilibria in Hydrocarbon Systems," Ind. Eng. Chem., 42(3), 534-9.
368		Canjar, L.N. and Manning, F.S., Thermodynamic Properties and Reduced Correlations for Gases, Gulf Publishing Co., Houston, Texas, 212 pp., 1967.
369		Reamer, H.H., Korpi, K.J., Sage, B.H., and Lacey, W.N., "Phase Equilibria in Hydrocarbons Systems," Ind. Eng. Chem., 39(2), 206-9, 1947.
370		Witonsky, R. and Miller, J.G., "Compressibility of Gases. IV. The Burnett Method Applied to Gas Mixtures at Higher Temperatures. The Second Virial Coefficients of the Helium-Nitrogen System from 175 to 475 Degrees," J. Am. Chem. Soc., 85, 282-6, 1963.
371	63037	Phillips, T.W. and Murphy, K.P., "Liquid Viscosity of Halogenated Refrigerants," ASHRAE Trans., 76, 146-56, 1970.
372	33096	Petker, I. and Mason, D., "Viscosity of the N_2O_4-NO_2 Gas System," J. Chem. Eng. Data, 9(2), 280-1, 1964.
373	13868	Trautz, M. and Heberling, R., "Viscosity, Heat Conductivity, and Diffusion in Gaseous Mixtures. XXV. Internal Viscosity of Xenon and Its Mixtures with Hydrogen and Helium," Ann. Physik, 20, 118-20, 1934.
374	30265	Agaev, N.A. and Golubev, I.F., "The Viscosities of Liquid and Gaseous n-Pentane at High Pressures at Different Temperatures," Gazovaya Prom., 8(5), 45-50, 1963.
375		Miller, J.E., Brandt, L.W., and Stroud, L., "Compressibility of Helium-Nitrogen Mixtures," J. Chem. Eng. Data, 5, 6-9, 1960.
376		Miller, J.E., Brandt, L.W., and Stroud, L., "Compressibility Factors for Helium and Helium-Nitrogen Mixtures," U.S. Bureau of Mines Rept. Invest., 5845, 11 pp., 1961.
377	5390	van Lierde, J., "Measurement of Thermal Diffusion and Viscosity of Certain Gas Mixtures at Low and Very Low Temperatures," Verhandel. Koninkl. Vlaam. Acad. Weten-Schap. Belg. Kl. Wetenschap., 9(24), 7-78, 1947.

Material Index

Material Index